AF663643

MATRIX: ALGEBRA, CALCULUS AND GENERALIZED INVERSE

MATRIX:

ALGEBRA, CALCULUS

AND

GENERALIZED INVERSE

(Vol. I)

A . K . HAZRA

New Delhi | Mumbai | Chennai | Kolkata | Bangalore | Hyderabad | Kochi

First Indian Edition 2009

VIVA BOOKS PRIVATE LIMITED

- 4737/23 Ansari Road, Daryaganj, New Delhi 110 002
  E-mail: vivadelhi@vivagroupindia.net, Tel. 42242200
- 76, Service Industries, Shirvane, Sector 1, Nerul, Navi Mumbai 400 706
  E-mail: vivamumbai@vivagroupindia.net, Tel. 27721273, 27721274
- Jamals Fazal Chambers, 26 Greams Road, Chennai 600 006
  E-mail: vivachennai@vivagroupindia.net, Tel. 28290304, 28294241
- B-103, Jindal Towers, 21/1A/3 Darga Road, Kolkata 700 017
  E-mail: vivakolkata@vivagroupindia.net, Tel. 22836381, 22816713
- 7, Sovereign Park Apartments, 56-58, K. R. Road, Basavanagudi, Bangalore 560 004
  E-mail: vivabangalore@vivagroupindia.net, Tel. 26607409, 26607410
- 101-102 Mughal Marc Apartments, 3-4-637 to 641, Narayanguda, Hyderabad 500 029
  E-mail: vivahyderabad@vivagroupindia.net, Tel. 27564481, 27564482
- Beevi Towers, First Floor, SRM Road, Kaloor, Kochi 682 018, Kerala
  E-mail: vivakochi@vivagroupindia.net, Tel: 0484-2403055, 2403056

www.vivagroupindia.com

Published by arrangement with
Cambridge International Science Publishing
7, Meadow Walk, Great Abington
Cambridge CB1 6AZ, UK

ISBN-10: 81-309-0952-9
ISBN-13: 978-81-309-0952-3

Published by Vinod Vasishtha for **Viva Books Private Limited, 4737/23, Ansari Road,** Daryaganj, New Delhi-110 002. **Printed and bound by Anand Sons, Delhi-110 092.**

Comments about the book

Professor S. Councilman, Prof. and Ex-Chairman, Department of Mathematics, California State University, Long Beach, U.S.A., writes "*Judging by the preface and table of contents, this looks like a very interesting and comprehensive text.*"

Professor M. Zuhair Nashed, Department of Mathematical Science, University of Delaware, New York, DE 19716, USA, writes, on the basis of the preface and table of contents "*It looks that a major undertaking in its scope and details of exposition. I would be happy to recommend the book to our library. In USA it is not clear where it could be used as a text since elementary linear algebra courses are different from your extensive coverage, and include applications and some computational methods. I hope it will be well received.*"

Professor W.B.R. Lickorish, Head of the Department of Pure Mathematics and Mathematical Statistics, University of Cambridge, UK, wishes the success to publication of the book .

ABOUT THE AUTHOR

A.K.Hazra (*Md. Abdul Khaleq Hazra*) is an Associate Professor of the Department of Mathematics, Bangladesh University of Engineering and Technology, Dhaka, Bangladesh, where he has been teaching undergraduate and graduate engineering students since October 1966. He started his professional career at the Saadat College (now it is Govt. Saadat University College), Karatia, Tangail, Bangladesh and served there from August 1965 to October 1966. He received his Diploma in Computer Application Technology from the Asian Institute of Technology (AIT), Bangkok, Thailand, in 1980 and M.Sc. in Applied Mathematics from University of Dhaka, Dhaka, Bangladesh, in 1965. He regularly contributes articles in mathematical conferences in Bangladesh and professional journals such as *Bangladesh Academy of Science*; *The Bangladesh Journal of Scientific Research*; *Ganit, Journal of Bangladesh Mathematical Society*; *Journal of Mathematics & Mathematical Sciences*, a publication of mathematics department, Jahangirnagar University, Savar, Dhaka; *BUET Studies*, a publication of Bangladesh University of Engineering and Technology, Dhaka; *Proceedings of Mathematical Conference of Bangladesh*, etc. His accepted paper '*Competition Coefficients in Population Dynamics*' for presentation in the conference '*A Research Conference on Mathematical Ecology*', Trieste, Italy, 1990, could not be presented due to lack of finance. But later it was later published in *Journal of Mathematics & Mathematical Sciences*. The Abstract of his paper '*Mathematics for Computerised Medical Diagnosis*' was printed in the proceedings of the '*First International Conference on Computer Applications in Science, Technology and Medicine*' held in the University of Isfahan, Isfahan, Iran, in 1991. His books have been published by national and international publishers. His published books are *Vector Analysis with Applications* (*3/e*) by Wiley Eastern Ltd (now it is New Age International (P) Limited, Publishers), New Delhi, India; *Integral Calculus with Applications* by Pragati Prakashan, Publisher, Meerut, India; *Integral Calculus and Elementary Differential Equations* (*Bengali medium*), 2/e by Bangla Academy, Dhaka, Bangladesh; the Bengali version of three volumes this book has been published by Bangla Academy, Dhaka, Bangladesh in 1998, 1999.

Dedicated

to

the sweet memories of those
whom I can't forget for a moment.
My primary school teacher.
I forgot his name but I didn't forget
his valuable contribution .
My University teacher late Professor
A.R. Khalifa,
a man of devotion to Mathematics.
My mother whom I lost forever
due to the cyclone of 9th May, '61
and my late father, a shade over me.

PREFACE

At the very outset it is pleasant to pay homage to those whose activities lay the general foundation of this work. Prof. Albert Einstein, the founder of the theory of relativity and Prof. S. Councilman of California State University belong to this group. These two persons in principle guided the author through their sayings and writings in selecting topics, illustrations and problems according to their ideal. Prof. Albert Einstein said,

"Everything should be made as simple as possible, but not simpler."

Prof. Councilman expressed his dissatisfaction and deepest sorrow about text books on linear algebra and put forward a constructive suggestion about the nature of this type of text. Of course this should be true generally for all kinds of texts. He writes in [11],

"*When teaching the topics of eigenvalues, eigenvectors and diagonalizability, it is quite desirable to have an abundant supply of examples and problems which illustrate all possible situations that can occur with respect to distinct or multiple eigenvalues and the existence of n or fewer linearly independent eigenvectors.*

Most linear algebra texts do a rather poor job in this regard. They fail to cover either all possible cases or, if they do, their examples and problems suffer from one or other of the two possible defects:

(i) *Computing the characteristic polynomial* det(A–λI) *is very tedious and finding its roots is somewhat difficult, or*

(ii) *matters are oversimplified by using only upper or lower triangular, or even diagonal matrices.*

The drawback in case (ii) *is that the eigenvalues are the diagonal entries, so the student is not required to compute the characteristic polynomial and find its roots. Furthermore, the system* $(A - \lambda_i I)\mathbf{x}^T = 0$ *is in "echelon form", so students get little practice in using Gaussian elimination.*"

The author left no stone unturned to fulfill the desire expressed by Prof. Councilman. Yet, nothing is complete in itself. There always remains something more.

Readers

The author prepared the major part of the text book during the ten months' sabbatical leave in 1994. Generally, it is designed for undergraduate and graduate students of Mathematics, Engineering and Social Sciences. Professionals and researchers of those fields may also use it as a reference and find the book useful. It is one of the major branches of Mathematics. The contents are arranged in such a way that a beginner can easily grasp the materials step by step. The theories are tried to make lucid through illustrated examples. No prerequisite is necessary to sequential study of the materials except some preliminary knowledge of calculus for the 'calculus part' of the book. The 'Generalized Inverse' is necessary for the study of unbalanced data of linear models in which the coefficient matrix is singular/rectangular and is essential for the undergraduate and graduate study in Statistics, Numerical Analysis, Engineering, Physics and other Social Sciences of the majority of the institutions of the world.

Synopsis

The book is divided into three volumes. Volume I includes chapter 1 to chapter 8, Volume II includes chapter 9 to chapter 13 and Volume III includes chapter 14 to chapter 18. Each chapter begins with chapter summary and ends with exercises and answers to each problem. Chapter one is on fundamental definitions. It interlinks matrix and linear algebra, tests definiteness of matrices and includes special definitions. Chapters two and three provide with 'Basic algebraic operations'. 'Square matrices' including some more square matrices normally not available in other texts comprise chapter three. Submatrices and partition matrices with distinction are in chapter four. Chapter five includes other matrix operations like Kronecker sums, products, etc., which are normally beyond the contents of many texts. Chapter six is on determinants and includes a new method of expansion of the author, comparison of methods, determinant of the product of square and rectangular matrices, factorization of determinants, definiteness tests of matrices, etc., as special features. Chapter seven is on simultaneous linear algebraic equations with analytic construction of magic squares. In chapter eight the author establishes the properties of the cofactor matrix, proofs of some properties of the adjoint matrix from the cofactor matrix and introduces classical methods of inversion of nonsingular matrices, and in sec.8.3 the author lists 17 inversion procedures with appropriate places of discussion. Chapter nine is on elementary operations and matrices similar to other texts with the collection of different symbols from different texts. The material of chapter ten is rank and nullity of a matrix with the development of Gerstein algorithm from 88's journal and introduces definiteness tests in terms of rank and signature. Chapter eleven and twelve introduce 'Vector spaces' with the idea of various vector spaces in practice and 'Linear Transformation' with best available collections of materials. Chapter thirteen and fourteen on 'Eigenvalues and Eigenvectors' and 'Canonical Forms and Matrix Factorization' include some results available in journals only and not in any texts. Chapter fifteen describes linear, bi-linear and quadratic forms in detail. Chapters sixteen to eighteen on 'Norms and Measures of Matrices', 'Matrix Calculus' and 'Generalized Inverse' mostly are dependent on journals and not available in texts of matrices and linear algebra.

Key features

* Most attractive feature of the book might be GRE sample tests.
* Most chapters include some definitions and ideas different from classical texts.
* Dozens of algorithm can easily be converted to computer programs.
* Formulation of the modal matrix in case of multiple eigenvalues in terms grand-parent, parent and daughter is perhaps a new and pleasant terminology.
* Most texts ignore the inverse of singular as well as rectangular matrices. As a result, misconception grows among learners that inverse of a matrix is only possible if it is nonsingular. This is removed at the start of the inverse matrix in sec.8.3 mentioning 17 procedures of inversion.
* Definiteness tests occupy appropriate places in appropriate contexts such as in secs.1.3, 3.10, 6.17, 10.11.1, 13.24.
* The submatrix and partition matrix have been dealt with in details in chapter 4 instead of finishing the idea with definition only.
* Chapter 5 on 'Other matrix operations' is a special feature of the book.

* Chapter 6 introduces a new method of expansion of the determinant in sec.6.7.3 and comparison of different methods in sec.6.7.6.
* Analytic construction of magic squares is a new idea in chapter 7.
* Chapter 8 introduces properties of cofactor matrices and avoids classical back-gearing process of proving the identity adj (AB) = adjB adjA .
* Text book in general is a mouth-piece of journals. It lucidly and understandably explains and popularizes the research articles. This text recovered and used many results such as Sylvester's dialytic elimination, different types of circulant matrices, properties of Helmertian matrices, properties and construction of Hadamard matrices, inversion of Vandermode matrices, developed Gerstein algorithms for the computing rank of a matrix, different types and procedures of generalized inverses, etc., in appropriate sections from the bondage of journals of different centuries [references].
* Norms & measure, matrix calculus and generalized inverse in independent chapters constitute a new flavour of the text.

Gratefulness

The author is indebted to researchers and authors of text books and papers listed in the references. The author remains grateful to them for ever.

To err is human. Errors of different types may have crept in. Any suggestion(s) for the improvement of the book will be cordially and gratefully accepted.

Acknowledgment

Heartiest thanks and gratitude to Professor S. Councilman, Prof. & Ex-Chairman, Department of Mathematics, California State University, Long Beach, U.S.A, and Professor M. Zuhair Nashed, Department of Mathematical Science, University of Delaware, New York, DE 19716, U.S.A for their valuable remarks on the books and sparing useful time & energy.

The author expresses his heartiest and sincere thanks to his colleagues Mr. Obaidullah, Mr. Abdul Maleque, Mr. Ilias, Prof. A.K.M. Kafi Majumder, Prof. Kaykobad, Prof. M. Monimul Huque for supplying him references while he was on sabbatical leave and not permitted to borrow books from his University Libraries. He is very much grateful to Dr. Nurul Islam, University of Western Sydney, Australia , Dr. G. K. Beg , King Fahd University of Petroleum and Minerals, Dhahran, Saudi Arabia, Dr. F. Kabir in Canada and Mr. Abdul Latif, Asstt. Librarian, Dhaka University, Dhaka for their cooperation in supplying many reference papers and books. Cooperation of the staff particularly ex-deputy librarian Mr. Kazi Abdul Hye, librarian M. Zahirul Islam, deputy librarian Mrs. Suraiya Begum, assistant librarians Dr. M. M. Rezaul Islam, Mr. S. M. Zakir Hossain and Mr. Abdul Ohab of BUET library is no less important for time to time helping the author in collecting papers from different sources.

Friend in need

Once the author did not have a PC of his own and was in real crisis with incomplete computer compose. At such a crucial moment Prof. A.K.M.A. Quader came forward with his kind and honourable sympathy and allowed the author to use his official PC. So, the author is really indebted to his act of kindness and remains ever grateful to such a friend in need.

Cordial thanks

Cordial thanks go to Mrs. Sabina Askar (Rupa) and to all Mr. Md. Ashfaque Iftekher Khan, Md. Tariqul Alam, M. Shafiqul Islam, M.A. Momin, etc. for their pains in computer typing of the manuscript, Prof. Mohammad Ali Choudhury, EEE department and Mr. Saidul Islam, CSE department for their technical support in computer problems. Final editing was performed by the author himself. His younger son, Mr. Ziaul Ahsan (Mithul) and nephew, Schachi of UK helped a lot during editing. So, the author is thankful for their sincere cooperation.

Cordial thanks to the management and associates of the Publisher for publishing the book timely and efficiently.

BUET, Dhaka. A. K. Hazra

Table of contents

CHAPTER ONE

FUNDAMENTAL DEFINITIONS

CHAPTER TWO

BASIC ALGEBRAIC OPERATIONS

CHAPTER THREE

SQUARE MATRIX AND ITS SPECIAL FORMS

CHAPTER FOUR

SUBMATRIX AND PARTITION MATRIX

CHAPTER FIVE

OTHER MATRIX OPERATIONS

CHAPTER SIX
DETERMINANTS

CHAPTER SEVEN

SYSTEM OF SIMULTANEOUS LINEAR ALGEBRAIC EQUATIONS

CHAPTER EIGHT
COFACTOR, ADJOINT AND INVERSE MATRICES

CHAPTER NINE

ELEMENTARY OPERATIONS AND MATRICES

CHAPTER TEN

RANK AND NULLITY OF A MATRIX

CHAPTER ELEVEN

VECTOR(LINEAR) SPACES

CHAPTER TWELVE

LINEAR TRANSFORMATION (OPERATORS)

CHAPTER THIRTEEN

EIGENVALUES AND EIGENVECTORS

CHAPTER ONE

FUNDAMENTAL DEFINITIONS

Chapter summary: *The first chapter begins with a historical note including few words about the relation between matrix theory and linear algebra. Next comes sequentially A B C of matrices, classification of matrices, definitions of scalar, row matrix, column matrix, analogy of row/column matrix with geometric vectors, alternative definition of matrix, principal diagonal, trace of a matrix, equality of matrices and its properties, inequality of matrices, zero matrix, s.u. vectors, one vector, J-matrix, matrix unit, trapezoidal matrices. Matrices in different disciplines.*

1.1 Historical note

Matrix theory and linear algebra are twin. Someone calls them synonymous. Matrix theory is an essential part of linear algebra and linear algebra may be thought as an application of matrix theory. These two branches of mathematics go hand in hand.

Linear algebra is the study of homomorphism [a transformation T from one algebraic system to another that preserves the operation(s)] of finite dimensional vector spaces, including the characterization of matrices representing the same homomorphism relative to different bases. The study of vector spaces & linear transformation gives a geometrical basis to matrix theory. So, the study of matrix theory provides the fundamentals of both topics.

The term "matrix" dates back to 1850 due to the famous English mathematician James Joseph Sylvester. He calls "His rectangular array of terms" matrix . Matrix means " Something that represents the point or location from which another thing originates or develops". Later Sir William Rowan Hamilton in 1853 and Arthur Cayley in 1858 make notable contributions to the subject. Normally discoverers remain ignorant about the applications of their discoveries. In this case, it is not an exception .

Till 1925 , matrix theory remains confined to pure mathematicians as an algebra of manipulation. In 1925 , physicists Heisenberg , Born and Jordan first introduce matrix in quantum mechanics and in 1934 aeronautical engineers W. J. Duncan , A. R. Collar apply matrix algebra to engineering sciences. Since then matrix become omnipresent in mathematical areas of applications including algebra, differential equations, calculus of several variables, probability and statistics, optimization and graph theory, biological sciences, economics and management sciences etc.

1.2 A B C of matrices

The matrix and the other related terms are defined and explained in this section.

1.2.1 Matrix : Matrix is an arrangement of some information in a rectangular / square *array* and obeys its own rules of addition and multiplication. Commonly, matrix is defined just as a rectangular *array* of numbers.

The word "array" in the definition bears special significance. Two meanings of array are

(i) a set of identical (or very similar) elements arranged in a *regular pattern*,

(ii) order of soldiers in a parade ground or battle field in which a soldier cannot change his position at his own will.

Both these meanings imply that any change either in the pattern or in the order of soldiers, results into a new matrix and hence the original significance of the matrix changes to a new one. Thus the words "matrix" and "array" together strictly emphasis that once an information (a soldier) is placed, it (he) is not supposed to change its (his) location in the given pattern.

The plural of matrix is matrices or matrixes. It is to be noted that a matrix is simply a two dimensional array, no value or number has been assigned to or associated with a given matrix; matrix as a whole, not the individual information in it, constitutes a mathematical object for mathematical analysis.

1.2.2 Elements : The information are individually called *elements* or *entities*. They may be positive or negative, real or complex numbers, algebraic expressions, matrices or matrix expressions even. The physical interpretations of those element is essential. So, matrices may arise from a variety of sources and in variety of forms. A change in *any one* of the elements *at least* produces a different matrix. Hence the matrix implies the idea of the elements as a ***whole***. It is something more than the arrays of their elements and separate rules of addition and multiplication for matrices are developed for their uses.

1.2.3 Rows and Columns : Elements along a horizontal line together constitutes a *row* while those along a vertical line together constitutes a *column* of a matrix. Location of an element is the point of intersection of a row and a column.

1.2.4 Interior elements : Elements that are neither in the first or last row (or column) are called *interior elements*.

1.2.5 Width and height : The number of elements in a row, i.e. number of columns, is called the *width* of the matrix while those in a column , i.e. number of rows,is called the *height* of the matrix.

1.2.6 Illustration : If a matrix consists of m rows and n columns (cols) or in other words, if a matrix has width n and height m (i.e. mn elements altogether) then it is called an $m{\times}n$ (read as m by n) matrix. Thus

$$A = \begin{bmatrix} a_{11} & a_{12} & \cdots & a_{1n} \\ a_{21} & a_{22} & \cdots & a_{2n} \\ \vdots & \vdots & \vdots & \vdots \\ a_{m1} & a_{m2} & \cdots & a_{mn} \end{bmatrix} = \left[a_{ij}\right]_{m\times n}, \tag{1.1}$$

is an $m{\times}n$

$$i = 1, 2, \cdots, m \quad \text{and}\; j = 1,2, \cdots, n.$$

matrix in double subscript notation. Note that among the two subscripts i and j, the first one, i defines row position from top to bottom along the height and the second one, j defines column position from left to right along the width. Both m and n are positive integers. The elements

a_{11}, a_{12} etc. are read as 'a' one-one, 'a' one-two etc. and *not* as 'a' eleven, 'a' twelve etc. The elements a_{ij} , $i \neq 1$, and $j \neq 1$, are interior elements.

Sometimes a comma is inserted between the numbers representing i and j if numbers / number consists of more than one digit such as $a_{2,12}$ which clearly signifies the elements in second row and twelfth column otherwise confusion may arise whether it belongs to 21st row and second column or 2nd row and 12th column if it is written simply as a_{212}.

Any of the above a_{ij}'s *cannot* be blank or haphazardly arranged in a matrix. Thus

$\begin{bmatrix} 2 & 1 \\ 3 & \end{bmatrix}$, $\begin{bmatrix} & 7 & \\ 8 & & 9 \end{bmatrix}$, $\begin{bmatrix} & 0 & \\ & 1 & \\ 5 & & 4 \\ & 2 & \end{bmatrix}$, months of calender , do not form matrices. But

incidentally the month of February , 1998 , forms a matrix shown below :

February , 1998

Sun	Mon	Tues	Wed	Thirs	Fri	Sat
1	2	3	4	5	6	7
8	9	10	11	12	13	14
15	16	17	18	19	20	21
22	23	24	25	26	27	28

1.2.7 Notations : (a) Generally a matrix is represented by

(i) roman capital letters A, B, C, D, P, X, Y etc., which do not show the elements;

(ii) the elements explicitly with square brackets as in (1.1). In compact form it is written as $[a_{ij}]$ as in (1.1) also where the subscripts $i = 1, 2, \cdots, m$; indicate number of rows and $j = 1, 2, \cdots, n$, indicate that of columns. The subscripts i and j of element a_{ij} together called the *address* or *position* or *location* of the element as the ith row and jth column. This is the important feature of matrix which provides a grid for storing the objects a_{ij}.

(iii) formula of the elements which exhibits the matrix such as

$$a_{ij} = \begin{cases} 0\,, & i = j \\ 1\,, & i,\ j \text{ both even or odd but not equal,} \\ -1\,, & \text{either } i \text{ or } j \text{ is odd.} \end{cases}$$

i.e. $$\left[a_{ij}\right]_{3\times 4} = \begin{bmatrix} 0 & -1 & 1 & -1 \\ -1 & 0 & -1 & 1 \\ 1 & -1 & 0 & -1 \end{bmatrix}.$$

Other symbols for enclosing the elements are a pair of parentheses,(:::), a pair of double vertical bars , || :::: || etc.

(b) Vectors are nomally represented by boldfaced small lĕtters **a, b, c,** $\cdots$,**x** , **y** , etc. while row and column vectors of a matrix A are represented by $\mathbf{A}_{i.}$ (or $\mathbf{A}_{i*}$) and $\mathbf{A}_{.j}$ (or $\mathbf{A}_{*j}$) (cf. sec.1.5, 1.8 etc.).

(c) Determinants , minors & cofactors are represented by capital italic letters A , B, $\cdots$, A_{ij} , while determinant, sets and transformations are represented by detA , S , T etc.

Note: *Better formula representing matrix under order below.*

1.2.8 Order: The order (dimension or size) of a matrix is the *ordered pair* of the number of rows and the number of columns present in the matrix. Thus the order of the matrix in (1.1) is the ordered pair (m, n) and is usually written as $m \times n$. The order of a matrix is usually displayed using subscripts. Any one of the following forms may be conveniently used for explicit display of order of a matrix. Following notations are used for representing an m×n matrix;

$$ {}_{m}^{n}\mathrm{A};\ {}_{m}\mathrm{A}_{n};\ \mathrm{A}_{m\times n} \equiv \left[a_{ij}\right]_{m\times n};\ \mathrm{A}_{(m,n)} \equiv \left[a_{ij}\right]_{m\times n}; $$

$$ \mathrm{A}_{(m,n)} \equiv \left(a_{ij}\right)_{(m,n)};\ \mathrm{A}_{(m,n)} \equiv \left\|a_{ij}\right\|_{(m,n)} \quad \text{etc.} $$

If $m = n$ then the above notations simplify to $\mathrm{A}_n = [a_{ij}]_n$ only and the matrix is then called a *square* matrix while others are *rectangular*. Algebraic formula involving i and j with order symbols m and n together define a matrix in short. Thus $\mathrm{A}_3 = \left(i^{j+1}\right)$ and $\mathrm{B}_{2\times 3} = (i+j-1)$ define the matrices

$$ \mathrm{A}_3 = \begin{bmatrix} 1 & 1 & 1 \\ 2^2 & 2^3 & 2^4 \\ 3^2 & 3^3 & 3^4 \end{bmatrix} \text{ and } \mathrm{B}_{2\times 3} = \begin{bmatrix} 1 & 2 & 3 \\ 2 & 3 & 4 \end{bmatrix} \text{ in short .} $$

The ith row of the matrix A is denoted by $\mathbf{A}_{i*}$ or $\mathrm{A}_{i.}$ and jth column by $\mathbf{A}_{*j}$ or $\mathbf{A}_{.j}$. Thus the second row of A_3 and $\mathrm{B}_{2\times 3}$ are

$$ \mathbf{A}_{2.} = [1^2\ 2^2\ 3^2] \text{ and } \mathbf{B}_{2.} = [2\ 3\ 4] $$

respectively and second column of those matrices are

$$ \mathbf{A}_{.2} = \begin{bmatrix} 1 \\ 2^3 \\ 3^3 \end{bmatrix} \text{ and } \mathbf{B}_{.2} = \begin{bmatrix} 2 \\ 3 \end{bmatrix} \text{ respectively.} $$

1×1 matrix is written as [a] or simply by 'a' which is technically incorrect. This causes no harm and is very convenient.

1.2.9 Equality of orders : Two matrices are said to be of the *same* or *equal order* if they have same number of rows and same number of columns. Thus, the matrices

$$ \mathrm{A} = \begin{bmatrix} 1 & 0 & 1 & 2 \\ -1 & 0 & 1 & 3 \\ 0 & 1 & -1 & 4 \end{bmatrix} \text{ and } \mathrm{B} = \begin{bmatrix} 2 & 3 & -1 & 4 \\ -2 & -3 & 1 & -4 \\ 0 & 0 & 0 & 0 \end{bmatrix} $$

are of same order both being 3×4 matrices. Equality of matrices does not mean equality of orders (cf. section 1.11).

1.3 Classification of matrices

Matrices can be classified into *five* major groups on the basis of

(i) order or size,

(ii) nature of elements,
(iii) number of zero elements,
(iv) number of rows/columns and
(v) sign of elements.

They are discussed below.

(i) Size : Matrices are classified into *rectangular* and *square* matrices on the basis of size. If the number m of rows and the number n of columns are unequal then the matrix is *rectangular.* If $m = n$, the matrix is *square*. Different special forms of square matrices are discussed in chapter three while special forms of rectangular matrices are mentioned in sec. 1.13 , 1.16 , 1.17, 1.18 of this chapter.

(ii) Nature of elements : On the basis of the nature of elements — real and complex, matrices are classified into R-matrix, C-matrix and purely imaginary matrix respectively. These matrices may be rectangular or square depending on size .

R-matrix : A matrix is an R-matrix (or a real matrix) if all the elements are real. Thus,

$$\begin{bmatrix} a & b & c \\ x & y & z \\ 1 & 0 & 1 \end{bmatrix}, \begin{bmatrix} 1 & 2 & 3 & 4 \\ 0 & 0 & 0 & 0 \end{bmatrix}, \begin{bmatrix} 1 & 1 & 1 \\ 1 & 1 & 1 \\ 1 & 1 & 1 \end{bmatrix} \text{ etc.}$$

are all examples of R-matrices, provided a, b, c, x, y, z in the first matrix are all real.

C-matrix : A matrix is a C-matrix (or a complex matrix) if all the elements are complex. Thus

$$\begin{bmatrix} 1+i & 0 & 1 \\ 2 & 2-i & i \end{bmatrix}, \begin{bmatrix} 1 & 2 & 3 \\ 0 & 0 & 0 \\ 1 & 0 & 1 \end{bmatrix} \text{ etc.}$$

are examples of C-matrices. *Note* that real numbers are complex numbers with zero imaginary part and so R-matrices are C-matrices but not the converse is true.

Purely imaginary matrix: A matrix is purely imaginary if all the elements of the matrix are purely imaginary. Thus

$$\begin{bmatrix} i & -i \\ -i & 2i \end{bmatrix} \text{ and } \begin{bmatrix} i & -2i & 3i \\ -2i & i & -4i \end{bmatrix}$$

are examples of purely imaginary matrices.

(iii) Number of zero elements: Depending on the number of zero elements, matrices are further classified into *sparse* and *dense* matrices. These matrices may be rectangular or square, real or complex.

Sparse matrix: A sparse matrix is one that has preponderance of zero elements i.e. most of the elements of the matrix are zero. It may be rectangular or square. This type of matrix mostly arises in models where local connection or local influence principle exists. Thus in bridge model, electric network, sociometric model, numerical analysis often given rise to sparse matrices. Examples of sparse matrices are

$$A = \begin{array}{c} \\ 1 \\ 2 \\ 3 \\ 4 \\ 5 \\ 6 \\ 7 \\ 8 \\ 9 \\ 10 \\ 11 \\ 12 \\ 13 \end{array} \begin{array}{c} \begin{array}{cccccccccccc} 1 & 2 & 3 & 4 & 5 & 6 & 7 & 8 & 9 & 10 & 11 & 12 \end{array} \\ \begin{bmatrix} 0 & 0 & 0 & 0 & 0 & 0 & 0 & 0 & 0 & 0 & 0 & 0 \\ 0 & 0 & 0 & 0 & 0 & 0 & 0 & 0 & 0 & 0 & 1 & 0 \\ 1 & 0 & 0 & 1 & 0 & 0 & 0 & 0 & 0 & 0 & 0 & 0 \\ 0 & 0 & 0 & 0 & 1 & 0 & 0 & 0 & 0 & 0 & 0 & 0 \\ 0 & 0 & 0 & 0 & 0 & 0 & 0 & 0 & 0 & 0 & 0 & 0 \\ 0 & 0 & 0 & 0 & 0 & 0 & 0 & 0 & 0 & 0 & 0 & 1 \\ 0 & 0 & 0 & 0 & 1 & 0 & 0 & 0 & 0 & 0 & 0 & 0 \\ 0 & 0 & 0 & 0 & 0 & 0 & 0 & 1 & 0 & 0 & 0 & 0 \\ 0 & 0 & 0 & 0 & 0 & 0 & 0 & 1 & 0 & 0 & 0 & 0 \\ 0 & 0 & 0 & 0 & 0 & 1 & 0 & 0 & 0 & 0 & 0 & 1 \\ 0 & 0 & 1 & 0 & 0 & 0 & 0 & 1 & 0 & 0 & 0 & 0 \\ 0 & 0 & 0 & 1 & 0 & 0 & 0 & 0 & 0 & 0 & 1 & 0 \\ 0 & 0 & 0 & 0 & 1 & 0 & 0 & 0 & 0 & 0 & 0 & 0 \end{bmatrix} \end{array}, \text{ diagonal matrices etc.}$$

In matrix A there are only *fifteen* 1's (encircled) out of 12×13 = 156 elements and in diagonal matrix there are only n nonzero elements out of n^2 elements. So, there exist preponderance of zero elements in those examples and hence they are sparse matrices.

Addition, subtraction and multiplication of sparse matrices are very easy. In addition and subtraction corresponding nonzero elements are required to be added & subtracted. From multiplication rule (cf. chapter two)

$$A = [a_{ik}]_{m\times n},\ B = [b_{kj}]_{n\times p},\ C = [c_{ij}]_{m\times p} = \left[\sum_{k=1}^{n} a_{ik} b_{kj}\right]_{m\times p}$$

It follows that nonzero elements of the kth column of A and that of kth rows of B produces nonzero elements of C.

Dense matrix: A dense matrix is one that has preponderance of nonzero element i.e. most of the elements of the matrix are nonzero. Algebraic operations of addition, subtraction and multiplication are laborious for dense matrices.

(iv) Number of rows/columns: On the basis of the finiteness or infiniteness of the number of rows or columns or both, matrices are again classified into *finite* and *infinite* matrices.

Finite matrix: A matrix with finite number of rows and columns is called a *finite matrix.* All the examples cited above are finite matrices.

Infinite matrix: A matrix is called an *infinite matrix* if its number of rows or its number of columns is infinite but the elements are finite.

Note that if an element of a matrix is infinite then the matrix does not exist.

For example the matrix

$$A = \begin{bmatrix} a_{11} & a_{12} & \cdots & a_{1n} \\ a_{21} & a_{22} & \cdots & a_{2n} \\ \vdots & \vdots & \vdots & \vdots \\ a_{i1} & a_{i2} & \cdots & a_{in} \\ \vdots & \vdots & \vdots & \vdots \end{bmatrix}$$

is an infinite matrix where $i = 1, 2, 3, \cdots, \infty$ and a_{ij} are all finite.

(v) Sign of elements: Matrices with nonnegative elements arise in many important applications such as probability theory for the investigation of Markov chains (Stochastic matrices). Theory of small oscillations of elastic systems (oscillation matrices), numerical analysis, elliptic partial differential equation, economics, operation research etc. Under these matrices "positive" and "nonnegative" matrices need discussion. These matrices are also used to describe "positive definite", "nonnegative definite" (positive indefinite) matrices (cf. sec.3.10, 6.17, 11.10.1, 15.4.5 etc.) which will not appear together with positive or nonnegative matrices. So, there is no danger of confusion. On the basis of sign of the elements, the following *four* types are encountered.

Positive matrix : A real rectangular $m \times n$ matrix A is *positive* if all its elements are positive i.e. if the elements $a_{ij} > 0$, $i = 1, 2, \cdots, m$; $j = 1, 2, \cdots, n$. It is denoted by $A > 0$. The matrix

$$A = \begin{bmatrix} 2 & 3 & \frac{1}{2} & 7 \\ 1 & 5 & \frac{1}{4} & 9 \end{bmatrix}$$

is a positive matrix. Because all its elements are positive and none is zero or negative.

Likewise a *positive vector* **x**, has all its components positive and is denoted by $\mathbf{x} > 0$. The vector $\mathbf{x} = \begin{bmatrix} 2 & 1 & 3 & 5 \end{bmatrix}$ is a positive vector.

Nonnegative matrix : A real rectangular $m \times n$ matrix A is *nonnegative* if all its elements are nonnegative i.e. if the elements $a_{ij} \geq 0$, $i = 1, 2, \cdots, m$; $j = 1, 2, \cdots, n$. It is denoted by $A \geq 0$. The matrix

$$A = \begin{bmatrix} 2 & 3 & \frac{1}{2} & 0 \\ 0 & 4 & 9 & 1 \\ 1 & 5 & \frac{1}{4} & 0 \end{bmatrix}$$

is a nonnegative matrix. Because each element is either greater than or equal to zero and none is negative.

Likewise a *nonnegative vector* **x**, has all its elements nonnegative and is denoted by $\mathbf{x} \geq 0$. The vector $\mathbf{x} = \begin{bmatrix} 2 & 1 & \frac{1}{2} & 0 \end{bmatrix}$ is nonnegative.

Semipositive matrix : It is a special case of nonnegative matrix. A nonnegative matrix is semipositive if $a_{ij} \geq 0$ and *at least* one element of every row and column is strictly positive. The matrix

$$A = \begin{bmatrix} 0 & 2 & 0 & 0 & 0 \\ 0 & 0 & 4 & 1 & 3 \\ 1 & 3 & 5 & 0 & 0 \end{bmatrix}$$

is a semipositive matrix. It is also known as *irreducible* (or *undecomposable*) matrix.

Mod A : If the elements of an $m \times n$ matrix $A = \left[a_{ij}\right]$ are replaced by the corresponding modulus or magnitude then the new positive or nonnegative matrix is called *modulus* of matrix A and is denoted by mod A.

Similarly mod **x** is obtained by replacing the components of the vector **x** by the corresponding modulus or absolute values of the components. *Do not* confuse with modulus of the vector $|\mathbf{x}| = \sqrt{x_1^2 + x_2^2 + \cdots + x_n^2}$, which is different from mod $\mathbf{x} = (|x_1|, |x_2|, \cdots, |x_n|)$.

1.4 Scalar

A scalar is an 1 × 1 ordered matrix and it is simply an element enclosed within braces such as (70). So, to distinguish it from an array or a matrix of higher order than 1 × 1, it is called a *scalar* in matrix theory. The scalar (70) is completely equivalent to the real number 70 as can be easily seen from the operations of matrices. This "scalar" and the "scalar matrix" (cf. sec.3.3) are completely different.

1.5 Column matrix

A column matrix is a matrix of order $m \times 1$. It consists of m rows and one column. It is also called *column vector* or simply a *vector* with m components expressed as m-vector. Thus

$$\mathbf{A}_{.1} = \begin{bmatrix} a_{11} \\ a_{21} \\ \vdots \\ a_{m1} \end{bmatrix}$$

is a column matrix or m-vector. Thus vector is a special kind of matrix.

1.6 Row matrix

A row matrix is a matrix of order $1 \times n$. It consists of one row and n columns. It is also called a *line matrix* or *row vector* with n components expressed as n-vector and

$$\mathbf{A}_{1.} = [a_{12} \quad a_{13} \quad \cdots \quad a_{1n}]$$

is an example of a row matrix or n-vector.

1.7 Analogy of column/row matrix with geometric vectors

When m and n are equal to 2 or 3 both column and row matrices become identical with vectors of vector analysis and coordinates of a point in geometry. So, they are termed as *geometric vectors.*

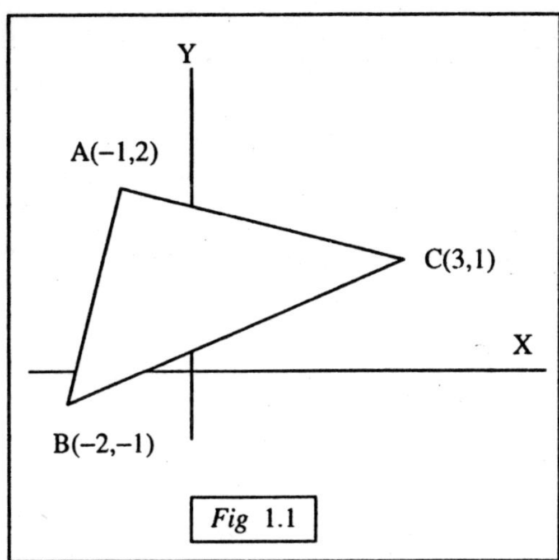

Fig 1.1

Thus a triangle ABC with coordinates A(−1,2), B(−2,−1) and C(3,1) in *figure* 1.1 can be represented either by a 2×3 matrix

$$\begin{bmatrix} -1 & -2 & 3 \\ 2 & -1 & 1 \end{bmatrix}$$

if coordinates correspond to columns, or by a 3×2 matrix

$$\begin{bmatrix} -1 & 2 \\ -2 & -1 \\ 3 & 1 \end{bmatrix}$$

if coordinates correspond to rows.

1.8 Alternate definition of matrix

The matrix in (1.1) is *either* a side by side collection or representation of n column vectors each with m components, *or* a stack of m row vectors each with n components enclosed by square brackets.

Thus matrix (1.1) can be written *either* as $A = \begin{bmatrix} \mathbf{x}_{.1} & \mathbf{x}_{.2} & \cdots & \mathbf{x}_{.j} \end{bmatrix}$, where $\mathbf{x}_{.j}$ are collection of n column vectors each with m components i.e.

$$\mathbf{x}_{.j} = \begin{bmatrix} a_{1j} \\ a_{2j} \\ \vdots \\ a_{mj} \end{bmatrix}, \quad j = 1, 2, \cdots, n \tag{1.2}$$

or as
$$A = \begin{bmatrix} \mathbf{y}_{1.} \\ \mathbf{y}_{2.} \\ \vdots \\ \mathbf{y}_{m.} \end{bmatrix},$$

where $\mathbf{y}_{i.}$ are stack of m vectors each with n components i.e.

$$\mathbf{y}_{i.} = \begin{bmatrix} a_{i1} & a_{i2} \cdots a_{in} \end{bmatrix},\ i = 1, 2, \cdots, m. \tag{1.3}$$

In (1.2) and (1.3) vectors $\mathbf{x}_{.j}$ and $\mathbf{y}_{i.}$ are expressed with single subscript and a dot "." which indicates that row/column is free. Hence every row of A is a row vector and every column of A is a column vector. It should be *noted* that matrices and vectors are conceptually quite different. But the identification as above is useful for manipulation.

Since $\mathbf{A}_{i.}$ and $\mathbf{A}_{.j}$ denote the *i*th row and *j*th column of a matrix (cf. sec.1.2.8). So for a fixed i, $1 \le i \le m$, from (1.3), $\mathbf{A}_{i.} = \begin{bmatrix} a_{i1} & a_{i2} & \cdots & a_{in} \end{bmatrix}$ and for a fixed $j, 1 \le j \le n$, from (1.2) by

$$\mathbf{A}_{.j} = \begin{bmatrix} a_{1j} \\ a_{2j} \\ \vdots \\ a_{nj} \end{bmatrix} = \mathbf{x}_{.j}\ .$$

Therefore, the matrix (1.1) can be written *either*

$$\mathrm{A} = \begin{bmatrix} \mathbf{A}_{.1} & \mathbf{A}_{.2} & \cdots & \mathbf{A}_{.n} \end{bmatrix} \ \ or \ \ \mathrm{A} = \begin{bmatrix} \mathbf{A}_{1.} \\ \mathbf{A}_{2.} \\ \vdots \\ \mathbf{A}_{m.} \end{bmatrix}.$$

1.9 Principal diagonal

Let A = $[a_{ij}]$ be an $m \times n$ matrix and let h be the smaller of m and n. The *principal* or *main diagonal* or *simply diagonal* of the matrix A is the line running through the equal subscripted elements $a_{11}, a_{22}, \cdots, a_{hh}$ from left top corner a_{11} down to the right most bottom elements a_{hh}. This diagonal does not coincide with the geometrical diagonal of a rectangle. The elements $a_{11}, a_{22}, \cdots, a_{hh}$ are called the *diagonal elements* and is sometimes expressed by the column matrix

$$\text{diag A} = \begin{bmatrix} a_{11} \\ a_{22} \\ \vdots \\ a_{hh} \end{bmatrix}$$

where diag A stands for "the diagonal of A".

Thus diagonal elements of $[a_{ij}]_{3\times4}$ are $a_{11}, a_{22}, \cdots$. Also the diagonal elements of

$$\mathrm{A} = \begin{bmatrix} 1 & 0 & 1 \\ 2 & 3 & 4 \\ 0 & 1 & 9 \end{bmatrix},\ \mathrm{B} = \begin{bmatrix} 2 & 3 \\ -2 & 1 \\ 6 & 4 \end{bmatrix},\qquad \mathrm{C} = \begin{bmatrix} 1 & 2 & 3 \end{bmatrix}$$

Diagonal Diagonal Diagonal

are 1, 3, 9 ; 2, 1 and 1 respectively and can be written as

$$\text{diag A} = \begin{bmatrix} 1 \\ 3 \\ 9 \end{bmatrix}, \quad \text{diag B} = \begin{bmatrix} 2 \\ 1 \end{bmatrix} \text{ and } \text{diag C} = [1]$$

respectively, while the elements with unequal subscripts a_{ij} , $i \neq j$ are called *off diagonal* elements.

1.10 Trace of a matrix

Trace or *spur/character* of an $m \times n$ matrix A is the sum of the diagonal elements and denoted by Tr (A), trA or Sp (A). If h is smaller of m and n then trace of a rectangular matrix

$$A_{(m,n)} = \left[a_{ij}\right]_{(m,n)} \text{ is } \operatorname{Tr}(A) = \sum_{i=1}^{h} a_{ii} = \left(a_{11} + a_{22} + \cdots + a_{hh}\right)$$

Therefore,

$$\operatorname{Tr}\left(\begin{bmatrix} 1 & 0 & 1 \\ 2 & 3 & 4 \\ 0 & 1 & 9 \end{bmatrix}\right) = 1 + 3 + 9 = 13\ ,$$

$$\operatorname{Tr}\left(\begin{bmatrix} 2 & 3 \\ -2 & 1 \\ 6 & 4 \end{bmatrix}\right) = 2 + 1 = 3,\ \operatorname{Tr}\left(\begin{bmatrix} 1 & 2 & 3 \end{bmatrix}\right) = 1.$$

More on trace of square matrix are available in section 3.1.

1.11 Equality of matrices

Two matrices are equal if

(i) they are of the same order and

(ii) the corresponding elements are equal.

Thus matrices $A = [a_{ij}]_{m \times n}$ and $B = [b_{ij}]_{p \times q}$ are equal i.e. $A = B$ if

(i) $m = p$ and $n = q$,

(ii) $a_{ij} = b_{ij}$ for all i and j

i.e. the elements satisfy mn equations between them. So, the equality of two matrices is the compact representation of mn scalar equations after the first condition (i) has been satisfied. Note that equal matrices are of equal order but equal order does not imply equality of matrices.

1.11.1 Properties of equal matrices: The above definition of equality of matrices and properties of equality in ordinary algebra together give the following *four* properties of equal matrices:

(i) Determinative : If A and B are any two matrices then either $A = B$ or $A \neq B$.

(ii) Reflexive : If A is any matrix, then $A = A$.

(iii) Symmetric : If $A = B$, then $B = A$.

(iv) Transitive : If $A = B$ and $B = C$, then $A = C$.

Thus equality of matrices is determinative, reflexive, symmetric & transitive. While a relation between pairs of mathematical objects which satisfy the above four properties is called *equivalence* relation. Thus the equivalence properties of ordinary algebra are applicable to matrices.

The examples of equal matrices are

$$\begin{bmatrix} 1 & 2 & -1 \\ 3+1 & 0 & 1+2 \end{bmatrix}, \quad \begin{bmatrix} \frac{4}{4} & 2 & -1 \\ 4 & 0 & 3 \end{bmatrix}.$$

Because both simplify to

$$\begin{bmatrix} 1 & 2 & -1 \\ 4 & 0 & 3 \end{bmatrix}.$$

1.12 Inequality of matrices

Inequality of matrices exists between real matrices. Because real numbers or variables can be compared as >, ≥, <, ≤, but not the imaginary numbers or variables. Therefore, a real matrix A is *greater than* (>) another real matrix B if and only if each element of A is greater than the corresponding element of B. Similarly, inequality relations for ≥, < or ≤ can be stated by changing "greater than" to "greater than or equal to", "less than" or "less than or equal to" in the above statement. Thus,

$$\text{if } A = \begin{bmatrix} 1 & 1 \\ 3 & 2 \end{bmatrix} \text{ and } C = \begin{bmatrix} -1 & 0 \\ 2 & 2 \end{bmatrix}, \text{ then } A \geq C.$$

Because each element of A is either greater than or equal to the corresponding element of C.

1.13 Zero / null matrix

A matrix is said to be a zero / null matrix if all the elements are zero. It is denoted by 0 (zero). It is of arbitrary order. A 3×2 zero matrix is

$$0_{3\times 2} = \begin{bmatrix} 0 & 0 \\ 0 & 0 \\ 0 & 0 \end{bmatrix}$$

and $A = [a_{ij}]_{m\times n}$ is a null matrix if $a_{ij} = 0$ for $i = 1, \cdots, m$ and $j = 1, \cdots, n$. It is to be noted that a null matrix is ***not equal*** to the scalar quantity zero unless it is a scalar i.e. 1×1 matrix.

1.14 Standard unit (s.u.) vectors

Standard unit (s.u.) vectors in different dimensions are the concern of this section

One dimensional : Usually numbers are pictorially represented along a straight line, called number line. The positive and the negative numbers are located at unit distances to the right and the left of the origin as follows (*figure* 1.2).

Fig. 1.2

The numbers 1, 2, 3 etc. are at distances 1, 2, 3 etc. units to the right of 0 (zero) or origin. Similarly, the numbers –1, –2, –3 etc. are at distances 1, 2, 3 etc. units to the left of origin. So, these numbers 1, 2, 3 etc. and –1, –2, –3 etc. possess both direction and magnitude. Hence they may be considered as one dimensional vectors. Also the unit of measurement of these numbers is clearly 1(one). Therefore, 1 is the one dimensional *unit vector*.

Two dimensional : The two dimensional vectors are represented in a two dimensional XOY plane (*figure* 1.3).

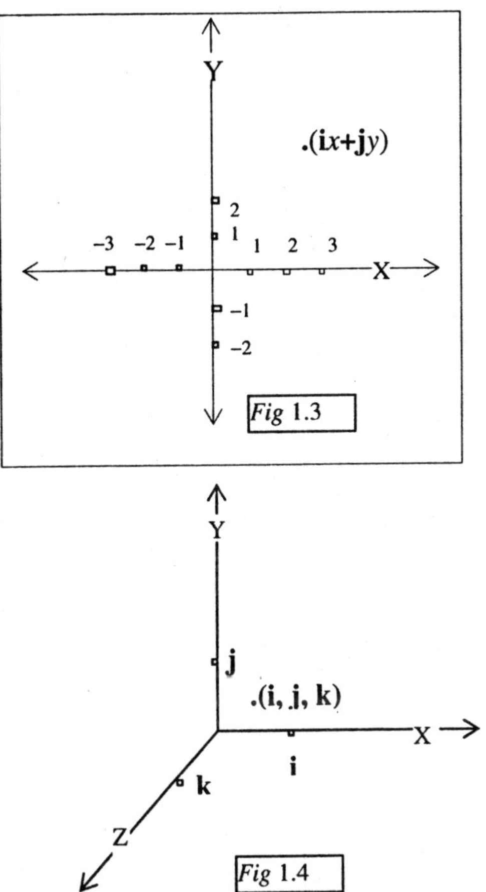

Fig 1.3

Fig 1.4

The unit vectors are **i** = (1 , 0) and **j** = (0 , 1).

Three dimensional : The three dimensional vectors are represented in a three dimensional orthogonal space (*figure* 1.4).

The unit vectors are **i** = (1 , 0 , 0) , **j** = (0 , 1 , 0) and **k** = (0 , 0 , 1).

***n*-dimensional :** From the analogy of 1, 2, 3-dimensional unit vectors, if $\mathbf{e}_i$ denotes the ith n-dimensional unit vector, then

$$\left.\begin{array}{l} \mathbf{e}_i = (0,\ 0,\ \cdots,\ \underset{\substack{\uparrow \\ i\text{th position}}}{1},\ 0,\ 0,\ \cdots) \end{array}\right\}. \text{ Specifically,}$$

$$\left.\mathbf{e}_1 = (\underset{\substack{\uparrow \\ \text{1st position}}}{1},\ 0,\ 0,\ \cdots\ \underset{\substack{\uparrow \\ n\text{th position}}}{,0)}\right\}, \qquad \left.\mathbf{e}_2 = (0,\ \underset{\substack{\uparrow \\ \text{2nd position}}}{1},\ 0,\ \cdots\ \underset{\substack{\uparrow \\ n\text{th position}}}{,0)}\right\},$$

$$\left.\mathbf{e}_i = (0,\ 0,\cdots\ \underset{\substack{\uparrow \\ i\text{th position}}}{1},\ \underset{\substack{\uparrow \\ n\text{th position}}}{\cdots,0)}\right\}, \qquad \left.\mathbf{e}_n = (0,\ 0,\ 0,\ \cdots\ ,0\cdots,\underset{\substack{\uparrow \\ n\text{th position}}}{1)}\right\}$$

are the standard unit vectors.

So one dimensional unit vector 1 in association with $(n-1)$ zeros produces other standard n-dimensional unit vectors and 1 moves according to the position of unit vectors among the $(n-1)$ zeros. Thus in ith unit vector 1 moves to ith position from the first position in $\mathbf{e}_1$ keeping $(i-1)$ zero to the left and $(n-i+1)$ zero to the right.

A caution : Unit vectors and standard unit vectors should not be confused. they are two different objects. A unit vector is expressible in terms of standard unit vectors but not the converse. Thus unit vector $\begin{bmatrix}\frac{3}{4} & \frac{4}{5}\end{bmatrix} = \frac{3}{4}\begin{bmatrix}1 & 0\end{bmatrix} + \frac{4}{5}\begin{bmatrix}0 & 1\end{bmatrix}$. The common property of these vectors is their magnitude and it is unity which can be easily seen.

1.15 One vector

The one vector is also a column vector with all its elements equal to 1(unity). Thus an n-dimensional one vector is

$$\mathbf{e} = \begin{bmatrix}1\\1\\\vdots\\1\end{bmatrix}_n.$$

Some authors denote $\mathbf{e}$ by $\mathbf{1}_n$. This one vector is the sum of n standard unit vectors $\mathbf{e}_1$, $\mathbf{e}_2$, $\cdots$, $\mathbf{e}_n$. Thus

$$\mathbf{e} = \begin{bmatrix}1\\1\\\vdots\\1\end{bmatrix}_{n\times 1} = \begin{bmatrix}1\\0\\\vdots\\0\end{bmatrix}_{n\times 1} + \begin{bmatrix}0\\1\\\vdots\\0\end{bmatrix}_{n\times 1} + \cdots + \begin{bmatrix}0\\0\\\vdots\\1\end{bmatrix}_{n\times 1} = \sum_{i=1}^{n}\mathbf{e}_i$$

which follows from addition of vectors in vector analysis.

1.16 Unity matrix

If every element of a matrix equals to 1, then the matrix is called a *unity matrix* and is denoted by

$$J_{m\times n} \text{ or } \mathbf{1}_{m\times n} \quad \text{i.e } J_{3\times 4} = \mathbf{1}_{3\times 4} = \begin{bmatrix} 1 & 1 & 1 & 1 & 1 \\ 1 & 1 & 1 & 1 & 1 \\ 1 & 1 & 1 & 1 & 1 \end{bmatrix}$$

It is also known as the J-matrix. The $m\times n$ J-matrix is the side by side collection of m-dimensional n one vector. Thus

$$J_{m\times n} = \begin{bmatrix} J^1_{m\times 1} & J^2_{m\times 1} & \cdots & J^n_{m\times 1} \end{bmatrix} = \begin{bmatrix} \mathbf{1}^1_m & \mathbf{1}^2_m & \cdots & \mathbf{1}^n_m \end{bmatrix}$$

where each superscript 1, 2, $\cdots$, n indicates the position of each vector.

1.17 Matrix unit

A sparse matrix with only one nonzero element 1 in the (i, j)th position and $(mn - 1)$ zero elements is called a *matrix unit*. This matrix is denoted by E_{ij}.
Examples are

$$E_{11} = \begin{bmatrix} 1 & 0 & 0 & \cdots & 0 \\ 0 & 0 & 0 & \cdots & 0 \\ \vdots & \vdots & \vdots & \vdots & \vdots \\ 0 & 0 & 0 & \cdots & 0 \end{bmatrix}_{m\times n}, \quad E_{14} = \begin{bmatrix} 0 & 0 & 0 & 1 \\ 0 & 0 & 0 & 0 \\ 0 & 0 & 0 & 0 \end{bmatrix}_{3\times 4} \quad \text{etc.}$$

It is also called *natural* or *fundamental matrix unit.* Justification of the term "unit" in matrix unit and s.u. vector will appear soon in section 2.8, where they are used as unit of measurement or unit of construction of other matrices and vectors respectively. Graham misnomered the matrix unit as an elementary matrix which is obtained by elementary operations (cf. chapter two).

1.18 Trapezoidal matrices

If the elements below (above) the principal diagonal of a rectangular (square) matrix are zero and those on principal diagonal are zero or not then the matrix is called a *trapezoidal matrix.* Different types of it are given below.

(i) Upper trapezoidal matrix : If the elements *below* the principal diagonal of a rectangular (square) matrix are zero i.e. $a_{ij} = 0$ for $i > j$, then the matrix is called an *upper trapezoidal matrix*. The following are some examples:

$$U_{3\times 4} = \begin{bmatrix} u_{11} & u_{12} & u_{13} & u_{14} \\ 0 & u_{22} & u_{23} & u_{24} \\ 0 & 0 & u_{33} & u_{34} \end{bmatrix}, \quad U_{4\times 3} = \begin{bmatrix} u_{11} & u_{12} & u_{13} \\ 0 & u_{22} & u_{23} \\ 0 & 0 & u_{33} \\ 0 & 0 & 0 \end{bmatrix},$$

$$U_3 = \begin{bmatrix} u_{11} & u_{12} & u_{13} \\ 0 & u_{22} & u_{23} \\ 0 & 0 & u_{33} \end{bmatrix}.$$

(ii) Strictly upper trapezoidal matrix : An upper trapezoidal matrix is *strictly upper trapezoidal* if its principal diagonal elements are zero i.e. $a_{ij} = 0$ for $i \geq j$. Examples are

$$\begin{bmatrix} 0 & u_{12} & u_{13} & u_{14} \\ 0 & 0 & u_{23} & u_{24} \\ 0 & 0 & 0 & u_{34} \end{bmatrix} \text{ and } \begin{bmatrix} 0 & u_{12} & u_{13} \\ 0 & 0 & u_{23} \\ 0 & 0 & 0 \\ 0 & 0 & 0 \end{bmatrix}.$$

(iii) Lower trapezoidal matrix : If the elements *above* the principle diagonal of a rectangular (square) matrix are zero i.e. $a_{ij} = 0$ for $i < j$, then the matrix is called a *lower trapezoidal matrix*. Examples are

$$L_{3\times4} = \begin{bmatrix} l_{11} & 0 & 0 & 0 \\ l_{21} & l_{22} & 0 & 0 \\ l_{31} & l_{32} & l_{33} & 0 \end{bmatrix}, \; L_{4\times3} = \begin{bmatrix} l_{11} & 0 & 0 \\ l_{21} & l_{22} & 0 \\ l_{31} & l_{32} & l_{33} \\ l_{41} & l_{42} & l_{43} \end{bmatrix}, \; L_3 = \begin{bmatrix} l_{11} & 0 & 0 \\ l_{21} & l_{22} & 0 \\ l_{31} & l_{32} & l_{33} \end{bmatrix}.$$

(iv) Strictly lower trapezoidal matrix : A lower trapezoidal matrix is *strictly lower trapezoidal* if its principal diagonal elements are zero i.e. $a_{ij} = 0$ for $i \leq j$. Examples are

$$\begin{bmatrix} 0 & 0 & 0 & 0 \\ l_{21} & 0 & 0 & 0 \\ l_{31} & l_{32} & 0 & 0 \end{bmatrix} \text{ and } \begin{bmatrix} 0 & 0 & 0 \\ l_{21} & 0 & 0 \\ l_{31} & l_{32} & 0 \\ l_{41} & l_{42} & l_{43} \end{bmatrix}.$$

1.19 Applications : matrices in different disciplines

The aim of this section is just to introduce the readers an idea about the occurrence of matrices in the real world. It is not possible to give an exhaustive listing of all applications. It is simply an attempt to motivate the learner showing how matrices are being used and how real life problems can be modeled in the language of mathematics of matrices. Some matrices those occur in different areas or disciplines, are also introduced here.

1.19.1 Matrices in linear algebra : Linear algebra originated in the study of linear equations play, on the other hand, a major part in the study of linear models mainly and nonlinear model partially. These models are studied with properties of matrices.

Matrices are primarily used for compact representation of simultaneous equations. This type of representation is known as *vector-matrix equation*, or simply *matrix equation*. Thus $A\mathbf{x} = \mathbf{b}$ is a vector-matrix equation where A is a matrix, called coefficient matrix and $\mathbf{x}$ and $\mathbf{b}$ are column vectors of variables and constants respectively. Justification of writing matrix equation from simultaneous system will be clear after the knowledge of matrix multiplication (cf. sec. 2.3.5).

(i) **Two types** of matrices that arise in the study of *simultaneous equations* are defined and explained below:

(a) Coefficient matrix : It is a matrix obtained from the coefficients of the unknowns in simultaneous linear algebraic, linear difference and differential equations. The elements a_{ij} of the matrix $[a_{ij}]_{m\times n}$ are the coefficients of the *j*th unkown in the *i*th equation. Thus, for the simultaneous equations

$$\left.\begin{array}{r} 3x_1 - 2x_2 + 3x_3 - x_4 = 1 \\ x_1 + x_2 \qquad\quad - x_4 = 0 \\ 5x_3 + 3x_4 = 9 \end{array}\right\} \tag{1.4}$$

the *coefficient matrix* is

$$A = \begin{bmatrix} 3 & -2 & 3 & -1 \\ 1 & 1 & 0 & -1 \\ 0 & 0 & 5 & 3 \end{bmatrix}$$

The way of writing the coefficient matrix is simply dropping the unknowns and enclosing their coefficients (no need to write '+' signs) within square brackets or like. The matrix A above displays the coefficients of the unknown in the pattern of the equations irrespective of the name of the unknown. Zero is used for missing coefficients.

(b) Augmented matrix : It is the coefficient matrix with an extra column of the elements obtained from the right-hand-side constants of the equations. Thus the augmented matrix of the equation (1.4) is

$$B = \left[\begin{array}{rrrr:r} 3 & -2 & 3 & -1 & 1 \\ 1 & 1 & 0 & -1 & 0 \\ 0 & 0 & 5 & 3 & 9 \end{array}\right]$$

The way of writing is just to delete all unknowns, plus sign and addition of a partition vertical dotted line instead of equal signs. Use zero for missing coefficients or constants.

These two matrices are useful in investigating solution of the system of simultaneous equations (cf. chapter seven).

(ii) **Linear (mapping) transformation** (cf. chapter twelve) is another vital topic in linear algebra. It can be represented and studied by the use of matrix properties. Let a mapping $f : R^3 \to R^3$ be defined by

$$f(x, y, z) = (a_1x + b_1y + c_1z,\ a_2x + b_2y + c_2z,\ a_3x + b_3y + c_3z)$$

where R^3 denote the set of all ordered triple of real numbers and a_i, b_i, c_i, $i = 1, 2, 3$ are constants. This mapping may be expressed as

$$f(\mathbf{v}) = A\mathbf{v} \quad \text{where} \quad \mathbf{v} = \begin{bmatrix} x \\ y \\ z \end{bmatrix} \text{ and } A = \begin{bmatrix} a_1 & b_1 & c_1 \\ a_2 & b_2 & c_2 \\ a_3 & b_3 & c_3 \end{bmatrix}.$$

Method of writing requires knowledge of matrix multiplication (cf. chapter two).

1.19.2 Matrices in Electrical network (Circuit) : In electrical network the symbol –||– represents a source of power (battery or generator) measured in volts and the symbol

represents a resistor measured in ohms. The capital letters represent nodes where three or more wires intersect and i's represent the currents from +ve to – ve by convention between the nodes. The current is determined in terms of resistance and voltages using the following *Kirchhoff's laws* (named after G. R. Kirchhoff, German, 1824-1887).

First law : The sum of the incoming currents is equal to the sum of the outgoing current at every node.

Second law : The algebraic sum of the voltage must equal to the algebraic sum of the voltage drops.

The *voltage drops* E emf for each resistance are given by *Ohm's law* : E = iR where i is the current in amperes and R is the resistance in Ohms in *figure* 1.5. In complicated circuits, the sum of voltage drops across the resistors must equal the *net* emf (the algebraic sum of emfs) i.e. opposite voltages are added and same voltages are subtracted. Let *figure* 1.5 be an electric network. According Kirchhoff's first law, we have

$$i_1 = i_2 + i_3 \quad \text{at node A}$$

Since i_1 is incoming and i_2, i_3 are outgoing,

i.e. $$i_1 - i_2 - i_3 = 0$$

$$i_1 = i_2 + i_3 \quad \text{same as at node A} \tag{1.5}$$

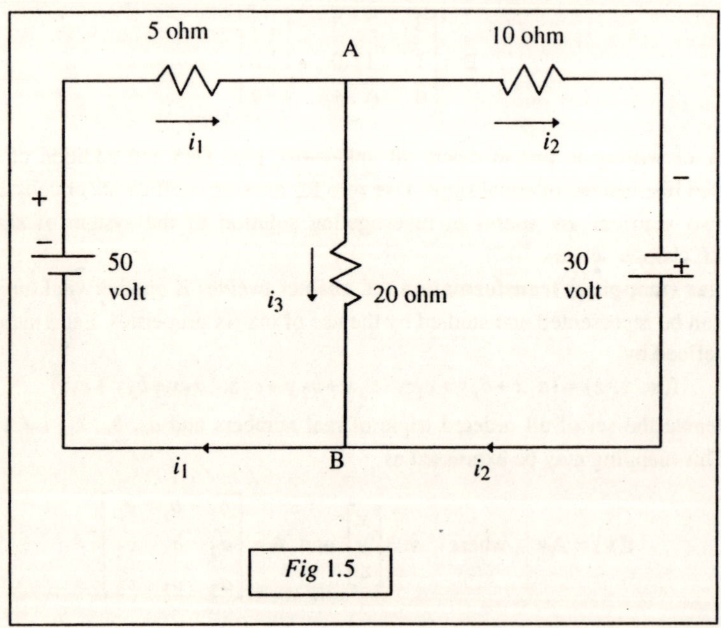

Fig 1.5

At B,

$$5i_1 + \quad 20i_3 = 50 \qquad \text{(left loop)} \qquad (1.6)$$

$$10i_2 - 20i_3 = 30 \qquad \text{(right loop)} \qquad (1.7)$$

$$5i_1 + 10i_2 = 50 + 30 = 80 \qquad \text{(outer loop)} \qquad (1.8)$$

Since voltages are of opposite signs.

The coefficient and augmented matrix corresponding to equations (1.5) to (1.8) are

$$\begin{bmatrix} 1 & -1 & 1 \\ 5 & 0 & 20 \\ 0 & 10 & -20 \\ 5 & 10 & 0 \end{bmatrix} \text{ and } \left[\begin{array}{ccc|c} 1 & -1 & -1 & 0 \\ 5 & 0 & 20 & 50 \\ 0 & 10 & -20 & 30 \\ 5 & 10 & 0 & 80 \end{array}\right] \text{ respectively.}$$

From the network these matrices can be written directly without the equations (1.5) to (1.8) just by noting the directions of currents in branches and loops.

1.19.3 Matrices in control theory : Electrical , mechanical , chemical engineers and other people in control theory deal with feedback control systems. Most of the contrl theory can be described by the matrix of differential equation

$$\dot{\mathbf{x}} = \mathbf{A}\mathbf{x} + \mathbf{B}\mathbf{u}, \quad \mathbf{y} = \mathbf{C}\mathbf{x}$$

where A, B, C are $n{\times}n$, $n{\times}m$, $p{\times}n$ matrices and **x**, **u**, **y** are column vectors of n, m, p elements respectively. The vectors **x**, **u** and **y** are known as *state* vector, *input* vector and *output* vector respectively. The matrices A, B, C are known as the *system matrix*, the *input distribution* matrix and the *measurement matrix* respectively.

1.19.4 Matrices in graph theory : The graph theory, a branch of mathematics, is originated from Euler's study in solving the famous problem of the seven Bridges of Konisburg. Konisburg is an old Prussian city. The present name of this city is Kaliningrad in Soviet Lithuania.

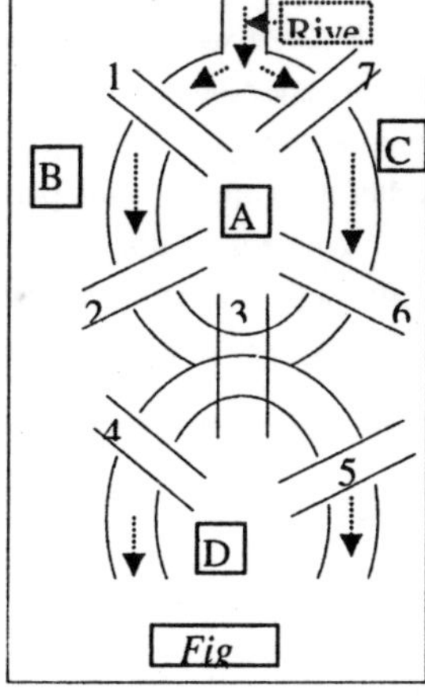

A river flows through the town consisting of two islands A and D and the main lands B and C. The problem runs as "Is it possible to walk through the city, cross each bridge *exactly once*, and return to the *starting point* ?" Euler in 1736 proves its impossibility while, on the other hand, his theorem on Euler cycle (not included in this book) gives birth to graph theory. In *figure* 1.6 A , B , C , D are lands of the city and 1 , 2 , 3 , 4 , 5 , 6 , 7 are the bridges.

Graph, in general, is a set of points called *vertices* or *nodes* connected by lines called *edges*. The graph theory requires the connection of points instead of their location, length of lines and geometry of arcs. That is, graph is "topological" rather than "geometrical". Thus an abstract graph of the seven Bridges of Konisburg is shown in *figure* 1.7, where edges (arcs) denote bridges and vertices denote lands.

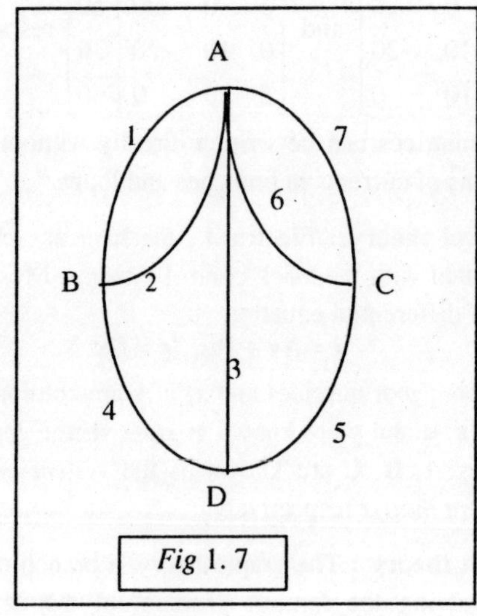

Fig 1. 7

Graphs may be a directed graph or a *digraph*, in which the arrow shows the direction of flow from the initial to terminal point, *undirected* which shows no direction of flow and *mixed* in which some edges are directed and some are not.

Graphs may be classified as *network* and *tree*. A network is a connected graph, i.e. a graph with each node connected to each other node along some path of edges.

So, all graphs are networks but the converse is not true. For example, the graph ABCDEFG in *figure* 1.8 is *not* a network but two separate networks ABCD and EFG because the graph ABCD is *not* connected to graph EFG by any arc (arcs). But the graph ABCDEFG in *figure* 1.9. is a network by definition.

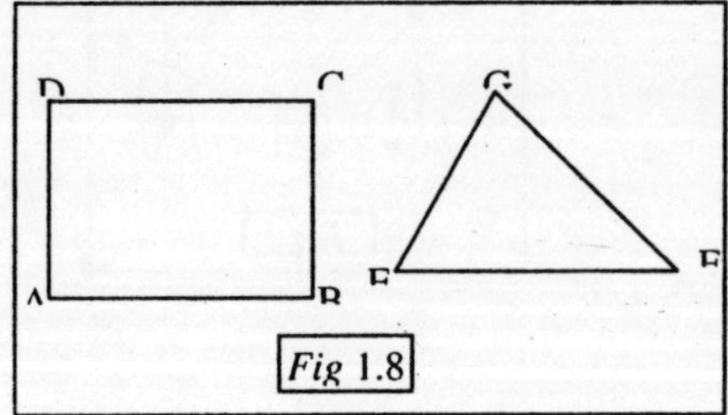

Fig 1.8

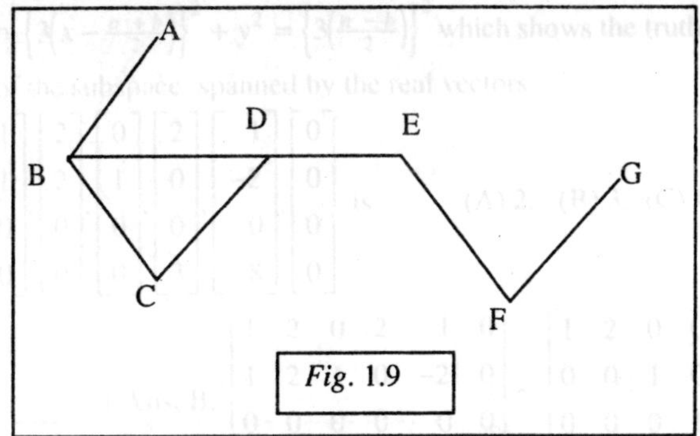

Fig. 1.9

A *tree* is a network without cycles. None of the above *figures* 1.8 & 1.9 is a tree. A network can be reduced to a tree by deleting arcs creating the cycle. Thus, if BD is removed in *Figure* 1.9, above then the cycle BCDB is lost and becomes a tree, *figure* 1.10.

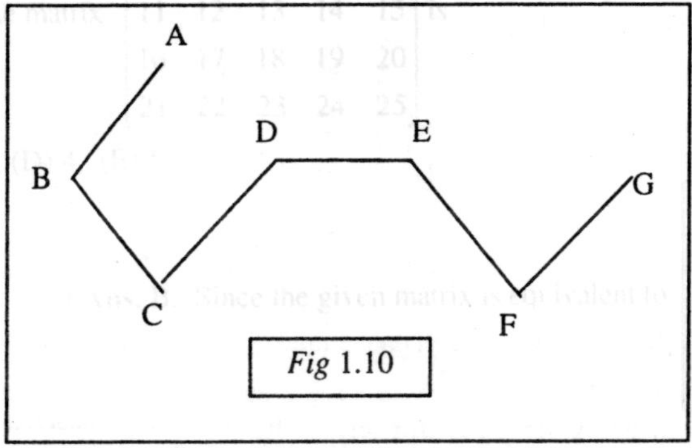

Fig 1.10

Graph theory is used in social sciences, business, communication, computer science, economics, geography, ecology etc.

Now our ultimate goal is to represent the graph by a matrix. Different matrices may be constructed from graphs. Let us consider the construction of a matrix called the *adjacency matrix* which consists of the elements 0 and 1 defined by

$$a_{ij} = \begin{cases} 1, & \text{if } i\text{th point (vertex or node) is connected to } j\text{th point by an edge.} \\ 0, & \text{if it does not connect points } i \text{ and } j. \end{cases}$$

Thus *adjacency matrix* of *figure* 1.9 is

$$[a_{ij}] = \begin{matrix} & A & B & C & D & E & F & G \\ A & 0 & 1 & 0 & 0 & 0 & 0 & 0 \\ B & 1 & 0 & 1 & 1 & 0 & 0 & 0 \\ C & 0 & 1 & 0 & 1 & 0 & 0 & 0 \\ D & 0 & 1 & 1 & 0 & 1 & 0 & 0 \\ E & 0 & 0 & 0 & 1 & 0 & 1 & 0 \\ F & 0 & 0 & 0 & 0 & 1 & 0 & 1 \\ G & 0 & 0 & 0 & 0 & 0 & 1 & 0 \end{matrix}.$$

In this matrix $a_{12} = 1$ as A is connected to B by an edge and $a_{31} = 0$ as C is *not* connected to A by an edge etc. Note that $a_{ii} = 0$ and two vertices are adjacent iff $a_{ij} = 1$.

Adjacency matrix for Multiple edge : A pair of vertices may be connected by more than one edges as *figure* 1.7 of the Konisburg bridge abstract graph. In such case the adjacency matrix needs modification showing the number of multiple edges. Thus, the *adjacency matrix* for the Konisburg bridge graph is

$$[a_{ij}] = \begin{matrix} & A & B & C & D \\ A & 0 & 2 & 2 & 1 \\ B & 2 & 0 & 0 & 1 \\ C & 2 & 0 & 0 & 1 \\ D & 1 & 1 & 1 & 0 \end{matrix}.$$

In this matrix, $a_{12} = a_{13} = a_{21} = a_{31} = 2$. Because land A (vertex) is connected to land B by two bridges etc.

Note that adjacency matrix is defined with respect to vertices (connected or not, directed or undirected) only and it is always a square matrix of order $n \times n$ for n vertices of the graph. The directed graph or digraph, on the other hand, only defines another matrix called the *incidence matrix* with respect to both vertices and edges and it is of order $m \times n$ if m is the number of vertices and n is the number of edges.

The elements of the incident matrix are defined by

$$a_{ij} = \begin{cases} 1, & \text{if arrow is towards } i\text{th vertex from } j\text{th vertex,} \\ & \text{i.e. incoming arrow from } i\text{th vertex from } j\text{th vertex.} \\ -1, & \text{if arrow is away from } i\text{th vertex to } j\text{th vertex} \\ 0, & \text{none of the above two.} \end{cases}$$

For example, the 5×6 *incidence matrix* of the digraph in *figure* 1.11 is

$$\begin{matrix} & E_1 & E_2 & E_3 & E_4 & E_5 & E_6 \\ A & -1 & 0 & 0 & 0 & -1 & 1 \\ B & 1 & -1 & 0 & 0 & 0 & 0 \\ C & 0 & 1 & -1 & 0 & 0 & -1 \\ D & 0 & 0 & 1 & -1 & 0 & 0 \\ E & 0 & 0 & 0 & 1 & 1 & 0 \end{matrix}$$

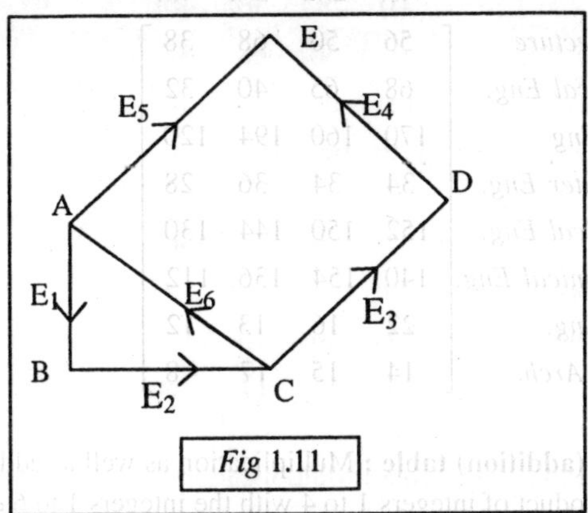

Fig 1.11

while the *adjacency matrix* of this digraph in *figure* 1.11 is a 5×5 matrix given by

$$\begin{array}{c} \\ A \\ B \\ C \\ D \\ E \end{array}\begin{array}{c} \begin{array}{ccccc} A & B & C & D & E \end{array} \\ \left[\begin{array}{ccccc} 0 & 1 & 1 & 0 & 1 \\ 1 & 0 & 1 & 0 & 0 \\ 1 & 1 & 0 & 1 & 0 \\ 0 & 0 & 1 & 0 & 1 \\ 1 & 0 & 0 & 1 & 0 \end{array}\right] \end{array}.$$

If the edge AC is deleted from the digraph in *figure* 1.11 then the *incidence matrix* is a 5×5 square matrix given by

$$\begin{array}{c} \\ A \\ B \\ C \\ D \\ E \end{array}\begin{array}{c} \begin{array}{ccccc} E_1 & E_2 & E_3 & E_4 & E_5 \end{array} \\ \left[\begin{array}{ccccc} -1 & 0 & 0 & 0 & -1 \\ 1 & -1 & 0 & 0 & 0 \\ 0 & 1 & -1 & 0 & 0 \\ 0 & 0 & 1 & -1 & 0 \\ 0 & 0 & 0 & 1 & 1 \end{array}\right] \end{array}.$$

1.19.5 Matrices in tabular information : Information may be stored in any of the three forms — written words, diagram or table of numbers. Any tabular information or data results into matrix. Thus, student enrollment, ordinary multiplication table, mileage chart between certain places of a city or country, scores of student in few tests, score board of a team, inventory or stock situation of an industry or a shop, traffic situation, etc. may be represented by matrices. The following are few examples.

(i) Enrollment of students : The number of students enrolled in an institution may be represented by a matrix. The following 8 (departments) × 4 (years) matrix shows the number of undergraduate students enrolled for the academic session 1990-91 in Bangladesh University of Engineering & Technology (BUET), Dhaka,

$$S_t = \begin{array}{l} \\ \textit{Architecture} \\ \textit{Chemical Eng.} \\ \textit{Civil Eng} \\ \textit{Computer Eng.} \\ \textit{Electrical Eng.} \\ \textit{Mechanical Eng.} \\ \textit{Met. Eng.} \\ \textit{Naval Arch.} \end{array} \begin{array}{c} \begin{array}{cccc} \textit{1st} & \textit{2nd} & \textit{3rd} & \textit{4th} \end{array} \\ \begin{bmatrix} 56 & 50 & 68 & 38 \\ 68 & 65 & 40 & 32 \\ 170 & 160 & 194 & 120 \\ 34 & 34 & 36 & 28 \\ 152 & 150 & 144 & 130 \\ 140 & 154 & 136 & 112 \\ 22 & 16 & 13 & 12 \\ 14 & 15 & 17 & 8 \end{bmatrix} \end{array}$$

(ii) Multiplication (addition) table : Multiplication as well as addition can be represented in matrix form. Thus product of integers 1 to 4 with the integers 1 to 6 can be written as

×	1	2	3	4	5	6
1	1	2	3	4	5	6
2	2	4	6	8	10	12
3	3	6	9	12	15	18
4	4	8	12	16	20	24

.

1.19.6 Matrices in economatrix : Leontief input-output model in economics: This model is used to analyze interdependence of industries. Let a_{ij} denote the amount of input of a certain commodity i to produce unit output of commodity j. The first subscript i, refers to input, the second j, to output. The elements a_{ij} are called *input coefficients* and the matrix $A = [a_{ij}]$ is called the *input matrix* which defines interdependence of industries. For example

$$A = \begin{bmatrix} a_{11} & a_{12} & a_{13} \\ a_{21} & a_{22} & a_{23} \\ a_{31} & a_{32} & a_{33} \end{bmatrix} = \begin{bmatrix} 0.25 & 0.40 & 0.40 \\ 0.35 & 0.10 & 0.20 \\ 0.20 & 0.30 & 0.10 \end{bmatrix}.$$

In this small model $a_{32} = 0.30$ means that 30% worth of commodity 3 is required to produce one dollar's worth of commodity 2 etc. Note that $\sum_{i=1}^{3} a_{ij} < 1,\ j = 1, 2, 3$ in each column and each element $0 < a_{ij} < 1$, $i, j = 1, 2, 3$. So, input matrix is a positive matrix.

1.19.7 Matrices in calculus and differential equations : Any one conversant in calculus of multivariables has heard the names of mathematicians J. Jacobi , Hess and Wornski. Jacobian, Hessian and Wornskian matrices are named after them. Each of them is important for the study of calculus of several variables — differential and integral.

Elements of the Jacobian matrix are the first partial derivatives

$$\frac{\partial f_i}{\partial f_j}, \text{ where } f_i(x_1, x_2, x_3, \cdots, x_n),\ i = 1, 2, 3, \cdots, n.$$

Elements of the Hessian matrix are the second partial derivatives

$$\frac{\partial^2 f}{\partial x_i \partial x_j}, \text{ where } f_i(x_1, x_2, x_3, \cdots, x_n),\ i = 1, 2, 3, \cdots, n$$

is a function of n real variables. It is used in maxima and minima problems of sec.15.6. A Wronskian is used in testing functional independence. Let $y^n + a_{n-1}y^{n-1} + \cdots + a_1 y' + a_0 y = 0$ be an nth order ordinary differential equation (ODE) with constant coefficients. Let $y_1, y_2, \cdots, y_n$ be n independent solutions of the stated ODE. Then Wronskian of $y_1, y_2, \cdots, y_n$ is a determinant of the matrix

$$W(y_1, y_2, \cdots, y_n) = \begin{bmatrix} y_1 & y_2 & \cdots & y_n \\ y_1' & y_2' & \cdots & y_n' \\ \vdots & \vdots & \vdots & \vdots \\ y_1^{n-1} & y_2^{n-2} & \cdots & y_n^{n-1} \end{bmatrix}.$$

1.19.8 Matrices in elastro-mechanics : Stress matrix : An idea of engineering applications of matrix theory is given below and a matrix, called *stress matrix*, has been constructed.

Stress is the force per unit area at a point of a rigid body due to an external force. Let O be the point of interest in the body and let AB be a cross section through O (*figure* 1.12).

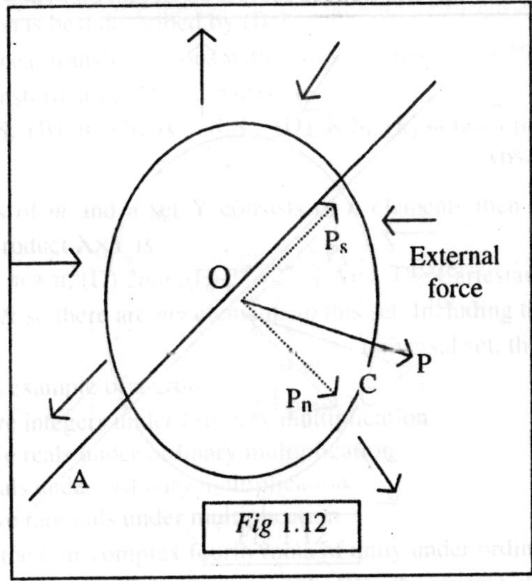

Fig 1.12

Let us consider the stress P exerted on the side C. Let P be decomposed into P_n, called *normal stress*, perpendicular to the plane of AB and P_s, called *shearing stress* parallel to the plane of AB. Thus

$$P = P_n + P_s.$$

The force P_n tends to elongate the body and the force P_s tends to cause the plane to slide on an adjacent plane.

Let three mutually perpendicular planes xy, xz and yz be considered at O. Let us first consider the xz plane (*figure*1.13).

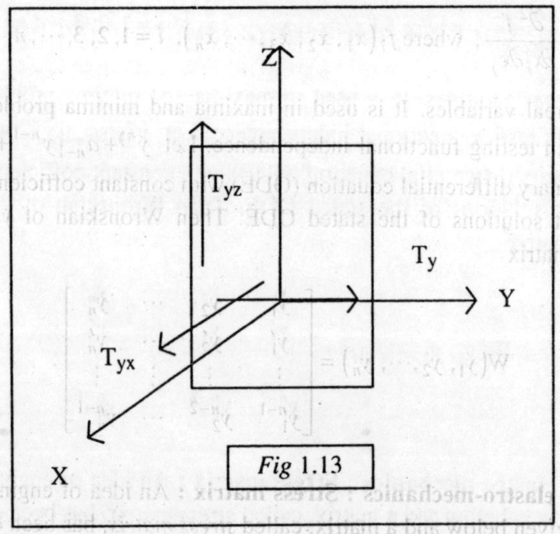

Fig 1.13

Using usual notations, T_y denotes the normal stresses in the direction of *y*-axis, T_{yx} and T_{yz} respectively denote the two components of shearing stresses in *x* and *z* directions. Thus, this stress in the *xz* plane is

$$T_{xy}\mathbf{i}+T_{zy}\mathbf{j}+T_{yz}\mathbf{k} \tag{1.9}$$

Similarly, stress in *xy* and *yz* planes are

$$T_{zx}\mathbf{i}+T_{zy}\mathbf{j}+T_z\mathbf{k} \tag{1.10}$$

$$T_x\mathbf{i}+T_{xy}\mathbf{j}+T_{xz}\mathbf{k} \tag{1.11}$$

These three equations (1.9),(1.10), (1.11) can be represented in a matrix as

$$T = \begin{matrix} & \begin{matrix} x & y & z \end{matrix} \\ \begin{matrix} x \\ y \\ z \end{matrix} & \begin{bmatrix} T_x & T_{xy} & T_{xz} \\ T_{yx} & T_y & T_{yz} \\ T_{zx} & T_{zy} & T_z \end{bmatrix} \end{matrix}$$

Exercises one

1. A matrix has 100 elements. What are its possible orders ?

[**Ans** . 1×100, 2×50, 4×25, 5×20, 10×10 and the reverse.]

2. A matrix has 17 elements. What are its possible orders ?

[**Ans.** 1×17, 17×1]

3. How many elements does 14×14 matrix have that are not on the main diagonal. How many interior elements does it have ?

[**Ans** . 14×13; 14×14 – 12×2 – 14×2]

4. Construct a 5×4 matrix whose elements a_{ij} are given by $a_{ij} = 2i - j$.

[Ans . $\begin{bmatrix} 1 & 0 & -1 & 2 \\ 3 & 2 & 1 & 0 \\ 5 & 4 & 3 & 2 \\ 7 & 6 & 5 & 4 \\ 9 & 8 & 7 & 6 \end{bmatrix}$]

5. Construct a 4×4 matrix whose elements a_{ij} = (minimum of i and j).

[Ans . $\begin{bmatrix} 1 & 1 & 1 & 1 \\ 1 & 2 & 2 & 2 \\ 1 & 2 & 3 & 3 \\ 1 & 2 & 3 & 4 \end{bmatrix}$]

6. Derive a formula for the elements of the matrix

$$\begin{bmatrix} 1 & 1 & 1 & 1 & 1 & \cdots \\ 1 & 2 & 3 & 4 & 5 & \cdots \\ 1 & 4 & 9 & 16 & 25 & \cdots \\ 1 & 8 & 27 & 64 & 125 & \cdots \\ \vdots & \vdots & \vdots & \vdots & \vdots & \vdots \end{bmatrix}$$

by inspection.

[Ans . $a_{ij} = j^{i-1}$]

7. Construct a 4×4 matrix that tabulates the mutual distances of the vertices of a square whose side is one unit long (*figure* 1.14).

[Hints :

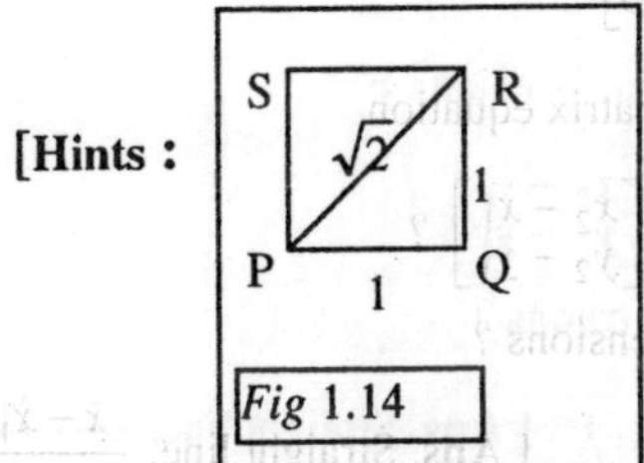

Fig 1.14

Ans . $\begin{array}{c} \\ P \\ Q \\ R \\ S \end{array}\begin{array}{c} \begin{array}{cccc} P & Q & R & S \end{array} \\ \begin{bmatrix} 0 & 1 & \sqrt{2} & 1 \\ 1 & 0 & 1 & \sqrt{2} \\ \sqrt{2} & 1 & 0 & 1 \\ 1 & \sqrt{2} & 1 & 0 \end{bmatrix} \end{array}$]

8. Is there any value of x and y so that the following two matrices are equal ?

$$A = \begin{bmatrix} 3x & 2 \\ 0 & 5y^2 \end{bmatrix} \text{ and } B = \begin{bmatrix} 9 & 1 \\ 2 & 125 \end{bmatrix}$$

[Ans . No, because $a_{12} \neq b_{12}$ and $a_{21} \neq b_{21}$]

9. Determine the values of x and y if A = B, where

$$A = \begin{bmatrix} 3x & 2 \\ 0 & 5y^2 \end{bmatrix} \text{ and } B = \begin{bmatrix} 9 & 1 \\ 2 & 125 \end{bmatrix}.$$

[Ans . $x = 3$, $y = \pm 5$]

10. Write out the 3×2 matrix units.

[Ans. $E_{11} = \begin{bmatrix} 1 & 0 \\ 0 & 0 \\ 0 & 0 \end{bmatrix}$, write remaining five matrices]

11. Construct a 3×5 matrix whose elements a_{ij} are given by $a_{ij} = i + j - 1$.

$$[\textbf{Ans. } A = \left[a_{ij}\right]_{3\times5} = \begin{bmatrix} 1 & 2 & 3 & 4 & 5 \\ 2 & 3 & 4 & 5 & 6 \\ 3 & 4 & 5 & 6 & 7 \end{bmatrix}\ ]$$

12. Construct the a 3×4 matrix such that $a_{ij} = ij$.

$$[\textbf{Ans}.\ \begin{bmatrix} 1 & 2 & 3 & 4 \\ 2 & 4 & 6 & 8 \\ 3 & 6 & 9 & 12 \end{bmatrix}\ ]$$

13. Which of the following statements are true?

$$(a, a) = (a, a, 0);\ (a - a, 0) = (0, 0);\ \begin{bmatrix} 5-5 & 2-2 \\ a-a & 6-6 \end{bmatrix} = 0.$$

[**Ans.** 2*nd* and 3*rd*]

14. If

$$\begin{bmatrix} x+y+z+w & x+y+2z+w \\ x+2y+z+w & 2x+y+z+w \end{bmatrix} = \begin{bmatrix} 7 & 6 \\ 6 & 6 \end{bmatrix},$$

find the values of *x, y, z, w*.

[**Ans.** $x = -1,\ y = -1,\ z = -1,\ w = 10$]

15. Show that it is impossible to hold the equality

$$\begin{bmatrix} x+y & x+2y \\ x-y & x-2y \end{bmatrix} = \begin{bmatrix} 5 & 6 \\ 3 & 0 \end{bmatrix}.$$

[**Ans.** *y* contradicts]

16. What is the geometrical significance of the matrix equation

$$\begin{bmatrix} x \\ y \end{bmatrix} = \begin{bmatrix} x_1 \\ y_1 \end{bmatrix} + t\begin{bmatrix} x_2 - x_1 \\ y_2 - y_1 \end{bmatrix} ?$$

What is the corresponding equation in three dimensions ?

[**Ans.** Straight line, $\dfrac{x - x_1}{x_2 - x_1} = \dfrac{y - y_1}{y_2 - y_1}$]

17. A matrix has 15 elements. What are its possible orders ?

[**Ans.** 1×15, 3×5, 15×1, 5×3]

18. A matrix is 3×5. How many interior elements of this matrix ?

[**Ans.** $15 - (5 + 5 + 1 + 1) = 3$]

Generalize this result for an *m*×*n* matrix.

[**Ans.** $mn - (2n + 2(m - 2)) = mn - 2(m + n - 2)$]

19. Write the 4 matrices of orders

$$m\times n,\ m\times(q - n),\ (p - m)\times n \text{ and } (p - m)\times(q - n)$$

contiguously to form a single matrix of order *p*×*q*.

$$[\textbf{Ans}.\ \begin{bmatrix} \left[a_{ij}\right]_{m\times n} & \left[a_{ij}\right]_{m\times(q-n)} \\ \left[a_{ij}\right]_{(p-m)\times n} & \left[a_{ij}\right]_{(p-m)\times(q-n)} \end{bmatrix}\ ]$$

20. What is the diagonal of $A_{2\times4} = [a_{ij}]_{2\times4} = (i - j)$? Of $A^{T}_{2\times4}$?

[**Ans** . $a_{11} = 0,\ a_{22} = 0$ for both]

21. Construct the matrix for the ordinary multiplication of 4 by 6.

$$\text{[Ans . } \begin{array}{c} \\ 1 \\ 2 \\ 3 \\ 4 \end{array}\begin{array}{c} \begin{matrix} 1 & 2 & 3 & 4 & 5 & 6 \end{matrix} \\ \begin{bmatrix} 1 & 2 & 3 & 4 & 5 & 6 \\ 2 & 4 & 6 & 8 & 10 & 12 \\ 3 & 6 & 9 & 12 & 15 & 18 \\ 4 & 8 & 12 & 16 & 20 & 24 \end{bmatrix} \end{array} \text{]}$$

22. Dominance means existence of relation between persons, nations, numbers or biological characteristics. In graph arrow from point i to point j denotes dominance of i over j and $a_{ij} = 1$ if i dominates j and $a_{ij} = 0$ if i does not dominate j. Draw the graph corresponding to the dominance matrix

$$\begin{array}{c} \\ \#1 \\ \#2 \\ \#3 \\ \#4 \end{array}\begin{array}{c} \begin{matrix} \#1 & \#2 & \#3 & \#4 \end{matrix} \\ \begin{bmatrix} 0 & 0 & 1 & 1 \\ 1 & 0 & 1 & 0 \\ 0 & 0 & 0 & 0 \\ 0 & 1 & 1 & 0 \end{bmatrix} \end{array}.$$

[Ans .

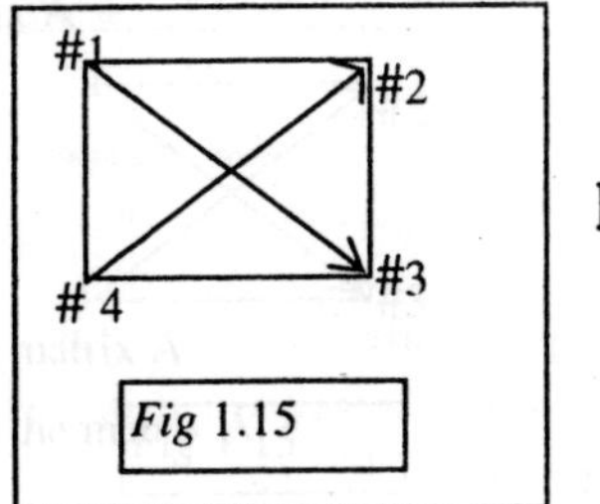

Fig 1.15

]

23. State the order of the matrix

$$\begin{bmatrix} 1 & 2 & 4 & 5 \\ 4 & 0 & 3 & 7 \end{bmatrix}.$$

What is the address of "3" ? [**Ans** . 2×4 and $a_{23} = 3$, 23 is the address of 3]

24. Does the array of dates of a month on a calendar form a matrix ? Why ?

[**Ans** . No. There are vacant spaces not occupied by numbers]

25. Find the values of the unknown from the following inequalities, if possible.

(i) $\begin{bmatrix} -1 & -2 \\ 4 & 3 \end{bmatrix} \le \begin{bmatrix} 0 & -1 \\ x & 3 \end{bmatrix}$;

(ii) $\begin{bmatrix} 3i & 4 \\ 3 & x \end{bmatrix} > \begin{bmatrix} i & 2 \\ 1 & 2 \end{bmatrix}$;

(iii) $\begin{bmatrix} 2 & 0 \\ 4 & -3 \end{bmatrix} < \begin{bmatrix} x & 2 \\ 4 & -2 \end{bmatrix}$

[**Ans.** (i) $x \ge 4$,
(ii) impossible as imaginary numbers are not ordered,
(iii) impossible as 4 $\nless$ 4 in the address 21]

CHAPTER TWO

BASIC ALGEBRAIC OPERATIONS

***Chapter summary** : This chapter two deals with three unary operations - complex conjugation, transposition and tranjugation, and four binary operations - addition, scalar multiplication, subtraction & matrix multiplication. Further, the algebra of these operations are stated and proved in details. Next s.u. vectors are used to construct matrix unit ; Kronecker delta; rows, columns and elements of a matrix. The chapter ends with application: random walks in crystal, simultaneous equations and their change of variables and quadratic in matrix form.*

2.1 Matrix operations

Matrix algebra possesses two main operations:

(i) Unary and

(ii) Binary.

Each one of them is described below separately.

2.2 Unary operations and derived matrices

In a number system, a unary operation is the operation of a number with itself producing a new number. Thus

cube of 2 is $2^3 = 8$,

magnitude of a vector $|(3, 0, 4)| = \sqrt{(3^2 + 0^2 + 4^2)} = 5$,

$\sin(\frac{\pi}{6}) = \frac{1}{2}$ are examples of unary operations.

Similarly, operation of a single matrix to produce a second matrix is called *unary operation* of the matrix. The *three* standard unary operations namely:

(i) complex conjugation,

(ii) transposition and

(iii) Hermitian conjugation or tranjugation

produce new matrices called *derived* matrices. Each of them is discussed and defined below:

2.2.1 Complex conjugation (conjugation): If the elements of a complex matrix $A = [a_{jk}]_{m \times n} = [p_{jk} + iq_{jk}]_{m \times n}$, i is the complex quantity, are replaced by the corresponding complex conjugate

$$\left[\bar{a}_{jk}\right]_{m \times n} = \left[p_{jk} - iq_{jk}\right]_{m \times n}$$

then the unary operation is called *complex conjugation* or simply *conjugation* and the derived matrix denoted by

$$A^c \text{ or } \bar{A} = \left[\bar{a}_{jk}\right]_{m \times n} = \left[p_{jk} - iq_{jk}\right]_{m \times n}$$

is called the *conjugate matrix*. Note that the order remains unchanged before and after conjugation. For example, if

$$A = \begin{bmatrix} 1+2i & 0 & x-iy \\ 3+2i & 1+i & 4-5i \end{bmatrix} \text{ then}$$

$$A^c \text{ or } \overline{A} = \begin{bmatrix} 1-2i & 0 & x+iy \\ 3-2i & 1-i & 4+5i \end{bmatrix}.$$

2.2.1.1 Necessary and sufficient condition for real matrix : The necessary and sufficient condition for real matrix is given by the next theorem.

Theorem. The matrix A is real if and only if $A = \overline{A}$.

Alternatively the necessary and sufficient condition for a matrix A to be real is $A = \overline{A}$.

Proof. Let the general element of A be $a_{jk} = \alpha_{jk} + i\beta_{jk}$, where $i = \sqrt{-1}$, the imaginary quantity, α_{jk} and β_{jk} are real. Then its conjugate $\overline{a}_{jk} = \alpha_{jk} - i\beta_{jk}$ which is a general element of the conjugate matrix $\overline{A}$.

If $A = \overline{A}$, then by definition of equality of matrices,

$$a_{jk} = \overline{a}_{jk}$$

i.e. $$\alpha_{jk} + i\beta_{jk} = \alpha_{jk} - i\beta_{jk}.$$

Therefore, equating imaginary parts $$\beta_{jk} = -\beta_{jk}$$

or $$2\beta_{jk} = 0$$

or $$\beta_{jk} = 0.$$

Therefore, $a_{jk} = \alpha_{jk} =$ purely imaginary. So, the matrix A is real.

Conversely, if A is real, $\beta_{jk} = 0$ and $a_{jk} = \alpha_{jk}$.

So, $$\overline{a}_{jk} = \alpha_{jk}.$$

Therefore, $$a_{jk} = \overline{a}_{jk}$$

and thus $$A = \overline{A}.$$

2.2.1.2 Identifying character of R-matrix and purely imaginary matrix : Conjugation provides with the identifying character of R-matrix and purely imaginary matrix:

(i) **R-matrix :** If the conjugate of a matrix equals the original matrix then the matrix is a R-matrix or real matrix i.e. if $[\overline{a}_{jk}] = [a_{jk}]$ then $[p_{jk} - iq_{jk}] = [p_{jk} + iq_{jk}]$. Hence $p_{jk} - iq_{jk} = p_{jk} + iq_{jk}$,

i.e. $$q_{jk} = 0.$$

Therefore, $a_{jk} = p_{jk}$ only i.e. elements are free of imaginary part and so the elements as well as the matrix are real. For example,

$$[a_{jk}]_{2\times3} = \begin{bmatrix} 1 & 2 & 3 \\ 0 & 1 & 4 \end{bmatrix} = [\overline{a}_{jk}]_{2\times3}$$

$$= [a_{jk}]_{2\times3} \text{ is a real matrix. 1}$$

(ii) **Purely imaginary matrix :** If the conjugate of a matrix equals the negative of the original matrix then the matrix is purely imaginary

i.e. if $[\overline{a}_{jk}] = -[a_{jk}]$

then $\left[p_{jk} - iq_{jk}\right] = -\left[p_{jk} + iq_{jk}\right]$.

Hence $p_{jk} - iq_{jk} = -p_{jk} - iq_{jk}$

i.e. $p_{jk} = 0.$

Therefore, $a_{jk} = q_{jk}$ only i.e. elements are free of real parts and so the elements are purely imaginary. So the matrix is purely imaginary.

Let $\left[a_{jk}\right]_{2\times3} = \begin{bmatrix} i & 2i & -3i \\ 0 & -i & 4i \end{bmatrix} = -\left[\overline{a}_{jk}\right]_{2\times3}$.

$\therefore$ $[a_{jk}]_{2\times3}$ is purely imaginary.

2.2.2 Transposition : The unary operation of interchanging rows (columns) into columns (rows) of a matrix produces a derived matrix called *transposed* matrix (*transpose* of a matrix). It is denoted by A′, A^T or $\widetilde{A}$ (~ is called *tilde* by physicists). The symbol A′ will not be used as calculus part uses A′ for derivative. So, a transposition of an $m\times n$ matrix A is an $n\times m$ matrix A^T. Hence in terms of elements , if $A = [a_{ij}]_{m\times n}$ then $A^T = [a^T_{ij}]_{m\times n}$, where $a^T_{ij} = a_{ji}$ i.e. (i, j)th element of A^T is the (j, i)th element of A. The order of A changes to $n\times m$ due to transposition. In the case of the square matrix transposition keeps the order unchanged. Thus, a column vector is the transposed of a row vector and vice-versa.

For example,

let $A = \begin{bmatrix} 1 & 2 & 3 & 0 \\ 0 & 1 & -2 & -3 \\ -1 & 5 & 4 & -5 \end{bmatrix}$ is a 3×4 matrix. Then $A^T = \begin{bmatrix} 1 & 0 & -1 \\ 2 & 1 & 5 \\ 3 & -2 & 4 \\ 0 & -3 & -5 \end{bmatrix}$ is a 4×3 matrix.

The scalar — a 1×1 matrix, is unaltered by transposition.

From the above discussions, transposition can be viewed *either* as a "*reflection*" of its elements in its diagonal which acts as a mirror *or* as a "*rotation*" of the elements a_{ij} , $i > j$ down to a_{ij} , $i < j$ about the diagonal and vice-versa.

2.2.3 Tranjugation (Hermitian conjugation) : If two unary operations — complex conjugation and transposition are performed one after another on a matrix then the unary operation is called *Hermitian conjugation* or *tranjugation* and the derived matrix is called *conjugate transposed matrix* or *tranjugate matrix* or *associate of a matrix* denoted by A^{CT} or A^* or $A^\theta = [\overline{a}_{ij}]_{m\times n}$.

This operation can be performed in two ways as follows:

$$\begin{array}{ccc} A = [a_{ij}]_{m\times n} & \xrightarrow{\text{Conjugation}} & [\bar{a}_{ij}]_{m\times n} \\ \downarrow & & \downarrow \\ \text{Transposition} & & \text{Transposition} \\ \downarrow & & \downarrow \\ [a_{ji}]_{n\times m} & \xrightarrow{\text{Conjugation}} & [\bar{a}_{ji}]_{n\times m} = A^*. \end{array}$$

Hence $A^* = \bar{A}^T = \left(\overline{A^T}\right)$. Physicists use $A^\dagger$ (called A dagger) for A^* or A^θ. The order of A^* or A^θ or $A^\dagger$ is $n \times m$ when the order of the original matrix is $m \times n$.

2.3 Binary operations

Algebraic operations like addition, subtraction, multiplication and division are binary operations. "Bi" means two. So binary operation operates between two objects. The first three operations are valid for matrices. But division does not valid for matrices as in vectors. The purpose of division is met using inverse which will be discussed in chapter seven.

2.3.1 Addition : Two matrices are comformable for addition if they are of same orders (dimensions). Let $A = [a_{ij}]_{m\times n}$ and $B = [b_{ij}]_{m\times n}$.
The sum or addition A + B is a third matrix $C = [c_{ij}]_{m\times n}$ defined by

$$C = [c_{ij}]_{m\times n} = A + B = [a_{ij} + b_{ij}]_{m\times n}$$

which adds corresponding elements of A and B. The idea of addition can be extended to any number of $m\times n$ matrices. Thus

$$A + B + \cdots + K = [a_{ij} + b_{ij} + \cdots + k_{ij}]_{m\times n}.$$

The sum A + B requires a total of mn additions or additive operations. Let

$$A = \begin{bmatrix} 1 & 0 & 3 \\ 2 & 5 & -1 \end{bmatrix},\ B = \begin{bmatrix} 2 & 5 & 1 \\ 3 & 0 & 7 \end{bmatrix},\ C = \begin{bmatrix} 1 & 2 \\ 3 & 4 \end{bmatrix}.4$$

Therefore,

$$A + B = \begin{bmatrix} 3 & 5 & 4 \\ 5 & 5 & 6 \end{bmatrix}$$

5whereas A + C as well as B + C are not defined. Note that addition of the null matrix, 0 to any one of A, B, C keeps each one of them unchanged. So, the null matrix is the *additive identity* for matrices.

Example. The undergraduate male and female students of BUET for the session 1990-91 in different years are given by the matrices M_s & F_s respectively. Find the matrix of the total number of students both male and female in each year in different departments during the session. Verify the result from the matrix in chapter1.

$$M_s = \begin{array}{c} \\ Chem. \\ Met. \\ Civil \\ Mech. \\ Naval \\ E.E. \\ CSE \\ Arch \end{array}\begin{array}{c} \begin{array}{cccc} 1st & 2nd & 3rd & 4th \end{array} \\ \begin{bmatrix} 48 & 56 & 32 & 30 \\ 18 & 16 & 12 & 10 \\ 149 & 150 & 162 & 104 \\ 132 & 146 & 129 & 101 \\ 14 & 15 & 17 & 8 \\ 138 & 137 & 122 & 118 \\ 28 & 31 & 34 & 26 \\ 37 & 32 & 50 & 23 \end{bmatrix} \end{array},$$

$$F_s = \begin{array}{c} \\ Chem. \\ Met. \\ Civil \\ Mech. \\ Naval \\ E.E. \\ CSE \\ Arch \end{array}\begin{array}{c} \begin{array}{cccc} 1st & 2nd & 3rd & 4th \end{array} \\ \begin{bmatrix} 20 & 9 & 8 & 2 \\ 4 & 0 & 1 & 2 \\ 21 & 10 & 32 & 16 \\ 8 & 8 & 7 & 11 \\ 0 & 0 & 0 & 0 \\ 14 & 13 & 22 & 12 \\ 6 & 3 & 2 & 2 \\ 19 & 18 & 18 & 15 \end{bmatrix} \end{array}$$

Adding the given matrices, we get the required result as follows:

$$M_s + F_s = \begin{array}{c} \\ Chem. \\ Met. \\ Civil \\ Mech. \\ Naval \\ E.E. \\ CSE \\ Arch \end{array}\begin{array}{c} \begin{array}{cccc} 1st & 2nd & 3rd & 4th \end{array} \\ \begin{bmatrix} 68 & 65 & 40 & 32 \\ 22 & 16 & 13 & 12 \\ 170 & 160 & 194 & 120 \\ 140 & 154 & 136 & 112 \\ 14 & 15 & 17 & 8 \\ 152 & 150 & 144 & 130 \\ 34 & 34 & 36 & 28 \\ 56 & 50 & 68 & 38 \end{bmatrix} \end{array}$$

2.3.2 Scalar multiplication : If k is a scalar real number and $A = [a_{ij}]_{m\times n}$ is a matrix then the scalar multiplication of A by k is another matrix

$$C = [c_{ij}]_{m\times n} = [ka_{ij}]_{m\times n}.$$

The matrix $C_{m\times n} = kA_{m\times n}$ is obtained by multiplying each element of A by k. So, the scalar multiplication requires a total of mn algebraic multiplications.

Let $$A = \begin{bmatrix} 1 & 0 & 3 \\ 2 & 5 & -1 \end{bmatrix}.$$

So $$C = 5A = 5\begin{bmatrix} 1 & 0 & 3 \\ 2 & 5 & -1 \end{bmatrix} = \begin{bmatrix} 5 & 0 & 15 \\ 10 & 25 & -5 \end{bmatrix}.$$

6Note the converse, if all the elements of a matrix has a common factor then this common factor can be taken out as a scalar multiplying a new matrix. Thus

$$C=\begin{bmatrix} 5 & 0 & 15 \\ 10 & 25 & -5 \end{bmatrix}$$

in which each element has a common factor 5. So,

$$C=5\begin{bmatrix} 1 & 0 & 5 \\ 2 & 5 & -1 \end{bmatrix}=5A\ 7.$$

2.3.3 Linear combination of matrices : Combination of the two rules addition and scalar multiplication, gives definition of linear combination as the sum of the scalar products of n matrices A, B, $\cdots$, K of the same order and n scalars $a_1, a_2, \cdots, a_n$. Thus

$$a_1A + a_2B + \cdots + a_nK = [a_1a_{ij} + a_2b_{ij} + \cdots + a_nk_{ij}].$$

2.3.4 Subtraction : Subtraction, A – B of two matrices A and B is defined by

$$A + (-1)\,B.$$

So, this operation is a combination of two binary operations — addition and scalar multiplication i.e. a linear combination of A and B with two scalars 1 and (–1). This is conformable if both A and B are of the same dimensions. Subtraction requires mn scalar multiplications and mn additions. So, total number of operations is $2mn$. Hence subtraction needs double computational time than addition.

Let
$$A=\begin{bmatrix} 3 & 5 & 6 \\ 7 & 0 & 5 \end{bmatrix},\quad B=\begin{bmatrix} -2 & 5 & -6 \\ 4 & -2 & 7 \end{bmatrix}.$$

Both A and B are of same dimensions 2×3. Therefore,

$$A-B=A+(-1)B=\begin{bmatrix} 3 & 5 & 6 \\ 7 & 0 & 5 \end{bmatrix}+\begin{bmatrix} 2 & -5 & 6 \\ -4 & 2 & -7 \end{bmatrix}$$

$$=\begin{bmatrix} 5 & 0 & 12 \\ 3 & 2 & -2 \end{bmatrix}$$

Note that $A + (-A) = A - A = 0$, a null matrix.

$\therefore$ (– A) is the additive inverse of the matrix A.

2.3.5 Matrix multiplication : Matrix multiplication, also called *Cayley product*, will be defined in *three* stages as follows:

Stage 1. The product of a row vector and a column vector

Let $\mathbf{u} = [\,u_1 \;\; u_2 \;\; \cdots \;\; u_n\,]$ be an n-dimensional row vector and

$\mathbf{v} = [\,v_1 \;\; v_2 \;\; \cdots \;\; v_n\,]^T$ be an n-dimensional column vector. Their product **uv** is called the *scalar* or *inner* product and is *defined* as

$$\mathbf{uv} = \begin{bmatrix} u_1 & u_2 & \cdots & u_n \end{bmatrix}_{1\times n} \begin{bmatrix} v_1 \\ v_2 \\ \vdots \\ v_n \end{bmatrix}_{n\times 1} = \begin{bmatrix} u_1v_1 + u_2v_2 + \cdots + u_nv_n \end{bmatrix}_{1\times 1} \tag{2.1}$$

Note that the number n representing both columns of $\mathbf{u}$ and rows of $\mathbf{v}$ has been absorbed in the product. This is same as the dot or scalar product **x.y** of two row vectors **x** & **y** in Vector Algebra.

Stage 2 . The product of an $m\times n$ matrix and an $n\times 1$ column vector

Let A be an $m\times n$ matrix with n-dimensional row vectors $\mathbf{A}_{1\cdot}, \mathbf{A}_{2\cdot}, \cdots, \mathbf{A}_{m\cdot}$ and $\mathbf{v}$ be an n-dimensional column vector. Then the product $A\mathbf{v}$ is defined by

$$A\mathbf{v} = \begin{bmatrix} \mathbf{A}_{1\cdot} \\ \mathbf{A}_{2\cdot} \\ \vdots \\ \mathbf{A}_{m\cdot} \end{bmatrix}_{m\times n} \mathbf{v}_{n\times 1} = \begin{bmatrix} \mathbf{A}_{1\cdot}\mathbf{v} \\ \mathbf{A}_{2\cdot}\mathbf{v} \\ \vdots \\ \mathbf{A}_{m\cdot}\mathbf{v} \end{bmatrix}_{m\times 1} \tag{2.2}$$

In terms of components

$$\mathbf{A}_{i.} = \begin{bmatrix} a_{i1} & a_{i2} & \cdots & a_{in} \end{bmatrix};\ i = 1, 2, \cdots, n.$$

and $\mathbf{v} = \begin{bmatrix} v_1 & v_2 & \cdots & v_n \end{bmatrix}^{\mathrm{T}}$, the above product is given by

$$A\mathbf{v} = \begin{bmatrix} a_{11} & a_{12} & \cdots & a_{1n} \\ a_{21} & a_{22} & \cdots & a_{2n} \\ \vdots & \vdots & \vdots & \vdots \\ a_{m1} & a_{m2} & \cdots & a_{mn} \end{bmatrix}_{m\times n} \begin{bmatrix} v_1 \\ v_2 \\ \vdots \\ v_n \end{bmatrix}_{n\times 1}$$

$$= \begin{bmatrix} a_{11}v_1 + a_{12}v_2 + \cdots + a_{1n}v_n \\ a_{21}v_1 + a_{22}v_2 + \cdots + a_{2n}v_n \\ \vdots \quad\quad \vdots \quad\quad \vdots \\ a_{m1}v_1 + a_{m2}v_2 + \cdots + a_{mn}v_n \end{bmatrix}_{m\times 1} = \begin{bmatrix} \sum_{i=1}^{n} a_{1i}v_i \\ \sum_{i=1}^{n} a_{2i}v_i \\ \vdots \\ \sum_{i=1}^{n} a_{mi}v_i \end{bmatrix}_{m\times 1} \tag{2.3}$$

Here also *note* that (i) common factor n of orders of A and $\mathbf{v}$ has been absorbed in the resulting matrix or vector, (ii) repeated use of (2.1) above; (iii) ith element in the product $A\mathbf{v}$ is the dot product of ith row of A with $\mathbf{v}$; (iv) the definition $A\mathbf{v}$ requires that the number of columns of A equals the number of elements of $\mathbf{v}$ and this number is n which disappears in the product, (v) order of the resulting product is $m\times 1$ where m is the number of rows in A .

Stage 3 . Product of an $m\times n$ matrix and an $n\times p$ matrix

Let A be an $m\times n$ matrix and B be an $n\times p$ matrix. Let the p columns of the matrix B be $\mathbf{B}_{.1}$, $\mathbf{B}_{..2}, \cdots, \mathbf{B}_{p}$. Each of them consists of n elements. Now the product AB is defined as

$$AB = A\,[\,\mathbf{B}_{.1} \quad \mathbf{B}_{.2} \quad \cdots \quad \mathbf{B}_{.p}\,] = [\,A\mathbf{B}_{.1} \quad A\mathbf{B}_{.2} \quad \cdots \quad A\mathbf{B}_{.p}\,],$$

which is the product of an $m\times n$ matrix A and $1\times p$ row vector, where $\mathbf{AB}_{.i}$ are defined by stage 2 above. Thus the ith column of the product AB is the product the matrix A with ith column of B. [This is an example of "*parallel processing*" in computer science in which the knowledge of A and the ith column of B only requires to find the ith column of AB independently].

Thus if $$A = \begin{bmatrix} \mathbf{A}_{1.} \\ \mathbf{A}_{2.} \\ \vdots \\ \mathbf{A}_{m.} \end{bmatrix}_{m\times 1}$$, then

$$C = AB = \begin{bmatrix} \mathbf{A}_{1.} \\ \mathbf{A}_{2.} \\ \vdots \\ \mathbf{A}_{m.} \end{bmatrix}_{m\times 1} \begin{bmatrix} \mathbf{B}_{.1} & \mathbf{B}_{.2} & \cdots & \mathbf{B}_{.p} \end{bmatrix}_{1\times p}$$

$$= \begin{bmatrix} \mathbf{A}_{1.}\mathbf{B}_{.1} & \mathbf{A}_{1.}\mathbf{B}_{.2} & \cdots & \mathbf{A}_{1.}\mathbf{B}_{.p} \\ \mathbf{A}_{2.}\mathbf{B}_{.1} & \mathbf{A}_{2.}\mathbf{B}_{.2} & \cdots & \mathbf{A}_{2.}\mathbf{B}_{.p} \\ \vdots & \vdots & \vdots & \vdots \\ \mathbf{A}_{m.}\mathbf{B}_{.1} & \mathbf{A}_{m.}\mathbf{B}_{.2} & \cdots & \mathbf{A}_{m.}\mathbf{B}_{.p} \end{bmatrix}_{m\times p} \qquad (2.4)$$

or in terms of components,

$$C = AB$$

$$= \begin{matrix} \mathbf{A}_{1.} \\ \mathbf{A}_{2.} \\ \vdots \\ \mathbf{A}_{i.} \\ \vdots \\ \mathbf{A}_{m.} \end{matrix} \begin{bmatrix} a_{11} & a_{12} & \cdots & a_{1n} \\ a_{21} & a_{22} & \cdots & a_{2n} \\ \vdots & \vdots & \vdots & \vdots \\ a_{i1} & a_{i2} & \cdots & a_{in} \\ \vdots & \vdots & \vdots & \vdots \\ a_{m1} & a_{m2} & \cdots & a_{mn} \end{bmatrix}_{m\times n} \begin{matrix} \mathbf{B}_{.1} & \mathbf{B}_{.2} & \cdots & \mathbf{B}_{.j} & \cdots & \mathbf{B}_{.p} \end{matrix} \begin{bmatrix} b_{11} & b_{12} & \cdots & b_{1j} & \cdots & b_{1p} \\ b_{21} & b_{22} & \cdots & b_{2j} & \cdots & b_{2p} \\ \vdots & \vdots & \vdots & \vdots & \vdots & \vdots \\ b_{n1} & b_{n2} & \cdots & b_{nj} & \cdots & b_{np} \end{bmatrix}_{n\times p} \qquad (2.5)$$

$$\text{or,} \quad C = \begin{bmatrix} \sum_{k=1}^{n} a_{1k}b_{k1} & \sum_{k=1}^{n} a_{1k}b_{k2} & \cdots & \sum_{k=1}^{n} a_{1k}b_{kj} & \cdots & \sum_{k=1}^{n} a_{1k}b_{kp} \\ \sum_{k=1}^{n} a_{2k}b_{k1} & \sum_{k=1}^{n} a_{2k}b_{k2} & \cdots & \sum_{k=1}^{n} a_{2k}b_{kj} & \cdots & \sum_{k=1}^{n} a_{2k}b_{kp} \\ \vdots & \vdots & \vdots & \vdots & \vdots & \vdots \\ \sum_{k=1}^{n} a_{ik}b_{k1} & \sum_{k=1}^{n} a_{ik}b_{k2} & \cdots & \sum_{k=1}^{n} a_{ik}b_{kj} & \cdots & \sum_{k=1}^{n} a_{ik}b_{kp} \\ \vdots & \vdots & \vdots & \vdots & \vdots & \vdots \\ \sum_{k=1}^{n} a_{mk}b_{k1} & \sum_{k=1}^{n} a_{mk}b_{k2} & \cdots & \sum_{k=1}^{n} a_{mk}b_{kj} & \cdots & \sum_{k=1}^{n} a_{mk}b_{kp} \end{bmatrix} \qquad (2.6)$$

Note that in all cases of multiplication common factor of the orders has been absorbed leaving the uncommon factors of the orders as the order of the resulting matrix.

Let C = AB and C = $[c_{ij}]$. Then in general

$$c_{ij} = \mathbf{A}_{i.}\mathbf{B}_{.j} = \sum_{k=1}^{n} a_{ik}b_{kj} = a_{i1}b_{1j} + a_{i2}b_{2j} + \ldots + a_{in}b_{nj}.$$

Number of operations : nmp multiplication + nmp addition = $2nmp$ operations.

Conformability : It is observed that matrix multiplication AB is conformable if the number of columns of the matrix A is equal to the number of rows of the matrix B. That is

$$(\text{an } m \times \not{n} \text{ matrix}) \times (\text{an } \not{n} \times p \text{ matrix}) = (\text{an } m \times p \text{ matrix})$$

or, $$(m \times \not{n}) \times (\not{n} \times p) = (m \times p)$$

Therefore the product AB has the same number of rows m as the matrix A on the left and same number of columns p as the matrix B on the right.

In the product AB, the matrix B is premultiplied by A (or multiplied by A by the left) called *premultiplier* and A is postmultiplied by B (or multiplied by B by the right) called *post multiplier*. This terminology is essential since the product BA may not be conformable as in ordinary algebra.

Ex. Let $A = \begin{bmatrix} 1 & 0 & -1 & 4 \\ 2 & 3 & 0 & 5 \\ -1 & 7 & -2 & 1 \end{bmatrix}_{3\times 4}$ and $B = \begin{bmatrix} 3 & -2 \\ -1 & 3 \\ 2 & 4 \\ 5 & 1 \end{bmatrix}_{4\times 2}$.

Illustrate the stages 2 and 3 respectively of matrix multiplication and hence obtain the product AB in each case. Use multiplication rule to obtain AB. Find the number operations required.

It is clear that number of columns 4 of A is equal to the number of rows 4 of B. Therefore, the product AB is conformable for matrix multiplication and BA is not. To illustrate stage 2 of multiplication, we compute the product of A with each of the two columns of B:

$$\mathbf{AB}_{.1} = \begin{bmatrix} 1 & 0 & -1 & 4 \\ 2 & 3 & 0 & 5 \\ -1 & 7 & -2 & 1 \end{bmatrix}_{3\times 4} \begin{bmatrix} 3 \\ -1 \\ 2 \\ 5 \end{bmatrix}_{4\times 1}$$

$$= \begin{bmatrix} 1\times 3 + 0\times -1 + (-1)\times 2 + 4\times 5 \\ 2\times 3 + 3\times -1 + \ 0 \ \times 2 + 5\times 5 \\ -1\times 3 + 7\times -1 + (-2)\times 2 + 1\times 5 \end{bmatrix}_{3\times 1} = \begin{bmatrix} 21 \\ 28 \\ -9 \end{bmatrix}$$

and

$$\mathbf{AB}_{.2} = \begin{bmatrix} 1 & 0 & -1 & 4 \\ 2 & 3 & 0 & 5 \\ -1 & 7 & -2 & 1 \end{bmatrix} \begin{bmatrix} -2 \\ 3 \\ 4 \\ 1 \end{bmatrix} = \begin{bmatrix} 1\times -2 + 0\times 3 + (-1)\times 4 + 4\times 1 \\ 2\times -2 + 3\times 3 + \ 0 \ \times 4 + 5\times 1 \\ -1\times -2 + 7\times 3 + (-2)\times 4 + 1\times 1 \end{bmatrix} = \begin{bmatrix} -2 \\ 10 \\ 16 \end{bmatrix}$$

$$\therefore\ \ AB=\begin{bmatrix}\mathbf{AB}_{.1} & \mathbf{AB}_{.2}\end{bmatrix}=\begin{bmatrix}21 & -2\\28 & 10\\-9 & 16\end{bmatrix}.\ \text{Now } A=\begin{bmatrix}\mathbf{A}_{1.} & \mathbf{A}_{2.} & \mathbf{A}_{3.}\end{bmatrix}^{T},$$

where $\mathbf{A}_{1.}=\begin{bmatrix}1 & 0 & -1 & 4\end{bmatrix}$, $\mathbf{A}_{2.}=\begin{bmatrix}2 & 3 & 0 & 5\end{bmatrix}$ and $\mathbf{A}_{3.}=\begin{bmatrix}-1 & 7 & -2 & 1\end{bmatrix}$.

Also, $B=\begin{bmatrix}\mathbf{B}_{.1} & \mathbf{B}_{.2}\end{bmatrix}$ where $\mathbf{B}_{.1}=\begin{bmatrix}3 & -1 & 2 & 5\end{bmatrix}^{T}$, $\mathbf{B}_{.2}=\begin{bmatrix}-2 & 3 & 4 & 1\end{bmatrix}^{T}$.

So, the elements of the matrix product AB are as follows (illustrated stages 3) :

$$\mathbf{A}_{1.}\mathbf{B}_{.1} = 1.3 + 0.(-1) + (-1).2 + 4.5 = 3 + 0 - 2 + 20 = 21$$
$$\mathbf{A}_{1.}\mathbf{B}_{.2} = 1.(-2) + 0.3 + (-1).4 + 4.1 = -2 + 0 - 4 + 4 = -2$$
$$\mathbf{A}_{2.}\mathbf{B}_{.1} = 2.3 + 3.(-1) + 0.2 + 5.5 = 6 - 3 + 0 + 25 = 28$$
$$\mathbf{A}_{2.}\mathbf{B}_{.2} = 2.(-2) + 3.3 + 0.4 + 5.1 = -4 + 9 + 0 + 5 = 10$$
$$\mathbf{A}_{3.}\mathbf{B}_{.1} = (-1).3 + 7.(-1) + (-2).2 + 1.5 = -3 - 7 - 4 + 5 = -9$$
$$\mathbf{A}_{3.}\mathbf{B}_{.2} = (-1).(-2) + 7.3 + (-2).4 + 1.1 = 2 + 21 - 8 + 1 = 16$$

$$\text{Hence } AB=\begin{bmatrix}\mathbf{A}_{1.}\\ \mathbf{A}_{2.}\\ \mathbf{A}_{3.}\end{bmatrix}\begin{bmatrix}\mathbf{B}_{.1} & \mathbf{B}_{.2}\end{bmatrix}=\begin{bmatrix}\mathbf{A}_{1.}\mathbf{B}_{.1} & \mathbf{A}_{1.}\mathbf{B}_{.2}\\ \mathbf{A}_{2.}\mathbf{B}_{.1} & \mathbf{A}_{2.}\mathbf{B}_{.2}\\ \mathbf{A}_{3.}\mathbf{B}_{.1} & \mathbf{A}_{3.}\mathbf{B}_{.2}\end{bmatrix}=\begin{bmatrix}21 & -2\\28 & 10\\-9 & 16\end{bmatrix}_{3\times2}$$

Alternatively, (multiplication rule)

$$AB=\begin{bmatrix}1 & 0 & -1 & 4\\2 & 3 & 0 & 5\\-1 & 7 & -2 & 1\end{bmatrix}_{3\times4}\begin{bmatrix}3 & -2\\-1 & 3\\2 & 4\\5 & 1\end{bmatrix}_{4\times2}$$

$$=\begin{bmatrix}(1.3+0.(-1)+(-1).2+4.5) & (1.(-2)+0.3+(-1).4+4.1)\\(2.3+3.(-1)+0.2+5.5) & (2.(-2)+3.3+0.4+5.1)\\((-1).3+7.(-1)+(-2).2+1.5) & ((-1).(-2)+7.3+(-2).4+1.1)\end{bmatrix}_{3\times2}=\begin{bmatrix}21 & -2\\28 & 10\\-9 & 16\end{bmatrix}_{3\times2}$$

Total number of operations = (number of multiplication) + (number of addition)
= 4.(3×2) + 4.(3×2) = 4.6 + 4.6 = 48.

2.3.6 Algorithm for checking multiplication : Let us begin with the matrix

$$C = AB$$

in the last line of stage 3 of sec. 2.3.5.

$$\text{Now, } (\text{sum of } i\text{th row of C}) = \Sigma\mathbf{C}_{i.} = \sum_{k=1}^{n} a_{ik}\left(b_{k1}+b_{k2}+\cdots+b_{kp}\right)$$

$$= \sum_{k=1}^{n} a_{ik}(\text{sum of } k\text{th row of B})$$

$$= \sum_{k=1}^{n} a_{ik}\,\Sigma\mathbf{B}_{k.} \tag{2.7}$$

This shows that if a new column is created with the sums, $\Sigma\mathbf{B}_{k.}$ and is multiplied by a_{ik} then the *sum* of the products namely that on the right of equation (2.7) is equal to the *sum*, $\Sigma\mathbf{C}_{i.}$. This result provides the following algorithm for checking matrix multiplication.

Algorithm

Step 1. Sum the rows of B i.e. obtain $\sum_{k=1}^{n} \mathbf{B}_{k.}$.

Step 2. Let the elements obtain in step1 constitute the column $\mathbf{B}_{.(p+1)}$ which is incorporated to the right of $\mathbf{B}_{.p}$.

Step 3. Now obtain the product $C_{m\times(p+1)} = A_{m\times n}B_{n\times(p+1)}$, instead of the given product $C_{m\times p} = A_{m\times n}B_{n\times p}$

Step 4. If (2.7) holds for all i, $i = 1, 2, \cdots, m$, then the multiplication done is correct in all respect and the required product $C_{n\times p}$ is obtained by rejecting $(p+1)$th column of $C_{n\times(p+1)}$.

Step 5. If (2.7) does not hold for any i, the rechecking of the multiplication is necessary for that i *only* and note step 4 again for correctness of the product as well as the result of the product.

Alternative of the example below illustrate checking procedure by introducing a sum row of A, obtained by summing columns of A.

Ex. Perform AB where,

$$A = \begin{bmatrix} 1 & 0 & -1 & 4 \\ 2 & 3 & 0 & 5 \\ -1 & 7 & -2 & 1 \end{bmatrix}_{3\times 4} \quad \text{and } B = \begin{bmatrix} 3 & -2 \\ -1 & 3 \\ 2 & 4 \\ 5 & 1 \end{bmatrix}_{4\times 2}$$

Check the product AB.

Let $C_{3\times 3} = A_{3\times 4}B_{4\times 3}$

$$= \begin{bmatrix} 1 & 0 & -1 & 4 \\ 2 & 3 & 0 & 5 \\ -1 & 7 & -2 & 1 \end{bmatrix} \left[\begin{array}{cc|c} 3 & -2 & 1 \\ -1 & 3 & 2 \\ 2 & 4 & 6 \\ 5 & 1 & 6 \end{array}\right]$$

$$= \underbrace{\left[\begin{array}{cc|c} 21 & -2 & 19 \\ 28 & 10 & 38 \\ -9 & 16 & 7 \end{array}\right]}_{C_{3\times 2}}$$

Therefore, the check is satisfied since

$$21 + (-2) = 19,\ 28 + 10 = 38 \text{ and } (-9) + 16 = 7$$

are the elements in the extra columns of C. The required product is

$$C_{3\times 2} = \begin{bmatrix} 21 & -2 \\ 28 & 10 \\ -9 & 10 \end{bmatrix}.$$

Alternatively, let

$$C_{4\times2} = A_{4\times4}B_{4\times2}$$

$$= \left[\begin{array}{cccc} 1 & 0 & -1 & 4 \\ 2 & 3 & 0 & 5 \\ -1 & 7 & -2 & 1 \\ \hdashline 2 & 10 & -3 & 10 \end{array}\right]\begin{bmatrix} 3 & -2 \\ -1 & 3 \\ 2 & 4 \\ 5 & 1 \end{bmatrix}$$

Sum row of A, obtained by summing columns of A.

Separation line

$$= \left[\begin{array}{cc} 21 & -2 \\ 28 & 10 \\ -9 & 16 \\ \hdashline 40 & 24 \end{array}\right] \Big\} C_{3\times2}$$

Separation line

Row for sum of column of $C_{3\times2}$

The check is satisfied since

$$21 + 28 + (-9) = 40,\ (-2) + 10 + 16 = 24$$

are the elements in the extra row of $C_{3\times2}$. The required product is

$$C_{3\times2} = \begin{bmatrix} 21 & -2 \\ 28 & 10 \\ -9 & 16 \end{bmatrix}.$$

2.3.7 Matrix multiplication – a linear combination : Matrix multiplication can be represented as a linear combination of the row (column) vectors $\mathbf{B}_{i.}$(or $\mathbf{A}_{.j}$) of the matrix B (or A) in which the scalars are the elements of rows (columns) of the matrix A (or B). The concept of addition is being used in this matrix multiplication. The product C = AB in (4) in section 2.3.5 written explicitly in terms of elements of both A and B may be compactly expressed in terms of row (column) vectors of B (or A) as follows:

$$C = AB = \begin{bmatrix} \sum_{k=1}^{n} a_{1k}\left(b_{k1}\ b_{k2} \cdots . b_{kj} \cdots b_{kp}\right) \\ \sum_{k=1}^{n} a_{2k}\left(b_{k1}\ b_{k2} \cdots b_{kj} \cdots b_{kp}\right) \\ \vdots \quad \vdots \quad \vdots \quad \vdots \\ \sum_{k=1}^{n} a_{mk}\left(b_{k1}\ b_{k2} \cdots b_{kj} \cdots . b_{kp}\right) \end{bmatrix} = \begin{bmatrix} \sum_{k=1}^{n} a_{1k}\mathbf{B}_{k.} \\ \sum_{k=1}^{n} a_{2k}\mathbf{B}_{k.} \\ \vdots \\ \sum_{k=1}^{n} a_{mk}\mathbf{B}_{k.} \end{bmatrix}$$

$$= \begin{bmatrix} a_{11}\mathbf{B}_{1\cdot} + a_{12}\mathbf{B}_{2\cdot} + \cdots + a_{1n}\mathbf{B}_{n\cdot} \\ a_{21}\mathbf{B}_{1\cdot} + a_{22}\mathbf{B}_{2\cdot} + \cdots . + a_{2n}\mathbf{B}_{n\cdot} \\ \vdots \quad \vdots \quad \vdots \quad \vdots \\ a_{m1}\mathbf{B}_{1\cdot} + a_{m2}\mathbf{B}_{2\cdot} + \cdots + a_{mn}\mathbf{B}_{n\cdot} \end{bmatrix}_{m\times p} \tag{2.8}$$

where $\mathbf{B}_{i.} = (b_{i1}\ b_{i2}\ \cdots\ b_{ip})$, $i = 1, 2, \cdots, n$, are rows of the matrix B. The matrix in (2.8) consists of m rows as a linear combination of the row vectors $\mathbf{B}_i$ of the matrix B and the scalar multipliers are the row elements of the matrix A. Apparently the matrix in (2.8) seems to be of order $m\times 1$. Actually it is of order $m\times p$ due to the presence of p columns in each $\mathbf{B}_i$.

Similarly, (2.4) can also be written as

$$C = AB = \left[\sum_{k=1}^{n} b_{k1} \begin{bmatrix} a_{1k} \\ a_{2k} \\ \vdots \\ a_{mk} \end{bmatrix} \quad \sum_{k=1}^{n} b_{k2} \begin{bmatrix} a_{1k} \\ a_{2k} \\ \vdots \\ a_{mk} \end{bmatrix} \quad \cdots \quad \sum_{k=1}^{n} b_{kp} \begin{bmatrix} a_{1k} \\ a_{2k} \\ \vdots \\ a_{mk} \end{bmatrix} \right]$$

$$= \left[\sum_{k=1}^{n} b_{k1}\mathbf{A}_{.k} \quad \sum_{k=1}^{n} b_{k2}\mathbf{A}_{.k} \quad \cdots \quad \sum_{k=1}^{n} b_{kp}\mathbf{A}_{.k} \right]$$

$$= \begin{bmatrix} \mathbf{A}_{.1}b_{11} + \mathbf{A}_{.2}b_{21} + \cdots + \mathbf{A}_{.n}b_{n1} \\ \mathbf{A}_{.1}b_{12} + \mathbf{A}_{.2}b_{22} + \cdots + \mathbf{A}_{.n}b_{n2} \\ \vdots \qquad \vdots \qquad \vdots \\ \mathbf{A}_{.1}b_{1p} + \mathbf{A}_{.2}b_{2p} + \cdots + \mathbf{A}_{.n}b_{np} \end{bmatrix}^{T}_{m\times p} \tag{2.9}$$

where $\mathbf{A}_{.j} = \begin{bmatrix} a_{1j} & a_{2j} & \cdots & a_{mj} \end{bmatrix}^{T}, j = 1, 2, \cdots, n,$

are columns of the matrix A. The matrix in (2.9) consists of p columns as a linear combination of the column vectors $\mathbf{A}_{.j}$ of the matrix A and the scalar multipliers are the column elements of the matrix B. Although the matrix in (2.9) appears as of order $1\times p$, it is actually of $m\times p$ order due the presence of m elements in each column vectors $\mathbf{A}_{.j}$.

Therefore, from (2.8) and (2.9) above it can be easily said that the rows (columns) of the product matrix C are the linear combinations of the rows (columns) of the post multiplier B (or the premultiplier A) and the scalar multipliers for linear combination are the corresponding row (column) elements of the premultiplier A (or the post multiplier B).

Ex. Find

$$AB = \begin{bmatrix} 1 & 0 & -1 & 4 \\ 2 & 3 & 0 & 5 \\ -1 & 7 & -2 & 1 \end{bmatrix} \begin{bmatrix} 3 & -2 \\ -1 & 3 \\ 2 & 4 \\ 5 & 1 \end{bmatrix}.$$

Let $\mathbf{B}_{1.} = [3 \quad -2]$, $\mathbf{B}_{2.} = [-1 \quad 3]$, $\mathbf{B}_{3.} = [2 \quad 4]$ and $\mathbf{B}_{4.} = [5 \quad 1]$.

Expressing each row of the product as a linear combination of the rows of B, we have,

$$AB = \begin{bmatrix} 1\mathbf{B}_{1.} + 0\mathbf{B}_{2.} + (-1)\mathbf{B}_{3.} + 4\mathbf{B}_{4.} \\ 2\mathbf{B}_{1.} + 3\mathbf{B}_{2.} + \ 0\mathbf{B}_{3.} \ + 5\mathbf{B}_{4.} \\ (-1)\mathbf{B}_{1.} + 7\mathbf{B}_{2.} + (-2)\mathbf{B}_{3.} + 1\mathbf{B}_{4.} \end{bmatrix}$$

$$= \begin{bmatrix} 1[3 \ \ -2] + 0[-1 \ \ 3] - [2 \ \ 4] + 4[5 \ \ 1] \\ 2[3 \ \ -2] + 3[-1 \ \ 3] + 0[2 \ \ 4] + 5[5 \ \ 1] \\ -[3 \ \ -2] + 7[-1 \ \ 3] - 2[2 \ \ 4] + [5 \ \ 1] \end{bmatrix}$$

$$= \begin{bmatrix} [3-0-2+20] & [-2+0-4+4] \\ [6-3+0+25] & [-4+9+0+5] \\ [-3-7-4+5] & [2+21-8+1] \end{bmatrix} = \begin{bmatrix} 21 & -2 \\ 28 & 10 \\ -9 & 16 \end{bmatrix}$$

Alternate method :

Let

$$\mathbf{A}_{.1} = \begin{bmatrix} 1 \\ 2 \\ -1 \end{bmatrix}, \mathbf{A}_{.2} = \begin{bmatrix} 0 \\ 3 \\ 7 \end{bmatrix}, \mathbf{A}_{.3} = \begin{bmatrix} -1 \\ 0 \\ -1 \end{bmatrix}, \mathbf{A}_{.4} = \begin{bmatrix} 4 \\ 5 \\ 1 \end{bmatrix}.$$

Expressing each column of the product as a linear combination the columns of A, we have,

$$\mathrm{AB} = \left[(3\mathbf{A}_{.1} - \mathbf{A}_{.2} + 2\mathbf{A}_{.3} + 5\mathbf{A}_{.4}) \quad (-2\mathbf{A}_{.1} + 3\mathbf{A}_{.2} + 4\mathbf{A}_{.3} + \mathbf{A}_{.4})\right]$$

$$= \left[\left(3\begin{bmatrix} 1 \\ 2 \\ -1 \end{bmatrix} - \begin{bmatrix} 0 \\ 3 \\ 7 \end{bmatrix} + 2\begin{bmatrix} -1 \\ 0 \\ -2 \end{bmatrix} + 5\begin{bmatrix} 4 \\ 5 \\ 1 \end{bmatrix}\right) \quad \left(-2\begin{bmatrix} 1 \\ 2 \\ -1 \end{bmatrix} + 3\begin{bmatrix} 0 \\ 3 \\ 7 \end{bmatrix} + 4\begin{bmatrix} -1 \\ 0 \\ -2 \end{bmatrix} + \begin{bmatrix} 4 \\ 5 \\ 1 \end{bmatrix}\right)\right]$$

$$= \begin{bmatrix} 21 & -2 \\ 28 & 10 \\ -9 & 16 \end{bmatrix}$$

N.B. Checking can be performed by introducing a column of row sum or a row of column sum.

2.3.8 Matrix division not defined : The knowledge of the binary operations of addition, subtraction and multiplication of matrices has no natural extension to division of matrices because matrix division produces *unlimited* answer as in vector division. This grave situation is discussed below in two ways —

(i) taking matrices with numerical elements and

(ii) taking one matrix with arbitrary elements and two matrices with numerical elements.

(i) The matrix products

$$\begin{bmatrix} 1 & 1 \\ 1 & 1 \end{bmatrix}\begin{bmatrix} 2 & 2 \\ 1 & 1 \end{bmatrix} = \begin{bmatrix} 3 & 3 \\ 3 & 3 \end{bmatrix} \qquad \text{(a)}$$

and

$$\begin{bmatrix} 2 & -1 \\ 2 & -1 \end{bmatrix}\begin{bmatrix} 2 & 2 \\ 1 & 1 \end{bmatrix} = \begin{bmatrix} 3 & 3 \\ 3 & 3 \end{bmatrix} \qquad \text{(b)}$$

are considered.

Using knowledge of ordinary algebra (a) and (b) reveal that the division the matrix

$\begin{bmatrix} 3 & 3 \\ 3 & 3 \end{bmatrix}$ by $\begin{bmatrix} 2 & 2 \\ 1 & 1 \end{bmatrix}$ produces two different matrices $\begin{bmatrix} 1 & 1 \\ 1 & 1 \end{bmatrix}$ and $\begin{bmatrix} 2 & -1 \\ 2 & -1 \end{bmatrix}$. So,

matrix division is not unique as in ordinary algebra. It produces *unlimited answer* if more trials are taken as before.

(ii) Let $\mathrm{A} = \begin{bmatrix} a_{11} & a_{12} \\ a_{21} & a_{22} \end{bmatrix}$, $\mathrm{B} = \begin{bmatrix} 2 \\ 1 \end{bmatrix}$ and $\mathrm{C} = \begin{bmatrix} 5 \\ 2 \end{bmatrix}$ be such that $\mathrm{A} = \dfrac{\mathrm{B}}{\mathrm{C}}$.

Therefore, $\mathrm{AC} = \mathrm{B}$.

or $$\begin{bmatrix} a_{11} & a_{12} \\ a_{21} & a_{22} \end{bmatrix}\begin{bmatrix} 5 \\ 2 \end{bmatrix} = \begin{bmatrix} 2 \\ 1 \end{bmatrix}$$

or $5a_{11} + 2a_{12} = 2$

and $5a_{21} + 2a_{22} = 1$

This system of simultaneous equations has *two* equations with *four* unknowns. The arbitrary choice of values of any *two* unknowns will give the value of the other two. Hence matrix A will *not* be unique.

[**N.B.** The case (ii) should be read after chapter seven]

From the above two considerations, it is clear that the matrix division is *not defined*. But the purpose of matrix division is carried out by matrix inversion discussed in chapter eight. The inversion is limited by two necessary conditions

(i) square and

(ii) nonsingular

until the study of generalized inverse of matrix in chapter eighteen.

2.4 Homogeneity and additivity

The meaning of homogeneity is indeed necessary for the better understanding of the algebraic properties of matrices discussed in section 2.5 next. So, in the following paragraph it is defined and explained.

Homogeneity : If the independent variable (input) of a function is multiplied by a constant '*a*' then the dependent variable (output) i.e. the original function gets multiplied by the same constant '*a*'. This property is called *homogeneity*. This is stated as "Effect is proportional". This is also stated as "Effect is proportional to cause". Mathematically,

$$f(ax) = af(x)$$

This is one of the two *properties of principle of superposition* (P.S.). The other is *additivity*.

Additivity : $f(x + y) = f(x) + f(y)$. Here effect is the sum of causes. The P.S. is the *characteristic of linearity* of a system. The two properties of P.S. may be combined mathematically as

$$f(ax + by) = af(x) + bf(y)$$

which can be extended to several variables.

2.5 Algebraic properties

The previous sections describe definitions of the operations on matrices. This section will be devoted to the algebra of those operations one after another.

2.5.1 Algebra of transposition : The algebra of transposition are given by—

(i) Let A and B be $m \times n$ matrices and k be a scalar. Then

(a) **Additivity** : $(A + B)^T = A^T + B^T$.

(b) **Homogeneity** : $(kA)^T = kA^T$.

(ii) **Reflexive** : For any $m \times n$ matrix A, $(A^T)^T = A$.

(iii) **Reversible rule** : Let $A_{m\times n}$ and $B_{n\times p}$ be two matrices conformable for multiplication then $$(AB)^T = B^T A^T$$

i.e. *transpose of a product is the transpose of the product taken in reverse order.*

Proof. (**i.a**) Let $A = [a_{ij}]_{m\times n}$ and $B = [b_{ij}]_{m\times n}$ be the matrices conformable for addition. So,

$$A + B = [a_{ij} + b_{ij}]_{m\times n} = [c_{ij}]_{m\times n} \text{ (say).}$$

Hence by definition of transpose,

$$(A+B)^T = [c_{ij}]_{n\times m} = [a_{ji} + b_{ji}]_{n\times m} = [a_{ji}]_{n\times m} + [b_{ji}]_{n\times m} = A^T + B^T$$

i.e. *transpose of the sum of matrices is equal to the sum of their transposes.*

(**i.b**) Let $A = [a_{ij}]_{m\times n}$ be a matrix and k be a scalar. Therefore, by definition of scalar multiplication of a matrix, $kA = [ka_{ij}]_{m\times n}$. Using definition of transpose,

$$(kA)^T = [ka_{ji}]_{n\times m} = k[a_{ji}]_{n\times m} = kA^T.$$

Therefore, transpose of a scalar product of a matrix is equal to the same scalar product of the transposed matrix.

(**ii**) Let $A = [a_{ij}]_{m\times n}$. Therefore, $A^T = [a_{ji}]_{n\times m}$. Hence $(A^T)^T = [a_{ij}]_{m\times n} = A$.

Therefore, transpose of the transpose matrix is the matrix itself.

(iii) Let $A = [a_{ij}]_{m\times n}$ and $B = [b_{jk}]_{n\times p}$ be two matrices and A is conformable to B for multiplication. Therefore, $AB = [c_{ik}]_{m\times p}$ where

$$c_{ik} = \sum_{j=1}^{n} a_{ij} b_{jk} = a_{i1}b_{1k} + a_{i2}b_{2k} + \cdots + a_{in}b_{nk}, i = 1, 2, \cdots, m$$

and $k = 1, 2, \cdots, p$. Therefore,

$$(AB)^T = [c_{ki}]_{p\times m} = \left[\sum_{j=1}^{n} a_{kj} b_{ji}\right]_{p\times m} = \left[a_{k1}b_{1i} + a_{k2}b_{2i} + \cdots + a_{kn}b_{ni}\right].$$

From which it is evident that $\begin{bmatrix} a_{k1} & a_{k2} & \cdots & a_{kn} \end{bmatrix}$ is the kth row of A and

$$\begin{bmatrix} b_{1i} \\ b_{2i} \\ \vdots \\ b_{ni} \end{bmatrix}$$

is the ith column of B. This means that $\begin{bmatrix} b_{1i} & b_{2i} & \cdots & b_{ni} \end{bmatrix}$ is the ith row of B^T and

$$\begin{bmatrix} a_{k1} \\ a_{k2} \\ \vdots \\ a_{kn} \end{bmatrix}$$

is the kth column of A^T.

Therefore, B^T is conformable to A^T for the (i, k)th element of $B^T A^T$ is

$$\left(b_{1i}a_{k1} + b_{2i}a_{k2} + \cdots + b_{ni}a_{kn}\right) = \left(a_{k1}b_{1i} + a_{k2}b_{2i} + \cdots + a_{kn}b_{ni}\right),$$

using commutative law of ordinary algebra. Hence the matrix

$$B^T A^T = \left[\sum_{j=1}^{n} a_{kj} b_{ji} \right]_{p\times m}.$$

$$\therefore \qquad (AB)^T = B^T A^T.$$

Hence *transpose of a matrix product is the matrix product of the transposes taken in reverse order.* This is known as the *reversible* rule of transposition. It is of great importance and can be extended to any number of matrices. Thus

$$(ABCD \cdots K)^T = K^T \cdots C^T B^T A^T.$$

2.5.2 Algebra of conjugation : If $u = p_1 + iq_1$ and $v = p_2 + iq_2$ be complex numbers then the conjugate of u is $\bar{u} = p_1 - iq_1$ and conjugate of the conjugate u is $\bar{\bar{u}} = p_1 + iq_1 = u$, the complex number itself. Also $\overline{u+v} = \bar{u} + \bar{v}$ and $\overline{uv} = \bar{u}\,\bar{v}$. These properties are equally applicable to the matrices as follows:

(i) **Reflexive :** If A is any matrix then $\bar{\bar{A}} = A$

(ii) **Additivity :** If A and B are two matrices conformable for addition, then *conjugate of a sum (difference) is equal to the sum (difference) of the conjugates*

i.e. $$\overline{A+B} = \bar{A} + \bar{B}.$$

(iii)**Homogeneity :** If A is a matrix and k is a complex scalar then $\overline{kA} = \bar{k}\,\bar{A}$.

(iv) **Product rule :** If A and B are matrices conformable for multiplication, then $\overline{AB} = \bar{A}\,\bar{B}$,

i.e. ***conjugate of a product is equal to the product of the conjugates.***

(v) **Transpose rule :** If A is a matrix then $(\bar{A})^T = \overline{A^T}$.

Proof.

(i) Let $A = [a_{jk}]_{m\times n}$ be a C-matrix such that $a_{jk} = p_{jk} + iq_{jk}$. Then $\bar{a}_{jk} = p_{jk} - iq_{jk}$ and $\bar{\bar{a}}_{jk} = p_{jk} + iq_{jk}$. Hence $\bar{\bar{A}} = \left[\bar{\bar{a}}_{jk}\right]_{m\times n} = \left[a_{jk}\right]_{m\times n} = A$.

(ii) Let $A = [a_{jk}]_{m\times n}$ and $B = [b_{jk}]_{m\times n}$ be two C-matrices such that

$$a_{jk} = p_{jk} + iq_{jk} \quad \text{and} \quad b_{jk} = r_{jk} + is_{jk}.$$

Then
$$a_{jk} + b_{jk} = (p_{jk} + r_{jk}) + i(q_{jk} + s_{jk}) \quad \text{and}$$
$$\overline{a_{jk} + b_{jk}} = (p_{jk} + r_{jk}) - i(q_{jk} + s_{jk})$$
$$= (p_{jk} - iq_{jk}) + (r_{jk} - is_{jk}) = \bar{a}_{jk} + \bar{b}_{jk}$$

Therefore,

$$\overline{A+B} = \left[\overline{a_{jk} + b_{jk}}\right]_{m\times n} = \left[\bar{a}_{jk} + \bar{b}_{jk}\right]_{m\times n} = \left[\bar{a}_{jk}\right]_{m\times n} + \left[\bar{b}_{jk}\right]_{m\times n} = \bar{A} + \bar{B}.$$

(iii) It is left for the readers.

(iv)Let $A = [a_{jk}]_{m\times n}$ and $B = [b_{kl}]_{n\times p}$ be two C-matrices conformable for multiplication such that $a_{jk} = p_{jk} + iq_{jk}$ and $b_{kl} = r_{kl} + is_{kl}$.

Then $$a_{jk}\, b_{kl} = (p_{jk} + iq_{kl})\,(r_{kl} + is_{kl})$$
$$= (p_{jk}r_{kl} - q_{jk}s_{kl}) + i(p_{jk}s_{kl} + q_{jk}r_{kl})$$

$\therefore$ $$\overline{a_{jk}b_{kl}} = (p_{jk}r_{kl} - q_{jk}s_{kl}) - i(p_{jk}s_{kl} + q_{jk}r_{kl})$$
$$= p_{jk}(r_{kl} - is_{kl}) - iq_{jk}(r_{kl} - is_{kl})$$
$$= (p_{jk} - iq_{jk})(r_{kl} - is_{kl}) = \bar{a}_{jk}\bar{b}_{kl}$$

Hence
$$\overline{\mathrm{A\,B}} = \left[\sum_{k=1}^{n}\overline{a_{jk}b_{kl}}\right]_{m\times p} = \left[\sum_{k=1}^{n}\bar{a}_{jk}\bar{b}_{kl}\right]_{m\times p} = \left[\bar{a}_{jk}\right]_{m\times n}\left[\bar{b}_{kl}\right]_{n\times p} = \overline{\mathrm{A}}\,\overline{\mathrm{B}}.$$

(v) Let $\mathrm{A} = [a_{jk}]_{m\times n}$ be a C-matrix such that $a_{jk} = p_{jk} + iq_{jk}$. Then $\bar{a}_{jk} = p_{jk} - iq_{jk}$ and $\bar{a}_{kj} = p_{kj} - iq_{jk}$. Therefore

$$\overline{\mathrm{A}} = \left[\bar{a}_{jk}\right]_{m\times n} \text{ and } \left(\overline{\mathrm{A}}\right)^{\mathrm{T}} = \left[\bar{a}_{kj}\right]_{n\times m} \tag{a}$$

Also $$\mathrm{A}^{\mathrm{T}} = \left[a_{kj}\right]_{m\times n} \text{ and } \overline{\mathrm{A}^{\mathrm{T}}} = \left[\bar{a}_{kj}\right]_{n\times m} \tag{b}$$

Therefore, from (a) and (b) we have $\left(\overline{\mathrm{A}}\right)^{\mathrm{T}} = \overline{\mathrm{A}^{\mathrm{T}}}$.

2.5.3 Algebra of tranjugation : If A and B are C-matrices of suitable orders and k be a complex scalar, then :

(i) **Reflexive** : $(\mathrm{A}^*)^* = \mathrm{A}$
(ii) **Additivity** : $(\mathrm{A}+\mathrm{B})^* = \mathrm{A}^* + \mathrm{B}^*$
(iii) **Homogeneity** : $(k\,\mathrm{A})^* = \bar{k}\,\mathrm{A}^*$, k is a complex constant.
(iv) **Reversible rule** : $(\mathrm{AB})^* = \mathrm{B}^*\mathrm{A}^*$
(v) $\mathrm{A\,A}^* = 0$ implies $\mathrm{A} = 0$
(vi) **Cancellation law** : a . $\mathrm{BA\,A}^* = \mathrm{CA\,A}^*$ implies $\mathrm{BA} = \mathrm{CA}$
b . $\mathrm{BA}^*\mathrm{A} = \mathrm{CA}^*\mathrm{A}$ implies $\mathrm{BA}^* = \mathrm{CA}^*$

Proof .

(i) $$\left(\mathrm{A}^*\right)^* = \left[\left(\overline{\mathrm{A}}\right)^{\mathrm{T}}\right]^* = \left(\overline{\mathrm{A}^{\mathrm{T}}}\right)^* = \left(\overline{\overline{\mathrm{A}^{\mathrm{T}}}}\right)^{\mathrm{T}} = \left(\mathrm{A}^{\mathrm{T}}\right)^{\mathrm{T}} = \mathrm{A}$$

(ii) $$\left(\mathrm{A}+\mathrm{B}\right)^* = \left(\overline{\mathrm{A}+\mathrm{B}}\right)^{\mathrm{T}} = \left(\overline{\mathrm{A}}+\overline{\mathrm{B}}\right)^{\mathrm{T}} = \left(\overline{\mathrm{A}}\right)^{\mathrm{T}} + \left(\overline{\mathrm{B}}\right)^{\mathrm{T}} = \mathrm{A}^* + \mathrm{B}^*$$

(iii) $$\left(k\mathrm{A}\right)^* = \left(\overline{k\mathrm{A}}\right)^{\mathrm{T}} = \left(\bar{k}\,\overline{\mathrm{A}}\right)^{\mathrm{T}} = \bar{k}\left(\overline{\mathrm{A}}\right)^{\mathrm{T}} = \bar{k}\mathrm{A}^*$$

(iv) $$(\mathrm{A\,B})^* = (\overline{\mathrm{A\,B}})^{\mathrm{T}} = (\overline{\mathrm{A}}\,\overline{\mathrm{B}})^{\mathrm{T}} = (\overline{\mathrm{B}})^{\mathrm{T}}(\overline{\mathrm{A}})^{\mathrm{T}} = \mathrm{B}^*\mathrm{A}^*$$

(v) Let $\mathrm{AA}^* = 0$, so tr $\mathrm{AA}^* = 0$, which implies that squares of the sum of the moduli of the elements of A are equal to zero. Hence each element of A must be zero. Therefore, matrix $\mathrm{A} = 0$.

(vi) a. Since $\mathrm{BAA}^* = \mathrm{CAA}^*$,
So $\mathrm{BAA}^* - \mathrm{CAA}^* = 0$
or $(\mathrm{BAA}^* - \mathrm{CAA}^*)\,(\mathrm{B}-\mathrm{C})^* = 0\,(\mathrm{B}-\mathrm{C})^* = 0$
or $(\mathrm{BAA}^* - \mathrm{CAA}^*)\,(\mathrm{B}^* - \mathrm{C}^*) = 0$ by (ii)
or $\mathrm{BAA}^*\mathrm{B}^* - \mathrm{CAA}^*\mathrm{B}^* - \mathrm{BAA}^*\mathrm{C}^* + \mathrm{CAA}^*\mathrm{C}^* = 0$ (by distributive law)

or $(BA - CA)(BA)^* - (BA - CA)(CA)^* = 0,$

or $(BA - CA)((BA)^* - (CA)^*) = 0$

or $(BA - CA)(BA - CA)^* = 0$

or $BA - CA = 0$ by (v)

or $BA = CA$

Hence the assertion (vi) b. It follows the same procedure as of proof of (vi)a.

2.6 Algebra of addition and subtraction

Algebraic laws of matrix addition are identical with those of ordinary algebra of numbers although there remains fundamental differences in addition procedures. The following algebraic laws of addition hold true for matrix addition if the matrices concerned are conformable for addition:

1. **Commutative law** : $A + B = B + A$
2. **Associative law** : $A + (B + C) = (A + B) + C$
3. **Additive identity** : $A + 0 = A$
4. **Additive inverse** : $A + (-)A = 0$
5. **Solving equation** : $A + C = B$ iff $C = B - A$
6. **Cancellation law for addition** : $A + C = B + C$ iff $A = B$

Proof.

1. Let $A = [a_{ij}]_{m\times n}$ and $B = [b_{ij}]_{m\times n}$ be two matrices conformable for addition. Therefore,

$$A + B = [a_{ij} + b_{ij}]_{m\times n} = [b_{ij} + a_{ij}]_{m\times n}$$

$$= [b_{ij}]_{m\times n} + [a_{ij}]_{m\times n},$$

by ordinary commutative law for algebra.

$$= B + A \quad \text{, by addition law.}$$

Thus matrix addition is commutative.

2. Let $A = [a_{ij}]_{m\times n}$, $B = [b_{ij}]_{m\times n}$ and $C = [c_{ij}]_{m\times n}$ be three matrices conformable for addition pairwise. Now $B + C = [b_{ij} + c_{ij}]_{m\times n}$, addition law.

$$\therefore \quad A + (B + C) = [a_{ij}]_{m\times n} + [b_{ij} + c_{ij}]_{m\times n}$$

$$= [a_{ij} + (b_{ij} + c_{ij})]_{m\times n}, \text{ addition law}$$

$$= [(a_{ij} + b_{ij}) + c_{ij}]_{m\times n}, \text{ associative law for ordinary algebra.}$$

$$= [a_{ij} + b_{ij}]_{m\times n} + [c_{ij}]_{m\times n}$$

$$= (A + B) + C$$

In this way it can be shown that all other permutations for adding are possible i.e.

$$(A + C) + B = A + (C + B)$$

$$B + (A + C) = (B + A) + C \text{ etc.}$$

Therefore, it is immaterial to use parenthesis in the addition, $A + B + C$.

3. Additive identity is a quantity when any thing added to it, it contributes nothing. Let $A = [a_{ij}]_{m\times n}$ be an arbitrary matrix and $0 = [0_{ij}]_{m\times n}$ be a zero matrix. Therefore, adding

$$A + 0 = [a_{ij} + 0_{ij}]_{m\times n}$$

$$= [a_{ij}]_{m\times n} \quad [\because 0 \text{ is an additive identity of ordinary algebra of numbers}]$$

$$= A$$

Hence 0 (zero) matrix is an additive identity in matrix algebra.

4. Additive inverse is a quantity which on adding to some other quantity produces the additive identity. Let $A = [a_{ij}]_{m\times n}$ be a matrix. Then by subtraction law,

$$A + (-)A = [a_{ij}]_{m\times n} + \{-[a_{ij}]_{m\times n}\}$$
$$= [a_{ij} - a_{ij}]_{m\times n} = [0_{ij}]_{m\times n} = 0, \text{ (zero) matrix.}$$

which is the additive identity. Therefore, negative of any matrix is the additive inverse of that matrix. *Alternatively*, this property can be stated as if $A + B = 0_{m\times n}$ for any $m\times n$ matrix A then B is the additive inverse of A.

5. Let $A = [a_{ij}]_{m\times n}$, $B = [b_{ij}]_{m\times n}$ and $C = [c_{ij}]_{m\times n}$. Therefore, A and C are conformable for addition. As $A + C = B$, so $a_{ij} + c_{ij} = b_{ij}$ or $c_{ij} = b_{ij} - a_{ij}$. Hence,

$$[c_{ij}]_{m\times n} = [b_{ij} - a_{ij}]_{m\times n} = [b_{ij}]_{m\times n} - [a_{ij}]_{m\times n}, \text{ by subtraction law.}$$

or $\quad C = B - A.$

Similarly, $A = B - C$. These can be proved using 4 as follows: adding the additive inverse of A i.e. $(-A)$ to both sides of the given equation. We get,

$$(-A) + A + C = B + (-A)$$

or $\quad (-A + A) + C = B - A,$ associative law

or $\quad 0 + C = B - A,$ additive inverse

or $\quad C = B - A,$ additive identity.

Converse can be established by adding the additive inverse A of $(-A)$ to both sides of $C = B - A$.

i.e. $\quad C + A = B + ((-A) + A)$
$\quad = B + 0 = B$

or $\quad A + C = B.$

6. Let $A = [a_{ij}]_{m\times n}$, $B = [b_{ij}]_{m\times n}$ and $C = [c_{ij}]_{m\times n}$ be matrices conformable for addition. Now $A + C = B + C$ gives $[a_{ij} + c_{ij}]_{m\times n} = [b_{ij} + c_{ij}]_{m\times n}$. From equality of two matrices,

$$a_{ij} + c_{ij} = b_{ij} + c_{ij}$$

or $\quad a_{ij} = b_{ij}$, by cancellation law of ordinary algebra.

$\therefore \quad [a_{ij}]_{m\times n} = [b_{ij}]_{m\times n}$

or $\quad A = B.$

Converse can be established easily by adding matrix C on both sides of $A = B$.

2.7 Algebra of multiplication

Algebra of multiplication consists of (i) scalar multiplication of matrices, (ii) scalar multiplication of vectors and (iii) multiplication of matrix with vectors.

2.7.1 Scalar multiplication of matrices : Algebraic laws for scalar multiplication are as follows:

1. **Distributive law of a scalar over matrix addition** : $u(A + B) = uA + uB$
2. **Distributive law of a matrix over scalar addition** : $(u + v)A = uA + vA$
3. **Associative law of scalar multiplication** : $u(vA) = (uv)A$
4. $uA = 0_{m\times n}$ if $u = 0$ in which both u and 0 are scalars or $A = 0_{m\times n}$.
5. **Cancellation law for scalar multiplication** :

   $uA = uB$ and $u \neq 0$ imply $A = B$.

6. **Scalar multiplication by unity, 1** : $(1)A = A$.
7. **Negation** : $(-1)A = -A$.
8. **Scalar multiplication of a zero matrix** :

   $a0_{m\times n} = 0_{m\times n}$, both 0's are zero matrices.

9. **Scalar multiplication by zero** : $0A = 0_{m\times n}$, the left hand 0 is zero scalar whereas right hand zero is the zero matrix.

Proof .

1. Let $A = [a_{ij}]_{m\times n}$, $B = [b_{ij}]_{m\times n}$ be conformable for addition and u be a scalar. Therefore, $A + B = [a_{ij} + b_{ij}]_{m\times n}$. Now from scalar multiplication rule,

$$u(A+B) = u[a_{ij}+b_{ij}]_{m\times n} = [ua_{ij}+ub_{ij}]_{m\times n}$$

$$= [ua_{ij}]_{m\times n} + [ub_{ij}]_{m\times n}, \text{ reverse of addition law}$$

$$= u[a_{ij}]_{m\times n} + u[b_{ij}]_{m\times n}, \text{ reverse of scalar multiplication}$$

$$= uA + uB$$

Therefore, scalar u is distributed over the sum of the matrices A and B.

2. Let $A = [a_{ij}]_{m\times n}$ be a matrix and u and v be two scalars. Then $u + v = k$ (say) another scalar.

So $\quad (u+v)A = k[a_{ij}]_{m\times n} = [ka_{ij}]_{m\times n}$

$$= [(u+v)a_{ij}]_{m\times n} = [ua_{ij}+va_{ij}]_{m\times n}$$

$$= [ua_{ij}]_{m\times n} + [va_{ij}]_{m\times n}$$

$$= u[a_{ij}]_{m\times n} + v[a_{ij}]_{m\times n} = uA + vB$$

Thus the matrix A is distributed over scalar sum of u and v.

3. Let $A = [a_{ij}]_{m\times n}$. So, by scalar multiplication rule ,

$$vA = [va_{ij}]_{m\times n} \text{ and}$$

$$u(vA) = u[va_{ij}]_{m\times n} = [(uv)a_{ij}]_{m\times n} = (uv)[a_{ij}]_{m\times n} = (uv)A.$$

4. Let $A = [a_{ij}]_{m\times n}$ be a matrix and u be a scalar. From the scalar multiplication rule, $uA = [ua_{ij}]_{m\times n}$. Since $uA = 0_{m\times n}$ or , $[ua_{ij}]_{m\times n} = 0_{m\times n}$.

$\therefore$ from the definition of equality of two matrices

$ua_{ij} = 0_{ij}$, 0_{ij} is a scalar element of the zero matrix.

$\therefore$ either $u = 0$ or $a_{ij} = 0_{ij}$.

Hence $u = 0$ or $A = [a_{ij}] = [0_{ij}] = 0_{m\times n}$.

6 , 7. Both follow directly from scalar multiplication rule.

8 , 9. Both can be easily seen to be true. Since in 8, a nonzero scalar multiplies 0(zero) elements producing zero elements. Whereas in 9 nonzero elements multiplied by a zero scalar produce zero elements.

2.7.2 Scalar multiplication of vectors: If **x**, **y**, **z** are vectors and k is a scalar then

i. **Commutative** : **x.y = y.x** ,

ii. **Dot product with zero vector** : **x.0 = 0**

iii. **Distributive** : **x.(y + z) = x.y + x.z**

iv. **Association with scalar,** k : **x.**$(k\mathbf{y}) = k(\mathbf{x.y})$

v. **Square of Euclidean norm** : $\mathbf{x.x} = x_1^2 + x_2^2 + \cdots + x_n^2$.

Proof.

Let $\mathbf{x} = [x_1 \ x_2 \ \cdots \ x_n]$, $\mathbf{y} = [y_1 \ y_2 \ \cdots \ y_n]$, $\mathbf{z} = [z_1 \ z_2 \ \cdots \ z_n]$ and $\mathbf{0} = [0 \ 0 \ \cdots \ 0]$.

i. By definition,

$\mathbf{x}.\mathbf{y} = x_1y_1 + x_2y_2 + \cdots + x_ny_n$ and $\mathbf{y}.\mathbf{x} = y_1x_1 + y_2x_2 + \cdots + y_nx_n$. But for numbers $x_i\ y_i = y_ix_i$, $i = 1, 2, \cdots, n$.

$\therefore \quad \mathbf{x}.\mathbf{y} = \mathbf{y}.\mathbf{x}$

ii. $\mathbf{x}.0 = x_1(0) + x_2(0) + \cdots + x_n(0) = 0$

iii. $\mathbf{y}+\mathbf{z} = \begin{bmatrix} y_1+z_1 & y_2+z_2 & \cdots & y_n+z_n \end{bmatrix}$

$$\therefore \quad \mathbf{x}.(\mathbf{y}+\mathbf{z}) = x_1(y_1+z_1) + x_2(y_2+z_2) + \cdots + x_n(y_n+z_n)$$
$$= (x_1y_1 + x_1y_2 + \cdots + x_ny_n) + (x_1z_1 + x_2z_2 + \cdots + x_nz_n)$$
$$= \mathbf{x}.\mathbf{y} + \mathbf{x}.\mathbf{z}$$

iv. $k\mathbf{y} = \begin{bmatrix} ky_1 & ky_2 & \cdots & ky_n \end{bmatrix}$

$$\mathbf{x}.(k\mathbf{y}) = x_1ky_1 + x_2ky_2 + \cdots + x_nky_n$$
$$= k(x_1y_1 + x_2y_2 + \cdots + x_ny_n) = k(\mathbf{x}.\mathbf{y})$$

v. Follows by definition (Also section 16.2.8).

2.7.3 Multiplication of matrix with vector : This algebra consists of two parts—

Part 1. If A is an arbitrary $m \times n$ matrix & **x**, **y** are $n \times 1$ column vectors, then for each scalar h & k,

(i) **Association of matrix with scalar times a vector** : $A(k\mathbf{x}) = kA\mathbf{x}$

(ii) **Distribution of matrix over vector addition** : $A(\mathbf{x}+\mathbf{y}) = A\mathbf{x} + A\mathbf{y}$

(iii) **Identity matrix as identity of vector** : $I\mathbf{x} = \mathbf{x}$

(iv) **Distribution of matrix over linear combination of vectors :**

$$A(h\mathbf{x} + k\mathbf{y}) = hA\mathbf{x} + kA\mathbf{y}.$$

Part 2. Multiplicative association of matrices with a vector as post multiplicant: If A is conformable for multiplication with B and **c** be a vector conformable for multiplication with AB, then A (B**c**) = (AB)**c**.

Proof .

Part 1. Recall stage 2 in section 2.3.5

(i) $$A(k\mathbf{x}) = \begin{bmatrix} \mathbf{A}_{1.} \\ \mathbf{A}_{2.} \\ \vdots \\ \mathbf{A}_{m.} \end{bmatrix}_{m \times n} \begin{bmatrix} kx_1 \\ kx_2 \\ \vdots \\ kx_n \end{bmatrix} = \begin{bmatrix} \mathbf{A}_{1.}k\mathbf{x} \\ \mathbf{A}_{2.}k\mathbf{x} \\ \vdots \\ \mathbf{A}_{m.}k\mathbf{x} \end{bmatrix} = k \begin{bmatrix} \mathbf{A}_{1.}\mathbf{x} \\ \mathbf{A}_{2.}\mathbf{x} \\ \vdots \\ \mathbf{A}_{m.}\mathbf{x} \end{bmatrix} = kA\mathbf{x}$$

(ii) $$A(\mathbf{x}+\mathbf{y}) = \begin{bmatrix} \mathbf{A}_{1.} \\ \mathbf{A}_{2.} \\ \vdots \\ \mathbf{A}_{m.} \end{bmatrix}_{m \times n} (\mathbf{x}+\mathbf{y}) = \begin{bmatrix} \mathbf{A}_{1.}(\mathbf{x}+\mathbf{y}) \\ \mathbf{A}_{2.}(\mathbf{x}+\mathbf{y}) \\ \vdots \\ \mathbf{A}_{m.}(\mathbf{x}+\mathbf{y}) \end{bmatrix} = \begin{bmatrix} \mathbf{A}_{1.}\mathbf{x} + \mathbf{A}_{1.}\mathbf{y} \\ \mathbf{A}_{2.}\mathbf{x} + \mathbf{A}_{2.}\mathbf{y} \\ \vdots \\ \mathbf{A}_{m.}\mathbf{x} + \mathbf{A}_{m.}\mathbf{y} \end{bmatrix}$$

$$= \begin{bmatrix} \mathbf{A}_{1.}\mathbf{x} \\ \mathbf{A}_{2.}\mathbf{x} \\ \vdots \\ \mathbf{A}_{m.}\mathbf{x} \end{bmatrix} + \begin{bmatrix} \mathbf{A}_{1.}\mathbf{y} \\ \mathbf{A}_{2.}\mathbf{y} \\ \vdots \\ \mathbf{A}_{m.}\mathbf{y} \end{bmatrix} = A\,\mathbf{x} + A\,\mathbf{y}$$

(iii) $$\mathbf{I}\mathbf{x} = \begin{bmatrix} \mathbf{e}_1^T \\ \mathbf{e}_2^T \\ \vdots \\ \mathbf{e}_n^T \end{bmatrix} \mathbf{x} = \begin{bmatrix} \mathbf{e}_1^T\mathbf{x} \\ \mathbf{e}_2^T\mathbf{x} \\ \vdots \\ \mathbf{e}_n^T\mathbf{x} \end{bmatrix} = \begin{bmatrix} x_1 \\ x_2 \\ \vdots \\ x_n \end{bmatrix} = \mathbf{x}$$

(iv) $$\begin{aligned} \mathrm{A}(h\mathbf{x} + k\mathbf{y}) &= \mathrm{A}(h\mathbf{x}) + \mathrm{A}(k\mathbf{y}) \quad \text{by (ii)} \\ &= h\mathrm{A}\,\mathbf{x} + k\mathrm{A}\,\mathbf{y} \quad \text{by (i)} \end{aligned}$$

Part 2. Let $\mathbf{c} = \begin{bmatrix} c_1 & c_2 & \cdots & c_n \end{bmatrix}$

$$= c_1\mathbf{e}_1 + c_2\mathbf{e}_2 + \cdots + c_n\mathbf{e}_n$$

$$\begin{aligned} \therefore \quad \mathrm{B}\mathbf{c} &= \mathrm{B}(c_1\mathbf{e}_1 + c_2\mathbf{e}_2 + \cdots + c_n\mathbf{e}_n) \\ &= c_1\mathrm{B}\mathbf{e}_1 + c_2\mathrm{B}\mathbf{e}_2 + \cdots + c_n\mathrm{B}\mathbf{e}_n \\ &= c_1\mathbf{B}_{.1} + c_2\mathbf{B}_{.2} + \cdots + c_n\mathbf{B}_{.n} \qquad [\because \mathrm{B}\,\mathbf{e}_i = \mathbf{B}_{.i}] \end{aligned}$$

$$\begin{aligned} \therefore \quad \mathrm{A}(\mathrm{B}\mathbf{c}) &= \mathrm{A}(c_1\mathbf{B}_{.1} + c_2\mathbf{B}_{.2} + \cdots + c_n\mathbf{B}_{.n}) \\ &= c_1\mathrm{A}\mathbf{B}_{.1} + c_2\mathrm{A}\mathbf{B}_{.2} + \cdots + c_n\mathrm{A}\mathbf{B}_{.n} \\ &= c_1(\mathrm{AB})\mathbf{e}_1 + c_2(\mathrm{AB})\mathbf{e}_2 + \cdots + c_n(\mathrm{AB})\mathbf{e}_n \\ &= (\mathrm{AB})(c_1\mathbf{e}_1 + c_2\mathbf{e}_2 + \cdots + c_n\mathbf{e}_n) = (\mathrm{AB})\mathbf{c} \end{aligned}$$

$$\left[\begin{aligned} \because\ &i\text{th column of AB} = (\mathrm{AB})\mathbf{e}_i \text{ and} \\ &i\text{th column of AB} = \text{A times } i\text{th column of B} \\ &\qquad = \mathrm{A}\mathbf{B}_{.i} = (\mathrm{AB})\mathbf{e}_i \end{aligned}\right]$$

2.7.4 Multiplication of matrix with matrix : Three marked *differences* are seen in matrix multiplication or Cayley product of two matrices with the multiplication in ordinary algebra.

(1) Matrix multiplication is not, in general, commutative and so this algebra of matrices are historically known as *noncommutative*.

(2) When the product AB = 0 but neither A nor B is zero then the factors A and B are called *divisor of zero* or *zero divisor*. Matrix product is zero divisor i.e. AB = 0 does not imply either the matrix A = 0 or the matrix B = 0.

(3) Cancellation law does not in general hold true in matrix multiplication i.e. AB = AC does not imply B = C even if A ≠ 0.

Algebraic laws true for Cayley product are embodied below with proofs fully or partly:

1. **Commutative law :** Does not true in general i.e. AB ≠ BA in general although two matrices A and B are conformable for multiplication, except in special cases.
2. **Zero divisor** : AB = 0 does not imply either A = 0 or B = 0
3. **Cancellation law** : AB = AC does not imply B = C.
4. **Associative law** : (AB)C = A(BC), provided concerned multiplications are conformable.
5. **Distributive laws :**

   (i) $(\mathrm{A}+\mathrm{B})\mathrm{C} = \mathrm{AC} + \mathrm{BC}$, right distribution

   (ii) $\mathrm{A}(\mathrm{B}+\mathrm{C}) = \mathrm{AB} + \mathrm{AC}$, left distribution,

provided respective conformability holds true.

6. **Multiplication by scalar zero :** $0\mathrm{A} = \mathrm{A}0 = 0_{m\times n}$.
7. **Multiplication by unit matrix :** $\mathrm{AI}_n = \mathrm{I}_m\mathrm{A} = \mathrm{A}$ for an $m\times n$ matrix A (cf. chapter 3).

8. Homogeneity : $u(\mathrm{A\,B}) = (u\mathrm{A})\mathrm{B} = \mathrm{A}(u\mathrm{B})$ for scalar u and matrices A and B conformable for multiplication.

9. Product of two scalar multiplications : $(u\mathrm{A})(v\mathrm{B}) = uv(\mathrm{AB})$ for scalars u and v and matrices A , B conformable for multiplication.

10. Product of double negation : $(-\mathrm{A})(-\mathrm{B}) = \mathrm{AB}$.

11. Product of single negation : $\mathrm{A}(-\mathrm{B}) = (-\mathrm{A})\mathrm{B} = -\mathrm{AB}$.

Proof .

1. The rule for matrix multiplication or Cayley product for two matrices is "*the number of columns* of premultiplicant must be equal to the *number* of *rows of post multiplicant*". Let A = $[a_{ij}]_{m\times n}$ and B = $[b_{jk}]_{n\times p}$ with $m \neq p$. Then the product AB is defined by the rule whereas the product BA is not defined. So, question of *commutative* does not arise at all. But if $m = p$ and the matrices are A = $[a_{ij}]_{m\times n}$, B = $[b_{ji}]_{n\times m}$ then both the products AB and BA are defined. Yet, AB may not be equal to BA. Thus

$$\mathrm{AB} = \left[\sum_{j=1}^{n} a_{ij} b_{ji}\right]_{m\times m}$$

which is an m-square matrix , with each element as a sum of n products and

$$\mathrm{BA} = \left[\sum_{i=1}^{m} b_{ji} a_{ij}\right]_{n\times n}$$

is an n-square matrix with each element as a sum of m products. Thus, AB $\neq$ BA.

Again if both A and B are square matrices in that case also AB may not be equal to BA as can be easily seen by checking elements in the above products when $m = n$. Thus the truth of the following statement is established:

matrix multiplication is not , in general , *commutative.*

2. In matrix theory, AB = 0 *does not imply that either* A = 0 *or* B = 0 as in ordinarily algebra. To illustrate the fact, let

$$\mathrm{A} = \begin{bmatrix} 1 & 0 & 3 \\ 2 & 0 & 4 \\ -1 & 0 & 7 \end{bmatrix} \text{ and } \mathrm{B} = \begin{bmatrix} 0 & 0 & 0 \\ 3 & -1 & 7 \\ 0 & 0 & 0 \end{bmatrix}$$

none of which is a zero matrix. But

$$\mathrm{AB} = \begin{bmatrix} 1 & 0 & 3 \\ 2 & 0 & 4 \\ -1 & 0 & 7 \end{bmatrix} \begin{bmatrix} 0 & 0 & 0 \\ 3 & -1 & 7 \\ 0 & 0 & 0 \end{bmatrix} = \begin{bmatrix} 0 & 0 & 0 \\ 0 & 0 & 0 \\ 0 & 0 & 0 \end{bmatrix},$$

a zero matrix. Thus there arises a definition "*Zero divisors*" for two such matrices.

If two nonzero matrices A and B are conformable for multiplication and their product AB = 0 then the pair of matrices A and B are called *zero divisors.*

This is a *special* character of matrix theory contradictory to ordinary algebra where $ab = 0$ means either $a = 0$ or $b = 0$ or both $a = b = 0$.

3. This is another contrast of the laws of matrix algebra with that of ordinary algebra. If A, B, C are three matrices such that AB = AC without having B = C i.e. A *cannot* be cancelled from the equation AB = AC although A ≠ 0. For example, let

$$A=\begin{bmatrix}1&2&0\\1&1&0\\-1&4&0\end{bmatrix},\ B=\begin{bmatrix}1&2&3\\1&1&-1\\2&2&2\end{bmatrix},\ C=\begin{bmatrix}1&2&3\\1&1&-1\\1&1&1\end{bmatrix}$$

Then $AB=\begin{bmatrix}3&4&1\\2&3&2\\3&2&-7\end{bmatrix}=AC$. But B ≠ C and A ≠ 0.

This verifies the truth that *law of cancellation does not hold good in matrix multiplication.*

4. Let $A=[a_{ij}]_{m\times n}$, $B=[b_{jk}]_{n\times p}$, $C=[c_{kl}]_{p\times q}$. Now using matrix multiplication rule, the product AB is conformable for multiplication. So, the (i, k)th element in the product is given by

$$a_{i1}b_{1k}+a_{i2}b_{2k}+\cdots+a_{in}b_{nk}=\sum_{j=1}^{n}a_{ij}b_{jk}$$

∴ the matrix $AB=\left[\sum_{j=1}^{n}a_{ij}b_{jk}\right]_{m\times p}$ which is again conformable for post multiplication

by

$C=[c_{kl}]_{p\times q}$. Hence,

$$(A\,B)C=\left[\sum_{k=1}^{p}\left(\sum_{j=1}^{n}a_{ij}b_{jk}\right)c_{kl}\right]_{m\times q}$$

$$=\left[\sum_{k=1}^{p}\left(\sum_{j=1}^{n}a_{ij}b_{jk}c_{kl}\right)\right],$$

obtained by multiplying each sum in parentheses by the factor c_{kl}.

$$=\left[\sum_{j=1}^{n}\left(\sum_{k=1}^{p}a_{ij}b_{jk}c_{kl}\right)\right],$$

since the order of summation is arbitrary in a finite sum.

$$\therefore\quad (A\,B)C=\left[\left(\sum_{j=1}^{n}a_{ij}\right)\left(\sum_{k=1}^{p}b_{jk}c_{kl}\right)\right]_{m\times q}$$

$$=\left[a_{ij}\right]_{m\times n}\left[\left(\sum_{k=1}^{p}b_{jk}c_{kl}\right)\right]_{n\times q},$$

by reverse of multiplication rule.

$$=A(BC).$$

∴ *associative property holds true for matrix multiplication.*

5.(i) Let A = $[a_{ij}]_{m\times n}$, B = $[b_{ij}]_{m\times n}$, C = $[c_{jk}]_{n\times p}$ be three matrices of which A, B are conformable for addition. Their sum (A + B) and C are conformable for multiplication. Also A, C and B, C are conformable for multiplications independently.

Now $A+B=[a_{ij}+b_{ij}]_{m\times n}$. So, the product

$$(A+B)C=[a_{ij}+b_{ij}]_{m\times n}[c_{jk}]_{n\times p}$$

$$=[(a_{i1}+b_{i1})c_{1k}+(a_{i2}+b_{i2})c_{2k}+\cdots+(a_{in}+b_{in})c_{nk}]_{m\times p}$$

$$=[(a_{i1}c_{1k}+a_{i2}c_{2k}+\cdots+a_{in}c_{nk})+(b_{i1}c_{1k}+b_{i2}c_{2k}+\cdots+b_{in}c_{nk})]_{m\times p}$$

using distributive and associative laws of ordinary algebra.

$$=\left[\sum_{j=1}^{n}a_{ij}c_{jk}\right]_{m\times p}+\left[\sum_{j=1}^{n}b_{ij}c_{jk}\right]_{m\times p}$$

$= [(i, j)$th element of the product $AC]_{m\times p}$ +

$[(i, j)$th element of the product $BC]_{m\times p}$

∴ $(A + B)C = AC + BC$, which establishes the right distribution of a matrix over a sum of matrices.

Similarly, letting A = $[a_{ij}]_{m\times n}$, B = $[b_{jk}]_{n\times p}$, C = $[c_{jk}]_{n\times p}$ left distribution can be easily established. Therefore combining the two distributions, we find that *matrix multiplication is distributive with respect to addition.*

6. If a nonzero matrix A is premultiplied or post multiplied by the scalar zero, the product is a zero matrix. This can be easily seen from the property of scalar multiplication of a matrix.

7. Let A = $[a_{ij}]_{m\times n}$ be an n square matrix, $I_n = [\delta_{ji}]_n$ and $I_m = [\delta_{ji}]_m$ be two square unit matrices. Now

$$AI_n=\left[\sum_{j=1}^{n}a_{ij}\delta_{ji}\right]_{m\times n}=\left[a_{i1}\delta_{1i}+a_{i2}\delta_{2i}+\cdots+a_{in}\delta_{ni}\right]_{m\times n}$$

$$=\left[a_{ij}\right]_{m\times n}=A.$$

Similarly, $I_mA=\left[\sum_{i=1}^{m}\delta_{ji}a_{ij}\right]_{m\times n}=\left[a_{ij}\right]_{m\times n}=A.$

∴ $AI_n = I_mA = A$ for any $(m\times n)$ matrix A.

8. Let $A = [a_{ij}]_{m\times n}$ and $B = [b_{jk}]_{n\times p}$ be two matrices conformable for matrix multiplication or Caley product and u be a scalar. From the law of matrix multiplication, let AB = C, where C = $[c_{ik}]_{m\times p}$ and

$$c_{ik}=a_{i1}b_{1k}+a_{i2}b_{2k}+\cdots+a_{in}b_{nk}.$$

Now from scalar multiplication,

$$(AB) = uC = u[c_{ik}]_{m\times p} = [uc_{ik}]_{m\times p}$$

$$= [ua_{i1}b_{1k} + ua_{i2}b_{2k} + \cdots + ua_{in}b_{nk}]_{m\times p}$$

$$= [(ua_{i1})b_{1k} + (ua_{i2})b_{2k} + \cdots + (ua_{in})b_{nk}]_{m\times p}$$

$$= \left[\sum_{j=1}^{n}(ua_{ij})b_{jk}\right]_{m\times p}, \text{ by associative law of ordinary algebra}$$

$$= \{(i,k)\text{th element of the product } (uA) \text{ and } B\}$$

$$= (uA)B$$

Similarly, $u(AB) = [a_{i1}(ub_{1k}) + a_{i2}(ub_{2k}) + \cdots + a_{in}(ub_{nk})]_{m\times p}$,

by associative law of ordinary algebra.

$$= \left[\sum a_{ij}(ub_{jk})\right]_{m\times p}$$

$$= \{(i,k)\text{th element of the product A and } (uB)\}$$

$$= A(uB)$$

$\therefore\ u(AB) = (uA)B = A(uB)$ is the required *homogeneity* law.

9. Let $A = [a_{ij}]_{m\times n}$ and $B = [b_{jk}]_{n\times p}$ be two matrices conformable for multiplication and u and v be two scalars. Then

$$(uA)(uB) = \left[\sum_{j=1}^{n}(ua_{ij})(vb_{jk})\right]_{m\times p} = \left[\sum_{j=1}^{n}(uv)(a_{ij}b_{jk})\right]_{m\times p}$$

$$= \left[(uv)\sum_{j=1}^{n}(a_{ij}b_{jk})\right]_{m\times p} = (uv)\left[\sum_{j=1}^{n}a_{ij}b_{jk}\right]_{m\times p}$$

$$= (uv)(AB)$$

10. This follows easily from 9 when $u = -1$, $v = -1$. By ordinary algebra, it is established that

$uv = (-1)(-1) = 1$ i.e. double negation becomes positive.

11. This also follows from 9 when $u = 1$ and $v = -1$ and vice versa.
Since $uv = (1)(-1) = (-1)(1) = (-1)$, then A (–B) = (–A)B = –(AB).

2.8 S.u. vectors in the construction of matrix unit, Kronecker delta, rows, columns and elements of a matrix

This section justifies the term "unit" of s.u. vectors through the construction of matrix unit , rows , columns and elements of a matrix. Finally, a matrix is expressed as a sum of s.u. vectors.

Matrix unit : The product of the s.u. vector $\mathbf{e}_i$ and the transpose of the s.u. vector $\mathbf{e}_j$ constructs the matrix unit E_{ij} and the order of E_{ij} is $m\times n$ where m and n are orders of $\mathbf{e}_i$ and $\mathbf{e}_j$ respectively, i.e. $\mathbf{e}_i\,\mathbf{e}_j^T = E_{ij}$.

For example, $\mathbf{e}_2\mathbf{e}_3^T = \begin{bmatrix}0\\1\\0\\0\end{bmatrix}_{4\times1} \begin{bmatrix}0 & 0 & 1 & 0\end{bmatrix}_{1\times4} = \begin{bmatrix}0&0&0&0\\0&0&1&0\\0&0&0&0\\0&0&0&0\end{bmatrix} = E_{23}$

$$\mathbf{e}_2\mathbf{e}_3^T = \begin{bmatrix}0\\1\\0\\0\end{bmatrix}_{4\times1} \begin{bmatrix}0 & 0 & 1 & 0 & 0\end{bmatrix}_{1\times5} = \begin{bmatrix}0&0&0&0&0\\0&0&1&0&0\\0&0&0&0&0\\0&0&0&0&0\end{bmatrix} = E_{23}$$

The order of the matrix units depends on the order of the s.u. vectors and it is chosen according to requirement.

Kronecker delta : The product of the transpose of the s.u. vector $\mathbf{e}_i^T$ and the s.u. vector $\mathbf{e}_j$ and vice-versa construct the Kronceker delta δ_{ij}, a scalar (an 1×1 matrix). That is $\mathbf{e}_i^T\mathbf{e}_j = \mathbf{e}_j^T\mathbf{e}_i = \delta_{ij}$ .For example

$$\delta_{11} = \mathbf{e}_1^T\mathbf{e}_1 = \begin{bmatrix}1 & 0 & 0 & 0\end{bmatrix}\begin{bmatrix}1\\0\\0\\0\end{bmatrix} = [1] \text{ and } \delta_{23} = \mathbf{e}_2^T\mathbf{e}_3 = \begin{bmatrix}0 & 1 & 0\end{bmatrix}\begin{bmatrix}0\\0\\1\end{bmatrix} = [0]$$

Thus in general, $\delta_{ij} = \begin{cases} 1, & \text{when } i = j \\ 0, & \text{when } i \neq j \end{cases}$

Column : Let the n columns of the $m\times n$ matrix A be denoted by $\mathbf{A}_{.1}, \mathbf{A}_{.2}, \cdots, \mathbf{A}_{.n}$. Hence

the jth column $= \mathbf{A}_{.j} = \begin{bmatrix}a_{1j}\\a_{2j}\\\vdots\\a_{mj}\end{bmatrix}, \quad j = 1, 2, \cdots, n$

$$= \begin{bmatrix} a_{11} & a_{12} & \cdots & a_{1j} & \cdots & a_{1n} \\ a_{21} & a_{22} & \cdots & a_{2j} & \cdots & a_{2n} \\ \vdots & \vdots & \vdots & \vdots & \vdots & \vdots \\ \vdots & \vdots & \vdots & \vdots & \vdots & \vdots \\ \vdots & \vdots & \vdots & \vdots & \vdots & \vdots \\ a_{m1} & a_{m2} & \cdots & a_{mj} & \cdots & a_{mn} \end{bmatrix}_{m\times n} \begin{bmatrix}0\\0\\\vdots\\1\\\vdots\\0\end{bmatrix}_{n\times1} \begin{matrix}\\\\\\\leftarrow j\text{th row}\\\\\\\end{matrix}$$

(jth column ↓ marks the column containing $a_{1j}, a_{2j}, \ldots, a_{mj}$)

$= \mathbf{Ae}_j$

i.e. jth column of A is the post multiplication of A by the ith s.u. vector. Therefore, an $m\times n$ matrix can be decomposed into m dimensional n column vectors.

Row : Let m rows of the $m\times n$ matrix A be denoted by $\mathbf{A}_{1.}, \mathbf{A}_{2.}, \cdots, \mathbf{A}_{n.}$. So that the ith row

$$= \mathbf{A}_{i.}^T = \begin{bmatrix}a_{i1} & a_{i2} & \cdots & a_{in}\end{bmatrix}, i = 1, 2, 3, \cdots, n$$

$$= \begin{bmatrix} 0 & 0 & \cdots & \overset{i\text{th column}}{\overset{\downarrow}{1}} & \cdots & 0 \end{bmatrix}_{1\times m} \begin{bmatrix} a_{11} & a_{12} & \cdots & a_{1n} \\ a_{21} & a_{22} & \cdots & a_{2n} \\ \vdots & \vdots & \vdots & \vdots \\ a_{i1} & a_{i2} & \cdots & a_{in} \\ \vdots & \vdots & \vdots & \vdots \\ a_{m1} & a_{m2} & \cdots & a_{mn} \end{bmatrix}_{m\times n} \leftarrow i\text{th row}$$

$$= \mathbf{e}_i^{\mathrm{T}} \mathrm{A}$$

$$\therefore \quad \mathrm{A}_{i.} = \mathrm{A}^{\mathrm{T}} \mathbf{e}_i$$

i.e. ith row of **A** is the post multiplication of transposed **A** by the ith s.u.vector. Therefore an $m\times n$ matrix can be decomposed into m row vectors each of dimension n.

Elements : Since jth column, $\mathbf{A}_{.j} = \mathrm{A}\mathbf{e}_j$ and ith row, $\mathbf{A}_{i.} = \mathrm{A}^{\mathrm{T}} \mathbf{e}_{i.}$

Hence (i, j)th element of can be written as

$$a_{ij} = \mathbf{e}_i^{\mathrm{T}} \ \mathbf{A}\mathbf{e}_j = \mathbf{e}_j^{\mathrm{T}} \mathbf{A}^{\mathrm{T}} \mathbf{e}_i$$

Hence matrix A can be expressed as a sum

$$\mathrm{A} = \sum_{i=1}^{m} \sum_{j=1}^{n} a_{ij} \mathrm{E}_{ij} = \sum_{i=1}^{m} \sum_{j=1}^{n} a_{ij} \mathbf{e}_i \mathbf{e}_j^{\mathrm{T}} \tag{2.10}$$

$$\therefore \text{ as a sum } j\text{th column} = \mathbf{A}_{.j} = \mathrm{A}\mathbf{e}_j = \left(\sum_{i=1}^{n} \sum_{j=1}^{m} a_{ij} \mathbf{e}_i \mathbf{e}_j^{\mathrm{T}} \right) \mathbf{e}_j \text{ by (2.10)}$$

$$= \sum_{i=1}^{n} \sum_{j=1}^{m} a_{ij} \mathbf{e}_i \left(\mathbf{e}_j^{\mathrm{T}} \mathbf{e}_j \right)$$

$$= \sum_{i=1}^{n} a_{ij} \mathbf{e}_i \qquad \left[\because \mathbf{e}_j^{\mathrm{T}} \mathbf{e}_j = 1 \right] \tag{2.11}$$

and ith row $= \mathbf{A}_{i.} = \mathrm{A}^{\mathrm{T}} \mathbf{e}_i$

$$= \sum_{i=1}^{n} \sum_{j=1}^{m} a_{ij} \mathbf{e}_i^{\mathrm{T}} \mathbf{e}_j \mathbf{e}_i = \sum_{i=1}^{n} \sum_{j=1}^{m} a_{ij} \mathbf{e}_j \left(\mathbf{e}_i^{\mathrm{T}} \mathbf{e}_i \right)$$

$$= \sum_{j=1}^{m} a_{ij} \mathbf{e}_j \qquad \left[\because \mathbf{e}_i^{\mathrm{T}} \mathbf{e}_i = 1 \right]$$

so that

$$\mathbf{A}_{i.}^{\mathrm{T}} = \sum_{j=1}^{m} a_{ij} \mathbf{e}_j^{\mathrm{T}} \tag{2.12}$$

Hence from (2.10) and (2.11) the matrix

$$\text{A as a sum of columns} = \sum \mathbf{A}_{.j} \mathbf{e}_j^{\mathrm{T}} \tag{2.13}$$

and from (2.10) and (2.11) the matrix

$$\text{A as a sum of rows} = \sum \mathbf{e}_i \mathbf{A}_i^{\mathrm{T}} \tag{2.14}$$

Ex.1 Write the matrix

$$\mathrm{A} = \begin{bmatrix} 2 & 3 \\ 4 & 5 \end{bmatrix}$$

as a sum of : (i) column vectors of A, (ii) row vectors of A.

The matrix $\mathrm{A}_{2\times 2}$ as a sum of column vectors of A

$$= \sum \mathbf{A}_{.j}\mathbf{e}_j^T = \mathbf{A}_{.1}\mathbf{e}_1^T + \mathbf{A}_{.2}\mathbf{e}_2^T = \begin{bmatrix}2\\4\end{bmatrix}[1 \;\; 0] + \begin{bmatrix}3\\5\end{bmatrix}[0 \;\; 1], \quad \text{using (2.13)}$$

Also the matrix $A_{2\times 2}$ as a sum of row vectors of A

$$= \sum \mathbf{e}_i \mathbf{A}_{i.}^T = \mathbf{e}_1\mathbf{A}_{1.}^T + \mathbf{e}_2\mathbf{A}_{2.}^T$$

$$= \begin{bmatrix}1\\0\end{bmatrix}[2 \;\; 3] + \begin{bmatrix}0\\1\end{bmatrix}[4 \;\; 5] \qquad \text{using (2.14)}$$

Ex.2 Prove that $A\mathbf{e}_i$ is the ith column of an arbitrary matrix A.

Since $\mathbf{e}_i$ is a standard unit vector of zeros and one in the ith position. Therefore, by stage 1 of matrix multiplications,

$$\mathbf{A}_{k.}\mathbf{e}_i = a_{k1}.0 + a_{k2}.0 + \cdots + a_{ik}.1 + \cdots + a_{kn}.0 = a_{ik}$$

$$\therefore \qquad A\mathbf{e}_i = \begin{bmatrix}\mathbf{A}_{1.}\mathbf{e}_i\\ \mathbf{A}_{2.}\mathbf{e}_i\\ \vdots\\ \mathbf{A}_{m.}\mathbf{e}_i\end{bmatrix} = \begin{bmatrix}a_{1i}\\ a_{2i}\\ \vdots\\ a_{mi}\end{bmatrix} = \mathbf{A}_{.i} = (i\text{th column of A})$$

Note : This example illustrates a fact that, by inspection it is possible to write a vector resulting from the post multiplication of matrix A by standard unit vectors.

$$\text{Thus} \qquad \begin{bmatrix}2 & 3 & 5\\ 0 & 1 & 9\\ -1 & 2 & 3\end{bmatrix}\begin{bmatrix}1\\0\\0\end{bmatrix} = A\mathbf{e}_1 = (\text{first column of A}) = \begin{bmatrix}2\\0\\-1\end{bmatrix}.$$

2.8.1 Relations between s.u. vectors and matrix units : Since $\mathbf{e}_i\,\mathbf{e}_j^T = E_{ij}$ and $\mathbf{e}_i^T\,\mathbf{e}_j = \delta_{ij}$, so

$$E_{ij}\mathbf{e}_r = \mathbf{e}_i\,\mathbf{e}_j^T\,\mathbf{e}_r = \mathbf{e}_i\,\delta_{jr}$$

and

$$\mathbf{e}_r^T\,E_{ij} = \mathbf{e}_r^T\,\mathbf{e}_i\mathbf{e}_j^T = \delta_{ri}\,\mathbf{e}_j^T$$

For example,

$$\mathbf{e}_1 = \begin{bmatrix}1\\0\\0\end{bmatrix}_{3\times 1}, \quad \mathbf{e}_2 = \begin{bmatrix}0\\1\\0\end{bmatrix}_{3\times 1}$$

$$\therefore \qquad \mathbf{e}_1\mathbf{e}_2^T = \begin{bmatrix}1\\0\\0\end{bmatrix}_{3\times 1}[0 \;\; 1 \;\; 0]_{3\times 1} = \begin{bmatrix}0 & 1 & 0\\ 0 & 0 & 0\\ 0 & 0 & 0\end{bmatrix}_{3\times 3} = E_{12}$$

$$\text{And} \qquad \mathbf{e}_1^T\mathbf{e}_2 = [1 \;\; 0 \;\; 0]_{1\times 3}\begin{bmatrix}0\\1\\0\end{bmatrix}_{3\times 1} = [0]$$

$$\therefore \qquad E_{12}\mathbf{e}_3 = \begin{bmatrix}0 & 1 & 0\\ 0 & 0 & 0\\ 0 & 0 & 0\end{bmatrix}_{3\times 3}\begin{bmatrix}0\\0\\0\end{bmatrix}_{3\times 1} = \begin{bmatrix}0\\0\\0\end{bmatrix}_{3\times 1}$$

$$\mathbf{e}_1\delta_{23} = \begin{bmatrix} 1 \\ 0 \\ 0 \end{bmatrix}_{3\times1} [0]_{1\times1} = \begin{bmatrix} 0 \\ 0 \\ 0 \end{bmatrix}_{3\times1}$$

$\therefore$ $E_{12}e_3 = \mathbf{e}_1\,\delta_{23}$ is verified . Similarly others one can be verified.

2.9 Properties of matrix units

Following are the properties of matrix unit:

(i) The matrix units satisfy the relation

$$E_{ij}E_{kl} = \delta_{jk}E_{il}$$

Since $E_{ij} = \mathbf{e}_i\,\mathbf{e}_j^{T}$

So, $E_{ij}E_{kl} = \mathbf{e}_i\mathbf{e}_j^{T}\;\mathbf{e}_k\,\mathbf{e}_l^{T} = \mathbf{e}_i\,\delta_{jk}\mathbf{e}_l^{T} = \delta_{jk}\mathbf{e}_i\,\mathbf{e}_l^{T}$

$= \delta_{jk}E_{il}$

In particular, if $j = k$, then $E_{ij}E_{jl} = E_{il}$ $\quad[\because \delta_{jj} = 1]$

More generally, $\quad E_{ij}\,E_{jk}\,E_{kl} = E_{ik}\,E_{kl} = E_{il}$

If $j \neq k$ then, $\quad E_{ij}\,E_{kl} = 0,\; j \neq k.$

$$\text{Thus}\quad E_{12}E_{23} = \begin{bmatrix} 0 & 1 & 0 \\ 0 & 0 & 0 \\ 0 & 0 & 0 \end{bmatrix}_{3\times3} \begin{bmatrix} 0 & 0 & 0 & 0 \\ 0 & 0 & 1 & 0 \\ 0 & 0 & 0 & 0 \end{bmatrix}_{3\times4} = \begin{bmatrix} 0 & 0 & 1 & 0 \\ 0 & 0 & 0 & 0 \\ 0 & 0 & 0 & 0 \end{bmatrix}_{3\times4}$$

$$= (1)\begin{bmatrix} 0 & 0 & 1 & 0 \\ 0 & 0 & 0 & 0 \\ 0 & 0 & 0 & 0 \end{bmatrix}_{3\times4} = \delta_{22}E_{13} = E_{13} \qquad [\because \delta_{22} = 1]$$

2.9.1 Matrix units in the construction of a matrix as a sum : A matrix can be constructed from the matrix units as well as from s.u. vectors as a sum. In other words, matrix units as well as s.u vectors can be used to represent a matrix as a sum i.e. a matrix is a linear combination of matrix units as well as s.u. vectors. Let

$$A = \left[a_{ij}\right]_{m\times n} = \begin{bmatrix} a_{11} & a_{12} & \cdots & a_{1n} \\ a_{21} & a_{22} & \cdots & a_{2n} \\ \vdots & \vdots & \vdots & \vdots \\ a_{m1} & a_{m2} & \cdots & a_{mn} \end{bmatrix}$$

$$= \begin{bmatrix} a_{11} & 0 & \cdots & 0 \\ 0 & 0 & \cdots & 0 \\ \vdots & \vdots & \vdots & \vdots \\ 0 & 0 & \cdots & 0 \end{bmatrix}_{m\times n} + \begin{bmatrix} 0 & a_{12} & \cdots & 0 \\ 0 & 0 & \cdots & 0 \\ \vdots & \vdots & \vdots & \vdots \\ 0 & 0 & \cdots & 0 \end{bmatrix}_{m\times n} + \ldots + \begin{bmatrix} 0 & \cdots & 0 & \cdots & 0 \\ \vdots & \vdots & \vdots & \vdots & \vdots \\ 0 & \cdots & a_{ij} & \cdots & 0 \\ \vdots & \vdots & \vdots & \vdots & \vdots \\ 0 & \cdots & 0 & \cdots & 0 \end{bmatrix}_{m\times n}$$

$$+ \cdots + \begin{bmatrix} 0 & 0 & \cdots & 0 \\ 0 & 0 & \cdots & 0 \\ \vdots & \vdots & \vdots & \vdots \\ 0 & 0 & \cdots & a_{mn} \end{bmatrix}_{m\times n},$$

which is a sum of mn sparse matrices with only one nonzero element a_{ij} in the (i, j)th position.

$\therefore \quad A = a_{11}E_{11} + a_{12}E_{12} + \cdots + a_{ij}E_{ij} + \cdots + a_{mn}E_{mn}$

$= \sum_{i=1}^{m}\sum_{j=1}^{n} a_{ij}E_{ij}$, which establishes the stated claim,

Again, $E_{ij} = \mathbf{e}_i\,\mathbf{e}_j^{T}$. Therefore,

$A = \sum_{i=1}^{m}\sum_{j=1}^{n} a_{ij}\mathbf{e}_i\,\mathbf{e}_j^{T}$, a linear combination in terms of s.u. vectors.

2.10 Applications

2.10.1 Random walks in crystals : In physics, an atom of a defective crystal may jump from a given site to an unoccupied neighbouring one. Another atom may jump to the site just vacated or even the same atom may jump back to its original site. This type of movement of atom(s) in a defective crystal is called a *random walk* of the atoms.

Mathematically this random walk is shown by graph in which points designate the positions to be occupied and the lines connect neighbouring points. An atom can move from one point to another if the two points are joined by line.

A ***walk of length*** n is a graph of a sequence $t_1, t_2, \cdots, t_{n+1}$ of points (not necessarily all different). The length and the number of such random walks are given by the following theorem without proof.

Theorem. Let A be the *adjacency* matrix of a graph having n points labeled $v_1, v_2, \cdots, v_n$. Then the number of distinct $v_i - v_j$ walks of length k is the ij entry of A^k, k is any positive integer.

Ex. Compute A^4 where A is the adjacency matrix of the graph (*figure* 2.1) and hence find the number of distinct walks of length 4 from v_1 to v_4 and from v_2 to v_4. From A^3, find the number of distinct walks of length 3 from v_2 to v_3 and from v_3 to v_4.

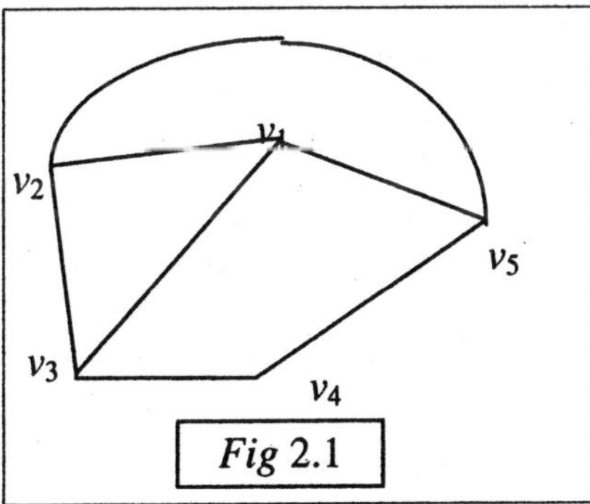

Fig 2.1

The Adjacency matrix is

$$A = \begin{array}{c} \\ v_1 \\ v_2 \\ v_3 \\ v_4 \\ v_5 \end{array}\begin{array}{c} \begin{array}{ccccc} v_1 & v_2 & v_3 & v_4 & v_5 \end{array} \\ \begin{bmatrix} 0 & 1 & 1 & 0 & 1 \\ 1 & 0 & 1 & 0 & 1 \\ 1 & 1 & 0 & 1 & 0 \\ 0 & 0 & 1 & 0 & 1 \\ 1 & 1 & 0 & 1 & 0 \end{bmatrix} \end{array}, \quad A^2 = \begin{bmatrix} 3 & 2 & 1 & 2 & 1 \\ 2 & 3 & 1 & 2 & 1 \\ 1 & 1 & 3 & 0 & 3 \\ 2 & 2 & 0 & 2 & 0 \\ 1 & 1 & 3 & 0 & 3 \end{bmatrix},$$

$$A^3 = \begin{bmatrix} 4 & 5 & 7 & 2 & 7 \\ 5 & 4 & 7 & 2 & 7 \\ 7 & 7 & 2 & 6 & 2 \\ 2 & 2 & 6 & 0 & 6 \\ 7 & 7 & 2 & 6 & 2 \end{bmatrix} \text{ and } A^4 = \begin{bmatrix} 19 & 18 & 11 & 14 & 11 \\ 18 & 19 & 11 & 14 & 11 \\ 11 & 11 & 20 & 4 & 20 \\ 14 & 14 & 4 & 12 & 4 \\ 11 & 11 & 20 & 4 & 20 \end{bmatrix}.$$

$\therefore$ number of distinct walks of length 4 from v_1 to $v_4 = 14$ (1)

and number of distinct walks of length 4 from v_2 to $v_4 = 14$ (2)

Also number of distinct walks of length 3 from v_2 to $v_3 = 7$ (3)

andnumber of distinct walks of length 3 from v_3 to $v_4 = 6$ (4)

The 14 paths in (1) are

$v_1\ v_3\ v_4\ v_3\ v_4$; $v_1\ v_3\ v_2\ v_3\ v_4$; $v_1\ v_5\ v_2\ v_3\ v_4$; $v_1\ v_2\ v_1\ v_3\ v_4$; $v_1\ v_3\ v_1\ v_3\ v_4$; $v_1\ v_5\ v_1\ v_3\ v_4$;

$v_1\ v_5\ v_1\ v_5\ v_4$; $v_1\ v_2\ v_1\ v_5\ v_4$; $v_1\ v_3\ v_1\ v_5\ v_4$; $v_1\ v_3\ v_4\ v_5\ v_4$; $v_1\ v_5\ v_4\ v_5\ v_4$; $v_1\ v_5\ v_2\ v_5\ v_4$;

$v_1\ v_3\ v_2\ v_5\ v_4$; $v_1\ v_3\ v_4\ v_3\ v_4$.

Similarly others can be read off from *figure* 2.1.

2.10.2 Simultaneous equations and their change of variables : Matrix multiplication provide compact notation for a system of simultaneous linear equations. Thus, the systems of m linear equations in n unknowns

$$\left.\begin{aligned} a_{11}x_1 + a_{12}x_2 + \cdots + a_{1n}x_n &= b_1 \\ a_{21}x_1 + a_{22}x_2 + \cdots + a_{2n}x_n &= b_2 \\ \vdots \qquad \vdots \qquad \vdots \qquad &\vdots \quad \vdots \\ a_{m1}x_1 + a_{m2}x_2 + \cdots + a_{mn}x_n &= b_m \end{aligned}\right\} \tag{2.15}$$

can be compactly written as

$$\mathbf{Ax} = \mathbf{b} \tag{2.16}$$

Where A = the $m \times n$ coefficient matrix = $\begin{bmatrix} a_{11} & a_{12} & \cdots & a_{1n} \\ a_{21} & a_{22} & \cdots & a_{2n} \\ \vdots & \vdots & \vdots & \vdots \\ a_{m1} & a_{m2} & \cdots & a_{mn} \end{bmatrix}$,

x = column matrix of the unknowns = $\begin{bmatrix} x_1 & x_2 & \cdots & x_n \end{bmatrix}^T$

and **b** = column matrix of the constants on the right = $\begin{bmatrix} b_1 & b_2 & \cdots & b_m \end{bmatrix}^T$.

Verification of (2.15) from (2.16) can be easily seen using equality of two matrices after multiplication.

Change of variables:

Let $n = 3$ and $m = 2$ in (2.15) and $\left.\begin{aligned} x_1 &= l_{11}u_1 + l_{12}v_1 \\ x_2 &= l_{21}u_1 + l_{22}v_1 \\ x_3 &= l_{31}u_1 + l_{32}v_1 \end{aligned}\right\}$

be substituted in (2.15). Therefore,

$$\begin{cases} a_{11}(l_{11}u_1 + l_{12}v_1) + a_{12}(l_{21}u_1 + l_{22}v_1) + a_{13}(l_{31}u_1 + l_{32}v_1) = b_1 \\ a_{21}(l_{11}u_1 + l_{12}v_1) + a_{22}(l_{21}u_1 + l_{22}v_1) + a_{23}(l_{31}u_1 + l_{32}v_1) = b_2 \end{cases}$$

or
$$\begin{cases} (a_{11}l_{11} + a_{12}l_{21} + a_{13}l_{31})u_1 + (a_{11}l_{12} + a_{12}l_{22} + a_{13}l_{32})v_1 = b_1 \\ (a_{21}l_{11} + a_{22}l_{21} + a_{23}l_{31})u_1 + (a_{21}l_{12} + a_{22}l_{22} + a_{23}l_{32})v_1 = b_2 \end{cases}$$

Now the new coefficient matrix is

$$\mathrm{C} = \begin{bmatrix} a_{11}l_{11} + a_{12}l_{21} + a_{13}l_{31} & a_{11}l_{12} + a_{12}l_{22} + a_{13}l_{32} \\ a_{21}l_{11} + a_{22}l_{21} + a_{23}l_{31} & a_{21}l_{12} + a_{22}l_{22} + a_{23}l_{32} \end{bmatrix}$$

$$= \begin{bmatrix} a_{11} & a_{12} & a_{13} \\ a_{21} & a_{22} & a_{23} \end{bmatrix} \begin{bmatrix} l_{11} & l_{12} \\ l_{21} & l_{22} \\ l_{31} & l_{32} \end{bmatrix}$$

$= (\text{coefficient matrix of the original eqns.})$

$(\text{coefficient matrix of substituted eqns.})$

$= \mathrm{A}_{2\times3}\mathrm{L}_{3\times2}$

This idea can be extended to all the m equations with substitution having t unknowns, not necessarily equal to the old n unknowns. Hence if a system of m linear equations in (2.15) with coefficient matrix A is subjected to a substitution

$$x_j = \sum_{k=1}^{t} l_{jk}u_k, \quad j = 1, 2, \cdots, n$$

with coefficient matrix L = $[l_{jk}]$, then the coefficient matrix of the resulting new systems is the product matrix AL.

2.10.3 Qudratic in matrix form : Let $f(x, y) = ax^2 + 2hxy + by^2$ be a quadratic in x,y. Now

$$f(x, y) = ax^2 + 2hxy + by^2 = x(ax + hy) + y(hx + by)$$

$$= \begin{bmatrix} x & y \end{bmatrix} \begin{bmatrix} a & h \\ h & b \end{bmatrix} \begin{bmatrix} x \\ y \end{bmatrix}$$

$\therefore$ $f(x, y)$ is a 1×1 matrix and is expressed as a product of three matrices.

Exercises two

1. Let $\mathbf{e}_j$ be an n-vector with 1 in the jth row, all other components being zero. Interpret each of the product

$$\mathbf{e}_j^{\mathrm{T}} \mathrm{A},\ \mathrm{A}\mathbf{e}_j,\ \mathbf{e}_j^{\mathrm{T}} \mathrm{A}\mathbf{e}_j,\ \mathbf{e}_j^{\mathrm{T}}\mathbf{e}_j,\ \mathbf{e}_j\mathbf{e}_j^{\mathrm{T}},$$

where A is an arbitrary matrix of order n.

[**Ans.** jth coln of A, jth row of A^{T}, element a_{ij}, δ_{ij}, δ_{ij}]

2. Calculate the following matrix products : AB, CA, DC, DCAB, $(\mathrm{AB})^2$,

$$\text{where } \mathrm{A}=\begin{bmatrix}1 & 2 & 3\\ 0 & 1 & -1\end{bmatrix},\ \mathrm{B}=\begin{bmatrix}0 & 2\\ -1 & 0\\ 1 & 0\end{bmatrix},\ \mathrm{C}=\begin{bmatrix}1 & -1\end{bmatrix},\ \mathrm{D}=\begin{bmatrix}1\\ 2\end{bmatrix}.$$

[**Ans.** $\mathrm{AB}=\begin{bmatrix}1 & 2\\ -2 & 0\end{bmatrix}$, $\mathrm{CA}=\begin{bmatrix}1 & 1 & 4\end{bmatrix}$, $\mathrm{DCAB}=\begin{bmatrix}3 & 2\\ 6 & 4\end{bmatrix}$, $(\mathrm{AB})^2=\begin{bmatrix}-3 & 2\\ -2 & -4\end{bmatrix}$]

3. If

$$\mathrm{A}=\begin{bmatrix}1 & 1 & 0\\ 0 & 1 & 2\\ 0 & 0 & 1\end{bmatrix} \text{ then prove that } \mathrm{A}^n=\begin{bmatrix}1 & n & n(n-1)\\ 0 & 1 & 2n\\ 0 & 0 & 1\end{bmatrix}.$$

4. Let A be $m\times n$ matrix and B an $n\times p$ matrix. Let $\mathbf{B}_1, \mathbf{B}_2, \cdots, \mathbf{B}_p$ be the columns of B. Show that $\mathrm{A}\mathbf{B}_1, \mathrm{A}\mathbf{B}_2, \cdots, \mathrm{A}\mathbf{B}_p$ are columns of AB.

5. Calculate the following:

(i) $\begin{bmatrix}1 & 4\\ 3 & 1\end{bmatrix}\begin{bmatrix}2 & 1 & 1\\ 3 & 4 & 0\end{bmatrix}$ [**Ans.** $\begin{bmatrix}14 & 17 & 1\\ 9 & 7 & 3\end{bmatrix}$]

(ii) $\begin{bmatrix}1 & 2\end{bmatrix}\begin{bmatrix}1 & 3\\ 3 & 1\end{bmatrix}$ [**Ans**. $\begin{bmatrix}8 & 5\end{bmatrix}$]

(iii) $\begin{bmatrix}2\\ 1\end{bmatrix}\begin{bmatrix}2 & 3\end{bmatrix}$ [**Ans**. $\begin{bmatrix}4 & 6\\ 2 & 3\end{bmatrix}$]

6. Decide which of the following pairs are orthogonal and orthonormal

$$\text{(i) } \begin{bmatrix}0\\ 1\end{bmatrix},\begin{bmatrix}0\\ 1\end{bmatrix};\ \text{(ii) } \begin{bmatrix}1\\ 1\end{bmatrix},\begin{bmatrix}1\\ -2\end{bmatrix};\ \text{(iii) } \begin{bmatrix}1+i\\ 1-i\end{bmatrix},\begin{bmatrix}1-i\\ 1+i\end{bmatrix}.$$

[**Ans**.(i) Orthonormal ; (ii) None; (iii) None]

7. Prove that

$$\begin{bmatrix}\cos\theta & \sin\theta\\ -\sin\theta & \cos\theta\end{bmatrix} \text{ is orthogonal for any real value of } \theta.$$

8. Which of the following matrices are orthogonal ?

$$\text{(i)} \begin{bmatrix} 1 & 0 \\ 0 & 1 \end{bmatrix}, \text{(ii)} \begin{bmatrix} 0 & 1 \\ 1 & 0 \end{bmatrix}, \text{(iii)} \begin{bmatrix} 1 & 1 \\ -1 & 1 \end{bmatrix}, \text{(iv)} \begin{bmatrix} \sin\theta & \cos\theta & 0 \\ -\sin\theta & \cos\theta & 0 \\ 0 & 0 & 1 \end{bmatrix}$$

[**Ans.** (i), (ii)]

9. Solve the matrix equation

$$\begin{bmatrix} i & 0 \\ 1 & -i \end{bmatrix} + \mathbf{X} = \begin{bmatrix} i & 2 \\ 3 & 4+i \end{bmatrix} - \mathbf{X}$$

[**Ans** . $\mathbf{X} = \begin{bmatrix} 0 & 1 \\ 1 & 2+i \end{bmatrix}$]

10. Show that if

$$\text{B} = \begin{bmatrix} 2 & 2 & 3 \\ 2 & 5 & 6 \\ 3 & 6 & 10 \end{bmatrix} \text{ and A} = \begin{bmatrix} 1 & 1 & 1 \\ 1 & 1 & 1 \\ -1 & -1 & -1 \end{bmatrix}, \text{ then BA} = \text{A}.$$

11. Prove the following rules of matrix algebra if the concerned operations are conformable and a, b are scalars.

(i) A (B – C) = AB – AC. (ii) (B – C) A = BA – CA.

(iii) a(B – C) = aB – aC. (iv) (a – b) C = aC – bC. (v) 0 – A = – A.

12. Show that

(i) A + (B + C) = (A + B) + C; (ii) (AB) C = A (BC)

(iii) (a + b) C = aC + bC; (iv) a(B – C) = a B – a C

where

$$\text{A} = \begin{bmatrix} -1 & 2 & 3 \\ 0 & 1 & 5 \\ 1 & -2 & 0 \end{bmatrix}, \text{B} = \begin{bmatrix} 2 & 3 & -1 \\ 5 & 1 & 0 \\ 0 & -2 & 1 \end{bmatrix}, \text{C} = \begin{bmatrix} 5 & 6 & 7 \\ -5 & -6 & -7 \\ 0 & 0 & 0 \end{bmatrix}, a = 5,\ b = -2.$$

13. For what real values of x is

$$\begin{bmatrix} x & 2 & 1 \end{bmatrix} \begin{bmatrix} 2 & 1 & 1 \\ -1 & 0 & 2 \\ 0 & 2 & 4 \end{bmatrix} \begin{bmatrix} x \\ 2 \\ 1 \end{bmatrix} > 0?$$

[**Ans** . for all x]

14. What must be true about a, b, c and d, if the matrices

$$\begin{bmatrix} a & b \\ c & d \end{bmatrix} \text{ and } \begin{bmatrix} 1 & 1 \\ -1 & 1 \end{bmatrix} \text{ are to commute ?}$$

[**Ans.** $a = d$, $b = -c$]

15. Prove that

(i) $(A^T B^T)^T = BA$

(ii) If A is $m \times n$ and B is a symmetric matrix of order m, then A^T BA is symmetric of order n.

(iii) If A and B are both symmetric and of the same order and if AB = BA then AB is

16. If

$$A = \begin{bmatrix} \cosh x & \sinh x \\ \sinh x & \cosh x \end{bmatrix} \text{ then show that } A^n = \begin{bmatrix} \cosh nx & \sinh nx \\ \sinh nx & \cosh nx \end{bmatrix}, n \text{ is integer.}$$

17. If

$$A = \begin{bmatrix} -1 & 0 & 2 \\ 3 & 4 & 5 \\ 1 & 2 & 0 \end{bmatrix} \text{ and } B = \begin{bmatrix} 1 & -1 \\ 3 & -2 \\ 3 & -2 \end{bmatrix}, \text{ determine the elements}$$

(i) in the 3rd row and 1st column of AB, **[Ans**. 7]

(ii) in the 2nd row and 2nd column of AB. **[Ans**. –31]

18. Let the following matrix A represent the prices of the grocery items in three different stores.

	store 1	store 2	store 3
Sugar (per kg.)	Tk. 30	Tk. 29	Tk. 31
Peaches (per can)	Tk. 25	Tk. 26	Tk. 27
A = Chicken (per kg.)	Tk. 80	Tk. 79	Tk. 81
Oil (per litre)	Tk. 49	Tk. 50	Tk. 49
Soap (per dozon)	Tk. 70	Tk. 72	Tk. 71

Let us purchase 5 kgs of sugar, 3 cans of peaches, 3 kgs of chicken, 5 litres of oil and 3 dozens of soap from each store during an average week. Determine bill in each store.

[**Hints:** Amounts purchased constitute a row vector $B = [5 \quad 3 \quad 3 \quad 5 \quad 3]$. Multiply BA.

Resulting row vector **C** = [Tk. 920 (store 1) Tk. 926 (store 2) Tk. 937 (store3)] is the answer.]

19. Compute JJ^T and (JJ)J, where J is a 3×4 J-matrix.

$$\textbf{[Ans. } \begin{bmatrix} 4 & 4 & 4 \\ 4 & 4 & 4 \\ 4 & 4 & 4 \end{bmatrix}, 12J.]$$

20. Prove that

(i) $J_{2\times 3}\, J_{3\times 4} = 3J_{2\times 4}$.

(ii) $J_{m\times 3}\, J_{3\times n} = 3J_{m\times n}$.

(iii) $J_{m\times l}\, J_{l\times n} = lJ_{m\times n}$.

21. For an arbitrary $m\times n$ matrix, find the following :

(i) $J_{l\times m}A$, (ii) $AJ_{n\times r}$ and (iii) $J_{l\times m}\, A\, J_{n\times r}$.

[**Ans.** (i) $l\times n$ matrix whose rows are the sum of rows of A

(ii) $m\times r$ matrix whose columns are the sum of columns of A

(iii) A matrix = (sum of elements of A) $J_{l\times r}$.]

22. Let a whole sale house supply apples and oranges to food stores A and B. Let the order for the first week be two bags of apples and three of oranges for store A and one bag of apples and four of oranges for store B. Display the order in matrix form. What is the cumulative order for the first two weeks ? Let the order for the 2nd week be displaced as follows:

$$N = \begin{array}{c} \\ \text{Apples} \\ \text{Oranges} \end{array} \begin{array}{c} \begin{array}{cc} \text{store A} & \text{store B} \end{array} \\ \begin{bmatrix} 4 & 3 \\ 2 & 2 \end{bmatrix} \end{array}$$

[**Ans**. $M = \begin{array}{c} \\ \text{Apples} \\ \text{Oranges} \end{array} \begin{array}{c} \begin{array}{cc} \text{store A} & \text{store B} \end{array} \\ \begin{bmatrix} 2 & 1 \\ 3 & 4 \end{bmatrix} \end{array}$, $M+N = \begin{array}{c} \\ \text{Apples} \\ \text{Oranges} \end{array} \begin{array}{c} \begin{array}{cc} \text{store A} & \text{store B} \end{array} \\ \begin{bmatrix} 6 & 4 \\ 5 & 6 \end{bmatrix} \end{array}$]

23. A contractor is building two different housing developments, one at Urban and one at Polis. At Urban he puts up three houses of type A and two houses of type B. At Polis he plans two houses of type A and four houses of type B. Now an A house requires four units of steel and three units of wood whereas a B house requires one unit of steel and two of wood. Find the amount of steel and wood required in those constructions.

[**Hints :** The matrix representation of the house in the two areas is

$$H = \begin{array}{c} \\ \text{Urban} \\ \text{Polis} \end{array} \begin{array}{c} \begin{array}{cc} A & B \end{array} \\ \begin{bmatrix} 3 & 2 \\ 2 & 4 \end{bmatrix} \end{array}$$

The matrix representation of the requirements of steel & wood in the two types of house is

$$R = \begin{array}{c} \\ A \\ B \end{array} \begin{array}{c} \begin{array}{cc} \text{Steel} & \text{Wood} \end{array} \\ \begin{bmatrix} 4 & 3 \\ 1 & 2 \end{bmatrix} \end{array}$$

$\therefore$ the product HR gives the records of the amount of steel and wood requirements.

$$HR = \begin{array}{c} \\ \text{Urban} \\ \text{Polis} \end{array} \begin{array}{c} \begin{array}{cc} A & B \end{array} \\ \begin{bmatrix} 3 & 2 \\ 2 & 4 \end{bmatrix} \end{array} \begin{array}{c} \\ A \\ B \end{array} \begin{array}{c} \begin{array}{cc} \text{Steel} & \text{Wood} \end{array} \\ \begin{bmatrix} 4 & 3 \\ 1 & 2 \end{bmatrix} \end{array} = \begin{array}{c} \\ \text{Urban} \\ \text{Polis} \end{array} \begin{array}{c} \begin{array}{cc} \text{Steel} & \text{Wood} \end{array} \\ \begin{bmatrix} 14 & 13 \\ 12 & 14 \end{bmatrix} \end{array}.]$$

24. If **a** is column (or row) vector then prove that $\mathbf{aa}^T$ (or $\mathbf{a}^T\mathbf{a}$) is a scalar and it is the sum of the squares of the elements of **a**.

25. Let J matrix be defined by the elements $a_{i,n-i+1} = 1,\ a_{ij} = 0,\ i+j \neq n+1$. If the matrix multiplications JA and AJ are conformable, then show that JA reverses the order of the rows of A where as AJ reverses the order of columns of A. What happens to the rectangular matrix A by the operations J_1AJ_2, where J_1 and J_2 are of the type J.

[**Ans.** Reverses the order of the elements of diagonals.]

26. Find all matrices B which satisfy the equation

$$\begin{bmatrix} 0 & 1 \\ 0 & 3 \end{bmatrix} \mathbf{B} = \begin{bmatrix} 0 & 0 \\ 0 & 0 \end{bmatrix}.$$

[**Ans.** (**Hints:** B must be 2×2 and $B = \begin{bmatrix} 0 & 0 \\ 0 & 0 \end{bmatrix}$).]

27. Find the matrices B such that

$$\begin{bmatrix} 0 & 1 \\ 0 & 3 \end{bmatrix} B = \begin{bmatrix} 0 & 0 & 1 \\ 0 & 0 & 3 \end{bmatrix}$$

[**Ans.** (**Hints:** B must be 2×3 and $B = \begin{bmatrix} 0 & 0 & 1 \\ 0 & 0 & 3 \end{bmatrix}$)]

28. Find matrices B such that they commute with

$$A = \begin{bmatrix} 0 & 1 \\ 0 & 3 \end{bmatrix}.$$

[**Ans.** (**Hints:** B must be 2×2)]

29. The table

	Production line 1	Production line 2
3 - speed bike	3	4
10 - speed bike	2	1

describes one hour's output in a factory and let

$P = \begin{bmatrix} 3 & 4 \\ 2 & 1 \end{bmatrix}$ be the corresponding matrix. Let $M = \begin{bmatrix} 7 \\ 8 \end{bmatrix}$ describes that line-1 and line-2 operate for 7 and 8 hours per day respectively. Compute PM and give its physical interpretation.

[**Ans.** $PM = \begin{bmatrix} 53 \\ 22 \end{bmatrix}$ which means that total output of line 1 and line 2 are 53 and 22 bikes per day respectively]

30. Let

$$A = \begin{bmatrix} 3 & -2 & 7 \\ 6 & 5 & 4 \\ 6 & 4 & 9 \end{bmatrix} \text{ and } B = \begin{bmatrix} 6 & -2 & 4 \\ 0 & 1 & 3 \\ 7 & 7 & 5 \end{bmatrix}.$$

Without finding the complete product of two matrices find:

(i) First row of AB. (ii) Third row of AB.
(iii) Second column of AB. (iv) First column of BA.
(v) Third row of AA. (vi) Third column of AA.

[**Ans.** (i) [67 41 41], (ii) [63 67 57], (iii) $\begin{bmatrix} 41 \\ 21 \\ 67 \end{bmatrix}$, (iv) $\begin{bmatrix} 6 \\ 6 \\ 63 \end{bmatrix}$, (v) [24 56 97], (vi) $\begin{bmatrix} 76 \\ 98 \\ 67 \end{bmatrix}$]

31. Let

$$A = \begin{bmatrix} 2 & 11 & 3 \\ 0 & 4 & -5 \\ 6 & 10 & 14 \\ 9 & -1 & 3 \end{bmatrix}, \ B = \begin{bmatrix} 2 & 11 & 6 \\ 5 & -3 & 4 \\ -9 & 2 & 0 \end{bmatrix}.$$

a. Find the trace of $A^T A$ and verify the equation Tr(AB) = Tr (BA).

b. Find the trace of the matrix B. [**Ans** . a. 598 , b. –1]

32. Let A and B be 4×5 matrices and C, D and E be 5×2, 4×2 and 5×4 matrrices respectively. Determine which of the following matrix expressions are defined and hence give their size.

(i) BA,(ii) AC + D, (iii) AE + B, (iv) AB + B, (v) E(A + B), (vi) E(AC)

[**Ans**. (i) Undefined, (ii) 4×2, (iii) undefined, (iv) Undefined, (v) 5×5, (vi) 5×2]

33. Determine a matrix B such that A + B = X, where $X = [x_{ij}]$, $x_{ij} = i + j$,

$$A = \begin{bmatrix} 2 & -1 & 4 \\ 3 & 2 & 0 \\ -1 & 4 & 4 \end{bmatrix} \text{ and } X = \begin{bmatrix} x_{11} & x_{12} & x_{13} \\ 0 & x_{22} & x_{23} \\ 0 & 0 & x_{33} \end{bmatrix}.$$

[**Ans**. $B = \begin{bmatrix} 0 & 4 & 0 \\ -3 & 2 & 5 \\ 1 & -4 & 2 \end{bmatrix}$]

34. Determine Q, so that $Q = \frac{1}{2}P$, $Q = 4P$,

$$\text{where } P = \begin{bmatrix} 6 & 4 & 5 & 0 \\ 8 & 1 & 6 & 1 \\ 3 & 0 & 2 & 2 \end{bmatrix}.$$

[**Ans.** $Q = \begin{bmatrix} 3 & 2 & \frac{5}{2} & 0 \\ 4 & \frac{1}{2} & 3 & \frac{1}{2} \\ \frac{3}{2} & 0 & 1 & 1 \end{bmatrix}$ and $Q = \begin{bmatrix} 24 & 16 & 20 & 0 \\ 32 & 4 & 24 & 4 \\ 12 & 0 & 8 & 8 \end{bmatrix}$]

35. Find all matrices B obeying the equations

(i) $\begin{bmatrix} 0 & 1 \\ 0 & 2 \end{bmatrix} B = \begin{bmatrix} 0 & 0 \\ 0 & 0 \end{bmatrix}.$ [**Ans.** $B = \begin{bmatrix} b_{11} & b_{12} \\ 0 & 0 \end{bmatrix}$]

(ii) $\begin{bmatrix} 0 & 1 \\ 0 & 2 \end{bmatrix} B = \begin{bmatrix} 0 & 0 & 1 \\ 0 & 0 & 2 \end{bmatrix}.$ [**Ans.** $B = \begin{bmatrix} b_{11} & b_{12} & b_{13} \\ 0 & 0 & 1 \end{bmatrix}$]

36. Find all matrices B which commute with

$$\begin{bmatrix} 0 & 1 \\ 0 & 2 \end{bmatrix}.$$

[**Ans**. $\begin{bmatrix} b_{11} & b_{12} \\ 0 & b_{11} + 2b_{12} \end{bmatrix}$]

37. Find the value of c if

$$P + 2Q + cR = \begin{bmatrix} 0 & 0 \\ -1 & 13 \end{bmatrix} \text{ and } P = \begin{bmatrix} 2 & 1 \\ -1 & 3 \end{bmatrix},\ Q = \begin{bmatrix} 0 & 1 \\ 2 & 5 \end{bmatrix},\ R = \begin{bmatrix} 2 & 3 \\ 4 & 0 \end{bmatrix}.$$

[**Ans.** $c = -1$]

38. If the 5th row of A is (1, –1, 0, 0), describe the 5th row of AB in terms of rows of B.

[**Ans.** $a_{51}\mathbf{B}_{1.} + a_{52}\mathbf{B}_{2.} + a_{53}\mathbf{B}_{3.} + a_{54}\mathbf{B}_{4.} = \mathbf{B}_{1.} - \mathbf{B}_{2.}$, where $\mathbf{B}_{i.}$ are rows of B.]

39. Show that every even power of the matrix

$$\begin{bmatrix} 1 & 0 & 0 \\ 0 & 1 & 0 \\ a & b & -1 \end{bmatrix},$$

where a and b are arbitrary scalars, equals the unit matrix and every odd power equals itself.

40. Find the cost (cost matrix) of the purchases of vegetables for a resident student hall in 3 weeks of a month where quantities purchased in different weeks and two sets of unit prices are given below in tabular form. What is the cost in week 2 at price set 1.

Quantities Q in kg					Prices P per kg				
	Potatoes	Carrots	Onions	Peas		Potatoes	Carrots	Onions	Peas
Week1	4	1	1	2	Price 1	15	32	24	18
Week2	3	2	1	2	Price 2	16	29	25	18
Week3	5	0	2	4					

[**Ans.** (i) Quantity matrix,
$$Q = \begin{matrix} \\ \text{wk1} \\ \text{wk2} \\ \text{wk3} \end{matrix}\begin{matrix} \text{Pot.} & \text{Car.} & \text{On.} & \text{Pea} \\ \left[\begin{matrix} 4 \\ 3 \\ 5 \end{matrix}\right. & \begin{matrix} 1 \\ 2 \\ 0 \end{matrix} & \begin{matrix} 1 \\ 1 \\ 2 \end{matrix} & \left.\begin{matrix} 2 \\ 3 \\ 4 \end{matrix}\right] \end{matrix},$$

(ii) Price matrix,
$$P = \begin{matrix} \\ \text{Pot.} \\ \text{Car.} \\ \text{On.} \\ \text{Pea.} \end{matrix}\begin{matrix} \text{Pr.1} & \text{Pr.2} \\ \left[\begin{matrix} 15 \\ 32 \\ 24 \\ 18 \end{matrix}\right. & \left.\begin{matrix} 16 \\ 29 \\ 25 \\ 18 \end{matrix}\right] \end{matrix}, \quad \text{Cost matrix} = QP = \begin{matrix} \\ \text{wk1} \\ \text{wk2} \\ \text{wk3} \end{matrix}\begin{matrix} \text{Pr.1} & \text{Pr.2} \\ \left[\begin{matrix} 152 \\ 187 \\ 195 \end{matrix}\right. & \left.\begin{matrix} 154 \\ 185 \\ 202 \end{matrix}\right] \end{matrix},$$

(ii) 187]

41. I_1, I_2 two islands, P_1, P_2, P_3 three ports of main land and T_1, T_2 two towns are connected by ferry and railways (*figure* 2.2).

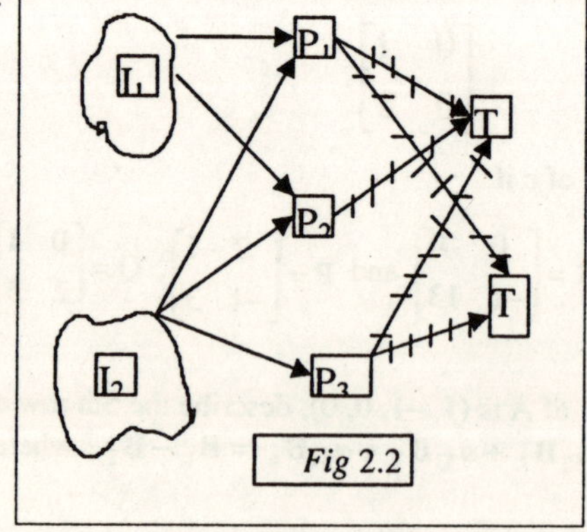

Fig 2.2

(i) Find the route matrix

(a) from I_1, I_2 to P_1, P_2, P_3;

(b) from P_1, P_2, P_3 to T_1, T_2.

Multiply these two matrices.

(ii) Use the results above in (i) to write down the number of routes there are from :

(a) $I_1 \to T_1$, (b) $I_2 \to T_1$, (c) $I_2 \to T_2$.

[**Ans** . (i) route matrix (a) $R_1 = \begin{matrix} & P_1 & P_2 & P_3 \\ I_1 & 1 & 1 & 0 \\ I_2 & 1 & 1 & 1 \end{matrix}$, (b) $R_2 = \begin{matrix} & T_1 & T_2 \\ P_1 & 1 & 1 \\ P_2 & 1 & 0 \\ P_3 & 1 & 1 \end{matrix}$,

$R_1R_2 = \begin{matrix} & T_1 & T_2 \\ I_1 & 2 & 1 \\ I_2 & 3 & 2 \end{matrix}$. (ii) (a) 2, (b) 3, (c) 2.]

42. Express that matrix

$$A = \begin{bmatrix} 3x^3 + 2x - 1 & 5x^2 - 3x + 2 & -2x^3 + 3x^2 - 5x + 1 \\ 9x^3 - 3 & 6x + 4 & 7x^2 - 4x \\ 2x^2 - 3x - 10 & x^3 + 6x^2 + x & 6x^3 - 4x^2 + 8 \end{bmatrix}$$

as a polynomial in x with matrix coefficients.

[**Ans.** $A = \begin{bmatrix} 3 & 0 & -2 \\ 9 & 0 & 0 \\ 0 & 1 & 6 \end{bmatrix} x^3 + \begin{bmatrix} 0 & 5 & 3 \\ 0 & 0 & 7 \\ 2 & 6 & -4 \end{bmatrix} x^2 + \begin{bmatrix} 2 & -3 & -5 \\ 0 & 6 & -4 \\ -3 & 1 & 0 \end{bmatrix} x + \begin{bmatrix} -1 & 2 & 1 \\ -3 & 4 & 0 \\ -10 & 0 & 8 \end{bmatrix}$]

43. Write the matrix

$$\begin{bmatrix} 1 & 2 \\ -1 & 3 \\ 2 & 0 \end{bmatrix}$$

as a sum of matrix units.

[**Ans.** $E_{11} + 2E_{12} - E_{21} + 3E_{22} + 2E_{22} + 2E_{31} + 0E_{32}$.]

44. Prove that, (a) $J_{2\times3}\, J_{3\times4} = 3J_{2\times4}$, (b) $J_{m\times4}\, J_{4\times n} = 4J_{m\times n}$. Generalize the result.

[**Ans.** $J_{m\times l}\, J_{l\times n} = lJ_{m\times n}$.]

45. For an arbitrary $m\times n$ matrix A, find (i) $J_{l\times m}A$, (ii) $AJ_{n\times r}$ and (iii) $J_{l\times m}AJ_{n\times r}$.

[**Ans.** (i) The $l\times n$ matrix each of whose rows is the sum of the rows of A.

(ii) The $m\times r$ matrix each of whose columns is the sum of the columns of A.

(iii) The $l\times r$ matrix each of whose elements is the sum of mn elements of A.]

46. (a) Under what conditions are both AB and BA defined ?

(b) When is $A^2 = AA$ defined ?

(c) When are both AB and A + B defined ?

[**Ans.** (a) If A is $m\times n$ and B is $n\times m$; (b) A is a square matrix;

(c) Both A and B are square matrices.]

47. The Pauli spin matrices are

$$\sigma_1=\begin{bmatrix}0&1\\1&0\end{bmatrix},\sigma_2=\begin{bmatrix}0&-i\\i&0\end{bmatrix};\sigma_3=\begin{bmatrix}1&0\\0&-1\end{bmatrix}.$$

Fill up the multiplication table

×	σ_1	σ_2	σ_3
σ_1	*	*	*
σ_2	*	*	*
σ_3	*	*	*

[**Ans.**

	σ_1	σ_2	σ_3
σ_1	I_2	$i\sigma_2$	$-i\sigma_2$
σ_2	$-i\sigma_3$	I_2	$i\sigma_1$
σ_3	$i\sigma_2$	$-i\sigma_1$	I_2

.]

48. Some of the isospin matrices used by physicists are called

$$\tau_1=\begin{bmatrix}0&1\\1&0\end{bmatrix},\tau_2=\begin{bmatrix}0&-i\\i&0\end{bmatrix},\ \tau_+=\begin{bmatrix}0&1\\0&0\end{bmatrix},\ \tau_-=\begin{bmatrix}0&0\\i&0\end{bmatrix}.$$

(i) Write τ_+ as a linear combination of τ_1 and τ_2.

(ii) Same for τ_-.

[**Ans.** $\tau_+=\frac{1}{2}\tau_1+\frac{1}{2}i\tau_2,\ \tau_-=\frac{1}{2}\tau_1-\frac{1}{2}i\tau_2$.]

49. Find the new coefficient matrix of the system of equations

$$\begin{cases}2x+3y+7z=1\\x-2y+5z=2\\4x-3y-z=4\end{cases}$$

by changing *x*, *y*, *z* to new variables *u*, *v* by the substitutions

$$\begin{cases}x=3u-4v\\y=5u+v\\z=u+8v\end{cases}$$

[**Ans.** $\begin{bmatrix}28&51\\-2&34\\-4&-27\end{bmatrix}$.]

50. Represent the quadratic $f(x,y,z)=ax^2+by^2+cz^2+2fxy+2gxz+2hyz$ in the matrix form.

[**Ans.** $[x\ \ y\ \ z]\begin{bmatrix}a&f&g\\f&b&h\\g&h&c\end{bmatrix}[x\ \ y\ \ z]$.]

51. Given that

$$B = \begin{bmatrix} 1 & 1 & 1 & 1 \\ 1 & 3 & 2 & 2 \\ 1 & -3 & -1 & -1 \\ 1 & -5 & -2 & -2 \end{bmatrix}, \quad A = \begin{bmatrix} a_1 \\ 1 \\ 1 \\ a_4 \end{bmatrix}.$$

Find a_1 and a_4 if $A^T B = \begin{bmatrix} 4 & -4 & 0 & 0 \end{bmatrix}$. [**Ans.** $a_1 = 1, a_4 = 1$]

CHAPTER THREE

SQUARE MATRIX AND ITS SPECIAL FORMS

Chapter summary *: The chapter starts with square matrix and its laws of exponents. Then follow traces and its properties, traces in terms of s.u. vectors; diagonals and diagonal dominance; diagonal matrix, its related theorems, scalar and unit matrix; tridiagonal and band matrix; triangular matrices; symmetric and Hermitian matrices and related theorems; orthogonal, normal, nilpotent, periodic, idempotent and involutory matrices; unitary, unimodular, proper and improper matrices; permutation matrix with propertics; Hadamard matrix with properties; Helmert matrices; circulant matrices with properties; Toeplitz matrix; Leslie, Jacobi, Vandermode, Heissenberg matrices; the chapter ends with square root of matrices and some applications.*

3.1 Square matrix

A matrix is said to be a *square matrix* if the number of rows and the number of columns are equal. Thus an $m\times n$ matrix becomes an n square matrix when $m = n$. For example,

$$A = \left[a_{ij}\right]_n = \begin{bmatrix} a_{11} & a_{12} & \cdots & a_{1n} \\ a_{21} & a_{22} & \cdots & a_{2n} \\ \vdots & \vdots & \ddots & \vdots \\ a_{n1} & a_{n2} & \cdots & a_{nn} \end{bmatrix}$$

and $[a_{11}]$ are square matrices of order n and 1 respectively.

Algebra of square matrices is the same as those discussed in chapter two, except the law of exponents, in sec.3.2 and in chapter 8 respectively.

3.2 Law of exponents for matrices

The product AA is defined for square matrix A and it is natural to abbreviate the product AA by A^2. As matrix multiplication is associative, so

$$AAA = (AA)A = A^2A \text{ and } AAA = A(AA) = AA^2.$$

The common product A^2A or AA^2 is abbreviated as A^3. In this way, we have the following definition for a positive power p of matrix A.

Definition : If p is a positive integer then A^p is defined by the product of p A's i.e. $A^p = \underbrace{A\,A\,A \cdots A}_{p \text{ A's}}$. If $A \neq 0$, the zeroth power of A is defined as $A^0 = I$. The matrix expression 0^0 is *not* defined. The following theorem gives the usual law of exponents for matrices:

Theorem . If p and q are positive or zero integers and A is a square matrix then (with exception 0^0 case)

(i) $(A^p)(A^q) = (A^q)(A^p) = A^{p+q}$

(ii) $(A^p)^q = (A^q)^p = A^{pq}$

(iii) $I^p = I$

(iv) $0^p = 0$, 0 indicating zero matrix.

Proof .

(i) $$(A^p)(A^q) = \left(\underbrace{AAA\ldots A}_{p\ A's}\right)\left(\underbrace{AAA\ldots A}_{q\ A's}\right)$$

$$= \underbrace{AA\ldots A\ AA\ldots A}_{(p+q)\ A's},$$

since grouping is immaterial by associative law

$$= A^{p+q}, \text{ by definition}$$

If $p = 0$, then $A^p = A^0 = I$. Therefore, $(A^p)(A^q) = IA^q = A^q = A^{p+q}$. Also if $q = 0$, $(A^p)(A^q) = A^pI = A^p = A^{p+q}$.

(ii) $$(A^p)^q = \left(\underbrace{AAA\cdots A}_{p\ A's}\right)^q$$

$$= \underbrace{\left(\underbrace{AAA\cdots A}_{p\ A's}\right)}_{\text{1st block}}\underbrace{\left(\underbrace{AAA\cdots A}_{p\ A's}\right)}_{\text{2nd block}}\cdots\underbrace{\left(\underbrace{AAA\cdots A}_{p\ A's}\right)}_{q\text{th block}}, \ q\text{blocks of } p \text{ A's}$$

$$= \left(\underbrace{AA\cdots A\ AA\cdots A\ AA\cdots A\ AA\cdots A}_{pq\ A's}\right) = A^{pq}$$

Similarly, $(A^q)^p = A^{pq}$

$\therefore \quad (A^p)^q = (A^q)^p = A^{pq}$

(iii) Since $I^2 = II = I$ etc. therefore,

$$I^p = \underbrace{II\cdots I}_{p\ I's} = I.$$

(iv) Here also $0^2 = 00 = 0$ etc. therefore, $0^p = \underbrace{000\cdots 0}_{p\ 0's} = 0$

Ex.1 If

$$A = \begin{bmatrix} \cos\theta & \sin\theta \\ -\sin\theta & \cos\theta \end{bmatrix}, \text{ prove that } A^2 = \begin{bmatrix} \cos 2\theta & \sin 2\theta \\ -\sin 2\theta & \cos 2\theta \end{bmatrix}.$$

$$\because A^2 = A.A = \begin{bmatrix} \cos\theta & \sin\theta \\ -\sin\theta & \cos\theta \end{bmatrix}\begin{bmatrix} \cos\theta & \sin\theta \\ -\sin\theta & \cos\theta \end{bmatrix}$$

$$= \begin{bmatrix} \cos^2\theta - \sin^2\theta & 2\sin\theta\cos\theta \\ -2\sin\theta\cos\theta & \cos^2\theta - \sin^2\theta \end{bmatrix}$$

$$= \begin{bmatrix} \cos 2\theta & \sin 2\theta \\ -\sin 2\theta & \cos 2\theta \end{bmatrix}.$$

Note that the result is identical to DeMoivre's theorem which states $(\cos\theta + i\sin\theta)^n = \cos n\theta + i\sin n\theta$, for positive integer n. This is because the given matrix A represents the complex number $\cos\theta + i\sin\theta$ (cf. sec. 3.25.1).

Ex.2 Verify $(A^2)(A^3) = (A^3)(A^2) = A^{2+3} = A^5$ for the matrix

$$A = \begin{bmatrix} 2 & 0 \\ 1 & -1 \end{bmatrix}.$$

$$A^2 = \begin{bmatrix} 2 & 0 \\ 1 & -1 \end{bmatrix}\begin{bmatrix} 2 & 0 \\ 1 & -1 \end{bmatrix} = \begin{bmatrix} 4 & 0 \\ 1 & 1 \end{bmatrix}.$$

$$A^3 = A^2A = \begin{bmatrix} 4 & 0 \\ 1 & 1 \end{bmatrix}\begin{bmatrix} 2 & 0 \\ 1 & -1 \end{bmatrix} = \begin{bmatrix} 8 & 0 \\ 3 & -1 \end{bmatrix}.$$

$$\therefore \quad (A^2)(A^3) = \begin{bmatrix} 4 & 0 \\ 1 & 1 \end{bmatrix}\begin{bmatrix} 8 & 0 \\ 3 & -1 \end{bmatrix} = \begin{bmatrix} 32 & 0 \\ 11 & -1 \end{bmatrix}.$$

Also $$(A^3)(A^2) = \begin{bmatrix} 8 & 0 \\ 3 & -1 \end{bmatrix}\begin{bmatrix} 4 & 0 \\ 1 & 1 \end{bmatrix} = \begin{bmatrix} 32 & 0 \\ 11 & -1 \end{bmatrix}.$$

and $$A^4 = A^3A = \begin{bmatrix} 8 & 0 \\ 3 & -1 \end{bmatrix}\begin{bmatrix} 2 & 0 \\ 1 & -1 \end{bmatrix} = \begin{bmatrix} 16 & 0 \\ 5 & 1 \end{bmatrix}.$$

$$A^5 = A^4A = \begin{bmatrix} 16 & 0 \\ 5 & 1 \end{bmatrix}\begin{bmatrix} 2 & 0 \\ 1 & -1 \end{bmatrix} = \begin{bmatrix} 32 & 0 \\ 11 & -1 \end{bmatrix}.$$

$$\therefore \quad (A^2)(A^3) = (A^3)(A^2) = A^5 = \begin{bmatrix} 32 & 0 \\ 11 & -1 \end{bmatrix}.$$

Ex.3 Prove that $(A^2)^3 = (A^3)^2 = A^{2.3} = A^6$ for A in Ex.2 above.

$$(A^2)^2 = (A^2)(A^2) = \begin{bmatrix} 4 & 0 \\ 1 & 1 \end{bmatrix}\begin{bmatrix} 4 & 0 \\ 1 & 1 \end{bmatrix} = \begin{bmatrix} 16 & 0 \\ 5 & 1 \end{bmatrix}.$$

$$(A^2)^3 = (A^2)^2A^2 = \begin{bmatrix} 16 & 0 \\ 5 & 1 \end{bmatrix}\begin{bmatrix} 4 & 0 \\ 1 & 1 \end{bmatrix} = \begin{bmatrix} 64 & 0 \\ 21 & 1 \end{bmatrix}.$$

Also $$(A^3)^2 = (A^3)(A^3) = \begin{bmatrix} 8 & 0 \\ 3 & -1 \end{bmatrix}\begin{bmatrix} 8 & 0 \\ 3 & -1 \end{bmatrix} = \begin{bmatrix} 64 & 0 \\ 21 & 1 \end{bmatrix}.$$

$$A^6 = A^5A = \begin{bmatrix} 32 & 0 \\ 11 & -1 \end{bmatrix}\begin{bmatrix} 2 & 0 \\ 1 & -1 \end{bmatrix} = \begin{bmatrix} 64 & 0 \\ 21 & 1 \end{bmatrix}.$$

$$\therefore \qquad (A^2)^3 = (A^3)^2 = A^{2.3} = A^6 = \begin{bmatrix} 64 & 0 \\ 21 & 1 \end{bmatrix}.$$

N.B. It is easy to verify that $I_2^2 = I_2$ and $0^3 = 0$.

3.3 The trace and its properties : The trace of a square $n \times n$ matrix $A=[a_{ij}]_{n\times n}$ denoted by TrA, trA or tr(A), is the sum of its diagonal elements:

$$\text{tr A} = \sum\sum_{i=1}^{n} a_{ii}$$

The trace of a matrix is a function of the diagonal elements.

Properties : The following are the properties of the trace:

(i) tr (A + B) = tr A + tr B

(ii) tr (λA) = λ tr A, λ is a scalar

(iii) tr (λ_1A + λ_2B) = λ_1 tr A + λ_2 tr B, λ_1, λ_2 are scalars

(iv) tr A = tr A^T

(v) tr (AB) = tr (BA)

(vi) tr (AA^T) = tr (A^TA)

(vii) tr (A^{-1}BA) = tr B

(viii) matrix norm, $\|A\| = \left(\text{tr}A^TA\right)^{\frac{1}{2}}$

(ix) tr (A⊗B) = tr A tr B

(x) $\mathbf{x}^T A\mathbf{x}$ = tr ($\mathbf{x}^T A\mathbf{x}$) = tr ($\mathbf{x}A\mathbf{x}^T$) for column vector **x** and matrix A.

Proof. The property (iii) is more general than (i) and (ii) if $\lambda_1 = \lambda_2 = 1$, then property (i) is obtained and if $\lambda_1 = \lambda$ and $\lambda_2 = 0$, then follows (ii). Similarly if B is replaced by A^T in (v) we obtain (vi). So proving (iii) and (v) is sufficient for the proofs of (i), (ii) and (vi).

Proof of (iii). $\lambda_1 A + \lambda_2 B = \lambda_1 \left[a_{ij}\right]_{n\times n} + \lambda_2 \left[b_{ij}\right]_{n\times n}$

$$= \left[\lambda_1 a_{ij}\right]_{n\times n} + \left[\lambda_2 b_{ij}\right]_{n\times n}$$

$$= \left[\lambda_1 a_{ij} + \lambda_2 b_{ij}\right]_{n\times n}$$

$$\therefore \qquad \text{tr}(\lambda_1 A + \lambda_2 B) = \sum_{i=1}^{n}(\lambda_1 a_{ii} + \lambda_2 b_{ii})$$

$$= \lambda_1 \sum_{i=1}^{n} a_{ii} + \lambda_2 \sum_{i=1}^{n} b_{ii}$$

$$= \lambda_1 \text{ tr A} + \lambda_2 \text{ tr B}$$

Therefore, trace of a linear combination of matrices is equal to the linear combination of the traces of matrices.

Proof of (iv). Since the diagonal elements a_{ii} of A and A^T remain invariant (i.e. same) or unaltered, so

$$\text{tr A} = \text{tr } A^T$$

Proof of (v). $\text{tr}(AB) = \sum_{i=1}^{n}\sum_{r=1}^{n} a_{ir}b_{ri} = \sum_{i=1}^{n}\sum_{r=1}^{n} b_{ri}a_{ir} = \text{tr}(BA)$

Proof of (vii). $\text{tr}\left(A^{-1}BA\right) = \text{tr}\left(A^{-1}(BA)\right) = \text{tr}\left((BA)A^{-1}\right)$

$$= \text{tr}\left(B\left(AA^{-1}\right)\right) = \text{tr } B$$

Proof of (viii). See chapter 16

Proof of (ix). See chapter 5

Proof of (x). Since the product $\mathbf{x}^T A\mathbf{x}$ is a scalar, so it is equal to its own trace. Hence

$$\mathbf{x}^T A\mathbf{x} = \text{tr}\left(\mathbf{x}^T A\mathbf{x}\right) = \text{tr}\left(\mathbf{x}A\mathbf{x}^T\right).$$

3.3.1 Trace in terms of s.u. vectors : Since $\text{tr } A = \sum a_{ii}$, where $A = \left[a_{ij}\right]_{n\times n}$, but $a_{ii} = \mathbf{e}_i^T A\mathbf{e}_i$.

$\therefore \quad \text{tr } A = \sum \mathbf{e}_i^T A\mathbf{e}_i = \sum \mathbf{e}_i^T \mathbf{A}_{.i}$

Hence trA is the sum of the product of transposed s.u. vector and the corresponding column of A. Thus, let

$$A = \begin{bmatrix} a_{11} & a_{12} & a_{13} \\ a_{21} & a_{22} & a_{23} \\ a_{31} & a_{32} & a_{33} \end{bmatrix}$$

Now, $\sum \mathbf{e}_i^T \mathbf{A}_{.i} = \mathbf{e}_1^T \mathbf{A}_{.1} + \mathbf{e}_2^T \mathbf{A}_{.2} + \mathbf{e}_3^T \mathbf{A}_{.3}$

$$= [1 \quad 0 \quad 0]\begin{bmatrix} a_{11} \\ a_{21} \\ a_{31} \end{bmatrix} + [0 \quad 1 \quad 0]\begin{bmatrix} a_{12} \\ a_{22} \\ a_{32} \end{bmatrix} + [0 \quad 0 \quad 1]\begin{bmatrix} a_{13} \\ a_{23} \\ a_{33} \end{bmatrix}$$

$$= \left[a_{11}\right] + \left[a_{22}\right] + \left[a_{33}\right] = a_{11} + a_{22} + a_{33} = \text{tr } A$$

$\therefore$ tr A can be written other way round as

$$\text{tr } A = \sum \mathbf{A}_{i.}\mathbf{e}_i.$$

3.4 Diagonals and diagonal dominance

This section defines different types of diagonals and the diagonal dominance.

Diagonals : A square matrix has two geometrical diagonals. The *main* or *principal* diagonal which contains the elements a_{11}, a_{22}, $\cdots$, a_{nn}, i.e. the elements with equal subscripts $i = j$ from top left corner down to bottom right corner. The other diagonal at right angle to the principal diagonal is called the *secondary* or *cross* diagonal starts from bottom left corner to top right corner. The elements with

$$i = n + j - 1 \text{ i.e. } a_{(n+j-1),j} \quad \text{i.e. } a_{n1}, a_{(n-1)2}, \ \cdots, a_{1n}$$

lie on the secondary diagonal.

Super and subdiagonals : The elements a_{ij} of A with $i < j$ constitutes the superdiagonals. Thus elements a_{12}, a_{23}, $\cdots$, $a_{(n-1)n}$ just above the principal diagonal constitute first superdiagonal. Similarly, elements a_{ij} with $i > j$ constitute the subdiagonals. Thus elements a_{21}, a_{32}, $\cdots$, $a_{n(n-1)}$ just below the principal diagonal belong to first subdiagonal. Thus

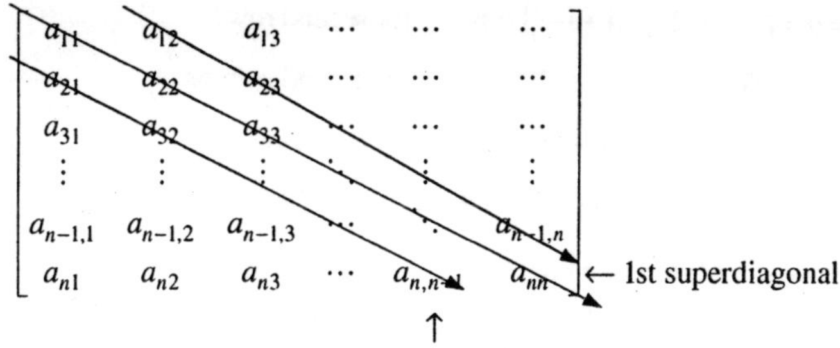

Off diagonal elements : The element a_{ij} such that $i \neq j$ are called *off diagonal* elements. Thus

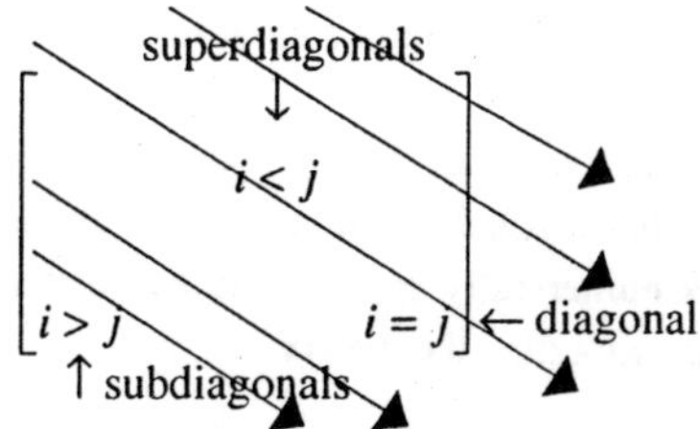

Diagonal dominance : An n-square real or complex matrix A is diagonally dominant or has diagonal dominance if

$$|a_{ii}| \geq \sum_{j \neq i} |a_{ij}|, \quad i = 1, 2, \cdots, n \tag{3.1}$$

In other words, the magnitude of any diagonal element a_{ii} must be greater than or equal to the sum of the magnitudes of other elements (i.e. off diagonal elements) in the corresponding row (column). That is, magnitudes of the diagonal elements "dominate" sum of the remaining elements in the corresponding row (column).

In terms of elements *either*

$$|a_{ii}| \geq |a_{i1}| + |a_{i2}| + \cdots + |a_{i,(i-1)}| + |a_{i,(i+1)}| + \cdots + |a_{in}|$$

for ith row.

or $$|a_{ii}| \geq |a_{1i}| + |a_{2i}| + \cdots + |a_{(i-1),i}| + |a_{(i+1),i}| + \cdots + |a_{ni}|$$

for ith column.

For example, the matrix

$$A = \begin{bmatrix} -5 & 0 & -1 & 1 & 2 \\ 3 & 6 & -2 & 0 & -1 \\ 1 & -2 & 12 & 4 & 3 \\ 0 & -1 & 3 & 7 & 2 \\ 1 & 2 & 3 & 4 & 11 \end{bmatrix}$$

is diagonally dominant because

$|a_{11}| = |-5| = 5 > (|0| + |-1| + |1| + |2|) = 4$ for first row

$|a_{11}| = |-5| = 5 = (|3| + |1| + |0| + |1|) = 5$ for first column

$|a_{22}| = 6 = (|3| + |-2| + |0| + |-1|) = 6$ for second row

$|a_{22}| = 6 > (|0| + |-2| + |-1| + |2|) = 5$ for second column

$|a_{33}| = 12 > (|1| + |-2| + |4| + |3|) = 10$ for third row

$|a_{33}| = 12 > (|-1| + |-2| + |3| + |3|) = 9$ for third column

$|a_{44}| = 9 > (|0| + |-1| + |3| + |-2|) = 6$ for fourth row

$|a_{44}| = 9 = (|1| + |0| + |4| + |4|) = 9$ for fourth column

$|a_{55}| = 11 > (|1| + |-2| + |3| + |4|) = 10$ for fifth row

$|a_{55}| = 11 > (|2| + |-1| + |3| + |2|) = 8$ for fifth column

Dominant diagonal matrices play important role in input-output analysis in economic theory and inter regional studies.

Strictly diagonal dominant : Diagonal dominance is *strict* if only inequality holds in (3.1). Thus the matrix A above is strictly diagonally dominant if the elements $a_{11} = 5$, $a_{22} = 6$ and $a_{44} = 9$ are replaced by some number numerically greater than 6. That is if $a_{11} = 5$, $a_{22} = 6$ and $a_{44} = 9$ are replaced by –6, –7 and 10 (say), then the matrix

$$A = \begin{bmatrix} -6 & 0 & -1 & 1 & 2 \\ 3 & -7 & -2 & 0 & -1 \\ 1 & -2 & 12 & 4 & 3 \\ 0 & -1 & 3 & 10 & 2 \\ 1 & 2 & 3 & 4 & 11 \end{bmatrix}$$

is strictly diagonally dominant.

3.5 Diagonal matrix

The square matrix in which off diagonal elements $a_{ij} = 0$, $i < j$ and $i > j$ and diagonal elements $a_{ii} \neq 0$ is called a *quasiscalar* or *diagonal* matrix. In short, $a_{ij} = \alpha_{ij}\delta_{ij}$, where $\delta_{ij} = 0$, $i \neq j$ and $\delta_{ij} = 1$, $i = j$, is the Kronecker delta. Thus

$$D_n = \begin{bmatrix} \alpha_{11} & 0 & 0 & \cdots & 0 \\ 0 & \alpha_{22} & 0 & \cdots & 0 \\ 0 & 0 & \alpha_{33} & \cdots & 0 \\ \vdots & \vdots & \vdots & \ddots & \vdots \\ 0 & 0 & 0 & \cdots & \alpha_{nn} \end{bmatrix} \text{ and } D_1 = [\alpha_{11}]$$

are n square and 1 square diagonal matrices. This diagonal square matrix is often expressed as

$$A = \text{diag}\,(a_{11}, a_{22}, \cdots, a_{nn}).$$

If the elements are *positive* then the matrix is called a *positive* diagonal matrix.

3.5.1 Theorems on diagonal matrix : The following three theorems are important to note for *diagonal* matrix.

Theorem 1 . Premultiplication of an $m \times n$ matrix A by an mth order diagonal matrix D multiplies each row of A by the corresponding diagonal element of D.

Proof . Let $A = [a_{kj}]_{m \times n}$ and $D = [\alpha_{ik}\,\delta_{ik}]_{m \times m}$ be the corresponding matrices. So,

$$DA = \left[\sum_{k=1}^{m} \alpha_{ik}\delta_{ik}a_{kj}\right]_{m\times m}$$

$$= \left[\alpha_{ii}a_{ij}\right]_{m\times n}, \quad \text{for } i = k, \delta_{ik} = 1 \text{ and } i \neq k, \delta_{ik} = 0$$

which shows that the elements of *i*th row of A gets multiplied by the (*i, i*)th diagonal elements of D. For example, let

$$A = \begin{bmatrix} 2 & 3 & 4 \\ -1 & 1 & 2 \end{bmatrix} \text{ and } D = \begin{bmatrix} 5 & 0 \\ 0 & 7 \end{bmatrix}$$

$$\therefore \qquad DA = \begin{bmatrix} 5 & 0 \\ 0 & 7 \end{bmatrix}\begin{bmatrix} 2 & 3 & 4 \\ -1 & 1 & 2 \end{bmatrix} = \begin{bmatrix} 10 & 15 & 20 \\ -7 & 7 & 14 \end{bmatrix}$$

Note that first row of A gets multiplied by (1,1)th diagonal element 5 of D and similarly the second row gets multiplied by (2, 2)th diagonal element 7.

Theorem 2 . Post multiplication of an $m\times n$ matrix A by an *n*th order diagonal matrix D multiplies each column of A by the corresponding diagonal element of D. Proof is left for the reader. Follow the proof of theorem 1 above. For example

$$AD = \begin{bmatrix} 2 & 3 & 4 \\ -1 & 1 & 2 \end{bmatrix}\begin{bmatrix} 5 & 0 & 0 \\ 0 & 7 & 0 \\ 0 & 0 & 2 \end{bmatrix} = \begin{bmatrix} 10 & 21 & 8 \\ -5 & 7 & 4 \end{bmatrix}.$$

Check the result.

Theorem 3 . If A and B are two *n*th order diagonal matrices, then their sum A + B, their product AB is also diagonal and AB = BA. (Product is commutative).

Proof . Part (i) Let

$$A = [a_{ij}\,\delta_{ij}]_{n\times n} \text{ and } B = [b_{ij}\,\delta_{ij}]_{n\times n}$$

be two diagonal matrices. Then

$$A + B = \begin{bmatrix} a_{11} & 0 & 0 & \cdots & 0 \\ 0 & a_{22} & 0 & \cdots & 0 \\ 0 & 0 & a_{33} & \cdots & 0 \\ \vdots & \vdots & \vdots & \ddots & \vdots \\ 0 & 0 & 0 & \cdots & a_{nn} \end{bmatrix} + \begin{bmatrix} b_{11} & 0 & 0 & \cdots & 0 \\ 0 & b_{22} & 0 & \cdots & 0 \\ 0 & 0 & b_{33} & \cdots & 0 \\ \vdots & \vdots & \vdots & \ddots & \vdots \\ 0 & 0 & 0 & \cdots & b_{nn} \end{bmatrix}$$

$$= \begin{bmatrix} (a_{11}+b_{11}) & 0 & 0 & \cdots & 0 \\ 0 & (a_{22}+b_{22}) & 0 & \cdots & 0 \\ 0 & 0 & (a_{33}+b_{33}) & \cdots & 0 \\ \vdots & \vdots & \vdots & \ddots & \vdots \\ 0 & 0 & 0 & \cdots & (a_{nn}+b_{nn}) \end{bmatrix}$$

$$= \text{diag}\left(a_{11} + b_{11},\ a_{22} + b_{22},\ \cdots,\ a_{nn} + b_{nn}\right)$$

Note that the diagonal elements of (A+B) is the sum of the corresponding diagonal elements of A and B.

Part(ii)

$$AB = \begin{bmatrix} a_{11}b_{11} & 0 & 0 & \cdots & 0 \\ 0 & a_{22}b_{22} & 0 & \cdots & 0 \\ 0 & 0 & a_{33}b_{33} & \cdots & 0 \\ \vdots & \vdots & \vdots & \ddots & \vdots \\ 0 & 0 & 0 & \cdots & a_{nn}b_{nn} \end{bmatrix}$$

$$= \text{diag}\left(a_{11}b_{11},\ a_{22}b_{22},\ \cdots,\ a_{nn}b_{nn}\right)$$

Note also that the diagonal elements of the product is the product of the corresponding diagonal elements of A and B . Again

$$BA = \begin{bmatrix} b_{11}a_{11} & 0 & 0 & \cdots & 0 \\ 0 & b_{22}a_{22} & 0 & \cdots & 0 \\ 0 & 0 & b_{33}a_{33} & \cdots & 0 \\ \vdots & \vdots & \vdots & \ddots & \vdots \\ 0 & 0 & 0 & \cdots & b_{nn}a_{nn} \end{bmatrix}$$

$$= \text{diag}\left(a_{11}b_{11},\ a_{22}b_{22},\ \cdots,\ a_{nn}b_{nn}\right)$$

$\therefore \qquad AB = BA$

3.6 Scalar and unit matrices

This section introduces scalar matrix and the unit matrix I_n of order n and I_s matrix, where s denotes secondary diagonal.

Scalar matrix : If $a_{ii} = a$, a scalar, in the diagonal matrix in sec. 3.5 then the matrix is called a *scalar matrix.* The elements $a_{ij} = \alpha\delta_{ij}$, where δ_{ij} is the Kronecker delta.

Thus, $$S_n = \begin{bmatrix} \alpha & 0 & 0 & \cdots & 0 \\ 0 & \alpha & 0 & \cdots & 0 \\ 0 & 0 & \alpha & \cdots & 0 \\ \vdots & \vdots & \vdots & \ddots & \vdots \\ 0 & 0 & 0 & \cdots & \alpha \end{bmatrix} \text{ and } S_3 = \begin{bmatrix} 3 & 0 & 0 \\ 0 & 3 & 0 \\ 0 & 0 & 3 \end{bmatrix}$$

are scalar matrices. In matrix multiplication a scalar matrix behaves like a scalar i.e. $S_m A_{m\times n} = A_{m\times n} S_n = \alpha A_{m\times n}$. Do not confuse "scalar matrix" with "scalar" of sec. 1.4.

Unit matrix : If $\alpha = 1$ in the scalar matrix then the matrix is called a *unit* or an *idem* or an *identity* matrix. It is denoted by I or I, suffix n indicates the order of the unit matrix. In terms of elements it is $[a_{ij}]_{n\times n} = [\delta_{ij}]_{n\times n}$, where δ_{ij} is the Kronecker delta. Thus,

$$I_1 = [1],\ I_2 = \begin{bmatrix} 1 & 0 \\ 0 & 1 \end{bmatrix} = [\delta_{ij}]_{2\times2},\ I_3 = \begin{bmatrix} 1 & 0 & 0 \\ 0 & 1 & 0 \\ 0 & 0 & 1 \end{bmatrix} = [\delta_{ij}]_{3\times3}$$

are unit matrices. It plays the role of unity as in ordinary algebra, and is the multiplicative identity in matrix theory. Thus $I_m A_{m\times n} = A_{m\times n} I_n = A_{m\times n}$ and scalar matrix $S_n = \alpha I_n$, $I_n^2 = I_n^3 = \cdots = I_n^n = I_n$, which justifies the name *unit*.

Unit matrix can be *constructed* from matrix unit as well as from *s.u. vectors*. Thus I_4 is the sum of four matrix units E_{11}, E_{22}, E_{33} and E_{44}. That is,

$$I_4 = E_{11} + E_{22} + E_{33} + E_{44} = \sum_{i=1}^{4} \mathbf{e}_i \mathbf{e}_i^T$$

Similarly , I_n is constructed from the sum of equal subscripted matrix units, i.e.

$$I_n = E_{11} + E_{22} + \cdots + E_{nn} = \sum_{i=1}^{n} \mathbf{e}_i \mathbf{e}_i^T$$

I_s matrix : It is a square matrix in which nonzero elements are units along the secondary diagonal. Thus, the elements are $a_{(n-j+1)j} = 1$, $a_{ij} = 0$, $i + j \neq n + 1$. For example, if $n = 3$, then

$$I_s = \begin{bmatrix} 0 & 0 & 1 \\ 0 & 1 & 0 \\ 1 & 0 & 0 \end{bmatrix},$$

the suffix "*s*" stands for secondary diagonal.

Note that $I_s^T = I_s$ and $I_s^2 = I$. Also note that $I_s A$ and AI_s for rectangular matrix A, reverse the rows and the columns respectively.

N.B. Some author calls the I_s matrix as J-matrix.

3.7 Tridiagonal and band matrices

Tridiagonal matrix : A square matrix having nonzero elements only along the *three* diagonals — main, first sub and first super while other elements $a_{ij} = 0$ for $|i-j| < 1$, then the matrix is called *tridiagonal* (*continuant*) matrix. Thus

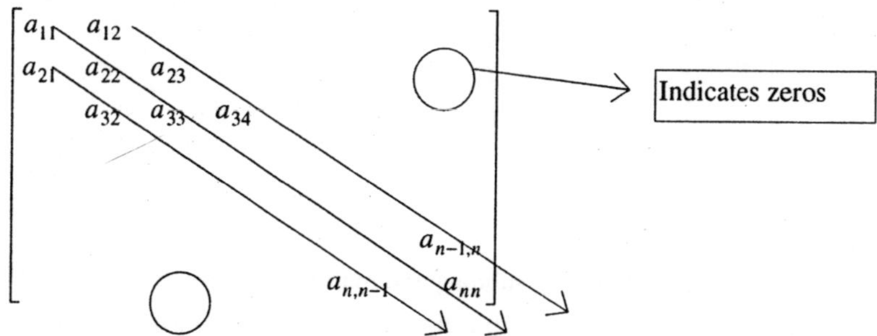

An easy theorem on the above definitions is the following (its proof is left for reader).

Theorem . The sum and the product of two *n*th order upper triangular (lower triangular or tridiagonal) matrices are again an *n*th order upper triangular (lower triangular or triadiagonal) matrices.

Band matrix : It is a sparse matrix but the nonzero elements are organised into bands instead of scattered in sparse. The zero elements occupy positions such that $a_{ij} = 0$, for $|i - j| >$ (bandwidth). The bandwidth is the distance between the main and the super (sub) diagonal. Thus tridiagonal matrix is an example of band marix with bandwidth $= |i - j| = 1$.

3.8 Triangular matrices

This section defines upper, strictly upper, lower and strictly lower triangular matrices.

Upper triangular matrix : A square upper trapezoidal matrix (i.e. $a_{ij} = 0$, $i > j$) is called an *upper triangular* matrix and is normally denoted by,

$$\mathrm{U} = \begin{bmatrix} u_{11} & u_{12} & u_{13} & \cdots & u_{1n} \\ 0 & u_{22} & u_{23} & \cdots & u_{2n} \\ 0 & 0 & u_{33} & \cdots & u_{3n} \\ \vdots & \vdots & \vdots & \ddots & \vdots \\ 0 & 0 & 0 & \cdots & u_{nn} \end{bmatrix}.$$

Strictly upper triangular matrix (upper or superdiagonal matrix) : An upper triangular matrix is *strictly upper triangular* (an *upper* or a *superdiagonal*) matrix if its diagonal elements are all zero. In general, a square matrix is strictly upper triangular if the elements $a_{ij} = 0$ for $i \geq j$. Thus

$$\mathrm{U}_s = \begin{bmatrix} 0 & u_{12} & u_{13} & \cdots & u_{1n} \\ 0 & 0 & u_{23} & \cdots & u_{2n} \\ \vdots & \vdots & \ddots & \ddots & \vdots \\ 0 & 0 & 0 & \ddots & u_{n-1,n} \\ 0 & 0 & 0 & \cdots & 0 \end{bmatrix},$$

suffix "*s*" stands for strictly.

Lower triangular matrix : A square lower trapezoidal matrix (i.e. $a_{ij} = 0$, $i < j$) is called a *lower triangular* matrix and is normally denoted by

$$\mathrm{L} = \begin{bmatrix} l_{11} & 0 & 0 & \cdots & 0 \\ l_{21} & l_{22} & 0 & \cdots & 0 \\ l_{31} & l_{32} & l_{33} & \cdots & 0 \\ \vdots & \vdots & \vdots & \ddots & \vdots \\ l_{n1} & l_{n2} & l_{n3} & \cdots & l_{nn} \end{bmatrix}.$$

Note that $\mathrm{U}^{\mathrm{T}} = \mathrm{L}$ and $\mathrm{L}^{\mathrm{T}} = \mathrm{U}$.

Strictly lower triangular matrix (lower or subdiagonal matrix) : A lower triangular matrix is *strictly lower triangular* (*lower* or *subdiagonal*) matrix if its diagonal elements are all zero. In general, a square matrix is strictly lower triangular if the elements $a_{ij} = 0$ for $i \leq j$. Thus

$$L_s = \begin{bmatrix} 0 & 0 & 0 & \cdots & \cdots & 0 \\ l_{21} & 0 & 0 & \cdots & \cdots & 0 \\ l_{31} & l_{32} & 0 & \cdots & \cdots & 0 \\ \vdots & \vdots & \ddots & \ddots & \vdots & \vdots \\ \vdots & \vdots & \vdots & \ddots & \ddots & \vdots \\ l_{n1} & l_{n2} & l_{n3} & \cdots & l_{n,n-1} & 0 \end{bmatrix},$$

suffix " s" stands for strictly.

Note that all diagonal matrices, scalar matrices, identity matrix and zero matrix are simultaneously upper and lower triangular matrices.

Conversely, every matrix which is both upper and lower triangular is diagonal.

Also *note* that any square matrix has a unique decomposition as a sum of *three* matrices—

strictly subdiagonal + diagonal + strictly superdiagonal.

For example,

$$\begin{bmatrix} a_{11} & a_{12} & a_{13} \\ a_{21} & a_{22} & a_{23} \\ a_{31} & a_{32} & a_{33} \end{bmatrix} = \begin{bmatrix} 0 & 0 & 0 \\ a_{21} & 0 & 0 \\ a_{31} & a_{32} & 0 \end{bmatrix} + \begin{bmatrix} a_{11} & 0 & 0 \\ 0 & a_{22} & 0 \\ 0 & 0 & a_{33} \end{bmatrix} + \begin{bmatrix} 0 & a_{12} & a_{13} \\ 0 & 0 & a_{23} \\ 0 & 0 & 0 \end{bmatrix}$$

3.9 Symmetric and Hermitian matrices

This section describes the symmetric matrix and its four other types, Hermitian and skew Hermitian, and theorems in relation to them.

3.9.1 Symmetric matrix : It is a square matrix that remains unchanged by unary operation of transposition. Clearly, speaking the elements $a_{ij} = a_{ji}$, $i \neq j$ i.e. $A = A^T$ or in other words, matrices that are their own transpose are symmetric. *Geometrically* speaking, the elements are "symmetric" about the main diagonal i.e. the elements coincide on folding about the main diagonal. In other words equidistant elements from the main diagonal are equal in all respects. Thus

$$\begin{bmatrix} a_{11} & a_{12} & a_{13} \\ a_{12} & a_{22} & a_{23} \\ a_{13} & a_{23} & a_{33} \end{bmatrix}, \quad \begin{bmatrix} 1 & 2 & 3 \\ 2 & 5 & 4 \\ 3 & 4 & 6 \end{bmatrix},$$

I_n and diagonal matrices are symmetric matrices.

Note that if A is a square matrix, then AA^T is symmetric. *Prove* the statement: A symmetric matrix of order $n \times n$ has $\frac{1}{2}n(n+1)$ distinct elements. So, it needs half computer storage and half the computation time and labour.

3.9.2 Skew (anti) symmetric matrix : It is a square matrix that *changes sign* on unary operation of transposition. Thus $A^T = -A$. In terms of elements $a_{ji} = -a_{ij}$. That is, equidistant elements from the main diagonal differ in sign but equal in magnitude. It follows from definition that $a_{ii} = -a_{ii}$, i.e. $2a_{ii} = 0$, i.e. $a_{ii} = 0$. So diagonal elements of a skew symmetric matrix are always zero.Thus

$$[0], \begin{bmatrix} 0 & -2 & 3 \\ 2 & 0 & 4 \\ -3 & -4 & 0 \end{bmatrix}$$

are skew symmetric matrices. A skew symmetric matrix of order $n \times n$, $n \neq 1$, has $\frac{1}{2}n(n-1)$ distinct elements if only magnitudes are concerned. If $n = 1$, it has one element only.

3.9.3 Crosssymmetric (persymmetric) matrix : It is a square matrix symmetric about the *secondary* or *cross diagonal* i.e. if $A = I_s A$. Examples are

$$A = \begin{bmatrix} a_{11} & a_{12} & a_{13} \\ a_{21} & a_{22} & a_{12} \\ a_{31} & a_{21} & a_{11} \end{bmatrix}, \; A = \begin{bmatrix} 2 & 0 & 1 \\ 8 & 1 & 0 \\ 5 & 8 & 2 \end{bmatrix}.$$

3.9.4 Skew Crosssymmetric matrix : It is a square matrix skew symmetric about the secondary or cross diagonal. Examples are

$$A = \begin{bmatrix} -a_{11} & a_{12} & a_{13} \\ -a_{21} & a_{22} & -a_{12} \\ a_{31} & a_{21} & a_{11} \end{bmatrix}, \; A = \begin{bmatrix} 2 & 0 & 1 \\ 8 & 1 & 0 \\ 5 & -8 & -2 \end{bmatrix}.$$

Centres

3.9.5 Centrosymmetric matrix : A square matrix which is symmetric about the centre of its array of elements is called *centrosymmetric*. Thus $A = [a_{ij}]_{n \times n}$ is centro-symmetric if $a_{ij} = a_{(n-i+1),(n-j+1)}$. Examples of centrosymmetric matrices are

$$\text{(i)} \begin{bmatrix} a_{11} & a_{12} & a_{13} \\ a_{21} & \boxed{a_{22}} & a_{23} \\ a_{31} & a_{32} & a_{33} \end{bmatrix} \quad \text{(ii)} \begin{bmatrix} 1 & 2 & 3 \\ 4 & \boxed{5} & 6 \\ 3 & 2 & 1 \end{bmatrix}, \quad \text{(iii)} \left[\begin{array}{cc|cc} a_{11} & a_{12} & a_{13} & a_{14} \\ a_{21} & a_{22} & a_{23} & a_{24} \\ \hline a_{24} & a_{23} & a_{22} & a_{21} \\ a_{14} & a_{13} & a_{12} & a_{11} \end{array}\right]$$

3.9.6 Hermitian matrix : A square complex matrix is said to be *Hermitian* if the unary operation tranjugation (Hermitian conjugation) keeps the matrix unchanged i.e. $A^* = A$ or in terms of elements $\overline{a}_{ji} = a_{ij}$. Therefore the diagonal elements $\overline{a}_{ii} = a_{ii}$. Let $a_{ii} = a + ib$. So $\overline{a}_{ii} = a - ib$.

Therefore

$$a + ib = a - ib$$

or $$2ib = 0$$

or $$b = 0.$$

Hence $a_{ii} = a$ = purely real. That is, diagonal elements of a Hermitian matrix is always real. Thus

$$\begin{bmatrix} a & a_1 + ib_1 & a_2 - ib_2 \\ a_1 - ib_1 & b & a_3 + ib_3 \\ a_2 + ib_2 & a_3 - ib_3 & c \end{bmatrix}, \begin{bmatrix} 4 & 1-i \\ 1+i & 2 \end{bmatrix}$$

are examples of Hermitian matrices for all real a, b, c, a_i and b_i , $i = 1, 2, 3$. The name "Hermitian matrix" bears the name of the French mathematician, Charles Hermite (1822 -1901). In 1855

Hermite proved that Hermitian matrices have real characteristic roots (cf. chapter 13). These matrices arise in applications, for example, in the theory of atomic physics.

3.9.7 Skew (anti) Hermitian matrix : A square complex matrix is said to be *skew* (*anti*) *Hermitian* if the unary operation tranjugation (Hermitian conjugation) changes the sign of the matrix i.e. $A^* = -A$ or in terms of elements $\bar{a}_{ji} = -a_{ij}$.

Therefore the diagonal elements $\bar{a}_{ii} = -a_{ii}$. Let $a_{ii} = a + ib$. So $\bar{a}_{ii} = a - ib$. Hence $a - ib = -a - ib$ or $a = 0$, which implies that diagonal elements of a skew Hermitian matrix are purely imaginary. Thus

$$\begin{bmatrix} ia & a_1 + ib_1 & a_2 - ib_2 \\ -(a_1 - ib_1) & ib & a_3 + ib_3 \\ -(a_2 + ib_2) & -(a_3 - ib_3) & ic \end{bmatrix}, \begin{bmatrix} 4 & 1-i \\ 1+i & 2 \end{bmatrix}$$

are examples of skew (anti) Hermitian matrix.

3.9.8 Theorem on symmetric and skew symmetric matrices : The following is a theorem relating symmetric and skew symmetric matrices.

Theorem . If A is square matrix then A = S + K where S is symmetric and K is skew symmetric. Moreover, this representation is unique.

Proof . By definition of symmetric matrix $A = A^T$. Therefore,

$$2S = A + A^T$$

or $$S = \tfrac{1}{2}(A + A^T).$$

Similarly, $$K = \tfrac{1}{2}(A - A^T).$$

We then have, $$A = S + K.$$

Now, $$S^T = \tfrac{1}{2}(A + A^T)^T = \tfrac{1}{2}(A^T + (A^T)^T) = \tfrac{1}{2}(A + A^T) = S$$

Therefore S is symmetric.

Also, $$K^T = \tfrac{1}{2}(A - A^T)^T = \tfrac{1}{2}(A^T - (A^T)^T) = -\tfrac{1}{2}(A - A^1) = -K$$

Therefore, K is skew symmetric.

Uniqueness : If possible let $A = S_1 + K_1$, where S_1 is symmetric and K_1 is skew symmetric. Now

$$A^T = S_1^T + K_1^T = S_1 - K_1 \quad \text{by definition.}$$

Therefore, $A + A^T = 2S_1$ and $A - A^T = 2K_1$.

or $$S_1 = \tfrac{1}{2}(A + A^T) \text{ and } K_1 = \tfrac{1}{2}(A - A^T).$$

Therefore, $S = S_1$ and $K = K_1$.

Hence the representation A = S + K is unique.

3.9.9 Theorems on symmetric and Hermitian matrices : The following are some important but easy theorems on symmetric (skew symmetric) and Hermitian (skew Hermitian) matrices and some of them brings relation between them under certain condition(s) :

Theorem 1. (i) A Hermitian (skew Hermitian) matrix is square and

(ii) a real Hermitian (skew Hermitian) matrix is symmetric (skew symmetric).

Proof. (i) By definition of Hermitian (skew Hermitian) matrix, $A^* = A$ (or $-A$). Then from the definition of equality of matrices, it follows that number of columns (or rows) of A becomes equal to the number of rows (or columns) of A^*. That is, if

$$A = [a_{ij}]_{m\times n} \text{ then } A^* = \left[a_{ji}\right]_{n\times m} \text{ and}$$

equality $A^* = A$ is only possible when (m of A) = (n of A^*) and vice-versa. Therefore, A^* must be a square matrix.

(ii) Since the matrix A is real Hermitian, so

$$A = A^* = \left(\overline{A}\right)^T, \text{ by definition.}$$

$= A^T$, conjugate $\overline{A}$ of a R-matrix is the matrix itself by identifying property of R-matrix which shows that A is symmetric i.e. a real Hermitian matrix is a symmetric matrix. Readers can verify statement with examples.

Theorem 2 . If the matrix A is Hermitian (skew Hermitian), then iA and ($-$ iA) are skew Hermitian (Hermitian). In other words the multiplication of Hermitian (skew Hermitian) matrix by the complex quantity i (or $-$ i) results into a skew Hermitian (Hermitian) matrix.

Proof . Part I : Let A be Hermitian.Therefore by defn. $A = A^*$ (3.2)

Now $(iA)^* = -\,iA^*$, as conjugate $i = \bar{i} = -i$, a complex scalar.

$= -\,iA = -\,(iA)$ by (3.2)

$\therefore$ A is skew Hermitian.

Part II : Let A be skew Hermitian.Therefore, by defn. $A^* = -A$ (3.3)

Now $(-\,iA)^* = iA^* = (-\,iA)$ by (3.3)

Therefore, ($-$ iA) is Hermitian.

Theorem 3 . If A is Hermitian, then A = M + iN, where M is real symmetric matrix and N is real skew symmetric matrix.

Proof . Since A is Hermitian, therefore , by defn. $A = A^*$.

Now, $(M + iN)^* = M^* + (iN)^* = M^* - iN^*$

$= M^T - iN^T$, as conjugates $\overline{M}, \overline{N}$ of R-matrices are the matrices themselves.

$= M^T + i(-N^T)$

$= M + iN$, as M is symmetric and N is skew symmetric.

Therefore, $A^* = A$, i.e. A is Hermitian.

Theorem 4 . If A is a square C-matrix, then $A + A^*$ is Hermitain and $A - A^*$ is skew Hermitian.

Proof . Let $A + A^* = H$. Then $H^* = (A + A^*)^* = A^* + A = A + A^* = H$.

Therefore , H is Hermitian.

Also let $(A - A^*) = H_s$. Therefore, $H_s^* = (A - A^*)^* = A^* - A$

$= -\,(A - A^*) = -\,H_s$.

Therefore, $H_s = A - A^*$ is skew Hermitian.

Theorem 5 . Any square C-matrix can be expressed as the sum of a Hermitian and a skew Hermitian.

Proof . It follows immediately from theorem 4 above. Here $H + H_s = A$ and hence proved.

Theorem 6 . If A is any C-matrix (not necessarily square) then A^*A and AA^* are both Hermitian.

Proof . Let $A^*A = H$. Then $H^* = (A^*A)^* = A^*(A^*)^* = A^*A = H$.

Therefore, H is Hermitian. Similarly, AA^* is Hermitian.

Theorem 7. If A is Hermitian, then so also is B^*AB for every comforbable C-matrix B.

Proof . A is (given) Hermitian i.e. $A = A^*$.

Now let $B^*AB = H$. Then $H^* = (B^*AB)^*$

$= (AB)^*(B^*)^* = B^*A^*B = B^*AB = H$.

Therefore, H is Hermitian.

Theorem 8 . If A is a Hermitian matrix of order n and if S is a C-matrix of order $n\times1$, then the scalar $\overline{S}^{T}AS$ is real.

Proof. $\overline{\overline{S}^{T}AS} = \left(\overline{\overline{S}^{T}AS}\right)^{T}$, as $\overline{S}^{T}AS$ is 1scalar i.e. single element matrix.

$$= \left(\overline{\overline{S}}^{T}\overline{AS}\right)^{T} = \left(S^{T}\overline{AS}\right)^{T} = \overline{S}^{T}\overline{A}^{T}S = \overline{S}^{T}AS \text{, as } A = \overline{A}^{T},$$

since conjugate of a real is real.

Therefore, $\overline{S}^{T}AS$ is real. ($\because A = \overline{A}$ for real)

3.10 Definite, semidefinite and indefinite matrices

These types of matrices arise in many different contexts such as obtaining minimum condition of functions of several variables, stability criteria in control theory etc. Positive definite matrices and positive real numbers are analogous and these matrices bear interesting relation with orthogonal matrices of section 3.11.

Definitions as well as tests for identifying these matrices are concern of this section. Note that $\mathbf{x}^{T}A\mathbf{x}$ is a scalar.

Definitions : An n square complex Hermitian (or real symmetric) matrix A is

(a) Positive definite (p.d.) if $\mathbf{x}^{*}A\mathbf{x} > 0$ (or $\mathbf{x}^{T}A\mathbf{x} > 0$) for all $\mathbf{x} \neq 0$;

(b) Positive semidefinite (p.s.d.) if $\mathbf{x}^{*}A\mathbf{x} \geq 0$ (or $\mathbf{x}^{T}A\mathbf{x} \geq 0$) for all $\mathbf{x}$;

(c) Negative definite (n.d) if $\mathbf{x}^{*}A\mathbf{x} < 0$ (or $\mathbf{x}^{T}A\mathbf{x} < 0$) for all $\mathbf{x} \neq 0$;

(d) Negative semidefinite (n.s.d.) if $\mathbf{x}^{*}A\mathbf{x} \leq 0$ (or $\mathbf{x}^{T}A\mathbf{x} \leq 0$) for all $\mathbf{x}$;

(e) Indefinite if none of the above holds.

Note that definitions serve the tests for indentification of the matrices.

In chapter fifteen $\mathbf{x}^{T}A\mathbf{x}$ represents a quadratic form. Here the knowledge of transpose and transjugation will be applied for testing these matrices. The fundamental procedure is the method of squaring of a quadratic expression which is a scalar in the sense of matrix. Other definitions (tests) based on principal minor (cf. sec.6.17), eigenvalue (cf. sec.13.48), rank and signature (cf. sec.10.11) will be encounted as we pass onto those topics one after another.

Ex. Identify the following matrices for positive (negative) definite (semidefinite) and indefinite.

$$\text{(i)} \begin{bmatrix} 1 & 2 & 4 \\ 2 & -3 & 0 \\ 4 & 0 & 1 \end{bmatrix}, \quad \text{(ii)} \begin{bmatrix} -1 & 3 & 0 \\ 3 & -10 & 1 \\ 0 & 1 & -4 \end{bmatrix}, \quad \text{(iii)} \begin{bmatrix} 10 & 2 & 3 \\ 2 & 1 & 1 \\ 3 & 1 & 4 \end{bmatrix},$$

$$\text{(iv)} \begin{bmatrix} 37 & -2 & -24 \\ -2 & 13 & -3 \\ -24 & -3 & 17 \end{bmatrix}, \text{(v)} \begin{bmatrix} -37 & 2 & 24 \\ 2 & -13 & 3 \\ 24 & 3 & -17 \end{bmatrix}.$$

(i) Let $\mathbf{x} = \begin{bmatrix} x_1 & x_2 & x_3 \end{bmatrix}^{T}$ be a nonzero vector. Then

$$\mathbf{x}^{T}A\mathbf{x} = \begin{bmatrix} x_1 & x_2 & x_3 \end{bmatrix}^{T} \begin{bmatrix} 1 & 2 & 4 \\ 2 & -3 & 0 \\ 4 & 0 & 1 \end{bmatrix} \begin{bmatrix} x_1 \\ x_2 \\ x_3 \end{bmatrix}$$

$$= \begin{bmatrix} x_1 & x_2 & x_3 \end{bmatrix} \begin{bmatrix} x_1 + 2x_2 + 4x_3 \\ 2x_1 - 3x_2 \\ 4x_1 + \quad x_3 \end{bmatrix}$$

$$= x_1^2 + 2x_1x_2 + 4x_1x_3 + 2x_1x_2 - 3x_2^2 + 4x_1x_3 + x_3^2$$
$$= x_1^2 - 3x_2^2 + x_3^2 + 4x_1x_2 + 8x_1x_3$$
$$= \tfrac{1}{2}\left(2x_1^2 - 6x_2^2 + 2x_3^2 + 8x_1x_2 + 16x_1x_3\right)$$
$$= \tfrac{1}{2}\left[(x_1 + x_2)^2 + (x_1 + 2x_3)^2 - \left(7x_2^2 + 2x_3^2\right) + 6x_1x_2 + 12x_1x_3\right]$$

x^TAx does not satisfy any of the conditions (a) through (d) of sec.3.10 for nonzero **x**. Therefore, A is indefinite.

(ii) $\mathbf{x}^T\mathbf{A}\mathbf{x} = \begin{bmatrix} x_1 & x_2 & x_3 \end{bmatrix} \begin{bmatrix} -1 & 3 & 0 \\ 3 & -10 & 1 \\ 0 & 1 & -4 \end{bmatrix} \begin{bmatrix} x_1 \\ x_2 \\ x_3 \end{bmatrix}$

$$= \begin{bmatrix} x_1 & x_2 & x_3 \end{bmatrix} \begin{bmatrix} -x_1 + 3x_2 \\ 3x_1 - 10x_2 + x_3 \\ x_2 - 4x_3 \end{bmatrix}$$
$$= -x_1^2 + 3x_1x_2 + 3x_1x_2 - 10x_2^2 + x_2x_3 + x_2x_3 - 4x_3^2$$
$$= -x_1^2 - 10x_2^2 - 4x_3^2 + 6x_1x_2 + 2x_2x_3$$
$$= -(x_1 - 3x_2)^2 - (x_2 - x_3)^2 - 3x_3^2$$
$$= -\left[(x_1 - 3x_2)^2 + (x_2 - x_3)^2 + 3x_3^2\right] < 0 \text{ always}$$

$\therefore$ A is negative definite.

(iii) $\mathbf{x}^T\mathbf{A}\mathbf{x} = \begin{bmatrix} x_1 & x_2 & x_3 \end{bmatrix} \begin{bmatrix} 10 & 2 & 3 \\ 2 & 1 & 1 \\ 3 & 1 & 4 \end{bmatrix} \begin{bmatrix} x_1 \\ x_2 \\ x_3 \end{bmatrix}$

$$= \begin{bmatrix} x_1 & x_2 & x_3 \end{bmatrix} \begin{bmatrix} 10x_1 + 2x_2 + 3x_3 \\ 2x_1 + x_2 + x_3 \\ 3x_1 + x_2 + 4x_3 \end{bmatrix}$$
$$= 10x_1^2 + 2x_1x_2 + 3x_1x_3 + 2x_1x_2 + x_2^2 + x_2x_3 + 3x_1x_3 + x_2x_3 + 4x_3^2$$
$$= 10x_1^2 + x_2^2 + 4x_3^2 + 4x_1x_2 + 6x_1x_3 + 2x_2x_3$$
$$= (2x_1 + x_2 + x_3)^2 + (x_1 + x_3)^2 + 5x_1^2 + 2x_3^2$$

which is always positive. Therefore, A is positive definite.

(iv) $\mathbf{x}^T\mathbf{A}\mathbf{x} = \begin{bmatrix} x_1 & x_2 & x_3 \end{bmatrix} \begin{bmatrix} 37 & -2 & -24 \\ -2 & 13 & -3 \\ -24 & -3 & 17 \end{bmatrix} \begin{bmatrix} x_1 \\ x_2 \\ x_3 \end{bmatrix}$

$$= \begin{bmatrix} x_1 & x_2 & x_3 \end{bmatrix} \begin{bmatrix} 37x_1 - 2x_2 - 24x_3 \\ -2x_1 + 13x_2 - 3x_3 \\ -24x_1 - 3x_2 + 17x_3 \end{bmatrix}$$

$$= 37x_1^2 - 2x_1x_2 - 24x_1x_3 - 2x_1x_2 + 13x_2^2 - 3x_2x_3 - 24x_1x_3 - 3x_2x_3 + 17x_3^2$$

$$= 37x_1^2 + 13x_2^2 + 17x_3^2 - 4x_1x_2 - 48x_1x_3 - 6x_2x_3$$

$$= (x_1 - 2x_2)^2 + (6x_1 - 4x_3)^2 + (3x_2 - x_3)^2$$

$$= (x_1 - 2x_2)^2 + 4(3x_1 - 2x_3)^2 + (3x_2 - x_3)^2 \geq 0,$$

Because the sum of three squares is always positive, provided $x_1 \neq 2x_2$, $3x_1 \neq 2x_3$ and $3x_2 \neq x_3$ and equals to zero when $x_1 = 2x_2$, $3x_1 = 2x_3$ and $3x_2 = x_3$.

∴ A is positive semidefinite.

(v) $\mathbf{x}^T\mathbf{A}\mathbf{x} = \begin{bmatrix} x_1 & x_2 & x_3 \end{bmatrix} \begin{bmatrix} -37 & 2 & 24 \\ 2 & -13 & 3 \\ 24 & 3 & -17 \end{bmatrix} \begin{bmatrix} x_1 \\ x_2 \\ x_3 \end{bmatrix}$

$$= -37x_1^2 - 13x_2^2 - 17x_3^2 + 4x_1x_2 + 48x_1x_3 + 6x_2x_3$$

$$= -\left[(x_1 - 2x_2)^2 + 4(3x_1 - 2x_3)^2 + (3x_2 - x_3)^2\right] \leq 0,$$

because of identical reasons of (iv). Therefore, A is negative semidefinite.

3.11 Orthogonal vectors and unit vectors

The following two definitions on vector product are useful for further study of vectors and matrices. Their definitions are given below in terms of row vectors as well as scalar product of vectors. Similar definitions can be given in terms of column vectors also.

Orthogonal vectors : If the scalar product of two vectors is zero, the vectors are orthogonal. Thus, $\begin{bmatrix} 2 & -1 & 4 \end{bmatrix} \begin{bmatrix} -1 \\ 2 \\ 1 \end{bmatrix} = 0$, implies that vectors $\begin{bmatrix} 2 & -1 & 4 \end{bmatrix}$ and $\begin{bmatrix} -1 \\ 2 \\ 1 \end{bmatrix}$ are orthogonal to each other.

Unit vector : If the product of a row vector and its transpose is unity, then the row vector is a unit vector. Thus, $\begin{bmatrix} \frac{3}{5} & \frac{4}{5} \end{bmatrix} \begin{bmatrix} \frac{3}{5} \\ \frac{4}{5} \end{bmatrix} = 1$. Therefore $\begin{bmatrix} \frac{3}{5} & \frac{4}{5} \end{bmatrix}$ is a unit vector. Note that unit vector is also be defined below in terms of length or norm. These can be defined using scalar (inner) product of vectors as follows:

Scalar (inner) products of vectors : Let $\mathbf{x}$ and $\mathbf{y}$ denote $n\times 1$ column vectors. Then $\mathbf{x}^T$ and $\mathbf{y}^T$ are $1\times n$ row vectors. The product

$$\mathbf{x}^T\mathbf{y} = \mathbf{y}^T\mathbf{x} = x_1y_1 + x_2y_2 + \cdots + x_ny_n$$

is a 1×1 matrix, called *scalar* (*inner*) product of $\mathbf{x}$ and $\mathbf{y}$. Similarly, for complex elements of $\mathbf{x}$ and $\mathbf{y}$, the product

$$\mathbf{x}^*\mathbf{y} = \mathbf{y}^*\mathbf{x} = \bar{x}_1 y_1 + \bar{x}_2 y_2 + \cdots + \bar{x}_n y_n$$

is called *complex scalar (inner) product* of **x** and **y**. Two column vectors **x** and **y** are called *orthogonal* if $\mathbf{x}^T\mathbf{y} = 0$; *complex orthogonal* if $\mathbf{x}^*\mathbf{y} = 0$.

The *length* or *norm* $\|\mathbf{x}\|$ of a vector $\mathbf{x} \neq 0$ is the positive quantity

$$\|\mathbf{x}\| = \sqrt{\mathbf{x}^*\mathbf{x}} = \sqrt{\mathbf{x}^T\mathbf{x}} = \sqrt{x_1^2 + x_2^2 + \cdots + x_n^2} > 0 \text{ if } \mathbf{x} \neq 0, \|0\| = 0.$$

Unit vector : A vector of unit length is called a *unit vector*. The length of the $\begin{bmatrix} \frac{3}{5} & \frac{4}{5} \end{bmatrix}$ is 1 and so it is an unit vector.

3.12 Orthogonal, normal, nilpotent, periodic, idempotent and involutory matrices

This section introduces and illustrates orthogonal, normal, nilpotent, periodic, idempotent and involutory matrices.

3.12.1 Orthogonal matrix : A square matrix of dimension n is called *orthogonal* if the Cayley product $A^TA = I_n$.

$$\text{Let} \qquad A = \begin{bmatrix} \frac{2}{3} & -\frac{2}{3} & \frac{1}{3} \\ \frac{2}{3} & \frac{1}{3} & -\frac{2}{3} \\ \frac{1}{3} & \frac{2}{3} & \frac{2}{3} \end{bmatrix}.$$

$$\text{Then, } A^TA = \begin{bmatrix} \frac{2}{3} & \frac{2}{3} & \frac{1}{3} \\ -\frac{2}{3} & \frac{1}{3} & \frac{2}{3} \\ \frac{1}{3} & -\frac{2}{3} & \frac{2}{3} \end{bmatrix} \begin{bmatrix} \frac{2}{3} & -\frac{2}{3} & \frac{1}{3} \\ \frac{2}{3} & \frac{1}{3} & -\frac{2}{3} \\ \frac{1}{3} & \frac{2}{3} & \frac{2}{3} \end{bmatrix} = \begin{bmatrix} 1 & 0 & 0 \\ 0 & 1 & 0 \\ 0 & 0 & 1 \end{bmatrix} = I_3.$$

Therefore, the matrix A is orthogonal. *Note* that rows (columns) of an orthogonal matrix form an orthogonal system of vectors, i.e. sum of products of any two rows is zero whereas any row with itself is unity.

Properties :

(i) If A is orthogonal, so is A^T.

(ii) If A is orthogonal, then det $A = \pm 1$ (cf. chapter 6).

(iii) If A is orthogonal, then $A^{-1} = A^T$ (cf. chapter 8). Hence A is nonsingular.

(iv) If A is orthogonal, so is A^{-1}.

(v) If A and B are orthogonal, so are AB and BA.

(vi) If an orthogonal matrix whose determinant equals 1, each elemant is equal in value to its cofactor.

Proofs . (i) and (v) follow directly from definition. For (v) $AA^T = I_n$, $BB^T = I_n$ (given). Now

$$(AB)(AB)^T = ABB^TA^T = AI_nA^T = AA^T = I_n$$

$\therefore$ AB is orthogonal. Similarly BA is orthogonal.

(ii), (vi) (cf. chapter 6) and (iii), (iv) (cf. chapter 8).

3.12.2 Normal matrix : A square matrix A is said to be a *normal matrix* if

$$AA^T = A^TA.$$

Examples of normal matrices include a fairly large class of matrices. Thus,

(i) symmetric matrices

(ii) diagonal matrices

(iii) orthogonal matrices

(iv) permutation matrices (cf. sec. 3.14)

(v) pseudopermutation matrices (i.e. If some or all 1's of a permutation matrix are changed to (–1)'s) are classes of normal matrices.

Let matrix $A = [a_{ij}]_{2\times2}$ be normal. Then by definition,

$$AA^T = A^TA$$

i.e.
$$\begin{bmatrix} a_{11} & a_{12} \\ a_{21} & a_{22} \end{bmatrix}\begin{bmatrix} a_{11} & a_{21} \\ a_{12} & a_{22} \end{bmatrix} = \begin{bmatrix} a_{11} & a_{21} \\ a_{12} & a_{22} \end{bmatrix}\begin{bmatrix} a_{11} & a_{12} \\ a_{21} & a_{22} \end{bmatrix}$$

or,
$$\begin{bmatrix} a_{11}^2 + a_{12}^2 & a_{11}a_{21} + a_{12}a_{22} \\ a_{11}a_{21} + a_{12}a_{22} & a_{21}^2 + a_{22}^2 \end{bmatrix} = \begin{bmatrix} a_{11}^2 + a_{21}^2 & a_{11}a_{12} + a_{21}a_{22} \\ a_{11}a_{12} + a_{22}a_{21} & a_{12}^2 + a_{22}^2 \end{bmatrix}$$

$\therefore$
$$a_{11}^2 + a_{12}^2 = a_{11}^2 + a_{21}^2$$
$$a_{11}a_{21} + a_{12}a_{22} = a_{11}a_{12} + a_{21}a_{22}$$
$$a_{21}^2 + a_{22}^2 = a_{12}^2 + a_{22}^2$$

imply that, $a_{12}^2 = a_{21}^2$ i.e. $a_{12} = -a_{21}$, because $a_{12} = a_{21}$ makes a_{11} and a_{22} indeterminate. Hence $a_{11} = a_{22}$. Therefore, normal 2×2 matrix has the form

$$\begin{bmatrix} a & b \\ -b & a \end{bmatrix}.$$

3.12.3 Nilpotent matrix : A square matrix A is called *nilpotent* if $A^k = 0$ for some positive integer *k*. If $A^k = 0$ but $A^{k-1} \neq 0$, then A is said to be nilpotent of order '*k*'. The strictly upper triangular and the strictly lower triangular matrices are examples of nilpotent matrices.

3.12.4 Periodic matrix : A square matrix A is called *periodic* of period *k* for some nonnegative integer *k* if

$$A^k = I,\ A^{k-1} \neq I$$

Periodic matrices are always invertible and $A^{-1} = A^{k-1}$ (cf. chapter 8).

3.12.5 Idempotent matrix : A square matrix A is called an *idempotent* matrix if $A^2 = A$. Thus $A^n = A$ for all positive integer, *n* are idempotent matrix A. Thus

$$\begin{bmatrix} 4 & -2 \\ 6 & -3 \end{bmatrix}^2 = \begin{bmatrix} 4 & -2 \\ 6 & -3 \end{bmatrix}.$$

$\therefore$
$$\begin{bmatrix} 4 & -2 \\ 6 & -3 \end{bmatrix}$$

is an idempotent matrix. The zero matrix and the identity matrix are both idempotent. The null matrix and the unit matrix are trivial examples of idempotent matrices.

3.12.6 Involutory matrix : A square matrix A is called an *involutory* matrix if $A^2 = I$, i.e. $A = A^{-1}$. Therefore, $A^2 - I^2 = 0$ or $(A + I)(A - I) = 0$ is the necessary condition for the involutory matrix.

Thus
$$A^2 = \begin{bmatrix} 1 & 0 \\ 0 & -1 \end{bmatrix}^2 = \begin{bmatrix} 1 & 0 \\ 0 & 1 \end{bmatrix} = I_2.$$

Therefore , $A = \begin{bmatrix} 1 & 0 \\ 0 & -1 \end{bmatrix}$ is an involutory matrix.

Construction of involutory matrix is given in chapter 7.

3.13 Unitary, unimodular, proper and improper matrices

This section describes unitary matrix with its theorem, unimodular, proper and improper matrices with illustrations.

3.13.1 Unitary matrix : An n-square complex matrix U_n is called *unitary* matrix iff

$$U_n^* U_n = I_n .$$

In other words, U_n is unitary iff its tranjugate U_n^* is its inverse

i.e. $U_n^* = U_n^{-1}$ or $U_n^* U_n = I_n$.

For the real matrix, $U_n^T = U_n^*$. Then U_n, an n-square real matrix, is *unitary* iff

$$U_n^T U_n = U_n U_n^T = I_n .\ A$$

the real unitary matrix is orthogonal.

3.13.2 Theorems on unitary matrix : Two theorems on unitary matrix are stated and proved below:

Theorem 1. A square matrix U_n is unitary if its columns (rows) are mutually orthogonal unit vectors.

Let $U_n = \begin{bmatrix} U_{.1} & U_{.2} & \cdots & U_{.n} \end{bmatrix}$ be an unitary matrix with column vectors $U_{.i}$, $i = 1, 2, \cdots, n$. Then

$$U_n^* = \begin{bmatrix} U_{.1}^* \\ U_{.2}^* \\ \vdots \\ U_{.n}^* \end{bmatrix} 1.$$

Now $$U_n^* U_n = \begin{bmatrix} U_{.1}^* \\ U_{.1}^* \\ \vdots \\ U_{.n}^* \end{bmatrix} \begin{bmatrix} U_{.1} & U_{.2} \cdots U_{.n} \end{bmatrix}$$

$$= \begin{bmatrix} U_{.1}^* U_{.1} & U_{.2}^* U_{.2} & \cdots & U_{.n}^* U_{.n} \end{bmatrix} \tag{3.4}$$

Since U_n is unitary, then $U_n^* U_n = I_n = [\delta_{ij}]_n$.

Therefore $U_{.i}^* U_{.j} = \delta_{ij}$

$$= \begin{cases} 0 \ , i \neq j \\ 1 \ , i = j \end{cases} \tag{3.5}$$

which proves that columns are mutually orthogonal unit vectors.

Conversely, as columns are mutually orthogonal unit vectors i.e. $\quad U^*_{.i} U_{.j} = \delta_{ij}$.

$\therefore \quad U^*_n U_n = \left[\delta_{ij}\right]_n = I_n$.

Similarly, $U_n U^*_n = I_n$ can be established. Hence the truth of the theorem.

N.B. *Readers are requested to read Theorem* 2, *secs.*3.13.3 , 3.13.4 , *Ex.*4 *below after chaptersix if determinant is quite unknown .*

Theorem 2. If U_n is unitary the absolute value of det $U_n = |U_n|$ is 1.

Proof . Since $\left|U^*_n\right| = \left|\overline{U}^T_n\right| = \left|\overline{U}_n\right| = \overline{\left|U_n\right|}$.2

Now $\quad \overline{\left|U_n\right|}\left|U_n\right| = \left|U_n\right|^2 = 1$,3 since U_n is unitary. Therefore absolute value of $|U_n|$ is 1.

If U_n is orthogonal, then $|U_n| = \pm 1$.

3.13.3 Unimodular matrix : If $|U_n| = 1$, the matrix U_n is called *unimodular.* That is, if the determinant of an unitary matrix is unity then the unitary matrix is *unimodular.*

3.13.4 Proper/Improper matrix : If the determinant of an orthogonal matrix U_n is unity then it is called *proper* matrix. If detU_n is (–1) then it is improper.

Ex.1 Under what conditions will the matrix

$$\begin{bmatrix} a+b & b-a \\ a-b & b+a \end{bmatrix} 4,$$

where *a* and *b* represent real numbers, be orthogonal ?

Let $U_2 = \begin{bmatrix} a+b & b-a \\ a-b & b+a \end{bmatrix}$. Therefore $U_2^T = \begin{bmatrix} a+b & a-b \\ b-a & b+a \end{bmatrix}$.

$$\text{Now} \quad U_2 U_2^T = \begin{bmatrix} (a+b)^2 + (b-a)^2 & \left(a^2 - b^2 + b^2 - a^2\right) \\ \left(a^2 - b^2 + b^2 - a^2\right) & (a-b)^2 + (b+a)^2 \end{bmatrix}$$

$$= \begin{bmatrix} 2\left(a^2 + b^2\right) & 0 \\ 0 & 2\left(a^2 + b^2\right) \end{bmatrix} = 2(a_2 + b_2)I_2$$

$$= I_2 , \quad \text{if } 2\left(a^2 + b^2\right) = 1$$

i.e. $\quad a^2 + b^2 = \frac{1}{2}$

Therefore, the given matrix is orthogonal if $a^2 + b^2 = \frac{1}{2}$.

Ex. 2 Show that

$$\begin{bmatrix} \frac{1}{\sqrt{14}} & \frac{3}{\sqrt{14}} & \frac{2}{\sqrt{14}} \\ \frac{3}{\sqrt{10}} & -\frac{1}{\sqrt{10}} & 0 \\ -\frac{1}{\sqrt{35}} & -\frac{3}{\sqrt{35}} & \frac{5}{\sqrt{35}} \end{bmatrix} \text{5is orthogonal.}$$

Let $U_3 = \begin{bmatrix} \frac{1}{\sqrt{14}} & \frac{3}{\sqrt{14}} & \frac{2}{\sqrt{14}} \\ \frac{3}{\sqrt{10}} & -\frac{1}{\sqrt{10}} & 0 \\ -\frac{1}{\sqrt{35}} & -\frac{3}{\sqrt{35}} & \frac{5}{\sqrt{35}} \end{bmatrix}$. Therefore, $U_3^T = \begin{bmatrix} \frac{1}{\sqrt{14}} & \frac{3}{\sqrt{10}} & \frac{1}{\sqrt{35}} \\ \frac{3}{\sqrt{14}} & -\frac{1}{\sqrt{10}} & -\frac{3}{\sqrt{35}} \\ \frac{2}{\sqrt{14}} & 0 & \frac{5}{\sqrt{35}} \end{bmatrix}$.

$$\therefore \; U_3U_3^T = \begin{bmatrix} 1 & 0 & 0 \\ 0 & 1 & 0 \\ 0 & 0 & 1 \end{bmatrix} = I_3$$

Therefore , U_3 is orthogonal as U_3 is a real matrix.

Ex.3 Deduce conditions under which the following "Scattering matrices"

$$\text{(i)} \begin{bmatrix} 0 & \alpha & 0 & i\beta \\ \alpha & 0 & i\beta & 0 \\ 0 & i\beta & 0 & \alpha \\ i\beta & 0 & \alpha & 0 \end{bmatrix}, \text{(ii)} \begin{bmatrix} 0 & 0 & \gamma & i\delta \\ 0 & 0 & i\delta & \gamma \\ \gamma & i\delta & 0 & 0 \\ i\delta & \gamma & 0 & 0 \end{bmatrix}, \text{(iii)} \begin{bmatrix} 0 & \eta & 0 & \zeta \\ \eta & 0 & \zeta & 0 \\ 0 & \zeta & 0 & \eta \\ \zeta & 0 & \eta & 0 \end{bmatrix}.$$

in microwave circuit theory are *unitary* if $\alpha, \beta, \gamma, \delta, \eta, \zeta$ are real numbers.

(i) Let U_4 denote the given matrix. It is a complex matrix. The tranjugate of U_4 is

$$U_4^* = \begin{bmatrix} 0 & \alpha & 0 & -i\beta \\ \alpha & 0 & -i\beta & 0 \\ 0 & -i\beta & 0 & \alpha \\ -i\beta & 0 & \alpha & 0 \end{bmatrix}$$

$$\text{Now} \quad U_4^*U_4 = \begin{bmatrix} \alpha^2+\beta^2 & 0 & 0 & 0 \\ 0 & \alpha^2+\beta^2 & 0 & 0 \\ 0 & 0 & \alpha^2+\beta^2 & 0 \\ 0 & 0 & 0 & \alpha^2+\beta^2 \end{bmatrix}$$

$$= \left(\alpha^2+\beta^2\right) \begin{bmatrix} 1 & 0 & 0 & 0 \\ 0 & 1 & 0 & 0 \\ 0 & 0 & 1 & 0 \\ 0 & 0 & 0 & 1 \end{bmatrix}$$

$$= (\alpha^2+\beta^2)I_4$$

Therefore, the required condition is $(\alpha^2 + \beta^2) = 1$.

(ii) Do yourself as in (i)

(iii) Let U_4 denote the given matrix. It is a real matrix.

The transpose of U_4 is

$$U_4^{\mathrm{T}} = \begin{bmatrix} 0 & \eta & 0 & \zeta \\ \eta & 0 & \zeta & 0 \\ 0 & \zeta & 0 & \eta \\ \zeta & 0 & \eta & 0 \end{bmatrix}.$$

$$\text{Now} \quad U_n U_n^{\mathrm{T}} = \begin{bmatrix} \eta^2+\zeta^2 & 0 & 2\eta\zeta & 0 \\ 0 & \eta^2+\zeta^2 & 0 & 2\eta\zeta \\ 2\eta\zeta & 0 & \eta^2+\zeta^2 & 0 \\ 0 & 2\eta\zeta & 0 & \eta^2+\zeta^2 \end{bmatrix}$$

$$= \left(\eta^2+\zeta^2\right) I_4, \quad \text{if } \frac{2\eta\zeta}{\eta^2+\zeta^2} = 0$$

$$= I_4 \qquad , \quad \text{if } \eta^2+\zeta^2 = 1$$

Therefore, if $\eta\zeta = 0$, then U is unitary i.e. either $\eta = 0$ or $\zeta = 0$, then U is unitary.

Ex.4 Given that ω is a complex cube root of unity, for what values of α, β, γ will the matrix be

$$\begin{bmatrix} \frac{1}{\sqrt{3}}\alpha & \frac{1}{\sqrt{3}}\beta & \frac{1}{\sqrt{3}}\gamma \\ \frac{1}{\sqrt{3}}\alpha & \frac{1}{\sqrt{3}}\beta\omega^2 & \frac{1}{\sqrt{3}}\gamma\omega \\ \frac{1}{\sqrt{3}}\alpha & \frac{1}{\sqrt{3}}\beta\omega & \frac{1}{\sqrt{3}}\gamma\omega^2 \end{bmatrix}$$

both unitary and unimodular ?

$$\text{Let} \qquad U_3 = \begin{bmatrix} \frac{1}{\sqrt{3}}\alpha & \frac{1}{\sqrt{3}}\beta & \frac{1}{\sqrt{3}}\gamma \\ \frac{1}{\sqrt{3}}\alpha & \frac{1}{\sqrt{3}}\beta\omega^2 & \frac{1}{\sqrt{3}}\gamma\omega \\ \frac{1}{\sqrt{3}}\alpha & \frac{1}{\sqrt{3}}\beta\omega & \frac{1}{\sqrt{3}}\gamma\omega^2 \end{bmatrix}, \text{ then}$$

$$U_3^* = \begin{bmatrix} \frac{1}{\sqrt{3}}\bar{\alpha} & \frac{1}{\sqrt{3}}\bar{\alpha} & \frac{1}{\sqrt{3}}\bar{\gamma} \\ \frac{1}{\sqrt{3}}\bar{\beta} & \frac{1}{\sqrt{3}}\bar{\beta}\bar{\omega}^2 & \frac{1}{\sqrt{3}}\bar{\beta}\bar{\omega} \\ \frac{1}{\sqrt{3}}\bar{\gamma} & \frac{1}{\sqrt{3}}\bar{\gamma}\bar{\omega} & \frac{1}{\sqrt{3}}\bar{\gamma}\bar{\omega}^2 \end{bmatrix}.$$

$$\text{Now } U_3U_3^* = \frac{1}{3}\begin{bmatrix} \left(\alpha\bar{\alpha}+\beta\bar{\beta}+\gamma\bar{\gamma}\right) & \left(\alpha\bar{\alpha}+\beta\bar{\beta}\bar{\omega}^2+\gamma\bar{\gamma}\bar{\omega}\right) & \left(\alpha\bar{\alpha}+\beta\bar{\beta}\bar{\omega}+\gamma\bar{\gamma}\bar{\omega}^2\right) \\ \left(\alpha\bar{\alpha}+\beta\bar{\beta}\omega^2+\gamma\bar{\gamma}\omega\right) & \left(\alpha\bar{\alpha}+\beta\bar{\beta}+\gamma\bar{\gamma}\right) & \left(\alpha\bar{\alpha}+\beta\bar{\beta}\omega+\gamma\bar{\gamma}\bar{\omega}^2\right) \\ \left(\alpha\bar{\alpha}+\beta\bar{\beta}\omega+\gamma\bar{\gamma}\omega^2\right) & \left(\alpha\bar{\alpha}+\beta\bar{\beta}\bar{\omega}+\gamma\bar{\gamma}\omega\right) & \left(\alpha\bar{\alpha}+\beta\bar{\beta}+\gamma\bar{\gamma}\right) \end{bmatrix}$$

$$= I_3, \quad \text{if } \left.\begin{aligned} \alpha\bar{\alpha}+\beta\bar{\beta}+\gamma\bar{\gamma} &= 3 \\ \alpha\bar{\alpha}+\beta\bar{\beta}\bar{\omega}^2+\gamma\bar{\gamma}\bar{\omega} &= 0 \\ \alpha\bar{\alpha}+\beta\bar{\beta}\bar{\omega}+\gamma\bar{\gamma}\bar{\omega}^2 &= 0 \end{aligned}\right\} \qquad (1)$$

which are the required conditions that U is unitary.

Again $\quad \det U_3 = \dfrac{\alpha\beta\gamma}{3\sqrt{3}}\det\begin{bmatrix} 1 & 1 & 1 \\ 1 & \omega^2 & \omega \\ 1 & \omega & \omega^2 \end{bmatrix} = \dfrac{\alpha\beta\gamma}{\sqrt{3}}(2\omega+1)$.

Therefore, U_3 will be unimodular if det U = 1,

i.e. $\quad \frac{1}{\sqrt{3}}\alpha\beta\gamma(2\omega+1) = 1$

i.e $\quad \frac{1}{\sqrt{3}}\alpha\beta\gamma\left(-1+i\sqrt{3}+1\right) = 1$

i.e. $\quad i\alpha\beta\gamma = 1 \quad$ i.e. $\alpha\beta\gamma = -i \qquad (2)$

Therefore , U_3 is both unitary and unimodular if (1) and (2) are simultaneously satisfied.

3.14 Permutation matrix with properties

The definition and properties of the permutation matrix with illustrations are the object of this section.

Permutation matrix : A permutation of n objects can be represented by an n-square matrix, called a *permutation* matrix $A_p = [a_{ij}]$, such that,

$$a_{ij} = 1, \text{ if } i\text{th object replaces } j\text{th object.}$$
$$= 0, \text{ if } i\text{th object does not replace } j\text{th object.}$$

For example, let ABCD be the original order and BCDA be its one permutation in which B replaces A, C replaces B, D replaces C and A replaces D. Then they can be represented by the following two respective matrices as

$$I_4 = \begin{array}{c} \\ A \\ B \\ C \\ D \end{array}\!\!\begin{array}{c} \begin{matrix} A & B & C & D \end{matrix} \leftarrow \text{replacement} \\ \begin{bmatrix} 1 & 0 & 0 & 0 \\ 0 & 1 & 0 & 0 \\ 0 & 0 & 1 & 0 \\ 0 & 0 & 0 & 1 \end{bmatrix} \end{array} \quad \text{and } A_p = \begin{array}{c} \\ A \\ B \\ C \\ D \end{array}\!\!\begin{array}{c} \begin{matrix} B & C & D & A \end{matrix} \\ \begin{bmatrix} 0 & 0 & 0 & 1 \\ 1 & 0 & 0 & 0 \\ 0 & 1 & 0 & 0 \\ 0 & 0 & 1 & 0 \end{bmatrix} \end{array}$$

$\uparrow \longleftrightarrow$ original unit matrix $\qquad$ Permutation matrix

The unit matrix above shows that each object replaces the same object while in permutation matrix $a_{14} = 1$ means 1st object A replaces 4th object D, i.e. A occupies the position of D etc.

The permutation matrix A_p obtained above by one permutation of ABCD can be stated in another *equivalent* way as follows:

It is square matrix whose each row and each column contains precisely one nonzero element 1 and 0 elsewhere. That is, each row (column) is a unit vector. It is obtained by the permutation or rearrangement of rows or columns of unit matrix. The above 4th order permutation matrix is formed from 4th order unit matrix, I_4 *either* by 4 permutations of $r_1\ r_2\ r_3\ r_4$ to $r_4\ r_1\ r_2\ r_3$ *or* by 4 column permutations of $c_1\ c_2\ c_3\ c_4$ to $c_2\ c_3\ c_4\ c_1$.

Properties :

(1) An n-square permutation matrix has n ones and $n(n-1)$ zeros.

(2) All permutation matrices are orthogonal. That is,

$$A_pA_p^T = \begin{bmatrix} 0 & 0 & 0 & 1 \\ 1 & 0 & 0 & 0 \\ 0 & 1 & 0 & 0 \\ 0 & 0 & 1 & 0 \end{bmatrix} \begin{bmatrix} 0 & 1 & 0 & 0 \\ 0 & 0 & 1 & 0 \\ 0 & 0 & 0 & 1 \\ 1 & 0 & 0 & 0 \end{bmatrix} = \begin{bmatrix} 1 & 0 & 0 & 0 \\ 0 & 1 & 0 & 0 \\ 0 & 0 & 1 & 0 \\ 0 & 0 & 0 & 1 \end{bmatrix} = I_3 \text{ i.e. } A_p^T = A_p^{-1}.$$

(3) If A_p and B_p are two permutation matrices conformable for multiplication then their product A_pB_p is also a permutation matrix. For example, let

$$A_p = \begin{bmatrix} 0 & 0 & 0 & 1 \\ 1 & 0 & 0 & 0 \\ 0 & 1 & 0 & 0 \\ 0 & 0 & 1 & 0 \end{bmatrix} \text{ and } B_p = \begin{bmatrix} 0 & 0 & 1 & 0 \\ 0 & 1 & 0 & 0 \\ 0 & 0 & 0 & 1 \\ 1 & 0 & 0 & 0 \end{bmatrix}.$$

$$\therefore \quad A_pB_p = \begin{bmatrix} 0 & 0 & 0 & 1 \\ 1 & 0 & 0 & 0 \\ 0 & 1 & 0 & 0 \\ 0 & 0 & 1 & 0 \end{bmatrix} \begin{bmatrix} 0 & 0 & 1 & 0 \\ 0 & 1 & 0 & 0 \\ 0 & 0 & 0 & 1 \\ 1 & 0 & 0 & 0 \end{bmatrix} = \begin{bmatrix} 1 & 0 & 0 & 0 \\ 0 & 0 & 1 & 0 \\ 0 & 1 & 0 & 0 \\ 0 & 0 & 0 & 1 \end{bmatrix}$$

= a permutation matrix by definition.

(4) Determinant of a permutation matrix is equal to +1 or (–1) according to the permutation is even or odd. So, permutation matrix is nonsingular (cf. chapter 6).

(5) The effect of pre or post multiplication of a matrix A by a permutation matrix permutes or interchanges the rows or columns of A according to the permutation involved in the given permutation matrix. For example

$$A_p = \begin{bmatrix} 0 & 1 & 0 & 0 \\ 1 & 0 & 0 & 0 \\ 0 & 0 & 1 & 0 \\ 0 & 0 & 0 & 1 \end{bmatrix}$$

be a permutation matrix having $r_2\ r_1\ r_3\ r_4$ (or $c_2\ c_1\ c_3\ c_4$) permutations of rows (columns). Let an arbitrary matrix be $A = [a_{ij}]_{4\times 3}$.

$$\therefore \quad A_pA = \begin{bmatrix} 0 & 1 & 0 & 0 \\ 1 & 0 & 0 & 0 \\ 0 & 0 & 1 & 0 \\ 0 & 0 & 0 & 1 \end{bmatrix} \begin{bmatrix} a_{11} & a_{12} & a_{13} \\ a_{21} & a_{22} & a_{23} \\ a_{31} & a_{32} & a_{33} \\ a_{41} & a_{42} & a_{43} \end{bmatrix} = \begin{bmatrix} a_{21} & a_{22} & a_{23} \\ a_{11} & a_{12} & a_{13} \\ a_{31} & a_{32} & a_{33} \\ a_{41} & a_{42} & a_{43} \end{bmatrix}$$

in which r_2 and r_1 interchanges as in A_p .

The similar effect of post multiplication can be easily verified.

3.15 Hadamard matrices with properties

This section is devoted to the definition, properties and different types of Hadamard matrices.

3.15.1 Hadamard matrix : An n square matrix H is called the *Hadamard* matrix if

(i) all its elements are +1 and –1,

(ii) $HH^T = nI_n$ where $n = 1, 2$ or a multiple of 4.

It is to be noted that the Hadamard matrix of order $n = 268 = 67\times4$, a multiple of 4 has not been constructed till 1977. This type of matrix has been applied in optimal design theory, information theory, graph theory etc. This matrix also arises in certain problem in geometry and in the theory of numbers and is applied in the construction of optimum code words in space communication. Examples are

$$H=[1],\ H=\begin{bmatrix}1&1\\1&-1\end{bmatrix},\ H=\begin{bmatrix}1&1&1&1\\1&1&-1&-1\\1&-1&1&-1\\1&-1&-1&1\end{bmatrix}.$$

In practice (+) and (–) signs are used for +1 and –1. So, the above matrices can be written as

$$[+],\ \begin{bmatrix}+&+\\+&-\end{bmatrix},\ \begin{bmatrix}+&+&+&+\\+&+&-&-\\+&-&+&-\\+&-&-&+\end{bmatrix}.$$

3.15.2 Properties of Hadamard matrix : Following are some properties of the Hadamard matrix.

(i) $HH^T = nI_n$ implies that the rows are pairwise orthogonal.

(ii) $H^TH = nI_n$ implies that the columns are pairwise orthogonal.

(iii) The above two properties imply that H is nonsingular (cf. definition chapter 8)

(iv) A matrix is Hadamard if its transpose is a Hadamard.

Properties (i) and (ii) may be stated otherwise as the elements in two different rows or columns coincide in exactly *half* the number of positions.

3.15.3 Nomenclature : Sylvester in 1867 discussed these types of matrices. However, in 1893 J. Hadamard (1865-1963), a French mathematician, proved that the absolute value of detA cannot exceed $M^n\left(\sqrt{n}\right)^3$, where $|a_{ij}| \le M$ for every i and j in $A = [a_{ij}]$.

$$\begin{aligned}\text{Since}\quad (\det H)^2 &= \det H^T \det H\\ &= \det(H^TH) = \det(nI_n) = n^n, \text{ (cf. chapter 6).}\end{aligned}$$

Therefore, for M = 1, a Hadamard matrix of order n attains the maximum determinant value. This is why the name "Hadamard matrix" has been used after the name of Hadamard.

3.15.4 Normalized Hadamard : A Hadmard matrix is said to be *normalized* if the first row and the first column are converted to +1's. This is done by

(i) negating the rows whose first element is (–1) and

(ii) negating the columns whose first element is (–1).

Ex. Normalize the Hadamard matrix

$$H=\begin{bmatrix}+&-&+&+\\-&+&+&+\\+&-&+&+\\+&+&-&+\end{bmatrix}.$$

Negating second rows which begins with (–) sign, we obtain

$$\begin{bmatrix} + & - & + & + \\ + & - & - & - \\ + & - & + & + \\ + & + & - & + \end{bmatrix} \qquad (1)$$

Negating the second column which begins with (–) sign, we obtain the normalized Hadamard matrix

$$\begin{bmatrix} + & + & + & + \\ + & + & - & - \\ + & + & + & + \\ + & - & - & + \end{bmatrix} \qquad (2)$$

Note that each step of conversion keeps the orthogonal property unchanged.

3.15.5 Seminormalized Hadamard : The matrix (1) is seminormalized because of its all positive signs in the first column.

3.15.6 Regular Hadamard : A Hadamard matrix is regular when all row sums are equal to a constant. For example

$$H = \begin{bmatrix} - & + & + & + \\ + & + & + & - \\ + & - & + & + \\ + & + & - & + \end{bmatrix}$$

is regular. Because each row sum = 2.

3.15.7 Symmetric Hadamard : If a Hadamard matrix is symmetric, then the matrix is a *symmetric Hadamard*. Example is

$$\begin{bmatrix} - & + & + & + \\ + & - & + & + \\ + & + & - & + \\ + & + & + & - \end{bmatrix}$$

3.15.8 Skew Hadamard : An n-th order skew Hadamard matrix is defined to be a Hadamard matrix of the form $H = S + I$, where S is skew symmetric and S satisfies $SS^T = (n-1)I$. Thus if

$$S = \begin{bmatrix} 0 & 1 & -1 & -1 \\ -1 & 0 & -1 & 1 \\ 1 & 1 & 0 & 1 \\ 1 & -1 & -1 & 0 \end{bmatrix}, \text{ then } SS^T = 3I_4 .$$

$$\therefore \quad H = S + I_4 = \begin{bmatrix} 1 & 1 & -1 & -1 \\ -1 & 1 & -1 & 1 \\ 1 & 1 & 1 & 1 \\ 1 & -1 & -1 & 1 \end{bmatrix}, \text{ is a skew Hadamard matrix.}$$

N.B. Note the difference between the skew Hadamard matrix and the real skew symmetric matrix of sec. 3.9.2.

3.15.9 Theorem on Hadamard matrix : Following is a theorem on Hadamard matrix.

Theorem . If A is an $n \times n$ Hadamard matrix, where $n > 2$, then n is a multiple of 4.

Proof . Method 1 : (an ingenious simple proof). Let $n \geq 3$ and set

$$H = \begin{bmatrix} a_1 & a_2 & a_3 & \cdots & a_n \\ b_1 & b_2 & b_3 & \cdots & b_n \\ c_1 & c_2 & c_3 & \cdots & c_n \\ \vdots & \vdots & \vdots & \vdots & \vdots \end{bmatrix}$$

where a's, b's, c's are either +1 or –1. Therefore by property (i) the inner product with the same row is n and that with a different row is zero.

Hence,

$$\left\{ \begin{aligned} & a_1^2 + a_2^2 + \cdots \cdots + a_n^2 = n \\ & a_1b_1 + a_2b_2 + \cdots + a_nb_n = 0 \\ & a_1c_1 + a_2c_2 + \cdots + a_nc_n = 0 \\ & b_1c_1 + b_2c_2 + \cdots + b_nc_n = 0 \text{ etc.} \end{aligned} \right. \tag{3.6}$$

But we have the algebraic identity,

$$(a_1+b_1)(a_1+c_1)+(a_2+b_2)(a_2+c_2)+\cdots+(a_n+b_n)(a_n+c_n)$$

$$=(a_1^2+a_2^2+\cdots+a_n^2)+(a_1c_1+a_2c_2+\cdots+a_nc_n)$$

$$+(a_1b_1+a_2b_2+\cdots+a_nb_n)+(b_1c_1+a_2c_2+\cdots+b_nc_n)$$

or $\quad (a_1+b_1)(a_1+c_1)+(a_2+b_2)(a_2+c_2)+\cdots+(a_n+b_n)(a_n+c_n)=n \qquad (3.7)$

[by first three equations in (3.6)]

Since a's, b's and c's are either +1 or –1, so the product $(a_1 + b_1)(a_1 + c_1)$ must be either 0 or 4 [i.e. if $a_1 = 1$, $b_1 = -1$, $c_1 = \pm 1$, then product is zero and if a_1, b_1, c_1 are of same sign then the product is 4]. The same is true for other products. Therefore, from (3.7), n, the sum of the product must be divisible by 4.

Method 2 : Let H denotes a Hadamard matrix of order n. Let the matrix be normalized. Then it is clear that except for the first row, the number of (+1)'s and (–1)'s in each of the remaining rows must be $\frac{1}{2}n$. Let the columns be permuted in such a way that first $\frac{1}{2}n$ elements in the second row are (+1)'s and the rest $\frac{1}{2}n$ elements in the second row are (–1)'s. Now let us examine the third row. Let the number of (+1)'s (and (–1)'s) in the first half of third row be

$$t \text{ [and } \tfrac{1}{2}n - t \text{]}$$

and that in the second half be

$$t_1 \text{ [and } \tfrac{1}{2}n - t_1 \text{]} .$$

Then $\qquad t + t_1 = \frac{1}{2}n$.

Since the dot product of the second and the third rows is 0, so

$$t + (\tfrac{1}{2}n - t_1) = \tfrac{1}{2}n$$

i.e. $t + t_1 = \frac{1}{2}n$

and $t - t_1 = 0$

$\therefore$ $t = \frac{1}{4}n$ and $t_1 = \frac{1}{4}n$.

Since t and t_1 must be integers, it follows that $n = 4t$ or $4t_1$

Therefore n is a multiple of 4, i.e. $n \equiv 0 \pmod 4$.

Ex. Determine all 2×2 Hadamard matrices.

By definition and properties, the 2×2 Hadamard matrices are

$$\begin{bmatrix}1 & 1\\1 & -1\end{bmatrix}, \begin{bmatrix}1 & -1\\1 & 1\end{bmatrix}, \begin{bmatrix}-1 & 1\\1 & 1\end{bmatrix}, \begin{bmatrix}1 & 1\\-1 & 1\end{bmatrix}, \begin{bmatrix}-1 & -1\\-1 & 1\end{bmatrix},$$

$$\begin{bmatrix}-1 & 1\\-1 & -1\end{bmatrix}, \begin{bmatrix}1 & -1\\-1 & -1\end{bmatrix}, \begin{bmatrix}-1 & -1\\1 & -1\end{bmatrix}$$

Each of these 8 (eight) 2×2 matrices satisfies the Hadamard matrix properties.

3.16 Helmert matrices

This section defines Helmert matrix and its variety with examples.

3.16.1 Helmertian matrix : Helmert or Helmertian matrix, H_e or H^e is a square orthogonal matrix having a triangle of zeros below the first row and above the principal diagonal. In other words, if $h_{ij} = 0$ for $j > i > 1$ in $H_e = [h_{ij}]_{n\times n}$, then H_e is a Helmertian matrix. It was introduced by F.R. Helmert in 1876 and hence the name. Since then, generalized forms of this type of matrix have found uses in statistics, in group theory and in the evaluation of the Jacobian of transformation from rectangular Cartesian to polar coordinates. The following are examples

$$H_e = \begin{bmatrix}\frac{1}{2} & \frac{1}{2} & \frac{1}{2} & \frac{1}{2}\\ \frac{1}{\sqrt{2}} & -\frac{1}{\sqrt{2}} & 0 & 0\\ \frac{1}{\sqrt{6}} & \frac{1}{\sqrt{6}} & -\frac{2}{\sqrt{6}} & 0\\ \frac{1}{\sqrt{12}} & \frac{1}{\sqrt{12}} & \frac{1}{\sqrt{12}} & -\frac{3}{\sqrt{12}}\end{bmatrix}, \quad H_e = \begin{bmatrix}1 & 0 & 0 & 0\\0 & 1 & 0 & 0\\0 & 0 & 1 & 0\\0 & 0 & 0 & 1\end{bmatrix}.$$

2×2 orthogonal matrices like

$$\begin{bmatrix}\cos\theta & \sin\theta\\ \sin\theta & -\cos\theta\end{bmatrix}6 \text{ and } \begin{bmatrix}\cos\theta & \sin\theta\\ -\sin\theta & \cos\theta\end{bmatrix}7$$

are standard Helmertians by *convention* and not by definition. This type of Helmertian is also called *standard Helmertian* to distinguish it from other variations discussed below. Note that $H_eH_e^T = I \neq H_e^TH_e$, i.e. standard Helmertians are orthogonal by rows.

3.16.2 Generalized Helmertian matrix : An n square orthogonal matrix is called a *generalized Helmertian* matrix if it can be transformed to a standard Helmertian by permutation of its rows and columns, by transposition and by change of signs of rows. Examples are permutation matrices, diagonal matrices with diagonal elements ± 1, transpose of standard Helmert matrices.

3.16.3 Helmertian in the strict sense : A Helmertian matrix is said to be *Helmertian in the strict sense* if it has the property

$$h_{1j} = +\sqrt{\omega_j}\,;\ \omega_j > 0;\ \sum_1^n \omega_j = 1\,.\ \text{Thus}$$

$$\begin{bmatrix} \frac{1}{\sqrt{3}} & \frac{1}{\sqrt{3}} & \frac{1}{\sqrt{3}} \\ \frac{1}{\sqrt{2}} & -\frac{1}{\sqrt{2}} & 0 \\ \frac{1}{\sqrt{6}} & \frac{1}{\sqrt{6}} & -\frac{2}{\sqrt{6}} \end{bmatrix}$$

is Helmertian in the strict sense. Because $\omega_j = \frac{1}{3} > 0,\ \sum_1^3 \omega_j = \frac{1}{3} + \frac{1}{3} + \frac{1}{3} = 1$ and $h_{1j} = +\sqrt{\omega_j}$. Similarly,

$$\begin{bmatrix} \frac{1}{\sqrt{5}} & \frac{1}{\sqrt{5}} & \frac{1}{\sqrt{5}} & \frac{1}{\sqrt{5}} & \frac{1}{\sqrt{5}} \\ \frac{1}{\sqrt{2}} & -\frac{1}{\sqrt{2}} & 0 & 0 & 0 \\ \frac{1}{\sqrt{6}} & \frac{1}{\sqrt{6}} & -\frac{1}{\sqrt{6}} & 0 & 0 \\ \frac{1}{\sqrt{12}} & \frac{1}{\sqrt{12}} & \frac{1}{\sqrt{12}} & -\frac{3}{\sqrt{12}} & 0 \\ \frac{1}{\sqrt{20}} & \frac{1}{\sqrt{20}} & \frac{1}{\sqrt{20}} & \frac{1}{\sqrt{20}} & -\frac{4}{\sqrt{20}} \end{bmatrix}$$

is a Helmertian in the strict sense. But the permutation matrices, unit matrices are not Helmertian in the strict sense. An alternative definition of this matrix will be found in chapter 8.

$H_e = D^{-1}H_{e0}$,

where $D^2 = \text{diag}\,[\,n.1,\ 1.2,\ 2.3,\ \cdots\ ,(n-1)n\,]$ and

$$H_{e0} = \begin{bmatrix} 1 & 1 & 1 & 1 & \cdots & 1 \\ 1 & -1 & 0 & 0 & \cdots & 0 \\ 1 & 1 & -2 & 0 & \cdots & 0 \\ 1 & 1 & 1 & -3 & \cdots & 0 \\ \vdots & \vdots & \vdots & \vdots & \vdots & \vdots \\ 1 & 1 & 1 & 1 & \cdots & -(n-1) \end{bmatrix}_{n\times n} .$$

3.16.4 Helmertian with weights : In 1942 J.O. Irwin modified the above definition of Helmertian. He introduced unequal weights ω_j. So, the above Helmertians become a special case of Irwin with equal weights. According to Irwin the defining conditions for elements h_{ij} are

$$h_{ij} = \sqrt{\omega_j}\,,\ h_{ij} = 0\ \text{ for } j > i > 1,$$

$$h_{ii} = -\left(\frac{W_{i-1}}{W_i}\right)^{\frac{1}{2}};\ h_{ij} = \left(\frac{\omega_i\omega_j}{W_iW_{i-1}}\right)^{\frac{1}{2}}\ \text{ for } i > j,\ i > 1$$

where $W_j = \omega_1 + \omega_2 + \cdots + \omega_j$, $W_n = 1$.

Let $\omega_1 = \frac{1}{2}, \omega_2 = \frac{1}{4},\ \omega_3 = \frac{1}{4}$ be the weights such that $W_3 = \omega_1 + \omega_2 + \omega_3 = 1$. Therefore,

$$h_{11} = \sqrt{\omega_1} = \frac{1}{\sqrt{2}},\ h_{12} = \sqrt{\omega_2} = \frac{1}{\sqrt{4}},\ h_{13} = \sqrt{\omega_3} = \frac{1}{\sqrt{4}},$$

$$h_{22} = -\sqrt{\frac{W_1}{W_2}} = -\sqrt{\frac{\omega_1}{\omega_1+\omega_2}} = -\sqrt{\frac{\frac{1}{2}}{\frac{1}{2}+\frac{1}{4}}} = -\frac{1}{\sqrt{3}}$$

$$h_{33} = -\sqrt{\frac{W_2}{W_3}} = -\sqrt{\frac{\omega_1+\omega_2}{\omega_1+\omega_2+\omega_3}} = -\sqrt{\frac{\frac{1}{2}+\frac{1}{4}}{1}} = -\sqrt{\frac{3}{4}}$$

$$h_{21} = \left(\frac{\omega_2\omega_1}{W_2W_1}\right)^{\frac{1}{2}} = \left(\frac{\omega_2\omega_1}{(\omega_1+\omega_2)\omega_1}\right)^{\frac{1}{2}} = \left[\frac{\frac{1}{4}\cdot\frac{1}{2}}{\left(\frac{1}{2}+\frac{1}{4}\right)\frac{1}{2}}\right]^{\frac{1}{2}} = \frac{1}{\sqrt{3}}$$

$$h_{31} = \left(\frac{\omega_3\omega_1}{W_3W_2}\right)^{\frac{1}{2}} = \left(\frac{\frac{1}{4}\cdot\frac{1}{2}}{1.\frac{3}{4}}\right)^{\frac{1}{2}} = \frac{1}{\sqrt{6}}$$

$$h_{32} = \left(\frac{\omega_3\omega_1}{W_3W_2}\right)^{\frac{1}{2}} = \left(\frac{\frac{1}{4}\cdot\frac{1}{4}}{1.\frac{3}{4}}\right)^{\frac{1}{2}} = \frac{1}{\sqrt{12}}\ .$$

$$\therefore \qquad \mathrm{H_e} = \begin{bmatrix} \frac{1}{\sqrt{2}} & \frac{1}{\sqrt{4}} & \frac{1}{\sqrt{4}} \\ \frac{1}{\sqrt{3}} & -\frac{1}{\sqrt{3}} & 0 \\ \frac{1}{\sqrt{6}} & \frac{1}{\sqrt{12}} & -\sqrt{\frac{3}{4}} \end{bmatrix}$$

which is a weighted Helmertian with weights $\frac{1}{2}, \frac{1}{4}, \frac{1}{4}$.

3.17 Vandermode matrix

The Vandermode matrix is a square matrix which arises as the coefficient matrix of a system of polynomial equations

$$\mathrm{P}(x_i) = y_i\ , \quad i = 1, 2, \cdots, (n-1)$$

obtained from the interpolating polynomial

$$\mathrm{P}(x) = a_0 + a_1x + a_2x^2 + \cdots + a_{n-1}x^{n-1}$$

in fitting a curve through n points $(x_0, y_0), (x_1, y_1), \cdots, (x_{n-1}, y_{n-1})$ for all distinct x_i.

An n square Vandermode matrix is

$$\mathrm{V} = \begin{bmatrix} 1 & x_0 & x_0^2 & \cdots & x_0^{n-1} \\ 1 & x_1 & x_1^2 & \cdots & x_1^{n-1} \\ 1 & x_2 & x_2^2 & \cdots & x_2^{n-1} \\ \vdots & \vdots & \vdots & \vdots & \vdots \\ 1 & x_n & x_n^2 & \cdots & x_n^{n-1} \end{bmatrix} = \left[v_{ij}\right]$$

where $v_{ij} = (x_i)^j$ for $i, j = 0, 1, 2, \cdots, (n-1)$.

This matrix has various applications. It is important in control theory and statistical and structure analysis. It is used in developing formulas in numerical analysis based on Lagrange interpolation.

Alternate forms : An alternate form of the Vandermode matrix is obtained by interchanging the first column with the nth one, the second one with the $(n-1)$th one, and so on. That is,

$$V = \begin{bmatrix} x_0^{n-1} & x_0^{n-2} & \cdots & x_0 & 1 \\ x_1^{n-1} & x_1^{n-2} & \cdots & x_1 & 1 \\ x_2^{n-1} & x_2^{n-2} & \cdots & x_2 & 1 \\ \vdots & \vdots & \vdots & \vdots & \vdots \\ x_{n-1}^{n-1} & x_{n-1}^{n-2} & \cdots & x_{n-1} & 1 \end{bmatrix}$$

Another alternate form is V^T i.e.

$$V = \begin{bmatrix} 1 & 1 & 1 & \cdots & 1 \\ x_0 & x_1 & x_2 & \cdots & x_n \\ x_0^2 & x_1^2 & x_2^2 & \cdots & x_n^2 \\ \vdots & \vdots & \vdots & \vdots & \vdots \\ x_0^{n-1} & x_1^{n-1} & x_2^{n-1} & \cdots & x_n^{n-1} \end{bmatrix}$$

In particular,

$$\begin{bmatrix} 1 & 1 & 1 \\ a & b & c \\ a^2 & b^2 & c^2 \end{bmatrix}$$

is a third order Vandermode matrix.

A Vandermode matrix is nonsingular if all x_i are distinct. So, the inverse (cf. chapter 8) exists for the matrix. Some methods for this inversion are given in chapter 8.

The $\det V = \prod_{i<j}(x_j - x_i)$, $i, j = 0, 1, \cdots, (n-1)$ (cf. chapter 6).

3.18 Circulant matrices with properties

These matrices have been studied since the time of Catalan (1846) and arise in the study of periodic or multiply symmetric dynamical systems, particularly in the theory of crystal structure.

3.18.1 Definition : A square matrix is called *circulant* or *circulant matrix* if each row after the first is obtained from its predecessor by a circular shift.

3.18.2 G-circulant matrix : A g-circulant matrix is an $n \times n$ square matrix of complex numbers, in which each row, except the first one, is obtained from the preceeding row by shifting the elements cyclically g columns to the right. Thus, the relationship of the elements a_{ij} of the i-th row and those of the preceding row is given by

$$a_{ij} = a_{i-1,j-g} \tag{3.8}$$

where indices are reduced to their least remainders modulo n.

If $g = 0$, the circulant matrix is the zero circulant matrix. In this matrix, all rows are same. The rank of the zero circulant is one (cf. chapter10).

If $g = 1$, i.e. the shift is one column to the right, then the circulant is *simple* or the matrix is 1-circulant. It is symmetrical about secondary diagonal i.e. persymmetric and consists of n distinct elements which constitute the first row. For example, the matrix

$$A = \begin{bmatrix} 2 & 5 & 8 & 9 & 6 \\ 6 & 2 & 5 & 8 & 9 \\ 9 & 6 & 2 & 5 & 8 \\ 8 & 9 & 6 & 2 & 5 \\ 5 & 8 & 9 & 6 & 2 \end{bmatrix} \tag{3.9}$$

is a simple circulant matrix.

Using (3.8) second row of the simple circulant matrix can be constructed as

$$a_{21} = a_{2-1,1-1} = a_{1,0} = a_{15} = 6 \qquad [\because \text{ modulo } 5 = 0, \text{ modulo } 4 = -1 \text{ etc.}]$$

$$a_{22} = a_{2-1,2-1} = a_{11} = 2$$

$$a_{23} = a_{2-1,3-1} = a_{12} = 5$$

$$\vdots \qquad \vdots \qquad \vdots \quad \vdots$$

$$a_{41} = a_{4-1,1-1} = a_{3,0} = a_{35} = 8 \text{ etc.}$$

If $g = 2$ i.e., shift to the right is by two columns then the second row can be constructed from the first row as follows using (3.8)

$$\left.\begin{aligned} a_{21} &= a_{2-1,1-2} = a_{1,-1} = a_{14} = 9 \\ a_{22} &= a_{2-1,2-2} = a_{10} = a_{15} = 6 \\ a_{23} &= a_{2-1,3-2} = a_{11} = 2 \\ a_{24} &= a_{2-1,4-1} = a_{12} = 5 \\ a_{25} &= a_{2-1,5-2} = a_{13} = 8 \end{aligned}\right\}.$$

In this way, 3rd row from 2nd, 4th row from 3rd and 5th row from 4th row may be constructed and the 2-*circulant* matrix is

$$\begin{bmatrix} 2 & 5 & 8 & 9 & 6 \\ 9 & 6 & 2 & 5 & 8 \\ 5 & 8 & 9 & 6 & 2 \\ 6 & 2 & 5 & 8 & 9 \\ 8 & 9 & 6 & 2 & 5 \end{bmatrix}.$$

Similarly, 3- *circulant* is

$$\begin{bmatrix} 2 & 5 & 8 & 9 & 6 \\ 8 & 9 & 6 & 2 & 5 \\ 6 & 2 & 5 & 8 & 9 \\ 5 & 8 & 9 & 6 & 2 \\ 9 & 6 & 2 & 5 & 8 \end{bmatrix} \text{ etc.}$$

3.18.3 Skew g-circulant matrix : As in skew symmetric matrix, the succeeding row of the skew g-circulant matrix from the preceding row, if the first row is given, is constructed by the equation

$$a_{ij} = -a_{i-1,\, j-g}$$

Other conditions remain as in the g-circulant. An example of the skew 1-circulant is

$$\begin{bmatrix} 2 & 5 & 8 & 9 & 6 \\ -6 & 2 & 5 & 8 & 9 \\ -9 & -6 & 2 & 5 & 8 \\ -8 & -9 & -6 & 2 & 5 \\ -5 & -8 & -9 & -6 & 2 \end{bmatrix}.$$

3.18.4 Properties of circulant matrix : Following are some properties of the circulant matrix .

(i) Circulant matrix is nonsingular (cf. chapter 6).

(ii) Conjugate transpose of the circulant matrix is circulant. In general, the transpose of the g-circulant is the $(g + 1)$-circulant for $g > 1$.

(iii) Circulant matrix is orthogonal i.e. $AA^* = A^*A$.

N.B. (iv) to (vi) (cf. chapter13)

(iv)If the first row of the circulant matrix X is $[x_0 \quad x_1 \quad \cdots \quad x_{n-1}]$, then eigenvalues become discrete Fourier expansion

$$\lambda_h = \sum_{j=0}^{n-1} \zeta^{h_j} x_j, \; 0 \le h \le n-1, \; \zeta = e^{2\pi j/n}.$$

(v) Diagonalized circulant matrix X with eigenvalues $\lambda_0, \; \lambda_1, \; \cdots, \; \lambda_{n-1}$ is $\mathrm{V\,diag}(\lambda_0 \quad \lambda_1 \quad \cdots \quad \lambda_{n-1})\mathrm{V}^{-1} = \mathrm{X}$, where V is the Vandermode matrix.

$$\mathrm{V} = \begin{bmatrix} 1 & 1 & 1 & \cdots & 1 \\ 1 & \zeta & \zeta^2 & \cdots & \zeta^{n-1} \\ \vdots & \vdots & \vdots & \vdots & \vdots \\ 1 & \zeta^{n-1} & \zeta^{2(n-1)} & \cdots & \zeta^{(n-1)^2} \end{bmatrix}.$$

(vi) $\mathrm{Det\, X} = \prod_{h=0}^{n-1}\left(\sum_{j=0}^{n-1} \zeta^{h_j} x_j\right)$.

3.19 Toeplitz matrix

A square matrix T is called a *Toeplitz* matrix if t_{ij}, the (i, j)th element of T is a function of $(i - j)$. Let $t_{ij} = b_{-(i-j)}$. Then the Toeplitz matrix is

$$\mathrm{T} = \begin{bmatrix} b_0 & b_1 & b_2 & \cdots & \cdots & b_{n-1} \\ b_{-1} & b_0 & b_1 & \cdots & \cdots & b_{n-2} \\ b_{-2} & b_{-1} & b_0 & \ddots & \cdots & b_{n-3} \\ \vdots & \vdots & \ddots & \ddots & \ddots & \vdots \\ b_{-n+2} & b_{-n+3} & b_{-n+4} & \ddots & & b_1 \\ b_{-n+1} & b_{-n+2} & b_{-n+3} & \cdots & b_{-1} & b_0 \end{bmatrix}_{n\times n}$$

For the sake of convenience, the single subscript notation is adopted for Toeplitz matrices. They are symmetric. If $b_i = b_{-i}$, each of the principal diagonal, super and subdiagonals consist of constant elements. In other words, the Toeplitz matrix is a square matrix with constant diagonals.

Toeplitz matrices mostly arise in time-series analysis of certain stochastic processes, discrete approximations of kernels $k(x,t)$ of integral equations when the kernels are functions of $(x - t)$. Examples are

$$\mathrm{T} = \begin{bmatrix} 1 & 2 & 3 & 4 & 5 \\ 6 & 1 & 2 & 3 & 4 \\ 7 & 6 & 1 & 2 & 3 \\ 8 & 7 & 6 & 1 & 2 \\ 9 & 8 & 7 & 6 & 1 \end{bmatrix},$$

1-circulant matrix, skew 1-circulant matrix, J-matrix, unit matrix, tridiagonal matrix, etc. A Toeplitz matrix is symmetric about the secondary diagonal, i.e. it is persymmetric. So, an $n \times n$ Toeplitz matrix consists of $(2n - 1)$ distinct elements. The above 5×5 Toeplitz matrix consists of $(2.5 - 1) = 9$ distinct elements 1 to 9.

The Toeplitz matrix differs from the circulant matrix in respect of

(i) number of elements — Toeplitz has $(2n-1)$ elements while circulant has n distinct elements.

(ii) each row begins with a distinct element not in the previous row as in the circulant.

3.20 Fröbenius (companion) matrix

A square matrix with nonzero elements in the first row and unit elements along the first subdiagonal and zero elsewhere is called a *Fröbenius* (*companion*) matrix. Thus

$$\mathrm{F} = \begin{bmatrix} a_1 & a_2 & a_3 & \cdots & a_{n-1} & a_n \\ 1 & 0 & 0 & \cdots & 0 & 0 \\ 0 & 1 & 0 & \cdots & 0 & 0 \\ \vdots & \vdots & \vdots & \vdots & \vdots & \vdots \\ 0 & 0 & 0 & \cdots & 1 & 0 \end{bmatrix}_{n \times n}, \quad \mathrm{F} = \begin{bmatrix} 6 & -11 & 6 \\ 1 & 0 & 0 \\ 0 & 1 & 0 \end{bmatrix}$$

are examples of Fröbenius matrices. If the nonzero elements of this matrice are positive, then the matrix is called the *nonnegative Fröbenius* matrix.

3.21 Leslie matrix

A square matrix with nonzero positive elements in the first row and positive real numbers along the first subdiagonal is called a *Leslie* matrix. This is frequently used in animal population ecology and human demography. A suitable example is

$$\begin{bmatrix} 0.427 & 0.849 & 0.123 \\ 0.992 & 0.0 & 0.0 \\ 0.0 & 0.982 & 0.0 \end{bmatrix}$$

3.22 Jacobi matrix

A tridiagonal square matrix with nonconstant nonzero elements along diagonal and first sub and super diagonals and the product of corresponding elements of sub & super diagonals is positive. That is, a matrix of the form

$$\begin{bmatrix} b_1 & c_1 & 0 & \cdots & 0 \\ a_1 & b_2 & c_2 & \cdots & 0 \\ 0 & a_2 & b_3 & \cdots & 0 \\ \vdots & \vdots & \vdots & \vdots & c_{n-1} \\ \cdots & \cdots & \cdots & a_{n-1} & b_n \end{bmatrix}$$

with $a_i\, c_i > 0$, $i = 1, 2, \cdots, (n-1)$.

3.23 Hessenberg matrix

A square matrix $A = \left[a_{ij}\right]_{n \times n}$ is an *upper Hessenberg* matrix if $a_{ij} = 0$ for $i > j - 1$.

Thus

$$A = \begin{bmatrix} x & x & x & x \\ x & x & x & x \\ 0 & x & x & x \\ 0 & 0 & x & x \end{bmatrix}$$

is a 4×4 upper Hessenberg matrix. It has zeros below the first subdiagonal. Similarly, a *lower Hessenberg* matrix has zeros above its first superdiagonal. A tridiagonal matrix is both the upper and lower Hesenberg matrix.

3.24 Square root of matrices

There is no general procedure for obtaining the square root of a square matrix A as in complex numbers except for the semipositive definite symmetric matrices which are analogous to positive real numbers and the method of constituent idempotents (cf. sec.17.2.17B), a more general procedure.

In complex numbers, $z^{\frac{1}{n}}$ has n distinct roots. But no analogous property exists for the matrix in general. The nth root of a square matrix A expresses the existence of a square matrix B such that $A = B^n$. But this matrix B is not always unique. When $n = 2$, there arises three cases:

(i) infinite numbers of square roots, (ii) no square roots

and (iii) finite numbers of square roots.

Calculations of square roots of matrices of higher order than second are unmanageable. So, the present discussion is limited to 2×2 matrices only for understanding the procedures.

Definition : A square matrix B is called a *square root* of a matrix A if $A = B^2$.

This definition calls for the following three facts:

I. Some square matrices possess the *infinite* number of square roots.

Let the square roots of

$$I_2 = \begin{bmatrix} 1 & 0 \\ 0 & 1 \end{bmatrix} \text{ be } B = \begin{bmatrix} a & b \\ c & d \end{bmatrix}.$$

Therefore, $\begin{bmatrix} a & b \\ c & d \end{bmatrix}^2 = \begin{bmatrix} 1 & 0 \\ 0 & 1 \end{bmatrix}.$

i.e.

$$a^2 + bc = 1 \quad \text{(i)}$$
$$b(a + d) = 0 \quad \text{(ii)}$$
$$c(a + d) = 0 \quad \text{(iii)}$$
$$bc + d^2 = 1 \quad \text{(iv)}$$

From (ii) and (iii) $b = 0$, $c = 0$ if $a + d \neq 0$. Hence from (i) and (iv) $a = \pm 1$, $d = \pm 1$. Therefore, the square roots of I_2 are

$$\begin{bmatrix} 1 & 0 \\ 0 & 1 \end{bmatrix} \text{ and } \begin{bmatrix} -1 & 0 \\ 0 & -1 \end{bmatrix}, \quad \text{i.e. } I_2 \text{ and } (-I_2).$$

If $a + d = 0$ and $b \neq 0$, $c \neq 0$, then system becomes either $bc + d^2 = 1$ or $a^2 + bc = 1$ which possesses the infinite number of solutions as there are three unknowns and one equation. Hence the square root of I_2 is infinite in number and their *general* form is

$$\begin{bmatrix} a & b \\ c & -a \end{bmatrix},$$

where a, b, c are any three numbers satisfying $a^2 + bc = 1$.

II. Some square matrices have *no* square roots:

Let the square roots of

$$\begin{bmatrix} 0 & 0 \\ 1 & 0 \end{bmatrix} \text{ be } \begin{bmatrix} a & b \\ c & d \end{bmatrix}.$$

Therefore, $\begin{bmatrix} a & b \\ c & d \end{bmatrix}^2 = \begin{bmatrix} 0 & 0 \\ 1 & 0 \end{bmatrix}.$

Hence

$$a^2 + bc = 0 \quad \text{(i)}$$
$$b(a + d) = 0 \quad \text{(ii)}$$
$$c(a + d) = 1 \quad \text{(iii)}$$
$$bc + d^2 = 0 \quad \text{(iv)}$$

From (ii) either $b = 0$ or $a + d = 0$. If $b = 0$, then from (i) and (iv) $a = 0$ and $d = 0$, which makes $a + d = 0$. If $a + d = 0$, then from (iii) $0 = 1$, an absurdity. Therefore, the system (i) through (iv) has no solution and hence the matrix

$$\begin{bmatrix} 0 & 0 \\ 1 & 0 \end{bmatrix}$$

has no square roots.

III. It can be shown by the following straight forward algebra that the square roots of the matrix

$$A = \begin{bmatrix} 3 & 1 \\ -4 & -1 \end{bmatrix}$$

are finite in number and they are

$$B = \pm \begin{bmatrix} 2 & \frac{1}{2} \\ -2 & 0 \end{bmatrix}.$$

Let the square roots of

$$A=\begin{bmatrix}3 & 1\\ -4 & -1\end{bmatrix} \text{ be } B=\begin{bmatrix}a & b\\ c & d\end{bmatrix}.$$

Then $B^2 = A.$

i.e.
$$\begin{bmatrix}a^2+bc & b(a+d)\\ c(a+d) & bc+d^2\end{bmatrix}=\begin{bmatrix}3 & 1\\ -4 & -1\end{bmatrix}$$

or
$$a^2+bc = 3 \qquad (i)$$
$$b(a+d) = 1 \qquad (ii)$$
$$c(a+d) = -4 \qquad (iii)$$
$$bc+d^2 = -1 \qquad (iv)$$

Dividing (ii) by (iii), $\dfrac{b}{c}=-\dfrac{1}{4}$ or $b=-\dfrac{c}{4}$ (v)

Using this value of b in (ii), $a+d=-\dfrac{4}{c}$ (vi)

(i) – (iv) gives $a^2-d^2=4$ or $(a+d)(a-d)=4$

or $(a-d)=-c$ by (vi) (vii)

From (vi) and (vii) $a=\frac{1}{2}\left(-\frac{4}{c}-c\right)$, $d=\frac{1}{2}\left(-\frac{4}{c}+c\right)$.

Using a, d, b, in (iv) $c=\pm 2$.

$\therefore$ $a=\mp 2$, $d=0$, 0 and $b=\mp\frac{1}{2}$.

$\therefore$
$$B=\pm\begin{bmatrix}2 & \frac{1}{2}\\ -2 & 0\end{bmatrix}.$$

3.25 Applications

Out of many applications of square matrices, some of them are described below:

3.25.1 Matrices to represent complex numbers : A complex number $a+ib$, $i=\sqrt{-1}$ can be represented by 2×2 matrices of the form

$$\begin{bmatrix}a & b\\ -b & a\end{bmatrix} \qquad (3.10)$$

where diagonal elements are same and those on the secondary diagonal are same but of opposite signs. Thus matrices like

$$\begin{bmatrix}4 & 2\\ -2 & 4\end{bmatrix}, \begin{bmatrix}-4 & -2\\ 2 & -4\end{bmatrix}, \begin{bmatrix}\frac{1}{2} & 0\\ 0 & \frac{1}{2}\end{bmatrix}, \begin{bmatrix}0 & 1\\ -1 & 0\end{bmatrix}$$

are of the form (3.10). The matrix (3.10) can be written as

$$\begin{bmatrix}a & b\\ -b & a\end{bmatrix}=\begin{bmatrix}a & 0\\ 0 & a\end{bmatrix}+\begin{bmatrix}0 & b\\ -b & 0\end{bmatrix}$$

$$= a\begin{bmatrix} 1 & 0 \\ 0 & 1 \end{bmatrix} + b\begin{bmatrix} 0 & 1 \\ -1 & 0 \end{bmatrix} = a\mathrm{I}_2 + bi \tag{3.11}$$

where $$i^2 = \begin{bmatrix} 0 & 1 \\ -1 & 0 \end{bmatrix}\begin{bmatrix} 0 & 1 \\ -1 & 0 \end{bmatrix} = \begin{bmatrix} -1 & 0 \\ 0 & -1 \end{bmatrix} = (-1)\mathrm{I}_2 = -\mathrm{I}_2$$

Therefore, (3.9) is of the form $a + ib,\ i^2 = -1$ with the matrix complex quantity

$$i = \begin{bmatrix} 0 & 1 \\ -1 & 0 \end{bmatrix}.$$

The real and imaginary parts of the complex number represented by (3.10) are

real part = same element on the diagonal

and *imaginary part = magnitude of the one in secondary diagonal*.

It can be shown that the matrix (3.10) representing complex number $a + ib$, obeys algebraic laws as the complex number, $a + ib$.

Ex. Find $(4 + 3i)(1 + 2i)$ using matrices.

$$(4 + 3i)(1 + 2i) = \begin{bmatrix} 4 & 3 \\ -3 & 4 \end{bmatrix}\begin{bmatrix} 1 & 2 \\ -2 & 1 \end{bmatrix} = \begin{bmatrix} -2 & 11 \\ -11 & -2 \end{bmatrix} = -2 + 11i.$$

3.25.2 Matrices to represent quaternions : Sir William R. Hamilton invented the quaternion in 1835. It is defined as a hypercomplex number, $Q = a + bi + cj + dk$, where a, b, c, d are real numbers and i, j, k are complex quantities defined by the equations

$$i^2 = j^2 = k^2 = -1$$

$$ij = k = -ji,\quad jk = i = -kj,\quad ki = j = -ik.$$

The conjugate of Q is $\overline{Q} = a - bi - cj - dk$. It may also be thought of as a 4-dimensional vectors with unit vectors $1,\ i,\ j, k$ with the following multiplication table along with *figure* 3.1

×	1	i	j	k
1	1	i	j	k
i	i	-1	k	$-j$
j	j	$-k$	-1	i
k	k	j	$-i$	-1

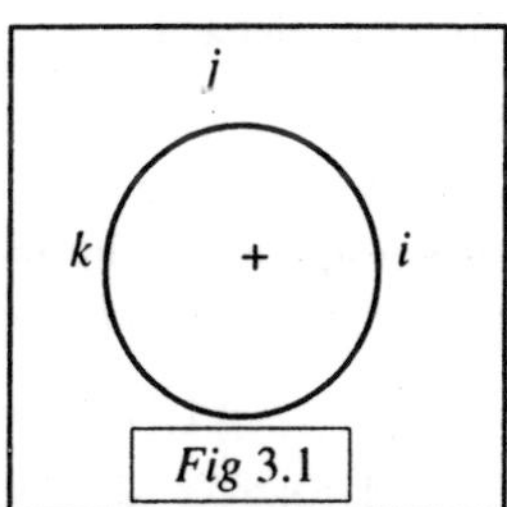

Fig 3.1

As the quaternion is a linear combination of *four* quantities $1,\ i\ , j\ , k$, hence the name quaternion is derived from Latin word *quarter* means four. Quaternions obey all the algebraic laws but they are non-commutative. A quaternion can be represented by a 4×4 matrix

$$\mathrm{Q} = \begin{bmatrix} a & b & c & d \\ -b & a & -d & c \\ -c & d & a & -b \\ -d & -c & b & a \end{bmatrix}$$

where the diagonal elements are same, the secondary diagonal elements are of the same magnitude with the alternate change of signs. Note that

$$Q = a\begin{bmatrix} 1 & 0 & 0 & 0 \\ 0 & 1 & 0 & 0 \\ 0 & 0 & 1 & 0 \\ 0 & 0 & 0 & 1 \end{bmatrix} + b\begin{bmatrix} 0 & 1 & 0 & 0 \\ -1 & 0 & 0 & 0 \\ 0 & 0 & 0 & -1 \\ 0 & 0 & 1 & 0 \end{bmatrix} + c\begin{bmatrix} 0 & 0 & 1 & 0 \\ 0 & 0 & 0 & 1 \\ -1 & 0 & 0 & 0 \\ 0 & -1 & 0 & 0 \end{bmatrix} + d\begin{bmatrix} 0 & 0 & 0 & 1 \\ 0 & 0 & -1 & 0 \\ 0 & 1 & 0 & 0 \\ -1 & 0 & 0 & 0 \end{bmatrix}$$

$$= aI_4 + b\mathtt{I} + c\mathtt{J} + d\mathtt{K}$$

Verify that I_4, $\mathtt{I}$, $\mathtt{J}$, $\mathtt{K}$ all satisfy the above multiplication table.

3.25.3 Matrices of magic squares : Formation of magic squares is an ancient mathematical recreation. The problem is "Fill up n^2 places of an $n\times n$ matrix by n^2 positive consecutive integers $1, 2, \cdots, n$ such that row, column and diagonal sums are equal and each sum is equal to $S = \frac{1}{2}n(n^2+1)$." Thus if $n = 4$ then $S = 34$ and positive (+) integers are 1, 2, $\cdots$,16. For an example :

M(4) =

15	10	3	6
4	5	16	9
14	11	2	7
1	8	13	12

Other examples with noninteger and nonconsecutive numbers are

M(3) =

$2+i$	$-1+i$	$2+i$
$1+i$	$1+i$	$1+i$
i	$3+i$	i

$\begin{cases} S = 3+3i \\ \text{complex magic square,} \\ \text{repetition of elements.} \end{cases}$

M(3) =

4	−1	0
−3	1	5
2	3	−2

$\begin{cases} S = 3 \\ \text{integers are not} \\ \text{consecutive.} \end{cases}$

The construction method was just a trial and error method. On that basis, some rules were created. A general procedure is developed in chapter 7.

Semimagic (Pseudomagic) square : In this magic square only the row and column sums will be equal. No consideration of diagonal sums is made. Examples are

$$A = M_s(3) = \begin{bmatrix} 1 & 8 & 6 \\ 5 & 3 & 7 \\ 9 & 4 & 2 \end{bmatrix}, \quad B = M_s(3) = \begin{bmatrix} 1 & 6 & 8 \\ 5 & 7 & 3 \\ 9 & 2 & 2 \end{bmatrix}$$

with *row sum* = *column sum* = 15 in both cases.

Algebra of magic squares : Addition of two semimagic squares is a semimagic square. Thus, addition of A and B above is

$$A + B = \begin{bmatrix} 2 & 14 & 14 \\ 10 & 10 & 10 \\ 18 & 6 & 4 \end{bmatrix}, \text{ a semimagic square.}$$

Scalar multiplication of a semimagic is again a semimagic. Thus

$$2A = 2\begin{bmatrix} 1 & 8 & 6 \\ 5 & 3 & 7 \\ 9 & 4 & 2 \end{bmatrix} = \begin{bmatrix} 2 & 16 & 12 \\ 10 & 6 & 14 \\ 18 & 8 & 4 \end{bmatrix} \text{ is a semimagic.}$$

Let A and B be two pseudomagic squares with common sums d_1 and d_2. Therefore,

$$(AB)J = A(BJ) = A(d_2J) = d_2(AJ) = d_2(d_1J) = d_1d_2J = d_1d_2J$$

and $\quad J(AB) = (JA)B = (d_1J)B = d_1(JB) = d_1d_2J$

$\therefore \quad (AB)J = J(AB) = d_1d_2J$

Hence the Cayley product AB of pseudomagic squares A and B is a pseudomagic square. Also

$$AB = \begin{bmatrix} 1 & 8 & 6 \\ 5 & 3 & 7 \\ 9 & 4 & 2 \end{bmatrix}\begin{bmatrix} 1 & 6 & 8 \\ 5 & 7 & 3 \\ 9 & 2 & 2 \end{bmatrix} = \begin{bmatrix} 89 & 74 & 56 \\ 83 & 65 & 77 \\ 47 & 86 & 92 \end{bmatrix} \text{ is a semimagic.}$$

Thus in general sum, scalar multiplication and matrix multiplication of two semimagic squares is again a semimagic square.

For magic squares, sum and scalar multiplication preserve the magic squares but matrix multiplication does not preserve the magic square. For example, let

$$A = \begin{bmatrix} 8 & 1 & 6 \\ 3 & 5 & 7 \\ 4 & 9 & 2 \end{bmatrix} \text{ and } B = \begin{bmatrix} 6 & 1 & 8 \\ 7 & 5 & 3 \\ 1 & 9 & 4 \end{bmatrix}.$$

Then $$AB = \begin{bmatrix} 67 & 67 & 91 \\ 67 & 91 & 67 \\ 91 & 67 & 67 \end{bmatrix} \text{ is not a magic square.}$$

Construction of magic squares : Magic squares have been constructed since the antiquity by the trial and error method and thus construction remained as a simple mathematical recreation. But Hazra developed a general method of construction (cf. chapter 7). The semimagic square can be constructed from the magic square by interchange of at least two columns (or rows).

3.25.4 Matrices of polynomial, ordinary differential equation and difference equations : The coefficients of monic polynomials (leading coefficient is unity in a polynomial), Ordinary differential equations and difference equations can be arranged in matrix form and this matrix is known as the *companion matrix*. It is a square matrix. All its elements are zero except for elements 1 along the first sub (supper) diagonal and negative of the coefficients of monic polynomial (ordinary differential or difference equation) in question along the last column (row). It appears from the definition that the transpose of the companion matrix corresponding to a polynomial or equation is the companion matrix of the same polynomial or equation. Let

$$f(x) = x^n + a_1x^{n-1} + a_2x^{n-2} + \cdots + a_n, a_0 = 1 \tag{3.12}$$

be the monic polynomial. Let $f(x) = 0$ and

$$\begin{cases} y_1 = x \\ y_2 = x^2 \\ \vdots \\ y_n = x^n = -a_n - a_{n-1}x - \cdots - a_2 x^{n-2} - a_1 x^{n-1} \end{cases}$$

be a system of equations. Therefore, introducing matrix notation, we have

$$\begin{bmatrix} y_1 \\ y_2 \\ y_3 \\ \vdots \\ y_n \end{bmatrix} = \begin{bmatrix} 0 & 1 & 0 & \cdots & 0 \\ 0 & 0 & 1 & \cdots & 0 \\ \vdots & \vdots & \vdots & \vdots & \vdots \\ 0 & 0 & 0 & \cdots & 1 \\ -a_n & -a_{n-1} & -a_{n-2} & \cdots & -a_1 \end{bmatrix} \begin{bmatrix} 1 \\ x \\ x^2 \\ \vdots \\ x^{n-1} \end{bmatrix}$$

$$\therefore \qquad \mathbf{y} = \mathbf{Ax}$$

where A is the companion matrix associated with the monic polynomial (3.12). The companion matrix of linear expression $f(x) = (x + a_1)$ is simply the scalar $[-a_1]$. This matrix is also evident from the system of equations corresponding to:

(i) the linear differential equation $f(D)x = 0$, $\left(D \equiv \dfrac{d}{dx}\right)$, and

(ii) the linear difference equation $f(E)y_n = 0$, $(E = 1 + \Delta)$.

Thus the differential equation $x''' + a_1 x'' + a_2 x' + a_3 x = 0$ can be replaced by the system

$$\begin{cases} x' = y \\ y' = z \text{ i.e. } x'' = z \text{ and } x''' = z' \\ z' = -a_3 x - a_2 y - a_1 z \end{cases}$$

Therefore, the matrix equation corresponding to this system is

$$\begin{bmatrix} x' \\ y' \\ z' \end{bmatrix} = \begin{bmatrix} 0 & 1 & 0 \\ 0 & 0 & 1 \\ -a_3 & -a_2 & -a_1 \end{bmatrix} \begin{bmatrix} x \\ y \\ z \end{bmatrix}$$

in which the coefficent matrix

$$\begin{bmatrix} 0 & 1 & 0 \\ 0 & 0 & 1 \\ -a_3 & -a_2 & -a_1 \end{bmatrix}$$

is a companion matrix with a cubic expression $\quad f(x) = x^3 + a_1 x^2 + a_2 x + a_1$.

Similarly, the difference equation

$$x_{n+3} + a_1 x_{n+2} + a_2 x_{n+1} + a_3 x_n = 0$$

can be replaced by the system

$$\begin{cases} x_{n+1} = y_n \\ y_{n+1} = z_n \quad \text{i.e. } x_{n+2} = y_{n+1} = z_n \\ z_{n+1} = -a_3 x_n - a_2 y_n - a_1 z_n \end{cases}$$

The matrix equation is

$$\begin{bmatrix} x_{n+1} \\ y_{n+1} \\ z_{n+1} \end{bmatrix} = \begin{bmatrix} 0 & 1 & 0 \\ 0 & 0 & 1 \\ -a_3 & -a_2 & -a_1 \end{bmatrix} \begin{bmatrix} x_n \\ y_n \\ z_n \end{bmatrix}.$$

The coefficient matrix is a companion matrix as before. The properties of the companion matrix concerning inverse, eigenvalue problems and rank are very easy and interesting (cf. chapter 8, 13 and 10). Polynomial equations can be solved using this matrix.

3.25.5 Matrices in probability and statistics : the matrix arises in probability theory and statistics. Some examples are cited below:

(i) Game theory : Many situations in business, economics, politics and the sciences call for decisions in a competitive basis among two or more people. These situations can be represented as *games*. These games are called *matrix games* and the matrix representing the games is called the *pay off* matrix. If competition takes place between two persons with constant total pay offs then the games are called a *two person constant sum* games, and if the constant is zero then it is called *two person zero sum* games. These games may be pure strategic and mixed strategic. Matrices and probability theory can be used in making decision in such situations.

Conventions of signs of elements of these matrices :

(a) positive element represents amounts the column player II pays to row player I and

(b) negative element represents amounts the row player I pays to column player II.

Note the following pay off matrix formed for the children's game "stone, scissors, paper". The game is: Each of two players writes down one of the words stone, scissors or paper. Stone beats scissors, scissors beat paper and paper beats stone. The winner receives Tk 1.00 from the loser and the pay off is Tk 0.00 in case of a tie. The pay off matrix is

		Player II		
		stone	scissors	paper
Player I	stone	0	1	−1
	scissors	−1	0	1
	paper	1	−1	0

This is a pure strategic game because player I does not know prehand what player II will write and a zero sum game because sum of pay offs by player I and II is zero. So, the game is two person zero sum pure strategic game.

To explain about the elements note that the diagonal elements are zero as both players write the same thing and tie condition (given) is satisfied. The element $a_{12} = 1$ because player I writes stone while play II writes scissors & by "stone beats scissors" condition player I wins and so player II pays Tk 1.00 to player I. By convention of signs it is positive. Similarly, others are formed.

(ii) Stochastic matrix : An n square matrix A with real elements is called a *stochastic matrix* if $a_{ij} \geq 0$ for all i, j and either $\sum_{i=1}^{n} a_{ij} = 1$ for each j (i.e. each column sum is unity) or $\sum_{j=1}^{n} a_{ij} = 1$ for each i (i.e. each row sum is unity). If both row and column sums are unity, the *stochastic matrix* A is called *doubly stochastic.* For example, let the probability of people staying in city be 0.95 and that moving from city to suburbs be 0.05. Again consider the reverse case, the probability of people staying in suburbs is 0.97 while those towards city from suburbs is 0.03. These informational data may be represented as

$$\mathrm{P} = \left[p_{ij}\right] = \begin{matrix} & \text{city} & \text{suburb} \\ \text{city} & 0.95 & 0.05 \\ \text{suburb} & 0.03 & 0.97 \end{matrix}$$

which is called a *row stochastic* matrix. Because each of row sum of P is unity. Note that P^{T} is a *column stochastic* matrix.

Markov chain and transition matrix : Let $\mathbf{x} = \left[x_1 \quad x_2 \quad \cdots \quad x_n\right]^{\mathrm{T}}$ be a column vector such that $\sum_{i=1}^{n} x_i = 1$ and $x_i \geq 0$ for each i and let A be a stochastic matrix. The sequence of vectors $\left\{\mathrm{A}^k \mathbf{x}\right\}$ is called a *finite Markov process* or *Markov chain.*

The matrix A is called the *transition matrix* of the process. The vector **x** is called the *initial state vector* and the vector $\mathrm{A}^k \mathbf{x}$ is called the state *vector after k steps* of the process.

Transition matrix : A nonnegative square matrix is called a *transition matrix* if each of the column sums to unity, 1. The condition implies that $0 \leq a_{ij} \leq 1$ for each i and j and

$$\sum_{i=1}^{n} a_{ij} = 1, \quad \text{for each } j = 1, 2, \cdots, n.$$

Examples are

$$\tfrac{1}{n}\mathrm{J}_n, \quad \frac{(\mathrm{J}_n - \mathrm{I}_n)}{(n-1)}, \quad \mathrm{I}, \mathrm{A} = \begin{bmatrix} 0 & \frac{2}{3} & 0 & 0 \\ \frac{1}{3} & 0 & \frac{2}{3} & 0 \\ 0 & \frac{1}{3} & 0 & 1 \\ \frac{2}{3} & 0 & \frac{1}{3} & 0 \end{bmatrix}, \quad \mathrm{A} = \begin{bmatrix} \frac{1}{2} & \frac{1}{4} & \frac{1}{3} \\ \frac{1}{2} & \frac{1}{4} & \frac{1}{3} \\ 0 & \frac{1}{2} & \frac{1}{3} \end{bmatrix}.$$

Note that $\mathbf{1}^{\mathrm{T}}\mathrm{A} = \mathbf{1}^{\mathrm{T}}$. In other words, the matrix A is a transition matrix if $\mathbf{1}^{\mathrm{T}}\mathrm{A} = \mathbf{1}^{\mathrm{T}}$.

Probability vector : P is a nonnegative column vector such that sum of components equal to 1.

Note that $\mathbf{1}^{\mathrm{T}}\mathrm{P} = 1$ is the criteria for P to be the probability vector.

Theorem . If A is a transition matrix and P is a probability vector, then AP is also a probability vector.

Proof . Clearly AP is a nonnegative vector.

Now $\qquad \mathbf{1}^{\mathrm{T}}(\mathrm{AP}) = (\mathbf{1}^{\mathrm{T}}\mathrm{A})\mathrm{P} = \mathbf{1}^{\mathrm{T}}\mathrm{P} = 1.$

$\therefore$ AP is a probability vector.

Regular stochastic matrix : The stochastic matrix A is *regular* if for some natural number k, $\mathrm{A}^k = \left[a_{ij}\right]_{n \times n}$ is positive i.e. all elements are greater than zero i.e. $t_{ij}^{(k)}$ for all i, j. In other words A^k has no zero element. Otherwise it is *irregular.* Thus, the stochastic matrix

$$A = \begin{bmatrix} \frac{1}{3} & \frac{2}{3} & 0 \\ \frac{1}{2} & 0 & \frac{1}{2} \\ 0 & \frac{1}{3} & \frac{2}{3} \end{bmatrix} \text{ is regular, because } A^2 = \begin{bmatrix} \frac{8}{18} & \frac{4}{16} & \frac{6}{18} \\ \frac{3}{18} & \frac{9}{18} & \frac{6}{18} \\ \frac{3}{18} & \frac{4}{18} & \frac{11}{18} \end{bmatrix} \text{ has no zero element. Here } k = 2.$$

Note that given A is a row stochastic matrix. Subsequent each A^k will remain row stochastic. This is a check in raising power.

3.25.6 Matrices in classical and quantum mechanics : Matrices arise in physics in various occasions. It is not possible to enumerate all such matrices. Some matrices those are encountered in classical and quantum mechanics, are mentioned below only to give a superficial idea of the uses of matrices in physics.

(i) Spin matrices : The following four matrices

$$I = \begin{bmatrix} 1 & 0 \\ 0 & 1 \end{bmatrix},\ \sigma_x = \begin{bmatrix} 0 & 1 \\ 1 & 0 \end{bmatrix},\ \sigma_y = \begin{bmatrix} 0 & -i \\ i & 0 \end{bmatrix},\ i = \sqrt{-1} \text{ and } \sigma_z = \begin{bmatrix} 1 & 0 \\ 0 & -1 \end{bmatrix}$$

are known as *spin* or *Pauli spin* matrices. They are used in Wolfgang Pauli's theory of electron spin in quantum mechanics to determine spin angular momentum operator **s** with components $s_x = \frac{1}{2}\hbar\sigma_x, s_y = \frac{1}{2}\hbar\sigma_y, s_z = \frac{1}{2}\hbar\sigma_z$, where $\hbar = \frac{1}{2\pi}h$ and h is Planck's constant.

(ii) Dirac matrices : The matrices

$$\beta = \begin{bmatrix} I_2 & 0 \\ 0 & -I_2 \end{bmatrix} = \left[\begin{array}{cc|cc} 1 & 0 & 0 & 0 \\ 0 & 1 & 0 & 0 \\ \hline 0 & 0 & -1 & 0 \\ 0 & 0 & 0 & -1 \end{array}\right]$$

$$\alpha_x = \begin{bmatrix} 0 & \sigma_x \\ \sigma_x & 0 \end{bmatrix},\ \alpha_y = \begin{bmatrix} 0 & \sigma_y \\ \sigma_y & 0 \end{bmatrix},\ \alpha_z = \begin{bmatrix} 0 & \sigma_z \\ \sigma_z & 0 \end{bmatrix}$$

are known as *Dirac matrices* in physics where σ_x , σ_y , σ_z are Pauli spin matrices.

(iii) JI-matrix : In physics it means an $2n\times2n$ square matrix

$$JI = \begin{bmatrix} 0 & I \\ -I & 0 \end{bmatrix}$$

where 0 is an $n\times n$ null matrix and I is an $n\times n$ unit matrix. Physicists use notation J for JI. the notation JI is introduced to avoid collision with J defined in ch.1. This is used for symmetric notation for Hamilton's equation. Here $JI^2 = -I$. It is orthogonal i.e. $(JI)^T(JI) = 1$ or $(JI)^T JI = 1$.

(vi) Symplectic matrix : If $M^TJM = J$ or $MJM^T = J$ then M is called the *symplectic matrix*. This J is the JI-matrix mentioned above.

Exercises three

1. A Markov process has the following transition matrix

$$A = \begin{bmatrix} 0 & .2 & 0 & 0 & 0 & 0 \\ 1 & 0 & .4 & 0 & 0 & 0 \\ 0 & .8 & 0 & .6 & 0 & 0 \\ 0 & 0 & .6 & 0 & .8 & 0 \\ 0 & 0 & 0 & .4 & 0 & 1 \\ 0 & 0 & 0 & 0 & .2 & 0 \end{bmatrix}$$

Show that A is not regular.

[**Hints** : Find a natural number k such that A^k will not be free of zero elements

$$A^2 = \begin{bmatrix} .2 & 0 & .08 & 0 & 0 & 0 \\ 0 & .2 & 0 & .24 & 0 & 0 \\ .8 & 0 & .68 & 0 & .48 & 0 \\ 0 & .48 & .36 & .68 & .32 & .8 \\ 0 & 0 & .24 & 0 & 52 & 0 \\ 0 & 0 & 0 & .08 & 0 & .2 \end{bmatrix}]$$

2. Show that following transition matrices in Markov process are regular:

$$\begin{bmatrix} 0 & 1 \\ \frac{1}{2} & \frac{1}{2} \end{bmatrix}, \begin{bmatrix} 0 & 1 & 0 \\ 0 & 0 & 1 \\ \frac{2}{5} & \frac{3}{5} & 0 \end{bmatrix}, \begin{bmatrix} 0 & \frac{2}{5} & 0 & \frac{3}{5} \\ 0 & 0 & 0 & 1 \\ 1 & 0 & 0 & 0 \\ 0 & 0 & 1 & 0 \end{bmatrix}.$$

3. Prove that

$$A^n = \begin{bmatrix} \cos n\theta & \sin n\theta \\ -\sin n\theta & \cos n\theta \end{bmatrix} \text{ if } A = \begin{bmatrix} \cos\theta & \sin\theta \\ -\sin\theta & \cos\theta \end{bmatrix}.$$

4. If

$$A = \begin{bmatrix} 1 & 1 \\ 0 & 1 \end{bmatrix}, \text{ find } A^n. \qquad \left[\textbf{Ans}. \begin{bmatrix} 1 & n \\ 0 & 1 \end{bmatrix}\right]$$

5. If

$$H = \begin{bmatrix} 1 & 0 & 0 & \cdots & 0 \\ 0 & 0 & 0 & \cdots & 0 \\ \vdots & \vdots & \vdots & & \vdots \\ 0 & 0 & 0 & \cdots & 0 \end{bmatrix}_{n\times n}, \text{ prove that } H^2 = H. \text{ Compute } H^{10,000}.$$

[**Hints:** $H^4 = H^2 = H$, $H^8 = H$, $H^{10} = H$, $H^{100} = (H^{10})^{10} = H^{10} = H$, $H^{1000} = (H^{100})^{10} = H^{10} = H$, $H^{10,000} = (H^{1000})^{10} = H^{10} = H$]

6. Find $J^m_{n\times n}$. [**Ans** . $n^{m-1}J$]

7. If

$$M = \begin{bmatrix} 1 & a & b & c \\ a^{-1} & 1 & a^{-1}b & a^{-1}c \\ b^{-1} & ab^{-1} & 1 & b^{-1}c \\ c^{-1} & ac^{-1} & bc^{-1} & 1 \end{bmatrix},$$

prove that $M^2 = 4M$. Hence prove that $M^n = 4^{n-1}M$.

8. If the numbers p, q, r satisfy $p^2 + q^2 + r^2 = 1$, show that the matrix

$$A = \begin{bmatrix} 0 & r & q \\ -r & 0 & p \\ -q & -p & 0 \end{bmatrix}$$

satisfy the matrix equation $A^3 + A = 0$.

9. Let a matrix A satisfy the matrix equation $A^2 - 2A + I = 0$. Show that $A^3 = 3A - 2I$ and $A^4 = 4A - 3I$. Obtain the expression for A^n.

[**Ans**. $nA - (n-1)I$]

10. Prove that a matrix A whose elements are given by the relations

$$a_{ij} = \begin{cases} (-1)^{j-1}\binom{j-1}{i-1} & i < j \\ (-1)^{i-1} & i = j \\ 0 & i > j \end{cases}$$

satisfies the relation $A^2 = I$. Here $\binom{n}{k}$ is the binomial coefficient $\dfrac{n!}{k!(n-k)!}$.

11. If $A^2 = A$, prove that $(I + A)^n = I + (2^n - 1)A$.

12. If $A^2 = I$, prove that $(I + A)^n = 2^{n-1}(I + A)$.

13. Prove that $(J + I)^n = \left(\dfrac{(p+1)^n - 1}{p}\right)J + I$, where p is the order of unity matrix , J.

14. Prove that an upper triangular matrix is normal iff it is a diagonal matrix.

15. Distinguish between Hessenberg and triangular matrices.

16. Let E_{ij} be an n square matrix with a 1 in the ith row and jth column and 0's elsewhere, evaluate the products $E_{ij}E_{jk}$, $E_{ij}E_{km}$, AE_{ij}, $E_{ij}A$, $E_{ij}AE_{jk}$,

$E_{ij}AE_{km}$. Show that $E_{ij} = \mathbf{e}_i\mathbf{e}^T{}_j$, $E_{jj}^2 = E_{jj}$ and $(E_{ij}E_{jk})^T = E_{ki} = E_{kj}E_{ji}$.

[**Ans**. E_{ik}, $\delta_{jk}E_{im}$, jth column of A, ith row of A, $a_{jj}E_{ik}$, $a_{jk}E_{im}$]

17. Show that, for arbitrary $n \times n$ matrices A and B,

$A = \sum_{i,j=1}^{n} a_{ij}E_{ij}$, $B = \sum_{j,k=1}^{n} b_{jk}E_{jk}$, $(AB)^T = B^TA^T$.

[**Hints** : use example 16]

18. Prove that symmetric matrices are square.

[**Hints** : Let $A_{m\times n}$ be symmetric. But by defn. $A = A^T$. Therefore order of A^T is $n\times m$. Also from equality of matrices order must be same i.e. $m\times n = n\times m$ i.e. $m = n$, which proves the assertion]

19. Show that every skew symmetric matrix must have zeros on the diagonal.

[**Hints** : Use defn]

20. Show that $A + A^T$ is symmetric whereas $A - A^T$ is skew symmetric for an arbitrary matrix A.

21. Describe the matrices that are both symmetric and skew symmetric.

[**Ans** . Zero matrix]

22. Prove that the matrix product $Q\overline{Q}$ is scalar where Q is a quaternion.

23. (i) If $A^2 = AA$, what condition must be imposed on A and B for $(A+B)^2$ to have meaning? [**Ans** : A and B must be comfortable for addition]

(ii) Does $(A+B)^2 = (A+B)(A+B)$?

(iii) Does $(A+B)^2 = A^2 + 2AB + B^2$?

(iv) Does $(A+B)(A-B) = A^2 - B^2$?

(v) Does $(A-B)\ (A+B) = A^2 - B^2$?

(vi) Does $(A+B)(A-B) = (A-B)\ (A+B)$?

[**Ans** . (ii) yes, if both A and B are square of the same order, (iii) not always true if AB = BA, (iv),(v),(vi) No]

24. Show that

$$A = \begin{bmatrix} 2 & 1 \\ -4 & -2 \end{bmatrix},\ A = \begin{bmatrix} 1 & 2 & 3 \\ 1 & 2 & 3 \\ -1 & -2 & -3 \end{bmatrix},\ A = \begin{bmatrix} -4 & 4 & -4 \\ 1 & -1 & 1 \\ 5 & -5 & 5 \end{bmatrix}$$

are nilpotent matrices each of index 2, while $B = \begin{bmatrix} 3 & 1 \\ 0 & 2 \end{bmatrix}$ is not nilpotent.

25. If A is symmetric or skew symmetric, show that both AA^T and A^2 are symmetric.

26. Show that

$$A = \begin{bmatrix} 2 & -2 & -4 \\ -1 & 3 & 4 \\ 1 & -2 & -3 \end{bmatrix} \text{ and } B = \begin{bmatrix} -1 & 2 & 4 \\ -1 & -2 & -4 \\ -1 & 2 & 4 \end{bmatrix}$$

are idempotent and that AB = 0.

27. If

$$A = \begin{bmatrix} 1 & 0 & 0 & 0 \\ 0 & 1 & 0 & 0 \\ a & b & -1 & 0 \\ c & d & 0 & -1 \end{bmatrix}, \text{ find } A^{18}.$$

-[**Ans** . I_4]

28. Let A and B be square matrices of the same order and such that

A and AB – BA

commute. Show that for any positive integer n,

$$A^n B - BA^n = n(AB - BA)A^{n-1}.$$

29. Show that $\mathbf{x}^T(\mathbf{y}\mathbf{x}^T)\mathbf{y}$ is a 1×1 matrix whose sole entry is $(\mathbf{x}.\mathbf{y})^2$.

30. Is 1×1 matrix upper triangular ? Lower triangular ? Diagonal ? A vector ? A row vector ? A symmetric matrix ?

[**Ans** . Diagonal matrix, a vector, symmetric matrix]

31. Verify that

$$A^2 = \begin{bmatrix} -1 & -2 \\ 0 & 1 \end{bmatrix}^2 = I_2.$$

Hence write the square root of I_2.

32. Under what conditions the binomial formula

$$(A+B)^p = \sum_{k=0}^{p} {}^kC_p A^k B^{p-k} \text{ holds true?}$$

[**Ans** . If A and B are square matrices of one and the same order and AB = BA, then it is true]

33. Let

$$P(A) = A^2 + \begin{bmatrix} 1 & -1 \\ 1 & 1 \end{bmatrix} A - \begin{bmatrix} 0 & 1 \\ 1 & 0 \end{bmatrix},$$

where A is a matrix of order 2. Find $P\begin{bmatrix} 0 & 0 \\ 1 & 0 \end{bmatrix}$. [**Ans** . $\begin{bmatrix} -1 & -1 \\ 0 & 0 \end{bmatrix}$]

34. Represent the matrix

$$A = \begin{bmatrix} 1 & -1 & 2 \\ -1 & 5 & 4 \\ 2 & 4 & 14 \end{bmatrix}$$

as a product of two triangular matrices U and L such that

(i) A = LU and (ii) A = UL.

[**Hints** : Let $U = \begin{bmatrix} 1 & u_{12} & u_{13} \\ 0 & 1 & u_{23} \\ 0 & 0 & 1 \end{bmatrix}$ and $L = \begin{bmatrix} l_{11} & 0 & 0 \\ l_{21} & l_{22} & 0 \\ l_{31} & l_{32} & l_{33} \end{bmatrix}$.

Obtain the unknowns from A = LU]

[**Ans** . (i) $U = \begin{bmatrix} 1 & -1 & 2 \\ 0 & 1 & \frac{3}{2} \\ 0 & 0 & 1 \end{bmatrix}, L = \begin{bmatrix} 1 & 0 & 0 \\ -1 & 4 & 0 \\ 2 & 6 & 1 \end{bmatrix}$ and

(ii) $U = \begin{bmatrix} 1 & -\frac{11}{27} & \frac{1}{7} \\ 0 & 1 & \frac{2}{7} \\ 0 & 0 & 1 \end{bmatrix},\ L = \begin{bmatrix} \frac{64}{22\cdot7} & 0 & 0 \\ -\frac{11}{7} & \frac{27}{7} & 0 \\ 2 & 4 & 14 \end{bmatrix}$]

35. Write the matrix

$$A = \begin{bmatrix} 3+2i & 1-i & 5 \\ -2+7i & 6i & -1+4i \\ 4-3i & 2+2i & 6+i \end{bmatrix}$$

as (i) the sum of a real and a purely imaginary matrix,

(ii) the sum of a symmetric and a skew symmetric matrix,

(iii) the sum of a Hermitian and a skew Hermitian matrix.

[**Ans** . (i) $A = \begin{bmatrix} 3 & 1 & 5 \\ -2 & 0 & -1 \\ 4 & 2 & 6 \end{bmatrix} + \begin{bmatrix} 2i & -i & 0 \\ 7i & 6i & 4i \\ -3i & 2i & i \end{bmatrix}$

(ii), (iii) $A = \frac{1}{2}\begin{bmatrix} 6+4i & -1+6i & 9-3i \\ -1+6i & 12i & 1+6i \\ 9-3i & 1+6i & 12+2i \end{bmatrix} + \frac{1}{2}\begin{bmatrix} 0 & 3-8i & 1+3i \\ -3+8i & 0 & -3+2i \\ -1-3i & 3-2i & 0 \end{bmatrix}$]

36. Show that the sum of two Hermitian matrices is Hermitian while the product of two Hermitian matrices is Hermitian if the two matrices commute.

[**Hints** : 2nd part : Let A and B be two Hermitian matrices such that $A^* = A$, $B^* = B$. Hence

$$(AB)^* = B^*A^* = BA \qquad (1)$$

Firstly let AB = BA. Then (1) reduces to $(AB)^* = AB$ and so AB is Hermitian. Secondly let AB is Hermitian i.e. $(AB)^* = AB$. Therefore by (1), AB =BA showing that A and B commute. Thus proves each of the problem]

37. For each of the following assertions concerning 2×2 matrices A and B, state whether it is true or false, justify your answer.

(i) If $A^2 = A$, then A is a singular matrix.

(ii) For all A, $A^2 - 5A + 6I = (A - 3I)(A - 2I)$ where I is the 2×2 unit matrix.

(iii) For all A and B, $A^2 - 5AB + 6B^2 = (A - 3B)(A - 2B)$.

(iv) If A and B are symmetric matrices, then AB is also a symmetric matrix.

(v) If

$$A = \begin{bmatrix} \cos\frac{\pi}{6} & -\sin\frac{\pi}{6} \\ \sin\frac{\pi}{6} & \cos\frac{\pi}{6} \end{bmatrix}, \text{ then } A^{12} = I.$$

[**Ans.** (i) $A = AA^{-1} = I$, I is nonsingular. So (i) is false.

(ii) True, can be easily verified.

(iii) False, AB may not be commutative.

(iv) $(AB)^T = B^TA^T = BA$.

So AB is symmetric and (iv) is true.

(v) False, A= complex number = $\cos\frac{\pi}{6} + i\sin\frac{\pi}{6}$.

$A^{12} = \cos\frac{\pi}{2} + i\sin\frac{\pi}{2}$, by DeMoivre 's theorem.

$$A^{12} = I\cos^6 \tfrac{\pi}{3} = \tfrac{1}{64} I \ ]$$

38. Find the square roots of

$$\text{(i)} \begin{bmatrix} 0 & 1 \\ 0 & 0 \end{bmatrix}, \quad \text{(ii)} \begin{bmatrix} 3 & -4 \\ 1 & -1 \end{bmatrix} \text{ and (iii) } I_3.$$

[**Ans.** (i) no, (ii) $\pm\begin{bmatrix} 2 & -2 \\ \frac{1}{2} & 0 \end{bmatrix}$, (iii) infinite]

39. Find the square roots of

$$\begin{bmatrix} p & 0 \\ 0 & q \end{bmatrix}.$$

[**Ans** . $\begin{bmatrix} \sqrt{p} & 0 \\ 0 & \sqrt{q} \end{bmatrix}$]

Hence or otherwise show that the square roots of the diagonal matrix

$$\begin{bmatrix} p & & & & O \\ & q & & & \\ & & r & & \\ & & & \ddots & \\ O & & & & z \end{bmatrix} \text{ are } \pm \begin{bmatrix} \sqrt{p} & & & & O \\ & \sqrt{q} & & & \\ & & \sqrt{r} & & \\ & & & \ddots & \\ O & & & & \sqrt{z} \end{bmatrix}.$$

40. Verify that

$$A^2 = \begin{bmatrix} -1 & -2 \\ 0 & 1 \end{bmatrix}^2 = I_2 .$$

Further verify that

$$\begin{bmatrix} -1 & -2 \\ 0 & 1 \end{bmatrix}$$

is one of the infinite number of square roots of I_2.

41. Show that the matrix

$$\begin{bmatrix} 0 & 1 \\ 1 & 0 \end{bmatrix}$$

has infinite number of square roots.

42. Find the square root of

$$\begin{bmatrix} 17 & 6 \\ 6 & 8 \end{bmatrix}.$$

[Ans. $\begin{bmatrix} \frac{9}{\sqrt{5}} & \frac{2}{\sqrt{5}} \\ \frac{2}{\sqrt{5}} & \frac{6}{\sqrt{5}} \end{bmatrix}$]

43. If A and B are transition matrices, prove that AB is also a transition matrix.

44. Verify that Pauli spin matrices satisfy $\sigma_x^2 = \sigma_y^2 = \sigma_z^2 = I$ and $\sigma_x\sigma_y + \sigma_y\sigma_x = 0$, $\sigma_y\sigma_z + \sigma_z\sigma_y = 0$, $\sigma_z\sigma_x + \sigma_x\sigma_z = 0$.

45. Prove that

(a) $\left[\sigma_x, \sigma_y\right] = 2i\sigma_x$, $\left[\sigma_x, \sigma_z\right] = 2i\sigma_x$, $\left[\sigma_z, \sigma_x\right] = 2i\sigma_y$,

(b) $\sigma^2 = \sigma_x^2 + \sigma_y^2 + \sigma_z^2 = 3I$

(c) $\left[\sigma^2, \sigma_i\right] = 0,\ i = x, y, z$

[**Hints** : $[X, Y] = XY - YX$ is called the commutator of X and Y]

46. Show that

$$\alpha_x^2 = \alpha_y^2 = \alpha_z^2 = \beta^2 = I$$

and $\{\alpha_x, \alpha_y\} = \{\alpha_y, \alpha_z\} = \{\alpha_z, \alpha_x\} = \{\alpha_x, \beta\} = \{\alpha_y, \beta\} = \{\alpha_z, \beta\}$

where $\{X, Y\} = XY + YX$ is called the *anti-commutator* of matrices X and Y.

47. Prove that the set of Pauli spin matrices possesses the following multiplication table:

Multi - plication	I	σ_x	σ_y	σ_z
I	I	σ_x	σ_y	σ_z
σ_x	σ_x	I	$i\sigma_z$	$-i\sigma_y$
σ_y	σ_y	$-i\sigma_z$	I	$i\sigma_x$
σ_z	σ_z	$i\sigma_y$	$-i\sigma_x$	I

CHAPTER FOUR
SUBMATRIX AND PARTITION MATRIX

Chapter summary: *The chapter begins with definition of submatrix. Derived formula for number of submatrices. Defined other submatrices and partitioned block matrix. Distinguished between submatrix and block matrix with example. Uses of partitioned matrix are noted. Supermatrices and bordered matrices are defined and explained. Partitioning has been used to determine a column /row of product and also to represent some matrices like Helmertian, etc. After the algebra of partitioned matrices, theorems determining submatrix and partitioned matrices of a product are included. The chapter ends with two sections on ascertaining order of blocks in partition product and, reducible and irreducible matrices with properties.*

4.1 Submatrix

A submatrix of a matrix of dimension $n \geq 1$ is obtained by deleting some rows or some columns or both (not necessarily consecutive) or neither. The matrix itself is its submatrix.

4.1.1 Number of submatrices : The concept of nonempty subset in set theory and the principle of combination are combinedly used for the "construction" and "calculation of the number" of submatrices of any matrix. Finally, from the analogy of the number of submatrices, a general formula for the number of submatrices of an $m \times n$ matrix is developed. The symbols rC and cC used below in the calculation denote combination of rows and columns respectively. Let

$$A = \begin{bmatrix} 3 & 2 & 0 \\ 4 & 6 & 9 \end{bmatrix}$$

be a 2×3 matrix.

Submatrices are calculated in the following steps:

(i) The matrix A itself a submatrix. It is a 2×3 submatrix and its number is 1.

(ii) Now the number of submatrices consisting of a single row out of two rows and three columns i.e. 1×3 submatrices is the same as the number of combinations of two rows taken one at a time and it is ${}^{2}rC_1 = 2$. They are [3 2 0], [4 6 9].

(iii) The number of submatrices consisting of two columns out of three columns and two rows i.e. 2×2 submatrices is same as the combination of three rows taken two at a time and it is ${}^{3}cC_2$. They are

$$\begin{bmatrix} 3 & 2 \\ 4 & 6 \end{bmatrix}, \begin{bmatrix} 3 & 0 \\ 4 & 9 \end{bmatrix}, \begin{bmatrix} 2 & 0 \\ 6 & 9 \end{bmatrix}.$$

(iv) Similarly, number of submatrices consisting of single column and two rows i.e. 2×1 submatrices is the same as the combination of three columns taken one at a time and it is ${}^{3}cC_1 = 3$. They are

$$\begin{bmatrix}3\\4\end{bmatrix},\begin{bmatrix}2\\6\end{bmatrix},\begin{bmatrix}0\\9\end{bmatrix}.$$

(v) Also number of submatrices consisting of one row and two columns i.e. 1×2 submatrices arising from (iii) is ${}^3cC_2$ times the combination of two rows taken one at time and it is ${}^3cC_2 \times {}^2rC_1 = 3\times 2 = 6$. They are

$$[3\ \ 2], [4\ \ 6], [3\ \ 0], [4\ \ 9], [2\ \ 0], [6\ \ 9].$$

(vi) Lastly, the number of submatrices consisting of one row and one column i.e. 1×1 submatrices is the total number of elements of the matrix and it is 2×3 = 6. They are [3] , [2] , [0] , [4] , [6] , [9] . Therefore, the total number of submatrices of the given 2×3 matrix

$$= (\text{i}) + (\text{ii}) + (\text{iii}) + (\text{iv}) + (\text{v}) + (\text{vi})$$
$$= 1 + {}^2rC_1 + {}^3cC_2 + {}^3cC_1 + \left({}^3cC_2 \times {}^3rC_1\right) + (2\times 3)$$
$$= 1 + 2 + 3 + 3 + 6 + 6$$
$$= 21 = 3\cdot 7 = \left(2^2-1\right)\left(2^3-1\right)$$

Now on, without going into so much details, the number of submatrices of 3×3 matrix are:

(i) No. of 3×3 submatrix = matrix itself = 1.

(ii) No. of 2×3 submatrices = ${}^3rC_2 = 3$.

(iii) No. of 1×3 submatrices = ${}^3rC_1 = 3$

(iv) No. of 3×2 submatrices = ${}^3cC_2 = 3$.

(v) No. of 3×1 submatrices = ${}^3cC_1 = 3$.

(vi) No. of 2×2 submatrices = 3. ${}^3cC_2 = 9$.

(vii) No. of 1×2 submatrices = 3. ${}^3rC_2 = 9$.

(viii) No. of 2×1 submatrices = 3. ${}^3rC_2 = 9$.

(ix) No. of 1×1 submatrices = 3×3 = 9.

Therefore, total number of submatrices of a 3×3 matrix

$$= 1+3+3+3+3+9+9+9+9 = 49$$
$$= 7.7 = (2^3-1)(2^3-1)$$

The number of submatrices of a 3×4 matrix are :

(i) No. of 3×4 submatrix = matrix itself = 1.

(ii) No. of 2×4 submatrices = ${}^3rC_2 = 3$.

(iii) No. of 1×4 submatrices = ${}^3rC_1 = 3$.

(iv) No. of 3×3 submatrices = ${}^4cC_3 = 4$.

(v) No. of 3×2 submatrices = ${}^4cC_2 = 6$.

(vi) No. of 3×1 submatrices = $c\,{}^4C_1 = 4$.

(vii) No. of 2×3 submatrices in (ii) = 3. ${}^4cC_3 = 12$.

(viii)No. of 2×2 submatrices in (ii) = 3 . ${}^4cC_2 = 18$.

(ix) No. of 2×1 submatrices in (ii) = 3. ${}^4cC_1 = 12$.

(x) No. of 1×3 submatrices in (iii) = 3. ${}^4cC_3 = 12$.

(xi) No. of 1×2 submatrices in (iii) = $3\,.\,{}^4cC_2 = 18$.

(xii) No. of 1×1 submatrices = 3×4 = 12.

Therefore , the total number of submatrices of a 3×4 matrix

$$= 1 + 3 + 3 + 4 + 6 + 4 + 12 + 18 + 12 + 12 + 18 + 12$$
$$= 105 = 7.15 = (2^3 - 1)(2^4 - 1)$$

Proceeding in this way, the formula for the number of submatrices of an m×n matrix $= (2^m - 1)(2^n - 1)$.

4.1.2 Principal submatrix : The submatrix of dimension $(n - r)$ obtained by deleting same r rows and columns of an n-square matrix A is called a *principal submatrix* of A.

The first order principal submatrices of third order matrix

$$A = \begin{bmatrix} a_{11} & a_{12} & a_{13} \\ a_{21} & a_{22} & a_{23} \\ a_{31} & a_{32} & a_{33} \end{bmatrix} \text{ are } [a_{11}], [a_{22}], [a_{33}]$$

and second order submatrices are

$$\begin{bmatrix} a_{11} & a_{12} \\ a_{21} & a_{22} \end{bmatrix}, \begin{bmatrix} a_{11} & a_{13} \\ a_{31} & a_{33} \end{bmatrix}, \begin{bmatrix} a_{22} & a_{23} \\ a_{32} & a_{33} \end{bmatrix}$$

obtained respectively by deleting the last row and column, second row and column and first row and column, while 3rd order matrix is the matrix itself. Note that in every submatrice the principal diagonal is a part of the principal diagonal of the original matrix.

4.1.3 Leading principal submatrix : The submatrix of dimension $(n - r)$ obtained by deleting last r rows and columns of an n-square matrix A is called the *leading principal submatrix*. The first order leading principal matrix is $[a_{11}]$ only, the second order leading principal submatrix of the above 3×3 matrix A is

$$\begin{bmatrix} a_{11} & a_{12} \\ a_{21} & a_{22} \end{bmatrix}$$

only. Note that the diagonals of both principal and leading principal submatrices are part of the diagonal of A.

4.1.4 Notation of submatrices : Sometimes submatrices of an m×n matrix A may be denoted by A(α, β) where $\alpha \subseteq \{1, 2, \cdots, m\}$ of rows and $\beta \subseteq \{1, 2, \cdots, n\}$ columns. Thus

$$\begin{bmatrix} 1 & 2 & 3 \\ 4 & 5 & 6 \\ 7 & 8 & 9 \end{bmatrix}(\{1, 3\}, \{1, 2, 3\}) = \begin{bmatrix} 1 & 2 & 3 \\ 7 & 8 & 9 \end{bmatrix}.$$

If $m = n$ and β = α, the submatrix A(α, α) is called a *principal submatrix*. For example,

$$\begin{bmatrix} 1 & 2 & 3 \\ 4 & 5 & 6 \\ 7 & 8 & 9 \end{bmatrix}(\{1\}, \{1\}) = [1],$$

a first order leading principal submatrix.

$$\begin{bmatrix} 1 & 2 & 3 \\ 4 & 5 & 6 \\ 7 & 8 & 9 \end{bmatrix}(\{1,2\},\{1,2\}) = \begin{bmatrix} 1 & 2 \\ 4 & 5 \end{bmatrix},$$

a second order leading principal submatrix.

$$\begin{bmatrix} 1 & 2 & 3 \\ 4 & 5 & 6 \\ 7 & 8 & 9 \end{bmatrix}(\{1,3\},\{1,3\}) = \begin{bmatrix} 1 & 3 \\ 7 & 9 \end{bmatrix},$$

a second order leading principal submatrix.

4.2 Partition matrix

If a matrix is divided or partitioned into smaller matrices, called *blocks* or *cells* with *consecutive* rows and columns by drawing dotted horizontal lines of full width between rows and vertical lines of full height between columns, then the matrix is called *partition matrix.* The elements of a partition matrix are smaller matrices. It is also called the *supermatrix* (cf. sec. 4.5). For example note the partitions and different cells /blocks of the matrix in sec. 4.2.1 and 4.3 etc.

Advantages of partitioning : The following advantages may be noted in partitioning a matrix A into blocks or cells :

i. The partitioning may simplify the writing or printing of A in compact form and thus saves space,

ii . It exhibits some smaller structure of A which is of interest,

iii. It simplifies computation.

Note these advantages while one proceeds on.

4.2.1 Conformal partition : Two matrices of same order are said to be *conformally* or *identically partitioned* if

(i) both the matrices are partitioned in such a way that the number of rows and number of columns of two partition matrices are same;

(ii) the corresponding blocks are of same order.

Thus, the matrices

$$\left[\begin{array}{cc:c} 0 & -2 & -3 \\ \hdashline 4 & -1 & 2 \\ 3 & 0 & 9 \end{array}\right] \text{ and } \left[\begin{array}{cc:c} 8 & 9 & 10 \\ \hdashline 11 & 12 & 13 \\ 14 & 15 & 16 \end{array}\right]$$

are conformally partitioned as the matrices are 3 square, partition matrices are 2×2 of the form

$$\begin{bmatrix} A & B \\ C & D \end{bmatrix} \text{ and } \begin{bmatrix} E & F \\ G & H \end{bmatrix}$$

and each block of both partition matrices are of same order. Row-wise orders of A and E , B and F, C and G , D and H are 1×2 , 1×1 , 2×2 , 2×1 respectively of 4 blocks.

4.3 Differences between a submatrix and a block

The most text books used the submatrix for block without making any differences. But the submatrices defined and explained above have fundamental differences with blocks of partition matrix. The differences are

(i) each element of the original matrix must belong to one and only one block whereas some element may belong to different submatrices;
(ii) a block is a submatrix but a submatrix may not be a block;
(iii) a matrix may be partitioned in many different ways as partitioning has no definite rule but it can be broken into submatrices uniquely;
(iv) partitioning lines go straight both in width and height in the partition matrix but not necessarily in submatrix division;
(v) block contains consecutive rows and columns but the submatrix may not contain consecutive rows and columns;
(vi) one system of partition retains all the elements of the matrix in compact form within blocks and can be put back to original matrix easily but unique division into submatrices splits the matrix into fragments which are difficult to put back into the original matrix.
(vii) all blocks in a given row/column must have equal number of rows/columns. But all submatrices do not have these properties. Note these criteria in the following illustrations.

The above points are *explained* below with examples.

Let

$$A = \left[\begin{array}{c|c|cc} 1 & 2 & 3 & 0 \\ -1 & -4 & 4 & 5 \\ \hline -2 & 6 & -3 & -5 \end{array}\right]$$

be partitioned by dotted horizontal line between the second and third row, and vertical lines between the first and second, second and third columns to divide the matrix into six smaller matrices or blocks —

$$A_{11} = \begin{bmatrix} 1 \\ -1 \end{bmatrix}, \ A_{12} = \begin{bmatrix} 2 \\ -4 \end{bmatrix}, \ A_{13} = \begin{bmatrix} 3 & 0 \\ 4 & 5 \end{bmatrix},$$

$$A_{21} = [-2], \ A_{22} = [6], \ A_{23} = [-3 \quad -5].$$

Note that each element of the matrix A belongs to one and only one of the six blocks. But the submatrix [1 2 3 0] obtained by deleting the second and third rows and the submatrix

$$\begin{bmatrix} 1 \\ -1 \\ -2 \end{bmatrix}$$

obtained by deleting the last three columns both containing the element $a_{11} = 1$. This explains (i) of differences.

The block $\begin{bmatrix} 3 & 0 \\ 4 & 5 \end{bmatrix}$ is a submatrix but the submatrix $\begin{bmatrix} 2 & 0 \\ -4 & 5 \end{bmatrix}$ is not a block; because the latter does not contain elements of consecutive columns. This explains the point (ii) of differences.

The above same matrix may be partitioned as

$$B = \left[\begin{array}{cc|cc} 1 & 2 & 3 & 0 \\ \hline -1 & -4 & 4 & 5 \\ -2 & 6 & -3 & -5 \end{array}\right]$$

into 4 blocks

$$B_{11} = \begin{bmatrix}1 & 2\end{bmatrix},\ B_{12} = \begin{bmatrix}3 & 0\end{bmatrix},\ B_{21} = \begin{bmatrix}-1 & -4\\ -2 & 6\end{bmatrix},\ B_{22} = \begin{bmatrix}4 & 5\\ -3 & -5\end{bmatrix}.$$

Other partitions are also possible but
the number of submatrices of this 3×4 matrix $=\left(2^3-1\right)\left(2^4-1\right) = 7.15 = 105$, which is unique. This clarifies the distintinctive point (iii).

To discuss the point (iv) above, note the division

$$\left[\begin{array}{c|ccc} 1 & 2 & 3 & 0\\ \hline -1 & -4 & 4 & 5\\ -2 & 6 & -3 & -5\end{array}\right]$$

of the above matrix into four submatrices in which horizontal lines do not go straight and as such submatrices

$$\begin{bmatrix}2 & 3 & 0\end{bmatrix} \text{ and } \begin{bmatrix}-4 & 4 & 5\\ 6 & -3 & -5\end{bmatrix}$$

are not blocks although both contains consecutive columns in the first and consecutive columns and rows in the second.
Examples of (v) and (vi) are obvious from the above discussions.

4.4 Uses of partition matrix

Techniques of partitioning is very useful when working with large matrix. The dimension i.e. width and height of a large matrix can be reduced by partitioning as each block is considered as an element of the partition matrix. It saves hand computing labour considerably as well as time and cost in computer calculation. If simplifies writing or printing the matrix. Many uses of partition matrix will be encountered in later study, particularly, in addition, multiplication and inverse of large matrices and in linear programming.

4.5 Supermatrix (block matrix)

Recall from sec. 4.2 that the matrix whose elements are blocks obtained by partitioning is called a *supermatrix* or *block matrix*.

$$\text{Thus } A = \left[\begin{array}{cc|cc} a_{11} & a_{12} & a_{13} & a_{14}\\ \hline a_{21} & a_{22} & a_{23} & a_{24}\\ a_{31} & a_{32} & a_{33} & a_{34}\end{array}\right] = \begin{bmatrix}P_{11} & P_{12}\\ P_{21} & P_{22}\end{bmatrix} \tag{4.1}$$

where $P_{11} = [a_{11}\ a_{12}]$, $P_{12} = [a_{13}\ a_{14}]$, $P_{21} = \begin{bmatrix}a_{21} & a_{22}\\ a_{31} & a_{32}\end{bmatrix}$ and $P_{22} = \begin{bmatrix}a_{23} & a_{24}\\ a_{33} & a_{34}\end{bmatrix}$.

The matrix

$$A = \begin{bmatrix}P_{11} & P_{12}\\ P_{21} & P_{22}\end{bmatrix}$$

is an example of supermatrix. Note that the subscripts have no relation with orders of blocks.

Transpose of supermatrix : The transpose of a supermatrix is the transpose of both blocks and constituents of blocks. Thus, from (4.1) $A^T = [P_{ij}]^T = [P^T_{ij}]$ i.e.

$$A^T = \begin{bmatrix} P_{11}^T & P_{12}^T \\ P_{21}^T & P_{22}^T \end{bmatrix}.$$

4.5.1 Diagonal blocks : The blocks along the diagonal of the supermatrix are called *diagonal blocks.* That is, the blocks P_{ij} for which $i = j$ are diagonal blocks. Thus P_{11} and P_{22} are diagonal blocks of the supermatrix

$$A = \begin{bmatrix} P_{11} & P_{12} \\ P_{21} & P_{22} \end{bmatrix}.$$

4.5.2 Square supermatrix : If the number of rows and the number of columns of blocks are equal then the matrix is called a *square supermatrix.* Thus, the partitioned matrix

$$A = \left[\begin{array}{cc|cc|cc} a_{11} & a_{12} & a_{13} & a_{14} & a_{15} & a_{16} \\ a_{21} & a_{22} & a_{23} & a_{24} & a_{25} & a_{26} \\ \hline a_{31} & a_{32} & a_{33} & a_{34} & a_{35} & a_{36} \\ a_{41} & a_{42} & a_{43} & a_{44} & a_{45} & a_{46} \end{array}\right] = \begin{bmatrix} A_{11} & A_{12} & A_{13} \\ A_{21} & A_{22} & A_{23} \end{bmatrix} \tag{4.2}$$

is a square supermatrix, because all A_{ij} are square blocks.

4.5.3 Regularly partition matrix : If the diagonal blocks of a square supermatrix are square matrices then the matrix is said to be *regularly partition* or *regular supermatrix.* Only the square matrix can be regularly partitioned.

4.5.4 Diagonal supermatrix : If a square supermatrix is such that the blocks $A_{ij} = 0$ for $i \neq j$, the matrix A is said to be a *diagonal supermatrix.* Note that matrix A needs not be square but it must be partitioned as a square supermatrix. Thus,

$$\begin{bmatrix} A_{11} & 0 & 0 \\ 0 & A_{22} & 0 \end{bmatrix}$$

is a diagonal supermatrix corresponding to (4.2).

4.5.5 Quasiupper triangular supermatrix : A square matrix A of order n is called the *quasiupper triangular supermatrix* if it can be partitioned into blocks A_{ij} such that the diagonal blocks A_{ii} are square superdiagonal blocks A_{ij} $(i < j)$ are *rectangular* and all the subdiagonal blocks A_{ij} $(i > j)$ are 0 (*zero*),

$$\text{i.e. } A = \begin{bmatrix} A_{11} & A_{12} & \cdots & A_{1r} \\ 0 & A_{22} & \cdots & A_{2r} \\ \vdots & \vdots & \ddots & \vdots \\ 0 & 0 & \cdots & A_{rr} \end{bmatrix}.$$

Similarly , the *quasilower triangular supermatrix* may be defined. For example,

$$A = \left[\begin{array}{c|cc|ccc} 3 & 9 & -7 & -6 & -5 & -4 \\ \hline 0 & 1 & 2 & 8 & 9 & 10 \\ 0 & 0 & 1 & 11 & 12 & 13 \\ \hline 0 & 0 & 0 & -1 & -3 & -4 \\ 0 & 0 & 0 & 0 & 2 & 5 \\ 0 & 0 & 0 & 7 & 1 & 9 \end{array}\right]_{6\times 6}$$

is a 6×6 square matrix.

Also note that diagonal blocks

$$A_{11} = [3],\ A_{22} = \begin{bmatrix} 1 & 2 \\ 0 & 1 \end{bmatrix} \text{ and } A_{33} = \begin{bmatrix} -1 & -3 & 4 \\ 0 & 2 & 5 \\ 7 & 1 & 9 \end{bmatrix}$$

are square, superdiagonal blocks

$$A_{12} = [9\ \ -7],\ A_{13} = [-6\ \ -5\ \ -4],\ A_{23} = \begin{bmatrix} 8 & 9 & 10 \\ 11 & 12 & 13 \end{bmatrix}$$

are all rectangular and subdiagonal blocks

$$A_{21} = \begin{bmatrix} 0 \\ 0 \end{bmatrix},\ A_{31} = \begin{bmatrix} 0 \\ 0 \\ 0 \end{bmatrix},\ A_{32} = \begin{bmatrix} 0 & 0 \\ 0 & 0 \\ 0 & 0 \end{bmatrix}$$

are all zero matrices. Therefore the matrix A is a quasiupper triangular supermatrix. It is sometimes called the *subdirect sum* of the diagonal blocks A_{ii}. It is a *direct sum* if $A_{ij} = 0$ for $i \neq j$. It reduces to simply upper triangular if all blocks A_{ij} are one dimensional.

4.5.6 Triangular supermatrix : If the square or rectangular blocks above (or below) the square diagonal blocks of a square supermatrix are all zero, then the matrix is said to be the *lower* (or *upper*) *triangular supermatrix*. Note that the original matrix need not be square but be partitioned as the square supermatrix.

Thus

$$A = \left[\begin{array}{cc|c|ccc} 1 & 2 & 0 & 0 & 0 & 0 \\ 1 & 4 & 0 & 0 & 0 & 0 \\ \hline 2 & 4 & 5 & 0 & 0 & 0 \\ 8 & 2 & 4 & 0 & 0 & 0 \\ 3 & 7 & 6 & 0 & 0 & 0 \end{array}\right]_{6\times 6} = \begin{bmatrix} A_{11} & 0 & 0 \\ A_{21} & A_{22} & 0 \end{bmatrix}$$

is a lower triangular supermatrix. Similarly,

$$A = \begin{bmatrix} A_{11} & A_{12} & \cdots & A_{1p} \\ 0 & A_{22} & \cdots & A_{2p} \\ \vdots & \vdots & \ddots & \vdots \\ 0 & 0 & \cdots & A_{pp} \end{bmatrix}$$

is an upper triangular supermatrix. Note that the first example is a rectangular one while the second is a square matrix.

4.5.7 Quasidiagonal supermatrix : It is a supermatrix whose diagonal blocks are square matrices of different order and off diagonal blocks are zero matrices. Thus

$$A = \begin{bmatrix} D_1 & 0 & \cdots & 0 \\ 0 & D_2 & \cdots & 0 \\ \vdots & \vdots & \ddots & \vdots \\ 0 & 0 & \cdots & D_s \end{bmatrix}$$

is a quasidiagonal matrix whose diagonal blocks $D_i = 1, 2, \cdots, 5$ are square matrices of different orders. $A = \text{diag}(D_1, D_2, \cdots, D_s)$ and $\det A = \det D_1 \det D_2 \cdots \det D_s$. The matrix A is also called *decomposable matrix* or *pseudodiagonal supermatrix* or *direct sum* of matrices $D_1, D_2, \cdots, D_s$ taken in this order. This direct sum is denoted by vertically elongated plus sign " ⧺ " or encircled plus sign $\oplus$ and the above matrix is shortly written as

$$(D_1 ⧺ D_2 ⧺ \cdots ⧺ D_s) = (D_1 \oplus D_2 \oplus \cdots \oplus D_s)$$

Algebra of this matrix is given in sec. 5.1.

4.6 Bordered matrix

If an $(n + 1)$ square matrix is constructed by adding a row at the top and a column at the left of an n square matrix $[a_{ij}]$ with a zero at the junction of the added row and column then the new matrix is called a *bordered matrix*. It looks like a partition matrix partitioned regularly under top row and right of left most column. For example, the matrix

$$B = \left[\begin{array}{c|cccc} 0 & u_1 & u_2 & \cdots & u_n \\ \hline v_1 & a_{11} & a_{12} & \cdots & a_{1n} \\ v_2 & a_{21} & a_{22} & \cdots & a_{2n} \\ \vdots & \vdots & \vdots & \vdots & \vdots \\ v_n & a_{n1} & a_{n2} & \cdots & a_{nn} \end{array}\right]$$

is a bordered matrix.

This can be compactly expressed by partition blocks as

$$B = \begin{bmatrix} 0 & R \\ C & A \end{bmatrix},$$

where $A = [a_{ij}]_{m \times n}$ square matrix , 0 = null matrix ,

$$R = \text{row matrix} = \begin{bmatrix} u_1 & u_2 & \cdots & u_n \end{bmatrix}, \text{ and } C = \text{column matrix} = \begin{bmatrix} v_1 \\ v_2 \\ \vdots \\ v_n \end{bmatrix}.$$

This idea can be extended to the construction of $(n + r)$ square bordered matrix in which case partitioned B will look like

$$A = \begin{bmatrix} 0_{r \times r} & R_{r \times n} \\ C_{n \times r} & A_{n \times n} \end{bmatrix},$$

where $0_{r \times r}$ = denotes a $r \times r$ zero or null block matrix
$R_{r \times n}$ = denotes a $r \times n$ rectangular block matrix
$C_{n \times r}$ = denotes a $n \times r$ rectangular block matrix
and $A_{n \times n}$ = denotes a $n \times n$ square block matrix.

So, any square matrix bordered by the top and the left to form a new higher ordered square matrix, the latter is a bordered matrix.

4.7 Finding a column / row of a product by partitioning

In some applications it is necessary to know a particular column / row of a product. The following theorem helps to find a particular column of the product of two matrices.

Theorem . If $A = [a_{ik}]_{m\times n}$ and $B = [b_{kj}]_{n\times p}$ such that $AB = C = [c_{ij}]_{m\times p}$ then the *j*th column of C is $AB_{j.}$,

$$\text{where } B_{j.} = \begin{bmatrix} b_{1j} \\ b_{2j} \\ \vdots \\ b_{nj} \end{bmatrix}, j = 1, 2, \cdots, p$$

are the column partitions of B.

Proof. For the purpose, let us partition an $n\times p$ matrix B into p column vectors ($n\times 1$ matrices) as

$$B = [B_{1.} \quad B_{2.} \quad \cdots \quad B_{j.} \quad \cdots \quad B_{p.}],$$

$$\text{where } B_{j.} = \begin{bmatrix} b_{1j} \\ b_{2j} \\ \vdots \\ b_{nj} \end{bmatrix}, j = 1, 2, \cdots, p$$

to find a column of the product AB. From the product rule of matrices, the elements of the product is known as $c_{ij} = \sum_{k=1}^{n} a_{ik}\, b_{kj}$, where $A = [a_{ik}]_{m\times n}$, $B = [b_{kj}]_{n\times p}$, $C = [c_{ij}]_{m\times p}$ and $C = AB$. Therefore, the product matrix is

$$C = [c_{ij}]_{m\times p} = \left[\sum_{k=1}^{n} a_{ik}\, b_{kj}\right]_{m\times p}, i = 1, 2, \cdots, m \text{ and } j = 1, 2, \cdots, p.$$

Therefore *j*th column of C is obtained by giving the values 1, 2 , $\cdots$, m to i and it is

$$C_{j.} = \begin{bmatrix} \sum_{k=1}^{n} a_{1k}\, b_{kj} \\ \sum_{k=1}^{n} a_{2k}\, b_{kj} \\ \vdots \\ \sum_{k=1}^{n} a_{mk}\, b_{kj} \end{bmatrix}$$

$$= \begin{bmatrix} a_{11} & a_{12} & \cdots & a_{1n} \\ a_{21} & a_{22} & \cdots & a_{2n} \\ \vdots & \vdots & \vdots & \vdots \\ a_{m1} & a_{m2} & \cdots & a_{mn} \end{bmatrix} \begin{bmatrix} b_{1j} \\ b_{2j} \\ \vdots \\ b_{nj} \end{bmatrix}, \; j = 1, 2, \cdots, p$$

$$= AB_{j.}, \; j = 1, 2, \cdots, p,$$ which proves the theorem.

Cor. (i) *j*th column of a product AB is obtained by the post multiplication of the matrix A by the *j*th column of B.

Cor. (ii) The complete product AB is row vector i.e.

$$AB = [AB_{1.} \quad AB_{2.} \quad \cdots \quad AB_{p.}], \text{ where } AB_{j.}, j = 1, 2, \cdots, p$$

are the column partitions of the product AB and each column partition is obtained by the post multiplication of the matrix by the respective columns of B.

Cor. (iii) Column partition of a matrix in terms of column partition of unit matrix i.e. in terms of unit vectors of the unit matrix. Since by column partitions the $m\times n$ matrix A is

$$[A_{1.}\ \ A_{2.}\ \cdots\ A_{n.}] = A = AI = [Ae_1\ \ Ae_2\ \cdots\ Ae_n]$$ by cor(ii) where

$$\mathbf{e}_j = \left(\begin{bmatrix}0\\ \vdots\\ 1\\ \vdots\\ 0\end{bmatrix} \leftarrow j\text{th row},\ j = 1, 2, \cdots, n\right) = \left[\delta_{ij}\right],\ i, j = 1, 2, \cdots, n.$$

Cor.(iv) Matrix in terms of rows of unit matrix:

By cor.(iii) $A^T = [A^T\mathbf{e}_1\ \ A^T\mathbf{e}_2\ \cdots\ A^T\mathbf{e}_m]$, where A is $m{\times}n$, matrix. Taking transpose of both sides,

$$A = \begin{bmatrix}\mathbf{e}_1^T A\\ \mathbf{e}_2^T A\\ \vdots\\ \mathbf{e}_m^T A\end{bmatrix},$$

which is also the row partition of A.

Cor.(v) The jth row vector of A is $\mathbf{e}_j^T A$ from cor.(iv).

Cor.(vi) Product matrix in terms of rows of unit matrix.

Let C = AB where A is an $m{\times}n$ matrix, B is an $n{\times}p$ matrix and consequently C is an $m{\times}p$ matrix. Then by cor. (iv)

$$C = \begin{bmatrix}\mathbf{e}_1^T C\\ \mathbf{e}_2^T C\\ \vdots\\ \mathbf{e}_m^T C\end{bmatrix} = \begin{bmatrix}(\mathbf{e}_1^T A)B\\ (\mathbf{e}_2^T A)B\\ \vdots\\ (\mathbf{e}_m^T A)B\end{bmatrix}.$$

Thus

$$AB = \begin{bmatrix}(\mathbf{e}_1^T A)\\ (\mathbf{e}_2^T A)\\ \vdots\\ (\mathbf{e}_m^T A)\end{bmatrix} B = \begin{bmatrix}(\mathbf{e}_1^T A)B\\ (\mathbf{e}_2^T A)B\\ \vdots\\ (\mathbf{e}_m^T A)B\end{bmatrix}.$$

4.8 Uses of partition

Use of partition is often advantageous for compact representation of some matrices. Some advantages are mentioned in sec. 4.2. The following are some more illustrations.

4.8.1 Helmertian in partition form : Helmertian in strict sense defined in sec 3.16.3 may be stated in compact form according to Helmert using row partition as

$$H^e = \begin{bmatrix}H^e_{\cdot 1}\\ \vdots\\ H^e_{\cdot 2}\end{bmatrix},$$

where $H^e_{\cdot 1}$ = (first row of H^e)= $\dfrac{1}{\sqrt{n.1}}1^T$ and $H^e_{\cdot 2}$ = last $(n-1)$ row of H^e,

in which rth row of

$$H^e_{.2} = \left[\frac{1}{\sqrt{r(r+1)}}1_r \quad \frac{-r}{\sqrt{r(r+1)}} \quad 0_{(n-r-1)\times 1}\right], \; r = 1, 2, \cdots, (n-1).$$

Thus

$$H^e_{4\times 4} = \begin{bmatrix} \frac{1}{\sqrt{4.1}} & \frac{1}{\sqrt{4.1}} & \frac{1}{\sqrt{4.1}} & \frac{1}{\sqrt{4.1}} \\ \frac{1}{\sqrt{1.2}} & \frac{-1}{\sqrt{1.2}} & 0 & 0 \\ \frac{1}{\sqrt{2.3}} & \frac{1}{\sqrt{2.3}} & \frac{-2}{\sqrt{2.3}} & 0 \\ \frac{1}{\sqrt{3.4}} & \frac{1}{\sqrt{3.4}} & \frac{1}{\sqrt{3.4}} & \frac{-3}{\sqrt{3.4}} \end{bmatrix} = \begin{bmatrix} H^e_{.1} \\ \vdots \\ H^e_{.2} \end{bmatrix},$$

$$\text{where } H^e_{.1} = \frac{1}{\sqrt{4.1}}1^T \text{ and } H^e_{.2} = \begin{bmatrix} \frac{1}{\sqrt{1.2}} & \frac{-1}{\sqrt{1.2}} & 0 & 0 \\ \frac{1}{\sqrt{2.3}} & \frac{1}{\sqrt{2.3}} & \frac{-2}{\sqrt{2.3}} & 0 \\ \frac{1}{\sqrt{3.4}} & \frac{1}{\sqrt{3.4}} & \frac{1}{\sqrt{3.4}} & \frac{-3}{\sqrt{3.4}} \end{bmatrix}.$$

4.8.2 Partitioning of centrosymmetric matrix : A centrosymmetric matrix (cf. sec. 3.9.5) of even and odd order can be partitioned, in general, into 4 blocks and 6 blocks respectively using the matrix I_s (sec. 3.6). Thus if A is an $2n\times 2n$ centrosymmetric matrix then partitioning of A is

$$A = \begin{bmatrix} P & Q \\ I_sQI_s & I_sPI_s \end{bmatrix}.$$

For example matrix (iii) above in sec. 3.9.5. if partitions under second row and after second column are drawn.

Also if A is an $(2n-1)\times(2n-1)$ centrosymmetric matrix then partitioning of A is

$$A = \begin{bmatrix} P & \mathbf{a} & Q \\ \mathbf{b}^T & a_{nn} & b^T I_s \\ I_sQI_s & I_sa & I_sPI_s \end{bmatrix},$$

where '**a**' is a column vector, '**b**' is a row vector, a_{nn} is the center element. Examples are (i) and (ii) in sec. 3.9.5 if two row and column partitions are drawn.

4.8.3 Block Toeplitz matrix : If the elements of the Toeplitz matrix are themselves matrices, then the matrix is called a *block Toeplitz* matrix.

Ex. Find the second column of the product AB without computing AB completely, where

$$A = \begin{bmatrix} 1 & 0 & 2 \\ -1 & 5 & 6 \end{bmatrix} \text{ and } B = \begin{bmatrix} -5 & -1 & 1 \\ 0 & 2 & -1 \\ -6 & 1 & 1 \end{bmatrix}.$$

By cor (i) theorem 2 above, $\mathbf{C}_{2\cdot}$ of the product $C = AB$ is

$$\mathbf{AB}_{.2} = \begin{bmatrix} 1 & 0 & 2 \\ -1 & 5 & 6 \end{bmatrix} \begin{bmatrix} -1 \\ 2 \\ 1 \end{bmatrix} = \begin{bmatrix} 1 \\ 17 \end{bmatrix}.$$

4.9 Block multiplication

Theorems . Let A be an $m \times n$ matrix and B be an $n \times p$ matrix. Let B (or A) be partitioned into two blocks by column partitioning only. Then the product AB is also partitioned into two blocks of same column (row) partitioning.

Proof . Let $B = [B_1 \quad B_2]$ where B_1 is $n \times t$ matrix and B_2 is an $n \times (p - t)$ matrix. Then

$$AB = A\Big[\overbrace{\mathbf{b}_{1.} \quad \mathbf{b}_{2.} \quad \cdots \quad \mathbf{b}_{t.}}^{B_1} \;\vdots\; \overbrace{\mathbf{b}_{(t+1).} \quad \cdots \quad \cdots \quad \mathbf{b}_{p.}}^{B_2}\Big]$$

where $\mathbf{b}_{j.} = \begin{bmatrix} a_{1j} \\ a_{2j} \\ \vdots \\ a_{nj} \end{bmatrix}$,

$$= \Big[\mathbf{Ab}_{1.} \quad \mathbf{Ab}_{2.} \quad \ldots \quad \mathbf{Ab}_{t.} \;\vdots\; \mathbf{Ab}_{(t+1).} \quad \cdots \quad \mathbf{Ab}_{p.}\Big], \quad j = 1, 2, \cdots, p$$

$$= [AB_1 \;\vdots\; AB_2]$$

4.10 Algebra of partition matrices

(i) **Addition and subtraction :** Two conformal matrices can be added or subtracted block by block as addition and subtracton of two matrices of the same dimensions. Thus if

$$A = \begin{bmatrix} A_{11} & A_{12} & \ldots & A_{1q} \\ A_{21} & A_{22} & \ldots & A_{2q} \\ \vdots & \vdots & \vdots & \vdots \\ A_{p1} & A_{p2} & \ldots & A_{pq} \end{bmatrix} \tag{4.3}$$

and

$$B = \begin{bmatrix} B_{11} & B_{12} & \ldots & B_{1q} \\ B_{21} & B_{22} & \ldots & B_{2q} \\ \vdots & \vdots & \vdots & \vdots \\ B_{p1} & B_{p2} & \ldots & B_{pq} \end{bmatrix} \tag{4.4}$$

are conformal then

$$A \pm B = \begin{bmatrix} A_{11} \pm B_{11} & A_{12} \pm B_{12} & \ldots & A_{1q} \pm B_{1q} \\ A_{21} \pm B_{21} & A_{22} \pm B_{22} & \ldots & A_{2q} \pm B_{2q} \\ \vdots & \vdots & \vdots & \vdots \\ A_{p1} \pm B_{p1} & A_{p2} \pm B_{p2} & \ldots & A_{pq} \pm B_{pq} \end{bmatrix} \tag{4.5}$$

(ii) Scalar multiplication : As in scalar multiplication of a matrix by a scalar, each block of partition matrix (4.3) is multiplied by a scalar α. Thus

$$\alpha A = \begin{bmatrix} \alpha A_{11} & \alpha A_{12} & \cdots & \alpha A_{1q} \\ \alpha A_{21} & \alpha A_{22} & \cdots & \alpha A_{2q} \\ \vdots & \vdots & \vdots & \vdots \\ \alpha A_{p1} & \alpha A_{p2} & \cdots & \alpha A_{pq} \end{bmatrix}$$

(iii) Multiplication of partition matrices : Let $A = [a_{ij}]_{m \times n}$ and $B = [b_{jk}]_{n \times p}$ be two matrices conformable for multiplication. Let A_{ij} and B_{jk} denote blocks of A and B. Let us partition the matrix A by n_i , $i = 1, 2, \cdots, r$, horizontal dotted lines and n_j , $j = 1, 2, \cdots, s$ vertical dotted lines as follows:

$$A = \begin{bmatrix} a_{11} & a_{12} & \cdots & a_{1n} \\ a_{21} & a_{22} & \cdots & a_{2n} \\ \vdots & \vdots & \vdots & \vdots \\ a_{m1} & a_{m2} & \cdots & a_{mn} \end{bmatrix} = \begin{array}{c} \begin{matrix} n_1 \text{ cols.} & n_2 \text{ cols.} & & n_s \text{ cols.} \end{matrix} \\ \left[\begin{array}{c:c:c:c} A_{11} & A_{12} & \cdots & A_{1s} \\ \hdashline A_{21} & A_{22} & \cdots & A_{2s} \\ \hdashline \vdots & \vdots & \vdots & \vdots \\ \hdashline A_{r1} & A_{r2} & \cdots & A_{rs} \end{array}\right] \end{array} \begin{matrix} m_1 \text{ rows} \\ m_2 \text{ rows} \\ \vdots \\ m_r \text{ rows} \end{matrix}$$

where $n_1 + n_2 + \cdots + n_s = n$ and $m_1 + m_2 + \cdots + m_r = m$.

Therefore each block A_{ij} consists of m_i rows and n_j columns of A, $i = 1, 2, \cdots, r$ and $j = 1, 2, \cdots, s$. Similarly, B is partitioned by n_j, $j = 1, 2, \cdots, s$ horizontal dotted lines and p_k , $k = 1, 2, \cdots, t$ vertical dotted lines as follows:

$$B = \begin{bmatrix} b_{11} & b_{12} & \cdots & b_{1p} \\ b_{21} & b_{22} & \cdots & b_{p} \\ \vdots & \vdots & \vdots & \vdots \\ b_{n1} & b_{n2} & \cdots & b_{np} \end{bmatrix} = \begin{array}{c} \begin{matrix} p_1 \text{ cols.} & p_2 \text{ cols.} & & p_t \text{ cols.} \end{matrix} \\ \left[\begin{array}{c:c:c:c} B_{11} & B_{12} & \cdots & B_{1t} \\ \hdashline B_{21} & B_{22} & \cdots & B_{2t} \\ \hdashline \vdots & \vdots & \vdots & \vdots \\ \hdashline B_{s1} & B_{s2} & \cdots & B_{st} \end{array}\right] \end{array} \begin{matrix} n_1 \text{ rows} \\ n_2 \text{ rows} \\ \vdots \\ n_r \text{ rows} \end{matrix}$$

where $p_1 + p_2 + \cdots + p_t = p$. So, each block B_{jk} consists of n_j rows and p_k columns, $j = 1, 2, \cdots, s$ and $k = 1, 2, \cdots, t$. Now considering each block of A and B as element, it is found that $A=[A_{ij}]_{r \times s}$ is conformable to $B = [B_{jk}]_{s \times t}$ for multiplication. Also each block

$$[A_{ij}]_{m_i \times n_j} \text{ is conformable to each block } [B_{jk}]_{n_j \times p_k}$$

for multiplication. These together constitute the rules for multiplication of partitioned matrices. Hence for multiplication AB by partition to be conformable, the number of column partitioning of A must be equal to that of row partitioning of B and in addition blocks must be comformbable for multiplication.

Therefore, using matrix multiplication rule,

$$AB = [C_{ik}]_{r \times t} = \left[\sum_{j=1}^{s} A_{ij} B_{jk}\right]_{r \times t}$$

which is again a partition matrix with r horizontal and t vertical dotted lines. Now performing multiplication and addition, we obtain

$$AB = \left[\begin{array}{c:c:c:c} C_{11} & C_{12} & \cdots & C_{1t} \\ \hdashline C_{21} & C_{22} & \cdots & C_{2t} \\ \hdashline \vdots & \vdots & \vdots & \vdots \\ \hdashline C_{r1} & C_{r2} & \cdots & C_{rt} \end{array}\right].$$

Hence removing the dotted lines , finally we get the product .

$$AB = \left[C_{jk}\right]_{m\times p} = \left[\sum_{j=1}^{n} a_{ij}b_{jk}\right]_{m\times p}$$

4.10.1 Addition and multiplication of quasidiagonal matrices : Addition and multiplication of quasidiagonal matrices are simple as diagonal matrices . Let

$$A = \left[\begin{array}{c:c:c} A_1 & & \\ \hdashline & \ddots & \\ & & \ddots \\ \hdashline & & A_s \end{array}\right] \quad \text{and} \quad B = \left[\begin{array}{c:c:c} B_1 & & \\ \hdashline & \ddots & \\ & & \ddots \\ \hdashline & & B_s \end{array}\right]$$

be two diagonal matrices such that A_i , B_i , $i = 1, 2, \cdots, s$ are of same order . Then

$$A \oplus B = \left[\begin{array}{c:c:c} A_1 + B_1 & & \\ \hdashline & \ddots & \\ & & \ddots \\ \hdashline & & A_s + B_s \end{array}\right] \quad \text{and } AB = \left[\begin{array}{c:c:c} A_1B_1 & & \\ \hdashline & \ddots & \\ & & \ddots \\ \hdashline & & A_sB_s \end{array}\right].$$

4.11 Submatrix and partition matrix of the product

This section discusses some theorems which ensures the structure of submatrix as well as partition blocks of a product.

4.11.1 Submatrix of product : The submatrix of the product AB can be determined from the following theorems.

Theorem . If AB = C then the submatrix containing rows $i_1, i_2, \cdots, i_r$ and columns $j_1, j_2, \cdots, j_s$ of C is equal to the product of the submatrix with these rows of A and the submatrix with these columns of B.

Proof . Let $A = [a_{ij}]_{m\times n}$ and $B = [b_{jk}]_{n\times p}$. Then $C = \left[c_{ik}\right]_{m\times p} = \left[\sum_{j=1}^{n} a_{ij}b_{jk}\right]_{m\times p}$, where $i = 1, 2, \cdots, m$ and $k = 1, 2, \cdots, p$. Now the submatrix C_{ik} of C with rows $i_1, i_2, \cdots, i_r$ and columns $j_1, j_2, \cdots, j_s$ is obtained by replacing row i and column k of C by these rows and columns. It is

$$C_{ik} = \left[\sum_{j=1}^{n} a_{ij}b_{jk}\right]_{r\times s}, \text{ where } i = i_1, i_2, \cdots, i_r \text{ and } j = j_1, j_2, \cdots, j_s. \qquad (4.6)$$

Again the product of the given submatrices A_i and B_k of A and B respectively is

$$A_iB_k = \begin{bmatrix} a_{i_1 1} & a_{i_1 2} & \cdots & a_{i_1 n} \\ a_{i_2 1} & a_{i_2 2} & \cdots & a_{i_2 n} \\ \vdots & \vdots & \vdots & \vdots \\ a_{i_r 1} & a_{i_r 2} & \cdots & a_{i_r n} \end{bmatrix}_{r\times n} \begin{bmatrix} b_{1j_1} & b_{1j_2} & \cdots & b_{1j_s} \\ b_{2j_1} & b_{2j_2} & \cdots & b_{2j_s} \\ \vdots & \vdots & \vdots & \vdots \\ b_{nj_1} & b_{nj_2} & \cdots & b_{nj_s} \end{bmatrix}_{n\times s}$$

$$= \begin{bmatrix} \sum_{j=1}^{n} a_{i_1 j}b_{jj_1} & \sum_{j=1}^{n} a_{i_1 j}b_{jj_2} & \cdots & \sum_{j=1}^{n} a_{i_1 j}b_{jj_s} \\ \vdots & \vdots & & \vdots \\ \sum_{j=1}^{n} a_{i_s j}b_{jj_1} & \sum_{j=1}^{n} a_{i_s j}b_{jj_2} & \cdots & \sum_{j=1}^{n} a_{i_s j}b_{jj_s} \end{bmatrix}$$

$$= \left[\sum_{j=1}^{n} a_{ij}b_{jk} \right]_{r\times s} \tag{4.7}$$

$i = i_1, i_2, \quad \cdots, i_r$ and $k = j_1, j_2, \quad \cdots, j_s$. Therefore (4.6) and (4.7) together complete the proof.

Ex. Find the product of the submatrix containing first, second and fifth row of A and the submatrix containing the second and fourth column of B where

$$A = \begin{bmatrix} 1 & 0 & 2 \\ 3 & 1 & -1 \\ 0 & 2 & 3 \\ 4 & 5 & -2 \\ -1 & 2 & -3 \end{bmatrix} \text{ and } B = \begin{bmatrix} 2 & -1 & 7 & 0 \\ -1 & -2 & 1 & -1 \\ 0 & 3 & 0 & -2 \end{bmatrix}.$$

Verify the result so obtained by direct multiplication of the matrices A and B.

Given:

$$\begin{matrix} & \text{submatrix of A} & \text{submatrix of B} & \\ & & j_1 = 2 \quad j_2 = 4 & \\ \begin{matrix} i_1 = 1 \\ i_2 = 2 \\ i_3 = 5 \end{matrix} & \begin{bmatrix} 1 & 0 & 2 \\ 3 & 1 & -1 \\ -1 & 2 & -3 \end{bmatrix}_{3\times 3} & \begin{bmatrix} -1 & 0 \\ -2 & -1 \\ 3 & -2 \end{bmatrix}_{3\times 2} & = \begin{bmatrix} 5 & -4 \\ -8 & 1 \\ -12 & 4 \end{bmatrix}_{3\times 2} \end{matrix} \tag{1}$$

Verification:

$$AB = \begin{bmatrix} 1 & 0 & 2 \\ 3 & 1 & -1 \\ 0 & 2 & 3 \\ 4 & 5 & -2 \\ -1 & 2 & -3 \end{bmatrix}_{5\times 3} \begin{bmatrix} 2 & -1 & 7 & 0 \\ -1 & -2 & 1 & -1 \\ 0 & 3 & 0 & -2 \end{bmatrix}_{3\times 4}$$

$$= \begin{bmatrix} 2 & 5 & 7 & -4 \\ 5 & -8 & 22 & 1 \\ -2 & 5 & 2 & -8 \\ 3 & -20 & 33 & -1 \\ -4 & -12 & -5 & 4 \end{bmatrix} = \mathrm{C} \quad (\text{say}).$$

Therefore , submatrix of C with first, second, fifth row and second, fourth column is

$$\begin{bmatrix} 5 & -4 \\ -8 & 1 \\ -12 & 4 \end{bmatrix}$$

which is equal to the right hand side of (1). Hence verified.

4.11.2 Partition of product : The following theorem decides the partitioning of matrix C, the product AB of partition matrices A and B.

Theorem 1. If the $m \times n$ matrix A is partitioned by consecutive groups of rows into blocks A_i, $i = 1, 2, \cdots, r$ and the $n \times p$ matrix B is partitioned by consecutive groups of columns into blocks B_j , $j = 1, 2, \cdots, s$ (the subscripts i and k have no meaning with rows of A and columns of B). Then the $m \times p$ product matrix AB = C is partitioned into blocks by row groups exactly as A and column groups exactly as B. The ikth block C_{ik} of C is given by $\mathrm{C}_{ik} = \mathrm{A}_i \mathrm{B}_k$.

Proof . The proof is an immediate consequence of the theorem in sec. 4.11.1. Let $i = i_1, i_2, \cdots, i_r$ be the consecutive rows of A and $k = j_1, j_2, \cdots, j_s$ be the consecutive columns of B. Then from (4.6)

$$\mathrm{C}_{ik} = \left[\sum_{j-1}^{n} a_{ij} b_{jk} \right]_{r \times s} \tag{4.8}$$

where $i = i_1, i_2, \cdots, i_r$ and $k = j_1, j_2, \cdots, j_s$ and from (4.7).

$$\mathrm{A}_i \mathrm{B}_k = \left[\sum_{j=1}^{n} a_{ij} b_{jk} \right]_{r \times s} \tag{4.9}$$

where $i = 1, 2, \cdots, r$ and $k = 1, 2, \cdots, s$.

From (4.8) and (4.9) we obtain our desired conclusion and hence $\mathrm{C}_{ik} = \mathrm{A}_i \mathrm{B}_k$.

Schematically,

$$\begin{bmatrix} \mathrm{A}_1 \\ \mathrm{A}_2 \\ \vdots \\ \mathrm{A}_i \\ \vdots \\ \mathrm{A}_r \end{bmatrix}_{r \times 1} \begin{bmatrix} \mathrm{B}_1 & \mathrm{B}_2 & \cdots & \mathrm{B}_j & \cdots & \mathrm{B}_s \end{bmatrix}_{1 \times s} = \left[\begin{array}{c:c:c:c:c:c} \mathrm{C}_{11} & \mathrm{C}_{12} & \cdots & \mathrm{C}_{1k} & \cdots & \mathrm{C}_{1s} \\ \hdashline \vdots & \vdots & & \vdots & & \vdots \\ \hdashline \mathrm{C}_{i1} & \mathrm{C}_{i2} & \cdots & \mathrm{C}_{ik} & \cdots & \mathrm{C}_{is} \\ \hdashline \vdots & \vdots & & \vdots & & \vdots \\ \hdashline \mathrm{C}_{r1} & \mathrm{C}_{r2} & \cdots & \mathrm{C}_{rk} & \cdots & \mathrm{C}_{rs} \end{array} \right]_{r \times s}$$

Converse of Theorem 1 :

Theorem 2. If the $m\times n$ matrix A is partitioned by consecutive groups of columns into blocks A_i, $i = 1, 2, \cdots, r$ with columns n_i, $i = 1, 2, \cdots, r$ respectively where $n_1 + n_2 + \cdots + n_r = n$ and the $n\times p$ matrix B is partitioned by consecutive groups of rows into blocks B_i, $i = 1, 2, \cdots, r$ then

$$C = AB = [A_1B_1 + A_2B_2 + \cdots + A_rB_r].$$

Proof . We have as in Theorem 1,

$$C_{ik} = \sum_{j=1}^{n} a_{ij}b_{jk} = \sum_{j=1}^{n_1} a_{ij}b_{jk} + \sum_{j=n_1+1}^{n_1+n_2} a_{ij}b_{jk} + \cdots + \sum_{j=n_1+\ldots+n_{r-1}+1}^{n} a_{ij}b_{jk}$$

$$= [A_1B_1]_{ik} + [A_2B_2]_{ik} + \cdots + [A_rB_r]_{ik}$$

$$= [A_1B_1 + A_2B_2 + \cdots + A_kB_k]_{ik}$$

$$\therefore \quad C = [C_{ik}]_{m\times p} = AB = [A_1B_1 + A_2B_2 + \cdots + A_rB_r]$$

Ex. Redo example under sec.4.11.1 using theorem 1 & 2 of sec.4.11.2. Check the result with previous one and note the partitions.

We have by theorem 1,

$$AB = \begin{matrix} A_1 \rightarrow \\ \\ A_2 \rightarrow \\ \\ A_3 \rightarrow \end{matrix} \left[\begin{array}{ccc} 1 & 0 & 2 \\ 3 & 1 & -1 \\ \hline 0 & 2 & 3 \\ 4 & 5 & -2 \\ \hline -1 & 2 & -3 \end{array}\right] \overset{\begin{matrix} B_1 & \quad B_2 & \quad B_3 \end{matrix}}{\left[\begin{array}{c|cc|c} 2 & -1 & 7 & 0 \\ -1 & -2 & 1 & -1 \\ 0 & 3 & 0 & -2 \end{array}\right]}$$

$$= \begin{bmatrix} A_1B_1 & A_1B_2 & A_1B_3 \\ A_2B_1 & A_2B_2 & A_2B_3 \\ A_3B_1 & A_3B_2 & A_3B_3 \end{bmatrix} = \left[\begin{array}{c|cc|c} 2 & 5 & 7 & -4 \\ 5 & -8 & 22 & 1 \\ \hline -2 & 5 & 2 & -8 \\ 3 & -20 & 33 & -1 \\ \hline -4 & -12 & -5 & -4 \end{array}\right]$$

Also we have by theorem 2

$$AB = \overset{\begin{matrix} A_1 & \quad A_2 \end{matrix}}{\left[\begin{array}{cc|c} 1 & 0 & 2 \\ 3 & 1 & -1 \\ 0 & 2 & 3 \\ 4 & 5 & -2 \\ -1 & 2 & -3 \end{array}\right]} \left[\begin{array}{cccc} 2 & -1 & 7 & 0 \\ -1 & -2 & 1 & -2 \\ \hline 0 & 3 & 0 & -2 \end{array}\right] \begin{matrix} B_1 \\ \\ B_2 \end{matrix}$$

$$= [A_1B_1 + A_2B_2]$$

$$= \left[\begin{bmatrix} 1 & 0 \\ 3 & 1 \\ 0 & 2 \\ 4 & 5 \\ -1 & 2 \end{bmatrix}_{5\times 2} \begin{bmatrix} 2 & -1 & 7 & 0 \\ -1 & -2 & 1 & -1 \end{bmatrix}_{2\times 4} + \begin{bmatrix} 2 \\ -1 \\ 3 \\ -2 \\ -3 \end{bmatrix}_{5\times 1} \begin{bmatrix} 0 & 3 & 0 & -2 \end{bmatrix}_{1\times 4}\right]_{5\times 4}$$

$$= \left[\begin{bmatrix} 2 & -1 & 7 & 0 \\ 5 & -5 & 22 & -1 \\ -2 & -4 & 2 & -2 \\ 3 & -14 & 33 & -5 \\ -4 & -3 & -5 & -2 \end{bmatrix}_{5\times 4} + \begin{bmatrix} 0 & 6 & 0 & -4 \\ 0 & -3 & 0 & 2 \\ 0 & 9 & 0 & -6 \\ 0 & -6 & 0 & 4 \\ 0 & -9 & 0 & 6 \end{bmatrix}_{5\times 4}\right]_{5\times 4}$$

$$= \begin{bmatrix} 2 & 5 & 7 & -4 \\ 5 & -8 & 22 & 1 \\ -2 & 5 & 2 & -8 \\ 3 & -20 & 33 & -1 \\ -4 & -12 & -5 & 4 \end{bmatrix}.$$

Checking the result with previous one follows immediately and noting partitions is left to readers..

Theorem 3. Let the number of column partitions and the number of columns in each such partition of the $m\times n$ matrix A be equal to that of row partitions and number of rows in each such partition of the $n\times p$ matrix B. Further, let the blocks A_{ij} and B_{ij} of A and B respectively be formed by arbitrary row group partitions of A and column group partitions of B where i and j denote row and column group occurrence respectively. Finally, let the product C = AB be partitioned according to the row partitioning of A and the column partitioning of B. Then the (i, j)th block of C is

$$C_{ij} = \sum_{k=1}^{n} A_{ik} B_{kj}.$$

N.B.(i) Combination of theorems in sec. 4.11.1 and theorem 2 leads to this remarkable theorem for block determination for a partitioned product.

(ii) Theorem 3 gives procedure for partitioning a matrix in product if the partitions of one is given.

Ex. Explain theorem3 for example in sec.4.11.1 with arbitrary partitions.

Let one column partition of A divides A into two groups — first and second column form one group and third column forms another group. To apply theorem 3, B is divided into two groups by row partition . So that first group contains first and second row and second group contains third row. Row partitions of A and column partitions of B are arbitrarily taken.

$$AB = \left[\begin{array}{cc|c} 1 & 0 & 2 \\ 3 & 1 & -1 \\ \hline 0 & 2 & 3 \\ \hline 4 & 5 & -2 \\ -1 & 2 & -3 \end{array}\right] \left[\begin{array}{cc|c|c} 2 & -1 & 7 & 0 \\ -1 & -2 & 1 & -1 \\ \hline 0 & 3 & 0 & -2 \end{array}\right] = \left[\begin{array}{c|c} A_{11} & A_{12} \\ \hline A_{21} & A_{22} \\ \hline A_{31} & A_{32} \end{array}\right] \left[\begin{array}{c|c|c} B_{11} & B_{12} & B_{13} \\ \hline B_{21} & B_{22} & B_{23} \end{array}\right]$$

$$= \left[\begin{array}{c:c:c} A_{11}B_{11}+A_{12}B_{21} & A_{11}B_{12}+A_{12}B_{22} & A_{11}B_{13}+A_{12}B_{23} \\ \hdashline A_{21}B_{11}+A_{22}B_{21} & A_{21}B_{12}+A_{22}B_{22} & A_{21}B_{13}+A_{22}B_{23} \\ \hdashline A_{31}B_{11}+A_{32}B_{21} & A_{31}B_{12}+A_{32}B_{22} & A_{31}B_{13}+A_{32}B_{23} \end{array}\right]$$

$$= \left[\begin{array}{c:c:c} \begin{bmatrix} 2 & -1 \\ 5 & -5 \end{bmatrix}+\begin{bmatrix} 0 & 6 \\ 0 & -3 \end{bmatrix} & \begin{bmatrix} 7 \\ 22 \end{bmatrix}+\begin{bmatrix} 0 \\ 0 \end{bmatrix} & \begin{bmatrix} 0 \\ -1 \end{bmatrix}+\begin{bmatrix} -4 \\ 2 \end{bmatrix} \\ \hdashline \begin{bmatrix} -2 & -4 \end{bmatrix}+\begin{bmatrix} 0 & 9 \end{bmatrix} & [2]+[0] & [-2]+[-6] \\ \hdashline \begin{bmatrix} 3 & -14 \\ -4 & -3 \end{bmatrix}+\begin{bmatrix} 0 & -6 \\ 0 & -9 \end{bmatrix} & \begin{bmatrix} 33 \\ -5 \end{bmatrix}+\begin{bmatrix} 0 \\ 0 \end{bmatrix} & \begin{bmatrix} -5 \\ -2 \end{bmatrix}+\begin{bmatrix} 4 \\ 6 \end{bmatrix} \end{array}\right]$$

$$= \left[\begin{array}{c:c:c} \begin{bmatrix} 2 & 5 \\ 5 & -8 \end{bmatrix} & \begin{bmatrix} 7 \\ 22 \end{bmatrix} & \begin{bmatrix} -4 \\ 1 \end{bmatrix} \\ \hdashline \begin{bmatrix} -2 & 5 \end{bmatrix} & [2] & [-8] \\ \hdashline \begin{bmatrix} 3 & -20 \\ -4 & -12 \end{bmatrix} & \begin{bmatrix} 37 \\ -5 \end{bmatrix} & \begin{bmatrix} -1 \\ 4 \end{bmatrix} \end{array}\right] = \left[\begin{array}{cc:c:c} 2 & 5 & 7 & -4 \\ 5 & -8 & 22 & 1 \\ \hdashline -2 & 5 & 2 & -8 \\ \hdashline 3 & -20 & 33 & -1 \\ -4 & -12 & -5 & 4 \end{array}\right] \text{ which verifies Theorem 3.}$$

4.12 How to ascertain order of blocks in a partitioned product

Each of the theorems in secs. 4.11.1 and 4.11.2 decides one way of ascertaining order of blocks in a partitioned product. Apart from those criterions corresponding to a given partition of one factor, the partitions of other factor can be ascertain on the basis of conformability of multiplication and addition of blocks. Thus let two $m\times n$ and $n\times p$ matrices A and B are partitioned as follows:

$$A = \begin{bmatrix} A_{11} & A_{12} \\ A_{21} & A_{22} \end{bmatrix}_{m\times n}, \quad B = \begin{bmatrix} B_{11} & B_{12} \\ B_{21} & B_{22} \end{bmatrix}_{n\times p}$$

Now, the problem is : what are orders of the blocks concerned ?

Here $$AB = \begin{bmatrix} A_{11}B_{11}+A_{12}B_{21} & A_{11}B_{12}+A_{12}B_{22} \\ A_{22}B_{11}+A_{22}B_{21} & A_{21}B_{12}+A_{22}B_{22} \end{bmatrix}$$

Let the orders of block A_{11} be $r\times s$. Then obviously order of A_{12} is $r\times(n-s)$, that of A_{21} and A_{22} are $(m-r)\times s$ and $(m-r)\times(n-s)$ respectively to satisfy the original order $m\times n$. Then to satisfy conformability for multiplication and addition, the orders of B_{ij} , $i, j = 1, 2$ can be easily determine. Therefore, order of B_{11} must be $s\times q$ (say) so that the order of the product $A_{11}B_{11}$ is $r\times q$; order of B_{21} must be $(n-s)\times q$, so that the order of the product $A_{12}B_{21}$ is $r\times q$. Hence addition of $A_{11}B_{11}$ and $A_{12}B_{21}$ is possible.

Similarly, order of B_{12} must be $s\times(p-q)$, so that the order of the product $A_{11}B_{12}$ is $r\times(p-q)$. Order of B_{22} must be $(n-s)\times(p-q)$, so that the order of the product $A_{12}B_{22}$ is $r\times(p-q)$. Hence the sum of $A_{11}B_{12}$ and $A_{12}B_{22}$ is admissible.

Therefore, tabulating the discussion, we get,

Blocks of A	Order	Blocks of B	Order
A_{11}	$r\times s$	B_{11}	$s\times q$
A_{12}	$r\times(n-s)$	B_{12}	$s\times(p-q)$
A_{21}	$(m-r)\times s$	B_{21}	$(n-s)\times q$
A_{22}	$(m-r)\times(n-s)$	B_{22}	$(n-s)\times(p-q)$

Ex. Partition B so that one can compute AB, if

$$A=\left[\begin{array}{cc|c|cc} 3 & 1 & 4 & 3 & 2 \\ 1 & 0 & 5 & -1 & 0 \\ \hline 0 & 0 & 2 & 0 & -2 \\ -3 & 1 & 1 & 0 & 0 \end{array}\right] \text{ and } B=\begin{bmatrix} 1 & 0 & 3 & -2 & 5 & 0 \\ 0 & 1 & 2 & 0 & 1 & -3 \\ 0 & 0 & 1 & 0 & -1 & 0 \\ 2 & 2 & -2 & 1 & 3 & 0 \\ 0 & 0 & 0 & 0 & 1 & 0 \end{bmatrix}$$

In this example, $B_{5\times6}$ is to be partitioned corresponding to the given six blocks of $A_{4\times5}$ for computing the product AB.

By theorem 3 of sec. 4.11.2, as A has two column partitions with first and second columns in one column division, third column in second column division and fourth and fifth columns in third column division, so first row division must contain first and second rows, second row division must contain third row and third row division must contain fourth and fifth rows of B.

Finally column division of B is the orders of the blocks in A are

4.13 Reducible and irreducible matrices

An n-square matrix $n \geq 2$ is said to be *redicible* (*decomposable*) if there is a permutation matrix P such that either

$$P^TAP = \begin{bmatrix} A_{11} & A_{12} \\ 0 & A_{22} \end{bmatrix} \text{ or } PAP^T = \begin{bmatrix} A_{11} & 0 \\ A_{21} & A_{22} \end{bmatrix} \quad (4.10)$$

where A_{11} and A_{12} are square blocks of order less than n. The matrix is *irreducible* (*indecomposable*) if it is not reducible i.e. there is no permutation matrix P such that P^TAP or PAP^T is any one of the form (4.10). Examples are (i) any matrix with nonzero elements is irreducible , (ii) positive matrices are reducible . The matrix

$$A=\begin{bmatrix} 1 & 1 & 2 \\ 0 & 1 & 0 \\ 2 & 3 & 1 \end{bmatrix} \text{ is decomposable if } P=\begin{bmatrix} 0 & 1 & 0 \\ 0 & 0 & 1 \\ 1 & 0 & 0 \end{bmatrix}. \text{ Then}$$

$$PAP^T = \begin{bmatrix} 0 & 1 & 0 \\ 0 & 0 & 1 \\ 1 & 0 & 0 \end{bmatrix}\begin{bmatrix} 1 & 1 & 2 \\ 0 & 1 & 0 \\ 2 & 3 & 1 \end{bmatrix}\begin{bmatrix} 0 & 0 & 1 \\ 1 & 0 & 0 \\ 0 & 1 & 0 \end{bmatrix}$$

$$=\begin{bmatrix}0 & 1 & 0\\0 & 0 & 1\\1 & 0 & 0\end{bmatrix}\begin{bmatrix}1 & 2 & 1\\1 & 0 & 0\\3 & 1 & 2\end{bmatrix}=\left[\begin{array}{c|cc}1 & 0 & 0\\ \hline 3 & 1 & 2\\1 & 2 & 1\end{array}\right]$$

Also $$P^TAP=\left[\begin{array}{cc|c}1 & 2 & 3\\2 & 1 & 1\\ \hline 0 & 0 & 1\end{array}\right].$$

4.13.1 Irreducibly diagonally dominant : An n square matrix A is irreducibly diagonally dominant if it is irreducible as well as eqn. (4.10) holds for strict inequality for *at least* one I.

4.13.2 Properties of irreducible matrices : Following are the properties of irreducible matrices :

1. If $A \geq 0$, is irreducible then for any a_{ij} there exists a positive integer k such that

   $a_{ij}^{(k)} > 0$, where $a_{ij}^{(k)}$ is the i, jth element of A^k.

2. For any $n\times n$ irreducible matrix $A \geq 0$, we have

$$(I+A)^{n-1} > 0 \text{ and } (I+A)^{n-1}\mathbf{x} > 0$$

where $\mathbf{x} \geq 0$ is a nonnegative vector.

Also $A(I+A)^{n-1} = (I+A)^{n-1}A$

Ex. Test whether the matrix

$$A=\begin{bmatrix}1 & 1 & 2\\0 & 1 & 1\\2 & 3 & 1\end{bmatrix}$$

is irreducible or not.

Test 1. The matrix

$$A=\begin{bmatrix}1 & 1 & 2\\0 & 1 & 1\\2 & 3 & 1\end{bmatrix} \geq 0 \text{ and } n = 3.$$

$$\therefore \quad A^2=\begin{bmatrix}1 & 1 & 2\\0 & 1 & 1\\2 & 3 & 1\end{bmatrix}\begin{bmatrix}1 & 1 & 2\\0 & 1 & 1\\2 & 3 & 1\end{bmatrix}=\begin{bmatrix}5 & 8 & 5\\2 & 4 & 2\\4 & 8 & 8\end{bmatrix}.$$

Here $a_{31} = 2$ and $a_{31}^2 = 4 = 2^2 > 0$.

$\therefore$ there exists an integer $k = 2$ such that $a_{31}^2 = 2^2 > 0$, an element of A^2.

$\therefore$ A is irreducible.

Test 2 :

$$(\mathrm{I}+\mathrm{A})^2 = \begin{bmatrix} 2 & 1 & 2 \\ 0 & 2 & 1 \\ 2 & 3 & 2 \end{bmatrix}^2 = \begin{bmatrix} 8 & 10 & 9 \\ 2 & 7 & 4 \\ 8 & 14 & 11 \end{bmatrix} > 0$$

$\therefore$ A is irreducible.

Exercises four

1(a). Show that an $m \times n$ matrix possesses $(2^m - 1)(2^n - 1)$ submatrices.

(b). How many submatrices does a 4×5 matrix possess ?

[**Hints** : (a) sec. 4.1.1; **Ans.** 465]

2. Show that

$$\left[\begin{array}{c|c} \mathrm{A} & \mathrm{B} \\ \hline \mathrm{C} & \mathrm{D} \end{array}\right]^{\mathrm{T}} = \left[\begin{array}{c|c} \mathrm{A}^{\mathrm{T}} & \mathrm{B}^{\mathrm{T}} \\ \hline \mathrm{C}^{\mathrm{T}} & \mathrm{D}^{\mathrm{T}} \end{array}\right]. \text{ Generalize.}$$

3. Partition

$$\begin{bmatrix} 2 & 1+i & 4 \\ 2 & 1 & i \\ -3 & 0 & 2 \end{bmatrix}$$

in such a way that one of the designated submatrices is a 2×2 R-matrix.

[**Ans .** $\left[\begin{array}{cc|c} 2 & 1+i & 4 \\ \hline 2 & 1 & i \\ -3 & 0 & 2 \end{array}\right]$ in which submatrix $\begin{bmatrix} 2 & 1 \\ -3 & 0 \end{bmatrix}$ is a 2×2 R-matrix]

4. Write all of the four 3×3 submatrices of

$$\begin{bmatrix} 3 & 4 & 0 & 9 \\ 4 & 3 & 2 & 1 \\ 4 & 0 & 2 & 1 \end{bmatrix}.$$

[**Ans.** Delete one column and get one 3×3 submatrix.
In this way obtain all of the remaining four 3×3 submatrices]

5. Write the principal submatrices and the leading principal submatrices of the matrix

$$A = \begin{bmatrix} 3 & 4 & 0 \\ 3 & 2 & 1 \\ 0 & 2 & 1 \end{bmatrix}.$$

[**Ans.** Principal submatrices : [3], [2], [1] along principal diagonal,

$\begin{bmatrix} 3 & 4 \\ 3 & 2 \end{bmatrix}, \begin{bmatrix} 3 & 0 \\ 0 & 1 \end{bmatrix}, \begin{bmatrix} 2 & 1 \\ 2 & 1 \end{bmatrix}$ and the given matrix;

leading principal submatrices :

[3] in (1, 1) position, $\begin{bmatrix} 3 & 4 \\ 3 & 2 \end{bmatrix}$ and the given matrix]

6. How many submatrices can be obtained from

$$\begin{bmatrix} 1 & 0 & -1 \\ 2 & 1 & 3 \end{bmatrix}.$$

Write all possible submatrices.

[**Ans.** 21, write all 21 submatrices by yourself]

7. Find the product AB using partition if

$$A = \begin{bmatrix} I_2 & 0 \\ 0 & A_1 \end{bmatrix}, \; B = \begin{bmatrix} B_1 & 0 & B_2 \\ 0 & I_1 & B_3 \end{bmatrix},$$

where $A_1 = [3]$, $B_1 = \begin{bmatrix} 1 & 2 \\ 3 & 4 \end{bmatrix}$, $B_2 = \begin{bmatrix} 6 \\ 5 \end{bmatrix}$, $B_3 = [7]$ and others with usual meaning.

[**Ans.** $AB = \begin{bmatrix} B_1 & 0 & B_2 \\ 0 & A_1 & A_1B_3 \end{bmatrix} = \left[\begin{array}{cc|c|c} 1 & 2 & 0 & 6 \\ 3 & 4 & 0 & 5 \\ \hline 0 & 0 & 3 & 21 \end{array}\right]$]

8. Write the submatrices indicated by

(i) $\begin{bmatrix} 1 & 3 & -2 & 5 \\ 0 & 1 & 2 & 4 \end{bmatrix}(\{1,2\},\{1,2,3\})$ [**Ans**. $\begin{bmatrix} 1 & 3 & -2 \\ 0 & 1 & 2 \end{bmatrix}$]

(ii) $\begin{bmatrix} 1 & 3 & -2 & 5 \\ 0 & 1 & 2 & 4 \end{bmatrix}(\{1\},\{2,3\})$ [**Ans.**[3 –2]]

(iii) $\begin{bmatrix} 1 & 3 & -2 & 5 \\ 0 & 1 & 2 & 4 \end{bmatrix}(\{2\},\{3\})$ [**Ans** . [2]]

9. If A is an $m\times n$ matrix and B is an $n\times p$ matrix whose columns are

$\mathbf{b}_{\cdot 1}, \mathbf{b}_{\cdot 2}, \cdots, \mathbf{b}_{\cdot p}$, prove that $AB = \begin{bmatrix} A\mathbf{b}_{\cdot 1} & A\mathbf{b}_{\cdot 2} & \cdots & A\mathbf{b}_{\cdot p} \end{bmatrix}$.

10. Partition B so that one can compute AB, if

$$A = \left[\begin{array}{cc|ccc} 2 & 4 & 0 & -3 & 7 \\ 1 & 0 & 0 & 5 & 7 \\ \hline -4 & -2 & 0 & 4 & 3 \end{array}\right] \text{ and } B = \begin{bmatrix} 0 & -1 & 2 & 5 \\ 1 & 0 & 0 & 8 \\ 2 & -3 & 0 & 4 \\ 5 & -2 & 3 & 4 \\ 1 & -1 & 1 & 1 \end{bmatrix}.$$

[**Ans.** $\left[\begin{array}{c|ccc} 0 & -1 & 2 & 5 \\ 1 & 0 & 0 & 8 \\ \hline 2 & -3 & 0 & 4 \\ 5 & -2 & 3 & 4 \\ 1 & -1 & 1 & 1 \end{array}\right], \left[\begin{array}{cc|cc} 0 & -1 & 2 & 5 \\ 1 & 0 & 0 & 8 \\ \hline 2 & -3 & 0 & 4 \\ 5 & -2 & 3 & 4 \\ 1 & -1 & 1 & 1 \end{array}\right], \left[\begin{array}{ccc|c} 0 & -1 & 2 & 5 \\ 1 & 0 & 0 & 8 \\ \hline 2 & -3 & 0 & 4 \\ 5 & -2 & 3 & 4 \\ 1 & -1 & 1 & 1 \end{array}\right], \left[\begin{array}{c|c|c|c} 0 & -1 & 2 & 5 \\ 1 & 0 & 0 & 8 \\ \hline 2 & -3 & 0 & 4 \\ 5 & -2 & 3 & 4 \\ 1 & -1 & 1 & 1 \end{array}\right]$]

11. A matrix is partitioned as

$$\left[\begin{array}{c|c|c} A & B & C \\ \hline D & E & F \end{array}\right].$$

Write down the complex conjugate, the transpose and the Hermitian conjugate of the matrix. [**Ans.** $\begin{bmatrix} A^* & B^* & C^* \\ D^* & E^* & F^* \end{bmatrix}, \begin{bmatrix} \tilde{A} & \tilde{D} \\ \tilde{B} & \tilde{E} \\ \tilde{C} & \tilde{F} \end{bmatrix}, \begin{bmatrix} A^+ & D^+ \\ B^+ & E^+ \\ C^+ & F^+ \end{bmatrix}$]

12. Let

$$M = \begin{bmatrix} A & 0 \\ X & B \end{bmatrix},$$

where A is of order $m \times m$, B of order $n \times n$, X of order $n \times m$ and 0 is the zero matrix of order $m \times n$. If M is unitary, show that X must be a zero matrix and A and B must be unitary.

13. Calculate AB in three ways : directly without partition; with the partition indicated; with a different partition of your own choosing :

$$A = \left[\begin{array}{cc|c|cc} 2 & 3 & 4 & 0 & 0 \\ 3 & 1 & 0 & 0 & 0 \\ \hline 1 & 0 & 1 & 0 & 4 \\ \hline -1 & 0 & 0 & 1 & 0 \\ 0 & -1 & 4 & 0 & 1 \end{array}\right], \quad B = \left[\begin{array}{c|cc} 1 & 0 & 0 \\ 3 & 0 & 0 \\ \hline 0 & 2 & 1 \\ \hline -1 & 0 & 0 \\ -1 & 0 & 0 \end{array}\right].$$

14. List all the 2×3 submatrices of

$$\begin{bmatrix} 3 & 4 & 2 & 1 \\ 0 & 2 & 1 & 4 \end{bmatrix}.$$

15. Partition

$$\begin{bmatrix} 1 & 1 & 0 \\ 2 & 2 & \sqrt{3} \end{bmatrix}$$

in such a way that a 2×2 real submatrix results.

$$\left[\textbf{Ans.}\begin{bmatrix} 1 & \vdots & 1 & 0 \\ 2 & \vdots & 2 & \sqrt{3} \end{bmatrix}\right]$$

16. Partition B and calculate AB in partitional form (with A partitioned as shown) if

$$A = \left[\begin{array}{cc|c|cc} 3 & 1 & 4 & 3 & 2 \\ 1 & 0 & 5 & -1 & 0 \\ \hline 0 & 0 & 2 & 0 & -2 \\ -3 & 1 & 1 & 0 & 0 \end{array}\right], \quad B = \begin{bmatrix} 1 & 0 & 3 & -2 & 5 & 0 \\ 0 & 1 & 2 & 0 & 1 & -3 \\ 0 & 0 & 1 & 0 & -1 & 0 \\ 2 & 2 & -2 & 1 & 3 & 0 \\ 0 & 0 & 0 & 0 & 1 & 0 \end{bmatrix} \quad \left[\textbf{Ans.}\ B = \begin{bmatrix} 1 \\ \hline 0 \\ \hline 0 \\ \hline 2 \\ \hline 0 \end{bmatrix}\right]$$

17. Formulate the method for adding partition matrices and verify your method by partitioning the matrices

$$A = \begin{bmatrix} 1 & 3 & -1 \\ 2 & 1 & 0 \\ 2 & -3 & 1 \end{bmatrix} \text{ and } B = \begin{bmatrix} 3 & 2 & 1 \\ -2 & 3 & 1 \\ 4 & 1 & 5 \end{bmatrix}$$

in two different ways and finding their sum.

$$\left[\textbf{Ans.}\ A + B = \left[\begin{array}{cc|c} 4 & 5 & 0 \\ 0 & 4 & 1 \\ \hline 6 & -2 & 6 \end{array}\right] \text{ is one possible answer}\right]$$

18. Let A and B be the following matrices

$$A = \begin{bmatrix} 2 & 1 & 3 & 4 & 2 \\ 1 & 2 & 3 & -1 & 4 \\ 2 & 3 & 2 & 1 & 4 \\ 5 & -1 & 3 & 2 & 6 \\ 3 & 1 & 2 & 4 & 6 \\ 2 & -1 & 3 & 5 & 7 \end{bmatrix} \text{ and } B = \begin{bmatrix} 1 & 2 & 3 & 4 & 1 \\ 2 & 1 & 3 & 2 & -1 \\ 1 & 5 & 4 & 2 & 3 \\ 2 & 1 & 3 & 5 & 7 \\ 3 & 2 & 4 & 6 & 1 \end{bmatrix}.$$

Find AB by partitioning A and B in two different ways and thus verify the result.

19. Distinguish between submatrix and block with examples.

20. Define a quasilower triangular matrix and illustrate by a 6×6 matrix. Is it all true that transpose of a quasilower triangular matrix is a quasiupper triangular matrix.

$$\left[\textbf{Ans.}\text{ Suggestion } \left[\begin{array}{c|cc|ccc} 3 & 0 & 0 & 0 & 0 & 0 \\ \hline 9 & 1 & 2 & 0 & 0 & 0 \\ -7 & 0 & 1 & 0 & 0 & 0 \\ \hline -6 & 8 & 11 & -1 & -3 & 4 \\ -5 & 9 & 12 & 0 & 2 & 5 \\ -7 & 10 & 13 & 7 & 1 & 9 \end{array}\right]_{6\times 6}, \text{ not necessary}\right]$$

21. Distinguish between quasitriangular and triangular upper matrices with examples.

22. Identify the following matrices

$$A = \left[\begin{array}{cc:cc:c} 1 & 2 & 6 & 1 & 1 \\ 3 & 2 & 3 & 2 & 0 \\ \hdashline 0 & 0 & 2 & 4 & 4 \\ 0 & 0 & 1 & 6 & 2 \\ \hdashline 0 & 0 & 0 & 0 & 3 \end{array}\right] \text{ and } B = \left[\begin{array}{ccc:c:cc} 2 & 1 & 3 & 0 & 0 & 0 \\ 4 & 2 & 1 & 0 & 0 & 0 \\ 6 & 0 & 9 & 0 & 0 & 0 \\ \hdashline 2 & 1 & 3 & 2 & 0 & 0 \\ \hdashline 1 & 6 & 9 & 2 & 3 & 4 \\ 3 & 2 & 4 & 4 & 6 & 2 \end{array}\right]$$

as lower or upper triangular supermatrices.

[**Ans.** A and B are upper and lower triangular matrices respectively]

23. Border the following matrix

$$A = \begin{bmatrix} 1 & 2 & 3 \\ 4 & 5 & 6 \\ 7 & 8 & 9 \end{bmatrix} \text{ by the row } \mathbf{U} = [3 \;\; 0 \;\; 5] \text{ and column } \mathbf{V} = [-1 \;\; -2 \;\; -3]^T.$$

$$\textbf{[Ans.} \left[\begin{array}{c|ccc} 0 & 3 & 0 & 5 \\ \hline -1 & 1 & 2 & 3 \\ -2 & 4 & 5 & 6 \\ -3 & 7 & 8 & 9 \end{array}\right]]$$

24. The matrices A and B are partitioned as follows :

$$A = \left[\begin{array}{cc:cc} 1 & 2 & 2 & 4 \\ -3 & 1 & 1 & -1 \end{array}\right] \text{ and } B = \left[\begin{array}{cc} 1 & 0 \\ \hdashline 2 & 1 \\ 0 & 1 \end{array}\right]. \text{ Find AB.}$$

$$\textbf{[Ans.} \begin{bmatrix} 9 & 6 \\ -6 & 0 \end{bmatrix}]$$

25. Obtain the direct sum of

$$A = \begin{bmatrix} 1 & 2 \\ 3 & 5 \end{bmatrix} \text{ and } B = \begin{bmatrix} 1 & 0 \\ 2 & 1 \end{bmatrix}.$$

$$\textbf{[Ans. } A \oplus B = \begin{bmatrix} 1 & 2 \\ 3 & 5 \end{bmatrix} \oplus \begin{bmatrix} 1 & 0 \\ 2 & 1 \end{bmatrix} = \begin{bmatrix} 1 & 2 & 0 & 0 \\ 3 & 5 & 0 & 0 \\ 0 & 0 & 1 & 0 \\ 0 & 0 & 2 & 1 \end{bmatrix}]$$

26. Verify that the sum and the product of two lower triangular matrices of the same order are lower triangular for the matrices

$$A = \begin{bmatrix} 1 & 0 & 0 \\ 3 & 2 & 0 \\ 1 & 1 & 1 \end{bmatrix} \text{ and } B = \begin{bmatrix} 3 & 0 & 0 \\ 2 & 1 & 0 \\ 6 & 0 & 4 \end{bmatrix}.$$

27. Determine which of the following matrices are diagonal, quasidiagonal, triangular (upper or lower) supermatrices.

$$A = \begin{bmatrix} 1 & 3 & 0 & 0 \\ 2 & 4 & 0 & 0 \\ 0 & 0 & 6 & 4 \\ 0 & 0 & 2 & 3 \end{bmatrix}, \quad B = \begin{bmatrix} 2 & 0 & 0 & 0 \\ 1 & 3 & 0 & 0 \\ 4 & 5 & 6 & 0 \\ 0 & 9 & 0 & 4 \end{bmatrix}, \quad C = \left[\begin{array}{ccc|ccc} 1 & 2 & 3 & 1 & 6 & 2 \\ 4 & 5 & 6 & 2 & 0 & 0 \\ 9 & 2 & 1 & 3 & 0 & 0 \\ \hline 0 & 0 & 0 & 3 & 0 & 0 \\ 0 & 0 & 0 & 0 & 4 & 0 \\ 0 & 0 & 0 & 0 & 2 & 2 \end{array}\right],$$

$$D = \begin{bmatrix} 3 & 2 & 1 & 0 & 0 \\ 3 & 4 & 6 & 0 & 0 \\ 0 & 0 & 0 & 2 & 6 \\ 0 & 0 & 0 & 3 & 9 \end{bmatrix}, \quad E = \begin{bmatrix} 0 & 0 & 0 & 0 \\ 0 & 0 & 0 & 0 \\ 0 & 0 & 0 & 0 \\ 0 & 0 & 0 & 1 \end{bmatrix}.$$

[**Ans.** A is diagonal supermatrix,
B is lower triangular supermatrix,
C is upper triangular supermatrix,
D is diagonal supermatrix,
E is diagonal supermatrix]

CHAPTER FIVE

OTHER MATRIX OPERATIONS

Chapter summary: This chapter deals with matrix sums and products of different kinds different from those belong to chapter two. Direct sum, Kronecker sum and product, Jordan product, Lie product, Khatri-Rao product and Hadamard product of matrices; cojoint, construction of permutation matrix from matrix units and Hadamard matrix; factorization of orthogonal matrix are the subject matters of this chapter. The chapter ends with the vec operator and its properties and commutation of Kronecker product.

5.1 Direct sum of matrices

Let $A_1, A_2, \cdots, A_r$ be square matrices of orders $m_1, m_2, \cdots, m_r$ respectively. The diagonal

$$\text{matrix diag}(A_1, A_2, \cdots, A_r) = \begin{bmatrix} A_1 & 0 & \cdots & 0 \\ 0 & A_2 & \cdots & 0 \\ \vdots & \vdots & \ddots & \vdots \\ 0 & 0 & \cdots & A_r \end{bmatrix}_{(m_1+m_2+\cdots+m_r)}$$

is called the *direct sum* of the square matrices $A_1, A_2, \cdots, A_r$ and is expressed by $(A_1 \oplus A_2 \oplus \cdots \oplus A_r)$ of order $(m_1 + m_2 + \cdots + m_r)$. It is also called *the block diagonalized form*. For example

$$A \oplus B = \begin{bmatrix} 1 & -2 & 0 & 0 & 0 \\ 2 & 3 & 0 & 0 & 0 \\ 0 & 0 & 1 & 0 & 1 \\ 0 & 0 & 2 & 3 & 0 \\ 0 & 0 & -1 & 1 & -1 \end{bmatrix}_{5\times 5},$$

where $A = \begin{bmatrix} 1 & -2 \\ 2 & 3 \end{bmatrix}$ and $B = \begin{bmatrix} 1 & 0 & 1 \\ 2 & 3 & 0 \\ -1 & 1 & -1 \end{bmatrix}$.

N.B. Authors who do not introduce both direct sum and Kronecker sum, use $\oplus$ for denoting either sum. Some of them use Kronecker sum for direct sum and vice-versa. So, this book uses "†" to denote Kronecker sum (sec. 5.2.1).

5.1.1 Properties of direct sum : Direct sum of matrices possesses the following algebraic properties.

1. Commutative property : Commutative property does *not hold* for direct sum of the square matrices A and B of orders m and n respectively. Because by definition.

$$A \oplus B = \begin{bmatrix} A & 0 \\ 0 & B \end{bmatrix} \text{ and } B \oplus A = \begin{bmatrix} B & 0 \\ 0 & A \end{bmatrix}$$

$\therefore\ A \oplus B \neq B \oplus A.$

2. Associative : $(A \oplus B) \oplus C = A \oplus (B \oplus C)$.

Since $A \oplus B = \begin{bmatrix} A & 0 \\ 0 & B \end{bmatrix} = D$ (say)

$$\therefore\ (A \oplus B) \oplus C = D \oplus C = \begin{bmatrix} D & 0 \\ 0 & C \end{bmatrix} = \begin{bmatrix} A & 0 & 0 \\ 0 & B & 0 \\ 0 & 0 & C \end{bmatrix}.$$

Similarly, $B \oplus C = \begin{bmatrix} B & 0 \\ 0 & C \end{bmatrix} = E$ (say)

$$\therefore\ A \oplus (B \oplus C) = A \oplus E = \begin{bmatrix} A & 0 \\ 0 & E \end{bmatrix} = \begin{bmatrix} A & 0 & 0 \\ 0 & B & 0 \\ 0 & 0 & C \end{bmatrix}.$$

$\therefore\ (A \oplus B) \oplus C = A \oplus (B \oplus C)$

3. Mixed sum : $(A + B) \oplus (C + D) = (A \oplus C) + (B \oplus D)$, if corresponding matrix additions are conformable.

By definition of direct sum and matrix addition,

$$(A + B) \oplus (C + D) = \begin{bmatrix} A + B & 0 \\ 0 & C + D \end{bmatrix}$$

$$= \begin{bmatrix} A & 0 \\ 0 & C \end{bmatrix} + \begin{bmatrix} B & 0 \\ 0 & D \end{bmatrix} = (A \oplus C) + (B \oplus D)$$

4. Matrix multiplication of direct sum : $(A \oplus B)(C \oplus D) = (AC) \oplus (BD)$ if each matrix multiplication is conformable.

$$(A \oplus B)(C \oplus D) = \begin{bmatrix} A & 0 \\ 0 & B \end{bmatrix} \begin{bmatrix} C & 0 \\ 0 & D \end{bmatrix},$$

since matrix multiplication is conformable.

$$= \begin{bmatrix} AC & 0 \\ 0 & BD \end{bmatrix} = (AC) \oplus (BD).$$

5. Transposition : $(A \oplus B)^T = A^T \oplus B^T$.

Since $A \oplus B = \begin{bmatrix} A & 0 \\ 0 & B \end{bmatrix}$

$$\therefore\ (A \oplus B)^T = \begin{bmatrix} A & 0 \\ 0 & B \end{bmatrix}^T = \begin{bmatrix} A^T & 0 \\ 0 & B^T \end{bmatrix} = A^T \oplus B^T$$

Let $A = \begin{bmatrix} a_{11} & a_{12} \\ a_{21} & a_{22} \end{bmatrix}$, $B = \begin{bmatrix} b_{11} & b_{12} \\ b_{21} & b_{22} \end{bmatrix}$.

$$\therefore \quad A \oplus B = \begin{bmatrix} a_{11} & a_{12} & 0 & 0 \\ a_{21} & a_{22} & 0 & 0 \\ 0 & 0 & b_{11} & b_{12} \\ 0 & 0 & b_{21} & b_{22} \end{bmatrix}$$

$$\therefore \quad (A \oplus B)^T = \begin{bmatrix} a_{11} & a_{21} & 0 & 0 \\ a_{12} & a_{22} & 0 & 0 \\ 0 & 0 & b_{11} & b_{21} \\ 0 & 0 & b_{12} & b_{22} \end{bmatrix} = \begin{bmatrix} A^T & 0 \\ 0 & B^T \end{bmatrix} = A^T \oplus B^T.$$

6. Inverse : If A and B are nonsingular, then so $A \oplus B$ and hence

$$(A \oplus B)^{-1} = A^{-1} \oplus B^{-1}$$

Since $|A| \neq 0 \neq |B|$. Therefore,

$$|A \oplus B| = \begin{vmatrix} A & 0 \\ 0 & B \end{vmatrix} = |A||B| \neq 0 \quad \text{(cf. chapter six)}$$

$\therefore$ $A \oplus B$ is nonsingular.

Now $\quad A \oplus B = \begin{bmatrix} A & 0 \\ 0 & B \end{bmatrix}$

$$\therefore \quad (A \oplus B)^{-1} = \begin{bmatrix} A & 0 \\ 0 & B \end{bmatrix}^{-1} = \begin{bmatrix} A^{-1} & 0 \\ 0 & B^{-1} \end{bmatrix} = A^{-1} \oplus B^{-1} \quad \text{(cf. chapter eight)}$$

This may extended to any finite number of direct sum of matrices such as

$$(A_1 \oplus A_2 \oplus \cdots \oplus A_k)^{-1} = A_1^{-1} \oplus A_2^{-1} \oplus \cdots \oplus A_k^{-1}$$

7. Rank : $R(A \oplus B) = R(A) \oplus R(B)$ i.e. rank of the direct sum of matrices is equal to the sum of the individual ranks (cf. chapter ten).

8. Let $M = A_1 \oplus A_2 \oplus \cdots \oplus A_k$ be a direct sum of k square matrices. Then it is easy to see that

a. $\det M = (\det A_1)(\det A_2) \cdots (\det A_k)$ (cf.chapter six)

b. $\operatorname{Tr} M = \operatorname{Tr} A_1 \oplus \operatorname{Tr} A_2 \oplus \cdots \oplus \operatorname{Tr} A_k$

5.2 Kronecker product of matrices

Kronecker product is also called *direct product* or *tensor product* or *outer product.* It is not restricted to conformability of matrix multiplication. It is a general procedure of multiplying any two matrices of any dimensions. It maps two matrices $A = [a_{ij}]$ and $B = [b_{ij}]$ into a matrix $C = [a_{ij}b_{ij}]$. It will be found useful in group theory, particle physics, stability theory of differential equations, multiple equation and regression estimation etc.

Let $A_{(m,n)} = [a_{ij}]_{(m,n)}$ and $B_{(p,q)} = [b_{ij}]_{(p,q)}$ be two rectangular matrices. Then the Kronecker product of A and B, denoted by $A \oplus B$ is defined as the partitioned matrix

$$A \otimes B = \begin{bmatrix} a_{11}B & a_{12}B & \cdots & a_{1n}B \\ a_{21}B & a_{22}B & \cdots & a_{2n}B \\ \vdots & \vdots & \vdots & \vdots \\ a_{m1}B & a_{m2}B & \cdots & a_{mn}B \end{bmatrix}_{(mp,nq)}$$

It has *mn* blocks. Thus (i, j)th block is the block $a_{ij}B$ of order (p, q).

Let $A = \begin{bmatrix} a_{11} & a_{12} \\ a_{21} & a_{22} \end{bmatrix}$, and $B = \begin{bmatrix} b_{11} & b_{12} \\ b_{21} & b_{22} \end{bmatrix}$ be two 2×2 matrices. Then by definition.

$$A \otimes B = \begin{bmatrix} a_{11}B & a_{12}B \\ a_{21}B & a_{22}B \end{bmatrix} = \begin{bmatrix} a_{11}\begin{bmatrix} b_{11} & b_{12} \\ b_{21} & b_{22} \end{bmatrix} & a_{12}\begin{bmatrix} b_{11} & b_{12} \\ b_{21} & b_{22} \end{bmatrix} \\ a_{21}\begin{bmatrix} b_{11} & b_{12} \\ b_{21} & b_{22} \end{bmatrix} & a_{22}\begin{bmatrix} b_{11} & b_{12} \\ b_{21} & b_{22} \end{bmatrix} \end{bmatrix}$$

$$= \left[\begin{array}{cc|cc} a_{11}b_{11} & a_{11}b_{12} & a_{12}b_{11} & a_{12}b_{12} \\ a_{11}b_{21} & a_{11}b_{22} & a_{12}b_{21} & a_{12}b_{22} \\ \hline a_{21}b_{11} & a_{21}b_{12} & a_{22}b_{21} & a_{22}b_{12} \\ a_{21}b_{21} & a_{21}b_{22} & a_{22}b_{21} & a_{22}b_{22} \end{array}\right]$$

Note the difference between the matrix product and Kronecker product. Matrix product AB requires equality of the number of columns in A and the number of rows in B while the Kronecker product is free from such restriction.

5.2.1 Kronecker sum (based on Kronecker product) : The Kronecker sum of two square matrices $A_{n \times n}$ and $B_{m \times m}$ is defined by

$A \dagger B = A \otimes I_m + I_n \otimes B$, which is an $nm \times nm$ matrix.

Thus if $A = \begin{bmatrix} 1 & -2 \\ 2 & 3 \end{bmatrix}$ and $B = \begin{bmatrix} 1 & 0 & 1 \\ 2 & 3 & 0 \\ -1 & 1 & -1 \end{bmatrix}$

Then $A \dagger B = \begin{bmatrix} 1 & -2 \\ 2 & 3 \end{bmatrix} \otimes \begin{bmatrix} 1 & 0 & 0 \\ 0 & 1 & 0 \\ 0 & 0 & 1 \end{bmatrix} + \begin{bmatrix} 1 & 0 \\ 0 & 1 \end{bmatrix} \otimes \begin{bmatrix} 1 & 0 & 1 \\ 2 & 3 & 0 \\ -1 & 1 & -1 \end{bmatrix}$

$$= \left[\begin{array}{ccc|ccc} 1 & 0 & 0 & -2 & 0 & 0 \\ 0 & 1 & 0 & 0 & -2 & 0 \\ 0 & 0 & 1 & 0 & 0 & -2 \\ \hline 2 & 0 & 0 & 3 & 0 & 0 \\ 0 & 2 & 0 & 0 & 3 & 0 \\ 0 & 0 & 2 & 0 & 0 & 3 \end{array}\right] + \left[\begin{array}{ccc|ccc} 1 & 0 & 1 & 0 & 0 & 0 \\ 2 & 3 & 0 & 0 & 0 & 0 \\ -1 & 1 & -1 & 0 & 0 & 0 \\ \hline 0 & 0 & 0 & 1 & 0 & 1 \\ 0 & 0 & 0 & 2 & 3 & 0 \\ 0 & 0 & 0 & -1 & 1 & -1 \end{array}\right]$$

$$= \begin{bmatrix} 2 & 0 & 1 & -2 & 0 & 0 \\ 2 & 4 & 0 & 0 & -2 & 0 \\ -1 & 1 & 0 & 0 & 0 & -2 \\ 2 & 0 & 0 & 4 & 0 & 1 \\ 0 & 2 & 0 & 2 & 6 & 0 \\ 0 & 0 & 2 & -1 & 1 & 2 \end{bmatrix}.$$

Note that Kronecker sum is *not commutative*. The definition is based on Kronecker product.

5.2.2 Kronecker product of two column vectors : Let $\mathbf{x}=\begin{bmatrix}x_1 & x_2 & \cdots & x_n\end{bmatrix}^{\mathrm{T}}$ and $\mathbf{y}=\begin{bmatrix}y_1 & y_2 & \cdots & y_n{}^{\mathrm{T}}\end{bmatrix}$ be two column vectors. Then by definition of the Kronecker product,

we have,
$$\mathbf{x}\otimes\mathbf{y}=\begin{bmatrix}x_1\mathbf{y}\\ x_2\mathbf{y}\\ \vdots\\ x_n\mathbf{y}\end{bmatrix}_{nm\times 1}=\begin{bmatrix}x_1y_1\\ \vdots\\ x_1y_m\\ x_2y_1\\ \vdots\\ x_2y_m\\ \vdots\\ x_ny_1\\ \vdots\\ x_ny_m\end{bmatrix}_{nm\times 1}$$

Theorem. If $\mathbf{x}$ and $\mathbf{y}$ are column vectors, not necessarily of the same order, then $\mathbf{x}^{\mathrm{T}}\otimes\mathbf{y}=\mathbf{y}\mathbf{x}^{\mathrm{T}}=\mathbf{y}\otimes\mathbf{x}^{\mathrm{T}}$.

Proof. Let $\mathbf{x}^{\mathrm{T}}=\begin{bmatrix}x_1 & x_2 & \cdots & x_n\end{bmatrix}$ and $\mathbf{y}=\begin{bmatrix}y_1\\ y_2\\ \vdots\\ y_n\end{bmatrix}$

$$\therefore\quad \mathbf{x}^{\mathrm{T}}\otimes\mathbf{y}=\begin{bmatrix}x_1\mathbf{y} & x_2\mathbf{y} & \cdots & x_n\mathbf{y}\end{bmatrix}$$

$$=\begin{bmatrix}x_1y_1 & x_2y_1 & \cdots & x_ny_1\\ x_1y_2 & x_2y_2 & \cdots & x_ny_2\\ \vdots & \vdots & \vdots & \vdots\\ x_1y_m & x_2y_m & \cdots & x_ny_m\end{bmatrix}$$ is an $m\times n$ matrix.

$$\mathbf{y}\otimes\mathbf{x}^{\mathrm{T}}=\begin{bmatrix}y_1\mathbf{x}^{\mathrm{T}}\\ y_2\mathbf{x}^{\mathrm{T}}\\ \vdots\\ y_m\mathbf{x}^{\mathrm{T}}\end{bmatrix}=\begin{bmatrix}y_1x_1 & y_1x_2 & \cdots & y_1x_n\\ y_2x_1 & y_2x_2 & \cdots & y_2x_n\\ \vdots & \vdots & \vdots & \vdots\\ y_mx_1 & y_mx_2 & \cdots & y_mx_n\end{bmatrix}$$
$$=\mathbf{x}^{\mathrm{T}}\otimes\mathbf{y}$$

Also
$$\mathbf{y}\mathbf{x}^{\mathrm{T}}=\begin{bmatrix}y_1\\ y_2\\ \vdots\\ y_m\end{bmatrix}_{m\times 1}\begin{bmatrix}x_1 & x_2 & \cdots & x_n\end{bmatrix}_{1\times n}=\begin{bmatrix}y_1x_1 & y_1x_2 & \cdots & y_1x_n\\ y_2x_1 & y_2x_2 & \cdots & y_2x_n\\ \vdots & \vdots & \vdots & \vdots\\ y_mx_1 & y_mx_2 & \cdots & y_mx_n\end{bmatrix}_{m\times n}$$

$$\therefore\quad \mathbf{x}^{\mathrm{T}}\otimes\mathbf{y}=\mathbf{y}\mathbf{x}^{\mathrm{T}}=\mathbf{y}\otimes\mathbf{x}^{\mathrm{T}}$$

Cor. 1 If **a** and **b** are $n\times 1$ and $1\times r$ vectors then $(\mathbf{a}\otimes\mathbf{b})^{\mathrm{T}}=\mathbf{a}\mathbf{b}^{\mathrm{T}}$.

5.2.3 Properties of Kronecker products : The Kronecker product possesses the following algebraic properties as in the ordinary product:

$$A\otimes 0=0\otimes B=0$$

$$1 \otimes 1 = 1$$

Anticommutative properties :

1. **Homogeneity :** If α is a scalar then $A \oplus (\alpha B) = \alpha(A \oplus B) = (\alpha A) \oplus B$.
2. **Commutative :** the Kronecker product is not commutative,

   i.e. $A \oplus B \neq B \oplus A$.
3. **Distributive :** If B and C are conformable for addition, then

   $A \otimes (B + C) = A \otimes B + A \otimes C$, (left distribution)

   $(B + C) \otimes A = B \otimes A + C \otimes A$, (right distribution)
4. **Associative :** $A \otimes B \otimes C = A \otimes (B \otimes C) = (A \otimes B) \otimes C$.
5. **Transposition :** $(A \otimes B)^T = A^T \otimes B^T$. This is different from matrix multiplication.
6. **Conjugation :** $\overline{A \otimes B} == \overline{A} \otimes \overline{B}$.
7. **Hermitian conjugation :** $(A \otimes B)^* = A^* \otimes B^*$, different from the matrix product.
8. **Mixed product rule :** If AC and BD exist, then

   $$(A \otimes B)(C \otimes D) = (AC) \otimes (BD).$$
9. **Inverse :** If A^{-1} and B^{-1} exist then $(A \otimes B)^{-1} = A^{-1} \otimes B^{-1}$.
10. **Rank :** $\text{Rank}(A \otimes B) = \text{Rank}(A)\text{Rank}(B)$ (cf. chapter ten).
11. **Trace :** $\text{Tr}(A \otimes B) = \text{Tr}A\, \text{Tr}B$.
12. **Two column vectors a and b, not necessarily of the same order :**

    $$\mathbf{a}^T \otimes \mathbf{b} = \mathbf{b}\mathbf{a}^T = \mathbf{b} \otimes \mathbf{a}^T.$$
13. **Power :** Let A be an $n \times k$ matrix. Then the Kronecker square is defined as

    $$A^{[2]} = A \otimes A \quad \text{and} \quad A^{[k]} = A \otimes A^{[k-1]}, \ k = 3, 4, \cdots$$

    $(AB)^{[k]} = A^{[k]} B^{[k]}$, if A and B are square matrices.
14. **Determinant :** $\det(A_{m \times m} \otimes B_{n \times n}) = (\det A)^m (\det B)^n$.

Proof.

1. The (i, j)th block of $A \otimes (\alpha B) = [a_{ij}(\alpha B)]$

$$= \alpha[a_{ij}B] = \alpha\,[(i, j)\text{th block of } A \otimes B].$$

$\therefore \qquad A \otimes (\alpha B) = \alpha(A \otimes B)$.

Also (i, j)th block of $\alpha(A \otimes B) = \alpha\,[(i, j)\text{th block of } A \otimes B]$

$$= \alpha[a_{ij}B] = [(\alpha a_{ij})B] = (i, j)\text{th block of } (\alpha a_{ij})B.$$

$\therefore \qquad \alpha(A \otimes B) = (\alpha A) \otimes B$

$\therefore \qquad A \otimes (\alpha B) = \alpha(A \otimes B) = (\alpha A) \otimes B$.

2. The (i, j)th block of $A \otimes B = [a_{ij} B]$.

The (i, j)th block of $B \otimes A = [b_{ij} A]$.

But $\qquad [a_{ij} B] \neq [b_{ij} A]$ for $A \neq B$.

$\therefore \qquad A \otimes B \neq B \otimes A$.

Hence the Kronecker product is not commutative.

3. The (i, j)th block of $A \otimes (B + C)$ is

$$[a_{ij}(B + C)] = [a_{ij} B] + [a_{ij} C]$$

$$= [(i, j)\text{th block of } A \otimes B] + [(i, j)\text{th block of } A \otimes C].$$

$\therefore\quad A \otimes (B + C) = (A \otimes B) + (A \otimes C)$

Similarly, $(B + C) \otimes A = B \otimes A + C \otimes A$. Note that in both distributions order of arrangements of matrices *must* be maintained.

4. The (i, j)th block of $A \otimes (B \otimes C)$ is

$$[a_{ij}(B \otimes C)] = [(a_{ij}B) \otimes C] = [a_{ij}(B \otimes C)]$$
$$= [(a_{ij}B) \otimes C]$$
$$= (k, l)\text{th block of } \{(a_{ij}B) \otimes C\}$$

$$= \left[a_{ij}b_{kl}C\right] \tag{5.1}$$

Now (i, j)th block of $A \otimes B = [a_{ij}\ B]$

$\therefore$ (k,l)th block of $(A \otimes B) \otimes C = \left[a_{ij}b_{kl}C\right]$ (5.2)

$\therefore$ matrices on R.H.S of (5.1) and (5.2) equal. Hence it is proved that

$$A \otimes (B \otimes C) = (A \otimes B) \otimes C\ .$$

5. The (i, j)th block of $(A \otimes B)^{-T}$ is $[a_{ji}\ B^T]$.

$\therefore$ $(A \otimes B)^T = [a_{ji}\ B^T] = A^T \otimes B^T$ by definition.

6,7. Left for reader. [Hints : Proceed as in 5.]

8. Let $A = [a_{ij}]_{m\times n}$, $B = [b_{ij}]_{p\times q}$, $C = [c_{ij}]_{n\times r}$, $D = [d_{ij}]_{q\times s}$.

The orders of matrices are chosen in such a way that the matrix product AC and BD are conformable. Now by definition of the Kronecker product,

$$P = A \otimes B = \begin{bmatrix} a_{11}B & a_{12}B & \cdots & a_{1n}B \\ \vdots & \vdots & \vdots & \vdots \\ a_{i1}B & a_{i2}B & \cdots & a_{in}B \\ \vdots & \vdots & \vdots & \vdots \\ a_{m1}B & a_{m2}B & \cdots & a_{mn}B \end{bmatrix}_{mp\times nq}$$

$$Q = C \otimes D = \begin{bmatrix} c_{11}D & c_{12}D & \cdots & c_{1r}D \\ \vdots & \vdots & \vdots & \vdots \\ c_{i1}D & c_{i2}D & \cdots & c_{ir}D \\ \vdots & \vdots & \vdots & \vdots \\ c_{n1}D & c_{n2}D & \cdots & c_{nr}D \end{bmatrix}_{nq\times rs}$$

Now the PQ matrix product is defined. So the (i, j)th element of PQ is

$$p_{ik}\,q_{kj} = (i\text{th row of P})\,(j\text{th column of Q})$$
$$= \left(a_{i1}c_{1j} + a_{i2}c_{2j} + \cdots + a_{in}c_{nj}\right)BD$$
$$= ((i, j)\text{th element of AC})\ BD$$
$$= (AC)_{ij}\,BD$$

$$\therefore\quad PQ = (A \otimes B)(C \otimes D)$$

$$=\begin{bmatrix}(AC)_{11}BD & (AC)_{21}BD & \cdots & (AC)_{1r}BD\\ \vdots & \vdots & \vdots & \vdots\\ (AC)_{m1}BD & (AC)_{m2}BD & \cdots & (AC)_{mr}BD\end{bmatrix}$$

$$=(AC)\otimes(BD) \quad \text{which is of order } mp\times rs.$$

9. See chapter eight.

10. See chapter ten.

11. Let A = $[a_{ij}]$ be an n square matrix. Since

$$A\otimes B=\begin{bmatrix}a_{11}B & a_{12}B & \cdots & a_{1n}B\\ a_{21}B & a_{22}B & \cdots & a_{2n}B\\ \vdots & \vdots & \vdots & \vdots\\ a_{n1}B & a_{n2}B & \cdots & a_{nn}B\end{bmatrix}.$$

Therefore, $\text{Tr}(A\otimes B) = \text{Tr}\, a_{11}B + \text{Tr}\, a_{22}B + \cdots + \text{Tr}\, a_{nn}B$

$$= (a_{11}+a_{22}+\cdots+a_{nn})\,\text{Tr B}$$

$$= \text{Tr A Tr B}.$$

$\therefore$ trace of a Kronecker product is equal to the product of traces of A and B taken in order.

12. See sec. 5.2.2

13. Kronecker square denoted by $A^{[2]}$ of a matrix $A_{m\times n}$ is defined from the Kronecker product as $A^{[2]} = A\otimes A$, which is of order $m^2\times n^2$. Therefore, the Kronecker cube is then

$$A^{[3]} = A\otimes A^{[2]} = A\otimes A\otimes A \text{ of order } m^3\times n^3.$$

Proceeding is this way,

$$A^{[k]} = A\otimes A^{[k-1]} = A\otimes A\otimes \cdots \otimes A \quad (k\text{times})$$

The matrix $A^{[k]}$ is of order $m^k\times n^k$.

Note that to distinguish between powers in the ordinary matrix product and that in the Kronecker product, the powers in the Kronecker product is written within square brackets.

Now, if A and B are such that the ordinary product AB is defined. Then kth power in the Kronecker product is $(AB)^{[k]} = A^{[k]}\,B^{[k]}$.

This can be established from property 8 after replacing B by A, C and D, both by B, then

$$(A\otimes B)(C\otimes D) = (AC)\otimes(BD)$$

becomes

$$(A\otimes A)(B\otimes B) = (AB)\otimes(AB)$$

or $$A^{[2]}B^{[2]} = (AB)^{[2]}$$

$\therefore$ $$(AB)^{[3]} = (AB)\otimes(AB)\otimes(AB)$$

$$= (AB)\otimes[(A\otimes A)(B\otimes B)], \text{ by prop. 8}$$

$$= (AB)\otimes(A^{[2]}B^{[2]})$$

$$= (A\otimes A^{[2]})(B\otimes B^{[2]}) = A^{[3]}B^{[3]}$$

Proceeding as above

$$(AB)^{[k]} = (A)^{[k]}(B)^{[k]}$$

14. See chapter thirteen.

5.3 Jordan product of matrices and its properties

The name of the product bears the name of physicist P. Jordan, a founder of modern quantum mechanics. The Jordan product of two square matrices is the arithmetic mean of the Cayley products AB and BA and is denoted by A*B; i.e.,

$$A*B=\tfrac{1}{2}(AB+BA).$$

Ex. Find

$$\begin{bmatrix}1&2&1\\3&4&-1\\0&0&2\end{bmatrix}*\begin{bmatrix}-1&-2&1\\0&3&4\\1&-1&0\end{bmatrix}.$$

Let $\quad A=\begin{bmatrix}1&2&1\\3&4&-1\\0&0&2\end{bmatrix}$ and $B=\begin{bmatrix}-1&-2&1\\0&3&4\\1&-1&0\end{bmatrix}.$

Now $\quad AB=\begin{bmatrix}1&2&1\\3&4&-1\\0&0&2\end{bmatrix}\begin{bmatrix}-1&-2&1\\0&3&4\\1&-1&0\end{bmatrix}=\begin{bmatrix}0&3&9\\-4&7&19\\2&-2&0\end{bmatrix}$

and $\quad BA=\begin{bmatrix}-1&-2&1\\0&3&4\\1&-1&0\end{bmatrix}\begin{bmatrix}1&2&1\\3&4&-1\\0&0&2\end{bmatrix}=\begin{bmatrix}-7&-10&3\\9&12&5\\-2&-2&2\end{bmatrix}$

By definition $\quad A*B=\tfrac{1}{2}(AB+BA)$

$$=\tfrac{1}{2}\begin{bmatrix}-7&-7&12\\5&19&24\\0&-4&2\end{bmatrix}=\begin{bmatrix}-\frac{7}{2}&-\frac{7}{2}&6\\\frac{5}{2}&\frac{19}{2}&12\\0&-2&1\end{bmatrix}$$

The product obeys commutative law and, in general, nonassociative. This product obeys the following laws:

1. Commutative : $\quad A*B=B*A.$

Proof. Since $A*B=\tfrac{1}{2}(AB+BA)=\tfrac{1}{2}(BA+AB)$

$$=B*A.$$

2. Homogeneity : $(kA)*B=A*(kB)=k(A*B)$ for scalar k.

Proof. $\quad (kA)*B=\tfrac{1}{2}\{(kA)B+B(kA)\}=\tfrac{1}{2}(k\{AB+BA\})$

$$=\tfrac{k}{2}(AB+BA)=k(A*B)$$

Similarly, $A*(kB)=\tfrac{1}{2}\{A(kB)+(kB)A\}$

$$=k(A*B)$$

3. Distributive :

(i) $A*(B+C)=(A*B)+(A*C)$

(ii) $(B+C)*A=B*A+C*A$, if B and C are conformable for addition.

Proof. (i) $A*(B+C)=\tfrac{1}{2}\{A(B+C)+(B+C)A\}$, by definition.

$$=\tfrac{1}{2}\{AB+AC+BA+CA\}$$

$$=\tfrac{1}{2}(AB+BA)+\tfrac{1}{2}(AC+CA)$$
$$=A*B+A*C$$

(ii) Similar to **(i)**.

4. Special associative law : $A*(B*A^2)=(A*B)*A^2$, where $A^2=A*A=AA$.

Proof. $A*(B*A^2)=\tfrac{1}{2}[A(B*A^2)+(B*A^2)A]$, by definition.

$$=\tfrac{1}{2}[A\tfrac{1}{2}(BA^2+A^2B)+\tfrac{1}{2}(BA^2+A^2B)A]$$
$$=\tfrac{1}{4}[ABA^2+A^3B+BA^3+A^2BA]$$

Now R.H.S. $=\tfrac{1}{2}[(A*B)A^2+A^2(AB)]$

$$=\tfrac{1}{2}[\tfrac{1}{2}(AB+BA)A^2+A^2\tfrac{1}{2}(AB+BA)]$$
$$=\tfrac{1}{4}[ABA^2+BA^3+A^3B+A^2BA]$$

$\therefore$ $A*(B*A^2)=(A*B)*A^2$.

5.4 Lie product (commutator) of matrices and its properties

This product again bears the name of Norwegian mathematician M. Sophus Lie (1842–1899) who studied continuous groups and so this product has its uses in Lie algebra of groups. It is defined by the difference of the two Cayley products AB and BA for two square matrices A and B. It is denoted by A×B ; i.e.

$$A\times B=AB-BA.$$

It is also known as *matrix cross product*. This product is noncommutative and nonassociative in general. This product appears in connection with the mathematical study of quantum theory and is well known as the *commutator* with notation [A, B]. The resulting matrix is also called the *commutating matrix*.

Properties : Let A, B, C be square C-matrices and k an arbitrary complex number, then the following properties can be easily established.

1. Noncommutativity : $A\times B=-B\times A$

2. Homogeneity : $(kA)\times B=A\times(kB)=k(A\times B)$

3. Distributive :

(i) $A\times(B+C)=(A\times B)+(A\times C)$

(ii) $(B+C)\times A=(B\times A)+(C\times A)$

4. Jacobi identity : $A\times(B\times C)+B\times(C\times A)+C\times(A\times B)=0$.

5. Skew symmetric : If A and B are skew symmetric, then A×B is skew symmetric.

Proof. Here $A=-A^T$ and $B=-B^T$ (given). Now

$$(A\times B)^T=(AB-BA)^T=(AB)^T-(BA)^T$$
$$=B^TA^T-A^TB^T=-(A^TB^T-B^TA^T)$$
$$=-A^T\times B^T=-A\times B$$

$\therefore$ A×B is skew symmetric.

Cor. Anticommutator : The sum of the Cayley products AB and BA is known as anticommutator of the square matrices A and B. This is denoted by

$$\{A,B\}=AB+BA.$$

6. $A\times A=0$.

7. $A\times I=I\times A=0$.

Ex. Prove that $[A,\{B,C\}] = \{[A,B],C\} + \{[A,C],B\}$.

Here, $[A,\{B,C\}] = A\{B,C\} - \{B,C\}A$

$= A(BC + CB) - (BC + CB)A$

$= ABC + ACB - BCA - CBA$

$= ABC - BAC + CAB - CBA + ACB - CAB + BAC - BCA$

$= (AB - BA)C + C(AB - BA) + (AC - CA)B + B(AC - CA)$

$= \{[A,B],C\} + \{[A,C],B\}$.

5.4.1 Relation between Jordan and Lie matrix product : For any square C-matrices A, B, C the following relations hold true for Jordan and Lie matrix products:

(i) $A \times (B * B) = 2[(A \times B) * B]$

(ii) $A\times (B \times C) = 4[(A * B) * C - (A * C) * B]$

Proof.

(i) L.H.S. $= A\times(B * B) = A(B * B) - (B * B)A = AB^2 - B^2A$

R.H.S. $= 2.\frac{1}{2}\{(A \times B)B + B(A \times B)\} = (AB - BA)B + B(AB - BA)$

$= AB^2 - BAB + BAB - B^2A = AB^2 - B^2A$.

$\therefore$ L.H.S. = R.H.S.

(ii) L.H.S. $= A\times(B\times C)$

$= A(B\times C) - (B\times C)\times A$

$= A(BC - CB) - (BC - CB)A$

$= ABC - ACB - BCA + CBA$

R.H.S. $= 4[(A * B) * C - (A * C) * B]$

$= 4\left[\frac{1}{2}\{(A * B)C + C(A * B)\} - \frac{1}{2}\{(A * C)B + B(A * C)\}\right]$

$= 2\left[\frac{1}{2}(AB + BA)C + C\frac{1}{2}(AB + BA) - \frac{1}{2}(AC + CA)B - B\frac{1}{2}(AC + CA)\right]$

$= ABC + BAC + CAB + CBA - ACB - CAB - BAC - BCA$

$= ABC + CBA - ACB - BCA$

$\therefore$ L.H.S. = R.H.S.

5.4.2 Relation among Cayley, Jordan and Lie matrix product : For square C-matrices A and B, $AB = A*B + \frac{1}{2}(A\times B)$.

Proof. By definition $A*B = \frac{1}{2}(AB + BA)$ (5.3)

and $A\times B = AB - BA$

or $\frac{1}{2}(A \times B) = \frac{1}{2}(AB - BA)$ (5.4)

Add (5.3) and (5.4), $AB = A*B + \frac{1}{2}(A\times B)$.

5.5 Khatri-Rao product and its properties

In 1968 C.G Khatri and C. R. Rao defined a special product known as the Khatri-Rao product.

Definition. If G is a $t\times u$ and F is a $q\times u$ matrix having same number of columns, then the Khatri-Rao product denoted by $F\Theta G$ is

$$F\Theta G = \left[F_{.1} \otimes G_{.1} : F_{.2} \otimes G_{.2} : \cdots : F_{.u} \otimes G_{.u}\right].$$

It is a $qt\times u$ matrix. "Encircled dot, $\odot$" is sometimes used to denote this product.

Properties :

1. $A\Theta(D^T \Theta F^T) = (A\Theta D^T)\Theta F^T$, where $A_{p\times q}$, $D_{q\times s}$, $F_{q\times n}$ matrices.

Proof. By definition, $D^T \Theta F^T = [D_{.1}^T \otimes F_{.1}^T : D_{.2}^T \otimes F_{.2}^T : \cdots : D_{.q}^T \otimes F_{.q}^T]_{su\times q}$

$$\therefore\ A\Theta(D^T \Theta F^T) = [A_{.1} \otimes (D_{.1}^T \otimes F_{.1}^T) : A_{.2} \otimes (D_{.2}^T \otimes F_{.2}^T) : \cdots : A_{.q} \otimes (D_{.q}^T \otimes F_{.q}^T)]_{psu\times q}$$

$$= [(A_{.1} \otimes D_{.1}^T) \otimes F_{.1}^T : (A_{.2} \otimes D_{.2}^T) \otimes F_{.2}^T : \cdots : (A_{.q} \otimes D_{.q}^T) \otimes F_{.q}^T]_{psu\times q}$$

$$= (A\Theta D^T)\Theta F^T$$

2. $L\Theta N = U_{n\times n}\ (N\Theta L)$ where $L_{n\times n}$ and $N_{n\times n}$ are n square matrices and $U_{n\times n}$ is a permutation matrix.

Proof. $L\Theta N = [L_{.1} \otimes N_{.1} : L_{.2} \otimes N_{.2} : \cdots : L_{.n} \otimes N_{.n}]_{n^2\times n}$

$= [U_{n\times n}N_{.1} \otimes L_{.1}U_{1\times 1} : U_{n\times n}N_{.2} \otimes L_{.2}U_{1\times 1} : \cdots : U_{n\times n}N_{.n} \otimes L_{.n}U_{1\times 1}]$, by sec. 5.5

$= U_{n\times n}[N_{.1} \otimes L_{.1} : N_{.2} \otimes L_{.2} : \cdots : N_{.n} \otimes L_{.n}]$ $\quad [\because U_{1\times 1}=1]$

$= U_{n\times n}\ (N\Theta L)$

Ex. Write the Khatri-Rao product of

$$F = \begin{bmatrix} f_{11} & f_{12} & f_{13} \\ f_{21} & f_{22} & f_{23} \end{bmatrix}_{2\times 3} \text{ and } G = \begin{bmatrix} g_{11} & g_{12} & g_{13} \\ g_{21} & g_{22} & g_{23} \\ g_{31} & g_{32} & g_{33} \end{bmatrix}.$$

By definition,

$$F\Theta G = \left[F_{.1} \otimes G_{.1} : F_{.2} \otimes G_{.2} : F_{.3} \otimes G_{.3}\right]_{2..3\times 3 = 6\times 3}$$

$$= \begin{bmatrix} f_{11}G_{.1} & f_{12}G_{.2} & f_{13}G_{.3} \\ f_{21}G_{.1} & f_{22}G_{.2} & f_{23}G_{.3} \end{bmatrix}_{6\times 3} = \begin{bmatrix} f_{11}g_{11} & f_{12}g_{12} & f_{13}g_{13} \\ f_{11}g_{21} & f_{12}g_{22} & f_{13}g_{23} \\ f_{11}g_{31} & f_{12}g_{32} & f_{13}g_{33} \\ f_{21}g_{11} & f_{22}g_{12} & f_{23}g_{13} \\ f_{21}g_{21} & f_{22}g_{22} & f_{23}g_{23} \\ f_{21}g_{31} & f_{22}g_{32} & f_{23}g_{33} \end{bmatrix}_{6\times 3}$$

5.6 Hadamard product

In literature the symbol for both the Khatri-Rao product and the Hadamard product is the encircled dot, " $\odot$ ". Sometimes " * " between matrices is used for the Hadamard product. But their definitions are basically different. To distinguish between these operations, the Hadamard product is denoted in this book by $A \bullet B$ matrices. The Hadamard matrix is widely used in statistical estimation.

Definition : If $A = [a_{ij}]$ and $B = [b_{ij}]$ are matrices of the same order ($m\times n$) then the Hadamard product of A and B is defined as

$$A \bullet B = [a_{ij}\, b_{ij}]_{m\times n}$$

where the order of original matrices remain same and the i, jth element of the product is the product of the corresponding elements (elementwise product) $a_{ij}b_{ij}$. Thus

$$A \bullet B = [a_{ij}] \bullet [b_{ij}] = [a_{ij}\, b_{ij}]_{m\times n}$$

$$= \begin{bmatrix} a_{11}b_{11} & a_{12}b_{12} & \cdots & a_{1n}b_{1n} \\ a_{21}b_{21} & a_{22}b_{22} & \cdots & a_{2n}b_{2n} \\ \vdots & \vdots & \vdots & \vdots \\ a_{m1}b_{m1} & a_{m2}b_{m2} & \cdots & a_{mn}b_{mn} \end{bmatrix}_{m\times n}$$

Note the differences between the Khatri-Rao product shown in Ex. of sec. 5.5 and the the Hadamard product written above.

The following *properties* are immediate consequence of the definition:

(Readers are advised to prove the properties)

1. **Commutative :** $A \bullet B = B \bullet A$
2. **Transposition :** $(A \bullet B)^T = A^T \bullet B^T$
3. **Associative :** $(A \bullet B) \bullet C = A \bullet (B \bullet C)$
4. $(A + B) \bullet (C + D) = A \bullet C + A \bullet D + B \bullet C + B \bullet D$
5. $A \bullet 0 = 0$
6. $A \bullet I = \text{diag } A$
7. $A \bullet J = A = J \bullet A$

5.7 Cojoint with properties

Alexander D. Wallace in 1934, denoted the cojoint of two n *square* matrices A and B by coj(A, B) and defined

$$\text{coj}(A, B) = \text{adj}(A + B) - \text{adj}A - \text{adj}B$$

It is useful in the geometry of a pair conics. It obeys the following properties:

(i) If A and B are $n\times n$ matrices then coj (A,B) = coj (B,A).

(ii) If A and B are $n\times n$ then $[\text{coj}(A, B)]^T = \text{coj}(A^T, B^T)$.

This is true since $(\text{adj } A)^T = \text{adj}(A^T)$.

(iii) If A and B are $n\times n$ matrices and k is a scalar, then

$$\text{coj}(kA, kB) = k^{n-1}\,\text{coj}(A, B).$$

Proof. $\text{coj}(kA, kB) = \text{adj}(kA + kB) - \text{adj}(kA) - \text{adj}(kB)$

$$= k^{n-1}\text{adj}(A+B) - k^{n-1}\text{adj } A - k^{n-1}\text{adj } B$$

$$= k^{n-1}\text{coj}(A, B). \qquad \because \text{adj } kA = k^{n-1}\text{adj } A$$

5.8 Construction of permutation matrix from matrix units

It can be easily checked that the particular fourth order permutation matrix is obtained as the sum of Kronecker products of 2×2 matrix units i.e.

$$U_{4\times 4} = \sum_{i=1}^{2}\sum_{j=1}^{2} E_{ij} \otimes E_{ji}$$

$$= E_{11} \otimes E_{11} + E_{12} \otimes E_{21} + E_{21} \otimes E_{12} + E_{22} \otimes E_{22}$$

$$= \begin{bmatrix} 1 & 0 \\ 0 & 0 \end{bmatrix} \otimes \begin{bmatrix} 1 & 0 \\ 0 & 0 \end{bmatrix} + \begin{bmatrix} 0 & 1 \\ 0 & 0 \end{bmatrix} \otimes \begin{bmatrix} 0 & 0 \\ 1 & 0 \end{bmatrix} + \begin{bmatrix} 0 & 0 \\ 1 & 0 \end{bmatrix} \otimes \begin{bmatrix} 0 & 1 \\ 0 & 0 \end{bmatrix} + \begin{bmatrix} 0 & 0 \\ 0 & 1 \end{bmatrix} \otimes \begin{bmatrix} 0 & 0 \\ 0 & 1 \end{bmatrix}$$

$$= \begin{bmatrix} 1 & 0 & 0 & 0 \\ 0 & 0 & 0 & 0 \\ 0 & 0 & 0 & 0 \\ 0 & 0 & 0 & 0 \end{bmatrix} + \begin{bmatrix} 0 & 0 & 0 & 0 \\ 0 & 0 & 1 & 0 \\ 0 & 0 & 0 & 0 \\ 0 & 0 & 0 & 0 \end{bmatrix} + \begin{bmatrix} 0 & 0 & 0 & 0 \\ 0 & 0 & 0 & 0 \\ 0 & 1 & 0 & 0 \\ 0 & 0 & 0 & 0 \end{bmatrix} + \begin{bmatrix} 0 & 0 & 0 & 0 \\ 0 & 0 & 0 & 0 \\ 0 & 0 & 0 & 0 \\ 0 & 0 & 0 & 1 \end{bmatrix}$$

$$= \begin{bmatrix} 1 & 0 & 0 & 0 \\ 0 & 0 & 1 & 0 \\ 0 & 1 & 0 & 0 \\ 0 & 0 & 0 & 1 \end{bmatrix}$$

In general, $U_{pq\times pq} = \sum_{i=1}^{p}\sum_{j=1}^{q} E_{ij} \otimes E_{ji} = \sum_{i=1}^{p}\sum_{j=1}^{q} E_{ij} \otimes E_{ij}^{T}$

The product of two permutation matrices is another permutation matrix. Multiplication of a full matrix A by a permutation matrix P as AP and AP rearranges rows and columns respectively of A in exactly the same pattern that P itself was obtained from the identity matrix.

Thus, multiplication of a 3×3 matrix A on the left by the permutation matrix

$$\begin{bmatrix} 0 & 0 & 1 \\ 0 & 1 & 0 \\ 1 & 0 & 0 \end{bmatrix}$$

interchanges the first and third rows of A and multiplication of the right interchanges the first and third columns of A.

5.9 Construction of the Hadamard matrix

The Kronecker product is used for the construction of Hadamard matrices. It is an early and easy method of construction. The method states that if H_1 is a Hadamard matrix of order m and H_2 is a Hadamard matrix of order n then $H_1 \otimes H_2$ is a Hadamard matrix of order mn.

For example,

$$\text{let } H_1 = \begin{bmatrix} + & + \\ + & - \end{bmatrix} \text{ and } H_2 = H_1^{T} = \begin{bmatrix} + & + \\ + & - \end{bmatrix}.$$

Therefore,

$$H_1 \otimes H_2 = \begin{bmatrix} + & + \\ + & - \end{bmatrix} \otimes \begin{bmatrix} + & + \\ + & - \end{bmatrix} = \begin{bmatrix} + & + & + & + \\ + & - & + & - \\ + & + & - & - \\ + & - & - & + \end{bmatrix} = H_4, \text{ (say)}$$

which is a 4th order Hadamard matrix constructed from 2nd order Hadamard matrix. Similarly $H_4 \otimes H_4^{T} = H_{16}$. Thus $4 = 2^2$, $16 = 2^4$ together imply that Hadamard matrix of order 2^k for every positive integer k can be constructed from kth order Hadamard matrix.

Various construction methods have been developed. Still there remains some Hadamard matrices of order 268 unconstructed.

Another example is

$$H_2 = \begin{bmatrix} + & + \\ + & - \end{bmatrix} \text{ and } H_4 = \begin{bmatrix} + & + & + & + \\ + & + & - & - \\ + & - & + & - \\ + & - & - & + \end{bmatrix}$$

$$\therefore \qquad H_2 \otimes H_4 = \begin{bmatrix} \begin{bmatrix} + & + & + & + \\ + & + & - & - \\ + & - & + & - \\ + & - & - & + \end{bmatrix} & \begin{bmatrix} + & + & + & + \\ + & + & - & - \\ + & - & + & - \\ + & - & - & + \end{bmatrix} \\ \begin{bmatrix} + & + & + & + \\ + & + & - & - \\ + & - & + & - \\ + & - & - & + \end{bmatrix} & \begin{bmatrix} - & - & - & - \\ - & - & + & + \\ - & + & - & + \\ - & + & + & - \end{bmatrix} \end{bmatrix}$$

$$= \begin{bmatrix} + & + & + & + & + & + & + & + \\ + & + & - & - & + & + & - & - \\ + & - & + & - & + & - & + & - \\ + & - & - & + & + & - & - & + \\ + & + & + & + & - & - & - & - \\ + & + & - & - & - & - & + & + \\ + & - & + & - & - & + & - & + \\ + & - & - & + & - & + & + & - \end{bmatrix}_{8\times 8}$$

5.10 Factorization of an orthogonal matrix in terms of standard Helmertian

The following provides us with a method of factoring any orthogonal matrix of order n into a product in which factors contain Helmertian.

Theorem . Any orthogonal matrix of order n can be factored into the product

$A=(1_{n-2} \oplus H_2^e)(1_{n-3} \oplus H_3^e) \cdots (1 \oplus H_{n-1}^e) H_n^e$,where $H_2^e, H_3^e, \cdots, H_{n-1}^e$,

H_n^e are standard Helmertians of orders 2, 3, $\cdots$, $(n-1)$, n.

5.11 Vec operation and its properties

The vec operator : "Vec" operator was introduced by Koopmans et al in 1950. It transforms an $m\times n$ matrix into an mn vector. Neudecker in 1969 used vec A, a vector valued function of an $m\times n$ matrix A as

$$\text{vec } A = \begin{bmatrix} A_{.1} \\ A_{.2} \\ \vdots \\ A_{.n} \end{bmatrix}_{mn\times 1},$$

which is a vector of mn components and is an unary operation. For example if

$$A = \begin{bmatrix} a_{11} & a_{12} & a_{13} \\ a_{21} & a_{22} & a_{23} \end{bmatrix},$$

$$\text{then vec A} = \begin{bmatrix} a_{11} \\ a_{21} \\ a_{12} \\ a_{22} \\ a_{13} \\ a_{23} \end{bmatrix},$$

which is a vector vecA of 2×3 = 6 components arranged column after column. Therefore, vec transforms an $m \times n$ matrix A into an $mn \times 1$ vector vecA by stacking columns of A one underneath the other, i.e. stacking first column then follows sequentially second, third, $\cdots$, nth columns.

Equality of vecs : VecA and vecB are equal only when matrices A and B are of the same order and can be written as vecA = vecB .

5.11.1 Vec on column vector : Let $\mathbf{a} = [a_1 \quad a_2 \quad \cdots \quad a_n]^T$ be a column vector. Then

$$\text{vec}\mathbf{a} = [a_1 \quad a_2 \quad \cdots \quad a_n]^T = \mathbf{a}.$$

On the other hand, $\mathbf{a}^T = [a_1 \quad a_2 \quad \cdots \quad a_n]$. So,

$$\text{vec } \mathbf{a}^T = \begin{bmatrix} a_1 \\ a_2 \\ \vdots \\ a_n \end{bmatrix},$$

by definition of vec operator. Therefore, vec $\mathbf{a}^T = \mathbf{a} =$ vec $\mathbf{a}$.

5.11.2 Relation between vec operator and Kronecker product : Let two vectors of orders n and m be

$$\mathbf{a} = \begin{bmatrix} a_1 \\ a_2 \\ \vdots \\ a_n \end{bmatrix} \text{ and } \mathbf{b} = \begin{bmatrix} b_1 \\ b_2 \\ \vdots \\ b_n \end{bmatrix}$$

respectively . Now

$$\mathbf{ab}^T = \begin{bmatrix} a_1 \\ a_2 \\ \vdots \\ a_n \end{bmatrix} [b_1 \quad b_2 \quad \cdots \quad b_m] = \begin{bmatrix} a_1b_1 & a_1b_2 & \cdots & a_1b_m \\ a_2b_1 & a_2b_2 & \cdots & a_2b_m \\ \vdots & \vdots & \vdots & \vdots \\ a_nb_1 & a_nb_2 & \cdots & a_nb_m \end{bmatrix}$$

$$\therefore \qquad \text{vec}\,\mathbf{ab}^{\mathrm{T}} = \begin{bmatrix} a_1b_1 \\ \vdots \\ a_nb_1 \\ a_1b_2 \\ \vdots \\ a_nb_2 \\ \vdots \\ a_1b_m \\ a_2b_m \\ \vdots \\ a_nb_m \end{bmatrix}$$

$$\text{Also} \quad \mathbf{b} \otimes \mathbf{a} = \begin{bmatrix} b_1 \\ b_2 \\ \vdots \\ b_m \end{bmatrix} \otimes \begin{bmatrix} a_1 \\ a_2 \\ \vdots \\ a_n \end{bmatrix} = \begin{bmatrix} b_1\mathbf{a} \\ b_2\mathbf{a} \\ \vdots \\ b_m\mathbf{a} \end{bmatrix} = \begin{bmatrix} b_1a_1 \\ b_1a_2 \\ \vdots \\ b_1a_n \\ \vdots \\ b_ma_1 \\ \vdots \\ b_ma_n \end{bmatrix}.$$

$$\therefore \qquad \text{vec}\,\mathbf{ab}^{\mathrm{T}} = \mathbf{b} \otimes \mathbf{a}.$$

This is the basic relation between vec operator and the Kronecker product when **a** and **b** are column vectors *not* necessarily of same order.

Note that there is no identical property in matrices i.e. vec AB^{T} is *not* equal to $\mathrm{B} \otimes \mathrm{A}$.

5.11.3 Other definitions & notations of vector valued functions of a matrix : Ransom and Sacks in 1976 defined a vector valued function vecd (M), a column vector by arranging the principal diagonal as column, i.e.,

$$\text{vecd}\,(\mathrm{M}) = \begin{bmatrix} m_{11} \\ m_{22} \\ \vdots \\ m_{nn} \end{bmatrix}$$

from the diagonal elements of an $m \times n$ matrix $\mathrm{M} = [m_{ij}]$. Barnet in 1973 defined a vector valued function $\mathbf{v}(\mathrm{X})$ a column vector stacking the rows of an $m \times n$ matrix $\mathrm{X} = [x_{ij}]$ taken in order i.e.

$$\mathbf{v}(\mathrm{X}) = \big[\underbrace{x_{11} \quad x_{12} \quad \cdots \quad x_{1n}}_{\leftarrow \text{ 1st row } \rightarrow} \quad \underbrace{x_{21} \quad \cdots\cdots \quad \cdots \quad x_{2n}}_{\leftarrow \text{ 2nd row } \rightarrow} \quad \cdots \quad \underbrace{x_{m1} \cdots\cdots \quad \cdots \quad x_{mn}}_{\leftarrow\ m\text{th row } \rightarrow}\big]^{\mathrm{T}}$$

with property $\mathbf{v}(\mathrm{CXD}^{\mathrm{T}}) = (\mathrm{C} \otimes \mathrm{D})\,\mathbf{v}(\mathrm{X})$, where C is $p \times m$, D is $q \times n$. Vetter in 1973 listed and used both the column and the row stacking definitions and used different symbols.

5.11.4 Properties of vec operators : If matrices A, D, H and H are of orders $p{\times}q$, $q{\times}s$, $p{\times}q$, $q{\times}q$, then

1. vec (A + H) = vec (A) + vec (H)
2. vec (ADB) = $(B^T \otimes A)$ vec D, B is of order $s{\times}t$.
3. vec $(A^T) = U_{p\times q}$ vec (A), $U_{p\times q}$ is a permutation matrix.
4. $\text{vec } A = \sum_{k}^{q}\left(\underset{(q)}{e_k}\otimes I_p\right)A\underset{(q)}{e_k}$, where (q) is the order of s.u. vector $\mathbf{e}_k$.
5. $\text{vec (AD)} = (I_s \otimes A)\,\text{vec(D)}$ (i)

   $= (D^T \otimes I_p)\,\text{vec(A)}$ (ii)

   $= (D^T \otimes A)(\,\text{vec}(I_q))$ (iii)
6. $\text{vec (AD)} = \sum_{k}^{q}\left(D^T\right)_{.k} \otimes A_{.k}$
7. vec (AVD) $= (D^T \ \Theta\ A)$ vecd (V), V is diagonal and $p{\times}q$.
8. tr. ADW = $(\text{vec } A^T)^T (I_p \otimes D)$ vec (W).
9. tr. ABCD $= (\text{vec } D^T)^T (C^T \otimes A)$ vec B

   $= (\text{vec } D)^T (A \otimes C^T)$ vec B^T.

Proof .

1. Let $A = [a_{ij}]_{p\times q}$ and $H = [h_{ij}]_{p\times q}$ be of the same order $p{\times}q$, so

$$A + H = (a_{ij} + h_{ij})_{p\times q}$$

$$\therefore \quad \text{vec (A + H)} = \begin{bmatrix} A_{.1} + H_{.1} \\ A_{.2} + H_{.2} \\ \vdots \\ A_{.q} + H_{.q} \end{bmatrix}$$

$$= \begin{bmatrix} A_{.1} \\ A_{.2} \\ \vdots \\ A_{.q} \end{bmatrix} + \begin{bmatrix} H_{.1} \\ H_{.2} \\ \vdots \\ H_{.q} \end{bmatrix}$$

$$= \text{vec}(A) + \text{vec}(H)$$

$\therefore$ vec operator on sum of conformable matrices is equal to the sum of vec operators on individual matrices.

2. It is a generalization of 5(iii). Neudeker has proved it in the following way :

AD is partitioned into columns and kth column of B is partitioned into a single element b_{ik}. Therefore, using rule for multiplication of partitioned matrices, the partition

$$(ADB)_{.k} = \sum_i (AD)_{.i} b_{ik} = \sum_i (b_{ik} A) D_{.i}$$

$$= \left[\left(B_{.k}\right)^T \otimes A\right] \text{vec D}$$

$$= \left[\left(\mathrm{B}^{\mathrm{T}}\right)_{k.} \otimes \mathrm{A} \right] \text{vec D}$$

$\therefore$ vec (ADB) $= (\mathrm{B}^{\mathrm{T}} \otimes \mathrm{A})$ vec D.

Cor. If **b** is $s\times1$ vector which replaces the matrix B above then

$$\mathrm{AD}\mathbf{b} = (\mathbf{b}^{\mathrm{T}} \otimes \mathrm{A}) \text{ vec D} = (\mathrm{A} \otimes \mathbf{b}^{\mathrm{T}}) \text{ vec } \mathrm{D}^{\mathrm{T}}.$$

3. Let $\mathrm{A} = [a_{ij}]_{p\times q}$

$= \sum_p \sum_q a_{ij} \mathrm{E}_{ij}$, where E_{ij} is matrix unit of order $p\times q$.

$\therefore$ $\mathrm{A}^{\mathrm{T}} = \sum_p \sum_q a_{ij} \mathrm{E}^{\mathrm{T}}{}_{ij}$

$\therefore$ vec $\mathrm{A}^{\mathrm{T}} = \sum_p \sum_q a_{ij}$ vec $\mathrm{E}^{\mathrm{T}}{}_{ij}$

$$= \left[\text{vec } \mathrm{E}_{11}^{\mathrm{T}} \mid \text{vec } \mathrm{E}_{21}^{\mathrm{T}} \mid \cdots \mid \text{vec } \mathrm{E}_{p1}^{\mathrm{T}} \mid \cdots \mid \text{vec } \mathrm{E}_{12}^{\mathrm{T}} \mid \cdots \mid \text{vec } \mathrm{E}_{pq}^{\mathrm{T}} \right] \begin{bmatrix} a_{11} \\ a_{21} \\ \vdots \\ a_{p1} \\ a_{12} \\ \vdots \\ a_{pq} \end{bmatrix}$$

$= \mathrm{U}_{p\times q}$ vec A, by sec. 5.8

4. Note that
$$\mathrm{A}\mathbf{e}_k = \begin{bmatrix} a_{11} & a_{12} & \cdots & a_{1k} & \cdots & a_{1q} \\ a_{21} & a_{22} & \cdots & a_{2k} & \cdots & a_{2q} \\ \vdots & \vdots & \vdots & \vdots & \vdots & \vdots \\ a_{p1} & a_{p2} & \cdots & a_{pk} & \cdots & a_{pq} \end{bmatrix} \begin{bmatrix} 0 \\ 0 \\ \vdots \\ 1 \\ 0 \\ \vdots \\ 0 \end{bmatrix}_{q\times1}$$

(*k*th column of A is marked "*k*th"; the 1 stands in the "*k*th row".)

$$= \begin{bmatrix} a_{1k} \\ a_{2k} \\ \vdots \\ a_{pk} \end{bmatrix} = \mathrm{A}_{.k}$$

$\therefore$ R.H.S. $= \sum_k^q \left(\underset{(q)}{\mathbf{e}_k} \otimes \mathrm{I}_p \right) \mathrm{A} \underset{(q)}{\mathbf{e}_k}$ = vec A , (q) denotes order of s.u. vector $\mathbf{e}_k$.

i.e. vec A is expressed as a *sum*.

5. (i) Replace B by I_s in 2,

$$\text{vec (AD)} = (\mathrm{I}_s^{\mathrm{T}} \otimes \mathrm{A}) \text{ vec D} = (\mathrm{I}_s \otimes \mathrm{A}) \text{ vec D}.$$

(ii) Replace B by D, A by I_p and D by A in 2,

$$\text{vec } (\mathrm{I}_p\mathrm{AD}) = (\mathrm{D}^{\mathrm{T}} \otimes \mathrm{I}_p) \text{ vec A}.$$

$\therefore$ vec (AD) = $(D^T \otimes I_p)$ vec (A).

(iii) Replace B by D and D by I_q,

$$\text{vec}\,(AI_qD) = (D^T \otimes A)\,\text{vec}\,(I_q)$$

i.e. $$\text{vec}\,(AD) = (D^T \otimes A)\,\text{vec}\,(I_q)$$

6. From 5(iii)

$$\text{vec}\,(AD) = (D^T \otimes A)\,\text{vec}\,(I_q)$$

$$= \left[\sum_i \left(D^T\right)_{.i} \mathbf{e}_i^T \otimes \sum_j A_{.j}\mathbf{e}_{.j}^T\right] \text{vec}\, I_q$$

$$= \sum_i\sum_j\left[\left(\left(D^T\right)_{.i} \otimes A_{.j}\right)\left(\mathbf{e}_i^T \otimes \mathbf{e}_j^T\right)\right] \text{vec}\, I_q$$

The product $\mathbf{e}_i^T \otimes \mathbf{e}_j^T$ is a row matrix having one unit element in the

$$[(i-1)n+j]\text{th}$$

column and zeros elsewhere. Hence the product

$$[(D^T)_{.i} \otimes A_{.j}]\,[\mathbf{e}_i^T \otimes \mathbf{e}_j^T]$$

is a matrix with $(B^T)_{.i} \otimes A_{.j}$ as its $[(i-1)n+j]$th column and zeros elsewhere. Since vec I is a column matrix having unity in the 1st, $(n+2)$, $(2n+3)$, $\cdots$, n^2 position and zeros elsewhere, the product of

$$[(D^T)_{.i} \otimes A_{.j}]\,[\mathbf{e}_i^T \otimes \mathbf{e}_j^T] \text{ and vec I}$$

is a column matrix whose elements are all zeros unless i and j satisfy

$$(i-1)n+j=1 \quad \text{or, } n+2 \text{ or, } 2n+3, \;\cdots \text{ or, } n^2,$$

i.e. $$i=j=1 \;\text{ or, }\; i=j=2 \;\text{ or, }\; i=j=3, \;\cdots\; \text{ or , } i=j=n.$$

in which case the column matrix is

$$(D^T)_{.i} \otimes A_{.j}\,,\; i=1,2,\;\cdots, n.$$

Hence the result.

7. From (6), $$\text{vec AVD} = \sum_k^q \left(D^T\right)_{.k} \otimes (AV)_{.k}$$

$$= \sum_k^q \left(D^T\right)_{.k} \otimes A_{.k}V_{kk}\,, \quad \text{as V is diagonal.}$$

Also from (6),

$$(D^T \Theta A)\,\text{vecd}\,(V) = \sum_k^q \left(D^T \Theta A\right)_{.k} V_{kk} = \sum_k^q \left(D^T\right)_{.k} \otimes A_{.k}V_{kk}\,,$$

using definition of the Khatri-Rao product.

5.11.5 Relation betweeen vec operator and trace : If matrices A and B are of the same order then $(\text{vec A})^T\,\text{vec B} = \text{tr. } A^TB$. This can be easily verified since both sides are equal to

$\sum_i\sum_j a_{ij}b_{ij}$. Extension of the result can be easily made. Thus tr. ADW = $(\text{vec } A^T)^T$ vec DW

where orders A, D and W are $p\times q$, $q\times s$, $s\times p$

$= (\text{vec } A^T)^T (I_p \otimes D)$ vec (W)by relation 5.(i) of sec. 5.11.5.

5.12 Commutation of the Kronecker product

In general, the Kronecker product does not possess the commutation property of algebra i.e.

$$A \otimes B \neq B \otimes A.$$

But under special circumstances, two matrices in a Kronecker product interchange.

Condition : $A \otimes B = U_1(B \otimes A)U_2$, if U_1 and U_2 are permutation matrices.

Proof. Let $AYB^T = \mathbf{x}$ (5.5)

Now taking vec operator on both sides, we get

$$\text{vec}(AYB^T) = \text{vec}\,\mathbf{x}$$

or $(B \otimes A)\,\text{vec}\,Y = \text{vec}\,\mathbf{x}$ (5.6)

Again taking transpose on (5.5),

$$BY^T A^T = \mathbf{x}^T$$

or taking vec operator

$$(A \otimes B)\,\text{vec}\,Y^T = \text{vec}\,\mathbf{x}^T \tag{5.7}$$

But vec property 3, $\text{vec}\,\mathbf{x}^T = U_1\,\text{vec}\,\mathbf{x}$ and $\text{vec}\,Y = U_2\,\text{vec}\,Y^T$ (5.8)

where U_1 and U_2 are two permutation matrices.

Now premultiplying (5.6) by U_1 and replacing vec Y in (5.6) by (5.8), we get

$$U_1(B \otimes A)U_2\,\text{vec}\,Y^T = U_1\,\text{vec}\,\mathbf{x} \tag{5.9}$$

Replacing vec $\mathbf{x}^T$ of (5.7) by (5.8) we have

$$(A \otimes B)\,\text{vec}\,Y^T = U_1\,\text{vec}\,\mathbf{x} \tag{5.10}$$

∴ from (5.10) and (5.9),

$$(A \otimes B)\,\text{vec}\,Y^T = U_1(B \otimes A)U_2\,\text{vec}\,Y^T$$

∴ $A \otimes B = U_1(B \otimes A)U_2$, by equating coefficients of vec Y^T.

In general, $A \otimes B = U_{s\times p}(A_{p\times q} \otimes B_{s\times t})U_{q\times t}$, (5.11)

where U's are permutation matrices.

Ex. Show that $(A \otimes I_3)^2 = A^2 \otimes I_3$ where

$$A = \begin{bmatrix} a_{11} & a_{12} \\ a_{21} & a_{22} \end{bmatrix}.$$

Generalize the result for arbitrary A and any integer power *k*.

$$A \otimes I_3 = \begin{bmatrix} a_{11}I_3 & a_{12}I_3 \\ a_{21}I_3 & a_{22}I_3 \end{bmatrix}$$

∴ $(A \otimes I_3)^2$

$$= \begin{bmatrix} a_{11} & 0 & 0 & a_{12} & 0 & 0 \\ 0 & a_{11} & 0 & 0 & a_{12} & 0 \\ 0 & 0 & a_{11} & 0 & 0 & a_{12} \\ a_{21} & 0 & 0 & a_{22} & 0 & 0 \\ 0 & a_{21} & 0 & 0 & a_{22} & 0 \\ 0 & 0 & a_{21} & 0 & 0 & a_{22} \end{bmatrix} \begin{bmatrix} a_{11} & 0 & 0 & a_{12} & 0 & 0 \\ 0 & a_{11} & 0 & 0 & a_{12} & 0 \\ 0 & 0 & a_{11} & 0 & 0 & a_{12} \\ a_{21} & 0 & 0 & a_{22} & 0 & 0 \\ 0 & a_{21} & 0 & 0 & a_{22} & 0 \\ 0 & 0 & a_{21} & 0 & 0 & a_{22} \end{bmatrix}$$

Also

$$A^2 = \begin{bmatrix} (a_{11}^2 + a_{12}a_{21}) & (a_{11}a_{12} + a_{12}a_{22}) \\ (a_{11}a_{21} + a_{22}a_{21}) & (a_{22}^2 + a_{12}a_{21}) \end{bmatrix}.$$

Therefore, $(A \otimes I_3)^2 = A^2 \otimes I_3$.

Proceeding in this way for arbitrary A and integer power k, we must have

$$(A \otimes I_m)^k = A^k \otimes I_m .$$

N.B. $(I_m \otimes A)^k = I_m \otimes A^k$. It is left for reader.

Ex. Solve the matrix equation

$$AX + XB = C$$

for the unknown matrix X where $A_{n\times n}$, $B_{m\times m}$ and $X_{n\times m}$.

Taking the vec operator,

$$\text{vec}\,(AX) + \text{vec}\,(XB) = \text{vec}\,C$$

or $(I_m \otimes A)\,\text{vec}\,X + (B^T \otimes I_n)\,\text{vec}\,X = \text{vec}\,C$, by 5(i),(ii) of sec. 5.11.4

or $(I_m \otimes A + B^T \otimes I_n)\,\text{vec}\,X = \text{vec}\,C$

or $(B^T \dagger A)\,\text{vec}\,X = \text{vec}\,C$, by definition of Kronecker sum.

or $\text{vec}\,X = (B^T \dagger A)^{-1}\,\text{vec}\,C$, if $(B^T \dagger A)$ is nonsingular.

N.B. This type of equation appears in system theory, stability theory, steady state analysis of stochastic dynamic systems, etc. An explicit solution can be obtained using eigenvalues (cf. chapter 13).

Exercises five

1. et A be an $m\times n$ matrix. Prove that

$\text{vec}\,A = (I_n \otimes A)\,\text{vec}\,I_n = (A^T \otimes I_m)\,\text{vec}\,I_m$.

2. If A, B and V are square matrices of the same order and $V = V^T$, prove that

$$(\text{vec}\,V)^T (A \otimes B)\,\text{vec}\,V = (\text{vec}\,V)^T (B \otimes A)\,\text{vec}\,V$$

[**Hints** : L.H.S. = $(\text{vec}\,V)^T (A \otimes B)\,\text{vec}\,V = (\text{vec}\,V)^T\,\text{vec}\,BVA^T$ by vec property 2

$= \text{tr.}\,V^T BVA^T$ by relation 5.11.5

R.H.S. = $(\text{vec}\,V)^T\,\text{vec}\,AVB^T$

$= \text{tr.}\,V^T AVB^T$]

3. Show that if A and B are unitary matrices then $A \otimes B$ is unitary.

[**Hints** : $(A \otimes B)^*(A \otimes B) = (A^*A) \otimes (B^*B) = I \otimes I = I$]

4. Show that if A and B are upper triangular matrices then $A \otimes B$ is upper triangular. Similarly prove for lower case.

5. If

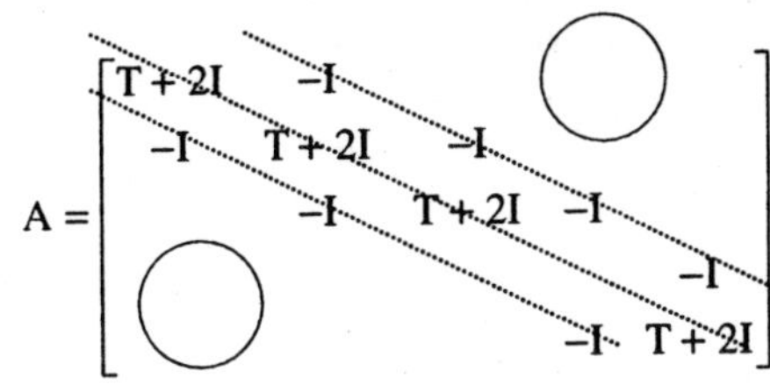

$$\text{where } T = \begin{bmatrix} 2 & -1 & & & \\ -1 & 2 & -1 & & \\ & -1 & 2 & -1 & \\ & & & & -1 \\ & & & -1 & 2 \end{bmatrix},$$

then show that $A = (I \otimes T) + (T \otimes I)$.

6. (a) If A = diag (A_1, A_2, $\cdots$, A_s) and B = diag (B_1, B_2, $\cdots$, B_s), where A_i and B_i have the same dimension for each $i = 1, 2, \cdots, s$, then show that

$$AB = \text{diag}\,(A_1B_1, A_2B_2, \cdots, A_sB_s).$$

(b) Compute

$$\begin{bmatrix} 2 & 0 & 0 & 0 & 0 \\ 0 & 1 & 2 & 0 & 0 \\ 0 & 0 & -1 & 0 & 0 \\ 0 & 0 & 0 & 2 & 3 \\ 0 & 0 & 0 & 1 & -2 \end{bmatrix}^2$$

[**Hints :** The given matrix A can be written as

$$A = \begin{bmatrix} A_1 & 0 & 0 \\ 0 & A_2 & 0 \\ 0 & 0 & A_3 \end{bmatrix}^2$$

which is the square of the direct sum of the matrices

$A_1 = \begin{bmatrix} 2 & 0 \\ 0 & 1 \end{bmatrix}$, $A_2 = \begin{bmatrix} 2 & 0 \\ -1 & 0 \end{bmatrix}$ and $A_3 = \begin{bmatrix} 2 & 3 \\ 1 & -2 \end{bmatrix}$. So

$$A = \begin{bmatrix} A_1^2 & 0 & 0 \\ 0 & A_2^2 & 0 \\ 0 & 0 & A_3^2 \end{bmatrix} = \begin{bmatrix} 4 & 0 & 0 & 0 & 0 \\ 0 & 1 & 4 & 0 & 0 \\ 0 & 0 & -2 & 0 & 0 \\ 0 & 0 & 0 & 7 & 0 \\ 0 & 0 & 0 & 0 & 7 \end{bmatrix}]$$

7. If $A_1, A_2, \cdots, A_k$ are all $m \times n$ matrices and $B_1, B_2, \cdots, B_k$ all $n \times m$ matrices then prove that $(A_1 \otimes B_1)(A_2 \otimes B_2)(A_3 \otimes B_3) \cdots (A_k \otimes B_k)$

$$= (A_1 A_2 \cdots A_k) \otimes (B_1 B_2 \cdots B_k).$$

8. Prove that $A_{m \times m} \otimes B_{n \times n} = (A_{m \times m} \otimes I_m)(I_n \otimes B_{n \times n})$.

9. Find $\mathbf{x} \otimes \mathbf{y}$ where $\mathbf{x} = [x_1 \quad x_2]^T$ and $\mathbf{y} = [y_1 \quad y_2 \quad y_3]^T$. Hence obtain

$$\begin{bmatrix} 1 \\ 2 \end{bmatrix} \otimes \begin{bmatrix} 3 \\ 0 \\ 5 \end{bmatrix}.$$

[**Ans.** $[3 \quad 0 \quad 5 \quad 6 \quad 0 \quad 10]^T$]

10. Can you obtain $[1 \quad 2]^T [3 \quad 0 \quad 5]^T$. [**Ans.** No, not defined]

11. Compare the Cayley matrix product and the Kronecker matrix product.

[**Hints :** Discuss order restriction, transposition, inverse]

12. Given

$$A = \begin{bmatrix} 2 & 1 \\ 0 & 1 \\ 1 & 1 \end{bmatrix} \text{ and } B = \begin{bmatrix} -1 & 1 \\ 2 & 0 \end{bmatrix}.$$

Calculate (i) $A \otimes B$, (ii) $B \otimes A$. Verify $(A \otimes B)^T = A^T \otimes B^T$.

[**Ans.** (i) $\begin{bmatrix} 2B & 1B \\ 0B & 1B \\ 1B & 1B \end{bmatrix} = \begin{bmatrix} -2 & 2 & -1 & 1 \\ 4 & 0 & 2 & 0 \\ 0 & 0 & -1 & 1 \\ 0 & 0 & 2 & 0 \\ -1 & 1 & -1 & 1 \\ 2 & 0 & 2 & 0 \end{bmatrix}$,

(ii) $\begin{bmatrix} -1A & 1A \\ 2A & 0A \end{bmatrix} = \begin{bmatrix} -2 & -1 & 2 & 1 \\ 0 & -1 & 0 & 1 \\ -1 & -1 & 1 & 1 \\ 4 & 2 & 0 & 0 \\ 0 & 2 & 0 & 0 \\ 2 & 2 & 0 & 0 \end{bmatrix}$]

13. Find Tr. $(A \otimes B)$, where A and B are given in 11 above.

[**Ans.** Tr (A) = 2 + 1 = 3, Tr (B) = –1. So, Tr $(A \otimes B) = 3(-1) = -3$]

14. Prove that if A and B are skew symmetric then A×B is skew symmetric.

15. Find the Lie product of

$$A = \begin{bmatrix} 1 & 2 \\ 3 & 4 \end{bmatrix} \text{ and } B = \begin{bmatrix} 0 & 1 \\ 3 & 2 \end{bmatrix}.$$

[**Ans.** $\begin{bmatrix} 3 & 1 \\ 3 & -3 \end{bmatrix}$]

16. Let

$$A = \begin{bmatrix} 0 & 1 & 0 \\ 1 & 0 & 1 \\ 0 & 1 & 0 \end{bmatrix}, B = \begin{bmatrix} 1 & 0 & 0 \\ 0 & 1 & 0 \\ 0 & 0 & 1 \end{bmatrix}, C = \begin{bmatrix} 1 & 0 & 0 \\ 0 & 0 & 0 \\ 0 & 0 & -1 \end{bmatrix}.$$

Show that A commutes with B and B commutes with C but that A does not commute with C.

[**Hints :** [A , B] = AB – BA = 0,

[B ,C] = BC – CB = 0,

[A , C] = AC – CA ≠ 0]

17. Find the Jordan product and the cross product of

$$A = \begin{bmatrix} 3 & 1 \\ 2 & 4 \end{bmatrix} \text{ and } B = \begin{bmatrix} 6 & 9 \\ 2 & 0 \end{bmatrix}.$$

[**Ans.** $\begin{bmatrix} 26 & \frac{69}{2} \\ 13 & 10 \end{bmatrix}$]

18. Find the Jordan product of

$$A = \begin{bmatrix} 1 & 2 & 1 \\ 3 & 4 & -1 \\ 0 & 0 & 2 \end{bmatrix} \text{ and } B = \begin{bmatrix} -1 & -2 & -1 \\ 0 & 3 & 4 \\ 1 & -1 & 0 \end{bmatrix}$$

[**Ans.** $\begin{bmatrix} -\frac{7}{2} & -\frac{9}{2} & 5 \\ \frac{5}{2} & \frac{19}{2} & 12 \\ 0 & -2 & 1 \end{bmatrix}$]

19. What are the union and the intersection of

$$\begin{bmatrix} 0 & 1 \\ 1 & 0 \end{bmatrix} \text{ and } \begin{bmatrix} 0 & 0 \\ 1 & 1 \end{bmatrix}?$$

[**Ans.** $\begin{bmatrix} 0 & 1 \\ 0 & 1 \end{bmatrix}$ and $\begin{bmatrix} 0 & 0 \\ 1 & 0 \end{bmatrix}$]

20. Show that $(A \otimes B)^* = A^* \otimes B^*$ and rank $(A \otimes B)$ = rank $A \otimes$ rank B.

[**Hints**: Do 2nd part after chapter 10]

21. Let **x** be an *n* vector and **y** is an *m* vector. Show that $\mathbf{x} \otimes \mathbf{y}^T = \mathbf{x}\mathbf{y}^T$.

22. If A is a partitional matrix,

$$A = \begin{bmatrix} A_{11} & A_{12} \\ A_{21} & A_{22} \end{bmatrix},$$

then show that $A \otimes B$ takes the form

$$A \otimes B = \begin{bmatrix} A_{11} \otimes B & A_{12} \otimes B \\ A_{21} \otimes B & A_{22} \otimes B \end{bmatrix}.$$

23. Prove that

$$(S^{-1} \otimes T^{-1})(A \otimes B)(S \otimes T) = L \otimes M, \text{ where } L = S^{-1}AS \text{ and } M = T^{-1}BT.$$

24. Verify that vecABC = $(C^{-1} \otimes B)$ vec B, where

$$A = \begin{bmatrix} 1 & 0 \\ 0 & 2 \end{bmatrix},\ B = \begin{bmatrix} 2 & -1 & 0 \\ 3 & 0 & -1 \end{bmatrix}, \text{ and } C = \begin{bmatrix} 1 & -1 \\ 0 & 0 \\ -1 & 1 \end{bmatrix}.$$

25. Establish each of the properties of the Haramard product.

26. Show that the commutators $[\sigma_x, \sigma_y]$, $[\sigma_y, \sigma_z]$ and $[\sigma_z, \sigma_x]$ are $2i\sigma_z$, $2i\sigma_x$, $2i\sigma_y$, where Pauli spin matrices

$$\sigma_x = \begin{bmatrix} 0 & 1 \\ 1 & 0 \end{bmatrix},\ \sigma_y = \begin{bmatrix} 0 & -i \\ i & 0 \end{bmatrix} \text{ and } \sigma_z = \begin{bmatrix} 1 & 0 \\ 0 & -1 \end{bmatrix}, i^2 = -1.$$

27. Verify $(A \otimes B)^T = A^T \otimes B^T$, if

$$A = \begin{bmatrix} 1 & 2 \\ 3 & 1 \end{bmatrix},\ B = \begin{bmatrix} 4 & 5 & 6 \\ 7 & 8 & 9 \end{bmatrix}.$$

28. Prove the properties of 6 and 7 of sec 5.2.3. **[Hints :** Proceed as in 5]

29. Commutator C = A×B = AB – BA is not necessarily nilpotent— establish the truth of the statement.

[Hints: $C^2 = CC = (AB - BA)(AB - BA)$

$= ABAB - ABBA - BAAB + BABA.$

If AB = BA, then $C^2 = (AB)^2 - (AB)^2 - (BA)^2 + (BA)^2 = 0$,

otherwise $C^2 \neq 0$. Hence the commutator is not necessarily nilpotent]

30. Find the commutator of the square matrices

$$A = \begin{bmatrix} 1 & 2 \\ 3 & 1 \end{bmatrix} \text{ and } B = \begin{bmatrix} 2 & 4 \\ 6 & 7 \end{bmatrix}.$$

What is $(A\times B)^T$ and $(A\times B)^{-1}$?

[Ans: & Hints: $AB - BA = \begin{bmatrix} 1 & 2 \\ 3 & 1 \end{bmatrix}\begin{bmatrix} 2 & 4 \\ 6 & 7 \end{bmatrix} - \begin{bmatrix} 2 & 4 \\ 6 & 7 \end{bmatrix}\begin{bmatrix} 1 & 2 \\ 3 & 1 \end{bmatrix}$

$$= \begin{bmatrix} 14 & 18 \\ 12 & 19 \end{bmatrix} - \begin{bmatrix} 14 & 8 \\ 27 & 19 \end{bmatrix} = \begin{bmatrix} 0 & 10 \\ -15 & 0 \end{bmatrix};$$

$B^TA^T - A^TB^T;\ B^{-1}A^{-1} - A^{-1}B^{-1}$]

CHAPTER SIX

DETERMINANTS

***Chapter summary**: After definition and introduction, the chapter discusses evaluation procedures of determinant by direct method, by Sarrus rule; theorems on properties; sum and product of determinants; expansion methods of Laplace, Chio, Gauss method, Hazra's simplification rule and expansion; determinant of sum and product of matrices and partitioned matrices; some special determinants; applications to vector algebra, elimination, geometry, electric network.*

6.1 Definition and introduction

Determinant A is a scalar valued function of a square matrix A. It is abbreviated as "det A" or D or $|A|$. The symbol used for a determinant is a pair of vertical bars $|\cdots|$ due to Cayley in 1841 and this has been continuing as standard symbol for more than a century. Thus the determinant of the *n square* matrix, $A = [a_{ij}]_{n\times n}$ is

$$\det A = \det (A) = |A| = D = |a_{ij}|_{n\times n}$$

and in terms of principal diagonal elements it is often written as

$$|a_{11} \quad a_{22} \quad a_{33} \quad \cdots \quad a_{nn}|$$

which is not convenient because this notation fails to distinguish between $|A|$ and $|A|^T$.

The double suffices bear the same meaning as in the matrix in earlier chapters. The *fundamental difference* between a matrix and a determinant is that the matrix is a tabular information/representation of a real world problem and the determinant is a single number denoting the scalar valued function of a square matrix. The determinant is always square whereas the matrix may be square as well as rectangular. Corresponding to every square matrix there exists a determinant although elements are not scalar constants.

The determinant arises in many connections such as scalar and vector triple products, inversion of matrices, solution of simultaneous equations, finding the area of a triangle and the volume of a parallelepiped, etc. It plays a significant role in many theoretical investigations as well as in the solution of many engineering problems such as in the determination of the nature of stability of structural mechanical and electrical systems, in the determination of amplitudes and modes of oscillations of a dynamic system, transient behaviour of electrical circuits and in solving many problems of machine theory.

Historically, the determinant was invented earlier than the matrix. But nowadays, the determinant is defined in terms of a matrix. It may have originated as a solution procedure of simultaneous algebraic equations, because the older name of the determinant is *eliminant*. In 1894, Cayley in reply to a question writes, "I certainly did not get the notion of a matrix in any way through quaternions: it was either directly from that of a determinant or as a convenient mode of expression of the equations

$$x_1 = ax + by$$
$$y_1 = cx + dy\text{.''}$$

This chapter is devoted to the evaluation procedures, properties and some applications of determinant. This knowledge is expected to be used in the further study of matrices in the following chapters and some previous sections of chapters 3 to 5.

6.2 Fundamental definition

As a matrix a determinant can be independently defined as a square array of n^2 elements a_{ij} bordered by vertical bars. The subscripts and the word array bear the same meaning as in the matrix in sec 1.1. Thus

$$D = \begin{vmatrix} a_{11} & a_{12} & \cdots & a_{1n} \\ a_{21} & a_{22} & \cdots & a_{2n} \\ \vdots & \vdots & \ddots & \vdots \\ a_{n1} & a_{n2} & \cdots & a_{nn} \end{vmatrix} = |a_{ij}|$$

is an nth order determinant, the n^2 element, may be number, variable, determinants etc.

6.3 Direct evaluation and definition

The title of the section means that the definition of the determinant is itself a procedure for evaluating the determinant to a single number. This is the oldest method of evaluation of the determinant. This procedure justifies the older name eliminant for determinant.

Let us begin with a single algebraic equation with a single unknown. Let

$$a_{11}x_1 = b_1. \qquad \text{Therefore, the solution } x_1 = \frac{b_1}{a_{11}},\ a_{11} \neq 0. \tag{6.1}$$

Again the 1×1 coefficient matrix for this single equation is $[a_{11}]$. Hence by definition, the determinant of this 1×1 matrix is

$$\det [a_{11}] = |a_{11}| = a_{11} \tag{6.2}$$

a number itself. This number is the denominator of the solution (6.1).

Let us consider two simultaneous algebraic equations

$$\left.\begin{aligned} a_{11}x_1 + a_{12}x_2 &= b_1 \\ a_{21}x_2 + a_{22}x_2 &= b_2 \end{aligned}\right\} \tag{6.3}$$

By elimination or by the method of cross multiplication, the solutions are

$$\left.\begin{aligned} x_1 &= \frac{(b_1a_{22} - b_2a_{12})}{(a_{11}a_{22} - a_{12}a_{21})} \\ x_2 &= \frac{(b_2a_{11} - b_1a_{21})}{(a_{11}a_{22} - a_{12}a_{21})} \end{aligned}\right\} \tag{6.4}$$

where $a_{11}a_{22} - a_{12}a_{21} \neq 0$. This nonzero quantity or number is actually obtained by the popular method of cross multiplication

$$\begin{matrix} a_{11} & & a_{12} & - \\ & \times & & = a_{11}a_{22} - a_{12}a_{21} \\ a_{21} & & a_{22} & + \end{matrix} \tag{6.5}$$

The downward arrow from left top to right down element is + ve whereas the upward arrow from left down to top right element is –ve. The 2×2 coefficient matrix of (6.3) is

$$\begin{bmatrix} a_{11} & a_{12} \\ a_{21} & a_{22} \end{bmatrix}.$$

So, the determinant of this matrix is

$$\det\begin{bmatrix} a_{11} & a_{12} \\ a_{21} & a_{22} \end{bmatrix} = \begin{vmatrix} a_{11} & a_{12} \\ a_{21} & a_{22} \end{vmatrix} \tag{6.6}$$

which is identical with the L.H.S. arrangement of (6.5).
Therefore, from R.H.S. of (6.5) and (6.6), we get

$$\begin{vmatrix} a_{11} & a_{12} \\ a_{21} & a_{22} \end{vmatrix} = a_{11}a_{22} - a_{12}a_{21} \tag{6.7}$$

= {(product of the main diagonal elements) – (product of secondary diagonal elements)} = (a single number) .

By the same method of elimination three nonhomogeneous equations

$$\left.\begin{aligned} a_{11}x_1 + a_{12}x_2 + a_{13}x_3 &= b_1 \\ a_{21}x_1 + a_{22}x_2 + a_{23}x_3 &= b_2 \\ a_{31}x_1 + a_{32}x_2 + a_{33}x_3 &= b_3 \end{aligned}\right\} \tag{6.8}$$

can be solved for x_1, x_2, x_3 with the nonzero denominator

$$a_{11}a_{22}a_{33} + a_{12}a_{23}a_{31} + a_{13}a_{21}a_{32} - a_{31}a_{22}a_{13} - a_{32}a_{23}a_{11} - a_{33}a_{21}a_{12} \tag{6.9}$$

$$= a_{11}(a_{22}a_{33} - a_{32}a_{23}) - a_{12}(a_{21}a_{33} - a_{23}a_{31}) + a_{13}(a_{21}a_{32} - a_{31}a_{22})$$

$$= a_{11}\begin{vmatrix} a_{22} & a_{23} \\ a_{32} & a_{33} \end{vmatrix} - a_{12}\begin{vmatrix} a_{21} & a_{23} \\ a_{31} & a_{33} \end{vmatrix} + a_{13}\begin{vmatrix} a_{21} & a_{22} \\ a_{31} & a_{32} \end{vmatrix} \tag{6.10}$$

$$= \begin{vmatrix} a_{11} & a_{12} & a_{13} \\ a_{21} & a_{22} & a_{23} \\ a_{31} & a_{32} & a_{33} \end{vmatrix} \tag{6.11}$$

= det {(coefficient matrix of equations (6.8)}

From (6.1), (6.5) and (6.9), it is observed that respective denominators consist of $n!$ terms and for $n > 1$, half of the terms are positive and half are negative. Each term consists of a product of n factors a_{ij} of which both row and column suffices of the positive terms maintain the natural order numbers (including anticlockwise) whereas those of negative terms either row or column suffices do not maintain the natural order and become inverted.

By inversion is meant

$$|A| = D = \Sigma(-1)^t\ a_{1j_1}\ a_{2j_2} \cdots a_{nj_n} \text{ (all permutation } j) \tag{6.12}$$

and $|A| = D = \sum (-1)^t \, a_{i_1 1} \, a_{i_2 2} \cdots a_{i_n n}$ (all permutation i) (6.13)

according to which inversion occurs in column and row suffices.

Thus, the determinant of an nth order matrix $A = [a_{ij}]_{n \times n}$ is defined to be the computed sum in (6.12) or (6.13) of $n!$ terms each being a product of n factors. This definition is due to Cauchy, Jacobi and others. The definition (6.12) is known as *row expansion* of det A and (6.13) is known as *column expansion* of det A. This definition is based on the evaluation of the determinant also and is known as *direct method* of evaluation.

Number of operations : The direct evaluation requires $n!$ additions and $n!(n-1)$ multiplication. The procedure becomes cumbersome for higher order determinants.

Sarrus rule : P.F Sarrus (1798-1861) developed a simple rule for the evaluation of third order determinant only by direct method. In this rule:

(i) 2 columns (2 rows) from the remotest to the nearest are written to the right or left (top or bottom) of the determinant;

(ii) product of 3 factors are taken downwards along and parallel to principal diagonal and upwards along and parallel to secondary diagonal.

(iii) downward terms are taken positive while upward terms are taken negative and finally

(iv) algebraic sum of these products gives the value of the determinant.

Ex . Evaluate the following determinant using the Sarrus rule

$$D = \begin{vmatrix} 3 & 7 & -1 \\ 4 & 0 & 2 \\ 5 & -2 & 3 \end{vmatrix}.$$

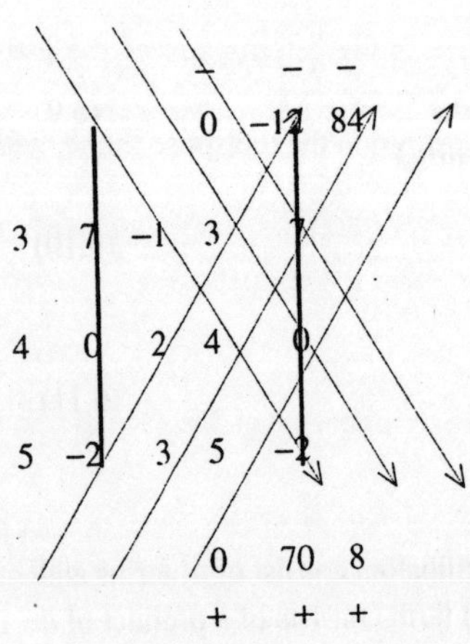

Procedure 1: Last two columns are written to the right of the determinant.

$\therefore D = (0+70+8)-(0-12+84)$

$= 78 - 72 = 6$

− − −
−12 84 0
7 −1 3 7 −1
0 2 4 0 2
−2 3 5 −2 3
70 8 0
+ + +

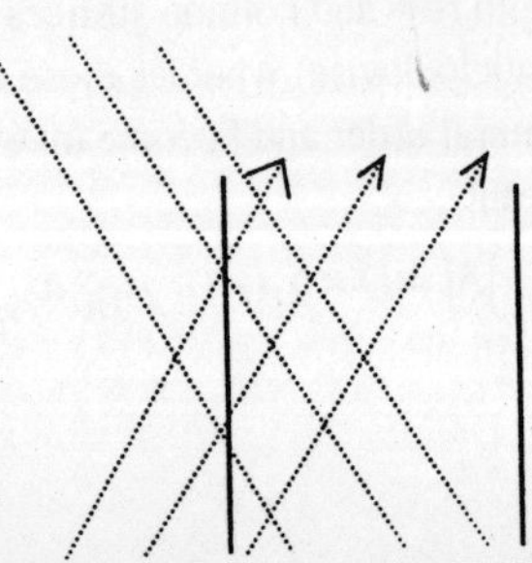

Procedure 2: First two columns are written to the left of the determinant.

$\therefore\ D = 78 - (-12 + 84)$

$= 78 - 72 = 6$

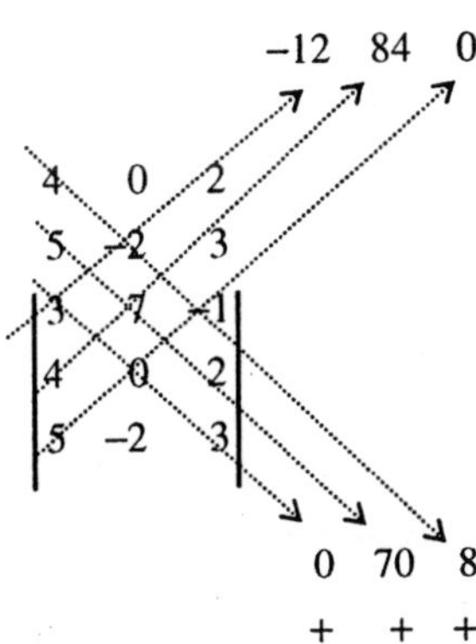

Procedure 3: Last two rows are written to top of the determinant.

$\therefore\ D = (0 + 70 + 8) - (0 - 12 + 84)$
$= 78 - 72 = 6$

Procedure 4: First two rows may be written at the bottom of the determinant. The result is the same.

6.4 Basic theorems on properties

Theorem 1 . The interchange of rows into columns and vice-versa of an *n square* matrix does not alter the value of the corresponding determinant. That is,

$$\det A^T = \det A.$$

In other words, the determinant of the transposed of a matrix is the determinant of the matrix itself.

Proof . Let $A = [a_{ij}]_{n\times n}$ be an *n square* matrix. Therefore, by definition the transpose of the matrix is $A^T = [a_{ji}]$. Now by equation (6.12),

$$\det A = \Sigma (-1)^t\ a_{1j_1}\, a_{2j_2}\ \cdots\ a_{nj_n} \tag{6.14}$$

and by equation (6.13)

$$\det A^T = \Sigma (-1)^t\ a_{i_1 1}\, a_{i_2 2}\ \cdots\ a_{i_n n} \tag{6.15}$$

The R.H.S. of (6.14) and (6.15) are actually equal on rearrangement of the factors. To be specific, let

$$A = \begin{bmatrix} a_{11} & a_{12} & a_{13} & a_{14} \\ a_{21} & a_{22} & a_{23} & a_{24} \\ a_{31} & a_{32} & a_{33} & a_{34} \\ a_{41} & a_{42} & a_{43} & a_{44} \end{bmatrix} \tag{6.16}$$

and
$$A^T = \begin{bmatrix} a_{11} & a_{21} & a_{31} & a_{41} \\ a_{12} & a_{22} & a_{32} & a_{42} \\ a_{13} & a_{23} & a_{33} & a_{43} \\ a_{14} & a_{24} & a_{34} & a_{44} \end{bmatrix} \tag{6.17}$$

Using (6.14) in (6.16) and (6.15) in (6.17), we have

$\det A = \Sigma(-1)^t\, a_{1j_1}\, a_{2j_2} a_{3j_3}\, a_{4j_4}$ = a sum of 4! terms = 24 terms (6.18) and

$\det A^T = \Sigma(-1)^t\, a_{j_1 1}\, a_{j_2 2}\, a_{j_3 3}\, a_{j_4 4}$ = a sum of 4! terms = 24 terms (6.19) Let us consider the permutation 2, 1, 4, 3 of j_r's, $r = 1, 2, 3, 4$.

From (6.18),

$$(-1)^{-1} a_{12}a_{21}a_{34}a_{43} = a_{12}a_{21}a_{34}a_{43} \tag{6.20}$$

and from (6.19),

$$(-1)^2\, a_{21}\, a_{12}\, a_{43}\, a_{34} = a_{21}\, a_{12}\, a_{43}\, a_{34} = a_{12}\, a_{21}\, a_{34}\, a_{43} \tag{6.21}$$

by rearrangement of factors.

Therefore (6.20) and (6.21) are equal in all respects.

Proceeding in this way, it can be shown by rearrangement among factors that R.H.S. of (6.18) and (6.19) are equal in all respects. Therefore, from (6.18) and (6.19) and hence from (6.14) and (6.15) we can conclude $\det A^T = \det A$.

Thus it proves the theorem about the interchange of rows and columns and hence establishes a fundamental distinction between a matrix and a determinant

i.e. $A^T \neq A$ but $\det A^T = \det A$.

Theorem 2. If every element of one row (column) of an *n square* matrix is zero, then det A = 0.

Proof. Let $A = [a_{ij}]_{n \times n}$ and let all the elements in the *i*th row are zero. That is,

$$A = \begin{bmatrix} a_{11} & a_{12} & \cdots & a_{1n} \\ a_{21} & a_{22} & \cdots & a_{2n} \\ \vdots & \vdots & \vdots & \vdots \\ 0_{i1} & 0_{i2} & \cdots & 0_{in} \\ \vdots & \vdots & \vdots & \vdots \\ a_{n1} & a_{n2} & \cdots & a_{nn} \end{bmatrix}, \leftarrow i\text{th row.}$$

Now by definition equation (6.12), we have,

$$|A| = \sum_{\substack{\text{all} \\ \text{permutation} \\ \text{over } j}} (-1)^t\, a_{1j_1}\, a_{2j_2} \cdots a_{nj_n} \tag{6.22}$$

Each term of which consists of exactly one factor of each row and each column. Since every element of the *i*th row is zero, so each term of *n* products must contain one element from the *i*th row and hence the term results into zero. Thus each of the *n*! terms in (6.20) becomes zero. Therefore, |A| = 0. Similar arguments can be given to prove the other case when the elements of a column are zero.

Alternatively, the theorem states that the determinant of a matrix with the zero row (column) is zero. The words zero row (column) means every element of a row (column) is zero.

Theorem 3. Interchange of two rows (columns) of a matrix changes the sign of the corresponding determinant.

Proof. Let A = $[a_{ij}]_{n\times n}$ be an *n square* matrix. Let interchange of *r*th row and *s*th row ($r < s$) result into a matrix B. Then obviously, $b_{rj} = a_{sj}$, $b_{sj} = a_{rj}$ and $b_{ij} = a_{ij}$ for $i \neq r, i \neq s$.

Now by definition in equation (6.12)

$$|B| = \sum_{\substack{\text{all}\\ \text{permutations}\\ \text{over } j}} (-1)^t\, b_{1j_1}\, b_{2j_2} \cdots b_{rj_r} \cdots b_{sj_s} \cdots b_{nj_n}$$

$$= \sum_{\substack{\text{all}\\ \text{permutations}\\ \text{over } j}} (-1)^t\, a_{1j_1}\, a_{2j_2} \cdots a_{sj_r} \cdots a_{rj_s} \cdots a_{nj_n}$$

$$= \sum_{\substack{\text{all}\\ \text{permutations}\\ \text{over } j}} -(-1)^t\, a_{1j_1}\, a_{2j_2} \cdots a_{rj_s} \cdots a_{sj_r} \cdots a_{nj_n}$$

This rearrangement of two factors creates a new permutation either with an even or odd inversion and hence its net effect on *t* is to multiply $(-1)^t$ by (–1). Therefore |B| = – |A|.

Now let B arise from A by an interchange of two columns of A. This is equivalent to "B^T arises from A^T by an interchange of two rows". So, $|B^T| = -|A^T|$. But $|B^T| = |B|$ and $|A^T| = |A|$. Hence |B| = |A|.

This property is analogous to that of definite integral, $\int_a^b f(x)dx = -\int_b^a f(x)dx$ and is known as the *alternating* property.

Theorem 4. If two rows (columns) of the *n square* matrix A are equal, then the corresponding determinant is zero, i.e. |A| = 0.

Proof. Let the *r*th and *s*th row of A be equal. Now interchanging *r*th and *s*th row of A let us have a matrix B. Therefore |B| = – |A|. Since *r*th and *s*th row of A are equal, so, B = A. Hence |A| = – |A| , i.e. 2|A| = 0, i.e. |A| = 0. Similar arguments hold for equal columns and the truth of the theorem can be established. **Geometrically,** this is the condition of coplanarity of vectors as the volume generated by |A| collapses due to the linear combination of side of the parallelotope of sec. 6.18.

Theorem 5. If a row (column) of an *n square* matrix A is multiplied by a real number *c* then determinant A gets multiplied by *c*.

Proof. Let A = $[a_{ij}]_{n\times n}$. Let *r*th row of A be multiplied by the real number *c* to obtain the matrix B = $[b_{ij}]_{n\times n}$. Then $b_{ij} = a_{ij}$ if $i \neq r$ and $b_{rj} = ca_{rj}$. Now definition of equation (6.12)

$$|B| = \sum_{\substack{\text{all}\\ \text{permutations}\\ \text{over } j}} (-1)^t\, b_{1j_1}\, b_{2j_2} \cdots b_{rj_r} \cdots b_{nj_n}$$

$$= \sum_{\substack{\text{all}\\ \text{permutations}\\ \text{over } j}} (-1)^t\, a_{1j_1}\, a_{2j_2} \cdots \left(ca_{rj_r}\right), \cdots a_{nj_n}$$

$$= c \sum_{\substack{\text{all}\\ \text{permutations}\\ \text{over } j}} (-1)^t\, a_{1j_1}\, a_{2j_2} \cdots a_{rj_r} \cdots a_{nj_n} \qquad = c\,|A|$$

Similar arguments hold for the column and the truth of the theorem can be easily established.

Cor. 1 If every element of an nth order matrix A is multiplied by scalar c, then

$$|cA| = c^n |A| \text{ and hence } |-A| = (-1)^n |A|.$$

Cor. 2 If two rows (columns) of an *n square* matrix A are proportional then $|A| = 0$.

Proof. Follows from above two theorems 2, 5.

Cor. 3 This is the linearity property of determinants.

Cor. 4 Computation $|A|$ can be simplified by this theorem. Any common factor in any row (column) can be taken out as a common factor. This theorem also helps in factorization of a determinant of a matrix. It is also known as the *factor theorem.*

For example

$$\begin{vmatrix} 2 & 4 & 6 \\ 1 & 2 & 3 \\ 4 & 5 & 6 \end{vmatrix} = 2\begin{vmatrix} 1 & 2 & 3 \\ 1 & 2 & 3 \\ 4 & 5 & 6 \end{vmatrix}$$

2 has been factored out from first row and then use of theorem 4 gives the value zero.

In the expression

$$|A| = \begin{vmatrix} 1 & a & a^2 \\ 1 & b & b^2 \\ 1 & c & c^2 \end{vmatrix} = \begin{vmatrix} 0 & a-b & a^2-b^2 \\ 0 & b-c & b^2-c^2 \\ 1 & c & c^2 \end{vmatrix},$$

theorem 6 has been used twice. From first and second row corresponding elements of second and third row have been subtracted. By theorem 5,

$$|A| = (a-b)(b-c)\begin{vmatrix} 0 & 1 & a+b \\ 0 & 1 & b+c \\ 1 & c & c^2 \end{vmatrix} = (a-b)(b-c)(c-a)$$

by cofactor expansion. This is one way of factorization of $|A|$.

Alternative method in sec. 6.16 on factorization.

Theorem 6. If to each elements of rth row (column) is added c times the corresponding elements of sth row, $r \neq s$ of an *n square* matrix then the value of the corresponding determinant remains unchanged.

An *equivalent statement* of the theorem is adding a multiple of one row (column) to another row (column) of an *n square* matrix yields a new *n square* matrix whose determinant is the same as that of the original matrix.

Proof . Let $A = [a_{ij}]_{n\times n}$ be an *n square* matrix. Let a new matrix $B = [b_{ij}]_{n\times n}$ be obtained from A by adding to each element of the rth row c times the corresponding element of the sth row, $r \neq s$, say $r < s$. Then obviously $b_{ij} = a_{ij}$ for $i \neq r$ and $b_{rj} = a_{rj} + ca_{sj}$, $r \neq s, r < s$.

Now using equation (6.13),

$$|B| = \sum_{\substack{\text{all} \\ \text{permutaions} \\ \text{over } j}} (-1)^t b_{1j_1}\, b_{2j_2} \cdots b_{rj_r} \cdots b_{sj_s} \cdots b_{nj_n}$$

$$= \sum_{\substack{\text{all} \\ \text{permutaions} \\ \text{over } j}} (-1)^t a_{1j_1}\, a_{2j_2} \cdots \left(a_{rj_r} + ca_{sj_r}\right) \cdots a_{sj_s} \cdots a_{nj_n}$$

$$= \sum_{\substack{\text{all} \\ \text{permutaions} \\ \text{over } j}} (-1)^t a_{1j_1}\ a_{2j_2} \cdots a_{rj_r} \cdots a_{sj_s} \cdots a_{nj_n} +$$

$$\sum_{\substack{\text{all} \\ \text{permutaions} \\ \text{over } j}} (-1)^t a_{1j_1}\ a_{2j_2} \cdots \left(ca_{sj_r}\right) \cdots a_{sj_s} \cdots a_{nj_n}$$

$$= |A| + c\begin{bmatrix} a_{11} & a_{12} & \cdots & \cdots & \cdots & \cdots & \cdots & a_{1n} \\ a_{21} & a_{22} & \cdots & \cdots & \cdots & \cdots & \cdots & a_{2n} \\ \vdots & \vdots & \ddots & \vdots & \vdots & \vdots & \vdots & \vdots \\ a_{s1} & a_{s2} & \cdots & \ddots & \cdots & \cdots & \cdots & a_{sn} \\ \vdots & \vdots & \vdots & \vdots & \ddots & \vdots & \vdots & \vdots \\ a_{s1} & a_{s2} & \cdots & \cdots & \cdots & \ddots & \cdots & a_{sn} \\ \vdots & \vdots & \vdots & \vdots & \vdots & \vdots & \ddots & \vdots \\ a_{n1} & a_{n2} & \cdots & \cdots & \cdots & \cdots & \cdots & a_{nn} \end{bmatrix} \begin{matrix} \\ \\ \\ \leftarrow \text{identical row} \\ \\ \leftarrow \text{identical row} \\ \\ \\ \end{matrix}$$

$$= |A| + c.0 = |A|$$

Similar argument holds for column.
In vector notation when rth row of A is replaced by {(rth row + c (sth row)}

$$|B| = \det\begin{bmatrix} \mathbf{y}_{1.} \\ \mathbf{y}_{2.} \\ \vdots \\ \mathbf{y}_{r.} + c\mathbf{y}_{s.} \\ \vdots \\ \mathbf{y}_{s.} \\ \vdots \\ \mathbf{y}_{n.} \end{bmatrix} = \det\begin{bmatrix} \mathbf{y}_{1.} \\ \mathbf{y}_{2.} \\ \vdots \\ \mathbf{y}_{r.} \\ \vdots \\ \mathbf{y}_{s.} \\ \vdots \\ \mathbf{y}_{n.} \end{bmatrix} + c\ \det\begin{bmatrix} \mathbf{y}_{1.} \\ \mathbf{y}_{2.} \\ \vdots \\ \mathbf{y}_{s.} \\ \vdots \\ \mathbf{y}_{s.} \\ \vdots \\ \mathbf{y}_{n} \end{bmatrix} \begin{matrix} \longrightarrow \\ \\ \longrightarrow \end{matrix} \left.\begin{matrix} r\text{th row} \\ \text{and} \\ s\text{th row are} \\ \text{identical} \end{matrix}\right\}$$

$$= \det\begin{bmatrix} \mathbf{y}_{1.} \\ \mathbf{y}_{2.} \\ \vdots \\ \mathbf{y}_{r.} \\ \vdots \\ \mathbf{y}_{n.} \end{bmatrix} = |A|\ .$$

If rth column of A is replaced by {(rth column + c (sth column)} then the vector notation

$$|B| = \det\left[\mathbf{x}_{.1}\ \ \mathbf{x}_{.2} \cdots \left(\mathbf{x}_{.r} + c\mathbf{x}_{.s}\right) \cdots\ \mathbf{x}_{.s} \cdots \mathbf{x}_{.n}\right]$$

$$= \det\left[\mathbf{x}_{.1}\ \ \mathbf{x}_{.2} \cdots\ \mathbf{x}_{.r} \cdots \mathbf{x}_{.s} \cdots\ \mathbf{x}_{.n}\right] + c\det\left[\mathbf{x}_{.1}\ \ \mathbf{x}_{.2} \cdots \mathbf{x}_{.r} \cdots \mathbf{x}_{.s} \cdots \mathbf{x}_{.n}\right]$$

rth ⋁ sth

columns are identical

$$= \det\left[\mathbf{x}_{.1}\ \mathbf{x}_{.2} \cdots \mathbf{x}_{.r} \cdots \mathbf{x}_{.s} \cdots \mathbf{x}_{.n}\right]$$

$$= |A|$$

6.5 Minor and cofactor

Minor and cofactor play an important role in the further study of the theory of matrices and determinants.

6.5.1 Minor : The minor determinant or simply minor of the element a_{ij} of the determinant of an *n square* matrix A = $[a_{ij}]_{n\times n}$, $n \geq 2$, is defined as the determinant of the $(n-1)\times(n-1)$ partition block obtained by deleting *i*th row and *j*th column of A. As there is no convention, different authors use different symbols for it or no symbol at all. Let M_{ij} denote the minor of a_{ij}. So,

$$M_{ij} = \begin{vmatrix} a_{11} & a_{12} & \cdots & a_{1j} & \cdots & a_{1n} \\ a_{21} & a_{22} & \cdots & a_{2j} & \cdots & a_{2n} \\ \vdots & \vdots & \ddots & \vdots & \vdots & \vdots \\ a_{i1} & a_{i2} & \cdots & a_{ij} & \cdots & a_{in} \\ \vdots & \vdots & \vdots & \vdots & \ddots & \vdots \\ a_{n1} & a_{n2} & \cdots & a_{nj} & \cdots & a_{nn} \end{vmatrix}$$

deleting *i*th row

deleting *j*th column

In particular, let $A = \begin{vmatrix} 2 & 3 & 4 & 5 \\ -2 & 1 & 0 & 7 \\ -1 & -3 & -4 & -5 \\ 8 & 9 & 10 & 11 \end{vmatrix}$

The minor of $a_{33} = -4$ in |A| is obtained by deleting the third row and the third column in which $a_{33} = -4$ belongs and it is

$$M_{33} = \begin{vmatrix} 2 & 3 & 5 \\ -2 & 1 & 7 \\ 8 & 9 & 11 \end{vmatrix} = \sum_{\substack{\text{all permutation} \\ \text{over } j}} (-1)^t a_{1j_1} a_{2j_2} a_{3j_3}, \text{ by equation 6.12}$$

$$= (-1)^0 a_{11}a_{21}a_{33} + (-1)^1 a_{11}a_{23}a_{32} + (-1)^2 a_{13}a_{21}a_{32} + (-1)^3 a_{13}a_{22}a_{31}$$
$$+ (-1)^1 a_{12}a_{21}a_{33} + (-1)^2 a_{12}a_{23}a_{31}$$

$$= 2.1.11 + 5(-2)9 + 3.7.8 - (2.7.9 + 3.-2.11 + 5.1.8)$$

$$= 22 - 90 + 168 - (126 - 66 + 40) = 100 - 100 = 0$$

By Sarrus rule, $M_{33} = (22 + 168 - 90) - (40 + 126 - 66) = 0$.

6.5.2 Principal and leading minors : Two more minors may be defined. They are *principal minor* (P.M.) and *leading principal minor* (L.P.M.). They are the determinants of principal and leading submatrices (block). Principal minors play an important role in determining positive, negative, semipositive and indefinite feature of a matrix as well as a quadratic form. The P.M.s of the matrix A = $[a_{ij}]_{n\times n}$ are the values D_k, $k = 1, 2, \cdots, n$ defined as

$$D_1 = \det[a_{11}],\ D_2 = \det\begin{bmatrix} a_{11} & a_{12} \\ a_{21} & a_{22}\end{bmatrix},\ D_3 = \det\begin{bmatrix} a_{11} & a_{12} & a_{13} \\ a_{21} & a_{22} & a_{23} \\ a_{31} & a_{32} & a_{33}\end{bmatrix},\ \cdots,\ D_n = \det A.$$

In other words P.M. is the determinant D_k in the top left hand corner $k{\times}k$ block of the matrix A.

6.5.3 Cofactor : The cofactor is useful in the computation of determinant and has provided a number of useful theorems. The cofactor can be defined in a simple way as the signed minor. The cofactor of a_{ij} is denoted by corresponding italic capital letter A_{ij}. The mechanical way of prefixing a sign to the minor of an element is to associate alternative (+) and (–) signs to each element of a determinant as follows

$$\begin{vmatrix} + & - & + & \cdots & (-1)^{1+n} \\ - & + & - & \cdots & (-1)^{2+n} \\ + & - & + & \cdots & (-1)^{3+n} \\ \vdots & \vdots & \vdots & \vdots & \vdots \\ (-1)^{n+1} & (-1)^{n+2} & (-1)^{n+3} & \cdots & (-1)^{n+n}\end{vmatrix}.$$

For a large determinant, this procedure is not suitable. To have a sound and mathematical basis, the following theorem is necessary.

Theorem 1 . The cofactor A_{ij} of the element a_{ij} of an *n square* matrix $A = [a_{ij}]_{n\times n}$ is $(-1)^{i+j}$ times the minor of a_{ij}. In other words,

$$A_{ij} = (-1)^{i+j} M_{ij}$$

Proof . Let us isolate the matrix $A = [a_{ij}]_{n\times n}$ by partitions. Let

$$\det A = \left|\begin{array}{c:c:c} P & Q & R \\ \hdashline S & a_{ij} & T \\ \hdashline U & V & W\end{array}\right|.$$

Now move a_{ij} to nn position by $(n-j)$ interchange of columns and by $(n-i)$ interchange of rows as follows:

$$\therefore \quad \det A = (-1)^{n-j}\left|\begin{array}{c:c:c} P & R & Q \\ \hdashline S & T & a_{ij} \\ \hdashline U & W & V\end{array}\right| = (-1)^{n-j}(-1)^{n-i}\left|\begin{array}{c:c:c} P & R & Q \\ \hdashline U & W & V \\ \hdashline S & T & a_{ij}\end{array}\right|$$

$$= (-1)^{2n-(i+j)}\left|\begin{array}{c:c:c} P & R & Q \\ \hdashline U & W & V \\ \hdashline S & T & a_{ij}\end{array}\right| = (-1)^{i+j}\left|\begin{array}{c:c:c} P & R & Q \\ \hdashline U & W & V \\ \hdashline S & T & a_{ij}\end{array}\right|$$

$$\because \quad (-1)^{2n-(i+j)} = (-1)^{2n}.(-1)^{-(i+j)} = (-1)^{i+j} \qquad \text{as } \left(\frac{1}{-1}\right) = (-1)$$

Also the minor M_{ij} of det A = $\begin{vmatrix} P & R \\ U & W \end{vmatrix}$.

Therefore, cofactor of a_{ij} in det A is given by

$$A_{ij} = (-1)^{i+j} \begin{vmatrix} P & R \\ U & W \end{vmatrix} = (-1)^{i+j} M_{ij} .$$

6.5.4 Expansion by cofactor : A determinant can be expanded or evaluated using cofactors of elements of any row (columns) of the determinant.

Theorem 1. If the elements of a row (column) of an *n square* matrix are multiplied by their corresponding cofactors then the sum of the products gives the expansion of the determinant.

Symbolically, if $A = [a_{ij}]_{n\times n}$ is the matrix, then

$$|A| = a_{1j}A_{1j} + a_{2j}A_{2j} + \cdots + a_{nj}A_{nj} \tag{6.23}$$

$$= \sum_{i=1}^{n} a_{ij}A_{ij}$$

is the expansion with respect to the jth column of A

and

$$|A| = a_{i1}A_{i1} + a_{i2}A_{i2} + \cdots + a_{in}A_{in} \tag{6.24}$$

$$= \sum_{j=1}^{n} a_{ij}A_{ij}$$

is the expansion with respect to the ith of A.

Proof . Since the cofactor of a_{ij} is given by

$A_{ij} = (-1)^{i+j} M_{ij}$

↓ jth column is missing

$$= (-1)^{i+j} \begin{vmatrix} a_{11} & a_{12} & \cdots & a_{1,j-1} & a_{1,j+1} & \cdots & a_{1n} \\ a_{21} & a_{22} & \cdots & a_{2,j-1} & a_{2,j+1} & \cdots & a_{2n} \\ \vdots & \vdots & \ddots & \vdots & \vdots & \vdots & \vdots \\ a_{i-1,1} & a_{i-1,2} & \cdots & a_{i-1,j-1} & a_{i-1,j+1} & \cdots & a_{i-1,n} \\ a_{i+1,1} & a_{i+1,2} & \cdots & a_{i+1,j-1} & a_{i+1,j+1} & \cdots & a_{i+1,n} \\ \vdots & \vdots & \vdots & \vdots & \vdots & \ddots & \vdots \\ a_{n1} & a_{n2} & \cdots & a_{n,j-1} & a_{n,j+1} & \cdots & a_{nn} \end{vmatrix}$$

← ith row is missing

by the definition of minor of a_{ij}. Now the expansion of the minor M_{ij} contains $(n-1)!$ Terms each being a product of $(n-1)$ factors. So by equation (6.12)

$$A_{ij} = (-1)^{i+j} \sum_{\substack{\text{all permutation} \\ \text{over } j\text{'s}}} (-1)^{t'} a_{1j_1} a_{2j_2} a_{3j_3} \cdots a_{i-1,j_{i-1}} a_{i+1,j_{i+1}} \cdots a_{nj_n} \tag{6.25}$$

Now to simplify the notation, let us consider $i = 1$ only. Then from (6.25),

$A_{1j} = (-1)^{1+j} M_{1j}$

$$= \begin{vmatrix} a_{21} & a_{22} & \cdots & a_{2,j-1} & a_{2,j+1} & \cdots & a_{2n} \\ a_{31} & a_{32} & \cdots & a_{3,j-1} & a_{3,j+1} & \cdots & a_{3n} \\ \vdots & \vdots & \vdots & \vdots & \vdots & \vdots & \vdots \\ a_{n1} & a_{n2} & \cdots & a_{n,j-1} & a_{n,j+1} & \cdots & a_{nn} \end{vmatrix}$$

(|← omission of *j*th column; ← omission of 1st row)

$$= (-1)^{1+j} \sum (-1)^{t'} a_{2j_2} a_{3j_3} \cdots a_{nj_n} \tag{6.26}$$

$$|A| = (-1)^{1+j} \sum_{\substack{\text{all permutation}\\ \text{over } j\text{'s}}} (-1)^{t} a_{1j_1} a_{2j_2} a_{3j_3} \cdots a_{nj_n} \tag{6.27}$$

Let us give $j_1 = 1, 2, \cdots, n$ and note that
there is no inversion for $j_1 = 1$, so inversion sign $= (-1)^0$
there is one inversion for $j_1 = 2$, so inversion sign $=(-1)^1$
$\vdots \qquad \vdots \qquad \vdots \qquad \vdots$
there is $(n-1)$ inversion for $j_1 = n$, so inversion sign $= (-1)^{n-1}$;
so Σ is taken over all permutations of $j_2, j_3, \cdots, j_n$. Hence

$$|A| = (-1)^0 \sum (-1)^{t'} a_{11} a_{2j_2} \cdots a_{nj_n} + (-1)^1 \sum (-1)^{t'} a_{12} a_{2j_2} \cdots a_{nj_n} + (-1)^2 \sum (-1)^{t'} a_{13} a_{2j_2} \cdots a_{nj_n} + \cdots + (-1)^{n-1} \sum (-1)^{t'} a_{1n} a_{2j_2} \cdots a_{nj_n}$$

$$= (-1)^{1+1} a_{11} \sum (-1)^{t'} a_{2j_2} \cdots a_{nj_n} + (-1)^{1+2} a_{12} \sum (-1)^{t'} a_{2j_2} \cdots a_{nj_n} + (-1)^{1+3} a_{13} \sum (-1)^{t'} a_{2j_2} \cdots a_{nj_n} + \cdots + (-1)^{1+n} a_{1n} \sum (-1)^{t'} a_{2j_2} \cdots a_{nj_n}$$

(multiplying each term by $(-1)^{1+1} = 1$)

$$= a_{11}A_{11} + a_{12}A_{12} + \cdots + a_{1n}A_{1n} \qquad \text{by equation (6.26)}$$

Proceeding in this way similar equations like (6.26) may be obtained with $i = 2, 3, \cdots, n$. Therefore, for any i, relation equivalent to (6.26) is

$$|A| = a_{i1}A_{i1} + a_{i2}A_{i2} + \cdots + a_{in}A_{in} = \sum_{j=1}^{n} a_{ij}A_{ij}.$$

Hence (6.24) is established.

Similarly (6.23) can be established using $j = 1$ in (6.25) and $i_1 = 1, 2, \cdots, n$ in the equation equivalent to (6.27).

Alternatively, the steps after equation (6.27) can explained with specific values of j_r's, $r = 1, 2, \cdots, n$. Let $r = 1, 2, 3$ only and each of j_r's ranges from 1 to 3.

Hence from (6.27)

$$|A| = (-1)^0 \sum (-1)^{t'} a_{11} a_{2j_2} a_{3j_3} + (-1)^1 \sum (-1)^{t'} a_{12} a_{2j_2} a_{3j_3} + (-1)^2 \sum (-1)^{t'} a_{13} a_{2j_2} a_{3j_3}$$

$$= (-1)^0 a_{11} \sum (-1)^{t'} a_{2j_2} a_{3j_3} + (-1)^1 a_{12} \sum (-1)^{t'} a_{2j_2} a_{3j_3} + (-1)^2 a_{13} \sum (-1)^{t'} a_{2j_2} a_{3j_3}$$

$$= a_{11}(a_{22}a_{33} - a_{23}a_{32}) - a_{12}(a_{21}a_{33} - a_{23}a_{31}) + a_{13}(a_{21}a_{32} - a_{22}a_{31})$$

$= a_{11}A_{11} + a_{12}A_{12} + a_{13}A_{13}$, for the third order determinant

Theorem 2. If each element of one row (column) of an *n square* matrix A is multiplied by the cofactor of the corresponding element of another row (column) of the same matrix A, then the sum of the products is zero.

Symbolically, $|A| = \begin{vmatrix} a_{11} & a_{12} \\ a_{21} & a_{22} \end{vmatrix} = \sum_{\substack{\text{permutation} \\ \text{over all } j}} (-1)^t \, a_{1j_1} a_{2j_2} = a_{11}a_{22} - a_{21}a_{12}$ is a number

$$\sum_{j=1}^{n} a_{ij}A_{kj} = a_{i1}A_{k1} + a_{i2}A_{k2} + \cdots + a_{in}A_{ki} = 0 \quad \text{for } i \neq k \tag{6.28}$$

$$\sum_{i=1}^{n} a_{ij}A_{ik} = a_{1j}A_{1k} + a_{2j}A_{2k} + \cdots + a_{nj}A_{nk} = 0 \quad \text{for } j \neq k \tag{6.29}$$

Proof. To make the proof easily understandable, let $i = 1$ and $k = 3$, i.e., the elements of the first row are being multiplied by the cofactors of the corresponding elements of the third row of the matrix

$$A = [a_{ij}]_{n \times n}.$$

Therefore, $A_{31} = (-1)^{3+1} M_{31} = M_{31}$; $A_{32} = (-1)^{3+2} M_{32} = -M_{32}$; $\cdots$;
$A_{3n} = (-1)^{3+n} M_{3n}$.

$$= \begin{vmatrix} a_{11} & a_{12} & \cdots & a_{1n} \\ a_{21} & a_{22} & \cdots & a_{2n} \\ a_{11} & a_{12} & \cdots & a_{1n} \\ \vdots & \vdots & \vdots & \vdots \\ a_{n1} & a_{n2} & \cdots & a_{nn} \end{vmatrix} \begin{matrix} \leftarrow \\ \\ \leftarrow \\ \\ \\ \end{matrix}$$

1st & 3rd rows are same

Hence $\sum_{j=1}^{n} a_{1j}A_{3j} = a_{11}A_{31} + a_{12}A_{32} + \cdots + a_{in}A_{3n}$

Cor. Kronecker delta combines theorem 1 and 2.

Since Kronecker delta is $\delta_{rs} = \begin{cases} 0, & r = s \\ 1, & r \neq s \end{cases}$

Therefore, the equation (6.23) , (6.29) and (6.24), (6.28) respectively can be combined into two equations only instead of four equations. These are

$$a_{i1}A_{k1} + a_{i2}A_{k2} + \cdots + a_{in}A_{kn} = \delta_{ik}\,|A| \tag{6.30}$$

$$a_{1j}A_{1k} + a_{2j}A_{2k} + \cdots + a_{nj}A_{nk} = \delta_{ik}\,|A| \tag{6.31}$$

6.5.5 Induction procedures for cofactor expansion: Cofactor expansion method for the evaluation or expansion of the determinant of an *n square* matrix can be achieved by the induction procedure.

Since the determinant of a 1×1 matrix is a number i.e.

if $A = [a_{11}]$, then $|A| = |a_{11}| = a_{11} = (-1)^{1+1}a_{11}$.

Again the determinant of a 2×2,

matrix $A = [a_{ij}]_{2\times 2} = \begin{bmatrix} a_{11} & a_{12} \\ a_{21} & a_{22} \end{bmatrix}$ is

$$= a_{11}(+a_{22}) + a_{12}(-a_{21}) = (-1)^{1+1}a_{11}(\text{minor of } a_{11}) + (-1)^{1+2}a_{12}(\text{minor of } a_{12})$$

$$= (-1)^{1+1}a_{11}M_{11} + (-1)^{1+1}a_{12}M_{12} = \sum_{j=1}^{2}(-1)^{1+j}a_{1j}M_{1j}.$$

where $(-1)^{1+j}$ ensures the proper sign of each term. So, a 2×2 determinant is expressed in terms of a 1×1 determinant.

Hence $|A| = \sum_{j=1}^{2} a_{1j}M_{1j}$.

Similarly, a determinant of 3×3 matrix can be expressed in terms of a 2×2 determinant. Let a 3×3 matrix be

$$A = [a_{ij}]_{3\times 3} = \begin{bmatrix} a_{11} & a_{12} & a_{13} \\ a_{21} & a_{22} & a_{23} \\ a_{31} & a_{32} & a_{33} \end{bmatrix}$$

Therefore,

$$|A| = \sum_{\substack{\text{permutation} \\ \text{over all } j\text{'s}}} (-1)^t a_{1j_1}a_{2j_2}a_{3j_3}$$

$$= (-1)^0 a_{11}a_{22}a_{33} + (-1)^1 a_{12}a_{21}a_{33} + (-1)^{1+1}a_{12}a_{23}a_{31} + (-1)^{2+0}a_{13}a_{21}a_{32} + (-1)^{2+1}a_{13}a_{22}a_{31} + (-1)^1 a_{11}a_{23}a_{32}$$

$$= a_{11}a_{22}a_{33} - a_{12}a_{21}a_{33} + a_{12}a_{23}a_{31} + a_{13}a_{21}a_{32} - a_{13}a_{22}a_{31} - a_{11}a_{23}a_{32}$$

$$= a_{11}(a_{22}a_{33} - a_{23}a_{32}) - a_{12}(a_{21}a_{33} - a_{23}a_{31}) + a_{13}(a_{21}a_{32} - a_{22}a_{31})$$

$$= a_{11}\begin{vmatrix} a_{22} & a_{23} \\ a_{32} & a_{33} \end{vmatrix} - a_{12}\begin{vmatrix} a_{21} & a_{23} \\ a_{31} & a_{33} \end{vmatrix} + a_{13}\begin{vmatrix} a_{21} & a_{22} \\ a_{31} & a_{32} \end{vmatrix}$$

$$= a_{11}\,(\text{minor of } a_{11}) - a_{12}\,(\text{minor of } a_{12}) + a_{13}\,(\text{minor of } a_{13})$$

$$= (-1)^{1+1}a_{11}M_{11} + (-1)^{1+2}a_{12}M_{12} + (-1)^{1+3}a_{13}M_{13}$$

$$= \sum_{j=1}^{3}(-1)^{1+j}a_{1j}M_{1j} = \sum_{j=1}^{3}a_{1j}A_{1j}$$

Therefore, the 3×3 determinant is expressed in terms of 2×2 determinants. Proceeding in this way, the determinant of the $n\times n$ matrix $A = [a_{ij}]_{n\times n}$ can be expanded in terms of $(n-1)\times(n-1)$ determinants.

Therefore, by definition (6.12),

$$|A| = \sum_{\substack{\text{permutations} \\ \text{over all j's}}} (-1)^{t'} a_{1j_1}\, a_{2j_2}\, a_{3j_3} \cdots a_{nj_n}$$

$$= (-1)^{1+1}a_{11}M_{11} + (-1)^{1+2}a_{12}M_{12} + \cdots + (-1)^{1+n}a_{1n}M_{1n},$$

where M_{ij}'s are $(n-1)\times(n-1)$ determinants.

$$= \sum_{j=1}^{n}(-1)^{1+j}a_{1j}M_{1j} = \sum_{j=1}^{n}a_{1j}A_{1j}$$

Therefore, an $n\times n$ determinant is expressed in terms of an $(n-1)\times(n-1)$ determinants. Hence, the theorem is found true for 1×1, 2×2, $(n-1)\times(n-1)$ and $n\times n$ matrices. If we replace n by $(n+1)$ it will be found true for all $(n+1)\times(n+1)$ square matrices of a finite positive order, n.

Hence the theorem is established with respect to the first row.

By taking out the common factors belonging to the second row, the theorem can be proved true for the second row. In this way, it can be proved for any arbitrary row and hence

$$|A| = |a_{ij}|_{n\times n} = \sum_{j=1}^{n} a_{ij} A_{ij} \ , \quad i = 1, 2, 3, \ \cdots, n.$$

Also in terms of columns, it is

$$|A| = |a_{ij}|_{n\times n} = \sum_{i=1}^{n} a_{ij} A_{ij} \ , \quad j = 1, 2, 3, \ \cdots, n.$$

N.B. The cofactor method of expansion is not the most efficient way (cf. table of comparison). But the method is of theoretical importance but not suitable for computer programming.

6.5.6 Alternate proofs of basic theorems in sec. 6.4 : The basic theorems of section 6.4 are being proved here using cofactors.

Theorem 1. det A^T = det A.

The theorem can be established by induction. It is true for 1×1 matrix. Since transpose of 1×1 matrix, $[a_{11}]^T$ is $[a_{11}]$, the matrix itself. So, det $[a_{11}]^T$ = det $[a_{11}]$.

Let $|A| = \begin{vmatrix} a_{11} & a_{12} \\ a_{21} & a_{22} \end{vmatrix}$ be a 2×2 matrix. So $|A| = a_{11}a_{22} - a_{12}a_{21}$, (6.32)

Then $|A|^T = \begin{vmatrix} a_{11} & a_{21} \\ a_{12} & a_{22} \end{vmatrix}$. Therefore, $|A|^T = a_{11}A_{11} + a_{21}A_{21}$, expansion by first row.

$$= a_{11}a_{22} + a_{21}(-a_{12}) = a_{11}a_{22} - a_{12}a_{21} \qquad (6.33)$$

Therefore, from (6.32) and (6.33), the theorem is found true for 2×2 matrix.

Proceeding in this way, the theorem can be proved true for n×n matrix. Now, if it is possible to show that the theorem is valid for $(n+1)$×$(n+1)$ matrix, then completes the proof by induction.

Let $A = [a_{ij}]_{(n+1)\times(n+1)}$ matrix. So $A^T = [a_{ji}]_{(n+1)\times(n+1)}$.

Now expansion of |A| in terms the elements of the first row and their cofactors of |A| is given by

$$|A| = \sum_{j=1}^{n+1} a_{1j} A_{1j} = a_{11}A_{11} + a_{12}A_{12} + \cdots + a_{1n}A_{1n} + a_{1,n+1}A_{1,n+1} \qquad (6.34)$$

Since the rows of A are identical to the columns of A^T, therefore, the cofactors of the elements of first column of $|A^T|$ are also identical to those of first row of |A|,

i.e. $\quad a_{1j} = a_{j1}$ and $A_{j1} = A_{1j}$ (6.35)

$\therefore \quad |A^T| = \sum_{j=1}^{n+1} a_{j1} A_{j1}$ by (6.23)

$\quad = \sum_{j=1}^{n+1} a_{1j} A_{1j}$ by (6.35)

$\quad = |A|$ by (6.34)

N.B. Let $A_{.j}$ denote column vectors of A.

Then $\quad$ det $[A_{.1} \ \ A_{.2} \ \ \cdots \ \ A_{.n}]^T$ = det $[A_{1.} \ \ A_{2.} \ \ \cdots \ \ A_{n.}]$

Theorem 2. |A| = 0 if all elements of one row (column) of the *n square* matrix A are zero.

Proof . Let the elements of ith row of the matrix A be a_{i}. Now expanding w.r.t. the ith row, we have,

$$|A| = \sum_{j=1}^{n} a_{ij}A_{ij} = \sum_{j=1}^{n} 0\, A_{ij} \qquad \because a_{ij} = 0, j = 1, 2, \cdots, n$$

$$= 0$$

Similarly, for the zero elements in jth column, the expansion w.r.t. the jth column gives the value of the determinant, $|A|$ zero.

Alternatively, since $|A^T| = |A|$, so jth column of A is the jth row of A^T. Hence expanding A^T w.r.t. the jth row, $|A^T| = 0$ as before. Thus the theorem is established.

N.B. Let $\mathbf{A}_{.j}$ and $0_{.j}$ denote nonzero column vectors and zero column vector respectively. Then

$$\det\begin{bmatrix}\mathbf{A}_{\cdot 1} & \mathbf{A}_{\cdot 2} & \cdots & \mathbf{A}_{\cdot i-1} & 0_{\cdot i} & \cdots & \mathbf{A}_{\cdot n}\end{bmatrix} = 0.$$

Theorem 3. $|B| = -|A|$ if B is obtained by the interchange of two rows (columns).

Proof . The theorem is proved by induction.

Let $\qquad A = [a_{ij}]_{2\times 2} = \begin{bmatrix} a_{11} & a_{12} \\ a_{21} & a_{22} \end{bmatrix}.$

Therefore, $\qquad |A| = a_{11}a_{22} - a_{12}a_{21}.$

Let $\qquad B = \begin{bmatrix} a_{21} & a_{22} \\ a_{11} & a_{12} \end{bmatrix}$ obtained from A by the interchange of rows.

Therefore, $|B| = a_{21}a_{12} - a_{11}a_{22} = -(a_{11}a_{22} - a_{21}a_{12}) = -|A|$. So, the theorem is true for 2×2 matrices.

Let $\qquad A = [a_{ij}]_{3\times 3} = \begin{bmatrix} a_{11} & a_{12} & a_{13} \\ a_{21} & a_{22} & a_{23} \\ a_{31} & a_{32} & a_{33} \end{bmatrix}.$

Therefore, $\qquad |A| = \sum_{j=1}^{3} a_{1j}A_{1j}$ expanded w.r.t. first row.

Let $\qquad B = \begin{bmatrix} a_{31} & a_{32} & a_{33} \\ a_{21} & a_{22} & a_{23} \\ a_{11} & a_{12} & a_{13} \end{bmatrix}$ obtained from A by interchanging the first and third rows.

$$\therefore \qquad |B| = a_{31}\begin{vmatrix} a_{22} & a_{23} \\ a_{12} & a_{13} \end{vmatrix} - a_{32}\begin{vmatrix} a_{21} & a_{23} \\ a_{11} & a_{13} \end{vmatrix} + a_{33}\begin{vmatrix} a_{21} & a_{22} \\ a_{11} & a_{12} \end{vmatrix}$$

$$= a_{31}(a_{22}a_{13} - a_{12}a_{23}) - a_{32}(a_{21}a_{13} - a_{23}a_{11}) + a_{33}(a_{21}a_{12} - a_{11}a_{22})$$

$$= -a_{11}(a_{22}a_{33} - a_{23}a_{32}) + a_{12}(a_{21}a_{33} - a_{23}a_{31}) - a_{13}(a_{21}a_{32} - a_{22}a_{31})$$

$$= -\sum_{j=1}^{3} a_{1j}A_{1j} = -|A|$$

Hence the theorem is true for the 3×3 matrix. Expansion can be performed w.r.t. any other rows. In this way it can be shown that the theorem is true for the $n\times n$ matrix. Now, if it is possible to prove the truth for $(n+1)\times(n+1)$ matrices then this completes the proof for the *n square* matrix.

Let $A = [a_{ij}]_{(n+1)\times(n+1)}$ and $B = [b_{ij}]_{(n+1)\times(n+1)}$ obtained from A by interchanging two rows. Expanding B in terms of the element of a row, i, other than those interchanged, so

$$|B| = \sum_{j=1}^{n+1} b_{ij} B_{ij}.$$

Here each b_{ij} , $j = 1, 2, \cdots, (n+1)$, is identical to the corresponding a_{ij} and each B_{ij}, $j = 1, 2, \cdots, (n+1)$ is obtained from the corresponding A_{ij} by interchanging two rows.

Hence $b_{ij} = a_{ij}$ and $B_{ij} = -A_{ij}, j = 1, 2, \cdots, (n+1)$, hence $|B| = -\sum_{j=1}^{n+1} a_{ij} A_{ij} = -|A|$.

A similar argument holds for the column. Alternatively, since $|A^T| = |A|$. Then rows of A^T are interchanged. Therefore $|A^T| = -|A^T| = -|A|$. This completes the proof.

In vector notation, the property can be represented in a better way. If the vector represents columns then

$$\det[\mathbf{A}_{\cdot 1} \quad \mathbf{A}_{\cdot 2} \quad \cdots \quad \underbrace{\mathbf{A}_{\cdot i} \quad \cdots \quad \mathbf{A}_{\cdot j}}_{\text{original position}} \quad \cdots \quad \mathbf{A}_{\cdot n}]$$

$$= -\det[\mathbf{A}_{\cdot 1} \quad \mathbf{A}_{\cdot 2} \quad \cdots \quad \underbrace{\mathbf{A}_{\cdot j} \quad \cdots \quad \mathbf{A}_{\cdot i}}_{\text{interchanged}} \quad \cdots \quad \mathbf{A}_{\cdot n}]$$

Theorem 4 . No alternative.

Theorem 5 . $|B| = c|A|$ if B is obtained by multiplying a row (column) of A by a nonzero scalar c.

Proof . Let $A = [a_{ij}]_{n\times n}$ and $B = [b_{ij}]_{n\times n}$ such that $b_{kj} = ca_{kj}$, $j = 1, 2, \cdots, n$. c is a nonzero scalar. Now expanding $|B|$ w.r.t. kth row,

$$|B| = \sum_{j=1}^{n} b_{kj} B_{kj} = \sum_{j=1}^{n} ca_{kj} A_{kj}.$$

As all other elements except ith row of B are identical to those of A.

$$= c\sum_{j=1}^{n} a_{kj} A_{kj}$$

$$= c|A| \qquad \because \quad |A| = \sum_{j=1}^{n} a_{kj} A_{kj} \text{ when expanded w.r.t. } k\text{th row.}$$

Similar argument holds for columns.

Alternatively, $\quad |A^T| = |A| \quad$ and $\quad |B^T| = |B| = c|A|$.

N.B. Let vector $\mathbf{A}_{\cdot j}$ denotes columns of A. Then

$$\det\left[\mathbf{A}_{\cdot 1} \quad \mathbf{A}_{\cdot 2} \quad \cdots \quad c\mathbf{A}_{\cdot k} \quad \cdots \quad \mathbf{A}_{\cdot n}\right] = c\det\left[\mathbf{A}_{\cdot 1} \quad \mathbf{A}_{\cdot 2} \quad \cdots \quad \mathbf{A}_{\cdot k} \quad \cdots \quad \mathbf{A}_{\cdot n}\right].$$

Cor. 1 If every element of an nth order matrix A is multiplied by scalar c,

then $\quad |cA| = c^n|A|$.

If $\quad c = -1$, then $|-A| = (-1)^n |A|$.

Cor. 2 If two rows (columns) of an nth order matrix A are proportional i.e. if $a_{ij} = ca_{kj}$, i.e., each element of ith row is equal c times the corresponding elements of kth row then $|A| = 0$.

Proof . Proof follows from theorem 2 and 5.

Theorem 6 . $|B| = |A|$ such that $B = [b_{kj}] = [a_{kj} + ca_{ij}]$

or $\quad B^T = b^T{}_{kj} = a^T{}_{kj} + ca^T{}_{ij}$, $i \neq k$, where $A = [a_{ij}]$.

Proof . Since B is obtained from A by replacing its kth row by $a_{kj} + ca_{ij}$, expanding $|B|$ w.r.t. its kth row, we have,

$$|B| = \sum_{j=1}^{n} b_{kj} B_{kj}$$

$$= \sum_{j=1}^{n} \left(a_{kj} + ca_{ij}\right) A_{kj} \quad ,$$

since B is identical to A except in kth row, so $B_{kj} = A_{kj}$.

$$= \sum_{j=1}^{n} a_{kj} A_{kj} + c \sum_{j=1}^{n} a_{ij} A_{kj} = |A| + c.0 \qquad \text{by theorem 4 in section 6.4}$$

$$= |A|$$

By similar argument, the theorem for the column may be established.

N.B. Let $\mathbf{A}_{.j}$ denote column vectors. Then

$$\det\left[\mathbf{A}_{\cdot 1} \quad \mathbf{A}_{\cdot 2} \quad \cdots \quad \mathbf{A}_{\cdot k} + c\mathbf{A}_{\cdot j} \quad \cdots \quad \mathbf{A}_{\cdot j} \quad \cdots \quad \mathbf{A}_{\cdot n}\right]$$

$$= \det\left[\mathbf{A}_{\cdot 1} \quad \mathbf{A}_{\cdot 2} \quad \cdots \quad \mathbf{A}_{\cdot k} \quad \cdots \quad \mathbf{A}_{\cdot j} \quad \cdots \quad \mathbf{A}_{\cdot n}\right]$$

6.6 Sum of determinants

Sometime it is necessary to express a determinant as a sum of determinants and vice-versa. The following theorem gives condition under which this is possible.

Theorem . If A = $[a_{ij}]_{n \times n}$ and B = $[b_{ij}]_{n \times n}$ are identical for all elements except for corresponding elements in the kth row (column) and if C = $[c_{ij}]$, then |A| + |B| = |C|, where

$$c_{ij} = \begin{cases} a_{ij} = b_{ij},\ i \neq k \\ a_{kj} + b_{kj},\ i = k \end{cases} \qquad \left(\text{or } c_{ij} = \begin{cases} a_{ij} = b_{ij},\ j \neq p \\ a_{ip} + b_{ip},\ i = p \end{cases}\right)$$

Proof . Since A = $[a_{ij}]_{n \times n}$ and B = $[b_{ij}]_{n \times n}$. Also $a_{ij} = b_{ij}$ when $i \neq k$ and $c_{ij} = a_{kj} + b_{kj}$ for $i = k$. So, the cofactors of the elements of kth row in |A| , |B|, |C| are same, i.e., $A_{kj} = B_{kj} = C_{kj}$.

Now expanding both |A| and |B| w.r.t. the kth row, we have,

$$|A| = \sum_{j=1}^{n} a_{kj} A_{kj} \quad \text{and} \quad |B| = \sum_{j=1}^{n} b_{kj} B_{kj} = \sum_{j=1}^{n} b_{kj} A_{kj}$$

$$\therefore \quad |A| + |B| = \sum_{j=1}^{n} (a_{kj} + b_{kj}) A_{kj}$$

$$= \sum_{j=1}^{n} c_{kj} C_{kj} = |C|$$

$$\therefore \quad c_{ij} = a_{ij} = b_{ij} \text{ for } i \neq k.$$

By similar arguments for column, it can be proved that |A| + |B| = |C|.

Ex. Verify that

$$\begin{vmatrix} 1 & 0 & -2 & 3 \\ 2 & 1 & 4 & 5 \\ -3 & -4 & 0 & 7 \\ -2 & 3 & 6 & 0 \end{vmatrix} + \begin{vmatrix} 1 & 0 & -2 & 3 \\ 5 & 4 & 1 & 2 \\ -3 & -4 & 0 & 7 \\ -2 & 3 & 6 & 0 \end{vmatrix} = \begin{vmatrix} 1 & 0 & -2 & 3 \\ 7 & 5 & 5 & 7 \\ -3 & -4 & 0 & 7 \\ -2 & 3 & 6 & 0 \end{vmatrix}$$

It can be found that the values of the three determinants are 498, 390, and 888 respectively from left to right. Hence 498 + 390 = 888. Verified.

N.B. Observe that addition of determinants and addition of matrices are *not the same.* For matrices corresponding elements are added in all position and they need not be square. For determinants two rows (columns) may be different but elements of the other remaining rows (columns) *must be the same.*

Converse Theorem. If $C = [c_{ij}]_{n\times n}$ such that $c_{kj} = a_{kj} + b_{kj}$, $i = k$ and $c_{ij} = c_{ij}$, $i \neq k$ then

$$|C| = |A| + |B| \text{ for } A = [a_{ij}] \text{ and } B = [b_{ij}]$$

where
$$a_{ij} = \begin{cases} c_{ij}, & i \neq k \\ a_{kj}, & i = k \end{cases} \quad \text{and } b_{ij} = \begin{cases} c_{ij}, & i \neq k \\ b_{kj}, & i = k \end{cases}.$$

Similar statement is for the kth column.

Proof. Expanding $|C|$ w.r.t. the kth row in which the elements occur as a combination only.

Therefore,
$$|C| = \sum_{j=1}^{n} c_{kj} C_{kj} = \sum_{j=1}^{n} (a_{kj} + b_{kj}) C_{kj}$$

Since $c_{ij} = a_{ij}$ and $c_{ij} = b_{ij}$ for $i \neq k$, so the cofactors $C_{kj} = A_{kj}$ and $C_{kj} = B_{kj}$.

Hence
$$|C| = \sum_{j=1}^{n} a_{kj} A_{kj} + \sum_{j=1}^{n} b_{kj} B_{kj} = |A| + |B|.$$

By similar arguments the converse theorem for columns can be established. Note the example below, in which the second column consists of a combination of two terms.

N.B. Writing $|C| = |A| + |B|$ explicitly, we get

$$\begin{vmatrix} c_{11} & c_{12} & \cdots & c_{1j} & \cdots & c_{1n} \\ c_{21} & c_{22} & \cdots & c_{2j} & \cdots & c_{2n} \\ \vdots & \vdots & \ddots & \vdots & \vdots & \vdots \\ (a_{k1}+b_{k1}) & (a_{k2}+b_{k2}) & \cdots & (a_{kj}+b_{kj}) & \cdots & (a_{kn}+b_{kn}) \\ \vdots & \vdots & \vdots & \vdots & \ddots & \vdots \\ c_{n1} & c_{n2} & \cdots & c_{nj} & \cdots & c_{nn} \end{vmatrix}$$

$$= \begin{vmatrix} c_{11} & c_{12} & \cdots & c_{1j} & \cdots & c_{1n} \\ c_{21} & c_{22} & \cdots & c_{2j} & \cdots & c_{2n} \\ \vdots & \vdots & \ddots & \vdots & \vdots & \vdots \\ a_{k1} & a_{k2} & \cdots & a_{kj} & \cdots & a_{kn} \\ \vdots & \vdots & \vdots & \vdots & \ddots & \vdots \\ c_{n1} & c_{n2} & \cdots & c_{nj} & \cdots & c_{nn} \end{vmatrix} + \begin{vmatrix} c_{11} & c_{12} & \cdots & c_{1j} & \cdots & c_{1n} \\ c_{21} & c_{22} & \cdots & c_{2j} & \cdots & c_{2n} \\ \vdots & \vdots & \ddots & \vdots & \vdots & \vdots \\ b_{k1} & b_{k2} & \cdots & b_{kj} & \cdots & b_{kn} \\ \vdots & \vdots & \vdots & \vdots & \ddots & \vdots \\ c_{n1} & c_{n2} & \cdots & c_{nj} & \cdots & c_{nn} \end{vmatrix}$$

If $\mathbf{C}_{i\cdot}$, $\mathbf{A}_{k\cdot}$, $\mathbf{B}_{k\cdot}$ denote rows, this can be written explicitly and shortly as

$$\det[\mathbf{C}_{1\cdot} \quad \mathbf{C}_{2\cdot} \quad \cdots \quad (\mathbf{A}_{k\cdot} + \mathbf{B}_{k\cdot}) \quad \cdots \quad \mathbf{C}_{n\cdot}]^{\mathrm{T}}$$
$$= \det[\mathbf{C}_{1\cdot} \quad \mathbf{C}_{2\cdot} \quad \cdots \quad \mathbf{A}_{k\cdot} \quad \cdots \quad \mathbf{C}_{n\cdot}]^{\mathrm{T}} +$$
$$\det[\mathbf{C}_{1\cdot} \quad \mathbf{C}_{2\cdot} \quad \cdots \quad \mathbf{B}_{k\cdot} \quad \cdots \quad \mathbf{C}_{n\cdot}]^{\mathrm{T}},$$

Ex. If

$$A = \begin{bmatrix} a_{11} & b_{11}+b_{12} & a_{13} \\ a_{21} & b_{21}+b_{22} & a_{23} \\ a_{31} & b_{31}+b_{32} & a_{33} \end{bmatrix}$$

then show that $|A| = \det[\mathbf{A}_{.1} \quad \mathbf{B}_{.1} \quad \mathbf{A}_{.3}] + \det[\mathbf{A}_{.1} \quad \mathbf{B}_{.2} \quad \mathbf{A}_{.3}]$, where $\mathbf{A}_{.j}$, $\mathbf{B}_{.j}$ denote column vectors.

$$|A| = \begin{vmatrix} a_{11} & b_{11}+b_{12} & a_{13} \\ a_{21} & b_{21}+b_{22} & a_{23} \\ a_{31} & b_{31}+b_{32} & a_{33} \end{vmatrix}$$

$$= (b_{11}+b_{12})A_{12} + (b_{21}+b_{22})A_{22} + (b_{31}+b_{32})A_{32},$$

by expansion w.r.t. second column

$$= (b_{11}A_{12} + b_{21}A_{22} + b_{31}A_{32}) + (b_{12}A_{12} + b_{22}A_{22} + b_{32}A_{32})$$

$$= \left\{-b_{11}\begin{vmatrix} a_{21} & a_{23} \\ a_{31} & a_{33} \end{vmatrix} + b_{21}\begin{vmatrix} a_{11} & a_{13} \\ a_{31} & a_{33} \end{vmatrix} - b_{31}\begin{vmatrix} a_{11} & a_{13} \\ a_{21} & a_{23} \end{vmatrix}\right\}$$

$$+ \left\{-b_{12}\begin{vmatrix} a_{21} & a_{23} \\ a_{31} & a_{33} \end{vmatrix} + b_{22}\begin{vmatrix} a_{11} & a_{13} \\ a_{31} & a_{33} \end{vmatrix} - b_{32}\begin{vmatrix} a_{11} & a_{13} \\ a_{21} & a_{23} \end{vmatrix}\right\}$$

$$= \begin{vmatrix} a_{11} & b_{11} & a_{13} \\ a_{21} & b_{21} & a_{23} \\ a_{31} & b_{31} & a_{33} \end{vmatrix} + \begin{vmatrix} a_{11} & b_{12} & a_{13} \\ a_{21} & b_{22} & a_{23} \\ a_{31} & b_{32} & a_{33} \end{vmatrix}$$

$$= \det[A_{.1} \quad B_{.1} \quad A_{.3}] + \det[A_{.1} \quad B_{.2} \quad A_{.3}].$$

6.7 Expansion of determinants

Apart from the expansion given in sec. 6.3 and by cofactor, mathematicians introduced several other expansion methods which are named after them.

6.7.1 Laplace's expansion : It is a generalization of cofactor method. In 1772 Laplace published a paper on expansion of determinant w.r.t. several rows or columns instead of a single row or column. To introduce this method some previous definitions need to be extended and redefined.

Minor determinant : Minor determinant or simply minor is obtained by deleting $(n-r)$ rows and $(n-r)$ columns from an n *square* matrix A. It is called a rth *order minor determinant* of A or an $(n-r)$th *minor determinant* of A. Note that the later term does not contain 'order' after $(n-r)$th.

For example

$$\begin{vmatrix} a_{21} & a_{23} \\ a_{31} & a_{33} \end{vmatrix}$$

is a 2nd order minor determinant of the 3rd order matrix

$$\begin{vmatrix} a_{11} & a_{12} & a_{13} \\ a_{21} & a_{22} & a_{23} \\ a_{31} & a_{32} & a_{33} \end{vmatrix}$$

obtained by deleting the first row and the second column. This example is not very interesting as it is the same as the minor of the a_{12} as in sec. 6.5.

Let us consider the 4 square matrix

$$A = \begin{bmatrix} 1 & 0 & 2 & 3 \\ -1 & -2 & 5 & 4 \\ -3 & -5 & -1 & 6 \\ -4 & -6 & 7 & 8 \end{bmatrix}.$$

Here

$$\begin{bmatrix} 1 & 2 \\ -3 & -1 \end{bmatrix}$$

is the minor determinant of A obtained by deleting 2nd & 4th rows and 2nd & 4th columns. It is different from the idea developed in sec. 6.5.

The number of rth order minor in an n *square* matrix is the number of combination of r different rows from n different rows i.e. in symbols it is

$${}^nC_r = \frac{n!}{r!\,(n-r)!} = {}^nC_{n-r}$$

i.e. the number of deleting $(n-r)$ rows from n rows. The same is true for columns too. So the total number of rth order minor obtained by deleting $(n-r)$ rows and $(n-r)$ columns is

$$\left({}^nC_r\right)^2 = \left(\frac{n!}{r!\,(n-r)!}\right)^2.$$

Thus, for a third order square matrix A, there are $\left({}^3C_2\right)^2 = 3^2 = 9$ second order minors and for a fourth order square matrix, there are $\left({}^4C_2\right)^2 = 6^2 = 36$ second order minors. These can be easily verified after listing all the minors. This is left to readers for saving time and space.

Representation of minors : Since the minor of a_{ij} is represented by M_{ij} in sec. 6.5, here also the minor obtained by deleting r rows $i_1, i_2, \cdots, i_r$ and r columns $j_1, j_2, \cdots, j_r$ from the n *square* matrix A is denoted by $M_{i_1 i_2 \cdots i_r,\, j_1 j_2 \cdots j_r}$.

Complementary set of indices : This is a new idea necessary for the next definition as well as for the main theorem of Laplace expansion.

Let $i_1\ i_2\ \cdots\ i_r$, $n > r \geq 1$ be any set S_r of r integers chosen form $1, 2, \cdots, n$ and arranged in *natural order* and $k_1\ k_2\ \cdots\ k_{n-r}$ be another set S_{n-r} of integers arranged also in *natural order* taken from the remaining $(n-r)$ integers. These two sets S_r and S_{n-r} are called *complementary sets of indices*. Thus

$$S_3 = 1\ 5\ 7 \text{ and } S_{8-3} = 2\ 3\ 4\ 6\ 8$$

are two complementary sets of indices of $S_8 = 1\ 2\ 3\ 4\ 5\ 6\ 7\ 8$.

Complementary minors : Let $i_1\ i_2\ \cdots\ i_r$ and $k_1\ k_2\ \cdots\ k_{n-r}$ be two complementary sets of row indices, and $j_1\ j_2\ \cdots\ j_r$ and $p_1\ p_2\ \cdots\ p_{n-r}$ be two complementary sets of column indices of an n *square* matrix A, then the two minor determinants $M_{i_1 i_2 \cdots i_r,\, j_1 j_2 \cdots j_r}$ and $M_{k_1 k_2 \cdots k_{n-r},\, p_1 p_2 \cdots p_{n-r}}$ are complementary minors of the matrix A. Thus for $n = 5$,

$$M_{135,124} = \begin{vmatrix} a_{23} & a_{25} \\ a_{43} & a_{45} \end{vmatrix} \text{ and } M_{24,35} = \begin{vmatrix} a_{11} & a_{12} & a_{14} \\ a_{31} & a_{32} & a_{34} \\ a_{51} & a_{52} & a_{54} \end{vmatrix}$$

are two complementary minors of a 5th order square matrix A.

Algebraic complement or cofactor : Since the cofactor $A_{ij} = (-1)^{i+j} M_{ij}$ in sec. 6.5, the quantity $A_{k_1k_2\cdots k_{n-r},\, p_1\, p_2 \cdots p_{n-r}} = (-)^{k_1+k_2+\cdots+k_{n-r}+p_1+p_2+\cdots+p_{n-r}} M_{k_1k_2\cdots k_{n-r},\, p_1p_2\cdots p_{n-r}}$ is called the *algebraic complement* or *cofactor* of the complementary minor $M_{i_1 i_2\cdots i_r,\, j_1 j_2 \cdots j_r}$.

Similarly, the quantity

$$A_{i_1i_2\cdots i_r,\, j_1j_2\cdots j_r} = (-1)^{i_1+i_2+\cdots+i_r+j_1+j_2+\cdots+j_r} M_{i_1i_2\cdots i_r,\, j_1j_2\cdots j_r}$$

is the algebraic complement or cofactor of the complementary minor

$$M_{k_1 k_2\cdots k_{n-r},\, p_1 p_2\cdots p_{n-r}}.$$

Adjacent transposition : It means interchange of adjacent integers in a permutation of natural numbers such as in the permutation 1 2 4 3 5 the adjacent numbers 3 and 4 are interchanged.

Main theorem on Laplace expansion : If an *n square* matrix A is any fixed set R_r of r row indices in the natural order of $i_1, i_2, \cdots, i_r$ and if $j_1, j_2, \cdots, j_r$ runs over every set C_r of r column indices in the natural order, then the expansion of the determinant A is the sum of the product of each of these complementary minor and its cofactor of the matrix, A. In other words, by the above definitions of complementary minor and cofactor

$$|A| = \sum_{(j)} \left(M_{k_1 k_2 \cdots k_{n-r},\, p_1 p_2 \cdots p_{n-r}} \right) \left(A_{i_1 i_2 \cdots i_r,\, j_1 j_2 \cdots j_r} \right)$$

The theorem is proved below after establishing of *three* more auxiliary theorems (*AT*).

AT_1: Determines the number of adjacent transposition. The permutation $i_1, i_2, \cdots, i_r, k_1, k_2, \cdots, k_{n-r}$ from the natural order $1, 2, \cdots, n$ for the complementary sets $i_1, i_2, \cdots, i_r$ and $k_1, k_2, \cdots, k_{n-r}$ is

$$\left[\sum_{s=1}^{r} i_s - \tfrac{1}{2} r(r+1) \right].$$

Proof. To put i_1 into the first position of the natural order $1, 2, \cdots, n$, it takes (i_1-1) adjacent transpositions. As $i_2 > i_1$, it also takes $(i_2 - 2)$ more transposition to put i_2 in the second position etc. Finally, to put i_r in the rth position, it takes $(i_r - r)$ adjacent transposition. So, the total number of adjacent transpositions is

$$\begin{aligned} &(i_1 - 1) + (i_2 - 2) + \cdots + (i_r - r) \\ &= (i_1 + i_2 + \cdots + i_r) - (1 + 2 + \cdots + r) \\ &= (i_1 + i_2 + \cdots + i_r) - \Sigma \tfrac{1}{2} r(r+1) \\ &= \left[\sum_{s=1}^{r} i_s - \tfrac{1}{2} r(r+1) \right] \end{aligned}$$

AT_2 : If an *n square* matrix A has $i_1, i_2, \cdots, i_r$ & $k_1, k_2, \cdots, k_{n-r}$ as complementary sets of row indices, and $j_1, j_2, \cdots, j_r$ & $p_1, p_2, \cdots, p_{n-r}$ as complementary sets of column indices, then a term in the expansion of det A is given by

$$(-1)^{\sum_{r=1}^{r} i_n \sum_{r=1}^{r} j_n} a_{i_1 j_1} a_{i_2 j_2} \cdots a_{i_r j_r} a_{k_1 p_1} a_{k_2 p_2} \cdots a_{k_{n-r} p_{n-r}} \qquad (6.36)$$

Proof . Since i_n's, k_n's and j_n's, p_n's are complementary sets of row and column subscripts, then every row and every column of A is represented once and only once in (6.36). Now there remains to show that the sign factor is correct. But by AT_1, the number of adjacent transpositions required by the row as well as column subscripts to be in natural order is

$$\left[\sum_{n=1}^{r} i_n - \frac{1}{2}r(r+1)\right] \text{ and } \left[\sum_{n=1}^{r} j_n - \frac{1}{2}r(r+1)\right]$$

respectively. Since the unscrambling of the row subscripts imposes an additional

$$\left[\sum_{n=1}^{r} i_n - \frac{1}{2}r(r+1)\right]$$

transposition on the column subscripts. So we can determine the sign from the permutation of column subscripts only, which is given by

$$(-1)^{\sum_{n=1}^{r} i_n - \frac{1}{2}r(r+1) + \sum_{n=1}^{r} j_n - \frac{1}{2}r(r+1)} = (-1)^{\sum_{n=1}^{r}(i_n+j_n) - r(r+1)} = (-1)^{\sum_{n=1}^{r}(i_n+j_n)}$$

$$\left[\because \frac{1}{(-1)^{r(r+1)}} = 1 \text{ as the order of a minor and either } r \text{ or } (r+1) \text{ is always even , so } r(r+1) \text{ is always even.}\right]$$

AT_3 : The expansion of det A of the *n square* matrix with complementary sets of row and column indices as in AT_2 contains a term as every product of a minor of A and its cofactor in the form

$$M_{k_1k_2 \cdots k_{n-r},\, p_1p_2 \cdots p_{n-r}} \; A_{i_1i_2 \cdots i_r,\, j_1j_2 \cdots j_r}$$

$$\equiv M_{k_1k_2 \cdots k_{n-r},\, p_1p_2 \cdots p_{n-r}} \; (-1)^{\sum_{n=1}^{r}(i_n+j_n)} \; M_{i_1i_2 \cdots i_{n-r},\, j_1j_2 \cdots j_{n-r}}$$

Proof . As in AT_2, the sign of each term is to be investigated as every row and column superscript will appear once and only once in each such term. A general term of the given product is

$$(-1)^{\sum_{n=1}^{r}(i_n+j_n)}\left[(-1)^{\mu_{j'}} a_{i_1j_1'}\, a_{i_2j_2'} \cdots a_{i_rj_r'}\right]\left[(-1)^{\mu_{p'}} a_{k_1p_1}\, a_{k_2p_2} \cdots a_{k_{n-r}p_{n-r}}\right]$$

where bracketed expressions are terms of the row expansion of two minors and $\mu_{j'}$ is the number of adjacent transpositions required to produce the permutation $j_1' j_2' \cdots j_r'$ from $j_1 j_2 \cdots j_r$ and similar explanation for $\mu_{p'}$.

The i's and k's in the product $a_{i_1j_1'}\, a_{i_2j_2'} \cdots a_{i_rj_r'} a_{k_1p_1'}\, a_{k_2p_2'} \cdots a_{k_{n-r}p_{n-r}'}$ are put back into the natural order 1, 2, $\cdots$, n by transposing factors. This can be performed by

$$\left[\sum_{n=1}^{r} i_n - \frac{1}{2}r(r+1)\right]$$

adjacent transpositions (cf. AT_1) and it produces further scrambling of the column scripts. Checking the natural order, the total number of column subscripts adjacent transposition is

$$\left[\sum_{n=1}^{r} j_n - \frac{1}{2}r(r+1)\right] + \mu_{j'} + \mu_{p'} + \left[\sum_{n=1}^{r} i_n - \frac{1}{2}r(r+1)\right].$$

Therefore, the required sign is given by

$$(-1)^{\sum_{n=1}^{r}(i_n+j_n)+\mu_{j'}+\mu_{p'}} = (-1)^{\sum_{n=1}^{r}(i_n+j_n)}, \text{[cf. last part of proof } AT_2\text{]}$$

Proof of the main theorem: Note that (i) Auxiliary theorems 1 and 2 proved that the summation contains only terms $|A|$; (ii) no two sets of subscripts $j_1 j_2 \cdots j_r$ are same, so the terms of $|A|$ are all distinct. Lastly, (iii) two minors under summation have $r!$ and $(n-r)!$ terms respectively. So $j_1 j_2 \cdots j_r$ from $1, 2, \cdots, n$ can be selected in

$$^nC_r = \frac{n!}{r!\,(n-r)!} \text{ ways.}$$

Therefore, the total number of terms in the expansion is

$$r!\,(n-r)!\frac{n!}{r!\,(n-r)!} = n!$$

which is equal to the number of terms in the expansion of $|A|$. These *three* facts together proved the assertion of the main theorem.

6.7.2 Gauss method of elimination : This method is discussed under sec. 7.6 of chapter seven with the number of operations needed for the purpose. This is mentioned here just to name a method and an example is cited below just to illustrate the method. Note that pivot elements are highest in magnitude in a column.

Ex. Compute the determinant of

$$A = \begin{bmatrix} 1 & 2 & -1 & 3 & -2 \\ 1 & 3 & 0 & 2 & 0 \\ 1 & 5 & -2 & 3 & 5 \\ 2 & 3 & -3 & 0 & -4 \\ 3 & 1 & -8 & 8 & 1 \end{bmatrix}$$

using Gauss pivotal condensation.

$$\det A = \begin{vmatrix} 1 & 2 & -1 & 3 & -2 \\ 1 & 3 & 0 & 2 & 0 \\ 1 & 5 & -2 & 3 & 5 \\ 2 & 3 & -3 & 0 & -4 \\ 3 & 1 & -8 & 8 & 1 \end{vmatrix}$$

$$= \begin{vmatrix} 3 & 1 & -8 & 8 & 1 \\ 1 & 2 & -1 & 3 & -2 \\ 1 & 3 & 0 & 2 & 0 \\ 1 & 5 & -2 & 3 & 5 \\ 2 & 3 & -3 & 0 & -4 \end{vmatrix}, \text{ by interchange of rows}$$

$$= \frac{81}{2}\begin{vmatrix} 3 & 1 & -8 & 8 & 1 \\ 0 & 5 & 5 & 1 & -7 \\ 0 & 8 & 8 & -2 & -1 \\ 0 & \boxed{14} & 2 & 1 & 14 \\ 0 & \frac{7}{2} & \frac{7}{2} & -8 & -7 \end{vmatrix} \begin{matrix} \\ 3r_2 - r_1 \\ 3r_3 - r_1 \\ 3r_4 - r_1 \\ \frac{3}{2}r_5 - r_1 \end{matrix}$$

pivot element in 2nd column

$$= \frac{81}{2}\begin{vmatrix} 3 & 1 & -8 & 8 & 1 \\ 0 & 14 & 2 & 1 & 14 \\ 0 & 5 & 5 & 1 & -7 \\ 0 & 8 & 8 & -2 & -1 \\ 0 & \frac{7}{2} & \frac{7}{2} & -8 & -7 \end{vmatrix}, \text{ by interchange of rows}$$

$$= \frac{81}{2}\left(\frac{5}{14}\cdot\frac{4}{7}\cdot\frac{1}{4}\right)\begin{vmatrix} 3 & 1 & -8 & 8 & 1 \\ 0 & 14 & 2 & 1 & 14 \\ 0 & 0 & \boxed{12} & \frac{9}{5} & -\frac{168}{5} \\ 0 & 0 & 12 & -\frac{9}{2} & -\frac{63}{4} \\ 0 & 0 & 12 & -33 & -42 \end{vmatrix} \begin{matrix} \\ \\ \frac{14}{5}r_3' - r_2' \\ \frac{7}{4}r_4' - r' \\ 4r_5' - r' \end{matrix}$$

pivot element in 3rd colum

$$= \frac{81}{2}\left(\frac{5}{14}\cdot\frac{4}{7}\cdot\frac{1}{4}\right)\begin{vmatrix} 3 & 1 & -8 & 8 & 1 \\ 0 & 14 & 2 & 1 & 14 \\ 0 & 0 & 12 & \frac{9}{5} & \frac{168}{5} \\ 0 & 0 & 0 & -\frac{63}{10} & \frac{357}{20} \\ 0 & 0 & 0 & \boxed{-\frac{174}{5}} & -\frac{42}{5} \end{vmatrix} \begin{matrix} \\ \\ \\ r_4'' - r_3'' \\ r_5'' - r_3'' \end{matrix}$$

pivot element in 4th column

$$= (-)\frac{81}{2}\left(\frac{5}{14}\cdot\frac{4}{7}\cdot\frac{1}{4}\right)\begin{vmatrix} 3 & 1 & -8 & 8 & 1 \\ 0 & 14 & 2 & 1 & 14 \\ 0 & 0 & 12 & \frac{9}{5} & -\frac{168}{5} \\ 0 & 0 & 0 & -\frac{174}{5} & -\frac{42}{5} \\ 0 & 0 & 0 & -\frac{63}{10} & \frac{357}{20} \end{vmatrix},$$

by interchange of rows.

$$= (-)\left(\frac{81}{2}\left(\frac{5}{14}\cdot\frac{4}{7}\cdot\frac{1}{4}\right)\frac{21}{116}\right)\begin{vmatrix} 3 & 1 & -8 & 8 & 1 \\ 0 & 14 & 2 & 1 & 14 \\ 0 & 0 & 12 & \frac{9}{5} & -\frac{168}{5} \\ 0 & 0 & 0 & -\frac{174}{5} & -\frac{42}{5} \\ 0 & 0 & 0 & 0 & 107 \end{vmatrix} \begin{matrix} \\ \\ \\ \\ \frac{116}{21}r_5''' - r_4 \end{matrix}$$

$$= (-1)^2\frac{81}{2}\left(\frac{5}{14}\cdot\frac{4}{7}\cdot\frac{1}{4}\right)\frac{21}{116}\cdot 3.\,14.\,12.\frac{174}{5}.\,107 = 428$$

N.B. More about pivot will be discussed in sec.7.6 in chapter seven.

6.7.3 Hazra's simplification rule for increasing the order and the expansion method: Hazra developed

i. a simplification rule for the difference of products to a higher order determinant and

ii. an expansion method which is independent of sign consideration of cofactors.

i. Simplification rule : The difference of the products of two pairs of second order determinants with the following properties simplifies to a product of the common element and a third order determinant.

Properties :

1. in the difference, the determinants possess a common element;
2. second row of prefactor of minuend is identical with first row of post factor of subtrahend;
3. first row of post factor of minuend is identical with second row of prefactor of subtrahend;
4. second column of the prefactor of minuend is identical with first column of the prefactor of subtrahend;
5. the first column of the post factor of minuend is identical with the second column of the post factor of subtrahend.

A prototype is

$$\begin{vmatrix} a & b \\ d & e \end{vmatrix}\begin{vmatrix} e & f \\ h & i \end{vmatrix} - \begin{vmatrix} b & c \\ e & f \end{vmatrix}\begin{vmatrix} d & e \\ g & h \end{vmatrix} = e\begin{vmatrix} a & b & c \\ d & e & f \\ g & h & i \end{vmatrix} \qquad (6.37)$$

where the common element 'e' is *encircled.*

In (6.37) 'e' is the common element in the difference of the second order products and occurs as a multiplier of the resulting third order determinant with 'e' in the centre. Higher order determinants can be reduced to (6.37) by the special partition principle stated below.

No such literature of order increasing simplification procedures is found, except Turnbull (1948) in which the order of a single determinant but not an expression of determinants, has been increased to some arbitrary higher order without disturbing the position of its elements and the extension has been made diagonally with the unit matrix, bordered on one side with zeros and on the other with arbitrary constants. For example,

$$\begin{vmatrix} a & b \\ c & d \end{vmatrix} = \begin{vmatrix} a & b & x \\ c & d & y \\ 0 & 0 & 1 \end{vmatrix} = \begin{vmatrix} a & b & x & z \\ c & d & y & t \\ 0 & 0 & 1 & 0 \\ 0 & 0 & 0 & 1 \end{vmatrix}$$

bordered by arbitrary constants

unit matrix

bordered by zeros

The relation (6.37) has been repeatedly applied by the author to derive Abel's identity for nth order ODE (cf. paper 24 of references).

Proof of simplification rule : Two proofs are given below one after another.

(a) A mechanical proof : The post factors of minued and subtrahend are moved down and each hangs at 'e' with prefactors taking the shape

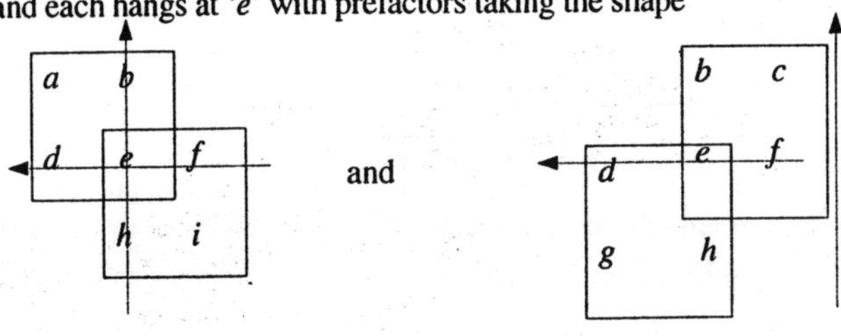

respectively. Finally these two move towards each other overlapping along the common vertical and horizontal elements forming the third order determinant with the one excess 'e' outside it , i.e.

$$e\begin{vmatrix} a & b & c \\ d & e & f \\ g & h & i \end{vmatrix}.$$

(b) Analytical proof :

$$\text{L.H.S. of (6.37)} = \begin{vmatrix} a & b \\ d & e \end{vmatrix}\begin{vmatrix} e & f \\ h & i \end{vmatrix} - \begin{vmatrix} b & c \\ e & f \end{vmatrix}\begin{vmatrix} d & e \\ g & h \end{vmatrix}$$

$$= (ae - bd)(ei - hf) - (bf - ce)(dh - ge)$$

$$= e[(aei + bfg + cdh) - (gec + hfa + idb)] \tag{6.38}$$

$$= e\begin{vmatrix} a & b & c \\ d & e & f \\ g & h & i \end{vmatrix}, \text{ by definition. (Sarrus rule) of expansion.}$$

Extension of simplification rule :

This rule can be extended using matrix partitioning technique and the definition of the complementary block of Hazra stated below:

Principle of partitioning : Let XY block or simply XY denote the partitioned block below/above X row and left / right of Y column and let T = below Top row, B = above Bottom row, R = **R**ight of first column and L = **L**eft of last column.

Complementary block : The TR and BL blocks are defined as the complementary blocks along principal diagonal while BR and TL blocks are the complementary blocks along the secondary diagonal of the determinant.

Thus using properties and above terminologies the following expression of third order determinants can be simplified to the form of (6.37).

$$\underset{\text{TR}}{\overset{\text{R}}{\text{T}\left|\begin{array}{c:cc} a & b & c \\ \hdashline e & f & g \\ i & j & k \end{array}\right|}}\ \underset{\text{BL}}{\overset{\text{L}}{\left|\begin{array}{cc:c} f & g & h \\ j & k & l \\ \hdashline n & o & p \end{array}\right|\text{B}}} - \underset{\text{TL}}{\overset{\text{L}}{\text{T}\left|\begin{array}{cc:c} b & c & d \\ \hdashline f & g & h \\ j & k & l \end{array}\right|}}\ \underset{\text{BR}}{\overset{\text{R}}{\left|\begin{array}{c:cc} e & f & g \\ i & j & k \\ \hdashline m & n & o \end{array}\right|\text{B}}} \tag{6.39}$$

$$= \begin{vmatrix} f & g \\ j & k \end{vmatrix} \underset{\text{R}}{\overset{\text{L}}{\left|\begin{array}{c:cc:c} a & b & c & d \\ \hdashline e & f & g & h \\ i & j & k & l \\ \hdashline m & n & o & p \end{array}\right|\begin{array}{c} \\ \text{T} \\ \text{B} \\ \\ \end{array}}} \tag{6.40}$$

$$\because \text{ group of common element} = \begin{vmatrix} f & g \\ j & k \end{vmatrix}.$$

Using double suffix notation, the following (n –1)th order expression has been simplified to an nth order determinant.

$$\underset{\text{TR}}{\left|a_{ij}\right|_{i,j=2\ldots n}}\ \underset{\text{BL}}{\left|a_{ij}\right|_{i,j=1\ldots(n-1)}} - \underset{\text{TL}}{\left|a_{ij}\right|_{\substack{i=2\ldots n\\ j=1,2\ldots(n-1)}}}\ \underset{\text{BR}}{\left|a_{ij}\right|_{\substack{i=1\ldots(n-1)\\ j=2\ldots n}}}$$

$$= \left|a_{ij}\right|_{i,j=2\ldots(n-1)} \left|a_{ij}\right|_{i,j=1\ldots n} \qquad (6.41)$$

where the common block $\left|a_{ij}\right|_{i,j=2\ldots(n-1)}$ lies among the four blocks.

(ii) Hazra's expansion method : This method states that the determinant is equal to the quotient of (a) the product of complementary blocks along secondary diagonal from that along principal diagonal and (b) the common block between complementary blocks. A prototype is

$$\begin{matrix} & \text{R} \ \ \text{L} \\ \begin{matrix}\text{T}\\ \\ \text{B}\end{matrix} & \begin{vmatrix} a & b & c \\ d & e & f \\ g & h & i \end{vmatrix}\end{matrix} = \left(\frac{1}{e}\right)\left\{\underset{\text{BL}}{\begin{vmatrix} a & b \\ d & e \end{vmatrix}}\ \underset{\text{TR}}{\begin{vmatrix} e & f \\ h & i \end{vmatrix}} - \underset{\text{BR}}{\begin{vmatrix} b & c \\ e & f \end{vmatrix}}\ \underset{\text{TL}}{\begin{vmatrix} d & e \\ g & h \end{vmatrix}}\right\},\ e \neq 0 \qquad (6.42)$$

which is a new method of expansion of determinant in terms difference of products of two pairs of determinants of one order lower than the original. The assumption $e \neq 0$ creates no loss of generality. Because, if 'e' were zero , any nonzero element can be brought to its position by interchanging rows or columns or both with necessary change in sign. This is the only property of the determinant necessary for the method for exceptional cases.

No literature of such expansion is found, except the Laplace expansion method of a determinant with respect to several rows (columns) as the sum of the products of some lower order determinants (cf. sec.6.7.1) and expansion of the determinant of a quasidiagonal matrix (cf.sec.4.5.7) in terms of square blocks as the product of determinants of those blocks. These two types of expansions are different from the one described here.

Proof . The expansion method follows directly from equations (6.38) and (6.39) and extension to higher order follows using the stated principle of partitioning. Therefore, given the determinant,

$$D = \left(\frac{1}{e}\right)\{(\text{TR})(\text{BL}) - (\text{TL})(\text{BR})\}$$

where e is the common block among four blocks of D.

Ex. Evaluate

$$\begin{vmatrix} -2 & 3 & 2 & -5 \\ 3 & -4 & -5 & 6 \\ 4 & -7 & -6 & 9 \\ -3 & 5 & 4 & -10 \end{vmatrix}$$

using Hazra's expansion method.

The determinant is partitioned as follows as per requirement

$$A=\begin{vmatrix} -2 & 3 & 2 & -5 \\ 3 & -4 & -5 & 6 \\ 4 & -7 & -6 & 9 \\ -3 & 5 & 4 & -10 \end{vmatrix} \begin{matrix} \\ T \\ \\ B \end{matrix}$$

(with columns labelled R, L) since, common block $=\begin{vmatrix} -4 & -5 \\ -7 & -6 \end{vmatrix} = -11 \neq 0$

$$= -\frac{1}{11}\left\{\underset{TR}{\begin{vmatrix} -4 & -5 & 6 \\ -7 & -6 & 9 \\ 5 & 4 & -10 \end{vmatrix}}\underset{BL}{\begin{vmatrix} -2 & 3 & 2 \\ 3 & -4 & -5 \\ 4 & -7 & -6 \end{vmatrix}} - \underset{TL}{\begin{vmatrix} 3 & -4 & -5 \\ 4 & -7 & -6 \\ -3 & 5 & 4 \end{vmatrix}}\underset{BR}{\begin{vmatrix} 3 & 2 & -5 \\ -4 & -5 & 6 \\ -7 & -6 & 9 \end{vmatrix}}\right\}$$

Now, $TR = -\frac{1}{6}\left\{\begin{vmatrix} -4 & -5 \\ -7 & -6 \end{vmatrix}\begin{vmatrix} -6 & 9 \\ 4 & -10 \end{vmatrix} - \begin{vmatrix} -5 & 6 \\ -6 & 9 \end{vmatrix}\begin{vmatrix} -7 & -6 \\ 5 & 4 \end{vmatrix}\right\} = 41$

$$BL = -\frac{1}{4}\left\{\begin{vmatrix} -2 & 3 \\ 3 & -4 \end{vmatrix}\begin{vmatrix} -4 & -5 \\ -7 & -6 \end{vmatrix} - \begin{vmatrix} 3 & 2 \\ -4 & -5 \end{vmatrix}\begin{vmatrix} 3 & -4 \\ 4 & -7 \end{vmatrix}\right\} = 6$$

$$TL = -\frac{1}{7}\left\{\begin{vmatrix} 3 & -4 \\ 4 & -7 \end{vmatrix}\begin{vmatrix} -7 & -6 \\ 5 & 4 \end{vmatrix} - \begin{vmatrix} -4 & -5 \\ -7 & -6 \end{vmatrix}\begin{vmatrix} 4 & -7 \\ -3 & 5 \end{vmatrix}\right\} = 3$$

$$BR = -\frac{1}{5}\left\{\begin{vmatrix} 3 & 2 \\ -4 & -5 \end{vmatrix}\begin{vmatrix} -5 & 6 \\ -6 & 9 \end{vmatrix} - \begin{vmatrix} 2 & -5 \\ -5 & 6 \end{vmatrix}\begin{vmatrix} -4 & -5 \\ -7 & -6 \end{vmatrix}\right\} = 16$$

$\therefore \quad A = -\frac{1}{11}(41.6 - 3.16) = -18\,.$

6.7.4 Chio's expansion (Pivotal condensation method) : F. Chio in 1853 devised an expansion method for the determinant. His procedure condensed the original larger determinant to a lower order one. Thus, the method of evaluation is known as Chio's pivotal condensation method.

Theorem. If each element a_{ij} of the minor A_{11} in *n square* matrix A = $[a_{ij}]$ such that $a_{11} \neq 0$ is replaced by

$$\begin{vmatrix} a_{11} & a_{1j} \\ a_{i1} & a_{ij} \end{vmatrix}$$

to form a new $(n-1)$ square matrix D then $|A| = \dfrac{|D|}{a_{11}^{n-2}}$. In other words

$$|A| = \frac{1}{a_{11}^{n-2}}\begin{vmatrix} \begin{vmatrix} a_{11} & a_{12} \\ a_{21} & a_{22} \end{vmatrix} & \begin{vmatrix} a_{11} & a_{13} \\ a_{21} & a_{23} \end{vmatrix} & \cdots & \begin{vmatrix} a_{11} & a_{1n} \\ a_{21} & a_{2n} \end{vmatrix} \\ \begin{vmatrix} a_{11} & a_{12} \\ a_{31} & a_{32} \end{vmatrix} & \begin{vmatrix} a_{11} & a_{13} \\ a_{31} & a_{33} \end{vmatrix} & \cdots & \begin{vmatrix} a_{11} & a_{1n} \\ a_{31} & a_{3n} \end{vmatrix} \\ \vdots & \vdots & \ddots & \vdots \\ \begin{vmatrix} a_{11} & a_{12} \\ a_{n1} & a_{n2} \end{vmatrix} & \begin{vmatrix} a_{11} & a_{13} \\ a_{n1} & a_{n3} \end{vmatrix} & \cdots & \begin{vmatrix} a_{11} & a_{1n} \\ a_{n1} & a_{nn} \end{vmatrix} \end{vmatrix}_{(n-1)\times(n-1)}$$

Proof. Since

$$|A| = \begin{vmatrix} a_{11} & a_{12} & a_{13} & \cdots & a_{1n} \\ a_{21} & a_{22} & a_{23} & \cdots & a_{2n} \\ a_{31} & a_{32} & a_{33} & \cdots & a_{3n} \\ \vdots & \vdots & \vdots & \ddots & \vdots \\ a_{n1} & a_{n2} & a_{n3} & \cdots & a_{nn} \end{vmatrix}, \quad a_{11} \neq 0 \tag{6.43}$$

$\therefore$ multiplying each row except first, by a_{11}, we have,

$$a_{11}^{n-1}|A| = \begin{vmatrix} a_{11} & a_{12} & a_{13} & \cdots & a_{1n} \\ a_{11}a_{21} & a_{11}a_{22} & a_{11}a_{23} & \cdots & a_{11}a_{2n} \\ a_{11}a_{31} & a_{11}a_{32} & a_{11}a_{33} & \cdots & a_{11}a_{3n} \\ \vdots & \vdots & \vdots & \ddots & \vdots \\ a_{11}a_{n1} & a_{11}a_{n2} & a_{11}a_{n3} & \cdots & a_{11}a_{nn} \end{vmatrix}$$

Now performing the row operations , we get

$$a_{11}^{n-1}|A| = \begin{vmatrix} a_{11} & a_{12} & a_{13} & \cdots & a_{1n} \\ 0 & \begin{vmatrix} a_{11} & a_{12} \\ a_{21} & a_{22} \end{vmatrix} & \begin{vmatrix} a_{11} & a_{13} \\ a_{21} & a_{23} \end{vmatrix} & \cdots & \begin{vmatrix} a_{11} & a_{1n} \\ a_{21} & a_{2n} \end{vmatrix} \\ 0 & \begin{vmatrix} a_{11} & a_{12} \\ a_{31} & a_{32} \end{vmatrix} & \begin{vmatrix} a_{11} & a_{13} \\ a_{31} & a_{33} \end{vmatrix} & \cdots & \begin{vmatrix} a_{11} & a_{1n} \\ a_{31} & a_{3n} \end{vmatrix} \\ \vdots & \vdots & \vdots & \ddots & \vdots \\ 0 & \begin{vmatrix} a_{11} & a_{12} \\ a_{n1} & a_{n2} \end{vmatrix} & \begin{vmatrix} a_{11} & a_{13} \\ a_{n1} & a_{n3} \end{vmatrix} & \cdots & \begin{vmatrix} a_{11} & a_{1n} \\ a_{n1} & a_{nn} \end{vmatrix} \end{vmatrix}$$

$$= a_{11} \begin{vmatrix} \begin{vmatrix} a_{11} & a_{12} \\ a_{21} & a_{22} \end{vmatrix} & \begin{vmatrix} a_{11} & a_{13} \\ a_{21} & a_{23} \end{vmatrix} & \cdots & \begin{vmatrix} a_{11} & a_{1n} \\ a_{21} & a_{2n} \end{vmatrix} \\ \begin{vmatrix} a_{11} & a_{12} \\ a_{31} & a_{32} \end{vmatrix} & \begin{vmatrix} a_{11} & a_{13} \\ a_{31} & a_{33} \end{vmatrix} & \cdots & \begin{vmatrix} a_{11} & a_{1n} \\ a_{31} & a_{3n} \end{vmatrix} \\ \vdots & \vdots & \ddots & \vdots \\ \begin{vmatrix} a_{11} & a_{12} \\ a_{n1} & a_{n2} \end{vmatrix} & \begin{vmatrix} a_{11} & a_{13} \\ a_{n1} & a_{n3} \end{vmatrix} & \cdots & \begin{vmatrix} a_{11} & a_{1n} \\ a_{n1} & a_{nn} \end{vmatrix} \end{vmatrix}$$

$= a_{11}|D|$, expansion w.r.t. first row, $|D|$ is $(n-1) \times (n-1)$ order.

$\therefore \quad |A| = \dfrac{|D|}{a_{11}^{n-2}}$, which proves the theorem.

Ex. Compute the determinant of

$$A = \begin{bmatrix} 1 & 2 & -1 & 3 & -2 \\ 1 & 3 & 0 & 2 & 0 \\ 1 & 5 & -2 & 3 & 5 \\ 2 & 3 & -3 & 0 & -4 \\ 3 & 1 & -8 & 8 & 1 \end{bmatrix}$$

using Chio's pivotal condensation method.

Here a_{11}=1 and n = 5. Therefore, by Chio's pivotal condensation method, the first condensed determinant is given by

$$\det A = \begin{vmatrix} \begin{vmatrix} 1 & 2 \\ 1 & 3 \end{vmatrix} & \begin{vmatrix} 1 & -1 \\ 1 & 0 \end{vmatrix} & \begin{vmatrix} 1 & 3 \\ 1 & 2 \end{vmatrix} & \begin{vmatrix} 1 & -2 \\ 1 & 0 \end{vmatrix} \\ \begin{vmatrix} 1 & 2 \\ 1 & 5 \end{vmatrix} & \begin{vmatrix} 1 & -1 \\ 1 & -2 \end{vmatrix} & \begin{vmatrix} 1 & 3 \\ 1 & 3 \end{vmatrix} & \begin{vmatrix} 1 & -2 \\ 1 & 5 \end{vmatrix} \\ \begin{vmatrix} 1 & 2 \\ 2 & 3 \end{vmatrix} & \begin{vmatrix} 1 & -1 \\ 2 & -3 \end{vmatrix} & \begin{vmatrix} 1 & 3 \\ 2 & 0 \end{vmatrix} & \begin{vmatrix} 1 & -2 \\ 2 & -4 \end{vmatrix} \\ \begin{vmatrix} 1 & 2 \\ 3 & 1 \end{vmatrix} & \begin{vmatrix} 1 & -1 \\ 3 & -8 \end{vmatrix} & \begin{vmatrix} 1 & 3 \\ 3 & 8 \end{vmatrix} & \begin{vmatrix} 1 & -2 \\ 3 & 1 \end{vmatrix} \end{vmatrix}.$$

$$= \begin{vmatrix} 1 & 1 & -1 & 2 \\ 3 & -1 & 0 & 7 \\ -1 & -1 & -6 & 0 \\ -5 & -5 & -1 & 7 \end{vmatrix}$$

$$\det A = \begin{vmatrix} \begin{vmatrix} 1 & 1 \\ 3 & -1 \end{vmatrix} & \begin{vmatrix} 1 & -1 \\ 3 & 0 \end{vmatrix} & \begin{vmatrix} 1 & 3 \\ 3 & 7 \end{vmatrix} \\ \begin{vmatrix} 1 & 1 \\ -1 & -1 \end{vmatrix} & \begin{vmatrix} 1 & -1 \\ -1 & -6 \end{vmatrix} & \begin{vmatrix} 1 & 2 \\ -1 & 0 \end{vmatrix} \\ \begin{vmatrix} 1 & 1 \\ -5 & -5 \end{vmatrix} & \begin{vmatrix} 1 & -1 \\ -5 & -1 \end{vmatrix} & \begin{vmatrix} 1 & 2 \\ -5 & 7 \end{vmatrix} \end{vmatrix}$$

$$= \begin{vmatrix} -4 & 3 & 1 \\ 0 & -7 & 2 \\ 0 & -6 & 17 \end{vmatrix}$$

$$= \frac{1}{(-4)^{3-2}} \begin{vmatrix} \begin{vmatrix} -4 & 3 \\ 0 & -7 \end{vmatrix} & \begin{vmatrix} -4 & 1 \\ 0 & 2 \end{vmatrix} \\ \begin{vmatrix} -4 & 3 \\ 0 & -6 \end{vmatrix} & \begin{vmatrix} -4 & 1 \\ 0 & 17 \end{vmatrix} \end{vmatrix} = \frac{1}{-4} \begin{vmatrix} 28 & -8 \\ 24 & -68 \end{vmatrix}$$

$$= \frac{1}{-4}(-1904 + 192) = 428$$

6.7.5 Operation count : The formulae for the number of operations required by different methods for the expansion of an $n \times n$ determinant are mentioned below for the purpose of comparison. Some of them are quoted from text books and others are derived.

The symbols A, S, M and D are used to denote addition, subtraction, multiplication and division respectively.

Sarrus method : It is only applicable for *third order* determinant. So, it is not of much interest for further study. It requires M = 12 and A-S = 5 operations.

Direct method : It is an expansion by definition. The number of $\text{M} = n!(n-1)$ and A-S = $(n!-1)$.

Laplace method : The method is valid for $n > 4$, otherwise it reduces to cofactor expansion with respect to a single row/column. The number of each of $m \times m$ and $(n - m) \times (n - m)$ determinant evaluations is

$$\frac{n!}{m!(n-m)!}$$

for an $n \times n$ determinant evaluation with respect to m rows/columns. The method is repeated if either $m > 4$ or $(n - m) > 4$ and ends if either m or $(n - m)$ equals to 2 or 3.

Each algebraic complement or cofactor needs $2m$ addition and $\Sigma(i_k + j_k)$ multiplication to find the proper sign by

$$(-1)^{(r_1+r_2+\cdots+r_m)+(c_1+c_2+\cdots+c_m)}$$

for expanding m rows.

The minimum operation count shown in table of sec. 6.7.6 is obtained if expanded w.r.t. first two rows because

(i) as m increases $(n - m)$ decreases and hence $\dfrac{n!}{m!(n-m)!}$ increases;

(ii) each 2×2 determinant needs M = 2 and A-S = 1, while a 3×3 needs M = 9 and A-S = 5.

Cofactor method: A special case of the Laplace method. Here

$$\text{M} = \sum_{k-1}^{n-1} \frac{n!}{k!} \text{ and A-S} = (n!-1).$$

Chio's pivotal condensation : Total number of 2×2 determinant evaluation is $\frac{1}{6}n(n-1)(2n-1)$ for an evaluation of $n \times n$ determinant. So, this requires $\frac{1}{3}n(n-1)(2n-1)$ multiplication and

$\frac{1}{3}n(n-1)(2n-1)$ multiplication and $\frac{1}{6}n(n-1)(2n-1)$ additions. Since the method reduces the determinant by one order in each step, the reduction of nth order to $(n-1)$th order needs $(n-2)$ multiplication and one division.

Therefore, number of this multiplication is

$$(n-2)+(n-3)+\cdots+2+1=\tfrac{1}{2}(n-1)(n-2)$$

and division $=1+1+\cdots+(n-2)$ times $=(n-2)$. Hence the grand total of

$$\begin{aligned}\text{M-D} &= \tfrac{1}{3}(n-1)(2n-1)+\tfrac{1}{2}(n-1)(n-2)+(n+2)\\ &= \tfrac{1}{3}n(n-1)(2n-1)+\tfrac{1}{2}(n-2)(n+1)\end{aligned}$$

Gauss method of elimination: It is also called Gauss pivotal condensation or upper triangular decomposition. It is a modification of Chio's pivotal condensation. But it keeps the order unchanged in each step. In this case

$$\text{M-D} = \tfrac{1}{3}(n-1)\left(n^2+n+3\right)$$

and $\qquad \text{A-S} = \frac{1}{6}n(n-1)(2n-1)$.

Hazra's method : The total number of operation required in this method is obtained as follows:

$$\text{A-S} = \tfrac{1}{3}(4^{n-1}-1)+\tfrac{1}{3}(4^{n-3}-1)=\tfrac{1}{3}\left(4^{n-1}+4^{n-3}-2\right)$$

$$\text{M} = \tfrac{2}{3}(4^{n-1}-1)+\tfrac{2}{3}(4^{n-3}-1)=\tfrac{2}{3}(4^{n-1}+4^{n-3}-2)$$

$$\text{D} = \tfrac{1}{3}(4^{n-2}-1)+\tfrac{1}{3}(4^{n-4}-1)=\begin{cases}\tfrac{1}{3}(4^{n-2}+4^{n-4}-2), & n>3\\ 1, & n=3\end{cases}$$

6.7.6 Comparison of different methods of expansions : The most important point is that Hazra's method is free from sign considerations while the Laplace method requires quite a good number of addition and multiplication to decide the proper sign of the cofactor. The cofactor and direct method also involve the consideration of signs. The direct method for higher orders is hopeless.

Hazra's method requires no knowledge of the properties of determinants, except the interchange property, as in the Gauss method. So, it is preferable for the beginners. It is quite a good method for cofactor expansion when $n>4$ (cf. table below).

The relative effectiveness of various methods of expanding a determinant is given in the following table which states the number of operations required by each depending on the order of determinant.

Table : Comparison of various methods in terms of number of operation.

Methods	Order									
	3		4		5		6		7	
	M-D	A-S	M-D	A-S	M-D	A-S	M-D	A-S	M-D	A-S
Direct method	12	5	72	23	480	119	3600	719	30240	5039
Laplace method	9	5	30	17	120	69	495	984	2853	1490
Cofactor expansion	9	5	40	23	205	119	1236	719	8659	5039
Chio's pivotal expansion	12	5	33	14	69	30	124	55	202	91
Gauss method	10	5	23	14	44	30	75	55	108	91
Hazra's method	11	5	49	22	202	90	814	362	3262	1450

From the table it appears that the Gauss method remains efficient and Hazra's method occupies the third efficient position in terms of number of operations required for the evaluation of determinants. Finally, efficiency may be judged from computer programming for a set of test problems of various types of determinants on the basis of operation counts as well as CPU time necessary for evaluation.

6.8 Determinant of a sum of matrices of same order

The following theorem determines the determinant of a sum of matrices of the same order.

Theorem. If A and B are two n *square* matrices, then $|A + B|$ is the sum of 2^n numbers of nth order determinants, jth column of each of the 2^n determinants, $j = 1, 2, \cdots, n$ is either the jth column of A or the jth column of B.

Proof . Let $A = [a_{ij}]_{n\times n}$ and $B = [b_{ij}]_{n\times n}$. Then $A+B=\left[a_{ij}+b_{ij}\right]_{n\times n}$ by definition of the addition of matrices.

$$\therefore\ |A+B|=\left|a_{ij}+b_{ij}\right|_{n\times n}=\begin{vmatrix} a_{11}+b_{11} & a_{12}+b_{12} & \cdots & a_{1n}+b_{1n} \\ a_{21}+b_{21} & a_{22}+b_{22} & \cdots & a_{2n}+b_{2n} \\ \vdots & \vdots & \ddots & \vdots \\ a_{n1}+b_{n1} & a_{n2}+b_{n2} & \cdots & a_{nn}+b_{nn} \end{vmatrix}.$$

Since each of the n columns is a sum of two elements and each column denotes the sum of two determinants,

$$|A+B|=\begin{vmatrix} a_{11} & a_{12} & \cdots & a_{1n} \\ a_{21} & a_{22} & \cdots & a_{2n} \\ \vdots & \vdots & \vdots & \vdots \\ a_{n1} & a_{n2} & \cdots & a_{nn} \end{vmatrix}+\begin{vmatrix} a_{11} & b_{12} & \cdots & a_{1n} \\ a_{21} & b_{22} & \cdots & a_{2n} \\ \vdots & \vdots & \vdots & \vdots \\ a_{n1} & b_{n2} & \cdots & a_{nn} \end{vmatrix}+\cdots+\begin{vmatrix} b_{11} & b_{12} & \cdots & b_{1n} \\ b_{21} & b_{22} & \cdots & b_{2n} \\ \vdots & \vdots & \vdots & \vdots \\ b_{n1} & b_{n2} & \cdots & b_{nn} \end{vmatrix}$$

= a sum of (2.2 $\cdots n$ times) determinants

= a sum of 2^n determinants in each of which jth, $j = 1, 2, \cdots, n$ columns belong to A or B.

Cor. If A , B, $\cdots$, K are m numbers of $n\times n$ matrices, than $|A + B + \cdots + K|$ is the sum of m^n numbers of nth order determinants, jth column of each of the m^n determinants , $j = 1, 2, \cdots, n$, is the jth column of some one of A, B, $\cdots$, or K.

Ex. If A and B are two 2×2 matrices, prove that $|A + B|$ is the sum of $2^2 = 4$ determinants in each of which one column belongs to either A or B. Find a condition such that $|A + B| = |A| + |B|$. Hence conclude that $|A + B| \neq |A| + |B|$ in general.

First part

Let $$A = \begin{bmatrix} a_{11} & a_{12} \\ a_{21} & a_{22} \end{bmatrix} \text{ and } B = \begin{bmatrix} b_{11} & b_{12} \\ b_{21} & b_{22} \end{bmatrix}$$

$$\therefore \quad A + B = \begin{bmatrix} a_{11} + b_{11} & a_{12} + b_{12} \\ a_{21} + b_{21} & a_{22} + b_{22} \end{bmatrix},$$

by definition of matrix addition

$$\therefore |A + B| = \begin{vmatrix} a_{11} + b_{11} & a_{12} + b_{12} \\ a_{21} + b_{21} & a_{22} + b_{22} \end{vmatrix} = \begin{vmatrix} a_{11} & a_{12} + b_{12} \\ a_{21} & a_{22} + b_{22} \end{vmatrix} + \begin{vmatrix} b_{11} & a_{12} + b_{12} \\ b_{21} & a_{22} + b_{22} \end{vmatrix},$$

by theorem in sec. 6.8

$$= \begin{vmatrix} a_{11} & a_{12} \\ a_{21} & a_{22} \end{vmatrix} + \begin{vmatrix} a_{11} & b_{12} \\ a_{21} & b_{22} \end{vmatrix} + \begin{vmatrix} b_{11} & a_{12} \\ b_{21} & a_{22} \end{vmatrix} + \begin{vmatrix} b_{11} & b_{12} \\ b_{21} & b_{22} \end{vmatrix}$$

= sum of $(4 = 2^2)$ determinants.

Observe that each of the 4 determinants contains columns from either A or B.

Second part :

If (1st column of A)$\propto$ (2nd column of B)

and (2nd column of A)$\propto$ (1st column of B)

then the second and the third determinants in $|A + B|$ vanish and satisfies

$$|A + B| = |A| + |B|.$$

Third part :

From the results of first and second parts it is evident that $|A + B| \neq |A| + |B|$ in general.

6.9 Determinant of a product of square matrices of same order

The theorem below determines the determinant of a product of square matrices of the same order.

Theorem . If A and B are $n\times n$ matrices then $|AB| = |A|\,|B|$. **In other words, determinant of the product of two** ***n square*** **matrices is equal to the product of the respective determinants of the matrices.**

This theorem is known as *product theorem* because of the involved products. It appears on November 30, 1812 before the French Academy of Sciences separately by **J. P. M. Binet (1786-1856)** and **A. L. Cauchy (1789-1857) in their masterful papers.** The theorem can be proved in three different ways. Of these proofs only one based on matrix multiplication is given below and rest will be given in chapters nine and ten. Geometry behind this product will be shown under transformation (chapter twelve).

Proof . Let A = $[a_{ij}]_{n\times n}$ and B = $[b_{ij}]_{n\times n}$. By matrix multiplication, the matrix

$$AB = \begin{bmatrix} \left(\sum_{k=1}^{n} a_{1k}b_{k1}\right) & \left(\sum_{k=1}^{n} a_{1k}b_{k2}\right) & \cdots & \left(\sum_{k=1}^{n} a_{1k}b_{kn}\right) \\ \vdots & \vdots & \ddots & \cdots \\ \left(\sum_{k=1}^{n} a_{nk}b_{k1}\right) & \left(\sum_{k=1}^{n} a_{nk}b_{k2}\right) & \cdots & \left(\sum_{k=1}^{n} a_{nk}b_{kn}\right) \end{bmatrix}$$

$$= \sum_{k=1}^{n} \begin{bmatrix} a_{1k}b_{k1} & a_{1k}b_{k2} & \cdots & a_{1k}b_{kn} \\ \vdots & \vdots & \ddots & \vdots \\ a_{nk}b_{k1} & a_{nk}b_{k2} & \cdots & a_{nk}b_{kn} \end{bmatrix}$$

$$\therefore \qquad |AB| = \sum \begin{vmatrix} a_{1i_1}b_{i_1 1} & a_{1i_1}b_{i_2 2} & \cdots & a_{1i_n}b_{i_n n} \\ \vdots & \vdots & \ddots & \vdots \\ a_{ni_1}b_{i_1 1} & a_{ni_2}b_{i_2 2} & \cdots & a_{ni_n}b_{i_n n} \end{vmatrix}$$

where each i_r = 1, 2, $\cdots$, n giving a sum of n^n determinants by cor. of theorem in sec. 6.8. Among these determinants some vanishes due to proportional columns. [Ex. for 2×2 matrices 2 matrices vanish shown below]. Eliminating these, we are left with

$$|AB| = \sum_{\substack{\text{permutation} \\ \text{over all } i\text{'s}}} \begin{vmatrix} a_{1i_1} & a_{1i_1} & \cdots & a_{1i_n} \\ \vdots & \vdots & \vdots & \vdots \\ a_{ni_1} & a_{ni_2} & \cdots & a_{ni_n} \end{vmatrix} \left\{ b_{i_1 1} \quad b_{i_2 2} \quad \cdots \quad b_{i_n n} \right\}$$

$$= \begin{vmatrix} a_{11} & a_{12} & \cdots & a_{1n} \\ a_{21} & a_{22} & \cdots & a_{2n} \\ \vdots & \vdots & \ddots & \vdots \\ a_{n1} & a_{n2} & \cdots & a_{nn} \end{vmatrix} \left\{ \sum_{\substack{\text{permutation} \\ \text{over all } i\text{''s}}} (-1)^t \, b_{i_1 1} \quad b_{i_2 2} \quad \cdots \quad b_{i_n n} \right\} = |A||B| \quad \text{by sec. 6.5}$$

Cor.1 The result can be extended as $|A \quad B \quad \cdots \quad K| = |A||B|\cdots|K|$ for n *square* matrices A, B , $\cdots$, K.

Cor.2 Multiplication of determinants : The theorem gives the method of multiplication of two determinants as |A||B| = |AB|. Since two conformable matrices are multiplied row by the column method, so is applicable to the corresponding determinants as the R.H.S. is the determinant of the product of two matrices.

Let $|A| = |a_{ij}|$ and $|B| = |b_{ij}|$ be two determinants of the same order. Then

$$|A||B| = |a_{ij}||b_{ij}| = \left|[a_{ij}][b_{ij}]\right|$$

$$= \left| \sum_{k=1}^{n} a_{ik}b_{kj} \right|, i, j = 1, 2, \cdots, n$$

This basic algorithm for the product of two determinants is differently explained in many text books which avoid the theorem. These explanations follow directly from the present theorem. Since $|A| = |A^T|$ and $|B| = |B^T|$, therefore row by the column method multiplication for determinants is equivalent to *three* more methods,

(i) column by column, (ii) row by row, (iii) column by row.

For example, let

$$A = \begin{bmatrix} a_{11} & a_{12} \\ a_{21} & a_{22} \end{bmatrix},\ B = \begin{bmatrix} b_{11} & b_{12} \\ b_{21} & b_{22} \end{bmatrix}.$$

Then
$$AB = \begin{bmatrix} (a_{11}b_{11} + a_{12}b_{21}) & (a_{11}b_{12} + a_{12}b_{22}) \\ (a_{21}b_{11} + a_{22}b_{21}) & (a_{21}b_{12} + a_{22}b_{22}) \end{bmatrix}$$

$$\therefore |AB| = \begin{vmatrix} a_{11}b_{11} + a_{12}b_{21} & a_{11}b_{12} + a_{12}b_{22} \\ a_{21}b_{11} + a_{22}b_{21} & a_{21}b_{12} + a_{22}b_{22} \end{vmatrix}$$

$$= \overset{i_1=1 \quad i_2=1 \;\to 0}{\begin{vmatrix} a_{11}b_{11} & a_{11}b_{12} \\ a_{21}b_{11} & a_{21}b_{12} \end{vmatrix}} + \overset{i_1=1 \quad i_2=2}{\begin{vmatrix} a_{11}b_{11} & a_{12}b_{22} \\ a_{21}b_{11} & a_{22}b_{22} \end{vmatrix}} + \overset{i_1=2 \quad i_2=1}{\begin{vmatrix} a_{12}b_{21} & a_{11}b_{12} \\ a_{22}b_{21} & a_{21}b_{12} \end{vmatrix}} + \overset{i_1=2 \quad i_2=2 \;\to 0}{\begin{vmatrix} a_{12}b_{21} & a_{12}b_{22} \\ a_{22}b_{21} & a_{22}b_{22} \end{vmatrix}}$$

$$= \begin{vmatrix} a_{11} & a_{12} \\ a_{21} & a_{22} \end{vmatrix} (b_{11}b_{22} - b_{21}b_{12}) = |A||B|$$

which verifies the statement in cor.2 .

6.10 Determinant of the product of rectangular matrices

Let $A = [a_{ij}]_{m\times n}$ and $[b_{ij}]_{n\times m}$ be two rectangular matrices. Our objective is to determine $|AB|$. Here arise two cases — (i) $m > n$, (ii) $m < n$.

Case (i) $m > n$:

Theorem 1 . If A is $m\times n$ and B is $n\times m$ matrices with $m > n$ then $|AB| = 0$.

Proof . In this case matrices are converted to square matrices by adjoining $(m - n)$ columns of zeros and $(m - n)$ rows of zeros to A and B respectively and determinant of the product of two square matrices can be applied. That is,

$$|AB| = \det \begin{bmatrix} a_{11} & a_{12} & \dots & a_{1n} \\ a_{21} & a_{22} & \dots & a_{2n} \\ \vdots & \vdots & \vdots & \vdots \\ a_{m1} & a_{m2} & \dots & a_{mn} \end{bmatrix} \begin{bmatrix} b_{11} & b_{12} & \dots & b_{1m} \\ b_{21} & b_{22} & \dots & b_{2m} \\ \vdots & \vdots & \vdots & \vdots \\ b_{n1} & b_{n2} & \dots & b_{nm} \end{bmatrix}$$

$$= \det\begin{bmatrix} a_{11} & a_{12} & \cdots & a_{1n} & 0 & \cdots & 0 \\ a_{21} & a_{22} & \cdots & a_{2n} & 0 & \cdots & 0 \\ \vdots & \vdots & \vdots & \vdots & \vdots & \vdots & \vdots \\ a_{m1} & a_{m2} & \cdots & a_{mn} & 0 & \cdots & 0 \end{bmatrix}_{m\times m} \begin{bmatrix} b_{11} & b_{12} & \cdots & b_{1m} \\ b_{21} & b_{22} & \cdots & b_{2m} \\ \vdots & \vdots & \vdots & \vdots \\ b_{n1} & b_{n2} & \cdots & b_{nm} \\ 0 & 0 & \cdots & 0 \\ \vdots & \vdots & \vdots & \vdots \\ 0 & 0 & \cdots & 0 \end{bmatrix}_{m\times m}$$

$(m-n)$ zero columns; $(m-n)$ zero rows

$= 0.0 = 0$ by theorem 2 of sec. 6.4 or 6.5.3

Case(ii) $m < n$ **:**

Theorem 2 . If A is $m\times n$ and B is $n\times m$ matrices with $m \le n$ then $|AB|$ is equal to the sum of the products of *corresponding majors* of A and B.

At the beginning, we need to define the *major determinant* and the *corresponding major* before we proceed to establish the theorem.

Definition 1. A *major determinant* of a $p\times q$ matrix is the determinant of any maximum order square partitioned block.

For example

$$\begin{vmatrix} a_{11} & a_{13} \\ a_{21} & a_{23} \end{vmatrix}$$

is one of the 3 major determinants of the rectangular 2×3 matrix

$$\begin{bmatrix} a_{11} & a_{12} & a_{13} \\ a_{21} & a_{22} & a_{23} \end{bmatrix}$$

but not of the 3×3 square matrix

$$\begin{bmatrix} a_{11} & a_{12} & a_{13} \\ a_{21} & a_{22} & a_{23} \\ a_{31} & a_{32} & a_{33} \end{bmatrix}$$

because the determinant of the maximum order square partitioned block is the determinant of the matrix itself.

Definition 2. Let $A = [a_{ij}]_{m\times n}$ and $B = [b_{ij}]_{n\times m}$ with $m \le n$. If *column indices* of the major determinant of A corresponds to the *rows indices* of the major determinant of B then these two majors are called *corresponding majors* of A and B. Thus if

$$A = \begin{bmatrix} a_{11} & a_{12} & a_{13} \\ a_{21} & a_{22} & a_{23} \end{bmatrix}_{2\times 3} \quad \text{and } B = \begin{bmatrix} b_{11} & b_{12} \\ b_{21} & b_{22} \\ b_{31} & b_{32} \end{bmatrix}_{3\times 2}$$

∴ Now the column indices of a major determinant of A,

$$\begin{vmatrix} a_{11} & a_{13} \\ a_{21} & a_{23} \end{vmatrix}$$

are 1 and 3 which corresponds to the row indices of a major determinant of B,

$$\begin{vmatrix} b_{11} & b_{12} \\ b_{31} & b_{32} \end{vmatrix}. \text{ So } \begin{vmatrix} a_{11} & a_{13} \\ a_{21} & a_{23} \end{vmatrix} \text{ and } \begin{vmatrix} b_{11} & b_{12} \\ b_{31} & b_{32} \end{vmatrix}$$

are two corresponding majors out of 3 such pair majors of A and B.

Proof . Let us first write the matrix A as n column vectors with m components and take the product AB i.e.

$$AB = \begin{bmatrix} \mathbf{A}_{.1} & \mathbf{A}_{.2} & \cdots & \mathbf{A}_{.n} \end{bmatrix} \begin{bmatrix} b_{11} & b_{12} & \cdots & b_{1m} \\ b_{21} & b_{22} & \cdots & b_{2m} \\ \vdots & \vdots & \vdots & \vdots \\ b_{n1} & b_{n2} & \cdots & b_{nm} \end{bmatrix}$$

$$= \Big[\underbrace{(b_{11}\mathbf{A}_{\cdot 1} + b_{21}\mathbf{A}_{\cdot 2} + \cdots + b_{n1}\mathbf{A}_{\cdot n})}_{\text{1st column}} \quad \underbrace{(b_{12}\mathbf{A}_{\cdot 1} + b_{22}\mathbf{A}_{\cdot 2} + \cdots + b_{n2}\mathbf{A}_{\cdot n})}_{\text{2nd column}} \quad \cdots \quad \underbrace{(b_{1m}\mathbf{A}_{\cdot 1} + b_{2m}\mathbf{A}_{\cdot 2} + \cdots + b_{nm}\mathbf{A}_{\cdot n})}_{n\text{th column}} \Big]_{1 \times n}$$

$$= \left[\sum_{j_1=1}^{n} \left(b_{j_1 1}\mathbf{A}_{.j_1}\right) \quad \sum_{j_2=1}^{n} \left(b_{j_2 2}\mathbf{A}_{.j_2}\right) \cdots \cdots \sum_{j_{m1}=1}^{n} \left(b_{j_m m}\mathbf{A}_{.j_m}\right) \right]_{1\times n}$$

Using converse theorem in sec. 6.6 each column vector of the above product can be decomposed into n determinants. As there are m such column vectors, so the total number of decomposed determinants is $(n.n \ \cdots m \text{ times}) = n^m$, which also includes some repetitions. Therefore,

$$|AB| = \sum_{j_1, j_2, \ldots, j_n=1}^{n} \det\left[\left(b_{j_1 1}\mathbf{A}_{.j_1}\right) \quad \left(b_{j_2 2}\mathbf{A}_{.j_2}\right) \cdots \left(b_{j_m m}\mathbf{A}_{.j_m}\right)\right]$$

The terms of the above sum in which $j_1, j_2, \ \cdots, j_n$ are not all distinct will vanish due to proportional columns. Thus the summation needs extension only over

$$^nP_m = \frac{n!}{(n-m)!}$$

permutations of $1, 2, \ \cdots, n$ taken m at a time. Denoting such a sum by $\Sigma'_{(j)}$ and factoring out the b's, we get

$$|AB| = \Sigma'_{(j)}\left(b_{j_1 1} \quad b_{j_2 2} \cdots b_{j_m m}\right)\det\left[\mathbf{A}_{.j_1} \quad \mathbf{A}_{.j_2} \cdots \mathbf{A}_{.j_m}\right]$$

$$= \sum_{(k)}\left(\sum_{(j)} \varepsilon^{k_1 k_2 \ldots k_m}_{j_1 j_2 \ldots j_m} b_{j_1 1}\, b_{j_2 2} \cdots b_{j_m m}\right)\det\left[A_{.k_1} \quad A_{.k_2} \cdots A_{.k_m}\right]$$

where (i) $k_1 \, k_2 \ \cdots k_m$ are m digits arranged in natural order

(ii) $\varepsilon^{k_1 k_2 \ldots k_m}_{j_1 j_2 \ldots j_m} = \pm 1$ depending on whether $j_1 \, j_2 \ \cdots \ j_m$ is an even or odd permutation of $k_1\ k_2\ \cdots k_m$.

$= \sum_{(k)}$ [(major determinant of B corresponding to that of A) (major determinant of A)]

= (sum of the corresponding major of A and B).

Thus the theorem is established.

Ex. Find the det AB where

$$A = \begin{bmatrix} 3 & 0 \\ 9 & 2 \\ 1 & 7 \end{bmatrix} \text{ and } B = \begin{bmatrix} 1 & 4 & 5 \\ 2 & 0 & 3 \end{bmatrix}.$$

Also find $|BA|$. In each case verify the method explained above.

First part : A is a 3×2 matrix and B is a 2×3 matrix. It is a case of finding determinant of the product of two rectangular matrices. In the first case $m > n$. So, by theory, its value must be zero.

Verification :

$$AB = \begin{bmatrix} 3 & 0 \\ 9 & 2 \\ 1 & 7 \end{bmatrix}\begin{bmatrix} 1 & 4 & 5 \\ 2 & 0 & 3 \end{bmatrix} = \begin{bmatrix} 3 & 12 & 15 \\ 13 & 36 & 51 \\ 15 & 4 & 26 \end{bmatrix}.$$

$$\therefore \quad |AB| = \begin{vmatrix} 3 & 12 & 15 \\ 13 & 36 & 51 \\ 15 & 4 & 26 \end{vmatrix} = 12\begin{vmatrix} 1 & 1 & 5 \\ 13 & 9 & 51 \\ 15 & 1 & 26 \end{vmatrix} = 12\begin{vmatrix} 1 & 0 & 0 \\ 13 & -4 & 6 \\ 15 & -14 & 21 \end{vmatrix}$$

$= 12(-84 + 84) = 0$, verified.

Second part m < n : \ |BA| must be nonzero.

$$BA = \begin{bmatrix} 1 & 4 & 5 \\ 2 & 0 & 3 \end{bmatrix}\begin{bmatrix} 3 & 0 \\ 9 & 2 \\ 1 & 7 \end{bmatrix} = \begin{bmatrix} 44 & 43 \\ 9 & 21 \end{bmatrix}. \quad \therefore \ |BA| = 537$$

Verification : Number of major determinant in each matrix

$$= \frac{n!}{n!(n-m)!} = \frac{3!}{2!(3-2)!} = 3,$$

which are shown below :

From columns 1, 2 of B and rows 1, 2 of A

$$|B_1| = \begin{vmatrix} 1 & 4 \\ 2 & 0 \end{vmatrix} = -8, \quad |A_1| = \begin{vmatrix} 3 & 0 \\ 9 & 2 \end{vmatrix} = 6$$

From columns 1, 3 of B and rows 1, 3 of A

$$|B_2| = \begin{vmatrix} 1 & 5 \\ 2 & 3 \end{vmatrix} = -7, \quad |A_2| = \begin{vmatrix} 3 & 0 \\ 1 & 7 \end{vmatrix} = 21$$

From columns 2, 3 of B and rows 2, 3 of A

$$|B_3| = \begin{vmatrix} 4 & 5 \\ 0 & 3 \end{vmatrix} = 12, \quad |A_3| = \begin{vmatrix} 9 & 2 \\ 1 & 7 \end{vmatrix} = 61$$

$$\therefore |AB| = |B_1||A_1| + |B_2||A_2| + |B_3||A_3| = (-8)(6) + (-7)(21) + (12)(61)$$
$$= -48 - 147 + 732 = 537. \quad \text{Verified.}$$

6.11 Special Determinants

Corresponding to each square matrix in chapter two, there exists one determinant. Some of them are mentioned in this section.

(1) Unit determinant : The determinant of a unit matrix is called *unit determinant.* It is denoted by 1, e.g.

$$1 = \begin{vmatrix} 1 & 0 & 0 \\ 0 & 1 & 0 \\ 0 & 0 & 1 \end{vmatrix} \text{ etc.}$$

It can be represented by the Kronecker delta, $1 = |\delta_{ij}|, \ \delta_{ij} = \begin{cases} 0, & i \neq j \\ 1, & i = j \end{cases}$

Properties :

(i) If zeros of one triangle are replaced by arbitrary elements, then the value of the determinant remains unchanged. $1 = \begin{vmatrix} 1 & 0 \\ x & 1 \end{vmatrix} = \begin{vmatrix} 1 & 0 & 0 \\ x & 1 & 0 \\ y & 1 & 1 \end{vmatrix}$ etc.

(ii) In raising the order of a lower order determinant to a higher order keeping original positions of elements, is performed by extending diagonally with the unit determinant, bordered on one side with zero and on the other with arbitrary elements.

(2) Vandermode determinant : the Vandermode determinant is the determinant corresponding to the Vandermode matrix.

$$\det V = \begin{vmatrix} 1 & x_0 & x_0^2 & \cdots & x_0^{n-1} \\ 1 & x_1 & x_1^2 & \cdots & x_1^{n-1} \\ 1 & x_2 & x_2^2 & \cdots & x_2^{n-1} \\ \vdots & \vdots & \vdots & \ddots & \vdots \\ 1 & x_{n-1} & x_{n-1}^2 & \cdots & x_{n-1}^{n-1} \end{vmatrix}$$

Now subtracting x_0 times the $(n-1)$th column from nth column, x_0 times the $(n-2)$th column from the $(n-1)$th column, $\cdots$, x_0 times the first column from the second column, it can be easily shown that

$$\det V = \prod_{i<j} (x_j - x_i), \quad i, j = 0, 1, 2, \cdots, (n-1).$$

Condition for singularity : V is singular if $x_i = x_j$ for any $i = j$.

6.12 Determinant of a triangular matrix

The theorem below determines the determinant of a triangular matrix as the product of the diagonal elements.

Theorem. If $A = [a_{ij}]_{n\times n}$ such that $a_{ij} = 0$ for $i > j$, then $\det A = a_{11}\, a_{22} \cdots a_{nn}$. In other words, the determinant of an upper triangular matrix is the product of the diagonal elements.

Proof. Since $a_{ij} = 0$ for $i > j$,

$$\text{then} \quad A = \begin{bmatrix} a_{11} & a_{12} & a_{13} & \cdots & a_{1n} \\ 0 & a_{22} & a_{23} & \cdots & a_{2n} \\ 0 & 0 & a_{33} & \cdots & a_{3n} \\ \vdots & \vdots & \vdots & \ddots & \vdots \\ 0 & 0 & 0 & \cdots & a_{nn} \end{bmatrix}$$

is an upper triangular matrix.

$$\therefore \quad |A| = \begin{vmatrix} a_{11} & a_{12} & a_{13} & \cdots & a_{1n} \\ 0 & a_{22} & a_{23} & \cdots & a_{2n} \\ 0 & 0 & a_{33} & \cdots & a_{3n} \\ \vdots & \vdots & \vdots & \ddots & \vdots \\ 0 & 0 & 0 & \cdots & a_{nn} \end{vmatrix} = a_{11} \begin{vmatrix} a_{22} & a_{23} & \cdots & a_{2n} \\ 0 & a_{33} & \cdots & a_{3n} \\ \vdots & \vdots & \ddots & \vdots \\ 0 & 0 & \cdots & a_{nn} \end{vmatrix}$$

Expanding w.r.t. first column.

$$= a_{11}a_{22} \begin{vmatrix} a_{33} & a_{34} & \cdots & a_{3n} \\ 0 & a_{44} & \cdots & a_{4n} \\ \vdots & \vdots & \ddots & \vdots \\ 0 & 0 & \cdots & a_{nn} \end{vmatrix} = a_{11}a_{22}a_{33} \begin{vmatrix} a_{44} & a_{45} & \cdots & a_{4n} \\ 0 & a_{55} & \cdots & a_{5n} \\ \vdots & \vdots & \ddots & \vdots \\ 0 & 0 & \cdots & a_{nn} \end{vmatrix}$$

Proceeding in this way, ultimately, we get $|A| = a_{11}\, a_{22} \cdots a_{nn}$ product of the diagonal elements.

Since $\det A = \det A^T$, so the determinant of the lower triangular matrix is equal to the product of the diagonal elements.

Cor. If any one of $a_{ii} = 0$, for any i, then the value of the determinant is zero.

6.13 Evaluation of determinant with computer

One efficient way of evaluating the determinant of a matrix with a computer is to use theorem 6 sec. 6.4 repeatedly to triangularize the given matrix. The use of theorem in sec. 6.12 gives the value of the determinant of the matrix as the product of the diagonal elements. The triangularization carried out by the following steps :

Step 1. Add multiple of 1st row to each of remaining $(n - 1)$ rows, choosing the appropriate multiple so that each of element in 1st column, except a_{11}, is zero.

Step 2. Next, add multiples of 2nd rows to each of $(n - 2)$ rows below the 2nd row, so that each element in 2nd column below the diagonal is zero.

⋮ ⋮ ⋮ ⋮

Step k. For $k = 3, 4, \cdots, (n-1)$ add multiples of kth row to each of the remaining $(n-k)$ rows below kth row, so that each element in kth column below the diagonal , is zero.

$\vdots \quad \vdots \quad \vdots \quad \vdots$

Step n. The resulting matrix will be upper triangular. Hence the desired determinant is the product of the diagonal elements of this matrix.

Cor. 1 During the triangularization process if any one of diagonal element $a_{ii} = 0$, for any i and the elements in ith column below a_{ii} are all zero then stop the procedures and conclude, by cor.1 in sec. 6.12, that the value of the determinant is zero.

Ex . Compute the determinant of the matrix

$$A = \begin{bmatrix} 1 & 2 & -1 & 3 & -2 \\ 1 & 3 & 0 & 2 & 0 \\ 1 & 5 & -2 & 3 & 5 \\ 2 & 3 & -3 & 0 & -4 \\ 3 & 1 & -8 & 8 & 1 \end{bmatrix}$$

by triangularization.

Step 1 .

$$\det A = \begin{vmatrix} 1 & 2 & -1 & 3 & -2 \\ 0 & 1 & 1 & -1 & 2 \\ 0 & 3 & -1 & 0 & 7 \\ 0 & -1 & -1 & -6 & 0 \\ 0 & -5 & -5 & -1 & 7 \end{vmatrix} \begin{matrix} \\ r_2 - r_1 \\ r_3 - r_1 \\ r_4 - 2r_1 \\ r_5 - 3r_1 \end{matrix}$$

Step 2 .

$$\det A = \begin{vmatrix} 1 & 2 & -1 & 3 & -2 \\ 0 & 1 & 1 & -1 & 2 \\ 0 & 0 & -4 & 3 & 1 \\ 0 & 0 & 0 & -7 & 2 \\ 0 & 0 & 0 & -6 & 17 \end{vmatrix} \begin{matrix} \\ \\ r_3' - 3r_2' \\ r_4' + r_2' \\ r_5' + 5r_2' \end{matrix}$$

Step 3.

$$\det A = \frac{1}{7}\begin{vmatrix} 1 & 2 & -1 & 3 & -2 \\ 0 & 1 & 1 & -1 & 2 \\ 0 & 0 & -4 & 3 & 1 \\ 0 & 0 & 0 & -7 & 2 \\ 0 & 0 & 0 & 0 & 107 \end{vmatrix} \begin{matrix} \\ \\ \\ \\ 7r_5'' - 6r_4'' \end{matrix}$$

Step 4 . $\det A = \frac{1}{7}(1.\ 1(-4)(-7).107) = 428$

N.B. Since 6 times fourth row has been subtracted from 7 times fifth row, so det A gets multiplied by 7 and to balance the value, the determinant is multiplied by $\frac{1}{7}$.

6.14 Determinant of partitioned matrix

The determinants of some particular forms of partitioned matrix are considered in this section.

(I) Let A and C be square blocks. Then det $\left[\begin{array}{c|c} A & 0 \\ \hline B & C \end{array}\right]$ = det A det C .

Proof . This is proved by induction. Since A and C are square blocks, the order of the null block, 0 and that of block B will adjust the orders of A and C so that the large matrix is square.

Case (i) Let A = $[a_{11}]$ and C = $[c_{22}]$ be 1×1 blocks.

Then $$\left[\begin{array}{c|c} A & 0 \\ \hline B & C \end{array}\right] = \left[\begin{array}{c|c} a_{11} & 0 \\ \hline b_{21} & c_{22} \end{array}\right]$$ is a 2×2 matrix.

∴ $$\det\left[\begin{array}{c|c} A & 0 \\ \hline B & C \end{array}\right] = \det\begin{bmatrix} a_{11} & 0 \\ b_{21} & c_{22} \end{bmatrix} = a_{11}c_{22} = |A|\,|C|$$

Case (ii) Let

$$A = \begin{bmatrix} a_{11} & a_{12} \\ a_{21} & a_{22} \end{bmatrix}$$ and $C = [c_{33}]$ be two 2×2 and 1×1 blocks respectively. Then

$$\det\left[\begin{array}{c|c} A & 0 \\ \hline B & C \end{array}\right] = \det\left[\begin{array}{cc|c} a_{11} & a_{12} & 0 \\ a_{21} & a_{22} & 0 \\ \hline b_{31} & b_{32} & c_{33} \end{array}\right]$$

$$= a_{11}A_{11} + a_{12}A_{12} + 0A_{13}$$

$$= a_{11}\begin{bmatrix} a_{22} & 0 \\ b_{32} & c_{33} \end{bmatrix} - a_{12}\begin{bmatrix} a_{21} & 0 \\ b_{31} & c_{33} \end{bmatrix}$$

$$= a_{11}a_{22}c_{33} - a_{12}a_{21}c_{33} = c_{33}(a_{11}a_{22} - a_{12}a_{21})$$

$$= \begin{vmatrix} a_{11} & a_{12} \\ a_{21} & a_{22} \end{vmatrix} |\,c_{33}\,| = |A|\,|C| .$$

Case (iii) Let

$$A = \begin{bmatrix} a_{11} & a_{12} \\ a_{21} & a_{22} \end{bmatrix}$$ and $$C = \begin{bmatrix} c_{33} & c_{34} \\ c_{43} & c_{44} \end{bmatrix}$$ be two 2×2 blocks. Then

$$\det\left[\begin{array}{cc|cc} a_{11} & a_{12} & 0 & 0 \\ a_{21} & a_{22} & 0 & 0 \\ \hline b_{31} & b_{32} & c_{33} & c_{34} \\ b_{41} & b_{42} & c_{43} & c_{44} \end{array}\right] = a_{11}A_{11} + a_{12}A_{12}$$

$$= a_{11}\begin{vmatrix} a_{22} & 0 & 0 \\ b_{32} & c_{33} & c_{34} \\ b_{42} & c_{43} & c_{44} \end{vmatrix} - a_{12}\begin{vmatrix} a_{21} & 0 & 0 \\ b_{31} & c_{33} & c_{34} \\ b_{41} & c_{43} & c_{44} \end{vmatrix}$$

$$= a_{11}a_{22}|C| - a_{12}a_{21}|C| = |A|\,|C|.$$

Proceeding in this way for all possibilities of square blocks A and C, the

$$\det\left[\begin{array}{c|c} A & 0 \\ \hline B & C \end{array}\right] = \det A \det C$$

= (Product of determinants of diagonal blocks).

Similarly,

$$\det\left[\begin{array}{c|c} A & B \\ \hline 0 & C \end{array}\right] = \det A \det C$$

= (Product of the determinants of diagonal blocks).

This idea can be applied to any matrix after reducing the given matrix to triangular block matrix. Since the determinant of a triangular matrix is the product of the diagonal elements, so the determinant of a triangular block matrix is the product of the determinants of the diagonal blocks.

Cor. (i) Let

$$A = \left[\begin{array}{ccc|cc} a_{11} & a_{12} & a_{13} & a_{14} & a_{15} \\ a_{21} & a_{22} & a_{23} & a_{24} & a_{25} \\ a_{31} & a_{32} & a_{33} & a_{34} & a_{35} \\ \hline 0 & 0 & 0 & a_{44} & a_{45} \\ 0 & 0 & 0 & a_{54} & a_{55} \end{array}\right] = \begin{bmatrix} A_1 & A_2 \\ 0 & A_3 \end{bmatrix}.$$

Then $\det A = \det A_1 \det A_3$.

Cor. (ii) Let

$$A = \begin{bmatrix} A_1 & \cdots & \cdots & \cdots & \cdots \\ 0 & A_2 & \cdots & \cdots\cdots & \cdots \\ 0 & 0 & A_3 & \cdots & \cdots \\ \vdots & \vdots & \vdots & \ddots & \cdots \\ 0 & 0 & 0 & 0 & A_5 \end{bmatrix}$$

be a block triangular matrix. Then

$$\det A = \det A_1 \det A_2 \cdots \det A_n$$

Ex. Compute the determinant of the partition matrix

$$\left[\begin{array}{c|cc|ccc} 0 & 0 & 0 & 3 & 2 & 1 \\ 0 & 0 & 0 & 2 & 1 & 0 \\ 0 & 0 & 0 & 1 & 0 & -1 \\ \hline 0 & 2 & 1 & 0 & 0 & 0 \\ 0 & 1 & 2 & 0 & 0 & 0 \\ \hline 4 & 0 & 0 & 0 & 0 & 0 \end{array}\right]$$

The given matrix is of the form

$$M = \left[\begin{array}{c|c|c} 0 & 0 & C \\ \hline 0 & A & 0 \\ \hline B & 0 & 0 \end{array}\right], \text{where } A = \begin{bmatrix} 2 & 1 \\ 1 & 2 \end{bmatrix}, B = [4], C = \begin{bmatrix} 3 & 2 & 1 \\ 2 & 1 & 0 \\ 1 & 0 & -1 \end{bmatrix}.$$

$\therefore \quad |M| = -|B|\,|A|\,|C| = -4.3.0 = 0 \quad [\because |C| = 0]$

6.15 Compound matrices

The array of all minors of a given size as in Laplace expansion from a given $m \times n$ matrix A is called a *compound* matrix of A. The kth compound matrix of A is denoted by C_k (A) or $A^{(k)}$ where

$k \leq \min\{m, n\}$. It is of order $\binom{m}{k} \times \binom{m}{k}$, where $\binom{m}{k}$ etc. denotes combination of m things

taken k at a time and α, β elements of C_k (A) are det A(α, β) (cf. chapter 4), taken lexicographically (dictionary order) i.e. minors of same group of rows (or columns) of A will occupy the same row (or column) of the compound matrix.

Ex. Obtain the 2nd compound matrix of

$$A = \begin{bmatrix} 7 & 2 & 3 \\ 4 & 5 & 2 \\ 8 & 0 & 9 \end{bmatrix}.$$

Here the order of 2nd compound matrix is

$$\binom{3}{2} \times \binom{3}{2} = 3 \times 3 \text{ and}$$

$\alpha = \{1, 2\}, \{1, 3\}, \{2, 3\}$; $\beta = \{1, 2\}, \{1, 3\}, \{2, 3\}$.

$$\therefore \quad C_2(A) = \begin{bmatrix} \begin{vmatrix} 7 & 2 \\ 4 & 5 \end{vmatrix} & \begin{vmatrix} 7 & 3 \\ 4 & 2 \end{vmatrix} & \begin{vmatrix} 2 & 3 \\ 5 & 2 \end{vmatrix} \\ \begin{vmatrix} 7 & 2 \\ 8 & 0 \end{vmatrix} & \begin{vmatrix} 7 & 3 \\ 8 & 9 \end{vmatrix} & \begin{vmatrix} 2 & 3 \\ 0 & 9 \end{vmatrix} \\ \begin{vmatrix} 4 & 5 \\ 8 & 0 \end{vmatrix} & \begin{vmatrix} 4 & 2 \\ 8 & 9 \end{vmatrix} & \begin{vmatrix} 5 & 2 \\ 0 & 9 \end{vmatrix} \end{bmatrix} = \begin{bmatrix} 27 & 2 & -11 \\ -10 & 39 & 18 \\ -40 & 20 & 45 \end{bmatrix}$$

Properties of compound matrix : The following are some properties of compound matrices:

(i) If A and B are $m \times k$ and $k \times n$ matrices, then

$$C_r(AB) = C_r(A)\, C_r(B), \; r \leq \min\{m, k, n\}.$$

This can be extended for several matrix products. Thus kth compound of a matrix product is equal to the product of the kth compounds.

(ii) $C_r(tA) = t\, C_r(A)$, t is a scalar.

(iii) For unit matrix I, $C_k(I) = I$, where I is of order $\binom{n}{k}$.

(iv) If A is an n *square* nonsingular matrix, then $C_k(A)^{-1} = C_k(A^{-1})$

(v) If A is an $m \times n$ matrix, then $C_k(A^T) = C_k(A)^T$, i.e. kth compound of the transpose of A is equal to the transpose of the kth compound of A.

(vi) If A is an $m \times n$ complex matrix, then $C_k(A^*) = C_k(A)^*$, i.e. kth compound of the Hermitian of A is equal to the Hermitian of its kth compound.

Ex. Prove that $C_k(A)^{-1}\, C_k(A) = I$,

L.H.S. $= C_k(A)^{-1} C_k(A) = C_k(A^{-1}) C_k(A)$ by (v)
$= C_k(A^{-1}A)$ by (i)
$= C_k(I) = I$ by (iii)

6.16 Factorization of symmetrical determinants

The determinant with nonnumerical elements is a polynomial of those nonnumerical elements, so the meaning of the symmetrical determinant is identical with that of symmetrical polynomial and thereby holds the properties of symmetrical polynomial. For ready reference, definition and properties of symmetrical polynomial are stated below from ordinary algebra.

6.16.1 Symmetrical polynomial : A homogeneous polynomial consisting of symbols a, b, c etc. is symmetrical if the following properties hold—

(i) The polynomial remains unchanged if 'a' is *replaced* by 'b', 'b' by 'c' etc.

(ii) The polynomial remains unchanged if 'a' and 'b' or 'b' and 'c' etc. are *interchanged*.

Examples are : $a+b+c$, $a^2+b^2+c^2$, $bc+ca+ab$, $a^3+b^3+c^3$, $abcd$ etc.

(iii) A linear combination (LC) of symmetrical polynomials of the same degree is again a symmetrical polynomial of that degree. Thus the LC

$$x(a^3+b^3+c^3) + y(b^2c+c^2a+a^2b+bc^2+ca^2+ab^2) + zabc$$

is a polynomial of three third degree symmetrical polynomials

$(a^3+b^3+c^3)$, $(b^2c+c^2a+a^2b+bc^2+ca^2+ab^2)$, (abc) of three symbols a, b, c.

6.16.2 Alternating polynomial : If the interchange of any two symbols changes the sign of the polynomial instead of remaining unchanged, the polynomial is called the *alternating polynomial.* Thus

$$(b-c)(c-a)(a-b) = -a^2(b-c) - b^2(c-a) - c^2(a-b)$$

changes sign on interchanging any two symbols b and c or c and a or a and b.

The product of symmetrical and alternating polynomials is an alternating polynomial.

6.16.3 Remainder theorem : The reader should remember the *remainder theorem* which states that if, in a polynomial containing $a, b, c, \cdots$ and replacement of a by b the polynomial vanishes then $(a-b)$ is a factor of the polynomial.

The properties of the symmetric polynomial, the remainder theorem and cor. 4, theorem 5 of sec.6.4 are together necessary for the factorization of symmetrical determinants.

6.16.4 Degree of symmetric determinant : The degree of the symmetric or alternate determinant is the degree of the product of the diagonal elements. Thus, the degree of

$$\begin{vmatrix} a & b & c \\ a^3 & b^3 & c^3 \\ a^5 & b^5 & c^5 \end{vmatrix}, \begin{vmatrix} a & c & b \\ b & a & c \\ c & b & a \end{vmatrix}$$

are the degree of the diagonal products ab^3c^5 and $a.a.a$ i.e. 9 and 3 respectively.

6.16.5 Procedure of factorization : The following steps are necessary to factor a determinant without expansion:

Step 1. Note the variables present and the degree of the determinant.

Step 2. Use the factor theorem, if possible.

Step 3. Replace one variable for the other and from the remainder theorem conclude the linear factors. *If r rows/columns become identical for a replacement of* $x = a$, *say, then* $(x-a)^{r-1}$ *is a factor of the determinant.*

Step 4. Repeat step 2 to obtain the maximum number of linear factors and note the degree of the product of the factors obtained so far.

Step 5. If the degree obtained in step 3 is less than that obtained in step 1 then adjust the degree by necessary forms of linear combinations shown in 6.11.1(iii).

Step 6. Finally, a constant multiple may be necessary to balance for adjustment.

Step 7. Obtain the numerical values of constants used in step 4 and step 5 by direct inspection or by giving special numerical values to the variables.

Ex. Factor the determinant

$$D = \begin{vmatrix} a & b & c \\ a^2 & b^2 & c^2 \\ a^5 & b^5 & c^5 \end{vmatrix}.$$

Step 1. Degree of D is 8 as the degree of diagonal product ab^2c^5 is 8 and variables are a, b, c.

Step 2. By cor. 4 of theorem 5 of sec. 6.4, remove common factors a, b, c from the 3 columns in D. Then

$$D = abc\begin{vmatrix} 1 & 1 & 1 \\ a & b & c \\ a^4 & b^4 & c^4 \end{vmatrix}.$$

Step 3, 4. On replacing a by b, b by c and c by a the determinant vanishes in each case. So, $(a-b)$, $(b-c)$ and $(c-a)$ are the factors of D.

Step 5, 6. No more factors can be obtained by replacement. So, the product of factors obtained is $abc(a-b)(b-c)(c-a)$ with degree equal to 6. Therefore, the remaining two degrees $(8-6)=2$ may be adjusted by a general second degree factor of a, b, c. Let

$$D = abc(a-b)(b-c)(c-a)\left[x\left(a^2+b^2+c^2\right)+y(bc+ca+ab)\right] \qquad (1)$$

Step 7. The constants x and y in (1) may be determined by inspection of various terms or by putting $a=1, b=-1, c=2$ and $a=1, b=2, c=3$ (values are chosen arbitrarily, so that, R.H.S. of D does not vanish). Using two sets of values in (1), we have,

$$\begin{vmatrix} 1 & -1 & 2 \\ 1 & 1 & 4 \\ 1 & -1 & 32 \end{vmatrix} = 1.-1.2(1+1)(-1-2)(2-1)\left[x\left(1^2+1^2+2^2\right)+y(-2+2-1)\right]$$

i.e. $\quad 60 = 12\,(6x-y)$

i.e. $\quad 5 = 6x-y \qquad (2)$

And
$$\begin{vmatrix} 1 & 2 & -2 \\ 1 & 4 & 4 \\ 1 & 32 & -32 \end{vmatrix} = 1.2.-2(1-2)(2+2)(-2-1)\left[x(1+4+4)+y(-4-2+2)\right]$$

i.e. $\quad -240 = -48\,[9x-4y]$

i.e. $\quad 5 = 9x-4y \qquad (3)$

Solving (2) and (3), we obtain, $x = 1$, $y = 1$.

$\therefore$ from (1) the required factors of D are give by,

$$D = abc(a-b)(b-c)(c-a)\left(a^2+b^2+c^2+bc+ca+ab\right)$$

6.17 Definite, semidefinite and infinite matrices

The positive (negative), definite (semidefinite) and indefinite matrices can be identified by the principal minor test. This is discussed and explained below one after another.

Positive definite matrix : A real symmetric (Hermitian) matrix A is positive definite iff each of the principal minors of A are positive.

For example, let

$$A = \begin{bmatrix} 1 & -1 & 2 & 0 \\ -1 & 2 & -1 & 1 \\ 2 & -1 & 6 & 2 \\ 0 & 1 & 2 & 3 \end{bmatrix}.$$

The pms are $D_1 = 1 > 0$, $D_2 = \begin{vmatrix} 1 & -1 \\ -1 & 2 \end{vmatrix} = 1 > 0$, $D_3 = \begin{vmatrix} 1 & -1 & 2 \\ -1 & 2 & -1 \\ 2 & -1 & 6 \end{vmatrix} = 5 > 0$,

$$D_4 = \begin{vmatrix} 1 & -1 & 2 & 0 \\ -1 & 2 & -1 & 1 \\ 2 & -1 & 6 & 2 \\ 0 & 1 & 2 & 3 \end{vmatrix} = \begin{vmatrix} 1 & 0 & 0 & 0 \\ -1 & 1 & 3 & 1 \\ 2 & 1 & 4 & 2 \\ 0 & 1 & 4 & 3 \end{vmatrix}$$

$$= \begin{vmatrix} 1 & 3 & 1 \\ 1 & 4 & 2 \\ 1 & 4 & 3 \end{vmatrix} = \begin{vmatrix} 1 & 3 & 1 \\ 0 & 1 & 1 \\ 0 & 1 & 2 \end{vmatrix} = \begin{vmatrix} 1 & 1 \\ 1 & 2 \end{vmatrix} = 1 > 0.$$

Since the pms are positive, so, A is positive definite.

Note that matrix A is positive definite if det A > 0. Then the matrix B^TB is positive definite, as det $(B^TB) = |B|^2 > 0$, assuming B is nonsingular.

Negative definite : A real symmetric matrix A is negative definite if the principal minors are alternatively negative and positive starting from principal minors of order one. In other words, principal minors of even orders are positive and those of odd orders are negative, i.e. $D_1 < 0$, $D_2 > 0$, $D_3 < 0$.

For example, let

$$A = \begin{bmatrix} -2 & 1 & -1 \\ 1 & -2 & 1 \\ -1 & 1 & -2 \end{bmatrix}. \text{ So } D_1 = -2 < 0,\ D_2 = \begin{vmatrix} -2 & 1 \\ 1 & -2 \end{vmatrix} = 3 > 0,$$

$$D_3 = \begin{vmatrix} -2 & 1 & -1 \\ 1 & -2 & 1 \\ -1 & 1 & -2 \end{vmatrix} = -2(3)-1(-1)-1(-1) = -6+1+1 = -4 < 0.$$

Since principal minors alternate –ve and +ve signs. Hence A is negative definite. Note that (–A) is a positive definite matrix while A is a negative definite matrix. That is $(-1)^n \ |A| > 0$.

Positive semidefinite : A real symmetric matrix A is positive semidefinite iff it is singular and all other principal minors are nonnegative. Thus if A = $[a_{ij}]_{n \times n}$ and D = |A| = 0 and $D_1 \geq 0, D_2 \geq 0, \ \cdots, D_n \geq 0$.

For example let

$$A = \begin{bmatrix} 1 & 2 & -2 \\ 2 & 4 & -4 \\ -2 & -4 & 4 \end{bmatrix}.$$

Here $$\det A = \begin{vmatrix} 1 & 2 & -2 \\ 2 & 4 & -4 \\ -2 & -4 & 4 \end{vmatrix} = 0,$$

as 2nd row is proportional to 1st row.

∴ A is singular.

Also the other pms are

$$D_1 = 1 > 0, D_2 = \begin{vmatrix} 1 & 2 \\ 2 & 4 \end{vmatrix} = 0,$$

which satisfy the criterion of positive semidefinite.

∴ the matrix A is positive semidefinite.

Negative semidefinite : A real symmetric matrix A is negative semidefinite iff A is singular and its other principal minors of even order are nonnegative and those of odd orders are nonpositive i.e. det A = 0 and $D_1 \leq 0, D_2 \geq 0, D_3 \leq 0$ etc.

For example, let

$$A = \begin{bmatrix} -1 & -2 & 2 \\ -1 & -4 & 4 \\ 2 & 4 & -4 \end{bmatrix}. \text{ Here } \det A = \begin{vmatrix} -1 & -2 & 2 \\ -2 & -4 & 4 \\ 2 & 4 & -4 \end{vmatrix} = 0. \qquad \therefore A \text{ is singular.}$$

Also the other pms are

$$D_1 = -1 < 0, \ D_2 = \begin{vmatrix} -1 & -2 \\ -2 & -4 \end{vmatrix} = 0.$$

∴ criterions for negative semidefinite are satisfied by A .

∴ A is negative semidefinite.

Indefinite : If a real symmetric matrix A fails to satisfy any of above four criteria then it is indefinite. For example, let

$$A = \begin{bmatrix} 1 & -1 & -1 \\ -1 & 2 & 4 \\ -1 & 4 & 6 \end{bmatrix}.$$

The pms are

$$D_1 = 1 > 0, D_2 = \begin{vmatrix} 1 & -1 \\ -1 & 2 \end{vmatrix} = 2 - 1 > 0,$$

$$D_3 = \det A = \begin{vmatrix} 1 & -1 & -1 \\ 0 & 1 & 3 \\ 0 & 3 & 5 \end{vmatrix} = 5 - 9 = -4 < 0.$$

Since D_1, D_2 are positive but D_3 is negative, so A is not positivedefinite. Also note that pms of odd orders are not all negative. So, A is not negative definite. Again A is not singular, so the case of semidefinite does not arise. Hence A is indefinite.

6.18 Applications of determinants

The determinant arises in many applications, as was mentioned in the beginning of the chapter. Some of them are mentioned below.

6.18.1 Vector Algebra : The vector cross product and the scalar triple product are expressed in terms of determinants (cf. any book on vector Analysis, particularly book by Hazra and Ashraf). The geometry of the scalar triple product is the volume of a parallelopiped. For vectors of higher dimension, this represents the volume of a parallelotope whose 2 and 3 dimensional corresponds are parallelogram and parallelopiped.

6.18.2 Elimination : It has already been mentioned that in solving simultaneous algebraic equations by the elimination method solution can be best expressed in terms of determinants. (cf. Cramer's method, chapter seven). Similarly the unknowns or variables of homogeneous simultaneous algebraic equations with equal numbers of unknowns are eliminated if the determinant of the coefficient matrix is equated to zero. Thus, the eliminant of x, y, z from the three simultaneous algebraic equations

$$\begin{cases} a_1x + b_1y + c_1z = 0 \\ a_2x + b_2y + c_2z = 0 \\ a_3x + b_3y + c_3z = 0 \end{cases}$$

is

$$\begin{vmatrix} a_1 & b_1 & c_1 \\ a_2 & b_2 & c_2 \\ a_3 & b_3 & c_3 \end{vmatrix} = 0 \text{ where } \begin{bmatrix} a_1 & b_1 & c_1 \\ a_2 & b_2 & c_2 \\ a_3 & b_3 & c_3 \end{bmatrix}$$

is the coefficient matrix of the given system.

Geometrically, the above eliminant expresses the condition for the planes represented by the three given systems which should have a common line of intersection.

6.18.3 Sylvester's dialytic elimination and bigradient determinant : Sylvester developed a method called *dialytic method* for the elimination of a single variable from two polynomial equations. This method is simple and useful when ordinary method of elimination fails. In this method the highest power of the variables in two equations are *noted*. Let they be m and n. Then $(m + n)$ equations are formed by multiplying one or both of the two equations by the variable to be eliminated so that both equations are raised to same degree. The eliminant is the ***bigradient*** determinant. It is the condition for having a common root of the given equations. More on the topic is in paper 25 in references.

Ex. Eliminate x between $x^2 + ax + b = 0$ and $x^3 + cx^2 + dx + e = 0$

Since $m = 2$ and $n = 3$, so 5 homogeneous equations

$$\begin{array}{lllllll} x^4 + & cx^3 + & dx^2 + & ex & & = & 0 \\ & x^3 + & cx^2 + & dx + & e & = & 0 \\ x^4 + & ax^3 + & bx^2 & & & = & 0 \\ & x^3 + & ax^2 + & bx & & = & 0 \\ & & x^2 + & ax + & b & = & 0 \end{array}$$

original eqns.

new eqns. formed

in $1, x, x^2, x^3, x^4$ are formed by multiplying 1st eqn. by $1, x, x^2$ and 2nd by $1, x$.

Therefore, the eliminant is

$$\begin{vmatrix} 1 & c & d & e & 0 \\ 0 & 1 & c & d & e \\ 1 & a & b & 0 & 0 \\ 0 & 1 & a & b & 0 \\ 0 & 0 & 1 & a & b \end{vmatrix} = 0$$

6.18.4 Electric circuit : Out of the many circuit laws of G.R. Kirchhoff published in 1847, the *matrix tree theorem* is one of these less well known theorem. This law gives a rule for writing a matrix from the labelled graph of an electric circuit and a number of spanning trees. The theorem is stated below without proof.

Theorem . If the vertices of the labelled graph are $1, 2, 3, \cdots, n$ then the matrix T of the graph is formed by the elements

$$t_{ii} = \text{number of lines into point } i,\ i = 1, 2, \cdots, n$$

and

$$t_{ij} = \begin{cases} -1, \text{ if there is a line between } i \text{ and } j = 1,2, \cdots, n \text{ and } i \neq j \\ 0, \text{ otherwise} \end{cases}$$

All the cofactors of $|T|$ are equal and their common value is the number of spanning trees.

Ex . Find the number of spanning trees in the following net work (*figure* 6.1).

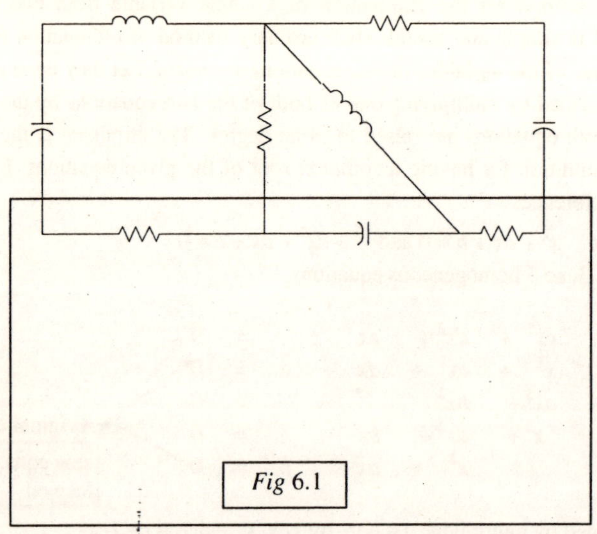

Fig 6.1

The labelled graph corresponding to the given network is

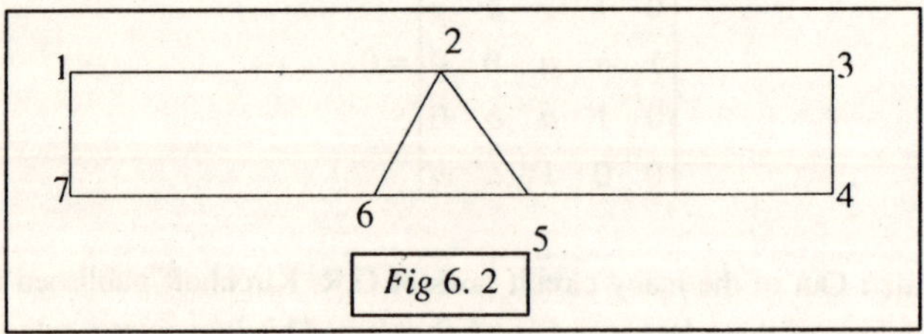

Fig 6. 2

The matrix is formed from the graph and the theorem. It is

$$T = \begin{array}{c} \\ 1 \\ 2 \\ 3 \\ 4 \\ 5 \\ 6 \\ 7 \end{array} \begin{array}{c} \begin{array}{ccccccc} 1 & 2 & 3 & 4 & 5 & 6 & 7 \end{array} \\ \begin{bmatrix} 2 & -1 & 0 & 0 & 0 & 0 & -1 \\ -1 & 4 & -1 & 0 & -1 & -1 & 0 \\ 0 & -1 & 2 & -1 & 0 & 0 & 0 \\ 0 & 0 & -1 & 2 & -1 & 0 & 0 \\ 0 & -1 & 0 & -1 & 3 & -1 & 0 \\ 0 & -1 & 0 & 0 & -1 & 3 & -1 \\ -1 & 0 & 0 & 0 & 0 & -1 & 2 \end{bmatrix} \end{array}$$

Note that $t_{22} = 4$ because there are 4 lines meet at vertex 2. Similarly $t_{46} = 0$ as there is no line from 4 to 6 directly. In this way all the elements have been written.

Find any cofactor of $|T|$. Let us find T_{45}.

$$T_{45} = (-1)^{4+5} M_{45} = -\begin{vmatrix} 2 & -1 & 0 & 0 & 0 & -1 \\ -1 & 4 & -1 & 0 & -1 & 0 \\ 0 & -1 & 2 & -1 & 0 & 0 \\ 0 & -1 & 0 & -1 & -1 & 0 \\ 0 & -1 & 0 & 0 & 3 & -1 \\ -1 & 0 & 0 & 0 & -1 & 2 \end{vmatrix} = 40,$$

which is the required number of spanning trees of the given network.

6.18.5 Applications to geometry : The geometric equations of line, plane, circle etc. can be suitably represented by determinants of different orders.

Equation of line : The equation of a straight line in a two dimensional space is

$$ax + by + c = 0 \qquad (6.43)$$

Let (x_1, y_1) and (x_2, y_2) be two points on line (6.43). So,

$$ax_1 + by_1 + c = 0 \qquad (6.44)$$

$$ax_2 + by_2 + c = 0 \qquad (6.45)$$

Eliminating a, b, c from (6.43) through (6.45), we obtain

$$\begin{vmatrix} x & y & 1 \\ x_1 & y_1 & 1 \\ x_2 & y_2 & 1 \end{vmatrix} = 0 \qquad (6.46)$$

The equation (6.46) is a first degree equation in x and y. So, it represents a *straight line* in determinantal form and passes through two points (x_1, y_1) and (x_2, y_2).

Equation of plane : The equation of a plane in a three dimensional space is

$$ax + by + cz + d = 0 \qquad (6.47)$$

If it passes through three noncollinear points (x_1, y_1), (x_2, y_2), (x_3, y_3), then from (6.47) we obtain

$$ax_1 + by_1 + cz_1 + d = 0 \qquad (6.48)$$

$$ax_2 + by_2 + cz_2 + d = 0 \qquad (6.49)$$

$$ax_3 + by_3 + cz_3 + d = 0 \qquad (6.50)$$

Elimination of a, b, c, d from (6.47) to (6.50), we obtain

$$\begin{vmatrix} x & y & z & 1 \\ x_1 & y_1 & z_1 & 1 \\ x_2 & y_2 & z_2 & 1 \\ x_3 & y_3 & z_3 & 1 \end{vmatrix} = 0 \qquad (6.51)$$

The equation (6.51) is a first degree equation in x, y, z. So, it represents a plane in determinantal form and passes through three noncollinear points (x_i, y_i), $i = 1, 2, 3$.

Equation of circle : Let the equation of a circle in a two dimensional space be

$$a(x^2 + y^2) + bx + cy + d = 0 \qquad (6.52)$$

If it passes through three noncollinear points (x_i, y_i), $i = 1, 2, 3$, then from equation (6.52) we have

$$a(x_i^2 + y_i^2) + bx_i + cy_i + d = 0, \; i = 1, 2, 3 \qquad (6.53)$$

Eliminating constants a, b, c, d, we obtain

$$\begin{vmatrix} x^2+y^2 & y & z & 1 \\ x_1^2+y_1^2 & y_1 & z_1 & 1 \\ x_2^2+y_2^2 & y_2 & z_2 & 1 \\ x_3^2+y_3^2 & y_3 & z_3 & 1 \end{vmatrix} = 0 \qquad (6.54)$$

Equation (6.54) is a second degree equation with no product term xy. So, it represents a *circle* in determinantal form through three noncollinear points (x_i, y_i), $i = 1, 2, 3$.

Equation of a sphere : Proceeding as above the equation of a *sphere* in determinantal from and passing through four noncoplanar points (x_i, y_i), $i = 1, 2, 3, 4$, is given by

$$\begin{vmatrix} x^2+y^2+z^2 & x & y & z & 1 \\ x_1^2+y_1^2+z_1^2 & x_1 & y_1 & z_1 & 1 \\ x_2^2+y_2^2+z_2^2 & x_2 & y_2 & z_2 & 1 \\ x_3^2+y_3^2+z_3^2 & x_3 & y_3 & z_3 & 1 \\ x_4^2+y_4^2+z_4^2 & x_4 & y_4 & z_4 & 1 \end{vmatrix} = 0 \qquad (6.55)$$

Area of a triangle : Let $A_1(x_1, y_1)$, $A_2(x_2, y_2)$ and $A_3(x_3, y_3)$ be the vertices of the triangle $A_1A_2A_3$. Let h be the altitude of the triangle from A_3 to the base A_1A_2 (*figure* 6.3). Since equation of A_1A_2 is

$$\begin{vmatrix} x & y & 1 \\ x_1 & y_1 & 1 \\ x_2 & y_2 & 1 \end{vmatrix} = 0$$

So,

$$h = \frac{\begin{vmatrix} x_3 & y_3 & 1 \\ x_2 & y_2 & 1 \\ x_1 & y_1 & 1 \end{vmatrix}}{\sqrt{(y_1-y_2)^2+(x_1-x_2)^2}},$$

[perpendicular distance of an external point from a line.]

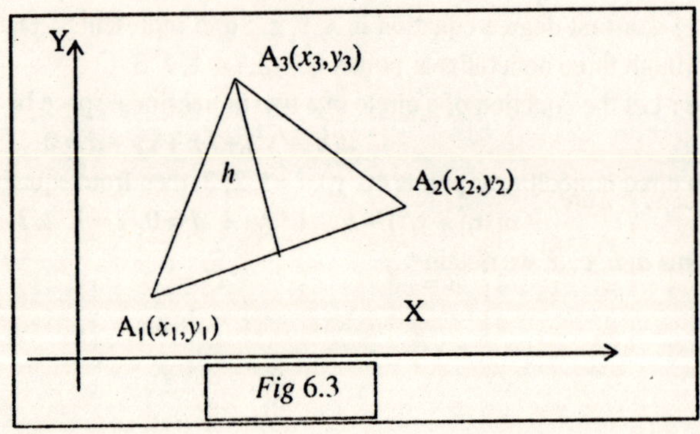

Fig 6.3

$$= \frac{\begin{vmatrix} x_1 & y_1 & 1 \\ x_2 & y_2 & 1 \\ x_3 & y_3 & 1 \end{vmatrix}}{A_1A_2}$$

Now area A of $\Delta A_1A_2A_3 = \frac{1}{2} \times$ base $\times$ altitude

i.e. $$A = \frac{1}{2} A_1A_2.h = \frac{1}{2}\begin{vmatrix} x_1 & y_1 & 1 \\ x_2 & y_2 & 1 \\ x_3 & y_3 & 1 \end{vmatrix}$$

points are taken in the anticlockwise direction.

$\therefore$ double the area $\Delta A_1A_2A_3$ is represented by the determinant

$$\begin{vmatrix} x_1 & y_1 & 1 \\ x_2 & y_2 & 1 \\ x_3 & y_3 & 1 \end{vmatrix},$$

where A_1, A_2, A_3 are considered in the anticlockwise direction in order to keep the sign of the area positive always.

Volume of a tetrahedron : Let $A_i(x_i, y_i, z_i)$, $i = 1, 2, 3, 4$, be the vertices of a tetrahedron (*figure* 6.4). The equation of the face (lateral plane) $A_1A_2A_3$ is

$$\begin{vmatrix} x & y & z & 1 \\ x_1 & y_1 & z_1 & 1 \\ x_2 & y_2 & z_2 & 1 \\ x_3 & y_3 & z_3 & 1 \end{vmatrix} = 0$$

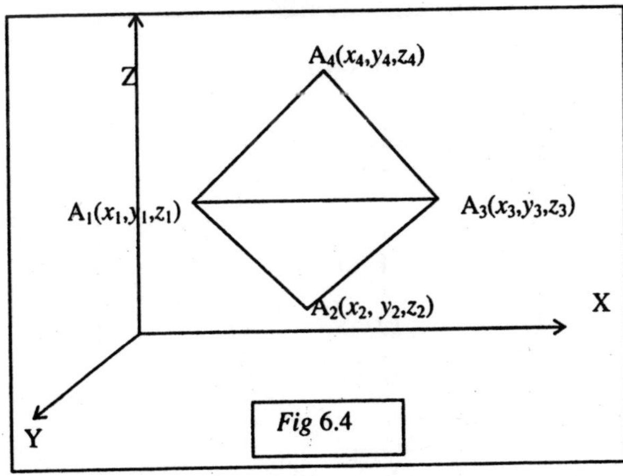

Fig 6.4

where coefficients of x, y, z are

$$A = \begin{vmatrix} y_1 & z_1 & 1 \\ y_2 & z_2 & 1 \\ y_3 & z_3 & 1 \end{vmatrix},\; B = -\begin{vmatrix} x_1 & z_1 & 1 \\ x_2 & z_2 & 1 \\ x_3 & z_3 & 1 \end{vmatrix},\; C = \begin{vmatrix} x_1 & y_1 & 1 \\ x_2 & y_2 & 1 \\ x_3 & y_3 & 1 \end{vmatrix}$$

which represent numerically double the areas of the projections of the triangle $A_1A_2A_3$ on the coordinate planes *YOZ, ZOX, XOY*. From analytic geometry

$$A^2 + B^2 + C^2 = 4\Delta^2, \quad \text{where } \Delta = \text{area of } A_1A_2A_3.$$

Now distance *d*, of A_4 from face $A_1A_2A_3$ is (taken with + ve or –ve sign)

$$d = \frac{\begin{vmatrix} x_4 & y_4 & z_4 & 1 \\ x_1 & y_1 & z_1 & 1 \\ x_2 & y_2 & z_2 & 1 \\ x_3 & y_3 & z_3 & 1 \end{vmatrix}}{\sqrt{A^2 + B^2 + C^2}}$$

$$= -\frac{1}{2\Delta}\begin{vmatrix} x_1 & y_1 & z_1 & 1 \\ x_2 & y_2 & z_2 & 1 \\ x_3 & y_3 & z_3 & 1 \\ x_4 & y_4 & z_4 & 1 \end{vmatrix}$$

Therefore, volume *V* of tetrahedron = $\pm\frac{1}{3}\Delta d$ i.e.

$$V = \pm\frac{1}{6}\begin{vmatrix} x_1 & y_1 & z_1 & 1 \\ x_2 & y_2 & z_2 & 1 \\ x_3 & y_3 & z_3 & 1 \\ x_4 & y_4 & z_4 & 1 \end{vmatrix}$$

which is the volume of the tetrahedron in the determinantal form. The sign ± depends on the sense attributed to the tetrahedron $A_1A_2A_3A_4$.

6.18.6 Power of a point with respect to (w.r.t.) a circle : The power *p* of a point P at a distance *d* from the centre O of the circle with radius *r* (*figure* 6.5) is

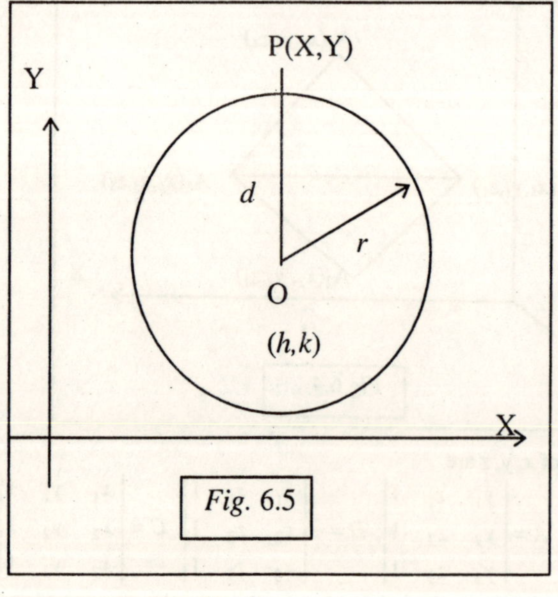

Fig. 6.5

$$p = d^2 - r^2 \tag{6.56}$$

$$= (X-h)^2 + (Y-k)^2 - r^2$$

$$= X^2 + Y^2 + AX + BY + C \tag{6.57}$$

(where $A = -2h$, $B = -2k$, $C = h^2 + k^2 - r^2$)

The expression on the right of (6.57) is in fact, the power of the point P(X,Y) w.r.t. the circle

$$x^2 + y^2 + Ax + By + c = 0 \tag{6.58}$$

i.e. the power of a point P(X , Y) is obtained from L.H.S. of (6.58) by replacing current coordinates x, y of the circle (6.58) by X, Y.

Let the equation of the circle through A_i (x_i, y_i), $i = 1, 2, 3$ be

$$\begin{vmatrix} x^2+y^2 & y & z & 1 \\ x_1^2+y_1^2 & y_1 & z_1 & 1 \\ x_2^2+y_2^2 & y_2 & z_2 & 1 \\ x_3^2+y_3^2 & y_3 & z_3 & 1 \end{vmatrix} = 0 \tag{6.59}$$

Now the coefficient of $(x^2 + y^2)$ in (6.59) is

$$\begin{vmatrix} x_1 & y_1 & 1 \\ x_2 & y_2 & 1 \\ x_3 & y_3 & 1 \end{vmatrix} \tag{6.60}$$

which represents the double area, $2A$ of $\Delta A_1A_2A_3$. Therefore, making the coefficients of $(x^2 + y^2)$ unity in (6.57), we have

$$\frac{1}{2A}\begin{vmatrix} x^2+y^2 & y & z & 1 \\ x_1^2+y_1^2 & y_1 & z_1 & 1 \\ x_2^2+y_2^2 & y_2 & z_2 & 1 \\ x_3^2+y_3^2 & y_3 & z_3 & 1 \end{vmatrix} = 0. \tag{6.61}$$

Power of P(x_4, y_4) w.r.t. circle (6.61) is

$$p = \frac{1}{2A}\begin{vmatrix} x_4^2+y_4^2 & y & z & 1 \\ x_1^2+y_1^2 & y_1 & z_1 & 1 \\ x_2^2+y_2^2 & y_2 & z_2 & 1 \\ x_3^2+y_3^2 & y_3 & z_3 & 1 \end{vmatrix}$$

or

$$-2Ap = \begin{vmatrix} x_1^2+y_1^2 & y_1 & z_1 & 1 \\ x_2^2+y_2^2 & y_2 & z_2 & 1 \\ x_3^2+y_3^2 & y_3 & z_3 & 1 \\ x_4^2+y_4^2 & y_4 & z_4 & 1 \end{vmatrix}, \tag{6.62}$$

by rearrangements of rows.

Geometric corollaries

Cor.1 The equation (6.52) can be written as

$$\begin{vmatrix} x_1^2+y_1^2 & -2x_1 & -2y_1 & 1 \\ x_2^2+y_2^2 & -2x_2 & -2y_2 & 1 \\ x_3^2+y_3^2 & -2x_3 & -2y_3 & 1 \\ x_4^2+y_4^2 & -2x_4 & -2y_4 & 1 \end{vmatrix} = -8Ap \tag{6.63}$$

Multiplying (6.63) by (6.62) row by row and using $l_{ij}^2 = (x_i - x_j)^2 + (y_i - y_j)^2$ it becomes,

$$-16A^2p^2 = \begin{vmatrix} 0 & l_{12}^2 & l_{13}^2 & l_{14}^2 \\ l_{12}^2 & 0 & l_{23}^2 & l_{24}^2 \\ l_{13}^2 & l_{23}^2 & 0 & l_{34}^2 \\ l_{14}^2 & l_{24}^2 & l_{34}^2 & 0 \end{vmatrix} \tag{6.64}$$

This is a relation between the area A of a triangle $A_1A_2A_3$, the power of the point A_4 with respect to circle circumscribed about that triangle and the six distances of the points A_1, A_2, A_3, A_4 taken two by two.

Cor.2 Multiplying columns 2, 3, 4 in the determinant (6.64) by $(l_{13}l_{14})^2$, $(l_{12}l_{14})^2$, $(l_{12}l_{13})^2$, respectively, we get

$$-16A^2p^2(l_{12}l_{13}l_{14})^2 = \begin{vmatrix} 0 & (l_{12}l_{13}l_{14})^2 & (l_{12}l_{13}l_{14})^2 & (l_{12}l_{13}l_{14})^2 \\ l_{12}^2 & 0 & (l_{12}l_{23}l_{14})^2 & (l_{12}l_{13}l_{24})^2 \\ l_{13}^2 & (l_{13}l_{14}l_{23})^2 & 0 & (l_{12}l_{13}l_{34})^2 \\ l_{14}^2 & (l_{13}l_{14}l_{24})^2 & (l_{12}l_{14}l_{34})^2 & 0 \end{vmatrix}$$

or $$-16A^2p^2(l_{12}l_{13}l_{14})^2 = (l_{12}l_{13}l_{14})^2 \begin{vmatrix} 0 & 1 & 1 & 1 \\ 1 & 0 & (l_{13}l_{14})^2 & (l_{13}l_{24})^2 \\ 1 & (l_{14}l_{23})^2 & 0 & (l_{12}l_{34})^2 \\ 1 & (l_{13}l_{24})^2 & (l_{12}l_{34})^2 & 0 \end{vmatrix}$$

or $$-16A^2p^2 = \begin{vmatrix} 0 & 1 & 1 & 1 \\ 1 & 0 & (l_{13}l_{14})^2 & (l_{13}l_{24})^2 \\ 1 & (l_{14}l_{23})^2 & 0 & (l_{12}l_{34})^2 \\ 1 & (l_{13}l_{24})^2 & (l_{12}l_{34})^2 & 0 \end{vmatrix} \tag{6.65}$$

Cor.3 Area in terms of sides.

Let A_4 coincide with the centre of the circle circumscribed by $\Delta A_1A_2A_3$. Then $d = 0$, $p = -R^2$ and $l_{14} = l_{24} = l_{34} = R$. So, (6.64) becomes

$$-16A^2R^2 = \begin{vmatrix} 0 & l_{12}^2 & l_{13}^2 & R^2 \\ l_{12}^2 & 0 & l_{23}^2 & R^2 \\ l_{13}^2 & l_{23}^2 & 0 & R^2 \\ R^2 & R^2 & R^2 & 0 \end{vmatrix} \tag{6.66}$$

$$16A^2 = -\begin{vmatrix} 0 & l_{12}^2 & l_{13}^2 & 1 \\ l_{12}^2 & 0 & l_{23}^2 & 1 \\ l_{13}^2 & l_{23}^2 & 0 & 1 \\ 1 & 1 & 1 & 0 \end{vmatrix} = \begin{vmatrix} 1 & 0 & l_{12}^2 & l_{13}^2 \\ 1 & l_{12}^2 & 0 & l_{23}^2 \\ 1 & l_{13}^2 & l_{23}^2 & 0 \\ 0 & 1 & 1 & 1 \end{vmatrix} \tag{6.67}$$

Let $l_{12} = a,\ l_{13} = b,\ l_{23} = c,$ then,

$$16A^2 = \begin{vmatrix} 1 & 0 & a^2 & b^2 \\ 1 & a^2 & 0 & c^2 \\ 1 & b^2 & c^2 & 0 \\ 0 & 1 & 1 & 1 \end{vmatrix} \qquad (6.68)$$

or $\qquad 16A^2 = (a+b+c)(a+b-c)(a+c-b)(b+c-a)$

Let $\qquad a+b+c = 2s$. So,

$$A^2 = s(s-a)(s-b)(s-c),$$

the well known formula for the area of a triangle.

Cor. 4 Comparing R.H.S. of (6.65) and (6.68),we obtain,

$$16A^2p^2 = (l_{23}l_{14} + l_{13}l_{24} + l_{12}l_{34})(l_{23}l_{14} + l_{13}l_{24} - l_{12}l_{34})$$
$$(l_{23}l_{14} + l_{12}l_{34} - l_{13}l_{24})(l_{13}l_{24} + l_{12}l_{34} - l_{23}l_{14}) \qquad (6.69)$$

Cor. 5 Let A_1, A_2, A_3, A_4 be on the same circle, then $p = 0$ and (6.66) reduces to

$$(l_{23}l_{14} + l_{13}l_{24} - l_{12}l_{34})(l_{23}l_{14} + l_{12}l_{34} - l_{13}l_{24})(l_{13}l_{24} + l_{12}l_{34} - l_{23}l_{14}) = 0 \qquad (6.70)$$

N.B. The above results can be extended to 3-dimension. The power of a point w.r.t. a sphere of radius R is $p = d^2 - R^2$, d is the distance of some point $A_5(x_5, y_5, z_5)$.

Exercises six

1. Evaluate determinant of the following matrices:

   i. [3], ii.[–5] **[Ans.** i. 3, ii. –5]

2. Find all zeros of the polynomial

$$f(t) = \det\begin{bmatrix} t-2 & 4 & -3 \\ 0 & t-3 & 2 \\ 0 & 0 & t \end{bmatrix}.$$

   [Ans. $f(t) = (t-2)(t-3)t = 0$ gives $t = 2, 3, 0$]

3. Without expanding the determinant, show that for any values of s, t and u

$$\det\begin{bmatrix} s & s+1 & s+2 \\ t & t+1 & t+2 \\ u & u+1 & u+2 \end{bmatrix} = 0.$$

4. Show by computation that

$$\begin{vmatrix} 5 & 6 & 1 \\ 5 & 2 & 3 \\ 11 & 5 & 0 \end{vmatrix} = \begin{vmatrix} 3 & 6 & 1 \\ 4 & 2 & 3 \\ 8 & 5 & 0 \end{vmatrix} + \begin{vmatrix} 2 & 6 & 1 \\ 1 & 2 & 3 \\ 3 & 5 & 0 \end{vmatrix}.$$

5. Consider the matrices

$$A = \begin{bmatrix} 2 & 4 & 5 \\ 1 & 1 & 2 \\ 0 & 6 & 8 \end{bmatrix} \text{ and } B = \begin{bmatrix} 2 & 2 & 9 \\ 6 & 1 & 6 \\ 5 & 1 & 0 \end{bmatrix}.$$

Show that

a. $|3A| = 3^3|A|$.

b. $|AB| = |A||B|$.

c. $|A \otimes B| = |A|^3|B|^3$.

6. Tell whether the following statements are true or false. Give your reasons. Do not expand determinants.

a. $\begin{vmatrix} 1 & 2 & 3 & 4 \\ 5 & 6 & 7 & 8 \\ 8 & 7 & 6 & 5 \\ 4 & 3 & 2 & 1 \end{vmatrix} = \begin{vmatrix} 2 & 1 & 4 & 3 \\ 6 & 5 & 8 & 7 \\ 7 & 8 & 5 & 6 \\ 3 & 4 & 1 & 2 \end{vmatrix}$.

b. $\begin{vmatrix} 1 & 3 & -4 \\ 2 & 8 & 3 \\ 0 & -2 & -5 \end{vmatrix} = \begin{vmatrix} 1 & 2 & 0 \\ 3 & 8 & -2 \\ -4 & 3 & -5 \end{vmatrix}$.

c. $\begin{vmatrix} 2x & 3x & 4x \\ 5x & 6x & 7x \\ 8x & 9x & 9x \end{vmatrix} = x\begin{vmatrix} 2 & 3 & 4 \\ 5 & 6 & 7 \\ 8 & 9 & 9 \end{vmatrix}$, if $x \neq 0$ and $x \neq \pm 1$.

[**Ans.** a. True, interchanges of 1st and 2nd columns, 3rd and 4th columns, keeps the sign unchanged.
b. True, interchanges of rows and columns keeps the value unchanged.
c. False, x^3 should be the common factor]

7. Without expanding, making use of two theorems, evaluate

$$\begin{vmatrix} 2 & 4 & 6 & 4 \\ 0 & 4 & 6 & 9 \\ 2 & 1 & 4 & 0 \\ 1 & 2 & 3 & 2 \end{vmatrix}.$$

[**Ans.** 1st row is proportional to the 4th row. Value = 0]

8. The value of the minor of the entry in the 13th row and 11th column of a 22nd order C-matrix is found 5. What is the cofactor of this entry ? Why ?

[**Ans.** $(-1)^{13+11}$ 4 = 4]

9. If det $[a_{ij}]_{3\times3} = 4$, find det $(3[a_{ij}]_{3\times3})$. Give reasons for your answer.

[**Ans.** $3^3.4 = 108$, by application of theorem 4 sec. 6.4, 3 times]

10. Without expanding, show that

$$\begin{vmatrix} x^2-y^2 & x+y & x \\ x-y & 1 & 1 \\ x-y & 1 & y \end{vmatrix} = 0.$$

Give reasons for your steps.

[**Ans.** $(x-y)$ is taken out common from column 1. Then 1st and 2nd column become identical]

11. Change the form, but not the value of

$$|A| = \begin{vmatrix} -1 & 1 & 2 & 0 \\ -2 & 1 & 3 & 1 \\ 1 & 0 & 2 & -1 \\ 2 & 1 & -1 & 3 \end{vmatrix},$$

so that zeros occur everywhere in the first column except in the third row.

$$\text{[Ans.} \quad \begin{matrix} r_1+r_3 \\ r_2+r_3 \\ \\ r_3+2r_1 \end{matrix} \begin{vmatrix} 0 & 1 & 4 & -1 \\ 0 & 2 & 2 & 3 \\ 1 & 0 & 2 & -1 \\ 0 & 3 & 3 & 2 \end{vmatrix} \text{]}$$

12. Evaluate the determinant

$$\begin{vmatrix} 3 & 5 & 2 & 4 \\ 1 & 1 & -1 & 6 \\ 2 & 3 & 5 & 1 \\ 2 & 1 & 4 & 8 \end{vmatrix}.$$ [**Ans.** 12]

(a) Laplace's expansion using two rows, by Laplace's expansion using two columns and by elements of a single row or column.

(b) By using the elementary operations to produce a number of zeros in rows and columns and then expanding.

13. Show that

$$\begin{vmatrix} a_1-x & a_2 & a_3 & a_4 \\ a_2 & a_3-x & a_4 & a_1 \\ a_3 & a_4 & a_1-x & a_2 \\ a_4 & a_1 & a_2 & a_3-x \end{vmatrix}$$

$$= [x_2-(a_1-a_3)^2-(a_2-a_4)^2]\,[x-a_1-a_2-a_3-a_4]\,[x-a_1+a_2-a_3+a_4]$$

without expanding the determinant.

14. Prove that

$$D = \begin{vmatrix} -1 & 0 & 0 & a \\ 0 & -a & 0 & b \\ 0 & 0 & -1 & c \\ x & y & z & -1 \end{vmatrix} = 1 - ax - by - cz \cdot$$

Hence $D = 0$ is the equation of a plane.

15. Find the determinant of the matrix whose column vectors are

(i) $\mathbf{X}_{\cdot 1} = [1\ \ 2\ \ 3]$, $\mathbf{X}_{\cdot 2} = [1\ \ 0\ \ -1]$, $\mathbf{X}_{\cdot 3} = [0\ 0\ 1]$ **[Ans.** -2]

(ii) $\mathbf{X}_{\cdot 1} = [0\ \ 1\ \ 0\ \ -1]$, $\mathbf{X}_{\cdot 2} = [1\ \ 0\ \ 1\ \ 0]$, $\mathbf{X}_{\cdot 3} = [0\ \ -1\ \ 0\ \ 1]$, $\mathbf{X}_{\cdot 4} = [1\ \ 0\ \ -1\ \ 0]$

[Ans. 0]

16. If A and B are square matrices, then show that

$$\det \begin{bmatrix} A & X \\ 0 & B \end{bmatrix} = \det A\ \det B$$

useful for large determinant evaluation.

17. Prove that $\det(A^*) = \overline{\det A}$ i.e. determinant of a conjugate transposed matrix (or tranjugate matrix) is the conjugate of the determinant of the original matrix. **[Hints:** Since $A^* = \overline{(A^T)}$ (cf. sec.2.2.4) and $\det A = \det A^T$,

$\therefore\ \det(A^*) = \overline{\det A}$]

18. Verify $|AB| = |A|\,|B|$ where

$$A = \begin{bmatrix} 3 & 2 & 4 \\ -1 & 1 & 0 \\ 1 & 2 & 1 \end{bmatrix},\ B = \begin{bmatrix} 2 & 1 & 3 \\ -5 & 0 & 2 \\ 1 & 3 & 2 \end{bmatrix}.$$

19. Eliminate a, b, c using determinant from the equations

$$p(b-c) = a\ ;\ \ q(c-a) = b\ ;\ \ r(a-b) = c.$$

[**Ans.** $pq + qr + rp + 1 = 0$]

20. Find the value of the following 10×10 determinant

(i) $$\det \begin{bmatrix} 0&0&0&0&0&1&0&0&0&0 \\ 0&0&0&0&0&0&2&0&0&0 \\ 0&0&0&0&0&0&0&3&0&0 \\ 0&0&0&0&0&0&0&0&4&0 \\ 5&0&0&0&0&0&0&0&0&5 \\ 0&6&0&0&0&0&0&0&0&0 \\ 0&0&7&0&0&0&0&0&0&0 \\ 0&0&0&8&0&0&0&0&0&0 \\ 0&0&0&0&9&0&0&0&0&0 \\ 0&0&0&0&0&10&0&0&0&0 \end{bmatrix};$$

(ii) $$\det\begin{bmatrix} 0 & 3 & 0 & 0 & 0 & 0 & 0 & 0 & 0 & 0 \\ 0 & 0 & 3 & 0 & 0 & 0 & 0 & 0 & 0 & 0 \\ 0 & 0 & 0 & 3 & 0 & 0 & 0 & 0 & 0 & 0 \\ 0 & 0 & 0 & 0 & 3 & 0 & 0 & 0 & 0 & 0 \\ 0 & 0 & 0 & 0 & 0 & 3 & 0 & 0 & 0 & 0 \\ 0 & 0 & 0 & 0 & 0 & 0 & 3 & 0 & 0 & 0 \\ 0 & 0 & 0 & 0 & 0 & 0 & 0 & 3 & 0 & 0 \\ 0 & 0 & 0 & 0 & 0 & 0 & 0 & 0 & 3 & 0 \\ 0 & 0 & 0 & 0 & 0 & 0 & 0 & 0 & 0 & 3 \\ 1 & 0 & 0 & 0 & 0 & 0 & 0 & 0 & 0 & 0 \end{bmatrix}.$$

[**Ans.** (i) 0 ; **Hints**: Partition into four 5×5 blocks or into suitable blocks ; (ii) -3^9]

21. Compute the determinants of the following matrices

(i) $$\left[\begin{array}{cc|ccc} 0 & 1 & 0 & 0 & 0 \\ 1 & 0 & 0 & 0 & 0 \\ \hline a & b & 0 & 1 & 2 \\ c & d & 2 & 0 & 1 \\ e & f & 1 & 2 & 0 \end{array}\right];$$ (ii) $$\left[\begin{array}{cc|cc|cc} 0 & 0 & 0 & 0 & 1 & 2 \\ 0 & 0 & 0 & 0 & 2 & 1 \\ \hline 0 & 0 & 1 & 2 & 0 & 0 \\ 0 & 0 & 2 & 1 & 0 & 0 \\ \hline a & b & c & d & e & f \\ \alpha & \beta & \gamma & \delta & g & h \end{array}\right]$$

[**Ans.** (i) 9,(ii) 9 $(a\beta - \alpha b)$]

22. Expand the determinant

$$\left|\begin{array}{ccccc} 2 & 0 & 1 & 0 & 1 \\ 1 & 0 & 1 & 0 & 2 \\ \hline 1 & 2 & 0 & 1 & 1 \\ 1 & 1 & 0 & 2 & 1 \\ 1 & 1 & 0 & 1 & 1 \end{array}\right|$$

w. r. t the first two rows. [**Ans.** –2]

23. Expand the determinant

$$\left|\begin{array}{ccc|cc} 0 & -2 & 4 & 1 & 1 \\ 1 & 2 & 1 & 1 & 2 \\ 1 & 2 & 1 & 2 & 2 \\ 0 & 1 & -2 & 3 & 3 \\ 1 & 0 & 1 & 4 & 4 \end{array}\right|$$

w. r. t the last two columns. [**Ans.** –30]

24. By means of a determinant, eliminate x, y, z from the equations

$$tx + a(y + z + w) = 0$$
$$ty + b(z + w + x) = 0$$
$$tz + c(w + x + y) = 0$$
$$tw + d(x + y + z) = 0$$

and show that the coefficients of t^2 in the result is —

$$(ab + ac + ad + bc + bd + cd).$$

25. Obtain as a determinant the result of eliminating x from the equations

$$ax^3 + bx^2 + cx + d = 0$$
$$a^1x^2 + b^1x + c^1 = 0$$

[**Hints :** Raise the degree of x to 4 and form 5 equations as

$$\begin{array}{lllllll} ax^4 + & bx^3 + & cx^2 + & dx & & = 0 & \leftarrow \text{new equations formed} \\ & ax^3 + & bx^2 + & cx + & d & = 0 & \\ a^1x^4 + & b^1x^3 + & c^1x^2 & & & = 0 & \\ & a^1x^3 + & b^1x^2 + & c^1x & & = 0 & \text{original equations} \\ & & a^1x^2 + & b^1x + & c^1 & = 0 & \leftarrow --- \end{array}$$

Ans. $\begin{vmatrix} a & b & c & d & 0 \\ 0 & a & b & c & d \\ a^1 & b^1 & c^1 & 0 & 0 \\ 0 & a^1 & b^1 & c^1 & 0 \\ 0 & 0 & a^1 & b^1 & c^1 \end{vmatrix}$]

26. Show that

$$\det\begin{bmatrix} x^2 & y^2 & z^2 \\ x & y & z \\ 1 & 1 & 1 \end{bmatrix} = (x-y)(x-z)(y-z).$$

27. Show that

$$\det\begin{bmatrix} a & a & \cdots & a & a \\ b & a & \cdots & a & a \\ 0 & b & \ddots & a & a \\ \vdots & \vdots & \ddots & \ddots & \vdots \\ 0 & 0 & \cdots & b & a \end{bmatrix}_{n\times n} = a\,(a-b)^{n-1}.$$

28. Show that

$$\det\begin{bmatrix} 0 & a & \cdots & a & a \\ a & 0 & \cdots & a & a \\ & & \ddots & & \\ \vdots & \vdots & \vdots & \ddots & \vdots \\ a & a & \cdots & a & b \end{bmatrix} = (-a)^{n-1}\,[n(a-b)-a+2b].$$

29. Factor the following determinants:

$$\text{(i)} \begin{vmatrix} 1 & 1 & 1 \\ a & b & c \\ a^4 & b^4 & c^4 \end{vmatrix}, \text{ (ii)} \begin{vmatrix} 1 & a^3 & a \\ a & a^5 & a^4 \\ a^4 & a^{11} & a^{13} \end{vmatrix}, \text{ (iii)} \begin{vmatrix} a & c & b \\ b & a & c \\ c & b & a \end{vmatrix},$$

$$\text{(iv)} \begin{vmatrix} (b+c)^2 & a^2 & a^2 \\ b^2 & (c+a)^2 & b^2 \\ c^2 & c^2 & (a+b)^2 \end{vmatrix}, \text{ (v)} \begin{vmatrix} \cos\alpha & \cos\beta & \cos\gamma \\ \sin 2\alpha & \sin 2\beta & \sin 2\gamma \\ \cos 3\alpha & \cos 3\beta & \cos 3\gamma \end{vmatrix},$$

$$\text{(vi)} \begin{vmatrix} a^2 & (s-a)^2 & (s-a)^2 \\ (s-b)^2 & b^2 & (s-b)^2 \\ (s-c)^2 & (s-c)^2 & c^2 \end{vmatrix}, \text{ (vii)} \begin{vmatrix} (s-a)^2 & a^2 & b^2 \\ b^2 & (s-b)^2 & b^2 \\ c^2 & c^2 & (s-c)^2 \end{vmatrix}$$

[**Ans.** (i) $(b-c)(c-a)(a-b)[(a^2+b^2+c^2)+(bc+ca+ab)]$,

(ii) $a^{10}(a+1)(a-1)^3(a^3+1)(a^2+a+1)$

(iii) $(a+b+c)[(a^2+b^2+c^2)-(bc+ca+ab)]$

(iv) $2abc\,(a+b+c)^2(a+b+c) = 2abc\,(a+b+c)^3$

(v) $-8\cos\alpha\cos\beta\cos\gamma(\sin\beta-\sin\gamma)(\sin\gamma-\sin\alpha)(\sin\alpha-\sin\beta)$

(vi) $2s^3(s-a)(s-b)(s-c)$

[**Hints:** $s = 0$ produces 3 column identical.]

(vii) $2abc\,s^3$. [**Hints:** $s = 0$ produces 3 columns identical,

Degree = (3 degrees of S) + (3 degrees of a, b, c) = 6]

30. Compute the determinants of the following matrices by triangularization:

$$\text{(i)} \begin{bmatrix} 1 & 3 & 2 & 1 & 0 \\ -2 & -4 & 0 & 1 & 3 \\ 3 & 0 & -9 & 1 & 2 \\ 1 & 2 & -1 & 2 & 5 \\ -1 & 2 & 3 & 4 & 7 \end{bmatrix}, \text{ (ii)} \begin{bmatrix} 1 & -2 & 3 & 0 \\ -2 & 4 & -4 & 1 \\ -3 & 5 & 1 & 1 \\ 1 & -3 & 0 & 2 \end{bmatrix}, \text{ (iii)} \begin{bmatrix} 1 & 1 & 2 & -1 \\ 2 & 3 & 7 & 0 \\ -1 & 0 & 1 & 4 \\ 1 & 1 & 2 & 0 \end{bmatrix},$$

$$\text{(vi)} \begin{bmatrix} 1 & 1 & 2 & -1 & 0 \\ 3 & 4 & 9 & -1 & 1 \\ 1 & 2 & 5 & 0 & 2 \\ 2 & 3 & 7 & 1 & -1 \\ 1 & 1 & 2 & 1 & 1 \end{bmatrix}, \text{ (v)} \begin{bmatrix} 1 & 0 & 2 & 1 \\ 2 & -1 & 1 & 0 \\ 1 & 0 & 0 & 3 \\ -1 & 0 & 2 & 1 \end{bmatrix}, \text{ (vi)} \begin{bmatrix} -1 & 1 & 2 & 1 \\ 1 & -1 & 3 & -1 \\ 2 & -2 & 3 & 1 \\ 1 & -1 & 0 & 1 \end{bmatrix},$$

(vii) $\begin{bmatrix} -1 & 2 & 0 & 1 \\ 1 & 1 & -1 & 0 \\ 2 & 1 & 1 & 0 \\ -1 & -1 & 0 & 1 \end{bmatrix}$

[**Ans.** (i) –204, (ii) 1,(iii) 0, (iv) 0, (v) 12, (vi) 0, (vii) –9]

31. Find the determinant of the following product of matrices:

(i) $\begin{bmatrix} 1 & 0 & 0 \\ 0 & 2 & 0 \\ 0 & 0 & 3 \end{bmatrix}\begin{bmatrix} 2 & 0 & 1 \\ 0 & 1 & 0 \\ 1 & 0 & 1 \end{bmatrix}\begin{bmatrix} 1 & 1 & \frac{1}{3} \\ 1 & \frac{1}{2} & 0 \\ 1 & 0 & 0 \end{bmatrix}$, [**Ans.** $6.1.\left(-\frac{1}{6}\right) = -1$]

(ii) $\begin{bmatrix} 1 & 0 & 0 & a \\ 0 & 1 & 0 & b \\ 0 & 0 & 1 & c \end{bmatrix}\begin{bmatrix} 1 & 0 & 0 \\ 0 & 1 & 0 \\ 0 & 0 & 1 \\ a & b & c \end{bmatrix}$, [**Ans.** $m = 3 < n = 4$, $(1 + a^2 + b^2 + c^2)$]

32. Given that

$$\begin{bmatrix} 1 & -2 & 0 \\ 2 & 1 & 0 \\ 0 & 0 & 1 \end{bmatrix} A \begin{bmatrix} 1 & 2 & 0 \\ -2 & 1 & 0 \\ 0 & 0 & 1 \end{bmatrix} = 5I,$$

what is det A?

[**Ans.** $5(\det A)5 = 5$ or $|A| = \frac{1}{5}$]

33. Compute the determinants of the following partition matrices:

(i) $\left[\begin{array}{cc|ccc} 0 & 1 & 0 & 0 & 0 \\ 1 & 0 & 0 & 0 & 0 \\ \hline a & b & 0 & 1 & 2 \\ c & d & 2 & 0 & 1 \\ e & f & 1 & 2 & 0 \end{array}\right]$, (ii) $\left[\begin{array}{c|cc|ccc} 0 & 0 & 0 & 3 & 2 & 1 \\ 0 & 0 & 0 & 2 & 1 & 0 \\ 0 & 0 & 0 & 1 & 0 & -1 \\ \hline 0 & 2 & 1 & 0 & 0 & 0 \\ 0 & 1 & 2 & 0 & 0 & 0 \\ \hline 4 & 0 & 0 & 0 & 0 & 0 \end{array}\right]$, (iii) $\left[\begin{array}{cc|cc|cc} 0 & 0 & 0 & 0 & 1 & 2 \\ 0 & 0 & 0 & 0 & 2 & 1 \\ \hline 0 & 0 & 1 & 2 & 0 & 0 \\ 0 & 0 & 2 & 1 & 0 & 0 \\ \hline a & b & c & d & e & f \\ \alpha & \beta & \gamma & \delta & g & h \end{array}\right]$

[**Ans.** (i) $\det\begin{bmatrix} 0 & 1 \\ 1 & 0 \end{bmatrix} \det\begin{bmatrix} 0 & 1 & 2 \\ 2 & 0 & 1 \\ 1 & 2 & 0 \end{bmatrix} = (-1)(9) = -9$

(ii) $-\det[4]\det\begin{bmatrix} 2 & 1 \\ 1 & 2 \end{bmatrix}\det\begin{bmatrix} 3 & 2 & 1 \\ 2 & 1 & 0 \\ 1 & 0 & -1 \end{bmatrix} = -4.3.0 = 0$

(iii) $-\det\begin{bmatrix} a & b \\ \alpha & \beta \end{bmatrix}\det\begin{bmatrix} 1 & 2 \\ 2 & 1 \end{bmatrix}\det\begin{bmatrix} 1 & 2 \\ 2 & 1 \end{bmatrix}$

$= (b\alpha - a\beta)(-3)(-3) = 9(b\alpha - a\beta)$]

34. Find the area of a triangle whose vertices are (–6, 0), (1, 1), (3, –4).

[**Ans.** $\frac{37}{2}$ sq. units]

35. Find the area of a quadrilateral whose vertices are at (1, 4), (1, 3), (–2, –3), (2, –2).
[**Ans.** $21\frac{1}{2}$ sq. units]

36. The vertices of a triangle are (0, t), (t, $2t$), ($3t$, 0). Find a formula for the area of the triangle. Is there any value of t for which the area is zero?

[**Ans.** $2t^2$ sq. units, $t = 0$]

37. Prove that if A and B are real, positive definite symmetric matrices of order n, then so is A + B.

38. Prove that if A_n is the $n \times n$ matrix

$$\begin{bmatrix} 2c & 1 & & & \bigcirc \\ 1 & 2c & 1 & & \\ & 1 & 2c & 1 & \\ & & \ddots & \ddots & \ddots \\ & & & \ddots & \ddots & \ddots \\ & & & & \ddots & \ddots & 1 \\ \bigcirc & & & & & 1 & 2c \end{bmatrix}.$$

$D_n = \det A_n$ and $c = \cos\theta$ then $D_n = \dfrac{\sin(n+1)\theta}{\sin\theta}$, and show how D_n may be evaluated if $|c| >$ 1.

39. Classify each of the following symmetric R-matrices according to whether they are positive definite, negative definite, positive semidefinite, negative semidefinite or indefinite:

$$\text{(i)} \begin{bmatrix} 2 & -2 \\ -2 & 5 \end{bmatrix}, \text{(ii)} \begin{bmatrix} 0 & 1 \\ 1 & 0 \end{bmatrix}, \text{(iii)} \begin{bmatrix} 1 & 2 \\ 2 & 1 \end{bmatrix}, \text{(iv)} \begin{bmatrix} 2 & 6 \\ 6 & -3 \end{bmatrix},$$

$$\text{(v)} \begin{bmatrix} 2 & 1 & 1 \\ 1 & 1 & 0 \\ 1 & 0 & 1 \end{bmatrix}, \text{(vi)} \begin{bmatrix} 1 & -2 & 0 \\ -2 & 4 & 0 \\ 0 & 0 & 3 \end{bmatrix}, \text{(vii)} \begin{bmatrix} 1 & 0 & 1 \\ 0 & 1 & 2 \\ 1 & 2 & 5 \end{bmatrix}, \text{(viii)} \begin{bmatrix} 2 & -1 & 1 \\ -1 & 2 & -1 \\ 1 & -1 & 2 \end{bmatrix}.$$

[**Ans.** (i) Positive definite, (ii) Indefinite, (iii) indefinite,
(iv) Indefinite, (v) Positive semidefinite, (vi) Indefinite,
(vii) Positive semidefinite, (viii) Negative semidefinite]

40. Examine for definiteness

$$\text{(i)} \begin{bmatrix} 1 & -1 & -1 \\ -1 & 2 & 4 \\ -1 & 4 & 6 \end{bmatrix}, \text{(ii)} \begin{bmatrix} 4 & 2 & -2 \\ 2 & 4 & 2 \\ -2 & 2 & 4 \end{bmatrix}, \text{(iii)} \begin{bmatrix} 1 & 2 & 0 \\ 2 & 1 & 3 \\ 0 & 3 & -3 \end{bmatrix},$$

$$\text{(iv)} \begin{bmatrix} 2 & 1 & 1 \\ 1 & 2 & 1 \\ 1 & 1 & \frac{2}{3} \end{bmatrix}, \text{(v)} \begin{bmatrix} x & 1 & 0 \\ 1 & x & 1 \\ 0 & 1 & x \end{bmatrix}$$

[**Ans.** (i) Indefinite, (ii) Indefinite, (iii) Indefinite, (iv) Indefinite, (v) Positive (negative) definite if $x > 0 \left(x < 0\right)$]

41. Which of the following matrices are positive definite? Negative definite? Indefinite?

$$\text{(i)} \begin{bmatrix} 3 & 2 \\ 2 & 2 \end{bmatrix}, \text{(ii)} \begin{bmatrix} 3 & 4 \\ 4 & 1 \end{bmatrix}, \text{(iii)} \begin{bmatrix} -2 & 0 & 1 \\ 0 & -1 & 0 \\ 1 & 0 & -2 \end{bmatrix}, \text{(iv)} \begin{bmatrix} 1 & 2 & 1 \\ 2 & 1 & 1 \\ 1 & 1 & 2 \end{bmatrix}.$$

[**Ans.** (i) Positive definite, (ii) Indefinite, (ii) Negative definite, (iv) Indefinite]

42. Show that an nth order determinant can be written as

$$|A| = \begin{vmatrix} a_{11} & a_{12} & \cdots & a_{1n} \\ a_{21} & a_{22} & \cdots & a_{2n} \\ \vdots & \vdots & \ddots & \vdots \\ a_{n1} & a_{n2} & \cdots & a_{nn} \end{vmatrix} = a_{11}A_{11} + \begin{vmatrix} 0 & a_{12} & \cdots & a_{1n} \\ a_{21} & a_{22} & \cdots & a_{2n} \\ \vdots & \vdots & \ddots & \vdots \\ a_{n1} & a_{n2} & \cdots & a_{nn} \end{vmatrix},$$

where A_{11} is the cofactor of a_{11}. Note that the cofactors $A_{12}, A_{13}, \cdots, A_{nn}$ contain the elements $a_{21}, a_{31}, \cdots, a_{n1}$ of the first column of A. Denote the cofactors of a_{i1} in A_{1j} by $A_{1j,\,i1}$ (this is a determinant of the partitioned matrix formed from A by crossing out rows 1, i and columns 1, j as does also $A_{11,ij}$ which, in turn, is the cofactor of a_{ij} in A_{11}. Thus $A_{11,ij}$ and $A_{1j,\,i1}$ can differ only in sign. Show that

$$A_{11,ij} = -A_{1j,i1}.$$

Now prove that | A | can be expanded in the form

$$|A| = a_{11}A_{11} - \sum_{i=2}^{n}\sum_{j=2}^{n} a_{1j}a_{i1}A_{11,ij}$$

This is famous *Cauchy expansion* of a determinant in terms of the first row and column. This is useful in statistics and in the study of quadratic surfaces.

43. By means of Cauchy method expand the following in terms of the first row and column:

$$\text{(i)} \begin{vmatrix} 1 & 2 & 3 & 4 \\ 2 & 0 & 7 & 1 \\ 5 & 9 & 0 & 6 \\ 6 & 1 & 1 & 2 \end{vmatrix}; \text{(ii)} \begin{vmatrix} 1 & 2 & 3 & 4 \\ 2 & -1 & 0 & 1 \\ -1 & 2 & -3 & 0 \\ -2 & -2 & 0 & 5 \end{vmatrix}; \text{(iii)} \begin{vmatrix} 1 & 2 & 3 & 4 \\ 2 & 3 & 4 & 5 \\ 3 & 4 & 5 & 6 \\ 4 & 5 & 6 & 7 \end{vmatrix}$$

[(i) **Ans. & Hints:**

$$|A| = a_{11}A_{11} - \sum_{i=2}^{4}\sum_{j=2}^{4} a_{1j}a_{i1}A_{11,ij}$$

$$= a_{11}A_{11} - \{a_{12}(a_{21}A_{11,22} + a_{31}A_{11,32} + a_{41}A_{11,42})$$

$$+ a_{13}(a_{21}A_{11,23} + a_{31}A_{11,33} + a_{41}A_{11,43}) + a_{14}(a_{21}A_{11,24} + a_{31}A_{11,34} + a_{41}A_{11,44})\}$$

$$= \begin{vmatrix} 0 & 7 & 1 \\ 9 & 0 & 6 \\ 1 & 1 & 2 \end{vmatrix} - \{2\left(2\begin{vmatrix} 0 & 6 \\ 1 & 2 \end{vmatrix} - 5\begin{vmatrix} 7 & 1 \\ 1 & 2 \end{vmatrix} + 6\begin{vmatrix} 7 & 1 \\ 0 & 6 \end{vmatrix}\right)$$

$$+3\left(-2\begin{vmatrix} 9 & 6 \\ 1 & 2 \end{vmatrix} + 5\begin{vmatrix} 0 & 1 \\ 1 & 2 \end{vmatrix} - 6\begin{vmatrix} 0 & 1 \\ 9 & 6 \end{vmatrix}\right) + 4\left(2\begin{vmatrix} 9 & 0 \\ 1 & 1 \end{vmatrix} - 5\begin{vmatrix} 0 & 7 \\ 1 & 1 \end{vmatrix} + 6\begin{vmatrix} 0 & 7 \\ 9 & 0 \end{vmatrix}\right)\}$$

$= -75 - \{350 + 75 - 1300\} = 800$; (ii) 216 ; (iii) 0]

44. Verify that

(i) $\begin{vmatrix} d_1 & b_{12} \\ b_{21} & d_2 \end{vmatrix} = \begin{vmatrix} 0 & b_{12} \\ b_{21} & 0 \end{vmatrix} + d_1 d_2$

(ii) $\begin{vmatrix} d_1 & b_{12} & b_{13} \\ b_{21} & d_2 & b_{23} \\ b_{31} & b_{32} & d_3 \end{vmatrix}$

$$= \begin{vmatrix} 0 & b_{12} & b_{13} \\ b_{21} & 0 & b_{23} \\ b_{31} & b_{32} & 0 \end{vmatrix} + d_1\begin{vmatrix} 0 & b_{23} \\ b_{32} & 0 \end{vmatrix} + d_2\begin{vmatrix} 0 & b_{13} \\ b_{31} & 0 \end{vmatrix} + d_3\begin{vmatrix} 0 & b_{12} \\ b_{21} & 0 \end{vmatrix} + d_1 d_2 d_3.$$

Hence show that if an arbitrary $n \times n$ matrix A = B + D, where

$$B = \begin{bmatrix} 0 & b_{12} & b_{13} & \cdots & b_{1n} \\ b_{21} & 0 & b_{23} & \cdots & b_{2n} \\ & & \ddots & & \\ \vdots & \vdots & \vdots & \ddots & \vdots \\ b_{n1} & b_{n2} & b_{n3} & \cdots & b_{nn} \end{bmatrix} \text{ and } D = \begin{bmatrix} d_1 & 0 & \cdots & 0 \\ 0 & d_2 & \cdots & 0 \\ \vdots & \vdots & \ddots & \vdots \\ 0 & 0 & \cdots & d_n \end{bmatrix},$$

then

$$\det A = \det B + \sum_{i=1}^{n} d_i B_{ii} + \sum_{1 \le i < j \le n} d_i d_j B_{ij,ij} + \sum_{1 \le i < j < k \le n} d_i d_j d_k B_{ijk.ijk} + \cdots + d_1 d_2 \cdots d_n$$

This is known as *diagonal expansion* of det A.

CHAPTER SEVEN

SYSTEM OF SIMULTANEOUS LINEAR ALGEBRAIC EQUATIONS

Chapter summary*: *Introducing system of simultaneous equations, their varieties and solutions, the chapter discusses equivalent system and different methods of solutions including Crammer's rule, Gaussian elimination, pivoting in Gaussian elimination, Gauss-Jordan elimination, Cholesky's/Crout's square-root; iterative methods of Jacobi, Gauss-Seidel; Evaluation of determinant by Gauss elimination; condition for unique solution of* **Ax = b** *and its solution by Chio's pivotal condensation; it discusses linear dependence and independence of vectors, matrices and functions, finally it discusses construction of magic squares .

7.1 Introduction and definitions

This section introduces the concept of linear algebraic equation, its geometrical significance and uses. It also introduces systems of simultaneous algebraic equations.

7.1.1 Linear algebraic equation : An equation with algebraic variables $x_1, x_2, \ldots x_n$, each with unit power is termed as *linear algebraic equation*. For example

$$a_1x_1 + a_2x_2 + \ldots + a_nx_n = b \tag{7.1}$$

in which the maximum power of each of the algebraic variables x_i is *one* and a_i are constants. Henceforth the word "equation" will stand for "linear algebraic equation".

Geometrically, it represents a hyperplane in an n-dimensional space. If $n = 3$, it represents a plane in 3-dimensional space and if $n = 2$, it represents a straight line in 2-dimension. This fact gives the name linear. It arises in varieties of mathematical representations of the real world problem called the *linear model.*

The main characteristic of a linear equation is that the equation contains no product or roots of the variables present. All variables occur to the first power and do not appear as arguments of trigonometric, logarithmic or exponential functions. The variation of these are the characteristics of nonlinear equations. Examples of nonlinear equations are

$$x_1 + \sqrt{x_2} + x_3 = 5, \quad x_1 + \sin x_2 = 0, \quad x_1 + x_2x_3 + x_4 = 1, \quad e^{x_1} + x_2 + x_3 = 5,$$

$$\log x_1 + x_2 + x_3 = 0, \quad {x_1}^2 + x_2 + x_3 = 7 \text{ etc.}$$

Linear equation arises in most linear models like economic and statistical regression models, electric circuit, mechanical vibrations and in many aspects of science and engineering problems. This linear equation is a fundamental need for linear algebra and hence the name linear for the algebra of vector spaces. This is necessary for further study of matrix theory such as ranks, eigenvalue and eigenvectors etc., systems of nonlinear equations, approximation of functions, and numerical analysis to study finite element method, finite difference method for the solution of ordinary and partial differential equations etc.

7.1.2 System of simultaneous linear equations: A set of equations in the variables $x_1, x_2, \cdots, x_n$ of the form (7.1) is termed as a system of *simultaneous linear equations* or simply the *linear system*.
For example

$$\left.\begin{array}{l} a_{11}x_1 + a_{12}x_2 + \cdots + a_{1n}x_n = b_1 \\ a_{21}x_1 + a_{22}x_2 + \cdots + a_{2n}x_n = b_2 \\ \vdots \quad\quad \vdots \quad\quad \vdots \quad\quad \vdots \\ a_{m1}x_1 + a_{m2}x_2 + \cdots + a_{mn}x_n = b_m \end{array}\right\} \tag{7.2}$$

is a system of m linear equations with n unknowns or variables or shortly $m \times n$ system. This is compactly written using matrix multiplication as

$$\mathbf{Ax} = \mathbf{b} \tag{7.3}$$

where $A = [a_{ij}]_{m\times n}$ is the coefficient matrix, $\mathbf{x} = [x_i]^{\mathrm{T}}_{n\times l}$ and $\mathbf{b} = [b_i]_{m\times l}$ are column vectors.

7.1.3 Homogeneous and Nonhomogeneous system : If the R.H. sides of all equations in (7.2) are zero then the system is a *homogeneous system.* Otherwise, if at least one of R.H sides of the equations in (7.2) is different from zero then the system is called a *nonhomogeneous system.* The homogenous system $\mathbf{Ax} = 0$ plays a fundamental role in science and engineering.

7.2 Variety of systems

Both homogeneous and nonhomogeneous system may occur in three different ways on the basis of number of equations and number of unknowns. Let the number of equations and that of unknowns in a system be m and n respectively. Then

1. a system may have equal numbers of equations and unknowns i.e. $m = n$;
2. a system may have more equations than unknowns i.e. $m > n$; is known as the *overdetermined system* ;
3. a system may have less equations than unknowns i.e. $m < n$; is known as the *underdetermined system.*

Most science and engineering problems belong to the *first* variety and gives unique solutions provided the coefficient matrix is nonsingular. An *n square* matrix A is *nonsingular* if $|A| \neq 0$ and *singular* if $|A| = 0$, and *ill conditioned* if $|A| \approx 0$. Accordingly an $n\times n$ system is *nonsingular* and *ill conditioned* if the corresponding determinant of the coefficient matrix is nonsingular and ill conditioned respectively.

Ex.1 Test the system

$$\begin{cases} 2x_1 - 3x_2 = 5 \\ \ \ x_1 - \ \ x_2 = 5 \end{cases}$$

for singular, nonsingular or ill condition.

The determinant of the coefficient matrix A is

$$|A| = \begin{vmatrix} 2 & -3 \\ 1 & -1 \end{vmatrix} = 1 \neq 0$$

so the system is nonsingular.

Ex.2 Test the system

$$\begin{cases} 2x_1 - 3x_2 = 5 \\ x_1 - 1.49x_2 = 5 \end{cases}$$

for singular, nonsingular or ill condition.

Here,

$$|A| = \begin{vmatrix} 2 & 3 \\ 1 & -1.49 \end{vmatrix} = -2.98 + 3 = 0.2 \approx 0$$

Therefore, the system is ill conditioned.

The *second* variety mostly occurs in experimental cases. No unique solution is possible. The *third* variety produces infinite solution and under this comes the eigenvalue and the optimization problems and the like.

A system may be *dense* if most of the coefficients of the unknowns are nonzero; while if most of the coefficients are zero, then the system is *sparse*.

7.3 Solution

Any set of n-tuple of numbers $\{s_i\}$, $i = 1, 2, \cdots, n$, that satisfies the system (7.2) or (7.3) is called a *solution set*. Solution interprets the validity of the model. Thus, depending on the number of equations and that of variables, a solution may exist uniquely, may not exist or may exist infinitely.

For example the equation $0.x + 0.y = 5$ and the system

$$\left.\begin{array}{l} x + y = 1 \\ x + y = 3 \end{array}\right\}$$

have *no solutions* as the equations express an absurdity $0 = 5$ and $1 = 3$.

The equation $0.x + 0.y = 0$ and the system

$$\left.\begin{array}{l} x \quad\;\; - z = 0 \\ \quad\;\; y - z = 1 \end{array}\right\}$$

are satisfied respectively for any pair (x, y) and any triple (k, k+1, k). So there exists an infinite solution. Lastly $x - 4 = 0$ and the system

$$\left.\begin{array}{l} x - y = 0 \\ x + y = 1 \end{array}\right\}$$

have unique solutions $x = 4$ and $(\frac{1}{2}, \frac{1}{2})$ respectively.

7.3.1 Consistent / inconsistent system : If the system of equations possesses at least one solution then the system is consistent; otherwise it is inconsistent i.e. no solution. Consistent equations may have a unique solution or may have infinite numbers of solutions.

Geometrically, a solution to a system of linear equations represents a point that lies in the intersection of the regions described by the individual equations. Now the *unique* solution means common points of intersection of the hyperplanes, *no solution* means no common points of the hyperplanes i.e. they are parallel and the *infinite solution* corresponds to coincident hyperplanes as the hyperplanes coincide in infinitely many points.

7.3.2 Trivial solution : The homogeneous system is always consistent and this solution is the zero solution, i.e. all the variables $x_i = 0$ satisfy the system. This solution is called a *trivial solution*. This solution is of theoretical interest only and not of practical interest to scientists and engineers.

7.3.3 Nontrivial solution : The homogeneous system possesses a nonzero solution, not all zero, other than trivial is called a *nontrivial solution* which has practical value to scientists and engineers.

7.3.4 A tree diagram on equations and solutions : Since the knowledge of both equations and solutions is essential in understanding linear modeling as well as the construction of the theory of vector spaces, the chapter one is devoted to the theory of linear equations as well as the existence and uniqueness of the solution and some solution methods.

The discussions of sec. 7.1 and 7.3 may be summarized in the following tree diagram.

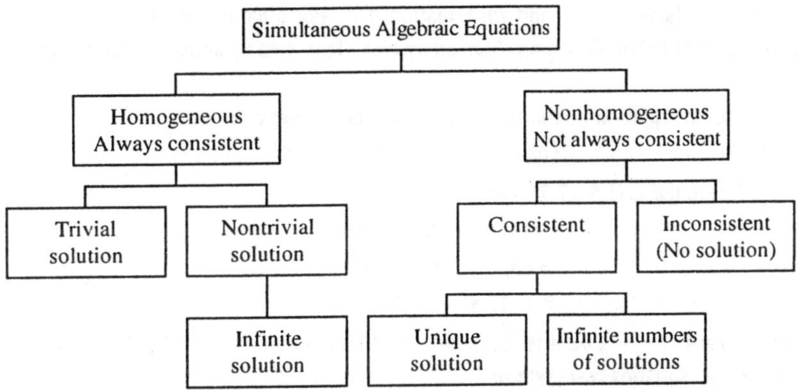

7.4 Equivalent system

If two systems of m linear equations in n variables x_i, $i = 1, 2, \cdots, n$, have the *same* solutions i.e. every solution of one system is also a solution of the other and vice-versa , then these two systems are said to be *equivalent system.* Certain simple procedure reduces one system to its equivalent system which is easier to solve than the original system. An equivalent system is obtained by a part or all of three elementary operations quoted below:

I. Interchange of any two equations in the system.

II. Multiplication of any equation by a nonzero scalar.

III. A multiple of one equation may be added to another equation.

For identical operations in matrix see sec.9.1. The following theorem guarantees for *invariance* of solution for two equivalent systems.

Theorem . If a linear system of equations is transformed into another linear system by any one of the three elementary operations on equations, then the two systems are *equivalent*.

Proof.

Let
$$\left.\begin{array}{l} a_{11}x_1 + a_{12}x_2 + \dots\dots\dots + a_{1n}x_n = b_1 \\ \vdots \\ a_{i1}x_1 + a_{i2}x_2 + \dots\dots\dots + a_{in}x_n = b_i \\ \vdots \\ a_{j1}x_1 + a_{j2}x_2 + \dots\dots\dots + a_{jn}x_n = b_j \\ \vdots \\ a_{m1}x_1 + a_{m2}x_2 + \dots\dots\dots + a_{mn}x_n = b_m \end{array}\right\} \tag{7.4}$$

where $m > 1$. Let

$$x_k = c_k,\ k = 1, 2, \ \cdots, n \tag{7.5}$$

be a set of solutions of the system (7.4), i.e. there exist identities

$$a_{ik}c_k = b_i,\ i = 1, 2, \ \cdots, m \text{ and } k = 1, 2, \ \cdots, n, \tag{7.6}$$

Now it is required to show that all the m identities in (7. 6) remains unaltered by the application of the elementary operations.

Case I. The interchange of ith and jth equation does not alter the m identities in (7.6) by the substitution of (7.5) in the new interchanged system. Thus two systems are satisfied by the same solution set.

Case II. Let the ith equation be multiplied by a nonzero scalar c. So,

$$c\left(a_{i1}x_1 + a_{i2}x_2 + \cdots + a_{in}x_n\right) = cb_i \tag{7.7}$$

Now using (7.5) in the L.H.S. (7.7) we have

$$c\left(a_{i1}c_1 + a_{i2}c_2 + \cdots + a_{in}c_n\right) \tag{7.8}$$
$$= cb_i \qquad \text{by (7.6)}$$
$$= \text{R.H.S. of (7.7)}$$

Hence, giving $i = 1, 2, \ \cdots, m$, it is found that all the equations are satisfied by (7.5) and thus the same solutions set satisfies two systems.

Case III. Now adding c times the terms of jth equation to the corresponding terms in ith equation, we have,

$$\left.\begin{array}{l} a_{11}x_1 + a_{12}x_2 + \cdots + a_{1n}x_n = b_1 \\ \vdots \\ (a_{i1} + ca_{j1})x_1 + (a_{i1} + ca_{j1})x_2 + \cdots + (a_{i1} + ca_{j1})x_1 = b_i + cb_j \\ \vdots \\ a_{ji}x_1 + a_{j2}x_2 + \cdots + a_{jn}x_n = b_j \\ \vdots \\ a_{m1}x_1 + a_{m2}x_2 + \cdots + a_{mn}x_n = b_m \end{array}\right\} \tag{7.9}$$

Note that in (7.9) ith equation is changed and others remain unchanged. Therefore, using (7.5) in ith equation,

$$\text{L.H.S.} = (a_{i1} + ca_{ji})c_1 + (a_{j2} + ca_{j2})c_2 + \cdots + (a_{in} + ca_{jn})c_n$$
$$= (a_{i1}c_1 + a_{i2}c_2 + \cdots + a_{in}c_n) + c(a_{j1} + a_{j2}c_2 + \cdots + a_{jn}c_n)$$
$$= b_i + cb_j$$
$$= \text{R.H.S. of (7.9)}$$

Therefore, solution set (7.5) satisfies the new system (7.9).

Hence the above three cases establish the requirement stated at the beginning. So the proof is complete.

Cor. There is no loss in generalizing the theorem by applications of elementary operations. Thus

(i) instead of interchanging only two equations, one may rearrange *all* equations in a system in *any order* of one's choice or necessity.

(ii) One may replace ith equation by the sum of c_i multiples, $i = 1, 2, \cdots, m$ of all m equations i.e.

$$\sum_{i=1}^{m} c_i(a_{i1}x_1 + a_{i2}x_2 + \cdots + a_{in}x_n) = \sum_{i=1}^{m} c_i b_i$$

may be used in place of ith equation. The resulting system will *remain equivalent* all the time.

7.5 Methods of solving

The system of simultaneous algebraic equations can be solved by various methods. These methods can be broadly divided into *two* groups - *exact* and *interative* methods. Some of the specific methods under these groups are discussed in the following subsections.

7.5.1 Exact methods : These require finite number of operations and include Crammer's rule, Gaussian elimination method, Gauss-Jordan elimination method, method of square roots, Cholesky's / Crout's method, Chio's pivotal condensation etc. These methods are useful for solving dense and up to moderate systems.

7.5.1.1 Crammer's rule : It was published in Geneva in 1750 by Swiss mathematician Gabriel Crammer (1704-1752) and was named after him. It is a determinantal formula for the solution of a nonsingular system of $n > 1$ simultaneous linear equations in n unknowns. In other words, this rule gives one of the applications of determinant for the solution of system of equations. This involves more computations and is thus attractive and useful for theoretical derivations. Of course, the efficiency of this rule is dependent on the methods by which the determinants in question are evaluated. The following theorem states the Crammer's rule:

Theorem. Let a system of n equations in n unknowns be

$$\sum_{j=1}^{n} a_{ij}x_j = c_i, \quad i = 1, 2, \cdots, n$$

such that $|A| = \det |a_{ij}| \neq 0$, then the unique solution is given by

$$x_j = \frac{|A_{\cdot j}|}{|A|}, \quad j = 1, 2, \cdots, n$$

where $|A_j|$ is obtained from $|A|$ by replacing the jth column of $|A|$ by the column vector $c_i, \ i = 1, 2, \cdots, n$.

Proof. Let $|A| \neq 0$ and let the column vector $\mathbf{x} = [x_1 \ x_2 \ \cdots x_j \cdots \ x_n]^T$ be a unique solution. Let

$$|A| = \det [A_{\cdot 1} \ \ A_{\cdot 2} \ \cdots A_{\cdot j} \ \cdots A_{\cdot n}]$$

where $A_{\cdot j}$, $j = 1, 2, \cdots, n$, be the column vectors of the n *square* matrix A.

Now
$$\begin{aligned} x_j|A| &= x_j \det [A_{\cdot 1} \ \ A_{\cdot 2} \ \cdots A_{\cdot j-1} \ \ A_{\cdot j} \ \ A_{\cdot j+1} \ \cdots A_{\cdot n}] \\ &= x_1 \det [A_{\cdot 1} \ \ A_{\cdot 2} \ \cdots A_{\cdot j-1} \ \ A_{\cdot j} \ \ A_{\cdot j+1} \ \cdots A_{\cdot n}] \\ &+ x_2 \det [A_{\cdot 1} \ \ A_{\cdot 2} \ \cdots A_{\cdot j-1} \ \ A_{\cdot j} \ \ A_{\cdot j+1} \ \cdots A_{\cdot n}] \\ &+ \cdots \quad \cdots \quad \cdots \quad \cdots \quad \cdots \quad \cdots \quad \cdots \\ &+ x_j \det [A_{\cdot 1} \ \ A_{\cdot 2} \ \cdots A_{\cdot j-1} \ \ A_{\cdot j} \ \ A_{\cdot j+1} \ \cdots A_{\cdot n}] \end{aligned}$$

$$+ \cdots \quad \cdots \quad \cdots \quad \cdots \quad \cdots$$

$$+ x_n \det[A_{\cdot 1} \quad A_{\cdot 2} \ \cdots . A_{\cdot j-1} \quad A_{\cdot j} \quad A_{\cdot j+1} \ \cdots A_{\cdot n}]$$

[As all the determinants, except the *j*th one on the right, are separately zero because of identical columns.]

$$= \det [A_{\cdot 1} \quad A_{\cdot 2} \ \cdots A_{\cdot j-1} \ (x_{\cdot 1} A_{\cdot 1} + x_{\cdot 2} A_{\cdot 2} + \ \cdots \ + x_{\cdot j} A_{\cdot j}) A_{\cdot j+1} \ \cdots \ A_n]$$

$= |A_j|$, by theorem 6, sec. 6.4.

$$\therefore \qquad x_j = \frac{|A_j|}{|A|}.$$

This division is only possible if and only if $|A| \neq 0$. Thus validity of Crammer's rule depends on

(i) $|A| \neq 0$ i.e. if the coefficient matrix A is *nonsingular* and

(ii) the number of equations and number of unknowns are equal, otherwise |A| cannot be formed.

N.B. Crammer's rule can also be established with the help of *inverse* (cf. sec. 8.9).

Ex. Solve the system of linear equations

$$\left.\begin{aligned} 2x_1 + x_2 - 5x_3 + x_4 &= 8 \\ x_1 - 3x_2 \qquad - 6x_4 &= 9 \\ 2x_2 - x_3 + 2x_4 &= -5 \\ x_1 + 4x_2 - 7x_3 + 6x_4 &= 0 \end{aligned}\right\}$$

using Crammer's rule.

The coefficient matrix is

$$A = \begin{bmatrix} 2 & 1 & -5 & 1 \\ 1 & -3 & 0 & -6 \\ 0 & 2 & -1 & 2 \\ 1 & 4 & -7 & 6 \end{bmatrix}.$$

$\therefore$ $\det A = |A| = 27 \neq 0$. Then A is nonsingular.

Since the solutions, $x_j = \dfrac{|A_j|}{|A|}, j = 1, 2, 3, 4.$

So,

$$x_1 = \frac{1}{|A|} \begin{vmatrix} 8 & 1 & -5 & 1 \\ 9 & -3 & 0 & -6 \\ -5 & 2 & -1 & 2 \\ 0 & 4 & -7 & 6 \end{vmatrix} = \frac{81}{27} = 3$$

$$x_2 = \frac{1}{|A|} \begin{vmatrix} 2 & 8 & -5 & 1 \\ 1 & 9 & 0 & -6 \\ 0 & -5 & -1 & 2 \\ 1 & 0 & -7 & 6 \end{vmatrix} = -\frac{108}{27} = -4$$

$$x_3 = \frac{1}{|A|}\begin{vmatrix} 2 & 1 & 8 & 1 \\ 1 & -3 & 9 & -6 \\ 0 & 2 & -5 & 2 \\ 1 & 4 & 0 & 6 \end{vmatrix} = -\frac{27}{27} = -1$$

$$x_4 = \frac{1}{|A|}\begin{vmatrix} 2 & 1 & -5 & 8 \\ 1 & -3 & 0 & 9 \\ 0 & 2 & -1 & -5 \\ 1 & 4 & -7 & 0 \end{vmatrix} = \frac{27}{27} = 1\,.$$

∴ the required solution is $x_1 = 3, x_2 = -4, x_3 = -1, x_4 = 1$.

7.5.1.2 Gaussian elimination method : It is the most common technique for the solution of linear system. The method is not restricted to nonsingular system as in Crammer's rule. So this is an advantage over Crammer's rule. Another advantage is that it is not also restricted to equal numbers of unknowns and equations. The Gauss reduction process or a process of successive elimination is known as the *Gaussian elimination method* or simply *Gaussian method*, after the name of K.F. Gauss (1777-1855) of Germany, although the procedure dates back from at least 250 B.C. It is a relatively simple and straight forward process. It consists of an algorithm of two steps and a series of operations to transform the original set of m equations with n unknowns to an equivalent system in Upper Trapezoidal Form (UTF), from which each unknown is determined by *back substitution*. This UTF is also known as *echelon form* (pronounced as esh'-uh-lon). The word "echelon" is a French word meaning "a series of steps". The necessary and sufficient condition for a system to be in echelon form is governed by the following two steps algorithm.

Algorithm

Step 1. The first nonzero coefficient of the first unknown x_1 in each equation is 1. If $a_{i1} = 0$, $i = 1, 2, \cdots, m$, use elementary operation I to satisfy the nonzero condition of first unknown x_1. Also if $a_{i1} \neq 0$ but $a_{i1} > 1$, then using elementary operation II divide throughout by a_{i1} to satisfy the condition of being unity.

Step 2. The number of leading zero coefficients in any equation except the first equation is *less* than that of those coefficients in the succeeding equation.

The method can be programmed for a computer according to the algorithm.

Let the system be

$$\left.\begin{aligned} a_{11}x_1 + a_{12}x_2 + \cdots + a_{1n}x_n &= b_1 \\ a_{21}x_1 + a_{22}x_2 + \cdots + a_{2n}x_n &= b_2 \\ \vdots \qquad \vdots \qquad\quad \vdots \quad &\quad \vdots \\ a_{m1}x_1 + a_{m2}x_2 + \cdots + a_{mn}x_n &= b_m \end{aligned}\right\} \tag{7.10}$$

with m equations in n unknow[illegible] Let $a_{11} \neq 0$ but >1.

To satisfy step 1 of algorit[illegible] *divide* the first equation of (7.10) by a_{11} giving

$$\left.\begin{array}{l} x_1 + a'_{12}x_2 + \cdots + a'_{1n}x_n = b'_1 \\ a_{21}x_1 + a_{22}x_2 + \cdots + a_{2n}x_n = b_2 \\ \vdots \qquad \vdots \qquad \vdots \qquad \vdots \\ a_{m1}x_1 + a_{m2}x_2 + \cdots + a_{mn}x_n = b_m \end{array}\right\}. \tag{7.11}$$

Now to use elementary operation III of sec. 7.4 to eliminate x_1 from all but the first equation in (7.11) *multiply* the first equation in (7.11) by $(-a_{i1})$, $i = 2, \cdots, m$ and add to the ith equation giving

$$\left.\begin{array}{r} x_1 + a'_{12}x_2 + \cdots + a'_{1n}x_n = b'_1 \\ a'_{22}x_2 + \cdots + a'_{2n}x_n = b'_2 \\ \vdots \qquad \vdots \qquad \vdots \qquad \vdots \\ a'_{m2}x_2 + \cdots + a'_{mn}x_n = b'_m \end{array}\right\}. \tag{7.12}$$

This process is repeated *until* the last equation is reached where these operations stop automatically. Now the first but the rest $(m-1)$ equations form a subsystem as (7.10). Continuing the process on the subsystem of $(m-1)$ equations, (7.12) becomes

$$\left.\begin{array}{r} x_1 + a'_{12}x_2 + a'_{13}x_3 + \cdots + a'_{1n}x_n = b'_1 \\ x_2 + a''_{23}x_3 + \cdots + a''_{2n}x_n = b_2 \\ a'_{33}x_3 + \cdots + a'_{3n}x_n = b'_3 \\ \vdots \qquad \vdots \qquad \vdots \\ a'_{m3}x_3 + \cdots + a_{mn}x_n = b'_m \end{array}\right\}. \tag{7.13}$$

Proceeding in this way finally we are left with the equivalent system for (7.10).

$$\left.\begin{array}{lllllll} x_1 + & b_{12}x_2 + & b_{13}x_3 + & \cdots & \cdots & \cdots +b_{1n}x_n & = c_1 \\ & x_2 + & b_{23}x_3 + & \cdots & \cdots & \cdots +b_{2n}x_n & = c_2 \\ & \vdots & \vdots & & \vdots & \vdots & \vdots \\ & & & x_r + & b_{r,r+1}x_{r+1} + & \cdots +b_{rn}x_n & = c_r \\ & & & & & 0 & = c_{r+1} \\ & & & \vdots & \vdots & \vdots & \vdots \\ & & & & & 0 & = c_n \end{array}\right\} \tag{7.14}$$

where $x_1, x_2, \ldots, x_{i_r}$ denote the leading variables with unit coefficients. This equivalent system is also known as *echelon form* of the given system. The variables $x_{i_{r+1}}, x_{i_{r+2}}, \cdots, x_{i_n}$ denoting remaining variables which *do not appear* at the beginning of any equations in the echelon form (7.14), are known as *free variables.*

Solution procedures:

The following points should be carefully noted for writing solutions from (7.14):

Nonhomogeneous :

(i) In the above process if an equation occurs as $0.x_1 + 0.x_2 + \cdots + 0.\,x_n = 0$ the equation is to be deleted. The deletion will not affect the solution.

(ii) If $r = m$, i.e. number of equations in echelon form (7.14) equals that in the original system (7.10) then there are no equations of the form $0 = c_j$.

(iii) If $r<m$, i.e. the number of equations in echelon form (7.14) is *less* than that of the original system (7.10) then the quantities $c_{r+1}, \cdots, c_m$ in (7.14) *may* or *may not* all be *zero*.
(iv) If *at least one* of the quantities $c_{r+1}, c_{r+2}, \cdots, c_m$ is *not* zero, then the given equations (7.10) imply a contradiction or absurdity and hence (7.10) is *inconsistent*.
(v) If all of c_{r+1}, c_{r+2}, (7.14), c_m are zero then (7.14) is *consistent* with $r<n$, i.e. the number of equations is less than the number of variables in echelon form (7.14). The solution is obtained by back substitution in terms of free variables. The number of free variables = $(n-r)$.

So, infinite numbers of solutions are obtained for infinite values of the free variables. Solutions in terms of free variables are known as the *general* solution. If particular values are assigned to free variables, then the solution is called the *particular* solution. If zero values are assigned to free variables then the particular solution is called the *basic* solution and the corresponding remaining variables other than free variables are called *basic* variables.
(vi) If $r = n$, i.e. number equations equals the number of variables in (7.14) after considering (i) and (ii) then the system (7.10) has a unique solution which is also obtained by back substitution.

Homogeneous :

In this case obviously $c_1, c_2, \cdots, c_r, \cdots, c_m$ in (7.14) are zero. Here also delete the equation $0.x_1 + 0.x_2 + \cdots + 0.x_n = 0$ if any, during the process. Two cases may arise:

(i) if in echelon form, $r = n$, the system (7.10) has trivial or zero solution;
(ii) if in echelon form $r<n$, the system (7.10) has infinite numbers of nontrivial or nonzero solutions in terms of $(n-r)$ free variables each one can take any value, as one likes.

Number of operations for $n \times n$ system: Let us calculate the number of operations for the ith equation. To make the coefficient of the first nonzero element unity we need $(n-i)$ divisions. For the elimination of x_i from equations $(i+1)$ to n requires $(n-i)(n-i+1)$ multiplications first and then $(n-i)(n-i+1)$ subtractions.

Therefore, total number of operations required in the elimination only for each $i = 1, 2, \cdots, (n-1)$ is given by

(a) Multiplications / divisions is

$$\sum_{i=1}^{n-1}(n-i)(n-i+2) = \sum_{i=1}^{n-1}[(n^2+2n)-2(n+1)i+i^2]$$

$$=(n^2+2n)\sum_{i=1}^{n-1}1-2(n+1)\sum_{i=1}^{n-1}i+\sum_{i=1}^{n-1}i^2$$

$$=(n^2+2n)(n-1)-2(n+1)\tfrac{1}{2}n(n-1)+\tfrac{1}{6}(n-1)(n)(2n-1)$$

$$=\tfrac{1}{6}(2n^3+3n^2-5n).$$

(b) Additions / subtractions is

$$\sum_{i=1}^{n-1}(n-i)(n-i+1) = \sum_{i=1}^{n-1}[(n^2+n)-(2n+1)i+i^2]$$

$$=(n^2+n)(n-1)-(2n+1)\tfrac{1}{2}n(n-1)+\tfrac{1}{6}(n-1)(n)(2n-1)$$

$$=\tfrac{1}{6}(n^3-n).$$

For back substitution: For the last equation, 1 division is needed to get x_n. Then $(n-i)$ multiplication and 1 division for each $i = n-1, \cdots, 2, 1$; and $(n-i-1)$ additions and 1 subtraction for each $i = n-1, \cdots, 2, 1$. Therefore, the total number of operations is given by

(c) Multiplications / divisions:

$$= 1 + \sum_{i=1}^{n-1}[(n-i)+1] = \tfrac{1}{2}(n^2+n).$$

(d) Additions / subtractions

$$= \sum_{i=1}^{n-1}[(n-i+1)+1] = \tfrac{1}{2}(n^2-n).$$

So, total number of operations in the process is, therefore,

Multiplications / divisions = (a) + (c) $= \tfrac{1}{6}(2n^3+3n^2-5n)+\tfrac{1}{2}(n^2+n)$

$$= \tfrac{1}{3}(n^3+3n^2-n) = \tfrac{1}{3}n(n^2+3n-1)$$

Additions / subtractions = (b) + (d) $= \tfrac{1}{6}(n^3-n)+\tfrac{1}{2}(n^2-n) = \tfrac{1}{6}(2n^3+3n^2-5n)$

$$= \tfrac{1}{6}n(2n^2+3n-5) = \tfrac{1}{6}n(n-1)(2n+5)$$

Checking Gauss elimination : Gauss elimination can be checked at each step by introducing a "check sums" denoted by Σ which sums the coefficients and the constant of each equation or each row of the augmented matrix. Operations that are performed on coefficients and constants are equally performed on Σ too. If the sum of coefficients and constants or rows of augmented matrix is equal to the corresponding element in Σ, then the step is correct. Finally if each element of Σ is equal to the $(1 + x_i)$ i.e. (1 + solution) then the solution is correct. In this way one can check intermediate steps as well as the final solution.

Ex. Solve the system

$$\left.\begin{aligned} 2x_1 - x_2 + x_3 &= -1 \\ 3x_1 + 3x_2 + 9x_3 &= 0 \\ 3x_1 + 3x_2 + 5x_3 &= 4 \end{aligned}\right\}$$

by the Gaussian elimination method showing the number of operations in each step and check the procedures.

The given system	The check sum (s_i)	No. of operations (ops.)
$2x_1 - x_2 + x_3 = -1$ (i)	$s_1 = 2-1+1-1 = 1$	
$3x_1 + 3x_2 + 9x_3 = 0$ (ii)	$s_2 = 3+3+9+0 = 15$	Nil
$3x_1 + 3x_2 + 5x_3 = 4$ (iii)	$s_3 = 3+3+5+4 = 15$	

Step 1 . Dividing eqn. (i) by 2, the system, the check sum and operations are

System	The check sum	Operations
$x_1 - \tfrac{1}{2}x_2 + \tfrac{1}{2}x_3 = -\tfrac{1}{2}$	$s_1 = 1 - \tfrac{1}{2} + \tfrac{1}{2} - \tfrac{1}{2} = \tfrac{1}{2}$	Div. 3
$3x_1 + 3x_2 + 9x_3 = 0$	$s_2 = 3+3+9+0 = 15$	
$3x_1 + 3x_2 + 5x_3 = 4$	$s_3 = 3+3+5+4 = 15$	

Step 2 . Eliminating x_1 from 2nd and 3rd eqns. we get,

System	The check sum	Operations
$x_1 - \frac{1}{2}x_2 + \frac{1}{2}x_3 = -\frac{1}{2}$	$\frac{1}{2}$	Mul. $3\times2 = 6$
$\frac{9}{2}x_2 + \frac{15}{2}x_3 = \frac{3}{2}$	$\frac{9}{2} + \frac{15}{2} + \frac{3}{2} = \frac{27}{2} = 15 - \frac{3}{2}$	Sub. $3\times2 = 6$
$\frac{9}{2}x_2 + \frac{7}{2}x_3 = \frac{11}{2}$	$\frac{9}{2} + \frac{7}{2} + \frac{11}{2} = \frac{27}{2} = 15 - \frac{3}{2}$	

Step 3 . Reducing coeff. of x_2 to unity in second eqn., we have,

System	The check sum	Operations
$x_1 - \frac{1}{2}x_2 + \frac{1}{2}x_3 = -\frac{1}{2}$	$\frac{1}{2}$	Div. 2
$x_2 + \frac{5}{3}x_3 = \frac{1}{3}$	$1 + \frac{5}{3} + \frac{1}{3} = \frac{27}{2} \times \frac{2}{9} = 3$	
$\frac{9}{2}x_2 + \frac{7}{2}x_3 = \frac{11}{2}$	$\frac{27}{2}$	

Step 4 . Eliminating x_2 from 3rd eqns. we obtain,

System	The check sum	Operations
$x_1 - \frac{1}{2}x_2 + \frac{1}{2}x_3 = -\frac{1}{2}$	$\frac{1}{2}$	
$x_2 + \frac{5}{3}x_3 = \frac{1}{3}$	3	Mul. 2
$-4x_3 = 4$	$-4 + 4 = 0 = \frac{27}{2} - 3.\frac{9}{2}$	Sub. 2

Step 5 . Reducing coeff. of x_3 unity in 3rd eqn., we have,

System	The check sum	Operations
$x_1 - \frac{1}{2}x_2 + \frac{1}{2}x_3 = -\frac{1}{2}$	$\frac{1}{2}$	
$x_2 + \frac{5}{3}x_3 = \frac{1}{3}$	3	Div. 1
$x_3 = -1$	$1 - 1 = 0 = \frac{0}{4}$	

Step 6 . Back substitution: Using 3rd eqn. of step 5 to 2nd and first equation, we obtain,

Solutions	The check sum	Operations
$x_3 = -1$	$1 - 1 = 0 = s_1$	
$x_2 = 2$	$1 + 2 = 3 = s_2$	Mul. $1 + 2 = 3$
$x_1 = 1$	$1 + 1 = 2 = s_3$	Add /Sub. $1 + 2 = 3$
	which verifies that $1 + x_i = s_i$	

$\therefore$ total number operations : Div / Mul.– $3 + 6 + 2 + 2 + 1 + 3 = 17$

and Add / Sub.– $6 + 0 + 2 + 0 + 3 = 11$

which verifies the operation count for $n = 3$.

Alternatively the above detail steps can be compactly expressed in tabular form as follows:

x_1	x_2	x_3	Constant term	Σ	No. of operations	Comments
2	−1	1	−1	1		
3	3	9	0	15	Nil	Given system
3	3	5	4	15		
1	$-\frac{1}{2}$	$\frac{1}{2}$	$-\frac{1}{2}$	$\frac{1}{2}$		
3	3	9	0	15	Div. 3	2 divides 1st row
3	3	5	4	15		
1	$-\frac{1}{2}$	$\frac{1}{2}$	$-\frac{1}{2}$	$\frac{1}{2}$	Mul. 3×2 = 6	x_1, x_2 are eliminated from
	$\frac{9}{2}$	$\frac{15}{2}$	$\frac{3}{2}$	$\frac{27}{2}$	Sub. 3×2 = 6	2nd and 3rd rows
	$\frac{9}{2}$	$\frac{7}{2}$	$\frac{11}{2}$	$\frac{27}{2}$		
1	$-\frac{1}{2}$	$\frac{1}{2}$	$-\frac{1}{2}$	$\frac{1}{2}$	Div. 2	$\frac{9}{2}$ divides 2nd row
	1	$\frac{5}{3}$	$\frac{1}{3}$	3		
	$\frac{9}{2}$	$\frac{7}{2}$	$\frac{11}{2}$	$\frac{27}{2}$		
1	$-\frac{1}{2}$	$\frac{1}{2}$	$-\frac{1}{2}$	$\frac{1}{2}$	Mul. 2	x_3 is eliminated from
	1	$\frac{5}{3}$	$\frac{1}{3}$	3	Sub. 2	3rd row
		−4	4	0		
1	$-\frac{1}{2}$	$\frac{1}{2}$	$-\frac{1}{2}$	$\frac{1}{2}$	Div. 1	−4 divides 3rd row and
	1	$\frac{5}{3}$	$\frac{1}{3}$	3		system is in echelon form
		1	−1	0		
		1	−1	0	Mul. 3	Back substitution
	1		2	3	Add/ Sub. 3	
1			1	2		

- total operations = 17 + 11 =28 .

7.5.1.3 Pivoting in Gaussian elimination : Before we pass on to the formal procedure, we need to know what do we mean by "*pivot*" and "*pivoting*"? The literal meaning of "pivot" is the "central point, something on which an argument etc. depends". In mathematics, the meaning of pivot is of no exception. How pivot arise in mathematics? In numerical solution of system of linear equations directly or by using matrix or determinant, division plays a major part, particularly in Gaussian elimination, triangularization and diagonalization of matrices etc. In all such cases, division of a bigger decimal number by a smaller one or vice-versa may result in an *overflow* or *underflow* in a computer. Also an error occurs due to *truncation* and *rounding* of the number as a result of such divisions. This error propagates creating a disaster against the true solution of the problem. To guard against this disaster, strategy known as "pivot" has been introduced.

Once it is understood for systems of equations, it is equally understandable for matrix and determinant as well and vice-versa. In the following paragraphs pivoting in Gaussian elimination has been defined and explained the necessity of its use and its classifications are treated in a nutshell.

The process of interchanging rows of the coefficient matrix of a linear system in Gaussian elimination so that the diagonal elements are nonzero, is called *pivoting* and the elements / coefficients used for the elimination are called *pivots.* A sufficient condition for theoretical correctness of Gaussian elimination is the use of nonzero pivots. But in practice one must be careful to have reliable results. Note the following example.

Ex. The system

$$\left.\begin{aligned} 0.0001x \quad +y &= 1 \\ x \quad +y &= 2 \end{aligned}\right\}$$

has the solution: $x = 1.0001$, $y = 0.9999$. (1)

Now to use Gaussian elimination, multiply the first equation by (–10,000) and add to the second, so that the equation reduces to

$$\left.\begin{aligned} 0.0001x \qquad\qquad +y &= 1 \\ -10{,}000y &= -10{,}000 \end{aligned}\right\}$$

and the solution becomes $x = 0.000, y = 1.000$ (2)

Solution (2) is a disaster against the true solution (1).

Let us *pivot* to have a larger coefficient in the diagonal instead of the smaller one. That is, by interchanging equations, we get,

$$\left.\begin{aligned} x + y &= 2 \\ 0.0001x + y &= 1 \end{aligned}\right\}$$

Now use again Gauss elimination, so that

$$\left.\begin{aligned} x + y &= 2 \\ 0.9999y &= 0.9998 \end{aligned}\right\} \quad \text{solution is} \quad \begin{cases} x = 1.00 \\ y = 1.00 \end{cases}$$

N.B. This example suggests not to avoid zero pivots only but to avoid relatively small ones also.

Kinds of pivoting : There are two standard kinds of pivoting.

(i) partial pivoting and

(ii) complete pivoting.

Partial pivoting: It takes place between rows or equations only. Let at the *k*th step, interchange of the rows of the coefficient matrix so that the *largest* remaining element in the *k*th column is used as the pivot. That is, after the pivoting

$$|a_{kk}| = \max|a_{ik}|, \quad \text{for } i = k, k+1, \cdots, n.$$

It serves *two* purposes in Gaussian elimination

(i) possibility of division by zero reduces,

(ii) accuracy in the method increases.

Complete pivoting: It takes place between both rows and columns. Let at the *k*th step, interchange of both the rows and the columns of the coefficient matrix so that the *largest* number in the remaining matrix is used as pivot. That is, after pivoting,

$$|a_{kk}| = \max|a_{ij}|, \text{ for } i = k, k+1, \cdots, n., \ j = k, k+1, \cdots, n.$$

and the unknowns are renumbered.

Complete pivoting is *safe*; errors are never *magnified* unreasonably. If two columns are interchanged, the corresponding variables must also be interchanged .

N.B. If pivoting cannot locate a nonzero element to place on the diagonal, the matrix must be singular.

Ex. Solve the system

$$0.00044x_1 + 0.0003x_2 - 0.0001x_3 = 0.00046$$
$$4x_1 + x_2 + x_3 = 1.5$$
$$3x_1 - 9.2x_2 - 0.5x_3 = -8.2$$

using Gaussian elimination with partial pivoting. Round off the result to 3 significant digits.

Given system / **Augmented matrix**

$$\begin{array}{l} 0.00044x_1 + 0.0003x_2 - 0.0001x_3 = 0.00046 \\ \rightarrow 4x_1 + x_2 + x_3 = 1.5 \\ 3x_1 - 9.2x_2 - 0.5x_3 = -8.2 \end{array} \qquad \left[\begin{array}{ccc|c} 0.00044 & 0.0003 & 0.0001 & 0.00046 \\ \rightarrow 4 & 1 & 1 & 1.5 \\ 3 & -9.2 & -0.5 & -8.2 \end{array}\right]$$

pivot entry   pivot entry

Step 1. By definition of partial pivoting, the largest coeff. of x_1 (element in 1st column) is 4 which belongs to 2nd eqn. (or row). Therefore, to bring pivot entry 4 in the top, we interchange 1st and 2nd equations (or rows), i.e.

$$\begin{array}{l} 4x_1 + x_2 + x_3 = 1.5 \\ 0.00044x_1 + 0.0003x_2 - 0.0001x_3 = 0.00046 \\ 3x_1 - 9.2x_2 - 0.5x_3 = -8.2 \end{array} \qquad \left[\begin{array}{ccc|c} 4 & 1 & 1 & 1.5 \\ 0.00044 & 0.0003 & -0.0001 & 0.00046 \\ 3 & -9.2 & -0.5 & -8.2 \end{array}\right]$$

Step 2. Dividing by the pivot entry 4, we obtain,

$$\begin{array}{l} x_1 + 0.25x_2 + 0.25x_3 = 0.375 \\ 0.00044x_1 + 0.0003x_2 - 0.0001x_3 = 0.00046 \\ 3x_1 - 9.2x_2 - 0.5x_3 = -8.2 \end{array} \qquad \left[\begin{array}{ccc|c} 1 & 0.25 & 0.25 & 0.375 \\ 0.00044 & 0.0003 & -0.0001 & 0.00046 \\ 3 & -9.2 & -0.5 & -8.2 \end{array}\right]$$

Step 3. Eliminating x_1 from 2nd and 3rd eqns. (or reducing a_{21} and a_{31} to zero), we are left with

$$\begin{array}{l} x_1 + 0.25x_2 + 0.25x_3 = 0.375 \\ \rightarrow -9.95x_2 - 1.25x_3 = -9.32 \\ 0.00019x_2 - 0.00021x_3 = 0.000295 \end{array} \qquad \left[\begin{array}{ccc|c} 1 & 0.25 & 0.25 & 0.375 \\ 0 & \rightarrow -9.95 & -1.25 & -9.32 \\ 0 & 0.00019 & -0.00021 & 0.000295 \end{array}\right]$$

pivot entry   pivot entry

Step 4. Dividing by the second pivot entry (– 9.95) yields

$$\begin{aligned} x_1 + 0.25x_2 + 0.25x_3 &= 0.375 \\ x_2 + 0.126x_3 &= 0.937 \\ 0.00019x_2 - 0.00021x_3 &= 0.000925 \end{aligned} \qquad \left[\begin{array}{ccc|c} 1 & 0.25 & 0.25 & 0.375 \\ 0 & 1 & 0.126 & 0.937 \\ 0 & 0.00019 & -0.00021 & 0.000295 \end{array}\right]$$

Step 5. Eliminating x_2 from 3rd eqn by 2nd eqn (or reducing a_{32} to zero by a_{22}) we obtain,

$$\begin{aligned} x_1 + 0.25x_2 + 0.25x_3 &= 0.375 \\ x_2 + 1.26x_3 &= 0.937 \\ -0.000234x_3 &= 0.000117 \end{aligned} \qquad \left[\begin{array}{ccc|c} 1 & 0.25 & 0.25 & 0.375 \\ 0 & 1 & 0.126 & 0.937 \\ 0 & 0 & -0.000234 & 0.000117 \end{array}\right]$$

Step 6. Dividing the 3rd eqn (or 3rd row) by (–0.000234) yields the echelon form (row echelon form)

$$\begin{aligned} x_1 + 0.25x_2 + 0.25x_3 &= 0.375 \\ x_2 + 1.26x_3 &= 0.937 \\ x_3 &= -0.5 \end{aligned} \qquad \left[\begin{array}{ccc|c} 1 & 0.25 & 0.25 & 0.375 \\ 0 & 1 & 0.126 & 0.937 \\ 0 & 0 & 1 & -0.5 \end{array}\right]$$

Step 7. Solving by back substitution rounding off the results to 3 significant digits, we have,

$$x_1 = 0.250, x_2 = 1.00, \quad x_3 = -0.500$$

Note : If the system is solved *without* pivoting then

$$x_1 = 0.245\ , x_2 = 1.01,\ x_3 = -0.492$$

while the *exact* solution is

$$x_1 = 0.25,\ \ x_2 = 1, \quad x_3 = -0.5$$

Ex. 2 Determine the values of 'a' so that the following system in unknowns x, y and z has:

$$\left.\begin{aligned} x + y - z &= 1 &(1) \\ 2x + 3y + az &= 3 &(2) \\ x + ay + 3z &= 2 &(3) \end{aligned}\right\}$$

(i) no solution, (ii) more than one solution, (iii) a unique solution.

Let us reduce the system to echelon form. The operations (2) – 2×(1), (3) – (1) give

$$\left.\begin{aligned} x + y - z &= 1 &(1') \\ y + (a+2)z &= 1 &(2') \\ (a-1)y + 4z &= 1 &(3') \end{aligned}\right\}$$

Subtraction $(a - 1)$ times (2') from (3') gives

$$\left.\begin{array}{llr} x+ \quad y- \quad z=1 & & (1'') \\ \quad y+ \quad (a+2)z=1 & & (2'') \\ \quad (3+a)(2-a)z=(2-a) & & (3'') \end{array}\right\}$$

$$[\because 4-(a-1)(a+2)=4-a^2-a+2=6-a-a^2=(3+a)(2-a)]$$

(i) There will be no solution if (3") is of the form $0.\,z=b,\ b\neq 0,$ i.e.

if $(3+a)=0$, i.e. $a=-3$.

$\therefore$ the required value of 'a' $=-3$.

(ii) There will be more than one solution if (3") is of the form $0.\,z=0$, i.e.

if $2-a=0$, i.e. $a=2$.

$\therefore$ the required value of 'a' $=2$.

(iii) There will be a unique solution if the coeff. of z in (3") is not zero i.e.

if $a\neq 2, -3$.

$\therefore$ the required value of 'a' are other than 2 and -3.

Ex. 3. Show that the following linear system has a unique solution for any choice of b_1, b_2, b_3 and b_4:

$$\left.\begin{array}{rrrrl} x_1 & & -\ x_3+ & 2x_4 & =b_1 \\ 2x_1+ & x_2+ & 2x_3- & x_4 & =b_2 \\ -2x_1+ & 2x_2- & x_3+ & 3x_4 & =b_3 \\ 3x_1+ & 4x_2- & x_3- & 2x_4 & =b_4 \end{array}\right\}$$

Steps reducing the system to echelon form are

$$\left.\begin{array}{rrrrl} x_1 & & -\ x_3+ & 2x_4 & =b_1 \\ & x_2+ & 4x_3- & 5x_4 & =b_2-2b_1 \\ & 2x_2- & 3x_3+ & 7x_4 & =b_3+2b_1 \\ & 4x_2+ & 2x_3- & 6x_4 & =b_4-3b_1 \end{array}\right\}$$

$$\Rightarrow \left.\begin{array}{rrrrl} x_1 & & -\ x_3+ & 2x_4 & =b_1 \\ & x_2+ & 4x_3- & 5x_4 & =b_2-2b_1 \\ & & -\ 11x_3+ & 17x_4 & =b_3+2b_1-2b_2+4b_1=b_3-2b_2+6b_1 \\ & & -\ 8x_3- & 14x_4 & =b_4-3b_1-4b_2+8b_1=b_4-4b_2+5b_1 \end{array}\right\}$$

$$\Rightarrow \left.\begin{array}{rrrrl} x_1 & & -\ x_3+ & 2x_4 & =b_1 \\ & x_2+ & 4x_3- & 5x_4 & =b_2-2b_1 \\ & & -\ 11x_3+ & 17x_4 & =b_3-2b_2+6b_1 \\ & & & 18x_4 & =11b_4-44b_2+55b_1-8b_3+16b_2-48b_1 \\ & & & & =11b_4-8b_3-28b_2+7b_1 \end{array}\right\}$$

which is the echelon form. There are 4 eqns. with 4 variables. So the system possesses a unique solution for any choice of b_1, b_2, b_3 and b_4.

Ex.4. Put the system of equations

$$\left.\begin{aligned} 5x_1 + 2x_2 - 7x_3 &= 1 \quad (1)\\ 7x_1 + 2x_2 - 7x_3 &= 1 \quad (2)\\ 2x_1 + 5x_2 - x_3 &= 1 \quad (3)\end{aligned}\right\}$$

into echelon form and hence solve the system.

Steps reducing the system to echelon form are (2)×5 – (1)×7 and (3)×5 – (1)×2 give

$$\left.\begin{aligned} 5x_1 + 2x_2 - 7x_3 &= 1 \quad (1')\\ -19x_2 + 59x_3 &= -7 \quad (2')\\ 21x_2 + 9x_3 &= 23 \quad (3')\end{aligned}\right\}$$

(3')×19 + (2')×21 gives

$$\left.\begin{aligned} 5x_1 + 2x_2 - 7x_3 &= 1 \quad (1'')\\ -19x_2 + 59x_3 &= -7 \quad (2'')\\ 1410x_3 &= 290 \quad (3'')\end{aligned}\right\}$$

which is in echelon form. The number of equations is equal to the number of unknowns. So, the system has a unique solution.

$\therefore$ by back substitution,

$x_3 = \frac{29}{141}$, $19x_2 = 7 + 59x_3$

=> $x_2 = \frac{142}{141}$, $x_1 = \frac{12}{141}$.

7.5.1.4 Gauss-Jordan elimination method : It is an extension of Gaussian method with *no pivoting* and requires more operations. The difference is that there is no need for back substitution. It uses three steps algorithm — two algorithms of Gaussian method plus a new one stated below :

Algorithm

Step 1 and 2. As in the Gaussian method.

Step 3. The variable appeared in the leading term of any equation and appears in no other equations. Up to equation (7.12) follow as in the Gaussian elimination method.

The UTF here is known as *Reduced Echelon Form* (REF). Use of this additional algorithm begins from equation (7.12). Divide the second equation of (7.12) by a'_{22}, then (7.12) becomes

$$\left.\begin{aligned} x_1 + a'_{12}x_2 + a'_{13}x_3 + \cdots + a'_{1n}x_n &= b'_1\\ x_2 + a''_{23}x_3 + \cdots + a''_{2n}x_n &= b'_2\\ a'_{32}x_2 + a'_{33}x_3 + \cdots + a'_{3n}x_n &= b'_3\\ \vdots \quad\quad \vdots \quad\quad \vdots \quad\quad \vdots \quad &\quad \vdots\\ a'_{m2}x_2 + a'_{m3}x_3 + \cdots + a'_{mn}x_n &= b'_m\end{aligned}\right\} \quad (7.13a)$$

Now multiply the second equation in (7.13a) by $(-a'_{12}),(-a'_{32}),\cdots,(-a'_{m2})$ successively and add each result with first equation and third to mth equation one after another, then (7.13a) becomes

$$\left.\begin{array}{llllll} x_1 + & & a''_{13}x_3 + & \cdots & +a''_{1n}x_n & = b''_1 \\ & x_2 + & a''_{23}x_3 + & \cdots & +a''_{2n}x_n & = b''_2 \\ & & a''_{33}x_3 + & \cdots & +a''_{3n}x_n & = b''_3 \\ & \vdots & \vdots & \vdots & \vdots & \vdots \\ & & a''_{m3}x_3 + & \cdots & +a''_{mn}x_n & = b''_m \end{array}\right\} \quad (7.14a)$$

in which x_2 belongs to the second equation of (7.14a) only and in no other equation. Note the difference with (7.14).

N.B. Equation numbers (7.13a) and (7.14a) are maintaining correspondence with (7.13) and (7.14).

Proceeding in this manner, the final REF is given by

$$\left.\begin{array}{llllll} x_1 & + b_{1,r+1}x_{i_{r+1}} & + b_{1,r+2}x_{i_{r+2}} & +\cdots & +b_{1n}x_{i_n} & = c_1 \\ \quad x_2 & + b_{2,r+1}x_{i_{r+1}} & + b_{2,r+1}x_{i_{r+2}} & +\cdots & b_{2,n}x_{i_n} & = c_2 \\ \vdots & \vdots & \vdots & \vdots & \vdots & \vdots \\ \quad x_r & + b_{r,r+1}x_{i_{r+1}} & + b_{r,r+1}x_{i_{r+2}} & +\cdots & b_{rn}x_{i_n} & = c_r \\ & & & & 0 & = c_{r+1} \\ & & & & \vdots & \\ & & & & 0 & = c_n \end{array}\right\} \quad (7.15)$$

In matrix notation, this is equivalent to

rth column (arrow); rth row (arrow)

$$\begin{bmatrix} 1 & 0 & \dots & 0 & b_{1,r+1} & \cdots & b_{1n} \\ 0 & 1 & \dots & 0 & b_{2,r+1} & \cdots & b_{2n} \\ \vdots & \vdots & \vdots & \vdots & & \vdots & \vdots \\ 0 & 0 & \dots & 1 & b_{r,r+1} & \cdots & b_{rm} \\ \vdots & \vdots & \vdots & \vdots & & \vdots & \vdots \\ 0 & 0 & \dots & 0 & 0 & \cdots & 0 \end{bmatrix} \begin{bmatrix} x_1 \\ x_2 \\ \vdots \\ x_{i_r} \\ \vdots \\ x_{i_n} \end{bmatrix} = \begin{bmatrix} c_1 \\ c_2 \\ \vdots \\ c_r \\ c_{r+1} \\ \vdots \\ c_m \end{bmatrix}$$

Solution procedures: As in sec. 7.5.1.2

Number of operations for an $n \times n$ system :

$$\textit{Multiplication / division} = \tfrac{1}{2}n^2(n+1)$$

$$\textit{Addition / substraction} = \tfrac{1}{2}(n-1)n(n+1)$$

Ex. Solve by Gauss-Jordan elimination:

$$\left.\begin{array}{rrrrrrl} x_1 + & 3x_2 - & 2x_3 & & + \;\; 2x_5 & & = 0 \\ 2x_1 + & 6x_2 - & 5x_3 - & 2x_4 + & 2x_5 - & 3x_6 & = -1 \\ & & 5x_3 + & 10x_4 + & & 15x_6 & = 5 \\ 2x_1 + & 6x_2 + & & 8x_4 + & 2x_5 + & 18x_6 & = 6 \end{array}\right\}$$

Step 1. This is a 4×6 system. Eliminating x_1 from 2nd to 4th eqns. by 1st eqn., gives

$$\left.\begin{array}{llllll} x_1 + & 3x_2 - & 2x_3 & & + \ 2x_5 & & = 0 \\ & & x_3 + & 2x_4 + & & 3x_6 & = 1 \\ & & 5x_3 + & 10x_4 + & & 15x_6 & = 5 \\ & & 4x_3 + & 8x_4 + & & 18x_6 & = 6 \end{array}\right\}$$

Step 2. Eliminating x_3 from 3rd and 4th eqns., by 3rd eqn., yields,

$$\left.\begin{array}{llllll} x_1 + & 3x_2 - & 2x_3 & & + \ 2x_5 & & = 0 \\ & & x_3 + & 2x_4 + & & 3x_6 & = 1 \\ & & & & & 0 & = 0 \\ & & & & & 30x_6 & = 10 \end{array}\right\}$$

Step 3. Discarding 3rd eqn. and dividing 4th eqn. by 6, the echelon form is

$$\left.\begin{array}{llllll} x_1 + & 3x_2 - & 2x_3 & & + \ 2x_5 & & = 0 \\ & & x_3 + & 2x_4 + & & 3x_6 & = 1 \\ & & & & & 0 & = 0 \\ & & & & & x_6 & = \frac{1}{3} \end{array}\right\}$$

Step 4 (Jordan part). Steps 1 to 3 are responsible for the Gaussian elimination. Now to eliminate the leading variables x_3 and x_6 from any of the equations, we eliminate x_6 from 2nd eqn. by the 3rd eqn. and hence we are left with

$$\left.\begin{array}{llllll} x_1 + & 3x_2 - & 2x_3 & & + \ 2x_5 & & = 0 \\ & & x_3 + & 2x_4 + & & 3x_6 & = 1 \\ & & & & & x_6 & = \frac{1}{3} \end{array}\right\}$$

Also to eliminate x_3 from 1st eqn. by 2nd eqn., we are left with

$$\left.\begin{array}{lllll} x_1 + & 3x_2 & & 4x_4 + \ 2x_5 & & = 0 \\ & & x_3 + & 2x_4 + & & = 0 \\ & & & & x_6 & = \frac{1}{3} \end{array}\right\}$$

Step 5. All eqns. in last part of step 4 are free from leading variables and the system is in echelon form with free variables x_2, x_4, x_5. Denoting these x_2, x_4, x_5 by r, s, t respectively, the solutions are

$$x_1 = -3r - 4s - 2t, \quad x_2 = r, \quad x_3 = -2s, \quad x_4 = s, \quad x_5 = t, \quad x_6 = \tfrac{1}{6}.$$

Note that no back substitution is needed here.

7.5.1.5 Cholesky's / Crout's method : This is a variation of Gauss or Gauss-Jordan elimination technique for a nonsingular $n \times n$ system. This can be readily adapted to computer usage and the corresponding flow chart or the software is available in many texts on numerical analysis. In this method the augmented or the coefficient matrix is factored into a lower triangular, L and a unit upper triangular T matrices. This factorization is known as LU factorization. The solution is obtained by back substitution using the unit upper triangular matrix only which saves computational work as well as the memory space of the computer. That is,

$$\mathrm{A} = \mathrm{LT} \text{ and } |\mathrm{A}| = |\mathrm{L}||\mathrm{T}| = |\mathrm{L}|, \quad \text{since } |\mathrm{T}| = 1 \text{ as } t_{ii} = 1$$

$$= \prod_{i=1}^{n} l_{ii}$$

from where *nonsingularity* and *ill conditioning* of the system can be easily checked. This is not possible in the Gauss elimination or Gauss-Jordan methods.

Let the original $n \times n$ system be

$$\mathbf{Ax} = \mathbf{b} \tag{7.16}$$

where coefficient matrix $A = [a_{ij}]_{n \times n}$, $\mathbf{x} = [x_1 \ x_2 \cdots x_n]^T$ and $\mathbf{b} = [b_1 \ b_2 \cdots b_n]^T$. For simplicity, let us consider a nonsingular 3×3 system where the augmented matrix is

$$\left[\begin{array}{ccc|c} a_{11} & a_{12} & a_{13} & b_1 \\ a_{21} & a_{22} & a_{23} & b_2 \\ a_{31} & a_{32} & a_{33} & b_3 \end{array}\right]$$

Let the required LU factorization be

$$\underbrace{\left[\begin{array}{ccc|c} a_{11} & a_{12} & a_{13} & b_1 \\ a_{21} & a_{22} & a_{23} & b_2 \\ a_{31} & a_{32} & a_{33} & b_3 \end{array}\right]}_{A \qquad\qquad \mathbf{b}} = \underbrace{\begin{bmatrix} l_{11} & 0 & 0 \\ l_{21} & l_{22} & 0 \\ l_{31} & l_{32} & l_{33} \end{bmatrix}}_{L} \underbrace{\left[\begin{array}{ccc|c} 1 & t_{12} & t_{13} & c_1 \\ 0 & 1 & t_{23} & c_2 \\ 0 & 0 & 1 & c_3 \end{array}\right]}_{U \qquad\qquad \mathbf{c}} \tag{7.17}$$

or, $$\left[a_{ij} \mid b_i\right] = \left[l_{ij}\right]\left[t_{ij} \mid c_i\right] \tag{7.18}$$

The equation (7.18) above contains 12 known quantities $a_{11}, \cdots, a_{33}, b_1, b_2$ and b_3 and 6 unknown quantities $l_{11}, l_{21}, l_{22}, l_{31}, l_{32}, l_{33}$ in the premultiplicant and $t_{12}, t_{13}, t_{23}, c_1, c_2, c_3$ in the post multiplicant on the right giving rise to a total of 12 and in general $n^2 + n$ equations instead of the 3 and in general n original equations. Yet the present equations are extremely simple to solve. Now multiplying the right side of (7.18), we get

$$\left[\begin{array}{ccc|c} a_{11} & a_{12} & a_{13} & b_1 \\ a_{21} & a_{22} & a_{23} & b_2 \\ a_{31} & a_{32} & a_{33} & b_3 \end{array}\right]$$

$$= \left[\begin{array}{ccc|c} l_{11} & l_{11}t_{12} & l_{11}t_{13} & l_{11}c_1 \\ l_{21} & (l_{21}t_{12} + l_{22}) & (l_{21}t_{13} + l_{22}t_{23}) & (l_{21}c_1 + l_{22}c_2) \\ l_{31} & (l_{31}t_{12} + l_{32}) & (l_{31}t_{13} + l_{32}t_{23} + l_{33}) & (l_{31}c_1 + l_{32}c_2 + l_{33}c_3) \end{array}\right] \tag{7.19}$$

Therefore, from the principle of equality of matrices, the 12 equations are

$$\left.\begin{array}{l} a_{11} = l_{11} \\ a_{21} = l_{21} \\ a_{31} = l_{31} \end{array}\right\} \quad (7.20) \qquad \left.\begin{array}{l} a_{12} = l_{11}t_{12} \\ a_{22} = l_{21}t_{12} + l_{22} \\ a_{32} = l_{31}t_{12} + l_{32} \end{array}\right\} \quad (7.21)$$

$$\left.\begin{aligned} a_{13} &= l_{11}t_{13} \\ a_{23} &= l_{21}t_{13} + l_{22}t_{23} \\ a_{33} &= l_{31}t_{13} + l_{32}t_{23} + l_{33} \end{aligned}\right\} \quad (7.22) \qquad \left.\begin{aligned} b_1 &= l_{11}c_1 \\ b_2 &= l_{21}c_1 + l_{22}c_2 \\ b_3 &= l_{31}c_1 + l_{32}c_2 + l_{33}c_3 \end{aligned}\right\} \quad (7.23)$$

Solving these 12 equations for the unknown, we get,

$$\begin{array}{l|l|l|} l_{11} = a_{11} & t_{12} = \dfrac{a_{12}}{l_{11}} & t_{13} = \dfrac{a_{13}}{l_{11}} \\ l_{21} = a_{21} & l_{22} = a_{22} - l_{21}t_{12} & t_{23} = \dfrac{1}{l_{22}}(a_{23} - l_{21}t_{13}) \\ l_{31} = a_{31} & l_{32} = a_{32} - l_{31}t_{12} & l_{33} = a_{33} - l_{31}t_{13} - l_{32}t_{23} \end{array} \quad (7.24)$$

$$\left.\begin{aligned} c_1 &= \frac{b_1}{l_{11}} \\ c_2 &= \frac{(b_2 - l_{21}c_1)}{l_{22}} \\ c_3 &= \frac{(b_3 - l_{31}c_1 - l_{32}c_2)}{l_{33}} \end{aligned}\right\} \quad (7.25)$$

Using (7.18) in (7.17), we have

$$\mathbf{LUx} = \mathbf{b} \quad (7.26)$$

or

$$\mathbf{Ly} = \mathbf{b} \quad (7.27)$$

$$\text{where } \mathbf{Ux} = \mathbf{y}, \quad \mathbf{y} = [y_1 \ \ y_2 \ \ y_3]^{\mathrm{T}} \quad (7.28)$$

Now (7.27) is equivalent to

$$\begin{aligned} l_{11}y_1 &= b_1 \\ l_{21}y_1 + l_{22}y_2 &= b_2 \\ l_{31}y_1 + l_{32}y_2 + l_{33}y_3 &= b_3 \end{aligned}$$

which by forward substitution and using (7.25) gives,

$$\left.\begin{aligned} y_1 &= \frac{b_1}{l_{11}} = c_1 \\ y_2 &= \frac{(b_2 - l_{21}y_1)}{l_{22}} = c_2 \\ y_3 &= \frac{(b_3 - b_{31}y_1 - l_{32}y_2)}{l_{33}} = c_3 \end{aligned}\right\} \quad (7.29)$$

Using (14) in (13) we have

$$\left.\begin{array}{r} {}_1+t_{12}x_2+t_{13}x_3=c_1 \\ x \qquad x_2+t_{23}x_3=c_2 \\ x_3=c_3 \end{array}\right\} \tag{7.30}$$

or
$$\begin{bmatrix} 1 & t_{12} & t_{13} \\ 0 & 1 & t_{23} \\ 0 & 0 & 1 \end{bmatrix}\begin{bmatrix} x_1 \\ x_2 \\ x_3 \end{bmatrix}=\begin{bmatrix} c_1 \\ c_2 \\ c_3 \end{bmatrix} \tag{7.31}$$

By backward substitution, (7.30) or (7.31) gives the solution set

$$\left.\begin{array}{l} x_3=c_3 \\ x_2=c_2-t_{23}x_3 \\ x_1=c_1-t_{12}x_2-t_{13}x_3 \end{array}\right\} \tag{7.32}$$

In practice, find $U=T=\left[t_{ij}\right]$, $t_{ii}=1$, $t_{ij}=0$ for $i>j$ and use (7.32) directly to get the solution.

For a set of $n\times n$ system, the solutions (7.24) and (7.25) are

$$l_{ij}=a_{ij}-\sum_{k=1}^{j-1} l_{ik}t_{kj}\,,\quad i\geq j,\ i=1,2,\cdots,n$$

$$t_{ij}=\{(a_{ij}-\sum_{k=1}^{i-1} l_{ik}t_{kj})/l_{ii}\},\quad i<j,\ j=2,3,\cdots,n$$

$$c_i=\{(b_i-\sum_{k=1}^{i-1} l_{ik}c_k)/l_{ii}\},\quad i=2,3,\cdots,n$$

$$t_{1j}=\frac{a_{1j}}{a_{11}},\quad j=2,3,\cdots,n$$

$$xl_{i1}=a_{i1}$$

$$c_1=\frac{b_1}{l_{11}}$$

The unknowns as in (7.32) are

$$x_n=c_n$$

$$x_j=c_j-\sum_{r=j+1}^{n} t_{jr}x_r,\quad j=1,2,\cdots,(n-1)$$

Ex. Solve the following set of equations by Cholesky's method:

$$\left.\begin{array}{rrrrl} 2x_1- & x_2+ & 3x_3- & 2x_4 & =1 \\ x_1+ & 2x_2- & x_3+ & x_4 & =6 \\ 3x_1+ & 3x_2+ & 4x_3- & 5x_4 & =1 \\ 5x_1- & 2x_2+ & 7x_3- & 6x_4 & =-2 \end{array}\right\}$$

Let the augmented matrix be

$$\left[\begin{array}{cccc|c} 2 & -1 & 3 & -2 & 1 \\ 1 & 2 & -1 & 1 & 6 \\ 3 & 3 & 4 & -5 & 1 \\ 5 & -2 & 7 & -6 & -2 \end{array}\right] = \begin{bmatrix} l_{11} & 0 & 0 & 0 \\ l_{21} & l_{22} & 0 & 0 \\ l_{31} & l_{32} & l_{33} & 0 \\ l_{41} & l_{42} & l_{43} & l_{44} \end{bmatrix} \left[\begin{array}{cccc|c} 1 & t_{12} & t_{13} & t_{14} & c_1 \\ 0 & 1 & t_{23} & t_{24} & c_2 \\ 0 & 0 & 1 & t_{34} & c_3 \\ 0 & 0 & 0 & 1 & c_4 \end{array}\right]$$

Multiplying the R.H.S. and equating like elements, we obtain,

$$l_{11} = a_{11} = 2$$
$$l_{21} = a_{21} = 1$$
$$l_{31} = a_{21} = 3$$
$$l_{41} = a_{41} = 5$$
$$l_{22} = a_{22} - l_{21}t_{12}$$
$$= 2 + \tfrac{1}{2} = \tfrac{5}{2}$$
$$l_{32} = a_{32} - l_{31}t_{12}$$
$$= 3 + \tfrac{3}{2} = \tfrac{9}{2}$$
$$l_{42} = a_{42} - l_{41}t_{12}$$
$$= -2 + \tfrac{5}{2} = \tfrac{1}{2}$$
$$l_{33} = a_{33} - (l_{13}t_{13} + l_{32}t_{23})$$
$$= 4 - (\tfrac{9}{2} - \tfrac{9}{2}) = 4$$
$$l_{43} = a_{43} - (l_{41}t_{13} + l_{42}t_{23})$$

$$t_{12} = \frac{a_{12}}{a_{11}} = -\tfrac{1}{2}$$
$$t_{13} = \frac{a_{13}}{a_{11}} = \tfrac{3}{2}$$
$$t_{14} = \frac{a_{14}}{a_{11}} = -1$$
$$t_{23} = \frac{1}{l_{22}}(a_{23} - l_{21}t_{13})$$
$$= \tfrac{2}{5}(-1 - \tfrac{3}{2}) = -1$$
$$t_{24} = \frac{1}{l_{22}}(a_{24} - l_{21}t_{14})$$
$$= \tfrac{2}{5}(1 + 1) = \tfrac{4}{5}$$
$$t_{34} = \frac{1}{l_{33}}[a_{34} - (l_{31}t_{14} + l_{32}t_{24})]$$

$$c_1 = \frac{b_1}{l_{11}} = \tfrac{1}{2}$$
$$c_2 = \frac{1}{l_{22}}(b_2 - l_{21}c_1)$$
$$= \tfrac{2}{5}(6 - c_1)$$
$$= \tfrac{2}{5}(6 - \tfrac{1}{2}) = \tfrac{11}{5}$$
$$c_3 = \frac{1}{l_{33}}(b_3 - (l_{31}c_1 + l_{32}c_2))$$
$$= \frac{1}{l_{33}}[1 - \tfrac{3}{2} + l_{32}.(\tfrac{11}{5})]$$
$$= \tfrac{1}{4}[1 - \tfrac{3}{2} + (\tfrac{9}{2}).(\tfrac{11}{5})]$$
$$= -\tfrac{13}{5}$$

$\therefore$ the solutions are

$$x_4 = c_4 = 4$$
$$x_3 = c_3 - t_{34}x_4 = -\tfrac{13}{5} + \tfrac{7}{5}.4 = 3$$
$$x_2 = c_2 - t_{23}x_3 - t_{24}x_4 = \tfrac{11}{5} + 3 - \tfrac{4}{5}.4 = 2$$
$$x_1 = c_1 - t_{12}x_2 - t_{13}x_3 - t_{14}x_4 = \tfrac{1}{2} + \tfrac{1}{2}.2 - \tfrac{3}{2}.3 + 4 = 1$$

7.5.1.6 Square root method: Banachiewicz derived a method similar to Crout's for an $n \times n$ system of equations with symmetrical coefficient matrix. This type of equation often occurs in science and engineering problems such as the numerical solution of ordinary and partial differential equations. The method saves memory space and reduces input data. It is by far the most efficient method for solving symmetrical systems.

Let the original $n \times n$ system be

$$\mathbf{Ax} = \mathbf{b} \tag{7.33}$$

where the coefficient matrix

$$A = [a_{ij}]_{n \times n} \text{ with } A^T = [a_{ji}]_{n \times n} = A \text{ such that}$$

$$|A| \neq 0, \text{ and } \mathbf{x} = [x_1 \quad x_2 \quad \cdots \quad x_n]^T \text{ and } \mathbf{b} = [b_1 \quad b_2 \quad \cdots \quad b_n]^T$$

since A is symmetrical, so A can be expressed as a product of transposed upper triangular and upper triangular matrices.

That is, let

$$A = T^T T \tag{7.34}$$

$$\text{where } T = \begin{bmatrix} t_{11} & t_{12} & \cdots & t_{1n} \\ 0 & t_{22} & \cdots & t_{2n} \\ \vdots & \vdots & \ddots & \vdots \\ 0 & 0 & \cdots & t_{nn} \end{bmatrix} \text{ and } T^T = \begin{bmatrix} t_{11} & 0 & \cdots & 0 \\ t_{12} & t_{22} & \cdots & 0 \\ \vdots & \vdots & \ddots & \vdots \\ t_{1n} & t_{2n} & \cdots & t_{nn} \end{bmatrix} \tag{7.35}$$

Now for simplicity, let us consider a 3×3 nonsingular symmetric system $\sum_{i,j=1}^{3} a_{ij}x_j = b_i$.

Therefore, let the augmented matrix be factored as

$$\underbrace{\begin{bmatrix} a_{11} & a_{12} & a_{13} \\ a_{12} & a_{22} & a_{23} \\ a_{13} & a_{23} & a_{33} \end{bmatrix}}_{A} \!\!\underbrace{\begin{matrix} \vdots \\ \vdots \\ \vdots \end{matrix}\begin{matrix} b_1 \\ b_2 \\ b_3 \end{matrix}}_{b} = \underbrace{\begin{bmatrix} t_{11} & 0 & 0 \\ t_{12} & t_{22} & 0 \\ t_{13} & t_{23} & t_{33} \end{bmatrix}}_{T^T} \underbrace{\begin{bmatrix} t_{11} & t_{12} & t_{13} \\ 0 & t_{22} & t_{23} \\ 0 & 0 & t_{33} \end{bmatrix}}_{T}\!\!\underbrace{\begin{matrix} \vdots \\ \vdots \\ \vdots \end{matrix}\begin{matrix} c_1 \\ c_2 \\ c_3 \end{matrix}}_{c} \tag{7.36}$$

$$\text{or} \qquad \left[a_{ij} \,\vdots\, b_i\right] = \left[t_{ij}\right]^T \left[t_{ij} \,\vdots\, c_i\right] \tag{7.37}$$

The equation (7.36) has 9 unknown $t_{11}, t_{12}, \cdots, t_{33}, c_1, c_2, c_3$ and 9 known quantities $a_{11}, a_{12}, \cdots, a_{33}, b_1, b_2, b_3$. Now multiplying the right side of (7.36) we get

$$\left[\begin{array}{ccc:c} a_{11} & a_{12} & a_{13} & b_1 \\ a_{12} & a_{22} & a_{23} & b_2 \\ a_{13} & a_{23} & a_{33} & b_3 \end{array}\right]$$

$$= \left[\begin{array}{ccc:c} t_{11}^2 & t_{11}t_{12} & t_{11}t_{13} & t_{11}c_1 \\ t_{11}t_{12} & (t_{12}^2 + t_{22}^2) & (t_{12}t_{13} + t_{22}t_{23}) & (t_{12}c_1 + t_{22}c_2) \\ t_{11}t_{13} & (t_{13}t_{12} + t_{22}t_{23}) & (t^2{}_{13} + t^2{}_{23} + t^2{}_{33}) & (t_{13}c_1 + t_{23}c_2 + t_{33}c_3) \end{array}\right] \tag{7.38}$$

Therefore, from the principle of equality of matrices, the 9 equations are

$$\left.\begin{aligned}
a_{11} &= t_{11}^2 \\
a_{12} &= t_{11}t_{12} \\
a_{13} &= t_{11}t_{13} \\
a_{22} &= t_{12}^2 + t_{22}^2 \\
a_{23} &= t_{13}t_{12} + t_{22}t_{23} \\
a_{33} &= t_{13}^2 + t_{23}^2 + t_{33}^2 \\
b_1 &= t_{11}c_1 \\
b_2 &= t_{12}c_1 + t_{22}c_2 \\
b_3 &= t_{13}c_1 + t_{23}c_2 + t_{33}c_3
\end{aligned}\right\} \quad (7.39) \quad \text{or} \quad \left.\begin{aligned}
t_{11} &= \sqrt{a_{11}} \\
t_{12} &= \frac{a_{12}}{t_{11}} \\
t_{13} &= \frac{a_{13}}{t_{11}} \\
t_{22} &= \sqrt{a_{22} - t_{12}^2} \\
t_{23} &= \frac{1}{t_{33}}(a_{23} - t_{13}t_{12}) \\
t_{33} &= \sqrt{a_{33} - t_{13}^2 - t_{23}^2} \\
c_1 &= \frac{b_1}{t_{11}} \\
c_2 &= \frac{1}{t_{22}}(b_2 - t_{12}c_1) \\
c_3 &= \frac{1}{t_{33}}(b_3 - t_{13}c_1 - t_{23}c_2)
\end{aligned}\right\} \quad (7.40)$$

Using (7.36) in (7.33), we have

$$\mathbf{T}^{\mathrm{T}}\mathbf{T}\mathbf{x} = \mathbf{b} \tag{7.41}$$

or

$$\mathbf{T}^{\mathrm{T}}\mathbf{y} = \mathbf{b} \tag{7.42}$$

where

$$\mathbf{T}\mathbf{x} = \mathbf{y}, \ \mathbf{y} = [y_1 \ \ y_2 \ \ y_3]^{\mathrm{T}} \tag{7.43}$$

Now (7.42) is equivalent to

$$\begin{aligned}
t_{11}y_1 \qquad\qquad\qquad &= b_1 \\
t_{12}y_1 + t_{22}y_2 \qquad\quad &= b_2 \\
t_{13}y_1 + t_{23}y_2 + t_{33}y_3 &= b_3
\end{aligned}$$

which by forward substitution and using last 3 equations of (7.40) yields

$$\left.\begin{aligned}
y_1 &= \frac{b_1}{t_{11}} = c_1 \\
y_2 &= \frac{1}{t_{22}}(b_2 - t_{12}y_2) = c_2 \\
y_3 &= \frac{1}{t_{33}}(b_3 - t_{13}y_1 - t_{23}y_2) = c_3
\end{aligned}\right\} \tag{7.44}$$

using (7.44) in (7.43) , we have,

$$\left.\begin{aligned} t_{11}x_1 + t_{12}x_2 + t_{13}x_3 &= c_1 \\ t_{22}x_2 + t_{23}x_3 &= c_2 \\ t_{33}x_3 &= c_3 \end{aligned}\right\} \tag{7.45}$$

which is equivalent to

$$\begin{bmatrix} t_{11} & t_{12} & t_{13} \\ 0 & t_{22} & t_{23} \\ 0 & 0 & t_{33} \end{bmatrix} \begin{bmatrix} x_1 \\ x_2 \\ x_3 \end{bmatrix} = \begin{bmatrix} c_1 \\ c_2 \\ c_3 \end{bmatrix} \tag{7.46}$$

By back substitution (7.45) or (7.46) gives the solutions

$$\left.\begin{aligned} x_3 &= \frac{c_3}{t_{33}} \\ x_2 &= \frac{1}{t_{22}}(c_2 - t_{23}x_3) \\ x_1 &= \frac{1}{t_{11}}(c_1 - t_{12}x_2 - t_{13}x_3) \end{aligned}\right\} \tag{7.47}$$

In practice, find T and use (7.47) directly to get the solution. The method gets its name, square root method, from the diagonal elements which involve square root. For a set of $n \times n$ system, the solution (7.40) takes the form

$$t_{ii} = \sqrt{a_{ii} - \sum_{k=1}^{i-1} t_{ki}^2}, \quad i = 2, \cdots, n$$

$$t_{ij} = \frac{1}{t_{ii}}(a_{ij} - \sum_{k=1}^{i-1} t_{ki}t_{kj}), \quad j = 2, \cdots, n$$

$$c_i = \frac{1}{t_{ii}}(b_i - \sum_{k=1}^{i-1} t_{ki}c_k), \quad i = 2, \cdots, n$$

$$t_{11} = \sqrt{a_{11}}$$

$$c_1 = \frac{b_1}{t_{11}}$$

$$t_{1j} = \frac{a_{1j}}{t_{11}}, \quad j = 2, \cdots, n$$

$$t_{ij} = 0, \quad i > j$$

and the unknowns in (7.47) are

$$x_n = \frac{c_n}{t_{nn}}$$

and $$x_i = \frac{1}{t_{ii}}(c_i - \sum_{k=i+1}^{n} t_{ik}x_k), \quad i = 1, 2, \cdots, (n-1)$$

Number of operations:

(a) n square roots on which depends the advantage of the method.

(b) $\frac{1}{6}(n^3 + 9n^2 + 2n)$ multiplication / division.

(c) $\frac{1}{6}\ (n^3 + 6n^2 - 7n)$ additions / subtractions.

Ex. Solve

$$\left.\begin{aligned} 2x_1 - x_2 + 3x_3 - 2x_4 &= 1 \\ -x_1 + 2x_2 + 3x_3 + x_4 &= 6 \\ 3x_1 + 2x_2 + 4x_3 - 5x_4 &= 1 \\ -2x_1 + x_2 - 5x_3 - 6x_4 &= -2 \end{aligned}\right\}$$

using the square root method.

In this method, let

$$\left[\begin{array}{cccc|c} 2 & -1 & 3 & -2 & 1 \\ -1 & 2 & 3 & 1 & 6 \\ 3 & 3 & 4 & -5 & 1 \\ -2 & 1 & -5 & -6 & -2 \end{array}\right] = \begin{bmatrix} t_{11} & 0 & 0 & 0 \\ t_{12} & t_{22} & 0 & 0 \\ t_{13} & t_{23} & t_{33} & 0 \\ t_{14} & t_{24} & t_{34} & t_{44} \end{bmatrix} \left[\begin{array}{cccc|c} t_{11} & t_{12} & t_{13} & t_{14} & c_1 \\ 0 & t_{22} & t_{23} & t_{24} & c_2 \\ 0 & 0 & t_{33} & t_{34} & c_3 \\ 0 & 0 & 0 & t_{44} & c_4 \end{array}\right]$$

Multiplying the R.H.S., equating like elements with L.H.S. and solving, we obtain,

$$t_{11} = \sqrt{a_{11}} = \sqrt{2}$$

$$t_{12} = \frac{a_{12}}{t_{11}} = -\frac{1}{\sqrt{2}}$$

$$t_{13} = \frac{a_{13}}{t_{11}} = \frac{3}{\sqrt{2}}$$

$$t_{14} = \frac{a_{14}}{t_{11}} = -\frac{2}{\sqrt{2}}$$

$$c_1 = \frac{1}{t_{11}} = \frac{1}{\sqrt{2}}$$

$$c_2 = \frac{(6 - t_{12}c_1)}{t_{22}} = \frac{13}{\sqrt{6}}$$

$$c_3 = \frac{(1 - t_{13}c_1 - t_{23}c_2)}{t_{33}} = \frac{20}{\sqrt{14}}i$$

$$c_4 = \frac{(-2 - t_{14}c_1 - t_{24}c_2 - t_{34}c_3)}{t_{44}}$$

$$= -\frac{13}{\sqrt{7.54}}i$$

$$t_{22} = \sqrt{a_{22} - t_{12}^2} = \sqrt{2 - \frac{1}{2}} = \sqrt{\frac{3}{2}}$$

$$t_{23} = \frac{(a_{23} - t_{13}t_{12})}{t_{22}} = \frac{\sqrt{2}}{\sqrt{3}}\left(3 + \frac{3}{\sqrt{2}}\cdot\frac{1}{\sqrt{2}}\right) = \frac{9}{\sqrt{6}}$$

$$t_{24} = \frac{(a_{24} - t_{12}t_{14})}{t_{22}} = \frac{\sqrt{2}}{\sqrt{3}}\left(1 - \frac{1}{\sqrt{2}}\cdot\frac{2}{\sqrt{2}}\right) = 0$$

$$t_{33} = \sqrt{a_{33} - t_{13}^2 - t_{23}^2} = \sqrt{4 - \frac{9}{2} - \frac{81}{6}} = \sqrt{14}$$

$$t_{34} = \frac{(a_{34} - t_{13}t_{14} - t_{23}t_{24})}{t_{33}} = \frac{2}{\sqrt{14}}i$$

$$t_{44} = \sqrt{a_{44} - t_{14}^2 - t_{24}^2 - t_{34}^2} = \sqrt{\frac{54}{7}}\ i$$

$\therefore$

$$x_4 = \frac{c_4}{t_{44}} = -\frac{13}{54},$$

$$x_3 = \frac{(c_3 - t_{34}x_4)}{t_{33}} = \frac{79}{54},$$

$$x_2 = \frac{(c_2 - t_{23}x_3 - t_{24}x_4)}{t_{22}} = -\frac{1}{18} \text{ and}$$

$$x_1 = \frac{(c_1 - t_{12}x_2 - t_{13}x_3 - t_{14}x_4)}{t_{11}} = -\frac{53}{27}.$$

7.5.1.7 Chio's pivotal condensation procedure to solve system of equations : As in the evaluation of determinants by Chio's pivotal condensation, the procedure consists of a series of evaluation of 2×2 determinants.

Let the n linear equations in n unknowns in matrix equation be

$\mathbf{Ax} = \mathbf{b}$ where $A = [a_{ij}]_{n\times n}$, $\mathbf{x} = [x_1\ x_2 \cdots x_n]^T$ and $\mathbf{b} = [b_1\ b_2 \cdots b_n]^T$.

Now the augmented matrix of the system is clearly an $n\times(n+1)$ matrix

$$\begin{bmatrix} a_{11} & a_{12} & \cdots & a_{1n} & b_1 \\ a_{21} & a_{22} & \cdots & a_{2n} & b_2 \\ \vdots & \vdots & \ddots & \vdots & \vdots \\ a_{n1} & a_{n2} & \cdots & a_{nn} & b_n \end{bmatrix}_{n\times(n+1)} \tag{7.48}$$

Then we form the reduced matrix of order $(n-1)\times n$ from (7.48)

$$\begin{bmatrix} c_{11} & c_{12} & \cdots & c_{1,n-1} & b_1' \\ c_{21} & c_{22} & \cdots & c_{2,n-1} & b_2' \\ \vdots & \vdots & \ddots & \vdots & \\ c_{n-1,1} & c_{n-1,2} & \cdots & c_{n-1,n-1} & b_{n-1}' \end{bmatrix} \tag{7.49}$$

where $c_{ij} = \begin{vmatrix} a_{11} & a_{1,j+1} \\ a_{j+1,1} & a_{i+1,j+1} \end{vmatrix}$, $b_i' = \begin{vmatrix} a_{11} & b_1 \\ a_{i+1,1} & b_{i+1} \end{vmatrix}$ and $c_{11} \neq 0$

The matrix (7.49) is the called the *first* reduction matrix of (7.48). In this way, continue the process of reduction $(n-1)$ times, so that the final reduction matrix is of order 1×2 and let the final reduced matrix take the form

$$\begin{bmatrix} g_{11} & b_1^{n-1} \end{bmatrix}, \quad g_{11} \neq 0 \tag{7.50}$$

This is the augmented matrix for *one* equation and the equation is

$$g_{11}x_n = b_1^{n-1}$$

$$\therefore \quad x_n = \frac{b_1^{n-1}}{g_{11}}, \quad g_{11} \neq 0.$$

Let the next to the last reduced matrix be

$$\begin{bmatrix} f_{11} & f_{12} & b_1^{n-2} \\ f_{21} & f_{22} & b_1^{n-2} \end{bmatrix}. \tag{7.51}$$

As x_n is known, x_{n-1} can be obtained from either of the equation

$$f_{11}x_{n-1} + f_{12}x_n = b_1^{n-2} \tag{7.52}$$

or the equation

$$f_{21}x_{n-1} + f_{22}x_n = b_1^{n-2}. \tag{7.53}$$

In this way of back calculation we can obtained all the x values. It is easy to consider the last row of each reduced equation to find an unknown.

Ex. Solve, using Chio's pivotal condensation procedure, the system of equations

$$\left.\begin{aligned} 3x_1 + x_2 + x_3 + x_4 &= 0 \\ 2x_1 - x_2 + 2x_3 - x_4 &= 4 \\ 2x_1 \quad\quad + 3x_3 + x_4 &= 1 \\ x_1 \quad\quad + 2x_3 + x_4 &= 3 \end{aligned}\right\}$$

The initial 4×5 augmented matrix is

$$\begin{bmatrix} 3 & 1 & 1 & 1 & 0 \\ 2 & -1 & 2 & -1 & 4 \\ 2 & 0 & 3 & 1 & 1 \\ 1 & 0 & 2 & 1 & 3 \end{bmatrix} \tag{1}$$

The first reduced 3×4 matrix obtained from (1) is given by

$$\begin{bmatrix} \begin{vmatrix} 3 & 1 \\ 2 & -1 \end{vmatrix} & \begin{vmatrix} 3 & 1 \\ 2 & 2 \end{vmatrix} & \begin{vmatrix} 3 & 1 \\ 2 & -1 \end{vmatrix} & \begin{vmatrix} 3 & 0 \\ 2 & 4 \end{vmatrix} \\ \begin{vmatrix} 3 & 1 \\ 2 & 0 \end{vmatrix} & \begin{vmatrix} 3 & 1 \\ 2 & 3 \end{vmatrix} & \begin{vmatrix} 3 & 1 \\ 2 & 1 \end{vmatrix} & \begin{vmatrix} 3 & 0 \\ 2 & 1 \end{vmatrix} \\ \begin{vmatrix} 3 & 1 \\ 1 & 0 \end{vmatrix} & \begin{vmatrix} 3 & 1 \\ 1 & 2 \end{vmatrix} & \begin{vmatrix} 3 & 1 \\ 1 & 1 \end{vmatrix} & \begin{vmatrix} 3 & 0 \\ 1 & 3 \end{vmatrix} \end{bmatrix} = \begin{bmatrix} -5 & 4 & -5 & 12 \\ -2 & 7 & 1 & 3 \\ -1 & 5 & 2 & 9 \end{bmatrix} \tag{2}$$

The second reduced 2×3 matrix is obtained from (2) as

$$\begin{bmatrix} \begin{vmatrix} -5 & 4 \\ -2 & 7 \end{vmatrix} & \begin{vmatrix} -5 & -5 \\ -2 & 1 \end{vmatrix} & \begin{vmatrix} -5 & 12 \\ -2 & 3 \end{vmatrix} \\ \begin{vmatrix} -5 & 4 \\ -1 & 5 \end{vmatrix} & \begin{vmatrix} -5 & -5 \\ 5 & 2 \end{vmatrix} & \begin{vmatrix} -5 & 12 \\ 2 & 9 \end{vmatrix} \end{bmatrix} = \begin{bmatrix} -27 & -15 & 9 \\ -21 & 15 & -69 \end{bmatrix} \tag{3}$$

The third reduced 1×2 matrix is obtained from (3) as

$$\begin{bmatrix} \begin{vmatrix} -27 & -15 \\ -21 & 15 \end{vmatrix} & \begin{vmatrix} -27 & 9 \\ -21 & -69 \end{vmatrix} \end{bmatrix} = \begin{bmatrix} -720 & 2052 \end{bmatrix} \tag{4}$$

From (4), $-720\,x_4 = 2052$

or, $x_4 = -2.85$

From (3), $-21x_3 + 15x_4 = -69$

or $$x_3 = \frac{-69-15(-2.85)}{-21} = 1.25$$

From (2), $-x_2 + 5x_3 + 2x_4 = 9$

or, $$x_2 = 5(1.25) + 2(-2.85) - 9 = -8.45$$

From (1), $x_1 + 2x_3 + x_4 = 3$

or $$x_1 = 3 - 2(-8.45) - (-2.85) = 3.35$$

N.B. Ref. Conkwright & J.D. Heide, The solution of simultaneous linear equations, The Mathematics Teacher, April 1945, P 177-180.

7.5.2 Iterative methods : These methods give approximate solution to a given accuracy by means of convergent infinite processes. These include the Jacobi method, Gauss-Seidel method, method of relaxation etc. Numerical approximate solution of a large and sparse system is obtained by these methods. In these methods solution starts with an "initial guess". The better the initial guess the sooner one gets a result within required accuracy. Some of the above methods are being discussed below.

Conditions for convergence of iteration:

(1) The necessary condition for the system is that the coefficient matrix must have a "*strong diagonal*" which means that it is greater than any other element in the row. In other words, diagonal elements dominate the rows.

(2) The sufficient condition for the system is that the absolute value of diagonal element of the coefficient matrix must be greater than the sum of the absolute values of all other elements in that row.

7.5.2.1 Jacobi's Method (Simultaneous displacement method) : This is one of the easiest iteration method. Let us consider a 3×3 system

$$\left.\begin{aligned} a_{11}x_1 + a_{12}x_2 + a_{13}x_3 &= b_1 \\ a_{21}x_1 + a_{22}x_2 + a_{23}x_3 &= b_2 \\ a_{31}x_1 + a_{32}x_2 + a_{33}x_3 &= b_3 \end{aligned}\right\} \tag{7.54}$$

with the strong diagonal of the coefficient matrix. The method begins with the solution of x_1, x_2, x_3 from 1st, 2nd and 3rd equations respectively. Iteration superscript k has been introduced with those solutions. Hence,

$$\left.\begin{aligned} x_1^{(k+1)} &= \frac{b_1}{a_{11}} - \frac{a_{12}}{a_{11}}x_2^{(k)} - \frac{a_{13}}{a_{11}}x_3^{(k)} \\ x_2^{(k+1)} &= \frac{b_2}{a_{22}} - \frac{a_{21}}{a_{22}}x_1^{(k)} - \frac{a_{23}}{a_{22}}x_3^{(k)} \\ x_3^{(k+1)} &= \frac{b_3}{a_{33}} - \frac{a_{31}}{a_{33}}x_1^{(k)} - \frac{a_{32}}{a_{33}}x_2^{(k)} \end{aligned}\right\} \tag{7.55}$$

Let the vector $\{x\}_k = \{x_1^{(k)}\ x_2^{(k)}\ x_3^{(k)}\}$ be an initial guess. On substitution in (7.55), a new vector $\{x\}_{k+1}$ is obtained. This process is continued until sufficient accuracy is achieved. That is, until

$$\{x\}_{k+1} \cong \{x\}_k, \quad k = 0, 1, \cdots, N.$$

where N is the total number of iterations performed.

The 3×3 system discussed above may be extended to an $n \times n$ system. Thus (7.55) is expressed as

$$x_i^{(k+1)} = \frac{1}{a_{ii}}\left(b_i - \sum_{\substack{j=1 \\ j \neq i}}^{n} a_{ij} x_j^{(k)} \right) \tag{7.56}$$

or in matrix form

$$\begin{bmatrix} x_1 \\ x_2 \\ \vdots \\ x_n \end{bmatrix}_{k+1} = \begin{bmatrix} B_1 \\ B_2 \\ \vdots \\ B_n \end{bmatrix} - \begin{bmatrix} 0 & A_{12} & \cdots & A_{1n} \\ A_{21} & 0 & \cdots & A_{2n} \\ \vdots & \vdots & \ddots & \vdots \\ A_{n1} & A_{n2} & \cdots & 0 \end{bmatrix} \begin{bmatrix} x_1 \\ x_2 \\ \vdots \\ x_n \end{bmatrix}_k \tag{7.57}$$

where $B_i = \dfrac{b_i}{a_{ii}}$, $i = 1, 2, \cdots, n$; $A_{ij} = \dfrac{a_{ij}}{a_{ii}}$, $i, j = 1, 2, \cdots, n$; $k = 0, 1, \cdots, N$; and N is the total number of iterations.

7.5.2.2 Gauss-Seidel Iteration (Successive displacements) : This method converges more rapidly than the Jacobi method under the same condition of diagonal dominance. This method does not guess initially all the unknowns. It requires the initial guessing of $x_2, x_3, \cdots, x_n$. Let the 3×3 system in the Jacobi method with the iteration constant k be

$$\left.\begin{aligned} x_1^{(k+1)} &= \frac{b_1}{a_{11}} - \frac{a_{12}}{a_{11}} x_2^{(k)} - \frac{a_{13}}{a_{11}} x_3^{(k)} \\ x_2^{(k+1)} &= \frac{b_2}{a_{22}} - \frac{a_{21}}{a_{22}} x_2^{(k+1)} - \frac{a_{23}}{a_{22}} x_3^{(k)} \\ x_3^{(k+1)} &= \frac{b_3}{a_{33}} - \frac{a_{31}}{a_{33}} x_2^{(k+1)} - \frac{a_{31}}{a_{33}} x_3^{(k+1)} \end{aligned}\right\} \tag{7.58}$$

Note the difference in the set (7.55) of the Jacobi method and this set (7.58). The initial guess required to find a new x_1 are x_2 and x_3 in the first equation of (7.58). This new value of x_1 (as $x_1^{(k+1)}$) and initial guess for x_3 are used in 2nd equation of (7.58). Finally, the two new values of both x_1 and x_2 are used to find x_3. Thus each newly calculated x_i replaces its previous value in subsequent calculations. The iteration continues until *all* the newly calculated x's converge to within a convergence criterion ε of their previous values.

In general, for an $n \times n$ system, the solution is given by

$$x_i^{(k+1)} = \frac{b_i}{a_{ii}} - \sum_{j=1}^{i-1} \frac{a_{ij}}{a_{ii}} x_j^{(k+1)} - \sum_{j=i+1}^{n} \frac{a_{ij}}{a_{ii}} x_j^{(k)} \tag{7.59}$$

$i, j = 1, 2, \cdots, n$, $k = 1, 2, \cdots, N$ and N is the number of iterations.

Ex 1. Solve

$$\left.\begin{aligned} 5x_1 - 2x_2 + x_3 &= 4 \\ x_1 + 4x_2 - 2x_3 &= 3 \\ x_1 + 2x_2 + 4x_3 &= 17 \end{aligned}\right\}$$

using the Jacobi method of iterations.

Let us check the diagonal dominance or strong diagonal of the coefficient matrix for the convergence of the method. Thus

$$|5| > |-2| + |\ 1| \Rightarrow 5 > 3$$

$$|4| > |\ 1| + |-2| \Rightarrow 4 > 3$$

$$|4| > |\ 1| + |\ 2| \Rightarrow 4 > 3$$

Therefore, the coefficient matrix has a strong diagonal. Obviously the iteration will converge. Therefore, expressing the given system in the form of (7.57) yields

$$\begin{bmatrix} x_1 \\ x_2 \\ x_3 \end{bmatrix}_{k+1} = \begin{bmatrix} 0.8 \\ 0.75 \\ 4.25 \end{bmatrix} - \begin{bmatrix} 0 & -0.4 & 0.2 \\ 0.25 & 0 & -0.5 \\ 0.25 & 0.50 & 0 \end{bmatrix} \begin{bmatrix} x_1 \\ x_2 \\ x_3 \end{bmatrix}_k \quad (1)$$

Let the initial iteration be $[x]_0 = [0]$. Using this initial guess to eqn (1) we obtain,

$$\begin{bmatrix} x_1 \\ x_2 \\ x_3 \end{bmatrix}_1 = \begin{bmatrix} 0.8 \\ 0.75 \\ 4.25 \end{bmatrix} - \begin{bmatrix} 0 & -0.4 & 0.2 \\ 0.25 & 0 & -0.5 \\ 0.25 & 0.50 & 0 \end{bmatrix} \begin{bmatrix} 0 \\ 0 \\ 0 \end{bmatrix}_0 = \begin{bmatrix} 0.8 \\ 0.75 \\ 4.25 \end{bmatrix}_1 \quad (2)$$

Using (2) in (1) gives the second approximation as

$$\begin{bmatrix} x_1 \\ x_2 \\ x_3 \end{bmatrix}_2 = \begin{bmatrix} 0.8 \\ 0.75 \\ 4.25 \end{bmatrix} - \begin{bmatrix} 0 & -0.4 & 0.2 \\ 0.25 & 0 & -0.5 \\ 0.25 & 0.50 & 0 \end{bmatrix} \begin{bmatrix} 0.8 \\ 0.75 \\ 4.25 \end{bmatrix}_1 = \begin{bmatrix} 0.25 \\ 2.675 \\ 3.675 \end{bmatrix}_2$$

This process is continued until successive values of each vector are sufficiently close in magnitude. The first 12 iterations are shown in the table below :

Successive approximation of solution (Jacobi's method)

Iterations

Var	1	2	3	4	5	6	7	8	9	10	11	12
x_1	0.8	0.25	1.14	1.24	1.02	0.92	0.98	1.02	1.01	0.99	0.99	1.00
x_2	0.75	2.68	2.53	1.89	1.79	1.99	2.07	2.02	1.98	1.99	2.01	2.01
x_3	4.25	4.25	2.85	2.70	2.99	3.10	3.02	2.97	2.98	3.01	3.01	2.94

- Note that the actual solutions are $x_1 = 1,\ x_2 = 2,\ x_3 = 3$.

Ex 2. Redo the example 1 using Gauss-Seidel method.

Since x_1 is need not to be initialised as in the Jacobi method, so let $x_2 = 0$ and $x_3 = 0$ be initial guess for x_2 and x_3. Therefore, eqn. (7.59) gives for $k = 0$,

$$x_1^{(1)} = 0.8 + 0 - 0 = 0.8$$

$$x_2^{(1)} = 0.75 - 0.25\ (0.8) + 0 = 0.55$$

$$x_3^{(1)} = 4.25 - 0.25\ (0.8) - 0.5\ (0.55) = 3.775$$

Process continues by using the new $x_2^{(1)}$ and $x_3^{(1)}$ into eqn (7.59), obtaining another set of values. First seven iterations are shown in the table below :

Successive approximation of solution (Gauss-Seidel method)

Iterations

Var	1	2	3	4	5	6	7
x_1	0.8	0.265	1.249	0.956	1.002	1.001	0.999
x_2	0.55	2.571	1.887	2.008	2.003	1.999	2.000
x_3	3.775	2.898	2.994	3.007	2.998	3.000	3.000

- Note that the results obtained in 7 iterations are more accurate than those obtained in 12 iterations in Jacobi's method.

7.6 Evaluation of determinant by Gaussian elimination

The Gaussian elimination is very *useful* in the evaluation or expansion of the determinant of the matrix. The elementary operations used in the Gaussian elimination are consistent with the properties of determinants (sec. 6.4). So the reduction of a matrix to the equivalent matrix by the procedures of Gauss elimination would not *alter* the value of the determinant of the matrix. Since the determinant of a triangular matrix is equal to the product of its diagonal elements; so the determinant converted to triangular form by Gauss elimination can be easily evaluated.

Number of operation :

$$\text{Multiplication / division} = \tfrac{1}{3}(n-1)(n^2+n+3)$$

$$\text{Addition / subtraction} = \tfrac{1}{6}n(n-1)(2n-1)$$

7.7 Condition for unique solution of Ax = b

The system of n equations in n unknowns can be compactly expressed by the matrix equation

$$\mathbf{Ax} = \mathbf{b}\,.$$

This section provides a condition under which the above system possesses a unique solution. The condition is embodied in the following theorem.

Theorem. The system of n equations in n variables, $\mathbf{Ax}=\mathbf{b}$ possesses a unique solution if det A $\neq 0$, i.e. A is nonsingular.

Proof. Since det A $\neq$ 0, then there exists an A^{-1} (existence condition of A^{-1}). Now premultiplying both sides of $\mathbf{Ax}=\mathbf{b}$ by A^{-1}, we have the solution $\mathbf{x} = A^{-1}\mathbf{b}$. Since A^{-1} is unique if A^{-1} exist, by uniqueness of A^{-1}, the solution $\mathbf{x} = A^{-1}\mathbf{b}$ must be unique.

Conversely, since the solution $\mathbf{x} = A^{-1}\mathbf{b}$ of $\mathbf{Ax}=\mathbf{b}$ is unique and A^{-1} is unique, then from existence condition of A^{-1} the det A must not be equate to zero, i.e. det A $\neq$ 0. Hence the theorem is proved.

Cor. The three statements (i) A is nonsingular, (ii)A^{-1} exists and (iii) det A $\neq 0$ are equivalent for square matrix A. Similarly, (I) A is singular, (ii) A^{-1} does not exist and (iii) det A = 0 are equivalent for the square matrix A. Equivalent statements can be used interchangeably wherever needed.

7.8 Linear dependence and independence of vector, matrices and functions

The notion of linear dependence and independence of vectors, matrices and functions are essential for a further study of the system of linear homogeneous and nonhomogeneous equations, ranks, vector space, linear transformations etc.

The theories of linear dependence and independence for vectors, matrices, functions and equations are same. If one theory is known for vectors, the replacement of vectors by matrices, functions or equations gives the equivalent theory for matrices, function or equations.

Linear combination : The sum of the products

$$c_1\mathbf{u}_1 + c_2\mathbf{u}_2 + \cdots + c_n\mathbf{u}_n \tag{7.60}$$

of n scalars $c_1, c_2, \cdots, c_n$ and n vectors $\mathbf{u}_1, \mathbf{u}_2, \cdots, \mathbf{u}_n$ multiplied in order is called a *linear combination* of the n vectors. Thus by trial $\mathbf{u}_1 = \frac{2}{3}\mathbf{u}_2 - \frac{1}{3}\mathbf{u}_3$, where $\mathbf{u}_1 = (1, -1, 1)$, $\mathbf{u}_2 = (2, 0, 1)$, $\mathbf{u}_3 = (1, 3, -1)$ expresses the linear combination of the vectors $\mathbf{u}_2$ & $\mathbf{u}_3$, and scalars $c_1 = \frac{2}{3}$ & $c_2 = \frac{1}{3}$. Trial method is time consuming. So a procedure is needed and that will be obtained soon after the definition of the linear dependence given below. The above linear combination may be written the other way round by simple algebraic manipulation. Thus, $\mathbf{u}_2 = \frac{2}{3}\mathbf{u}_1 + \frac{1}{2}\mathbf{u}_3$, or, $\mathbf{u}_3 = 2\mathbf{u}_2 - 3\mathbf{u}_1$ are two more linear combinations of the same three vectors. Geometrically, these show that one vector lies in the plane of the other two.

Linearly dependent : If scalars $c_1, c_2, \cdots, c_n$ are not all zero but the linear combination (7.60) is zero. Then vectors $\mathbf{u}_i$, $i = 1, 2, \cdots, n$, are called *linearly dependent* i.e.

$$c_1\mathbf{u}_1 + c_2\mathbf{u}_2 + \cdots + c_n\mathbf{u}_n = 0 \tag{7.61}$$

In such case, any one vector can be represented as a linear combination of others. Hence the *alternate* definition of linear dependence is the representation of one of the vectors as a linear combination of the remaining vectors. Note the theorem in the following subsection.

Ex.1 Test whether $\mathbf{u}_1 = (1, -1, 1)$, $\mathbf{u}_2 = (2, 0, 1)$, $\mathbf{u}_3 = (1, 3, -1)$ are linearly dependent or not.

By defn. (7.61), let c_1, c_2, c_3 be three scalars not all zero, such that

$$c_1\mathbf{u}_1 + c_2\mathbf{u}_2 + c_3\mathbf{u}_3 = 0 \tag{1}$$

for linear dependence.

$\therefore$ $$c_1(1, -1, 1) + c_2(2, 0, 1) + c_3(1, 3, -1) = (0, 0, 0)$$

or $$(c_1, -c_1, c_1) + (2c_2, 0, c_2) + (c_3, 3c_3, -c_3) = (0, 0, 0)$$

by scalar multiplication of vectors or row matrices.

or $$(c_1 + 2c_2 + c_3, -c_1 + 3c_3, c_1 + c_2 - c_3) = (0, 0, 0)$$

by addition rule, like components are added.

From the principle of equality of vectors,

$$\left.\begin{aligned} c_1 + 2c_2 + c_3 &= 0 \\ -c_1 \qquad + 3c_3 &= 0 \\ c_1 + c_2 - c_3 &= 0 \end{aligned}\right\} \tag{2}$$

This is a system of 3 homogeneous linear equations. Reducing the system to the echelon form

$$\left.\begin{aligned} c_1+2c_2+\ c_3&=0\\ 2c_2+4c_3&=0\\ c_2+2c_3&=0 \end{aligned}\right\}$$

or
$$\left.\begin{aligned} c_1+2c_2+\ c_3&=0\\ c_2+2c_3&=0 \end{aligned}\right\} \qquad (3)$$

which is in the echelon form and has two equations in three unknowns. So, the system has a nonzero solution, i.e. not all of c_1, c_2, c_3 are zero. Hence the given vectors are linearly dependent. So, any one of them can be expressed as a linear combination of the other two.

Here c_3 is a free variable. So, $c_2 = -2c_3$ and $c_1 = -2c_2 - c_3 = 4c_3 - c_3 = 3c_3$. Therefore, the linear combination is obtained substituting c_1 and c_2 in (1) i.e.

$$3c_3\mathbf{u}_1 - 2c_3\mathbf{u}_2 + c_3\mathbf{u}_3 = 0$$

or $\quad 3\mathbf{u}_1 - 2\mathbf{u}_2 + \mathbf{u}_3 = 0$, dividing by $c_3 \neq 0$ for the nontrivial solution.

Now by algebraic manipulation, either

$$\mathbf{u}_1 = \tfrac{2}{3}\mathbf{u}_2 - \tfrac{1}{3}\mathbf{u}_3 \text{ or, } \mathbf{u}_2 = \tfrac{3}{2}\mathbf{u}_1 + \tfrac{1}{2}\mathbf{u}_3 \text{ or, } \mathbf{u}_3 = 2\mathbf{u}_2 - 3\mathbf{u}_1$$

i.e. any one is expressed as a linear combination of the remaining.

Thus definition of linear dependence provides a *procedure* for expressing one of the dependent vectors in terms of the others.

Observation : If the scalars c_1, c_2, c_3 are eliminated from (2) then

$$\begin{vmatrix} 1 & 2 & 1\\ -1 & 0 & 3\\ 1 & 1 & -1 \end{vmatrix} = 0$$

is obtained. Now it can be said that if $\mathrm{A} = [\mathbf{u}_1 \quad \mathbf{u}_2 \quad \mathbf{u}_3]$ be matrix $\mathbf{u}_i$, $i = 1, 2, 3$, as column vectors then $\det \mathrm{A} = \det \mathrm{A}^{\mathrm{T}} = 0$ is the condition for the linear dependence of the vectors $\mathbf{u}_i$, $i = 1, 2, 3$. Hence in general, if $\mathbf{u}_i$, $i = 1, 2, 3, \cdots, n$, be n vectors each with n components then the condition for their linear dependence is the vanishing of the determinant of the matrix $\mathrm{A} = [\mathbf{u}_1 \quad \mathbf{u}_2 \quad \cdots \quad \mathbf{u}_n]$, i.e. det A = 0, where $\mathrm{A} = [\mathbf{u}_1 \quad \mathbf{u}_2 \quad \cdots \quad \mathbf{u}_n]$.

Ex. 2 Show that equations

$$\left.\begin{aligned} 2x+3y\qquad &=4\\ x-\ y+3z&=1\\ 3x+7y-3z&=7 \end{aligned}\right\}$$

are linearly dependent.

Let

$$\begin{cases} \mathbf{u}_1 \equiv 2x+3y \quad -4 = 0 \\ \mathbf{u}_2 \equiv \ x- \ y+3z-1 = 0 \\ \mathbf{u}_3 \equiv 3x+7y-3z-7 = 0 \end{cases}$$

Now for the linear dependence, let there exist scalars c_1, c_2, c_3 not all zero such that

$$c_1\mathbf{u}_1 + c_2\mathbf{u}_2 + c_3\mathbf{u}_3 = 0$$

or $$c_1(2x+3y-4) + c_2(x-y+3z-1) + c_3(3x+7y-3z-7) = 0$$

This is an identity. So equating coefficients of x, y, z and constants to zero, we have a system of simultaneous equations in c_1, c_2, c_3 as

$$\left.\begin{aligned} 2c_1 + c_2 + 3c_3 &= 0 \\ 3c_1 - c_2 + 7c_3 &= 0 \\ 3c_2 - 3c_3 &= 0 \\ -4c_1 - c_2 - 7c_3 &= 0 \end{aligned}\right\} \quad (1)$$

This is a system of 4 equations with 3 unknowns.
The echelon form of (1) is

$$2c_1 + c_2 + 3c_3 = 0$$

$$c_2 - c_3 = 0$$

with 2 equations and 3 unknowns. So, the system has a nonzero solution and hence the given system is linearly dependent. Therefore, any one equation can be represented as a linear combination of the other two. One such linear combination is
$2\mathbf{u}_1 = \mathbf{u}_2 + \mathbf{u}_3$ i.e. $2(2x+3y-4) = (x-y+3z-1) + (3x+7y-3z-7)$.

Linearly independent : If all the scalars are zero to make the linear combination (7.60) zero then the vectors $\mathbf{u}_1, \mathbf{u}_2, \cdots, \mathbf{u}_n$ are linearly independent.

In this case no vector can be expressed as a linear combination of the others as in the linear dependence of vectors.

Ex. 3 Show that the matrices A , B , C are linearly independent if

$$\mathrm{A} = \begin{bmatrix} 2 & 0 \\ 1 & 3 \end{bmatrix}, \quad \mathrm{B} = \begin{bmatrix} 0 & 1 \\ 1 & 0 \end{bmatrix}, \quad \mathrm{C} = \begin{bmatrix} 1 & 1 \\ 0 & 1 \end{bmatrix}.$$

Let scalars c_1, c_2, c_3 exist such that

$$c_1\mathrm{A} + c_2\mathrm{B} + c_3\mathrm{C} = 0$$

or $$c_1\begin{bmatrix} 2 & 0 \\ 1 & 3 \end{bmatrix} + c_2\begin{bmatrix} 0 & 1 \\ 1 & 0 \end{bmatrix} + c_3\begin{bmatrix} 1 & 1 \\ 0 & 1 \end{bmatrix} = 0$$

or $$\begin{bmatrix} (2c_1 + c_3) & (c_2 + c_3) \\ (c_1 + c_2) & (3c_1 + c_3) \end{bmatrix} = 0$$

which implies that

$$\left.\begin{aligned} 2c_1 + c_3 &= 0 \\ c_2 + c_3 &= 0 \\ c_1 + c_2 &= 0 \\ 3c_1 + c_3 &= 0 \end{aligned}\right\}$$

The echelon form of the system is

$$\left.\begin{aligned} 2c_1 \quad + c_3 &= 0 \\ c_2 + c_3 &= 0 \\ c_3 &= 0 \end{aligned}\right\}$$

This echelon system of homogeneous equation has equal number of equations and unknowns. So, the system has a trivial solution $c_1 = c_2 = c_3 = 0$. Hence the given matrices are linearly independent.

7.8.1 Theorem on linear dependence / independence : The following theorem gives the characterization of linear dependence / independence of vectors discussed above.

Theorem . The vectors $\mathbf{u}_1, \mathbf{u}_2, \cdots, \mathbf{u}_n$ each having n components are linearly dependent if and only if the vector equation $c_1\mathbf{u}_1 + c_2\mathbf{u}_2 + \cdots + c_n\mathbf{u}_n = 0$ has a solution with at least one scalar $c_i \neq 0, i = 1, 2, \cdots, n$.

Proof .

Case of $n = 1$, the vector equation $c_1\mathbf{u}_1 = 0$ has a nonzero solution for c_1 iff $\mathbf{u}_1 = 0$ as the product is zero.

Case of $n > 1$, let the vectors $\mathbf{u}_1, \mathbf{u}_2, \cdots, \mathbf{u}_n$ are linearly dependent. Then one vector say $\mathbf{u}_1$ can be expressed as a linear combination of the others i.e. $\mathbf{u}_1 = k_2\mathbf{u}_2 + k_3\mathbf{u}_3 + \cdots + k_n\mathbf{u}_n$ for some scalars $k_i, i = 2, 3, \cdots, n$. This implies that

$$1\mathbf{u}_1 - k_2\mathbf{u}_2 - k_3\mathbf{u}_3 - \cdots - k_n\mathbf{u}_n = 0$$

Then by letting $c_1 = 1,\ c_i = -k_i,\ i > 1,$ we have a solution of the given vector equation with at least one scalar $c_i \neq 0$ namely $c_1 = 1$.

Conversely, let the vector equation

$$c_1\mathbf{u}_1 + c_2\mathbf{u}_2 + \cdots + c_n\mathbf{u}_n = 0$$

has a solution with at least one $c_i \neq 0$, say , $c_1 \neq 0$. Then

$$\mathbf{u}_1 = -\frac{c_2}{c_1}\mathbf{u}_2 - \frac{c_3}{c_1}\mathbf{u}_3 - \cdots - \frac{c_n}{c_1}\mathbf{u}_n.$$

Thus one vector $\mathbf{u}_1$ is expressed as a linear combination of others. Hence the given vectors are linearly dependent.

Cor. The vectors $\mathbf{u}_1, \mathbf{u}_2, \cdots, \mathbf{u}_n$ are linearly dependent iff the vector equation

$$c_1\mathbf{u}_1 + c_2\mathbf{u}_2 + \cdots + c_n\mathbf{u}_n = 0$$

has only *nontrivial* solution.

Cor. A vector can be represented as a linear combination of the others if its associated scalar is nonzero.

7.9 Decomposition of a matrix into two triangular matrices

The following theorem gives condition for decomposing a matrix into triangular matrices and thereby a procedure for the purpose.

Theorem . Any square matrix $A = [a_{ij}]_{n\times n}$ with nonzero principal diagonal minors $\Delta_1 = a_{11} \neq 0$,

$$\Delta_2 = \begin{vmatrix} a_{11} & a_{12} \\ a_{21} & a_{22} \end{vmatrix} \neq 0, \;\cdots,\; \Delta_n = |A| \neq 0,$$

may be decomposed into two triangular matrices of different structure (lower and upper) if the diagonal elements of the triangular matrices are taken fixed 1's before hand.

Without going into general proof (cf. chapter 14), the method is discussed below.

Let $$A = T_1T_2 \qquad (7.62)$$

$$\text{where } T_1 = [b_{ij}],\; b_{ij} = 0 \text{ when } j > i \qquad (7.63)$$

is a lower triangular matrix of order n ;

$$T_2 = [c_{ij}],\; c_{ij} = 0 \text{ when } i > j \text{ and } c_{ii} = 1 \text{ (say)} \qquad (7.64)$$

Using (7.63) and (7.64) in (7.62), it can be easily solved for elements.

Ex. Decompose the matrix

$$A = \begin{bmatrix} 1 & 2 & 0 \\ 3 & 5 & 6 \\ -2 & 0 & -1 \end{bmatrix}$$

into two triangular matrices (upper and lower) and verify the result by multiplication.

Let

$$A = T_1T_2 = \begin{bmatrix} t_{11} & 0 & 0 \\ t_{21} & t_{22} & 0 \\ t_{31} & t_{32} & t_{33} \end{bmatrix}\begin{bmatrix} 1 & c_{12} & c_{13} \\ 0 & 1 & c_{23} \\ 0 & 0 & 1 \end{bmatrix}$$

or $$\begin{bmatrix} 1 & 2 & 0 \\ 3 & 5 & 6 \\ -2 & 0 & -1 \end{bmatrix} = \begin{bmatrix} t_{11} & t_{11}c_{12} & t_{11}c_{13} \\ t_{21} & t_{21}c_{12}+t_{22} & t_{21}c_{13}+t_{22}c_{23} \\ t_{31} & t_{31}c_{12}+t_{32} & t_{31}c_{13}+t_{32}c_{23}+t_{33} \end{bmatrix}$$

which implies that,

$t_{11} = 1$, $\quad t_{21} = 3,\; t_{31} = -2,\; t_{11}c_{12} = 2$

i.e. $c_{12} = 2$, $\quad t_{11}c_{13} = 0$

i.e. $c_{13} = 0$, $\quad t_{21}c_{12} + t_{22} = 5$

i.e. $t_{22} = 5 - (3)(2) = -1$, $\quad t_{21}c_{13} + t_{22}c_{23} = 6$

i.e. $c_{23} = -6$, $\quad t_{31}c_{12} + t_{32} = 0$

i.e. $t_{32} = 4,\; t_{31}c_{13} + t_{32}c_{23} + t_{33} = -1$

i.e. $t_{33} = 23$.

Therefore, $$T_1 = \begin{bmatrix} 1 & 0 & 0 \\ 3 & -1 & 0 \\ -2 & 4 & 23 \end{bmatrix} \text{ and } T_2 = \begin{bmatrix} 1 & 2 & 0 \\ 0 & 1 & -6 \\ 0 & 0 & 1 \end{bmatrix}.$$

Verification :

$$T_1T_2 = \begin{bmatrix} 1 & 0 & 0 \\ 3 & -1 & 0 \\ -2 & 4 & 23 \end{bmatrix}\begin{bmatrix} 1 & 2 & 0 \\ 0 & 1 & -6 \\ 0 & 0 & 1 \end{bmatrix} = \begin{bmatrix} 1 & 2 & 0 \\ 3 & 5 & 6 \\ -2 & 0 & -1 \end{bmatrix} = A.$$

$\therefore$ verified.

7.10 Analytic construction of magic squares

Constructions of magic squares based on trial and error methods have been surveyed in Ball and Coxeter's "Mathematical Recreations and Essays". But no analytic method for the purpose is known to the author except the one of his own described here. The notation M(n) means magic square of dimension n.

Magic (2) . It consists of four equal numbers.

Proof . Let magic (2) = $[a_{ij}]$, i, j = 1, 2. Let s be the common sum. By definition there are six simultaneous equations:

Row sum $\longrightarrow$ $a_{i1} + a_{i2} = s,\ i = 1, 2.$

Column sum $\longrightarrow$ $a_{1j} + a_{2j} = s,\ j = 1, 2.$

Diagonal sums $\longrightarrow$ $\begin{cases} \sum\limits_{i=1}^{2} a_{ii} = s \\ \sum\limits_{J=1}^{2} a_{2-j+1,\, j} = s \end{cases}$

with four unknowns. So, a unique solution exists and the solution is

$$a_{11} = a_{12} = a_{21} = a_{22} = \tfrac{s}{2}.$$

That is, magic (2) consists of four equal numbers and it is

$$\text{Magic (2)} = \begin{bmatrix} \frac{s}{2} & \frac{s}{2} \\ \frac{s}{2} & \frac{s}{2} \end{bmatrix} = \tfrac{s}{2} J_2 ,$$

for all s — real or complex where J_2 is a second order matrix with unit elements only. This is why, the classical assertion is that magic (2) of distinct four positive consecutive integers 1, 2, 3, 4 does not exists and is not interesting at all.

Magic (3). It consists of nine numbers with a specified common sum. Let s be the common sum and magic (3) = $[a_{ij}]$, i , j = 1 , 2 , 3. By definition, there are 8 simultaneous equations with 9 unknowns as:

Row sum $\longrightarrow$ $a_{i1} + a_{i2} + a_{i3} = s,\ i = 1, 2, 3.$

Column sum $\longrightarrow$ $a_{1j} + a_{2j} + a_{3j} = s,\ j = 1, 2, 3.$

Diagonal sums $\longrightarrow$ $\begin{cases} \sum\limits_{i=1}^{3} a_{ii} = s \\ \sum\limits_{J=1}^{3} a_{3-j+1,\, j} = s \end{cases}$

There exists no unique solution as number of unknown is greater than that of equations. The echelon form of the system is

$$\left.\begin{array}{l}
a_{11}+a_{12}+a_{13}=0\\
a_{12}+a_{13}-a_{21}-a_{31}=0\\
a_{13}-a_{21}-a_{22}-a_{32}=-s\\
a_{21}-a_{22}+a_{31}-a_{33}=0\\
2a_{22}+a_{23}+a_{32}+2a_{33}=2s\\
3a_{23}-2a_{31}+a_{32}+4a_{33}=2s\\
a_{31}+a_{32}+a_{33}=s
\end{array}\right\}$$

The echelon form consists of 7 equations with 9 unknowns taken in row-wise. Therefore, a_{32} and a_{33} are two free variables. These free variables are shown in magic (3) as

$$\text{Magic}(3)=\begin{array}{|c|c|c|}\hline a_{11} & a_{12} & a_{13}\\ \hline a_{21} & a_{22} & a_{23}\\ \hline a_{31} & a_{32} & a_{33}\\ \hline\end{array}$$

Hence the solution in terms of free variables is

$a_{11}=\frac{2}{3}s-a_{33}$, $\quad a_{12}=\frac{2}{3}s-a_{32}$, $\quad a_{13}=-\frac{1}{3}s+a_{32}+a_{33}$,

$a_{21}=-\frac{2}{3}s+(a_{32}+2a_{33})$, $\quad a_{22}=\frac{1}{3}s$, $\quad a_{23}=\frac{4}{3}s-a_{32}-2a_{33}$,

$a_{31}=s-(a_{32}+a_{33})$.

Therefore,

$$\text{Magic }(3)=\begin{array}{|c|c|c|}\hline \frac{2}{3}s-a_{33} & \frac{2}{3}s-a_{32} & -\frac{1}{3}s+s_1\\ \hline -\frac{2}{3}s+s_2 & \frac{1}{3}s & \frac{4}{3}s-s_2\\ \hline s-s_1 & a_{32} & a_{33}\\ \hline\end{array}$$

where $s_1=a_{32}+a_{33}$ and $s_2=a_{32}+2a_{33}$.

Particular cases :

(i) Any number system : The magic (3) obtained above is true for any number systems— rational, irrational, real or complex. For example if $s=3$, $a_{32}=2$ and $a_{33}=3$, then

$$\text{Magic (3)}=\begin{bmatrix}-1 & 0 & 4\\ 6 & 1 & -4\\ -2 & 2 & 3\end{bmatrix}.$$

Similarly, if $s=2$, $a_{32}=a_{33}=-1$ and $s=3+3i$, $a_{32}=0$, $a_{33}=i$, then the respective magic squares are

$$\text{Magic(3)}=\begin{bmatrix}\frac{7}{3} & \frac{7}{3} & -\frac{8}{3}\\ -\frac{13}{3} & \frac{2}{3} & \frac{17}{3}\\ 4 & -1 & -1\end{bmatrix}\quad\text{and}\quad \text{Magic(3)}=\begin{bmatrix}2+i & 2+2i & -1\\ -2 & 1+i & 4+2i\\ 3+2i & 0 & i\end{bmatrix}.$$

(ii) Distinct positive consecutive integers : Conventional magic squares of distinct and positive consecutive integers 1 to 9 with common sum $s=15$ may be formed by solving eight simultaneous equations. A computer program was developed and it was found that the number of such magic squares is 8 as stated in Ball and Coxeter.

(iii) Nonconsecutive positive integers : Arbitrary choice of s divisible by 3 and integer values to a_{32} and a_{33} using conditions in (ii) above, there can be an infinite number of magic (3). For example,

$$s = 24,\ a_{32} = 5,\ a_{33} = 7 \text{ and } s = 30,\ a_{32} = 9,\ a_{33} = 7$$

produce

$$\text{Magic}(3) = \begin{bmatrix} 11 & 9 & 4 \\ 1 & 8 & 15 \\ 12 & 7 & 5 \end{bmatrix} \quad \text{and} \quad \text{Magic}(3) = \begin{bmatrix} 13 & 11 & 6 \\ 3 & 10 & 17 \\ 14 & 9 & 7 \end{bmatrix}$$

which consists of nonconsecutive positive integers 1, 4, 5, 7, 8, 9, 11, 12, 15 and 3, 6, 7, 9, 10, 11, 13, 14, 17 respectively. Note that the numbers are *not* continuously consecutive.

Magic (4) : It consists of 16 numbers with specified common sums . Let Magic (4) = $[a_{ij}]$, $i, j = 1, 2, 3, 4$. By definition, there are 10 equations with 16 unknowns. They are

Row sum $\to$ $\quad a_{i1} + a_{i2} + a_{i3} + a_{i4} = s, \ i = 1, 2, 3, 4.$

column sum $\to$ $\quad a_{1j} + a_{2j} + a_{3j} + a_{4j} = s, \ j = 1, 2, 3, 4.$

$$\text{Diagonal sums} \to \quad \begin{cases} \displaystyle\sum_{i=1}^{4} a_{ii} = s \\ \displaystyle\sum_{J=1}^{4} a_{4-j+1,\, j} = s \end{cases}$$

The echelon form of this system has 7 free variables a_{24}, a_{3i}, a_{4i}, $i = 2, 3, 4$ and the magic square is

Magic (4)

$$= \begin{bmatrix} -s + s_3 & -s + s_3 + s_4 - s_6 & 2s - s_3 - s_4 + s_7 & s - s_5 \\ a_{32} + a_{33} - a_{24} & 2s - s_3 - s_4 + a_{44} & -s + s_3 + a_{44} - a_{32} & a_{24} \\ s - s_1 & a_{32} & a_{33} & a_{34} \\ s - s_2 & a_{42} & a_{43} & a_{44} \end{bmatrix} \text{ where } s_1 =$$

$a_{32} + a_{33} + a_{34}$, $s_2 = a_{42} + a_{43} + a_{44}$, $s_3 = a_{24} + a_{34} + a_{42} + a_{43} + a_{44}$, $s_4 = a_{33} + 2a_{44}$,

$s_5 = a_{24} + a_{34} + a_{44}$, $s_6 = a_{32} + a_{42} + a_{44}$, $s_7 = a_{32} - a_{43} + a_{44}$

A computer program was developed for positive consecutive numbers 1 to 16 with the common sum 34 and it is found that the number of such magic squares is 7040 as stated in Ball and Coxeter.

Proceeding in this way, the number of free variables for magic (n) is found to be $\{n^2 - (2n + 1)\}$ and magic squares, in general, of different orders may be constructed.

Exercises seven

1. Show that the following has a unique solution for all $k \neq -1$, but has no solution for $k = -1$:

$$\left.\begin{aligned} x + \ y + \ kz &= 1 \\ x - \ y - \ \ z &= 2 \\ 2x + \ y - \ 2z &= 3 \end{aligned}\right\}.$$

2. Reduce the following system of equations to an echelon system and find all solutions:

$$\left.\begin{aligned} x_1 - x_2 + 2x_3 &= 5 \\ 2x_1 + x_2 - x_3 &= 2 \\ 2x_1 - x_2 - x_3 &= 4 \\ x_1 + 3x_2 + 2x_3 &= 1 \end{aligned}\right\}.$$

3. Express the following system of equations in the echelon form and solve them:

$$\left.\begin{aligned} x + y \quad &= -3 \\ 2x - 2y - z &= -8 \\ 4x \quad - z &= -14 \\ x - 3y - z &= -5 \end{aligned}\right\}.$$

[**Ans** . $x = -\frac{7}{2} + \frac{a}{4}$, $y = \frac{(2-a)}{4}$, $z = a$]

4. Investigate for what values of λ, μ the system of simultaneous equations

$$\left.\begin{aligned} x + y + z &= 6 \\ x + 2y + 3z &= 1 \\ x + 2y + \lambda z &= \mu \end{aligned}\right\}$$

has (i) no solution, (ii) a unique solution and (iii) an infinite number of solutions.

[**Ans.** (i) $\lambda = 3$, $\mu \neq 10$; (ii) $\lambda \neq 3$, μ may have any value; (iii) $\lambda = 3$, $\mu = 10$]

5. For what values of η the equations

$$\begin{aligned} x + y + z &= 1 \\ x + 2y + 4z &= \eta \\ x + 4y + 10z &= \eta^2 \end{aligned}$$

have a solution and solve them completely in each case. [**Ans.** $\eta = 1$, $\eta = -2$.]

6. Show that the system

$$\left.\begin{aligned} 2x_1 + 3x_2 - x_3 + 3x_4 &= 1 \\ 5x_1 + x_2 + 7x_3 &= -2 \\ -x_1 + 5x_2 - 9x_3 + 6x_4 &= b \end{aligned}\right\}$$

is consistent when $b = 4$ and inconsistent otherwise.

7. Solve the following systems by Gaussian elimination with pivoting. Round off the solutions to 3 significant digits.

(i) $\begin{cases} 0.21x_1 + 0.33x_2 = 0.54 \\ 0.70x_1 + 0.24x_2 = 0.94 \end{cases}$ (ii) $\begin{cases} 0.11x_1 - 0.13x_2 + 0.20x_3 = -0.02 \\ 0.10x_1 - 0.01x_2 + 0.45x_3 = 0.25 \\ 0.50x_1 - 0.01x_2 + 0.30x_3 = -0.70 \end{cases}$

[**Ans** . (i) $x_1 = 0.997$, $x_2 = 1.00$; (ii) $x_1 = -2.00$, $x_2 = 0.00$, $x_3 = 1.00$]

8. Solve by Gauss-Jordan elimination:

(i) $\begin{cases} x_1 - 2x_2 + x_3 - 4x_4 = 1 \\ x_1 + 3x_2 + 7x_3 + 2x_4 = 2 \\ x_1 - 12x_2 - 11x_3 - 16x_4 = 5 \end{cases}$ (ii) $\begin{cases} 3x_1 + 2x_2 - x_3 = -15 \\ 5x_1 + 3x_2 + 2x_3 = 0 \\ 11x_1 + 7x_2 = -30 \end{cases}$

[**Ans** . (i)$x_1 = -4,\ x_2 = 2,\ x_3 = 7$,(ii) Inconsistent]

9. For any value of x_2 and x_3, show that

$$(4 - 2x_2 - 3x_3, x_2, x_3, 5 + x_2 - 2x_3, 6 - 4x_2 - x_3)$$

satisfies the system—

$$\begin{cases} 4x_1 + 2x_2 + 15x_3 + 2x_4 - x_5 = 20 \\ -2x_1 + 3x_2 - 2x_3 + x_4 + 2x_5 = 9 \\ x_1 - 7x_2 + 3x_3 + x_4 - 2x_5 = -3 \end{cases}$$

10. Solve the following system $\mathbf{A}\mathbf{x} = \mathbf{b}$ by Gaussian elimination without pivoting. Check that A = LU.

$$A = \begin{bmatrix} 1 & 1 & -1 \\ 1 & 2 & -2 \\ -2 & 1 & 1 \end{bmatrix}, \quad \mathbf{b} = \begin{bmatrix} 1 \\ 0 \\ 1 \end{bmatrix}.$$

$$[\textbf{Ans}: L = \begin{bmatrix} 1 & 0 & 0 \\ 1 & 1 & 0 \\ -2 & 3 & 1 \end{bmatrix}, U = \begin{bmatrix} 1 & 1 & -1 \\ 0 & 1 & -1 \\ 0 & 0 & 2 \end{bmatrix}, x = \begin{bmatrix} 2 \\ 2 \\ 3 \end{bmatrix}]$$

11. Consider the linear system

$$\begin{cases} 6x_1 + 2x_2 + 2x_3 = -2 \\ 2x_1 + \frac{2}{3}x_2 + \frac{1}{3}x_3 = 1 \\ x_1 + 2x_2 - x_3 = 0 \end{cases}$$

and verify that its solution is $x_1 = 2.6,\ x_2 = -3.8,\ x_3 = -5.0$.

Solve the above system by Gaussian elimination

(a) without pivoting ,

(b) with partial pivoting.

Comment on the procedures.

[**Ans.** (a) $x_1 = 1.335,\ x_2 = 0,\ x_3 = -5.003$;
(b) $x_1 = 2.602,\ x_2 = -3.801,\ x_3 = -5.003$]

12. The system $\mathbf{A}\mathbf{x} = \mathbf{b}$ where

$$A = \begin{bmatrix} 4 & -1 & 0 & -1 & 0 & 0 \\ -1 & 4 & -1 & 0 & -1 & 0 \\ 0 & -1 & 4 & 0 & 0 & -1 \\ -1 & 0 & 0 & 4 & -1 & 0 \\ 0 & -1 & 0 & -1 & 4 & -1 \\ 0 & 0 & -1 & 0 & -1 & 4 \end{bmatrix}, \quad \mathbf{b} = \begin{bmatrix} 2 \\ 1 \\ 2 \\ 2 \\ 1 \\ 2 \end{bmatrix}$$

has the solution $x = [1\ \ 1\ \ 1\ \ 1\ \ 1\ \ 1]^T$. Solve the system using the Jacobi iteration method and then solve it again using the Gauss-Seidel method. Comment on the number of iterations required for the best approximation in two cases.

13. Using the Cholesky method, calculate the decomposition $A = TT^T$ for

$$A = \begin{bmatrix} 2.25 & -3.0 & 4.5 \\ -3.0 & 5.0 & -10.0 \\ 4.5 & -10.0 & 34.0 \end{bmatrix} \qquad \textbf{[Ans. } T = \begin{bmatrix} 1.5 & -2.0 & 3.0 \\ 0 & 1.0 & 4.0 \\ 0 & 0 & 3.0 \end{bmatrix} \textbf{]}$$

14. Try the exercise 13 taking $t_{ii} = 1$, instead of $c_{ii} = 1$, $i = 1, 2, 3$ in sec. 7.9 and verify as exercise 13 also.

15. Represent the following matrix as a product of two triangular matrices T_1 and T_2 (upper and lower).

$$A = \begin{bmatrix} -1 & -1 & 2 \\ -1 & 5 & 4 \\ 2 & 4 & 14 \end{bmatrix}$$

$$\textbf{[Ans. } T_1 = \begin{bmatrix} 1 & 0 & 0 \\ -1 & 4 & 0 \\ 2 & 6 & 1 \end{bmatrix}, \; T_2 = \begin{bmatrix} 1 & -1 & 2 \\ 0 & 1 & \frac{3}{2} \\ 0 & 0 & 1 \end{bmatrix} \textbf{]}$$

16. Show that (1 , 0 , 0), (0 , 1 , 0) and (1 , 1 , 0) are linearly dependent.

17. Show that (1 , 2 , –3) , (2 , 0 , –1) , (7 , 6 , –11) are linearly dependent.

18. Find k, given that (k ,0, 0), (0, k, 0), (0, 0, k) are linearly independent.

[Ans. $k = 0$]

19. Show that (3, 0, 1, –1), (2, –1, 0, 1), (1, 1, 1, –2) are linearly dependent.

20. Show that there are 8 magic (3) of positive numbers 1 to 9 with common sum 15.

21. Are the following vectors linearly dependent
u = (5, –7, 0 , 1, –1) , **v** = (1, –3, –2, 3, –1) & **w** = (4, 0, 7, –9, 2) ? If dependent, find a relation among them. **[Ans.** yes, 3**u** = 7 + 2**w]**

22. Obtain the linear relation between the vectors **u** = (1, 1, –1), **v** = (1, –1, 1), **w** = (–1,1,1), **x** =(4,2 , –3). **[Ans. 6u + v – w –2x = 0]**

23. Determine whether the following sets of constants are linearly dependent or not:

(i) A = {5 , 2 , 1 , 3 , 4}, B = {0 , 3 , 0 , 0 , 8}, C = {15 , 7 , 3 , 9 , 7}. **[Ans.** No**]**

(ii) A = {1 , 0 , 0 , 5}, B = {1 , 2 , 6 , 7} , C = {3 , 1 , 3 , 16}.

[Ans. yes, 5A + B –2C = 0**]**

CHAPTER EIGHT

COFACTOR, ADJOINT AND INVERSE MATRICES

Chapter summary: *Cofactor matrix is defined in classical manner. Developed properties of cofactor matrix. Adjoint matrix and its properties are introduced. Inversion, its existence and uniqueness, algebra of inverses, inverse of rectangular matrix have been dealt with. Inverse by partitioning and by bordering; inverse when an element of a matrix changes are treated. It also include solution of simultaneous equations, Kaufman's inversion of Vandermode matrix, inversion by binomial expansion, Cramer's rule, Helmertian in the strict sense and Leontief input-out open model.*

8.1 Cofactor matrix

Let $A = [a_{ij}]_{n\times n}$, $n > 1$, be a square matrix. Then cofactor matrix of A denoted by cofA is a square matrix obtained by replacing a_{ij} in A by their cofactors A_{ij}. That is,

$$\text{cof A} = \begin{bmatrix} A_{11} & A_{12} & \cdots & A_{1n} \\ A_{21} & A_{22} & \cdots & A_{2n} \\ \vdots & \vdots & \ddots & \vdots \\ A_{n1} & A_{n2} & \cdots & A_{nn} \end{bmatrix} = \left[A_{ij}\right]_{n\times n}. \tag{8.1}$$

Note that if A is 1×1 , i.e., $A = [a_{11}]$, then cof A = [1].

8.1.1 Properties of cofactor matrix : The properties of the cofactor matrix are stated and proved below one after another.

Property 1. $\text{cof A}^{\text{T}} = (\text{cof A})^{\text{T}}$,

i.e. the cofactor matrix of a transposed matrix is the transpose of the cofactor matrix.

Proof. Since $\text{A}^{\text{T}} = \left[a_{ji}\right]$

$\therefore \text{cof A}^{\text{T}} = \left[A_{ji}\right] = (\text{cof A})^{\text{T}}$ by (8.1)

Ex. Find

$$\text{cof}\begin{bmatrix} 2 & -1 & 0 \\ 3 & -2 & 1 \\ -4 & 5 & 7 \end{bmatrix}. \text{ Hence find det}\begin{bmatrix} 2 & -1 & 0 \\ 3 & -2 & 1 \\ -4 & 5 & 7 \end{bmatrix}.$$

Let $$A = \begin{bmatrix} 2 & -1 & 0 \\ 3 & -2 & 1 \\ -4 & 5 & 7 \end{bmatrix}.$$

∴ $$A_{11} = (-1)^{1+1}\begin{vmatrix} -2 & 1 \\ 5 & 7 \end{vmatrix} = -14-5 = -19$$

$$A_{12} = (-1)^{1+2}\begin{vmatrix} 3 & 1 \\ -4 & 7 \end{vmatrix} = -(21+4) = -25$$

$$A_{13} = (-1)^{1+3}\begin{vmatrix} 3 & -2 \\ -4 & 5 \end{vmatrix} = 15-8 = 7$$

$$A_{21} = (-1)^{2+1}\begin{vmatrix} -1 & 0 \\ 5 & 7 \end{vmatrix} = -(-7-0) = 7$$

$$A_{22} = (-1)^{2+2}\begin{vmatrix} 2 & 0 \\ -4 & 7 \end{vmatrix} = (14-0) = 14$$

$$A_{23} = (-1)^{2+3}\begin{vmatrix} 2 & -1 \\ -4 & 5 \end{vmatrix} = -(10-4) = -6$$

$$A_{31} = (-1)^{3+1}\begin{vmatrix} -1 & 0 \\ -2 & 1 \end{vmatrix} = -1+0 = -1$$

$$A_{32} = (-1)^{3+2}\begin{vmatrix} 2 & 0 \\ 3 & 1 \end{vmatrix} = -(2-0) = -2$$

$$A_{33} = (-1)^{3+3}\begin{vmatrix} 2 & -1 \\ 3 & -2 \end{vmatrix} = -4+3 = -1$$

Hence $$\text{cof } A = \begin{bmatrix} -19 & -25 & 7 \\ 7 & 14 & -6 \\ -1 & -2 & -1 \end{bmatrix} \text{ and}$$

$$|A| = 2(-19) - 1(-25) + 0(7) = -38 + 25 = -13.$$

N.B. Verify from the above problem that (i) $\text{cof}(A^T) = (\text{cof } A)^T$,

(ii) $A \text{ cof } A = \begin{bmatrix} -45 & -54 & 20 \\ -62 & -105 & 32 \\ 104 & 156 & -51 \end{bmatrix}$, (iii) $|\text{cof } A| = (-13)^2 = |A|^2$.

Property 2. $A^T \text{ cof } A = |A|\, I$.

Proof. ⟶ ↓

$$A^T \text{ cof } A = \begin{bmatrix} a_{11} & a_{21} & \cdots & a_{n1} \\ a_{12} & a_{22} & \cdots & a_{n2} \\ \vdots & \vdots & \ddots & \vdots \\ a_{1n} & a_{2n} & \cdots & a_{nn} \end{bmatrix} \begin{bmatrix} A_{11} & A_{12} & \cdots & A_{1n} \\ A_{21} & A_{22} & \cdots & A_{2n} \\ \vdots & \vdots & \ddots & \vdots \\ A_{n1} & A_{n2} & \cdots & A_{nn} \end{bmatrix}$$

$$= \begin{bmatrix} |A| & 0 & \cdots & 0 \\ 0 & |A| & \cdots & 0 \\ \vdots & \vdots & \ddots & \vdots \\ 0 & 0 & \cdots & |A| \end{bmatrix}_{n\times n} = |A|I = \text{ a scalar matrix.}$$

N.B. (i) $(\text{cof } A^T) = |A|\, I$, (ii) $A \text{ cof } A \neq 0$ but something more.

Property 3. $|\text{cof } A| = |A|^{n-1}$

Proof.Since $A^T \text{ cof } A = |A|\, I$.

$\therefore$ taking determinant of both sides,

$$\left|A^T \text{ cof } A\right| = |A|^n$$

or $$\left|A^T\right|\left|\text{cof } A\right| = |A|^n$$

or $$|A|\left|\text{cof } A\right| = |A|^n$$

or $$\left|\text{cof } A\right| = |A|^{n-1}$$

Property 4. cof AB = cof A cof B, assuming that AB is defined.

Proof.This is established by induction. For the purpose, begin with 3×3 matrices. Let $A = [a_{ik}]_{3\times3}$ and $B = [b_{kj}]_{3\times3}$ be two three square matrices. Let C = AB.

Therefore, $[c_{ij}]_{3\times3} = \left[\sum_{k=1}^{3} a_{ik} b_{kj}\right]$

Now, $$\text{cof } C = [C_{ij}]_{3\times3} = \begin{bmatrix} C_{11} & C_{12} & C_{13} \\ C_{21} & C_{22} & C_{23} \\ C_{31} & C_{32} & C_{33} \end{bmatrix} \text{ where } C_{11} = \begin{vmatrix} c_{22} & c_{23} \\ c_{32} & c_{33} \end{vmatrix} \text{ etc.}$$

and $\text{cof } A = [A_{ik}]$, $\text{cof } B = [B_{kj}]$.

$$\therefore\ C_{11} = \begin{vmatrix} \sum_{k=1}^{3} a_{2k} b_{k2} & \sum_{k=1}^{3} a_{2k} b_{k3} \\ \sum_{k=1}^{3} a_{3k} b_{k2} & \sum_{k=1}^{3} a_{3k} b_{k3} \end{vmatrix}$$

$$= \left(\sum_{k=1}^{3} a_{2k} b_{k2} \sum_{k=1}^{3} a_{3k} b_{k3} - \sum_{k=1}^{3} a_{3k} b_{k2} \sum_{k=1}^{3} a_{2k} b_{k3} \right)$$

$$= (a_{21}b_{12} + a_{22}b_{22} + a_{23}b_{32})(a_{31}b_{13} + a_{32}b_{23} + a_{33}b_{33})$$
$$- (a_{31}b_{12} + a_{32}b_{22} + a_{33}b_{32})(a_{21}b_{13} + a_{22}b_{23} + a_{23}b_{33})$$

$$= \left(a_{21}b_{12}\left(a_{33}b_{33}+a_{32}b_{23}\right)+a_{22}b_{22}\left(a_{33}b_{33}+a_{31}b_{13}\right)+a_{23}b_{32}\left(a_{32}b_{23}+a_{31}b_{13}\right)\right)$$
$$-\left(a_{31}b_{12}\left(a_{22}b_{23}+a_{23}b_{33}\right)+a_{32}b_{22}\left(a_{21}b_{13}+a_{23}b_{33}\right)+a_{33}b_{32}\left(a_{21}b_{13}+a_{22}b_{23}\right)\right)$$

$$= \left(a_{22}a_{33}-a_{23}a_{32}\right)\left(b_{22}b_{33}-b_{23}b_{32}\right)+\left(a_{21}a_{33}-a_{23}a_{31}\right)\left(b_{12}b_{33}-b_{23}b_{32}\right)$$
$$+\left(a_{21}a_{32}-a_{22}a_{31}\right)\left(b_{12}b_{23}-b_{13}b_{22}\right)$$

$$= \begin{vmatrix} a_{22} & a_{23} \\ a_{32} & a_{33} \end{vmatrix}\begin{vmatrix} b_{22} & b_{23} \\ b_{32} & b_{33} \end{vmatrix} + \begin{vmatrix} a_{21} & a_{23} \\ a_{31} & a_{33} \end{vmatrix}\begin{vmatrix} b_{12} & b_{23} \\ b_{32} & b_{33} \end{vmatrix} + \begin{vmatrix} a_{21} & a_{22} \\ a_{31} & a_{32} \end{vmatrix}\begin{vmatrix} b_{12} & b_{13} \\ b_{22} & b_{23} \end{vmatrix}$$

$$= A_{11}B_{11}+A_{12}B_{21}+A_{13}B_{31} = \sum_{K=1}^{3} A_{1k}B_{k1}$$

Similarly other C_{ij}'s, $i, j = 2, 3$, except C_{11} which is already shown, can be calculated.

$$\therefore \quad \left[C_{ij}\right] = \left[\sum_{k=1}^{3} A_{ik}B_{kj}\right]$$
$$= \left[A_{ik}\right]\left[B_{kj}\right]$$

i.e. cof C = cof A cof B

or, cof AB = cof A cof B

i.e. cofactor matrix of the product of two matrices is the product of two cofactor matrices taken in order and in general cofactor matrix of the product of matrices is the product of cofactor matrices.

i.e. cof (A B ⋯ K) = cof A cof B ⋯ cof K.

Property 5. cof cof A = | A $|^{n-2}$A.

Proof. This is established by induction. First, it will be established for a 3×3 matrix, then for a 4×4 matrix and finally the result is obtained by analogy.

For a 3×3 matrix $A = [a_{ij}]_{3\times 3}$, cof $A = [A_{ij}]_{3\times 3}$

$$\text{Now (cofactor of } A_{11}) = \begin{vmatrix} A_{22} & A_{23} \\ A_{32} & A_{33} \end{vmatrix} = A_{22}A_{33} - A_{23}A_{32}$$
$$= a_{11}|A|, \text{ on simplification.}$$

Similarly, cofactors $A_{ij} = a_{ij}\,|A|$, $i, j = 1, 2, 3$

$\therefore$ cof cof A = | A | $[a_{ij}]_{3\times 3}$ = | A | A.

Derivation of this property for matrices of higher order greater than 3 use the following Hazra's technique of simplification of a difference of the product of two pairs of determinants to a higher order determinant (cf. sec. 6.7.3):

$$\begin{vmatrix} a & b \\ d & e \end{vmatrix}\begin{vmatrix} e & f \\ h & i \end{vmatrix} - \begin{vmatrix} b & c \\ e & f \end{vmatrix}\begin{vmatrix} d & e \\ g & h \end{vmatrix} = e\begin{vmatrix} a & b & c \\ d & e & f \\ g & h & i \end{vmatrix} \tag{8.2}$$

Let $A = [a_{ij}]_{4\times 4}$ and cof $A = [A_{ij}]_{4\times 4}$. Then the cofactor of

$$A_{11} = \begin{vmatrix} A_{22} & A_{23} & A_{24} \\ A_{32} & A_{33} & A_{34} \\ A_{42} & A_{43} & A_{44} \end{vmatrix}$$

$$= A_{22}\left(A_{33}A_{44} - A_{34}A_{43}\right) + A_{32}\left(A_{24}A_{43} - A_{23}A_{44}\right) + A_{42}\left(A_{23}A_{34} - A_{24}A_{33}\right)$$

Now $\quad A_{33}A_{44} - A_{34}A_{43}$

$$= \begin{vmatrix} a_{11} & a_{12} & a_{14} \\ a_{21} & a_{22} & a_{24} \\ a_{41} & a_{42} & a_{44} \end{vmatrix} \begin{vmatrix} a_{11} & a_{12} & a_{13} \\ a_{21} & a_{22} & a_{23} \\ a_{31} & a_{32} & a_{33} \end{vmatrix} - \begin{vmatrix} a_{11} & a_{12} & a_{13} \\ a_{21} & a_{22} & a_{23} \\ a_{41} & a_{42} & a_{43} \end{vmatrix} \begin{vmatrix} a_{11} & a_{12} & a_{14} \\ a_{21} & a_{22} & a_{24} \\ a_{31} & a_{32} & a_{34} \end{vmatrix}$$

$$= \left|\begin{array}{c|cc} a_{44} & a_{41} & a_{42} \\ \hline a_{14} & a_{11} & a_{12} \\ a_{24} & a_{21} & a_{22} \end{array}\right| \left|\begin{array}{cc|c} a_{11} & a_{12} & a_{13} \\ a_{21} & a_{22} & a_{23} \\ \hline a_{31} & a_{32} & a_{33} \end{array}\right| - \left|\begin{array}{cc|c} a_{41} & a_{42} & a_{43} \\ \hline a_{11} & a_{12} & a_{13} \\ a_{21} & a_{22} & a_{23} \end{array}\right| \left|\begin{array}{c|cc} a_{14} & a_{11} & a_{12} \\ a_{24} & a_{21} & a_{22} \\ \hline a_{34} & a_{31} & a_{33} \end{array}\right|$$

$$= \begin{vmatrix} a_{11} & a_{12} \\ a_{21} & a_{22} \end{vmatrix} |\,\mathrm{A}\,|, \text{by } (8.2)$$

Similarly, $A_{24}A_{43} - A_{23}A_{44} = \begin{vmatrix} a_{11} & a_{12} \\ a_{31} & a_{32} \end{vmatrix} |\,\mathrm{A}\,|$ and

$$A_{23}A_{34} - A_{24}A_{33} = \begin{vmatrix} a_{11} & a_{12} \\ a_{41} & a_{42} \end{vmatrix} |\,\mathrm{A}\,|.$$

$\therefore$(cofactor of A_{11})

$$= |\,\mathrm{A}\,| \left\{ A_{22}\left(a_{11}a_{22} - a_{21}a_{12}\right) + A_{32}\left(a_{11}a_{32} - a_{31}a_{12}\right) + A_{42}\left(a_{11}a_{42} - a_{12}a_{41}\right) \right\}$$

$$= |\,\mathrm{A}\,| \left\{ \begin{array}{l} a_{11}\left(A_{12}a_{12} + A_{22}a_{22} + A_{32}a_{32} + A_{42}a_{42}\right) \nearrow 0 \\ \qquad\qquad - a_{12}\left(A_{12}a_{11} + A_{22}a_{21} + A_{32}a_{31} + A_{42}a_{41}\right) \end{array} \right\}$$

$$= |\,\mathrm{A}\,| a_{11} |\,\mathrm{A}\,| = a_{11} |\,\mathrm{A}\,|^2$$

Similarly,

$$(\text{cofactors } A_{ij}) = |\,\mathrm{A}\,|^2 a_{ij}; \quad i, j = 1,2,3,4\,.$$

$$\therefore \ \text{cof cof A} = |\,\mathrm{A}\,|^2 \left[a_{ij}\right]_{4\times4} = |\,\mathrm{A}\,|^2 \mathrm{A}.$$

Proceeding in this way it can be easily established that cof cof A = $|\,\mathrm{A}\,|^{n-2}$A for an n square matrix A.

8.2 Adjugate / adjoint matrix

Let A = $[a_{ij}]_{n\times n}$, $n > 1$, be a square matrix. The adjoint / adjugate matrix is the square matrix obtained by the transpose of the matrix cof A and is denoted by adj A or A^{A}. That is,

$$\text{adj A} = (\text{cof A})^{\mathrm{T}} = \left[A_{ji}\right]$$

The corresponding determinant $|\text{adj A}| = |A_{ji}|$ is called the *adjoint determinant*.

N.B. Some books on determinants or higher algebra use the untransposed adjoint. But for consistency and unification of theory transposed definition for the [illegible]

Ex. Find adj A where

$$A = \begin{bmatrix} 2 & -1 & 0 \\ 3 & -2 & 1 \\ -4 & 5 & 7 \end{bmatrix}.$$

By the previous example in sec. 8.1.1,

$$\text{cof } A = \begin{bmatrix} -19 & -25 & 7 \\ 7 & 14 & -6 \\ -1 & -2 & -1 \end{bmatrix}.$$

Hence $\quad \text{adj} A = (\text{cof} A)^T = \begin{bmatrix} -19 & 7 & -1 \\ -25 & 14 & -2 \\ 7 & -6 & -1 \end{bmatrix}.$

8.2.1 Properties of adjoint matrix : The properties of adjoint matrix A can be embodied by the following theorem.

Theorem . Let adj A be the adjoint of the square matrix A of order $n > 1$. Then for scalar k and for matrices B, C conformable to A for multiplication

(i) $A\,(\text{adj } A) = (\text{adj } A)\,A = |A|\,I$

(ii) $(\text{adj } A)^T = \text{adj } A^T$

(iii) $\text{adj } (kA) = k^{n-1}\,(\text{adj } A)$

(iv) $\text{adj } (AB) = \text{adj } B \;\; \text{adj } A$

$\text{adj } (ABC) = \text{adj } C \;\; \text{adj } B \;\; \text{adj } A$

i.e. adjoint of a product is the product of the adjoints in the reversed order.

(v) $|\text{adj } A| = |A|^{n-1}$

(vi) $\text{adj } (\text{adj } A) = |A|^{n-2} A$

(vii) $|\text{adj } (\text{adj } A)| = |A|^{(n-1)^2}$

Proof .

(i) Since $A = [a_{ij}]_{n\times n}$, therefore, by definition of adj A,

$$A\,(\text{adj } A) = \begin{bmatrix} a_{11} & a_{12} & \cdots & \cdots & \cdots & a_{1n} \\ a_{21} & a_{22} & \cdots & \cdots & \cdots & a_{2n} \\ \vdots & \vdots & \ddots & \vdots & \vdots & \vdots \\ a_{i1} & a_{i2} & \cdots & \ddots & \cdots & a_{in} \\ \vdots & \vdots & \vdots & \vdots & \ddots & \vdots \\ a_{n1} & a_{n2} & \cdots & \cdots & \cdots & a_{nn} \end{bmatrix} \begin{bmatrix} A_{11} & A_{21} & \cdots & A_{j1} & \cdots & A_{n1} \\ A_{12} & A_{22} & \cdots & A_{j2} & \cdots & A_{n2} \\ \vdots & \vdots & \ddots & \vdots & \vdots & \vdots \\ \vdots & \vdots & \vdots & \ddots & \vdots & \vdots \\ \vdots & \vdots & \vdots & \vdots & \ddots & \vdots \\ A_{1n} & A_{2n} & \cdots & A_{jn} & \cdots & A_{nn} \end{bmatrix}$$

$\therefore$ (i, j)th element of the product A (adj A)

$$= a_{i1}A_{j1} + a_{i2}A_{j2} + \cdots + a_{in}A_{jn}$$

$$= \begin{cases} |A|, & \text{if } i = j \\ 0, & \text{if } i \neq j \end{cases}$$

i.e. $$A\,(\text{adj } A) = \begin{bmatrix} |A| & 0 & 0 & \cdots & 0 \\ 0 & |A| & 0 & \cdots & 0 \\ 0 & 0 & |A| & \cdots & 0 \\ \vdots & \vdots & \vdots & \ddots & \vdots \\ 0 & 0 & 0 & \cdots & |A| \end{bmatrix}_{n\times n} = |A|\, I$$

Similarly, {(*i*, *j*)th element of the product (adj A)A}

$$= A_{1i}a_{1j} + A_{2i}a_{2j} + \cdots + A_{ni}a_{nj}$$

$$= \begin{cases} |A|, & \text{if } i = j \\ 0, & \text{if } i \neq j \end{cases}$$

Hence (adj A) A $= |A|\, I$

$\therefore$ A (adj A) = (adj A) A $= |A|\, I =$ a scalar matrix whose diagonal elements are $|A|$.

Ex. Verify the property (i) if

$$A = \begin{bmatrix} 2 & -1 & 0 \\ 3 & -2 & 1 \\ -4 & 5 & 7 \end{bmatrix}.$$

Since $$\text{adj } A = \begin{bmatrix} -19 & 7 & -1 \\ -25 & 14 & -2 \\ 7 & -6 & -1 \end{bmatrix}.$$

Therefore, $$A\,(\text{adj } A) = \begin{bmatrix} 2 & -1 & 0 \\ 3 & -2 & 1 \\ -4 & 5 & 7 \end{bmatrix}\begin{bmatrix} -19 & 7 & -1 \\ -25 & 14 & -2 \\ 7 & -6 & -1 \end{bmatrix}$$

$$= \begin{bmatrix} -13 & 0 & 0 \\ 0 & -13 & 0 \\ 0 & 0 & -13 \end{bmatrix} = (-13)\, I = |A|\, I.$$

Similarly (adj A) A = (−13) I $= |A|\, I$

Hence A (adj A) = (adj A) A $= |A|\, I$ is verified.

(ii) Let $A = [a_{ij}]_{n\times n}$ be a square matrix. Then by definition,

$\text{adj } A = [A_{ji}]$

$\therefore$ $(\text{adj } A)^T = [A_{ij}]$ (8.3)

Also, $A^T = [a_{ji}]_{n\times n}$.

Now by definition, $\text{adj}\,(A^T) = \left[A_{ij}\right] = (\text{adj}\,A)^T$

Hence proved.

(iii) Let $A = \left[a_{ij}\right]_{n\times n}$ be a square matrix. Taking scalar multiplication with k which multiplies each element of the matrix A. Therefore,

$$kA = \left[ka_{ij}\right]_{n\times n} \tag{8.4}$$

To be specific, let $n = 3$, then

$$kA = \begin{bmatrix} ka_{11} & ka_{12} & ka_{13} \\ ka_{21} & ka_{22} & ka_{23} \\ ka_{31} & ka_{32} & ka_{33} \end{bmatrix}$$

Now cofactor of ka_{11} is $\begin{vmatrix} ka_{22} & ka_{23} \\ ka_{32} & ka_{33} \end{vmatrix} = k^2 A_{11} = k^{3-1} A_{11}$

Similarly, the other 8 cofactors can be found and they are $k^2 A_{12}$, $k^2 A_{13}$, $k^2 A_{21}$, $k^2 A_{22}$, $k^2 A_{23}$, $k^2 A_{31}$, $k^2 A_{32}$, and $k^2 A_{33}$. Therefore,

$$\text{adj}\,(kA) = k^2\ \text{adj}\,A$$

In general for (8.4), $\text{adj}\,(kA) = k^{n-1}\text{adj}\,A$.

(iv) $\text{adj}\,AB = (\text{cof}\,AB)^T = \text{cof}\,(AB)^T = \text{cof}\,(B^T A^T)$

$$= \text{cof}\,B^T\ \text{cof}\,A^T \text{ as } \text{cof}\,AB = \text{cof}\,A\ \text{cof}\,B$$

$$= (\text{cof}\,B)^T\ (\text{cof}\,A)^T = \text{adj}\,B\ \text{adj}\,A$$

i.e. adjoint of a product is the product of the adjoints in the reverse order,
i.e. $\text{adj}\,(A\,B\,C \cdots K) = \text{adj}\,K \cdots \text{adj}\,C\ \text{adj}\,B\ \text{adj}\,A$.

Alternate proof of (iv) sec. 8.3.6 under inverse.

(v) By (i) we have $A\,(\text{adj}\,A) = (\text{adj}\,A)\,A = |A|\,I$.

Taking the determinant,

$$|A||\text{adj}\,A| = |\text{adj}\,A||A| = \begin{vmatrix} |A| & 0 & \cdots & 0 \\ 0 & |A| & \cdots & 0 \\ \vdots & \vdots & \ddots & \vdots \\ 0 & 0 & \cdots & |A| \end{vmatrix}_{n\times n} = |A|^n$$

$$\therefore \qquad |\text{adj}\,A| = |A|^{n-1}.$$

(vi) $\text{adj}\,(\text{adj}\,A) = \text{adj}\,(\text{cof}\,A)^T = \text{adj}\,(\text{cof}\,A^T)$

$$= (\text{cof}\,(\text{cof}\,A^T))^T = \text{cof}\,(\text{cof}\,A^T)^T$$

$$= \text{cof}\,\text{cof}\,(A^T)^T = \text{cof}\,\text{cof}\,A = |A|^{n-2} A \text{, by property (iv) of cof A.}$$

(vii) $\left|\text{adj}\,(\text{adj}\,A)\right| = \left\| A \right|^{n-2} A\right| = \left(\left| A \right|^{n-2}\right)^n \left| A \right| = \left| A \right|^{(n-1)^2}$

8.3 Matrix inversion and its methods

Once the matrix inversion is possible, it will be analogous with *reciprocals* or *inverses* of ordinary algebra. Since the reciprocal of $x \neq 0$ is $\frac{1}{x} = x^{-1}$ and $xx^{-1} = 1$ in ordinary algebra. This leads to the definition of matrix inverse of a square matrix and a rectangular matrix.

Methods of matrix Inversion

(1) Inversion of nonsingular matrices using adjoint (cf. sec. 8.3.2).
(2) Inversion of rectangular matrices (cf. sec. 8.3.7).
(3) Inversion from tringular factorization (cf. sec. 8.3.10).
(4) Inversion by partitioning (cf. sec. 8.4).
(5) Inversion by successive bordering with nonsingular blocks (cf. sec. 8.5).
(6) Inversion when an element of a matrix changes (cf. sec. 8.6).
(7) Inversion by binomial expansion (cf. sec.8.9).
(8) Inversion by elementary operations (cf. sec.9.3).
(9) Inversion by Gauss-Jordan elimination (cf. sec. 9.8).
(10) Inversion by bordering with singular blocks (cf. sec. 9. 11).
(11) Inversion by Cayley-Hamilton theorem (cf. sec. 13.29).
(12) Inversion of Vandermode matrix by Tou's method (cf. sec. 13.30).
(13) Inversion of Vandermode matrix by Kaufman's algorihm (cf. sec. 8.8.1).
(14) Inversion by minimal polynomial (cf. sec. 13.4.5).
(15) Inversion by Lagrange interpolation method (cf. sec.17.2.17A).
(16) Inversion by method of costituent idempotents (cf. sec.17.2.17B).
(17) Generalized inverse (cf. chapter 18).

8.3.1. Inverse of a square matrix : A square matrix B is called the *inverse* of another given *square* matrix A and vice-versa if (i) matrices A, B and I are of the same order and (ii) AB = I and BA = I.

Ex. Verify that

$$\begin{bmatrix} -6 & 7 \\ 7 & -8 \end{bmatrix} \text{ is the inverse of } \begin{bmatrix} 8 & 7 \\ 7 & 6 \end{bmatrix}.$$

Let
$$A = \begin{bmatrix} 8 & 7 \\ 7 & 6 \end{bmatrix} \text{ and } B = \begin{bmatrix} -6 & 7 \\ 7 & -8 \end{bmatrix}.$$

Now
$$AB = \begin{bmatrix} 8 & 7 \\ 7 & 6 \end{bmatrix} \begin{bmatrix} -6 & 7 \\ 7 & -8 \end{bmatrix} = \begin{bmatrix} 1 & 0 \\ 0 & 1 \end{bmatrix} = I$$

and
$$BA = \begin{bmatrix} -6 & 7 \\ 7 & -8 \end{bmatrix} \begin{bmatrix} 8 & 7 \\ 7 & 6 \end{bmatrix} = \begin{bmatrix} 1 & 0 \\ 0 & 1 \end{bmatrix} = I.$$

$\therefore$ the equations of definition of inverse are satisfied. So, the problem is verified.

8.3.2 Singular and nonsingular matrices : Since 0A = 0 for all conformable A, so the zero matrix has no inverse. In ordinary algebra, 0 has no inverse. But in matrix algebra, there are matrices though nonzero, have no inverse in the sense of sec. 8.3.1. These matrices are called *singular* matrices. While those possess inverse are called nonsingular or invertible. *Note that* generalized inverse in chapter 18 determines inverses of all matrices— nonsingular, singular and rectangular.

Criteria for singularity :

(i) a row (column) of zeros

(ii) two rows (columns) are proportional.

In other words, if $|A| = 0$, then A is singular. Otherwise it is nonsingular.

8.3.3 Existence of inverse : Let us determine necessary and sufficient condition under which the inverse of a square matrix exists.

Necessary condition : For the purpose, let us consider a second order square matrix and the definition of inverse. Let the second order matrix be

$$A = \begin{bmatrix} a_{11} & a_{12} \\ a_{21} & a_{22} \end{bmatrix}$$

and its inverse be $\quad B = \begin{bmatrix} x & y \\ z & t \end{bmatrix}.$

Now by definition of inverse AB = I. That is,

$$\begin{bmatrix} a_{11} & a_{12} \\ a_{21} & a_{22} \end{bmatrix} \begin{bmatrix} x & y \\ z & t \end{bmatrix} = \begin{bmatrix} 1 & 0 \\ 0 & 1 \end{bmatrix},$$

which breaks up into system of equations

$$\left.\begin{aligned} a_{11}x + a_{12}z &= 1 \\ a_{21}x + a_{22}z &= 0 \\ a_{11}y + a_{12}t &= 0 \\ a_{21}y + a_{22}t &= 1 \end{aligned}\right\} \tag{8.5}$$

The equations are consistent if and only if $\Delta = a_{11}a_{22} - a_{12}a_{21} \neq 0$ which appears as coefficients of the variables x, y, z, t during the ordinary process of elimination. So, assuming, $\Delta = a_{11}a_{22} - a_{12}a_{21} \neq 0$ the solutions are found to be

$$x = \frac{a_{22}}{\Delta}, y = \frac{-a_{12}}{\Delta}, z = \frac{-a_{21}}{\Delta}, t = \frac{a_{11}}{\Delta}.$$

Therefore, the required inverse is

$$B = \begin{bmatrix} \frac{a_{22}}{\Delta} & -\frac{a_{12}}{\Delta} \\ -\frac{a_{21}}{\Delta} & \frac{a_{11}}{\Delta} \end{bmatrix} = \frac{1}{\Delta}\begin{bmatrix} a_{22} & -a_{12} \\ -a_{21} & a_{11} \end{bmatrix} = \frac{B_1}{\Delta} \text{ (say)}$$

To verify the result, .

$$AB = \frac{1}{\Delta}\begin{bmatrix} a_{11} & a_{12} \\ a_{21} & a_{22} \end{bmatrix} \begin{bmatrix} a_{22} & -a_{12} \\ -a_{21} & a_{11} \end{bmatrix}$$

$$= \frac{1}{\Delta}\begin{bmatrix} a_{11}a_{22} - a_{12}a_{21} & 0 \\ 0 & a_{11}a_{22} - a_{12}a_{21} \end{bmatrix} = \begin{bmatrix} 1 & 0 \\ 0 & 1 \end{bmatrix} = I$$

and $$BA = \frac{1}{\Delta}\begin{bmatrix} a_{22} & -a_{12} \\ -a_{21} & a_{11} \end{bmatrix} \begin{bmatrix} a_{11} & a_{12} \\ a_{21} & a_{22} \end{bmatrix}$$

$$= \frac{1}{\Delta}\begin{bmatrix} a_{11}a_{22} - a_{12}a_{21} & 0 \\ 0 & a_{11}a_{22} - a_{12}a_{21} \end{bmatrix} = \begin{bmatrix} 1 & 0 \\ 0 & 1 \end{bmatrix} = I$$

which satisfy the definition of inverse.

Note that Δ is the determinant of the matrix A, i.e.. As $\Delta \neq 0$, the matrix A is nonsingular. Also note that the matrix B_1 is the adjoint of A,

i.e. B_1= adj A and hence $B = \dfrac{\text{adj}A}{|A|}$.

Thus it is clear that if $\Delta = \det A \neq 0$, then A^{-1} exists. (8.6)

$\therefore$ the necessary condition for the square matrix A to have inverse A is "A must be nonsingular".

Sufficient condition: Conversely if A^{-1} exists, then by definition

$$A A^{-1} = I_2$$

or $\det(A A^{-1}) = \det I_2 = 1$

or $\det A \det A^{-1} = 1$

$\because$ both det A and det A^{-1} are numbers, then the product of two numbers is nonzero. So each is nonzero,i.e. det $A \neq 0$, det $A^{-1} \neq 0$. Thus if A^{-1} exists, then

$$\det A \neq 0 \qquad (8.7)$$

$\therefore$ (8.5) and (8.6) together constitute the necessary and sufficient condition for the existence of inverse of A. These observationscan be extended to higher order square matrices and lead to the following theorem.

Theorem . The square matrix A has an inverse if and only if it is nonsingular and the inverse is given by

$$A^{-1} = \frac{\text{adj}A}{|A|}.$$

Note : (i) A^{-1} is square, because adj A is square and A is square.

(ii) Since det A det $A^{-1} = 1$, then det $A^{-1} = \dfrac{1}{\det A} = (\det A)^{-1}$. Here carefully note the two uses of supercript (–1). The left one denotes the determinant of the inverse of a matrix while the right one denotes inverse (reciprocal) of a scalar, the determinant in this case.

(iii) This method of inversion is inefficient because it requires a large number of determinant evaluations.

8.3.4 Uniqueness of inverse : Let the matrix A has an inverse B. Then by definition,

$$AB = I \quad \text{and} \quad BA = I \qquad (8.8)$$

If possible, suppose A has another inverse C. Then by definition of inverse,

$$AC = I \text{ and } CA = I \qquad (8.9)$$

Now post multiplication of the second equation of (8.8) by C we get,

$$(BA)C = IC = C$$

or $B(AC) = C$

or $BI = C$ by first equation of (8.9).

or $\qquad B = C$.

$\therefore$ the two inverses become equal. Hence *a nonsingular square matrix possesses a unique inverse*. In other words, *a nonsingular matrix has exactly one inverse.*

8.3.5 Algebra of inverses : The algebraic properties of inverses are stated below and sequentially proved thereafter.

(i) $(I_n)^{-1} = I_n$.

(ii) If A has an inverse, A^{-1} then inverse of the inverse is the matrix itself i.e. $(A^{-1})^{-1} = A$.

(iii) If A has an inverse and $c \neq 0$ is a scalar then $(cA)^{-1} = c^{-1} A^{-1}$.

(iv) If A and B have inverses and A is conformable to B for multiplication then the product AB has an inverse and it is the product of the inverses taken in reverse order. Thus $(AB)^{-1} = B^{-1} A^{-1}$.

(v) If A has an inverse, so does A^T and $(A^T)^{-1} = (A^{-1})^T$.

(vi) If A and B are $n \times n$ matrices and either A or B is singular, then AB and B A are singular.

(vii) Cancellation law : AB = AC implies B = C if A^{-1} exists.

(viii) If A^{-1} exists and AB = 0 then B = 0.

(ix) For nonsingular matrices A and B, $(A \otimes B)^{-1} = A^{-1} \otimes B^{-1}$.

Proofs .

(i) Note that $\qquad I_n I_n = I_n$.

$\therefore \qquad (I_n)^{-1} = I_n$ by definition of inverse.

(ii) Let $A = [a_{ij}]_{n \times n}$ be a nonsingular matrix. Let A_{ij} denote the cofactors of a_{ij}. Then

$$\text{adj } A = \left[A_{ji}\right]_{n \times n}.$$

Hence by theorem in sec. 8.3.3, $\qquad A^{-1} = \dfrac{\text{adj}A}{|A|}$

$$\therefore \qquad AA^{-1} = \frac{1}{|A|}\left[a_{ij}\right]_{n \times n}\left[A_{ji}\right]_{n \times n} = \frac{1}{|A|}\begin{bmatrix} |A| & 0 & 0 & \cdots & 0 \\ 0 & |A| & 0 & \cdots & 0 \\ \vdots & \vdots & \vdots & \vdots & \vdots \\ 0 & 0 & 0 & \cdots & |A| \end{bmatrix} = I_n .$$

Similarly, $A^{-1}A = I_n$, thus the two equations for the definition of inverse are satisfied by A^{-1} and hence the inverse of (A^{-1}) is A i.e. in symbol $(A^{-1})^{-1} = A$, which proofs the assertion.

(iii) Let $A = [a_{ij}]_{n \times n}$ be a nonsingular square matrix and 'c' be a scalar. Let A_{ij} denote the cofactors of a_{ij}. Then

$$cA = c\left[a_{ij}\right]_{n \times n} \text{ and } \text{adj}(cA) = c^{n-1}\left[A_{ji}\right]_{n \times n}. \text{ Also } |cA| = c^n|A|.$$

Therefore,

$$(cA)^{-1} = \frac{\text{adj}(cA)}{|cA|} = \frac{c^{n-1}\left[A_{ji}\right]_{n \times n}}{c^n|A|} = \frac{1}{c}\frac{\left[A_{ji}\right]_{n \times n}}{|A|} = c^{-1}\frac{\text{adj}A}{|A|} = c^{-1}A^{-1}.$$

(iv) Let $A = [a_{ij}]_{n \times n}$ and $B = [b_{jk}]_{n \times n}$ be two nonsingular square matrices of the same order n. Then A is conformable to B for multiplication and the product AB is a nonsingular square matrix of order n. So, by definition of inverse,

$$(AB)(B^{-1}A^{-1}) = A(BB^{-1})A^{-1} = AIA^{-1} = AA^{-1} = I$$

and $(B^{-1}A^{-1})(AB) = B^{-1}(A^{-1}A)B = B^{-1}IB = B^{-1}B = I$

$\therefore$ $B^{-1}A^{-1}$ is the inverse of AB

i.e. symbolically,

$$(AB)^{-1} = B^{-1}A^{-1} = \text{(product of the inverses in reverse order)}.$$

Similarly, this can be extended to any finite number of matrices. Thus

$$(ABC \cdots K)^{-1} = K^{-1} \cdots C^{-1}B^{-1}A^{-1}.$$

(v) Using definition of inverse, we have, $AA^{-1} = I$. Taking transpose of both sides,

$$(AA^{-1})^T = I \text{ or } (A^{-1})^T A^T = I \tag{8.10}$$

Similarly, $A^{-1}A = I$ or $(A^{-1})^T A^T = I$ (8.11)

From (8.10) and (8.11) it is clear that the inverse of A^T is $(A^{-1})^T$,

i.e. in symbol, $(A^T)^{-1} = (A^{-1})^T$ which proves the identity.

(vi) Either A or B is singular means, by definition, $|A| = 0$ or $|B| = 0$.

Now $|AB| = |A||B| = 0$ as either $|A| = 0$ or $|B| = 0$. So, AB is singular. Similarly BA is singular.

(vii) $A^{-1}(AB) = A^{-1}(AC)$

or $(A^{-1}A)B = (A^{-1}A)C$

or $IB = IC$

or $B = C$.

Thus cancellation law holds true if the inverse of common factor exists.

(ix) $A^{-1}(AB) = A^{-1}B$,

or $IB = 0$,

or $B = 0$.

(x) Let $A = [a_{ij}]_{n\times n}$ and $B = [b_{ij}]_{m\times m}$. Also let $A^{-1} = [\hat{a}_{ij}]$.

Now

$$(A^{-1} \otimes B^{-1})(A \otimes B) = (\hat{a}_{ij} B^{-1})(a_{ij}B)$$

$$= \sum_{k=1}^{n} \hat{a}_{ik} a_{kj} B^{-1}B$$

$$= \sum_{k=1}^{n} \hat{a}_{ik} a_{kj} I_n = I_{mn}.$$

Hence $A^{-1} \otimes B^{-1} = (A \otimes B)^{-1}$. Note the difference with (iv).

Alternatively, $(A\otimes B)(A^{-1} \otimes B^{-1}) = AA^{-1} \otimes BB^{-1}$

$$= I_n \otimes I_m = I_{mn}.$$

$\therefore$ $(A\otimes B)^{-1} = A^{-1} \otimes B^{-1}$.

8.3.6 Laws of exponent (negative powers) : In sec.3.2 laws of exponents for nonnegative integers were discussed. In this section laws of negative exponents will be discussed. Precondition for the purpose is that the matrix must be nonsingular. The following theorem describes the laws of exponent for negative integers.

Theorem . If A has an inverse and if *m* and *n* are integers (either positive, negative or zero), then

I. $A^m A^n = A^{m+n}$

II. $(A^m)^n = (A^n)^m = A^{mn}$

III. $(A^T)^m = (A^m)^T$

Cor. (i) In II, let $n = -1$,

then $(A^m)^{-1} = (A^{-1})^m = A^{-m}$ for any m (+ ve , – ve or zero).

Cor. (ii) In I, let $m = \pm 1$ and $n = \pm 1$, then

$$A^{\pm}A^{\mp} = A^0 = 1 .$$

Cor. (iii) From III, $(\overline{A})^{-1} = \overline{(A^{-1})}$

and $(A^*)^{-1} = (A^{-1})^*$

which can be easily proved.

Cor. (iv) If $BAA^* = CAA^*$ then $BA = CA$.

Proof. Since $BAA^* = CAA^*$, so postmultiplying both sides by $(A^*)^{-1}$ we have ,

$$BAA^*(A^*)^{-1} = CAA^*(A^*)^{-1}$$

or $BAA^*(A^{-1})^* = CAA^*(A^{-1})^*$ by cor. (iii)

or $BA(AA^{-1})^* = CA(AA^{-1})^*$

or $BAI^* = CAI^*$

or $BA = CA$.

8.3.7 Inverse of a rectangular matrix : In case of rectangular matrix, two inverses are defined .

(i) Left inverse : If A is an $m \times n$ matrix and if A_L is an $n \times m$ matrix such that $A_L A = I_n$, then A_L is called a *left inverse* of A.

(ii) Right inverse: If A is an $m \times n$ matrix and if A_R is $n \times m$ matrix such that $AA_R = I_m$, then A_R is called a *right inverse* of A.

Note that rectangular matrix may not have both right and left inverse. The procedures for finding these inverses are illustrated below with examples.

Ex. Find the inverse of the matrix

$$\begin{bmatrix} 2 & 1 & 1 \\ 1 & 2 & 1 \end{bmatrix}.$$

The given matrix is a 2×3 rectangular matrix. Let the left inverse of A be $A_L = [a_{ij}]$, $i = 1$, 2 , 3 and $j = 1, 2$. Therefore, by definition

$$A_L A = I_3$$

i.e.
$$\begin{bmatrix} a_{11} & a_{12} \\ a_{21} & a_{22} \\ a_{31} & a_{32} \end{bmatrix} \begin{bmatrix} 2 & 1 & 1 \\ 1 & 2 & 1 \end{bmatrix} = \begin{bmatrix} 1 & 0 & 0 \\ 0 & 1 & 0 \\ 0 & 0 & 1 \end{bmatrix}$$

or
$$2a_{11} + a_{12} = 1,\ a_{11} + 2a_{12} = 0\ , a_{11} + a_{12} = 0 \quad \text{(a)}$$
$$2a_{21} + a_{22} = 0,\ a_{21} + 2a_{22} = 1,\ a_{21} + a_{22} = 0 \quad \text{(b)}$$
$$2a_{31} + a_{32} = 0\ ,\ a_{31} + 2a_{32} = 0\ ,\ a_{31} + a_{32} = 1 \quad \text{(c)}$$

From (a) , $a_{11} = -a_{12}$; $\therefore a_{12} = -1$

$$a_{11} = -2a_{12}\ , \quad \therefore a_{12} = -\tfrac{1}{2}$$

which is absurd because this proves the absurdity $-1 = -\frac{1}{2}$. Therefore, the nine equations in (a), (b), (c) have no solutions. Thus A has no left inverse.

Now we can try for right inverse. Let

$$A_R = \begin{bmatrix} b_{11} & b_{12} \\ b_{21} & b_{22} \\ b_{31} & b_{32} \end{bmatrix}.$$

So, by definition, $AA_R = I_2$, $\begin{bmatrix} 2 & 1 & 1 \\ 1 & 2 & 1 \end{bmatrix} \cdot \begin{bmatrix} b_{11} & b_{12} \\ b_{21} & b_{22} \\ b_{31} & b_{32} \end{bmatrix} = \begin{bmatrix} 1 & 0 \\ 0 & 1 \end{bmatrix}$.

or $\left.\begin{aligned} 2b_{11} + b_{21} + b_{31} &= 1 \\ b_{11} + 2b_{21} + b_{31} &= 0 \end{aligned}\right\}$ and $\left.\begin{aligned} 2b_{12} + b_{22} + b_{32} &= 0 \\ b_{12} + 2b_{22} + b_{32} &= 1 \end{aligned}\right\}$

or $\left.\begin{aligned} 2b_{11} + b_{21} + b_{31} &= 1 \\ -3b_{21} - b_{31} &= 1 \end{aligned}\right\}$ and $\left.\begin{aligned} 2b_{12} + b_{22} + b_{32} &= 0 \\ -3b_{22} - b_{32} &= -2 \end{aligned}\right\}$

or $\left.\begin{aligned} 2b_{11} + b_{21} + b_{31} &= 1 \\ 3b_{21} + b_{31} &= -1 \end{aligned}\right\}$ and $\left.\begin{aligned} 2b_{12} + b_{22} + b_{32} &= 0 \\ 3b_{22} + b_{32} &= 2 \end{aligned}\right\}$

which are the echelon forms for the set of equations. So, here b_{31} and b_{32} are free variables for the sets. So, any value of b_{31} and b_{32} satisfying the above sets will yield a right inverse. Thus, A has an infinite number of right inverses. Taking $b_{31} = b_{32} = 0$, and $b_{31} = 0$, $b_{32} = 1$, two of the right inverses are respectively

$$A_R = \begin{bmatrix} \frac{2}{3} & -\frac{1}{3} \\ -\frac{1}{3} & \frac{2}{3} \\ 0 & 0 \end{bmatrix} \text{ and } A_R = \begin{bmatrix} \frac{2}{3} & -\frac{2}{3} \\ -\frac{1}{3} & \frac{1}{3} \\ 0 & 1 \end{bmatrix}.$$

Verifications :

(i) $$AA_R = \begin{bmatrix} 2 & 1 & 1 \\ 1 & 2 & 1 \end{bmatrix} \begin{bmatrix} \frac{2}{3} & -\frac{1}{3} \\ -\frac{1}{3} & \frac{2}{3} \\ 0 & 0 \end{bmatrix} = \begin{bmatrix} 1 & 0 \\ 0 & 1 \end{bmatrix}.$$

(ii) $$AA_R = \begin{bmatrix} 2 & 1 & 1 \\ 1 & 2 & 1 \end{bmatrix} \begin{bmatrix} \frac{2}{3} & -\frac{2}{3} \\ -\frac{1}{3} & \frac{1}{3} \\ 0 & 1 \end{bmatrix} = \begin{bmatrix} 1 & 0 \\ 0 & 1 \end{bmatrix}.$$

Procedures :

(1) Set equations $AA_R = I_m$ and $A_LA = I_n$.

(2) Reduce the corresponding simultaneous equations in echelon form.

(3) If the equations are consistent, then the requisite inverse exists.

(4) Find some of the inverses with some chosen value of the free variable(s).

Theorem . If an $m \times n$ matrix A has both a left inverse A_L and a right inverse A_R, then

(i) $A_L = A_R$

(ii) A_L is unique.

Proof. By definition, $A_L A = I_n$ (8.12)

$$A A_R = I_m \quad (8.13)$$

Post multiplying both sides of (8.12) by A_R, yields

$$(A_L A)A_R = I_n A_R$$

or $A_L(AA_R) = A_R$, using associative law on the left.

or $A_L I_m = A_R$, by (8.13)

or $A_L = A_R$ (8.14)

∴ (i) is proved.

Now if there is another left inverse, say, $\hat{A}_L$ then proceeding as above,

$$\hat{A}_L = A_R \quad (8.15)$$

will be obtained.

∴ combining (8.14) and (8.15), $\hat{A}_L = A_L$

Hence A_L is unique. Thus (ii) is proved.

N.B. If a square matrix A has both A_L & A_R then the common matrix is unique and it is A^{-1}.

8.3.8 Alternate proofs : some properties of adjoint matrix : Alternate proofs of properties (iv), (vi) and (vii) of adjoint matrix in sec. 8.2.1. are given with the help of inverse and this is nothing but a back gearing process because the inverse matrix is defined in terms of the adjoint.

(a) **Proof of (iv)** of sec.8.2.1

Since $(AB)^{-1} = B^{-1} A^{-1}$,

then by definition of inverse

$$\frac{\text{adj}AB}{|AB|} = \frac{\text{adj}B}{|B|}\frac{\text{adj}A}{|A|}$$

or $\text{adj}(AB) = \text{adj}B \ \text{adj}A$, $\because \ |AB| = |A||B| = |B||A|$

i.e. adjoint of a product of matrices is the product of the adjoints of the individual matrices in reverse order.

i.e. $\text{adj}(A\,B\,C \cdots K) = \text{adj}K \cdots \text{adj}C \ \text{adj}B \ \text{adj}A$.

(b) **Proof of (vi)** of sec.8.2.1

Since $(A^{-1})^{-1} = A$

$$\therefore \quad \left(\frac{\text{adj}A}{|A|}\right)^{-1} = A$$

or $\frac{1}{|A|^{-1}}(\text{adj}A)^{-1} = A$ as $(cA)^{-1} = c^{-1}A^{-1}$

or $(\text{adj}A)^{-1} = |A|^{-1} A$

or $\frac{\text{adj}(\text{adj}A)}{|\text{adj}A|} = |A|^{-1} A$

or $\text{adj}(\text{adj}A) = |\text{adj}A||A|^{-1} A = |A|^{n-2} A$, by (v) of sec .8.2.1.

(c) **Proof of (vii)** of sec.8.2.1

The determinant of (vi) is given by

$$|\text{adj}(\text{adj}A)| = |A|^{n-2}|A| \qquad \text{as} \quad |kA| = k^n|A| = |A|^{(n-1)^2}$$

8.3.9 Left and right identity : If $AB = B$, but $A \neq I$, then A is called the *left identity* for B and if $BC = B$, $C \neq I$, then C is called the *right identity* for B.

Ex. Show that the matrix

$$B = \begin{bmatrix} 0 & 0 & 0 \\ 2 & -1 & 3 \\ 4 & 1 & 6 \end{bmatrix}$$

does not have an inverse and that every matrix

$$A = \begin{bmatrix} a & 0 & 0 \\ b & 1 & 0 \\ c & 0 & 1 \end{bmatrix}$$

is a left identity for B. Find a right identity matrix C for B that is different from I_3.

First part :

Here

$$|B| = \det\begin{bmatrix} 0 & 0 & 0 \\ 2 & -1 & 3 \\ 4 & 1 & 6 \end{bmatrix} = 0.$$

So, matrix B is singular and hence B has no inverse.

Second part :

Now

$$AB = \begin{bmatrix} a & 0 & 0 \\ b & 1 & 0 \\ c & 0 & 1 \end{bmatrix}\begin{bmatrix} 0 & 0 & 0 \\ 2 & -1 & 3 \\ 4 & 1 & 6 \end{bmatrix} = \begin{bmatrix} 0 & 0 & 0 \\ 2 & -1 & 3 \\ 4 & 1 & 6 \end{bmatrix} = B, \text{ (given).}$$

$\therefore \quad A \neq I_3$ is a left identity for B, by definition .

Third part :

Let

$$C = \begin{bmatrix} c_{11} & c_{12} & c_{13} \\ c_{21} & c_{22} & c_{23} \\ c_{31} & c_{32} & c_{33} \end{bmatrix}.$$

Let BC = B for right identity for B.

Therefore,

$$\begin{bmatrix} 0 & 0 & 0 \\ 2 & -1 & 3 \\ 4 & 1 & 6 \end{bmatrix}\begin{bmatrix} c_{11} & c_{12} & c_{13} \\ c_{21} & c_{22} & c_{23} \\ c_{31} & c_{32} & c_{33} \end{bmatrix} = \begin{bmatrix} 0 & 0 & 0 \\ 2 & -1 & 3 \\ 4 & 1 & 6 \end{bmatrix},$$

or

$$\begin{cases} 2c_{11} - c_{21} + 3c_{31} = 2 & \text{(i)} \\ 2c_{12} - c_{22} + 3c_{32} = -1 & \text{(ii)} \\ 2c_{13} - c_{23} + 3c_{33} = 3 & \text{(iii)} \end{cases}$$

$$\begin{cases} 4c_{11} + c_{21} + 6c_{31} = 2 & \text{(iv)} \\ 4c_{12} + c_{22} + 6c_{32} = 1 & \text{(v)} \\ 4c_{13} + c_{23} + 6c_{33} = 3 & \text{(vi)} \end{cases}$$

There are six nonhomogeneous simultaneous linear equations with nine variables. So, solution will contain three arbitrary constants or free variables .Thus a_{31}, a_{32}, and a_{33} are chosen as free variables respectively for three pairs of equations (i), (iv); (ii), (v); and (iii),(vi) in echelon form. Thus choosing $a_{31} = 2$, $a_{32} = 0$ and $a_{33} = 1$, other elements can be easily obtained from each pairs of equations and thus

$$C = \begin{bmatrix} -2 & 0 & 0 \\ 0 & 1 & 0 \\ 2 & 0 & 1 \end{bmatrix} \neq I_3$$

is the required one of the right identities.

8.3.10 Inverse from LU factorization : If a matrix A can be factored into a lower triangular matrix L and an upper triangular matrix U, then inverse A^{-1} can be obtained easily. Let

$$A = LU$$

$$\therefore \quad A^{-1} = U^{-1}L^{-1}$$

Ex. Find the inverse of

$$A = \begin{bmatrix} 1 & -1 & 2 \\ -1 & 5 & 4 \\ 2 & 4 & 14 \end{bmatrix}$$

using LU factorization.

Let $A = LU$

$$= \begin{bmatrix} 1 & 0 & 0 \\ -1 & 4 & 0 \\ 2 & 6 & 1 \end{bmatrix} \begin{bmatrix} 1 & -1 & 2 \\ 0 & 1 & \frac{3}{2} \\ 0 & 0 & 1 \end{bmatrix}.$$

$$\therefore \quad A^{-1} = U^{-1}L^{-1} = \begin{bmatrix} 1 & 1 & -\frac{7}{2} \\ 0 & 1 & -\frac{3}{2} \\ 0 & 0 & 1 \end{bmatrix} \begin{bmatrix} 1 & 0 & 0 \\ \frac{1}{4} & \frac{1}{4} & 0 \\ -\frac{7}{2} & -\frac{3}{2} & 1 \end{bmatrix} = \begin{bmatrix} \frac{27}{2} & \frac{11}{2} & -\frac{7}{2} \\ \frac{11}{2} & \frac{5}{2} & -\frac{3}{2} \\ -\frac{7}{2} & -\frac{3}{2} & 1 \end{bmatrix}.$$

Verification :

$$A^{-1}A = \begin{bmatrix} \frac{27}{2} & \frac{11}{2} & -\frac{7}{2} \\ \frac{11}{2} & \frac{5}{2} & -\frac{3}{2} \\ -\frac{7}{2} & -\frac{3}{2} & 1 \end{bmatrix} \begin{bmatrix} 1 & -1 & 2 \\ -1 & 5 & 4 \\ 2 & 4 & 14 \end{bmatrix} = \begin{bmatrix} 1 & 0 & 0 \\ 0 & 1 & 0 \\ 0 & 0 & 1 \end{bmatrix} = I$$

8.4 Computation of inverse by partitioning

The inversion of nonsingular matrix can be performed effectively using partitioning technique. Variations of this method have been used in numerical procedures for inverting matrix on digital computers.

In this method an nth order square matrix is partitioned into *four* blocks as

$$R = \left[\begin{array}{c|c} A & B \\ \hline C & D \end{array}\right] \tag{8.16}$$

where A and D are square matrices of orders p and q respectively such that $p + q = n$. Then evidently B and C are rectangular matrices of orders $p \times q$ and $q \times p$ respectively.

Let the inverse matrix R^{-1} be also in partitioned form as

$$R^{-1} = \left[\begin{array}{c|c} K & L \\ \hline M & N \end{array}\right] \tag{8.17}$$

where K and N are square matrices of orders p and q respectively. Obviously L and M are of dimensions $p \times q$ and $q \times p$ respectively. We have to find K, L, M, N in terms of A, B, C, D whence R^{-1} will be known.

Let D has an inverse D^{-1}, which is known. Since $RR^{-1} = I_n$, then from (8.16) and (8.17)

$$\left[\begin{array}{c|c} A & B \\ \hline C & D \end{array}\right] \left[\begin{array}{c|c} K & L \\ \hline M & N \end{array}\right] = \left[\begin{array}{c|c} I_p & 0 \\ \hline 0 & I_q \end{array}\right] \tag{8.18}$$

or

$$\left[\begin{array}{c|c} AK + BM & AL + BN \\ \hline CK + DM & CL + DN \end{array}\right] = \left[\begin{array}{c|c} I_p & 0 \\ \hline 0 & I_q \end{array}\right]$$

Equating blocks of two equal partitioned matrices, we have

$$AK + BM = I_p \tag{8.19}$$

$$AL + BN = 0 \tag{8.20}$$

$$CK + DM = 0 \tag{8.21}$$

$$CL + DN = I_q \tag{8.22}$$

Since by hypothesis, D^{-1} exists and is known, so from (8.21)

$$M = -D^{-1}CK \tag{8.23}$$

Using (8.23) in (8.19), we get

$$AK - B\,D^{-1}CK = I_p$$

or

$$(A - B\,D^{-1}C)K = I_p$$

or

$$K = (A - BD^{-1}C)^{-1}, \tag{8.24}$$

by definition of inverse,

Also from (8.22),

$$DN = I_q - CL$$

or
$$N = D^{-1} - D^{-1}CL, \tag{8.25}$$
as D^{-1} exists by hypothesis,

Now from (8.20),
$$AL + B(D^{-1} - D^{-1}CL) = 0$$
or
$$AL + BD^{-1} - BD^{-1}CL = 0$$
or
$$(A - BD^{-1}C)L = -BD^{-1}$$
or
$$L = -(A - BD^{-1}C)^{-1} BD^{-1}$$
$$= -KBD^{-1}, \quad \text{by (8.24)}, \tag{8.26}$$

$\therefore$ using (8.23), (8.24), (8.25), (8.26) we have

$$R^{-1} = \left[\begin{array}{c|c} (A - BD^{-1}C)^{-1} & -KBD^{-1} \\ \hline -D^{-1}CK & D^{-1} - D^{-1}CL \end{array}\right] \tag{8.27}$$

The formula (8.27) shows that the partitioning method is conveniently applied when any diagonal block is easily inverted.

Note that the inverse of

$$R = [a_{11}] \text{ is } R^{-1} = \left[\frac{1}{a_{11}}\right], \; a_{11} \neq 0.$$

Similarly, the inverse of a 2×2 matrix is easy to find. If D is *not so easy* then find D^{-1} separately using the above procedure after partitioning D into 4 blocks and then compute R^{-1}. This procedure may be repeated for a large matrix any number of times.

Ex. Compute by partitioning the inverse of the matrix

$$\begin{bmatrix} 4 & 2 & 1 & 8 \\ 7 & 9 & 4 & 3 \\ 1 & 0 & 5 & 2 \\ 6 & 6 & 1 & 7 \end{bmatrix}.$$

The only possible partitioning to keep the main diagonal blocks square is

$$R = \left[\begin{array}{cc|cc} 4 & 2 & 1 & 8 \\ 7 & 9 & 4 & 3 \\ \hline 1 & 0 & 5 & 2 \\ 6 & 6 & 1 & 7 \end{array}\right] = \left[\begin{array}{c|c} A & B \\ \hline C & D \end{array}\right].$$

Now,
$$D = \begin{bmatrix} 5 & 2 \\ 1 & 7 \end{bmatrix}, \quad \therefore \operatorname{cof} D = \begin{bmatrix} 7 & -1 \\ -2 & 5 \end{bmatrix}$$

and
$$\operatorname{adj} D = [\operatorname{cof} D]^{T} = \begin{bmatrix} 7 & -2 \\ -1 & 5 \end{bmatrix}. \text{ Also } |D| = 33 \neq 0.$$

$$\therefore \quad D^{-1} = \frac{\operatorname{adj} D}{|D|} = \begin{bmatrix} \frac{7}{33} & -\frac{2}{33} \\ -\frac{1}{33} & \frac{5}{33} \end{bmatrix}. \quad \text{Let} \quad R^{-1} = \left[\begin{array}{c|c} K & L \\ \hline M & N \end{array}\right].$$

Now $K = \left(A - BD^{-1}C\right)^{-1} = \left[\begin{bmatrix} 4 & 2 \\ 7 & 9 \end{bmatrix} - \begin{bmatrix} 1 & 8 \\ 4 & 3 \end{bmatrix}\begin{bmatrix} \frac{7}{33} & -\frac{2}{33} \\ -\frac{1}{33} & \frac{5}{33} \end{bmatrix}\begin{bmatrix} 1 & 0 \\ 6 & 6 \end{bmatrix}\right]^{-1}$

$$= \left[\begin{bmatrix} 4 & 2 \\ 7 & 9 \end{bmatrix} - \begin{bmatrix} \frac{227}{33} & \frac{228}{33} \\ \frac{67}{33} & \frac{42}{33} \end{bmatrix}\right]^{-1} = \begin{bmatrix} -\frac{95}{33} & -\frac{162}{33} \\ \frac{164}{33} & \frac{255}{33} \end{bmatrix}^{-1} = \frac{1}{71}\begin{bmatrix} 255 & 162 \\ -164 & -95 \end{bmatrix}$$

$\therefore$ $M = -D^{-1}CK = -\begin{bmatrix} \frac{7}{33} & -\frac{2}{33} \\ -\frac{1}{33} & \frac{5}{33} \end{bmatrix}\begin{bmatrix} 1 & 0 \\ 6 & 6 \end{bmatrix}\begin{bmatrix} \frac{255}{33} & \frac{162}{33} \\ -\frac{164}{71} & -\frac{95}{71} \end{bmatrix}$

$$= -\begin{bmatrix} \frac{21}{71} & \frac{10}{71} \\ \frac{75}{71} & \frac{57}{71} \end{bmatrix} = -\frac{1}{71}\begin{bmatrix} 21 & 10 \\ 75 & 56 \end{bmatrix}$$

Now $L = -KBD^{-1}$

$$= -\frac{1}{71}\begin{bmatrix} 255 & 162 \\ -164 & -95 \end{bmatrix}\begin{bmatrix} 1 & 8 \\ 4 & 3 \end{bmatrix}\frac{1}{33}\begin{bmatrix} 7 & -2 \\ -1 & 5 \end{bmatrix} = \frac{1}{71}\begin{bmatrix} -115 & -328 \\ 67 & 209 \end{bmatrix}$$

and $N = D^{-1} - D^{-1}CL$

$$= \frac{1}{33}\begin{bmatrix} 7 & -2 \\ -1 & 5 \end{bmatrix} - \frac{1}{71}\frac{1}{33}\begin{bmatrix} 7 & -2 \\ -1 & 5 \end{bmatrix}\begin{bmatrix} 1 & 0 \\ 6 & 6 \end{bmatrix}\begin{bmatrix} -115 & -328 \\ 67 & 209 \end{bmatrix} = \frac{1}{71}\begin{bmatrix} 22 & 22 \\ 38 & 109 \end{bmatrix}$$

$$\therefore \quad R^{-1} = \left[\begin{array}{cc|cc} \frac{255}{71} & \frac{162}{71} & -\frac{115}{71} & -\frac{328}{71} \\ -\frac{164}{71} & -\frac{95}{71} & \frac{67}{71} & \frac{209}{71} \\ \hline -\frac{21}{71} & -\frac{10}{71} & \frac{22}{71} & \frac{22}{71} \\ -\frac{75}{71} & -\frac{56}{71} & \frac{38}{71} & \frac{109}{71} \end{array}\right]$$

Note that $RR^{-1} = \left[\begin{array}{cc|cc} 4 & 2 & 1 & 8 \\ 7 & 9 & 4 & 3 \\ \hline 1 & 0 & 5 & 2 \\ 6 & 6 & 1 & 7 \end{array}\right]\frac{1}{71}\left[\begin{array}{cc|cc} 255 & 162 & -115 & -328 \\ -164 & -95 & 67 & 209 \\ \hline -21 & -10 & 22 & 22 \\ -75 & -56 & 38 & 109 \end{array}\right]$

$$= \frac{1}{71}\begin{bmatrix} 71 & 0 & 0 & 0 \\ 0 & 71 & 0 & 0 \\ 0 & 0 & 71 & 0 \\ 0 & 0 & 0 & 71 \end{bmatrix} = I_4$$

and is verified.

A special case of the above general method:

Let the matrix (8.16) be of the form

$$R = \begin{bmatrix} I & P \\ 0 & Q \end{bmatrix} \tag{8.28}$$

where I is the unit matrix, 0 is the null matrix and Q^{-1} exists and is known.

Then the equations (8.23), (8.24), (8.25), (8.26) become

$$M = -\,Q^{-1}0K = 0 \tag{8.29}$$

$$K = (I - P\ Q^{-1}0)^{-1} = I \tag{8.30}$$

$$N = QQ^{-1} - Q^{-1}0L = Q^{-1} \tag{8.31}$$

$$L = -KP\ Q^{-1} = -IP\ Q^{-1} = -P\ Q^{-1} \tag{8.32}$$

$$\therefore \quad R^{-1} = \begin{bmatrix} I & -PQ^{-1} \\ 0 & Q^{-1} \end{bmatrix}. \tag{8.33}$$

8.5 Matrix inversion by the successive bordering method

This section discusses a particular case of the previous partitioning method of 4 partition block. In this case the partitioned lines to construct a bordered matrix are always fixed. The horizontal line is above the last row and the vertical line is on the left of the right most column. Thus n square matrix is bordered as

$$A = A_n = \left[\begin{array}{cccc|c} a_{11} & a_{12} & \cdots & a_{1,n-1} & a_{1n} \\ a_{21} & a_{22} & \cdots & a_{2,n-1} & a_{2n} \\ \vdots & \vdots & \ddots & \vdots & \vdots \\ a_{n-1,1} & a_{n-2,2} & \cdots & a_{n-1,n-1} & a_{n-1,n} \\ \hline a_{n1} & a_{n2} & \cdots & a_{n,n-1} & a_{nn} \end{array}\right] = \begin{bmatrix} A_{n-1} & \mathbf{u}_n \\ \mathbf{v}_n & a_{nn} \end{bmatrix} \tag{8.34}$$

where A_{n-1} is a $(n-1)$th square block of the given matrix $A = A_n$ and $\mathbf{u}_n = \begin{bmatrix} a_{1n} & a_{2n} & \cdots & a_{n-1,n} \end{bmatrix}^T$ and $\mathbf{v}_n = \begin{bmatrix} a_{1n} & a_{2n} & \cdots & a_{n-1,n} \end{bmatrix}$. The inverse of A_{n-1} is supposed exist. This type of bordering continues until a simple A_{n-1} is obtained so that its inverse is easily known. All the A_n's considered here are nonsingular. For any one or more A_n is singular (cf. sec.9.11 and chapter 18).

Let the inverse of A in bordering form be

$$A^{-1} = A_n^{-1} = \begin{bmatrix} P_{n-1} & \mathbf{r}_n \\ \mathbf{q}_n & \frac{1}{\alpha_n} \end{bmatrix} \tag{8.35}$$

where P_{n-1} is a matrix, $\mathbf{r}$ a column vector, $\mathbf{q}$ a row vector and $\frac{1}{\alpha_n}$ a number, all of which are to be determined from the 4 equations formed as in sec. 8.4 from definition of inverse.

We know $\quad AA^{-1} = I_n$

or
$$\left[\begin{array}{c|c} A_{n-1} & \mathbf{u}_n \\ \hline \mathbf{u}_n & a_{nn} \end{array}\right] \left[\begin{array}{c|c} P_{n-1} & \mathbf{r}_n \\ \hline q_n & \frac{1}{\alpha_n} \end{array}\right] = \left[\begin{array}{c|c} I_{n-1} & 0 \\ \hline 0 & 1 \end{array}\right] \tag{8.36}$$

$$\therefore \quad A_{n-1}P_{n-1} + \mathbf{u}_n\mathbf{q}_n = I_{n-1} \tag{8.37}$$

$$A_{n-1}\mathbf{r}_n + \mathbf{u}_n \frac{1}{\alpha_n} = 0 \tag{8.38}$$

$$\mathbf{v}_nP_{n-1} + a_{nn}\mathbf{q}_n = 0 \tag{8.39}$$

$$\mathbf{v}_n\mathbf{r}_n + a_{nn}\frac{1}{\alpha_n} = 0 \tag{8.40}$$

From (8.38), $$\mathbf{r}_n = -\frac{A_{n-1}^{-1}\mathbf{u}_n}{\alpha_n} \tag{8.41}$$

Using (8.41) in (8.40),

$$-\frac{\mathbf{v}_n \mathrm{A}_{n-1}^{-1} \mathbf{u}_n}{\alpha_n} + a_{nn}\frac{1}{\alpha_n} = 1$$

or $$\alpha_n = a_{nn} - \mathbf{v}_n \mathrm{A}_{n-1}^{-1} \mathbf{u}_n \qquad (8.42)$$

From (8.37),

$$\mathrm{A}_{n-1}\mathrm{P}_{n-1} = \mathrm{I}_{n-1}\mathbf{u}_n\mathbf{q}_n$$

or $$\mathrm{P}_{n-1} = \mathrm{A}_{n-1}^{-1} - \mathrm{A}_{n-1}^{-1}\mathbf{u}_n\mathbf{q}_n \qquad (8.43)$$

Using (8.43) in (8.39), we have,

$$\mathbf{v}_n\left(\mathrm{A}_{n-1}^{-1} - \mathrm{A}_{n-1}^{-1}\mathbf{u}_n\mathbf{q}_n\right) + a_{nn}\mathbf{q}_n = 0$$

or $$\mathbf{v}_n\mathrm{A}_{n-1}^{-1} - \mathbf{v}_n\mathrm{A}_{n-1}^{-1}\mathbf{u}_n\mathbf{q}_n + a_{nn}\mathbf{q}_n = 0$$

or $$\mathbf{v}_n\mathrm{A}_{n-1}^{-1} - (a_{nn} - \alpha_n)\mathbf{q}_n + a_{nn}\mathbf{q}_n = 0, \quad \text{by} \qquad (8.42)$$

or $$\mathbf{v}_n\mathrm{A}_{n-1}^{-1} + \alpha_n\mathbf{q}_n = 0$$

or $$\mathbf{q}_n = -\frac{\mathbf{v}_n\mathrm{A}_{n-1}^{-1}}{\alpha_n} \qquad (8.44)$$

Now from (8.37),

$$\mathrm{P}_{n-1} = \mathrm{A}_{n-1}^{-1} - \mathrm{A}_{n-1}^{-1}\mathbf{u}_n\mathbf{q}_n$$

$$= \mathrm{A}_{n-1}^{-1} + \frac{\mathrm{A}_{n-1}^{-1}\mathbf{u}_n\mathbf{v}_n\mathrm{A}_{n-1}^{-1}}{\alpha_n}, \text{ by (8.44)} \qquad (8.45)$$

∴ (8.41), (8.42), (8.44) and (8.45) together give

$$\mathrm{A}^{-1} = \left[\begin{array}{c|c} \mathrm{A}_{n-1}^{-1} + \dfrac{\mathrm{A}_{n-1}^{-1}\mathbf{u}_n\mathbf{v}_n\mathrm{A}_{n-1}^{-1}}{\alpha_n} & -\dfrac{\mathrm{A}_{n-1}^{-1}\mathbf{u}_n}{\alpha_n} \\ \hline -\dfrac{\mathbf{v}_n\mathrm{A}_{n-1}^{-1}}{\alpha_n} & \dfrac{1}{\left(a_{nn} - \mathrm{A}_{n-1}^{-1}\mathbf{u}_n\mathbf{q}_n\right)} \end{array}\right] \qquad (8.46)$$

Ex. Find the inverse of

$$\mathrm{A} = \begin{bmatrix} 4 & 2 & 1 & 8 \\ 7 & 9 & 4 & 3 \\ 1 & 0 & 5 & 2 \\ 6 & 6 & 1 & 7 \end{bmatrix}$$

using the bordering method.

Let the first bordering be

$$\mathrm{A} = \mathrm{A}_4 = \left[\begin{array}{ccc|c} 4 & 2 & 1 & 8 \\ 7 & 9 & 4 & 3 \\ 1 & 0 & 5 & 2 \\ \hline 6 & 6 & 1 & 7 \end{array}\right] = \left[\begin{array}{c|c} \mathrm{A}_3 & \mathbf{u}_4 \\ \hline \mathbf{v}_4 & a_{44} \end{array}\right]$$

where $A_3 = \begin{bmatrix} 4 & 2 & 1 \\ 7 & 9 & 4 \\ 1 & 0 & 5 \end{bmatrix}$, $\mathbf{u}_4 = [8 \quad 3 \quad 2]^T$, $\mathbf{v}_4 = [6 \quad 6 \quad 1]$ and $a_{44} = 7$.

Using bordering second time, let

$$B = A_3 = \left[\begin{array}{cc|c} 4 & 2 & 1 \\ 7 & 9 & 4 \\ \hline 1 & 0 & 5 \end{array}\right] = \left[\begin{array}{c|c} A_2 & \mathbf{u}_3 \\ \hline \mathbf{v}_3 & a_{33} \end{array}\right]$$

where $A_2 = \begin{bmatrix} 4 & 2 \\ 7 & 9 \end{bmatrix}$, a second order matrix, $\mathbf{u}_3 = [1 \quad 4]^T$, $\mathbf{v}_3 = [1 \quad 0]$ and $a_{33} = 5$.

As $A_2 = \begin{bmatrix} 4 & 2 \\ 7 & 9 \end{bmatrix}$, therefore, $|A_2| = 22 \neq 0$

and

$$\text{cof } A_2 = \begin{bmatrix} 9 & -7 \\ -2 & 4 \end{bmatrix}, \text{ adj } A_2 = \begin{bmatrix} 9 & -2 \\ -7 & 4 \end{bmatrix},$$

$$\therefore \quad A_2^{-1} = \frac{\text{adj } A_2}{|A_2|} = \frac{1}{22}\begin{bmatrix} 9 & -2 \\ -7 & 4 \end{bmatrix}$$

Also, $\alpha_3 = a_{33} - \mathbf{v}_3 A_2^{-1} \mathbf{u}_3$ from (8.42)

$$= 5 - [1 \quad 0]\tfrac{1}{22}\begin{bmatrix} 9 & -2 \\ -7 & 4 \end{bmatrix}\begin{bmatrix} 1 \\ 4 \end{bmatrix} = 5 - \tfrac{1}{22}[1] = \tfrac{109}{22}$$

From (8.41), $\mathbf{r}_3 = -\dfrac{A_2^{-1}\mathbf{u}_3}{\alpha_3} = -\tfrac{22}{109}\tfrac{1}{22}\begin{bmatrix} 9 & -2 \\ -7 & 4 \end{bmatrix}\begin{bmatrix} 1 \\ 4 \end{bmatrix} = \tfrac{1}{109}\begin{bmatrix} -1 \\ -9 \end{bmatrix}$

From (8.44), $\mathbf{q}_3 = -\dfrac{\mathbf{v}_3 A_2^{-1}}{\alpha_3} = -\tfrac{22}{109}[1 \quad 0]\tfrac{1}{22}\begin{bmatrix} 9 & -2 \\ -7 & 4 \end{bmatrix} = -\tfrac{1}{109}[9 \quad -2] = \tfrac{1}{109}[-9 \quad 2]$

From (8.45), $P_2 = A_2^{-1} - A_2^{-1}\mathbf{u}_3\mathbf{q}_3$

$$= \tfrac{1}{22}\begin{bmatrix} 9 & -2 \\ -7 & 4 \end{bmatrix} - \tfrac{1}{22}\begin{bmatrix} 9 & -2 \\ -7 & 4 \end{bmatrix}\begin{bmatrix} 1 \\ 4 \end{bmatrix}\tfrac{1}{109}[-9 \quad 2]$$

$$= \tfrac{1}{22}\begin{bmatrix} 9 & -2 \\ -7 & 4 \end{bmatrix} + \tfrac{1}{22.109}\begin{bmatrix} 9 & -2 \\ 81 & -18 \end{bmatrix} = \tfrac{1}{109}\begin{bmatrix} 45 & -10 \\ -31 & 19 \end{bmatrix}$$

$$\therefore \quad B^{-1} = A_3^{-1} = \tfrac{1}{109}\left[\begin{array}{cc|c} 45 & -10 & -1 \\ -31 & 19 & -9 \\ \hline -9 & 2 & 22 \end{array}\right].$$

Now find A^{-1} with the help of A_3^{-1}, let $A^{-1} = A_4^{-1} = \left[\begin{array}{c|c} P_3 & \mathbf{r}_4 \\ \hline \mathbf{q}_4 & \frac{1}{\alpha_4} \end{array}\right]$

From (8.42), $\alpha_4 = a_{44} - \mathbf{v}_4 A_3^{-1} \mathbf{u}_4$

$$= 7 - [6 \quad 6 \quad 1]\tfrac{1}{109}\begin{bmatrix} 45 & -10 & -1 \\ -31 & 19 & -9 \\ -9 & 2 & 22 \end{bmatrix}\begin{bmatrix} 8 \\ 3 \\ 2 \end{bmatrix} = 7 - \tfrac{692}{109} = \tfrac{71}{109}$$

From (8.41), $\mathbf{r}_4 = -\dfrac{A_3^{-1}\mathbf{u}_4}{\alpha_4} = -\tfrac{109}{71}\tfrac{1}{109}\begin{bmatrix} 45 & -10 & -1 \\ -31 & 19 & -9 \\ -9 & 2 & 22 \end{bmatrix}\begin{bmatrix} 8 \\ 3 \\ 2 \end{bmatrix} = \tfrac{1}{71}\begin{bmatrix} -328 \\ 209 \\ 22 \end{bmatrix}$

From (8.44) $\mathbf{q}_4 = -\dfrac{\mathbf{v}_4 A_3^{-1}}{\alpha_4} = -\tfrac{109}{71}[6 \quad 6 \quad 1]\tfrac{1}{109}\begin{bmatrix} 45 & -10 & -1 \\ -31 & 19 & -9 \\ -9 & 2 & 22 \end{bmatrix}$

$$= \tfrac{1}{71}[-75 \quad -56 \quad 38]$$

From (8.45),

$$P_3 = A_3^{-1} - A_3^{-1}\mathbf{u}_4\,\mathbf{q}_4$$

$$= \tfrac{1}{109}\begin{bmatrix} 45 & -10 & -1 \\ -31 & 19 & -9 \\ -9 & 2 & 22 \end{bmatrix} - \tfrac{1}{109.71}\begin{bmatrix} 45 & -10 & -1 \\ -31 & 19 & -9 \\ -9 & 2 & 22 \end{bmatrix}\begin{bmatrix} 8 \\ 3 \\ 2 \end{bmatrix}[-75 \quad -56 \quad 38]$$

$$= \tfrac{1}{109.71}\left\{\begin{bmatrix} 3195 & -710 & -71 \\ -2201 & 1349 & -639 \\ -639 & 142 & 1562 \end{bmatrix} - \begin{bmatrix} -24600 & -18368 & 12464 \\ 15675 & 11704 & -7942 \\ 1650 & 1232 & -836 \end{bmatrix}\right\}$$

$$= \tfrac{1}{71}\begin{bmatrix} 255 & 162 & -115 \\ -164 & -95 & 67 \\ -21 & -10 & 22 \end{bmatrix}$$

$$\therefore \quad A^{-1} = \tfrac{1}{71}\left[\begin{array}{ccc|c} 255 & 162 & -115 & -328 \\ -164 & -95 & 67 & 209 \\ -21 & -10 & 22 & 22 \\ \hline -75 & -56 & 38 & 109 \end{array}\right]$$

8.6 Computation of the inverse matrix when an element of a given matrix changes

This section discusses the computational procedure of the inverse of a matrix which differs from another given matrix with the known inverse by a single element. Sherman and Morison derived this procedure in 1949 and thus it eliminates the necessity of computing the new inverse from the beginning.

Let $A = [a_{ij}]$ be a given nonsingular matrix whose inverse is $A^{-1} = [b_{ij}]$, where $b_{ij} = \dfrac{A_{ji}}{|A|}$. Let $C = [c_{ij}]$ be a nonsingular matrix obtained from A with a change of an element a_{rs} of A by an amount Δa_{rs}. Obviously

$$c_{rs} = a_{rs} + \Delta a_{rs}\,.$$

Let $C^{-1} = [d_{ij}]$ be the inverse of C. Here we want to compute d_{ij} from b_{ij} without computing C^T and adj C etc.

$$\text{Now} \quad |C| = |c_{ij}| = \begin{vmatrix} a_{11} & a_{12} & \cdots & a_{1s} & \cdots & a_{1n} \\ a_{21} & a_{22} & \cdots & a_{2s} & \cdots & a_{2n} \\ \vdots & \vdots & \ddots & \vdots & \vdots & \vdots \\ a_{r1} & a_{r2} & \cdots & a_{rs} + \Delta a_{rs} & \cdots & a_{rn} \\ \vdots & \vdots & & \ddots & \vdots & \vdots \\ a_{n1} & a_{n2} & \cdots & a_{ns} & \cdots & a_{nn} \end{vmatrix}$$

$$= |A| + \Delta a_{rs}(\text{cofactor of } a_{rs})$$

$$= |A|\,(1 + \Delta a_{rs} b_{sr}) \text{ where } b_{sr} = \frac{A_{rs}}{|A|}$$

By a straight forward calculation, it can be proved that cofactor of $b_{sr} = \frac{a_{rs}}{|A|}$. That is, a_{rs} the elements of the original matrix is proportional to b_{rs}, the elements of its inverse matrix and the constant of proportionality is $|A|$. This relation can be easily established for a 3 square matrix.

Now to find the elements of C^{-1} in terms of b_{ij}, let us consider a 3 square matrix for the sake of convenience and let c_{rs} be $c_{23} = a_{23} + \Delta a_{23}$ in particular. Therefore, $|C| = |A|(1 + \Delta a_{23} b_{32})$ and we assume that $1 + \Delta a_{23} b_{32} \neq 0$.

$$\therefore \quad C^{-1} = [d_{ij}] = \frac{\text{adj } C}{|C|}$$

$$= \frac{1}{|C|}\begin{bmatrix} C_{11} & C_{21} & C_{31} \\ C_{12} & C_{22} & C_{32} \\ C_{13} & C_{23} & C_{33} \end{bmatrix},$$

$$\text{where} \quad C = [c_{ij}] = \begin{bmatrix} a_{11} & a_{12} & a_{13} \\ a_{21} & a_{22} & (a_{23} + \Delta a_{23}) \\ a_{31} & a_{32} & a_{33} \end{bmatrix}$$

$$\text{Now} \quad d_{11} = \frac{C_{11}}{|C|} = \frac{a_{22}a_{33} - a_{32}(a_{32} + \Delta a_{23})}{|C|}$$

$$= \frac{(a_{22}a_{33} - a_{32}a_{23}) - a_{32}\Delta a_{23}}{|C|} = \frac{A_{11} - a_{32}\Delta a_{23}}{|C|}$$

$$= \frac{|A|b_{11} - \Delta a_{23}(\text{cofactor of } b_{23})|A|}{|C|}$$, replacing a_{32} by the product of the cofactor of b_{23}, an element of original inverse and $|A|$.

$$= \frac{|A|b_{11} + \Delta a_{23}(b_{11}b_{32} - b_{31}b_{13})|A|}{|A|(1 + \Delta a_{23}b_{32})}$$

$$= \frac{b_{11}(1+\Delta a_{23}b_{32}) - \Delta a_{23}b_{31}b_{13}}{(1+\Delta a_{23}b_{32})}$$

$$= b_{11} - \frac{\Delta a_{23}b_{31}b_{12}}{1+\Delta a_{23}b_{32}} \tag{8.47}$$

In this way,

$$d_{12} = \frac{C_{21}}{|C|} = \frac{-a_{12}a_{33} + a_{32}a_{13}}{|C|} = \frac{b_{12}|A|}{|A|(1+\Delta a_{23}b_{32})}$$

$$= \frac{b_{12}}{(1+\Delta a_{23}b_{32})} \tag{8.48}$$

$$d_{13} = \frac{a_{12}(a_{23} + \Delta a_{23}) - a_{22}a_{13}}{|C|} = \frac{|A|b_{13} + a_{12}\Delta a_{23}}{|C|}$$

$$= \frac{|A|b_{13} - (b_{12}b_{33} - b_{32}b_{13})\Delta a_{23}}{|C|}$$

$$= b_{13} - \frac{b_{12}b_{33}\Delta a_{23}}{1+\Delta a_{23}b_{32}} \tag{8.49}$$

$$d_{21} = \frac{C_{12}}{|C|} = \frac{-a_{21}a_{33} + a_{31}(a_{23} + \Delta a_{23})}{|C|} = \frac{|A|\,b_{21} + a_{31}\Delta a_{23}}{|C|}$$

$$= \frac{|A|\,b_{21} + (b_{21}b_{32} - b_{31}b_{22})\Delta a_{23}|A|}{|C|}$$

$$= \frac{|A|b_{21} - b_{31}b_{22}\Delta a_{23}|A|}{|C|} = b_{21} - \frac{b_{31}b_{22}\Delta a_{23}}{1+\Delta a_{23}b_{32}} \tag{8.50}$$

$$d_{22} = \frac{C_{22}}{|C|} = \frac{a_{11}a_{33} - a_{31}a_{13}}{|C|} = \frac{b_{22}}{1+\Delta a_{23}b_{32}} \tag{8.51}$$

$$d_{23} = \frac{C_{32}}{|C|} = \frac{-a_{11}(a_{23} + \Delta a_{23}) + a_{21}a_{13}}{|C|} = \frac{b_{23}|A| - a_{11}\Delta a_{23}}{|C|}$$

$$= \frac{b_{23}|A| - (b_{22}b_{33} - b_{32}b_{23})\Delta a_{23}}{|C|}$$

$$= \frac{b_{23}\,|A|\,(1+b_{32}\Delta a_{23}) - b_{22}b_{33}\Delta a_{23}}{|C|}$$

$$= b_{23} - \frac{b_{22}b_{33}\Delta a_{23}}{1+\Delta a_{23}b_{32}} \tag{8.52}$$

$$d_{31} = \frac{C_{13}}{|C|} = \frac{a_{21}a_{32} - a_{31}a_{22}}{|C|} = \frac{b_{31}}{1+\Delta a_{23}b_{32}} \tag{8.53}$$

$$d_{32} = \frac{C_{23}}{|C|} = \frac{-(a_{11}a_{32} - a_{31}a_{12})}{|C|} = \frac{b_{32}}{1 + \Delta a_{23} b_{32}} \tag{8.54}$$

$$d_{33} = \frac{C_{33}}{|C|} = \frac{a_{11}a_{22} - a_{21}a_{12}}{|C|} = \frac{b_{33}}{1 + \Delta a_{23} b_{32}} \tag{8.55}$$

From (8.47), (8.49), (8.50), (8.52) above it follows that

$$d_{Rj} = b_{Rj} - \frac{\Delta a_{23} b_{R2} b_{3j}}{1 + \Delta a_{23} b_{32}}, \quad R = 1,2,3 \; ; \; j = 1,2,3 \; ; \text{ with } r = 2, \; s = 3. \tag{8.56}$$

and from (8.53), (8.54), (8.55) when $R = s = 3$

$$d_{3j} = \frac{b_{3j}}{1 + \Delta a_{23} b_{32}}, \quad j = 1, \; 2, \; 3 \; ; \text{ with } s = 3, \; r = 2 \tag{8.57}$$

Also from (8.48), (8.51), (8.54) when $j = r$

$$d_{2j} = \frac{b_{2j}}{1 + \Delta a_{23} b_{32}}, \quad j = 1, \; 2, \; 3 \; ; \text{ with } r = 2, \; s = 3 \tag{8.58}$$

The equations (8.56) to (8.58) then calculate the elements of the new inverse corresponding to a change Δa_{23} in the element a_{23} of the 3 square matrix A. These equations can be easily converted for a change in any other element of A by changing the subscripts 2 and 3 to arbitrary subscripts r and s respectively as follows:

$$d_{Rj} = b_{Rj} - \frac{\Delta a_{rs} b_{Rr} b_{sj}}{1 + \Delta a_{rs} b_{sr}},$$

$$d_{sj} = \frac{b_{sj}}{1 + \Delta a_{rs} b_{sr}},$$

$$d_{rj} = \frac{b_{rj}}{1 + \Delta a_{rs} b_{sr}},$$

where $R = 1, \; 2, \; 3$; $j = 1, 2, 3$ and r, s = any fixed value from 1, 2, 3 and provided $1 + \Delta a_{rs} b_{sr} \neq 0$.

These three formulae can be extended for any higher order n square matrix by just extending the ranges of subscripts involved.

If two or more elements of the matrix change the new inverse can be obtained by successive applications of the method.

N.B. This is the inverse of a perturbed matrix.

Ex. Find the inverse of

$$\begin{bmatrix} 2 & -1 & 0 \\ 3 & -2 & 1.09 \\ -4 & 5 & 7 \end{bmatrix}, \text{ if } \begin{bmatrix} 2 & -1 & 0 \\ 3 & -2 & 1 \\ -4 & 5 & 7 \end{bmatrix}^{-1} = -\frac{1}{13}\begin{bmatrix} -19 & 7 & -1 \\ -25 & 14 & -2 \\ 7 & -6 & -1 \end{bmatrix}.$$

Let $A = \begin{bmatrix} 2 & -1 & 0 \\ 3 & -2 & 1 \\ -4 & 5 & 7 \end{bmatrix}$ and $C = \begin{bmatrix} 2 & -1 & 0 \\ 3 & -2 & 1.09 \\ -4 & 5 & 7 \end{bmatrix}$.

Comparing A and C, it is seen that they differ in (2, 3)th element.

Thus $\quad c_{23} = a_{23} + \Delta a_{23}$, where $\Delta a_{23} = \frac{9}{100}$. Let $C^{-1} = [d_{ij}]$ and $A^{-1} = [b_{ij}]$.

Now $\quad 1 + \Delta a_{23} b_{32} = 1 + \frac{9}{100} \cdot \frac{6}{13} = \frac{1354}{1300} \neq 0$.

Therefore,

$$d_{11} = b_{11} - \frac{\Delta a_{23} b_{31} b_{12}}{1 + \Delta a_{23} b_{32}} = \frac{19}{13} - \frac{\left(\frac{9}{100}\right)\left(-\frac{7}{13}\right)\left(-\frac{7}{13}\right).1300}{1354} = \frac{25285}{17602}$$

$$d_{12} = \frac{b_{12}}{1 + \Delta a_{23} b_{32}} = \frac{\left(-\frac{7}{13}\right)(1300)}{1354} = -\frac{9100}{17602}$$

$$d_{13} = b_{13} - \frac{\Delta a_{23} b_{12} b_{33}}{1 + \Delta a_{23} b_{32}} = \frac{1}{13} - \frac{\left(\frac{9}{100}\right)\left(-\frac{7}{13}\right)\left(\frac{1}{13}\right).1300}{1354} = \frac{1417}{17602}$$

$$d_{21} = b_{21} - \frac{\Delta a_{23} b_{31} b_{22}}{1 + \Delta a_{23} b_{32}} = \frac{25}{13} - \frac{\left(\frac{9}{100}\right)\left(-\frac{7}{13}\right)\left(-\frac{14}{13}\right).1300}{1354} = \frac{32968}{17602}$$

$$d_{22} = \frac{b_{22}}{1 + \Delta a_{23} b_{32}} = -\frac{14}{13} \cdot \frac{1300}{1354} = -\frac{18200}{17602}$$

$$d_{23} = b_{23} - \frac{\Delta a_{23} b_{22} b_{33}}{1 + \Delta a_{23} b_{32}} = \frac{2}{13} - \frac{\left(\frac{9}{100}\right)\left(-\frac{14}{13}\right)\left(-\frac{1}{13}\right).1300}{1354} = \frac{2834}{17602}$$

$$d_{31} = \frac{b_{31}}{1 + \Delta a_{23} b_{32}} = -\frac{7}{13} \cdot \frac{1300}{1354} = -\frac{9100}{17602}$$

$$d_{32} = \frac{b_{32}}{1 + \Delta a_{23} b_{32}} = \frac{6}{13} \cdot \frac{1300}{1354} = \frac{7800}{17602}$$

$$d_{33} = \frac{b_{33}}{1 + \Delta a_{23} b_{32}} = \frac{1}{13} \cdot \frac{1300}{1354} = \frac{1300}{17602}$$

$\therefore$ the required inverse in question is

$$C^{-1} = \frac{1}{17602} \begin{bmatrix} 25285 & -9100 & 1417 \\ 32968 & -18200 & 2834 \\ -9100 & 7800 & 1300 \end{bmatrix}$$

Note that $CC^{-1} = \frac{1}{17602} \begin{bmatrix} 2 & -1 & 0 \\ 3 & -2 & \frac{109}{100} \\ -4 & 5 & 7 \end{bmatrix} \begin{bmatrix} 25285 & -9100 & 1417 \\ 32968 & -18200 & 2834 \\ -9100 & 7800 & 1300 \end{bmatrix} = \begin{bmatrix} 1 & 0 & 0 \\ 0 & 1 & 0 \\ 0 & 0 & 1 \end{bmatrix} = I_3$

8.7 Solution of a single system of simultaneous nonhomogeneous algebraic equations using the inverse matrix

The existence of the inverse of an n square matrix provides a means for the solution of a system of simultaneous nonhomogeneous n equations with n variables. Let the given system be written in matrix form as $\mathbf{Ax} = \mathbf{b}$ which is a simple matrix. This matrix equation can be solved using the following theorem.

Theorem. The equation $\mathbf{Ax} = \mathbf{b}$ has a unique solution $\mathbf{x} = \mathrm{A}^{-1}\mathbf{b}$.

Proof. The proof consists of two parts

(i) $\mathbf{x} = \mathrm{A}^{-1}\mathbf{b}$ satisfies the equation,

(ii) The solution $\mathbf{x} = \mathrm{A}^{-1}\mathbf{b}$ is unique.

Proof of (i). To prove this, we substitute $\mathbf{x} = \mathrm{A}^{-1}\mathbf{b}$ in L.H.S of the equation. Therefore,

$$\text{L.H.S.} = \mathbf{Ax} = \mathrm{A}(\mathrm{A}^{-1}\mathbf{b}) = (\mathrm{AA}^{-1})\mathbf{b} = \mathbf{Ib} = \mathbf{b} = \text{R.H.S.}$$

Thus, the equation is satisfied by the given solution.

Proof of (ii). If possible let the equation has another solution **c**. Then this **c** must satisfy the equation, i.e. $\mathbf{Ac} = \mathbf{b}$. Now premultiplying both sides by the square matrix A^{-1}, we get

$$\mathrm{A}^{-1}(\mathrm{A}\mathbf{c}) = \mathrm{A}^{-1}\mathbf{b}$$

or $$(\mathrm{A}^{-1}\mathrm{A})\mathbf{c} = \mathrm{A}^{-1}\mathbf{b}$$

or $$\mathbf{Ic} = \mathrm{A}^{-1}\mathbf{b}$$

or $$\mathbf{c} = \mathbf{A}^{-1}\mathbf{b}$$

$\therefore$ **c** coincides with the given solution. So, there cannot be more than one solution. Hence the solution is unique.

Algorithm for solving systems of n equations with n unknowns

Step 1. Write the system as $\mathbf{Ax} = \mathbf{b}$ where A is the n square coefficient matrix, **x** and **b** are column matrices.

Step 2. Test whether $|\mathrm{A}| = 0$ or not. If $|\mathrm{A}| \neq 0$, then conclude that the coefficient matrix A is nonsingular and follow the next steps, otherwise stop procedure by the conclusion that the system has no solution i.e. the system is inconsistent.

Step 3. Find A^{-1}.

Step 4. Write $\mathbf{x} = \mathrm{A}^{-1}\mathbf{b}$ as the solution in matrix form.

Step 5. Multiply A^{-1} and **b**. The result is a $n\times 1$ column matrix, say **c**.

Step 6. Now column matrices **x** and **c** are equal. So equate corresponding elements and these are the required solutions.

Ex. Solve :

$$\begin{cases} 3x - y + z = 2 \\ 2x + y - z = 1 \\ x + 5y - 3z = 3 \end{cases}.$$

Step 1. The given system in matrix form is $\mathbf{Ax} = \mathbf{b}$ where the coefficient matrix

$$A = \begin{bmatrix} 3 & -1 & 1 \\ 2 & 1 & -1 \\ 1 & 5 & -3 \end{bmatrix} \text{ and } \mathbf{x} = \begin{bmatrix} x \\ y \\ z \end{bmatrix}, \mathbf{b} = \begin{bmatrix} 2 \\ 1 \\ 3 \end{bmatrix} \text{ are column matrices.}$$

Step 2 . Now

$$|A| = \begin{vmatrix} 3 & -1 & 1 \\ 2 & 1 & -1 \\ 1 & 5 & -3 \end{vmatrix} = |10| \neq 0$$

Hence the coefficient matrix A is nonsingular.

Step 3 . Now to find A^{-1} let us calculated the cofactors of the elements,

$$A_{11} = 2,\ A_{12} = 5,\ A_{13} = 9,\ A_{21} = 2,\ A_{22} = -10,$$
$$A_{23} = -16, A_{31} = 0, A_{32} = 5, A_{33} = 5.$$

$$\therefore \qquad \text{cof } A = \begin{bmatrix} 2 & 5 & 9 \\ 2 & -10 & -16 \\ 0 & 5 & 5 \end{bmatrix}.$$

$$\therefore \quad \text{adj } A = (\text{cof } A)^T = \begin{bmatrix} 2 & 2 & 0 \\ 5 & -10 & 5 \\ 9 & -16 & 5 \end{bmatrix}$$

$$\text{Hence } A^{-1} = \frac{\text{adj } A}{|A|} = \frac{1}{10}\begin{bmatrix} 2 & 2 & 0 \\ 5 & -10 & 5 \\ 9 & -16 & 5 \end{bmatrix} = \begin{bmatrix} \frac{1}{5} & \frac{1}{5} & 0 \\ \frac{1}{2} & -1 & \frac{1}{2} \\ \frac{9}{10} & -\frac{8}{5} & \frac{1}{2} \end{bmatrix}.$$

Step 4 . Since $\mathbf{x} = A^{-1}\mathbf{b}$,

$$\therefore \qquad \begin{bmatrix} x \\ y \\ z \end{bmatrix} = \begin{bmatrix} \frac{1}{5} & \frac{1}{5} & 0 \\ \frac{1}{2} & -1 & \frac{1}{2} \\ \frac{9}{10} & -\frac{8}{5} & \frac{1}{2} \end{bmatrix} \begin{bmatrix} 2 \\ 1 \\ 3 \end{bmatrix}$$

Step 5 .

$$\begin{bmatrix} x \\ y \\ z \end{bmatrix} = \begin{bmatrix} \frac{3}{5} \\ \frac{3}{2} \\ \frac{17}{10} \end{bmatrix}$$

Step 6 . Therefore, $x = \frac{3}{5},\ y = \frac{3}{2},\ z = \frac{17}{10}$.

8.7.1. Solution of several systems of simultaneous nonhomogeneous algebraic equations with common coefficient matrix (matrix solution of a system of matrix equations): The solution procedures for a single system can be extended to a several systems if the number of unknowns and the number of equations in all systems are equal and each equal to *n*. Also the coefficient matrix must be *same*. The necessary modifications in the agorithm of sec.8.7 are

Step 1 . Write all the systems as $Ax = b_1$, $Ax = b_2$, $\cdots$, $Ax = b_k$; where A is the common coefficient matrix and **x** is the column matrix for all the k systems, and the column matrices b_1, b_2, $\cdots$, b_n are distinct.

Steps 2 and 3 . Remain same.

Step 4 . Write $x = A^{-1}b$, $r = 1, 2, \cdots, k$.

Step 5 . Carry out multiplications of A^{-1} and k matrices b_1, b_2, $\cdots$, b_n and obtain the products $c_1, c_2, \cdots, c_k$.

Step 6 . Equate $x = c_1$, $x = c_2$, $\cdots$, $x = c_k$ and get the solutions accordingly.

Ex. Solve the following three systems of equations:

$$\text{(i)} \left.\begin{array}{r} x+y+z+t=3 \\ 2x-y+3z \quad =3 \\ 2y \quad +3t=1 \\ -x \quad +2z+t=0 \end{array}\right\}, \quad \text{(ii)} \left.\begin{array}{r} x+y+z+t=0 \\ 2x-y+3z \quad =-1 \\ 2y \quad +3t=0 \\ -x \quad +2z+t=-3 \end{array}\right\},$$

$$\text{(iii)} \left.\begin{array}{r} x+y+z+t=4 \\ 2x-y+3z \quad =4 \\ 2y \quad +3t=5 \\ -x \quad +2z+t=2 \end{array}\right\}$$

Note that the common coefficient matrix for all the 3 systems is

$$A = \begin{bmatrix} 1 & 1 & 1 & 1 \\ 2 & -1 & 3 & 0 \\ 0 & 2 & 0 & 3 \\ -1 & 0 & 2 & 1 \end{bmatrix}.$$

Onl, differences in the 3 systems are the right hand sides. Let

$$b_1 = \begin{bmatrix} 3 \\ 3 \\ 1 \\ 0 \end{bmatrix}, \; b_2 = \begin{bmatrix} 0 \\ -1 \\ 0 \\ -3 \end{bmatrix} \text{ and } b_3 = \begin{bmatrix} 4 \\ 4 \\ 5 \\ 2 \end{bmatrix}$$

Step 1. $Ax = b_1$, $Ax = b_2$ and $Ax = b_3$.

Step 2.

$$|A| = \begin{bmatrix} 1 & 1 & 1 & 1 \\ 2 & -1 & 3 & 0 \\ 0 & 2 & 0 & 3 \\ -1 & 0 & 2 & 1 \end{bmatrix} = 14 \neq 0.$$

Step 3. To find A^{-1} let us calculate the cofactors:

$A_{11} = 0,\quad A_{12} = 21,\quad A_{13} = 7,\quad A_{14} = -14$

$A_{21} = 4,\quad A_{22} = -9,\quad A_{23} = -1,\quad A_{24} = 6$

$A_{31} = 2,\quad A_{32} = -8,\quad A_{33} = -4,\quad A_{34} = 10$

$A_{41} = -6,\quad A_{42} = 3,\quad A_{43} = 5,\quad A_{44} = -2$

$$\therefore \qquad \text{cof } A = \begin{bmatrix} 0 & 21 & 7 & -14 \\ 4 & -9 & -1 & 6 \\ 2 & -8 & -4 & 10 \\ -6 & 3 & 5 & -2 \end{bmatrix}.$$

$$\therefore \qquad \text{adj } A = (\text{cof } A)^T = \begin{bmatrix} 0 & 4 & 2 & -6 \\ 21 & -9 & -8 & 3 \\ 7 & -1 & -4 & 5 \\ -14 & 6 & 10 & -2 \end{bmatrix}$$

$$\therefore \qquad A^{-1} = \frac{\text{adj } A}{|A|} = \begin{bmatrix} 0 & \frac{2}{7} & \frac{1}{7} & -\frac{3}{7} \\ \frac{3}{2} & -\frac{9}{14} & -\frac{4}{7} & \frac{3}{14} \\ \frac{1}{2} & -\frac{1}{14} & -\frac{2}{7} & \frac{5}{14} \\ -1 & \frac{3}{7} & \frac{5}{7} & -\frac{1}{7} \end{bmatrix}.$$

Step 4. Since $\mathbf{x} = A^{-1}\mathbf{b}$, $r = 1, 2, 3$ and

Step 5. (i)

$$\mathbf{x} = \begin{bmatrix} 0 & \frac{2}{7} & \frac{1}{7} & -\frac{3}{7} \\ \frac{3}{2} & -\frac{9}{14} & -\frac{4}{7} & \frac{3}{14} \\ \frac{1}{2} & -\frac{1}{14} & -\frac{2}{7} & \frac{5}{14} \\ -1 & \frac{3}{7} & \frac{5}{7} & -\frac{1}{7} \end{bmatrix} \begin{bmatrix} 3 \\ 3 \\ 1 \\ 0 \end{bmatrix} = \begin{bmatrix} 1 \\ 2 \\ 1 \\ -1 \end{bmatrix}$$

(ii)

$$\mathbf{x} = \begin{bmatrix} 0 & \frac{2}{7} & \frac{1}{7} & -\frac{3}{7} \\ \frac{3}{2} & -\frac{9}{14} & -\frac{4}{7} & \frac{3}{14} \\ \frac{1}{2} & -\frac{1}{14} & -\frac{2}{7} & \frac{5}{14} \\ -1 & \frac{3}{7} & \frac{5}{7} & -\frac{1}{7} \end{bmatrix} \begin{bmatrix} 0 \\ -1 \\ 0 \\ -3 \end{bmatrix} = \begin{bmatrix} 1 \\ 0 \\ -1 \\ 0 \end{bmatrix}$$

(iii)

$$\mathbf{x} = \begin{bmatrix} 0 & \frac{2}{7} & \frac{1}{7} & -\frac{3}{7} \\ \frac{3}{2} & -\frac{9}{14} & -\frac{4}{7} & \frac{3}{14} \\ \frac{1}{2} & -\frac{1}{14} & -\frac{2}{7} & \frac{5}{14} \\ -1 & \frac{3}{7} & \frac{5}{7} & -\frac{1}{7} \end{bmatrix} \begin{bmatrix} 4 \\ 4 \\ 5 \\ 2 \end{bmatrix} = \begin{bmatrix} 1 \\ 1 \\ 1 \\ 1 \end{bmatrix}$$

Step 6. Therefore, the required solutions are

$$\text{(i)} \left.\begin{matrix} x = 1 \\ y = 2 \\ z = 1 \\ t = -1 \end{matrix}\right\}, \quad \text{(ii)} \left.\begin{matrix} x = 1 \\ y = 0 \\ z = -1 \\ t = 0 \end{matrix}\right\}, \quad \text{(iii)} \left.\begin{matrix} x = 1 \\ y = 1 \\ z = 1 \\ t = 1 \end{matrix}\right\}.$$

8.8 Inversion of Vandemode matrix

Different methods for the inversion of Vandermode matrix have been developed by different researchers. These are Reis's method, Kaufman's method, Tou's method, Wertz method, Temple University Problem Solving Group's method etc., Kaufman's method will be discussed here and Tou's method will be discussed in chapter thirteen.

8.8.1 Kaufman's algorithm for inversion of Vandermode matrix : In 1969 Kaufman gave a simple algorithm for the determination of the inverse of Vandermode matrix as a *product* of two matrices.

Let the Vandermode matrix be

$$V = \begin{bmatrix} 1 & 1 & \cdots & 1 \\ x_1 & x_2 & \cdots & x_n \\ x_1^2 & x_2^2 & \cdots & x_n^2 \\ \vdots & \vdots & \ddots & \vdots \\ x_1^{n-1} & x_2^{n-1} & \cdots & x_n^{n-1} \end{bmatrix}$$

and let it be defined as

$$f(x) = \prod_{i=0}^{n}(x - x_i) = \sum_{i=0}^{n} a_i x_{n-1} \tag{8.59}$$

where x_i belongs to V.

Kaufman illustrated f(x) with a third order Vandermode matrix that

$$V^{-1} = D^{-1} Q \tag{8.60}$$

$$\text{where } Q = \begin{bmatrix} q_{n-1}(x_1) & \cdots & q_2(x_1) & q_1(x_1) & 1 \\ q_{n-1}(x_2) & \cdots & q_2(x_2) & q_1(x_2) & 1 \\ \vdots & \vdots & \vdots & \vdots & \vdots \\ q_{n-1}(x_n) & \cdots & q_2(x_n) & q_1(x_n) & 1 \end{bmatrix} \tag{8.61}$$

$$\text{and} \quad D^{-1} = \begin{bmatrix} \frac{1}{d_1} & & & O \\ & \frac{1}{d_2} & & \\ & & \ddots & \\ O & & & \frac{1}{d_n} \end{bmatrix} = \text{diag}\left(\frac{1}{d_1}\frac{1}{d_2}\cdots\frac{1}{d_n}\right) \tag{8.62}$$

The elements in Q are given by Bareiss in 1967 as the recursion formula

$$q_{k+1}(x_i) = q_k(x_i)x_i + a_{k+1}, \tag{8.63}$$

$$i = 1, 2, \cdots, n; \quad k = 0, 1, \cdots, (n-2)$$

$$q_0(x_i) = 1 \tag{8.64}$$

while the elements of matrix D are

$$d_i = \frac{d}{dx} f(x)\bigg|_{x=x_i} = \prod_{j \neq i} \left(x_i - x_j\right) \tag{8.65}$$

Illustration : Let the third order Vandermode matrix be

$$V = \begin{bmatrix} 1 & 1 & 1 \\ x_1 & x_2 & x_3 \\ x_1^{\ 2} & x_2^{\ 2} & x_3^{\ 2} \end{bmatrix} \text{ and}$$

$$f(x) = \prod_{i=1}^{3} (x - x_i) = (x - x_1)(x - x_2)(x - x_3)$$

$$= x^3 + a_1 x^2 + a_2 x + a_3$$

where $a_1 = -(x_1 + x_2 + x_3)$

$a_2 = x_1 x_2 + x_2 x_3 + x_3 x_1$

$a_3 = -x_1 x_2 x_3$

Now from (8.64) and (8.65),

$$q_1(x_1) = q_0(x_1) x_1 + a_1 = x_1 + a_1$$

$$q_2(x_1) = q_1(x_1) x_1 + a_2 = x_1^{\ 2} + a_1 x + a_2$$

Similarly,

$$q_1(x_2) = x_2 + a_1 \quad \text{and} \quad q_2(x_2) = x_2^{\ 2} + a_1 x_2 + a_2$$

$$q_1(x_3) = x_3 + a_1 \quad \text{and} \quad q_2(x_3) = x_3^{\ 2} + a_1 x_3 + a_2$$

$$\therefore \qquad Q = \begin{bmatrix} q_2(x_1) & q_1(x_1) & 1 \\ q_2(x_2) & q_1(x_2) & 1 \\ q_2(x_3) & q_1(x_3) & 1 \end{bmatrix} = \begin{bmatrix} x_1^{\ 2} + a_1 x_1 + a_2 & x_1 + a_1 & 1 \\ x_2^{\ 2} + a_1 x_2 + a_2 & x_2 + a_1 & 1 \\ x_3^{\ 2} + a_1 x_3 + a_2 & x_3 + a_1 & 1 \end{bmatrix}$$

Hence $$QV = \begin{bmatrix} 3x_1^2 + 2a_1 x_1 + a_2 & 0 & 0 \\ 0 & 3x_2^2 + 2a_1 x_2 + a_2 & 0 \\ 0 & 0 & 3x_3^2 + 2a_1 x_3 + a_2 \end{bmatrix}$$

or $\quad QV = D, \quad$ since, $f(x) = 3x^2 + 2a_1 x + a_2$

i.e. $\quad V^{-1} = D^{-1} Q$

8.9 Inversion of some special matrices by binomial expansion

The well known binomial expansion $(1-x)^{-1}$ and the like may be used for the inversion of certain special types of matrices.The types are (i) time independent matrices (TIM) and (ii) its special form tridiagonal matrices. Time independent matrices are again particular upper triangular matrices in which the principal diagonal & each of the supper diagonals are constant, i.e. consist of the same element. The pattern of these matrices is

$$A = \begin{bmatrix} a_1 & a_2 & a_3 & a_4 & \cdots & \cdots & a_n \\ 0 & a_1 & a_2 & a_3 & \cdots & \cdots & \cdots \\ 0 & 0 & a_1 & a_2 & \cdots & \cdots & \cdots \\ \vdots & \vdots & \vdots & \ddots & \ddots & \vdots & \vdots \\ \vdots & \vdots & \vdots & \vdots & \ddots & \ddots & \vdots \\ 0 & 0 & 0 & 0 & 0 & a_1 & a_2 \\ 0 & 0 & 0 & 0 & 0 & 0 & a_1 \end{bmatrix}$$

The sum, product and scalar multiplication of time independent matrices are again time independent.

Inverse of TIM: If $a_1 \neq 0$, the inverse of the TIM is the TIM

$$B = \begin{bmatrix} b_1 & b_2 & b_3 & b_4 & \cdots & \cdots & \cdots \\ 0 & b_1 & b_2 & b_3 & \cdots & \cdots & \cdots \\ 0 & 0 & b_1 & b_2 & \cdots & \cdots & \cdots \\ \vdots & \vdots & \vdots & \ddots & \ddots & \vdots & \vdots \\ \vdots & \vdots & \vdots & \vdots & \ddots & \ddots & \vdots \\ 0 & 0 & 0 & 0 & 0 & b_1 & b_2 \\ 0 & 0 & 0 & 0 & 0 & 0 & b_1 \end{bmatrix}$$

where $b_1, b_2, \cdots$ have been determined from the numbers $a_1, a_2, \cdots$ by the equation

$$b_1 + b_2 x + b_3 x^2 + \cdots = \frac{1}{a_1 + a_2 x + a_3 x^2 + \cdots} \tag{8.66}$$

Verification : If the equation (8.66) holds, then

$(a_1 + a_2x + a_3x^2 + \cdots)(b_1 + b_2x + b_3x^2 + \cdots) = 1$

This implies that

$$a_1b_1 = 1$$
$$a_1b_2 + a_2b_1 = 0$$
$$a_1b_3 + a_2b_2 + a_3b_1 = 0 \text{ etc.}$$

Now

$$AB = BA = \begin{bmatrix} 1 & 0 & 0 & \cdots & 0 \\ 0 & 1 & 0 & \cdots & 0 \\ 0 & 0 & 1 & \cdots & 0 \\ \vdots & \vdots & \vdots & \ddots & \vdots \\ 0 & 0 & 0 & \cdots & 1 \end{bmatrix} = I.$$

$\therefore$ B is the inverse of A and vice-versa.

N.B. The tridiagonal matrix is a special case of TIM. So, the inversion procedure remains same.

Ex.1 Find the inverse of

$$A = \begin{bmatrix} \frac{1}{2} & \frac{1}{4} & \frac{1}{8} & \frac{1}{16} & \cdots & \cdots & \cdots \\ 0 & \frac{1}{2} & \frac{1}{4} & \frac{1}{8} & \ddots & \cdots & \cdots \\ 0 & 0 & \frac{1}{2} & \frac{1}{4} & \ddots & \ddots & \cdots \\ \vdots & \vdots & \vdots & \ddots & \ddots & \ddots & \vdots \\ 0 & 0 & 0 & 0 & \frac{1}{2} & \frac{1}{4} & \frac{1}{8} \\ 0 & 0 & 0 & 0 & 0 & \frac{1}{2} & \frac{1}{4} \\ 0 & 0 & 0 & 0 & 0 & 0 & \frac{1}{2} \end{bmatrix}$$

This is a TIM in which

$a_1 = \frac{1}{2}, a_2 = \frac{1}{4}, a_3 = \frac{1}{8}, a_4 = \frac{1}{16}$ etc. Let A^{-1} be

$$B = \begin{bmatrix} b_1 & b_2 & b_3 & b_4 & \cdots & \cdots & \cdots \\ 0 & b_1 & b_2 & b_3 & \cdots & \cdots & \cdots \\ 0 & 0 & b_1 & b_2 & \cdots & \cdots & \cdots \\ \vdots & \vdots & \vdots & \ddots & \ddots & \vdots & \vdots \\ \vdots & \vdots & \vdots & \vdots & \ddots & \ddots & \vdots \\ 0 & 0 & 0 & 0 & 0 & b_1 & b_2 \\ 0 & 0 & 0 & 0 & 0 & 0 & b_1 \end{bmatrix}$$

where $b_1 + b_2 x + b_3 x^2 + \cdots = \dfrac{1}{a_1 + a_2 x + a_3 x^2 + \cdots}$

i.e. $a_1 b_1 = 1$

$a_1 b_2 + a_2 b_1 = 0$

$a_1 b_3 + a_2 b_2 + a_3 b_1 = 0$

$a_1 b_4 + a_2 b_3 + a_3 b_2 + a_4 b_1 = 0$ etc.

or $\frac{1}{2} b_1 = 1$ i.e. $b_1 = 2$

$\frac{1}{2} b_2 + \frac{1}{4} b_1 = 0$ i.e. $b_2 = -1$

$\frac{1}{2} b_3 + \frac{1}{4} b_2 + \frac{1}{8} b_1 = 0$ i.e. $b_3 = 0$

$\frac{1}{2} b_4 + \frac{1}{4} b_3 + \frac{1}{8} b_2 + \frac{1}{16} b_1 = 0$ i.e. $b_4 = 0$ etc.

$\therefore$ the required inverse is

$$A^{-1} = B = \begin{bmatrix} 2 & -1 & 0 & 0 & \cdots & \cdots & \cdots \\ 0 & 2 & -1 & 0 & \cdots & \cdots & \cdots \\ 0 & 0 & 2 & -1 & \cdots & \cdots & \cdots \\ \vdots & \vdots & \vdots & \ddots & \ddots & \vdots & \vdots \\ \vdots & \vdots & \vdots & \vdots & \ddots & \ddots & \vdots \\ 0 & 0 & 0 & 0 & 0 & 2 & -1 \\ 0 & 0 & 0 & 0 & 0 & 0 & 2 \end{bmatrix}$$

Ex.2 Find the inverse of the tridiagonal matrix

$$A = \begin{bmatrix} 1 & -2 & 1 & 0 & \cdots & \cdots & \cdots \\ 0 & 1 & -2 & 1 & \cdots & \cdots & \cdots \\ 0 & 0 & 1 & -2 & \ddots & \cdots & \cdots \\ 0 & 0 & 0 & \ddots & \ddots & \ddots & \cdots \\ \vdots & \vdots & \vdots & \vdots & \ddots & \ddots & \ddots \\ 0 & 0 & 0 & 0 & 0 & 1 & -2 \\ 0 & 0 & 0 & 0 & 0 & 0 & 1 \end{bmatrix}$$

Comparing with standard TIM we have $a_1 = 1$, $a_2 = -2$, $a_3 = 1$, $a_4 = \cdots = 0$. Let the inverse A^{-1} be

$$B = \begin{bmatrix} b_1 & b_2 & b_3 & b_4 & \cdots & \cdots & \cdots \\ 0 & b_1 & b_2 & b_3 & \cdots & \cdots & \cdots \\ 0 & 0 & b_1 & b_2 & \ddots & \cdots & \cdots \\ \vdots & \vdots & \vdots & \ddots & \ddots & \ddots & \vdots \\ \vdots & \vdots & \vdots & \vdots & \ddots & \ddots & \ddots \\ 0 & 0 & 0 & 0 & 0 & b_1 & b_2 \\ 0 & 0 & 0 & 0 & 0 & 0 & b_1 \end{bmatrix}$$

such that

$$b_1 + b_2 x + b_3 x^2 + \cdots\cdots = \frac{1}{a_1 + a_2 x + a_3 x^2 + \cdots\cdots}$$

i.e. $a_1 b_1 = 1$ or $b_1 = \frac{1}{a_1} = 1$

$a_1 b_2 + a_2 b_1 = 0$ or $b_2 = 2$

$a_1 b_3 + a_2 b_2 + a_3 b_1 = 0$ or $b_3 = 3$

$a_1 b_4 + a_2 b_3 + a_3 b_2 + a_4 b_1 = 0$ or $b_4 = 4$ etc.

$\therefore$ the required inverse is

$$A^{-1} = B = \begin{bmatrix} 1 & 2 & 3 & 4 & 5 & \cdots & \cdots \\ 0 & 1 & 2 & 3 & 4 & \cdots & \cdots \\ 0 & 0 & 1 & 2 & 3 & \ddots & \cdots \\ \vdots & \vdots & \vdots & \ddots & \ddots & \ddots & \ddots \\ 0 & 0 & 0 & 0 & 1 & 2 & 3 \\ 0 & 0 & 0 & 0 & 0 & 1 & 2 \\ 0 & 0 & 0 & 0 & 0 & 0 & 1 \end{bmatrix}.$$

8.10 Cramer's rule

For statement see section 7.5.1.1.

Proof (Using inverse): The system of equations can be compactly written in matrix form as

$$\mathbf{Ax} = \mathbf{b} \tag{8.68}$$

where $A = [a_{ij}]$, coefficient matrix, $\mathbf{x} = [x_j]^T$ and $\mathbf{b} = [b_j]^T$ are column matrices.

Since $|A| \neq 0$, so A is invertible i.e. there exists an A^{-1}. Therefore, from (8.68)

$$\mathbf{x} = A^{-1}\mathbf{b}$$

$$= \frac{\text{adj } A}{|A|}\mathbf{b}$$

$$= \frac{1}{|A|}\begin{bmatrix} A_{11} & A_{21} & \cdots & \cdots & \cdots & A_{n1} \\ A_{12} & A_{22} & \cdots & \cdots & \cdots & A_{n2} \\ \vdots & \vdots & \ddots & \vdots & \vdots & \vdots \\ A_{1j} & A_{2j} & \cdots & \ddots & \cdots & A_{nj} \\ \vdots & \vdots & \vdots & \vdots & \ddots & \vdots \\ A_{1n} & A_{2n} & \cdots & \cdots & \cdots & A_{nn} \end{bmatrix}\begin{bmatrix} b_1 \\ b_2 \\ \vdots \\ \vdots \\ \vdots \\ b_n \end{bmatrix}$$

$$= \frac{1}{|A|}\begin{bmatrix} (b_1 A_{11} + b_2 A_{21} + \cdots + b_n A_{n1}) \\ (b_1 A_{12} + b_2 A_{22} + \cdots + b_n A_{n2}) \\ \vdots \\ (b_1 A_{1j} + b_2 A_{2j} + \cdots + b_n A_{nj}) \\ \vdots \\ (b_1 A_{1n} + b_2 A_{2n} + \cdots + b_n A_{nn}) \end{bmatrix}$$

$$\therefore \quad (\text{the } j\text{th row of } \mathbf{x}) = \mathbf{x}_j = \frac{b_1 A_{1j} + b_2 A_{2j} + \cdots + b_n A_{nj}}{|A|} \tag{8.69}$$

Again by hypothesis,

← jth column

$$A_j = \begin{bmatrix} a_{11} & a_{12} & \cdots & a_{1,j-1} & b_1 & a_{1,j+1} & \cdots & a_{1n} \\ a_{21} & a_{22} & \cdots & a_{2,j-1} & b_2 & a_{2,j+1} & \cdots & a_{2n} \\ a_{31} & a_{32} & \ddots & a_{3,j-1} & b_3 & a_{3,j+1} & \cdots & a_{3n} \\ \vdots & \vdots & \vdots & \ddots & \vdots & \vdots & \vdots & \vdots \\ \vdots & \vdots & \vdots & \vdots & \ddots & \vdots & \vdots & \vdots \\ \vdots & \vdots & \vdots & \vdots & \vdots & \ddots & \vdots & \vdots \\ \vdots & \vdots & \vdots & \vdots & \vdots & \vdots & \ddots & \vdots \\ a_{n1} & a_{2n} & \cdots & a_{n,j-1} & b_n & a_{n,j+1} & \cdots & a_{nn} \end{bmatrix}$$

Since the matrices A and A_j differ only in jth column, so the cofactors of the elements in the jth column of both matrices must be the same. Therefore, expanding $|A_j|$ w.r.t. the jth column of $|A_j|$,

$$|A_j| = b_1 A_{1j} + b_2 A_{2j} + \cdots + b_n A_{nj} \tag{8.70}$$

Therefore, using (8.70) in (8.69), we get

$$x_j = \frac{|A_j|}{|A|}, \quad j = 1, 2, \cdots, n$$

Hence the proof.

Theorem . The system of equations $\mathbf{Ax} = \mathbf{b}$ has a unique solution if A is nonsingular.

Alternate proof . If A is nonsingular then $\mathbf{x} = A^{-1}\mathbf{b}$ is the only solution of $\mathbf{Ax} = \mathbf{b}$. Conversely, let $\hat{\mathbf{x}}$ be the unique solution of $\mathbf{Ax} = \mathbf{b}$. Let A is singular. Then $\mathbf{Ax} = 0$ has a nontrivial solution $\mathbf{z}$. Let

$$\mathbf{y} = \hat{\mathbf{x}} + \mathbf{z}\text{ . Clearly } \hat{\mathbf{x}} \neq \mathbf{z} \text{ and}$$

$$\mathbf{Ay} = A(\hat{\mathbf{x}} + \mathbf{z}) = A\hat{\mathbf{x}} + A\mathbf{z} = \mathbf{b} + 0 = \mathbf{b}$$

$\therefore$ $\mathbf{y} = \hat{\mathbf{x}} + \mathbf{z}$ is a solution of $\mathbf{Ax} = \mathbf{b}$. That is a contradiction. Therefore if $\mathbf{Ax} = \mathbf{b}$ has unique solution then A must be nonsingular.

8.11 Helmertian in the strict sense

This is an alternative definition of the Helmertian in the strict sense already defined in sec. 3.16.3. The Helmertian in the strict sense is defined as

$$H^e = D^{-1}H_0$$

$$= D^{-1}\begin{bmatrix} 1 & 1 & 1 & 1 & \cdots & 1 \\ 1 & -1 & 0 & 0 & \cdots & 0 \\ 1 & 1 & -2 & 0 & \cdots & 0 \\ 1 & 1 & 1 & -3 & \cdots & 0 \\ \vdots & \vdots & \vdots & \vdots & \ddots & \vdots \\ 1 & 1 & 1 & 1 & \cdots & -(n-1) \end{bmatrix}$$

where D^2 is a diagonal matrix.,

$$D^2 = \text{diag}\,(n.1 \quad 1.2 \quad 2.3 \quad \cdots \quad (n-1)n)$$

The matrix H^e is orthogonal, since rows of H_0 are mutually orthogonal, i.e. dot product of any two rows is zero. Also in H_0 the sum of the first row elements is n and that of other individual rows is zero. Therefore, the above H^e simplifies to

$$H^e = \begin{bmatrix} \frac{1}{\sqrt{n.1}} & 0 & 0 & \cdots & 0 \\ 0 & \frac{1}{\sqrt{1.2}} & 0 & \cdots & 0 \\ 0 & 0 & \frac{1}{\sqrt{2.3}} & \cdots & 0 \\ \vdots & \vdots & \vdots & \ddots & \vdots \\ 0 & 0 & 0 & \cdots & \frac{1}{\sqrt{(n-1)n}} \end{bmatrix} \begin{bmatrix} 1 & 1 & 1 & 1 & \cdots & 1 \\ 1 & -1 & 0 & 0 & \cdots & 0 \\ 1 & 1 & -2 & 0 & \cdots & 0 \\ 1 & 1 & 1 & -3 & \cdots & 0 \\ \vdots & \vdots & \vdots & \vdots & \ddots & \vdots \\ 1 & 1 & 1 & 1 & \cdots & -(n-1) \end{bmatrix}$$

$$= \begin{bmatrix} \frac{1}{\sqrt{n.1}} & \frac{1}{\sqrt{n.1}} & \frac{1}{\sqrt{n.1}} & \frac{1}{\sqrt{n.1}} & \cdots & \frac{1}{\sqrt{n.1}} \\ \frac{1}{\sqrt{1.2}} & -\frac{1}{\sqrt{1.2}} & 0 & 0 & \cdots & 0 \\ \frac{1}{\sqrt{2.3}} & \frac{1}{\sqrt{2.3}} & -\frac{2}{\sqrt{2.3}} & 0 & \cdots & 0 \\ \frac{1}{\sqrt{3.4}} & \frac{1}{\sqrt{3.4}} & \frac{1}{\sqrt{3.4}} & \frac{1}{\sqrt{3.4}} & \cdots & 0 \\ \vdots & \vdots & \vdots & \vdots & \ddots & \vdots \\ \frac{1}{\sqrt{(n-1)n}} & \frac{1}{\sqrt{(n-1)n}} & \frac{1}{\sqrt{(n-1)n}} & \frac{1}{\sqrt{(n-1)n}} & \cdots & -\frac{(n-1)}{\sqrt{(n-1)n}} \end{bmatrix}$$

which is an nth order square Helmertian matrix. Giving different integral values to n, Helmertians of different orders can be constructed.

8.12 Leontief input-output open model

Input-output models may be of two types — closed and open. In a closed model input commodities must be identified at least once with output commodities and there is no source of inputs other than current production and no use for outputs except as inputs. The model may be open in various ways.

Let x_i be the total output of industry i required to meet the demand of the open sector and all n industries. Then

$$x_i = a_{i1}x_1 + a_{i2}x_2 + \cdots + a_{in}x_n + d_i, \quad i = 1, 2, \cdots, n \tag{8.71}$$

where d_i denotes the demand of the open sector from the ith industry and $a_{ij}x_j$ represent input requirements of the jth industry from the ith. The equation (8.71) can be written as a system of n equations:

$$\left.\begin{array}{lllll} (1-a_{11})x_1 - & a_{12}x_2 - & \cdots & -a_{1n}x_n & = d_1 \\ -a_{21}x_1 + & (1-a_{22})x_2 - & \cdots & -a_{2n}x_n & = d_2 \\ \vdots & \vdots & \ddots & \vdots & \vdots \\ \vdots & \vdots & \vdots & \vdots & \vdots \\ -a_{n1}x_1 & -a_{n2}x_2 & \cdots & (1-a_{nn})x_n & = d_n \end{array}\right\} \tag{8.72}$$

or

$$\begin{bmatrix} 1-a_{11} & -a_{12} & -a_{13} & \cdots & -a_{1n} \\ -a_{21} & 1-a_{22} & -a_{23} & \cdots & -a_{2n} \\ \vdots & \vdots & \ddots & \vdots & \vdots \\ \vdots & \vdots & \vdots & \ddots & \vdots \\ -a_{n1} & -a_{n2} & -a_{n3} & \cdots & 1-a_{nn} \end{bmatrix} \begin{bmatrix} x_1 \\ x_2 \\ \vdots \\ x_n \end{bmatrix} = \begin{bmatrix} d_1 \\ d_2 \\ \vdots \\ d_n \end{bmatrix} \tag{8.73}$$

or

$$(\mathbf{I} - \mathbf{A})\,\mathbf{x} = \mathbf{d} \tag{8.74}$$

The matrix A in (8.74) is nonnegative and each column adds up to less than unity. So, the matrix $(\mathrm{I} - \mathrm{A})$, called the *technology matrix*, is nonsingular. Hence from (8.74),

$$\mathbf{x} = (\mathbf{I} - \mathbf{A})^{-1}\,\mathbf{d} \tag{8.75}$$

For large A, the inverse in (8.75) can be obtained in the following way when $\lim_{m \to \infty} \mathrm{A}^{m+1} = 0$

for some large positive integer m. Now

$$(I - A)(I + A + A^2 + \cdots + A^m)$$
$$= (I + A + A^2 + \cdots + A^m) - A(I + A + A^2 + \cdots + A^m)$$
$$= I - A^{m+1} \tag{8.76}$$

Using $\lim_{m \to \infty} A^{m+1} = 0$, R.H.S. (8.76) reduces to I,

i.e. $$(I - A)(I + A + A^2 + \cdots + A^m) = I$$

or $$(I + A + A^2 + \cdots + A^m) = (I - A)^{-1} \tag{8.77}$$

Using (8.77) in (8.75), $$\mathbf{x} = (I + A + A^2 + \cdots + A^m)\mathbf{d} \tag{8.78}$$

Exercises Eight

1. Prove that

$$\text{cof cof } A = \begin{bmatrix} 26 & -13 & 0 \\ 39 & -26 & 13 \\ -52 & 65 & 91 \end{bmatrix}, \text{ when } A = \begin{bmatrix} 2 & -1 & 0 \\ 3 & -2 & 1 \\ -4 & 5 & 7 \end{bmatrix}.$$

[Hints: Use cof cof A = $|A|^{n-2}A$ for *n* square matrix A]

2. Show that

$$\text{cof} \begin{bmatrix} 1 & 0 & -1 \\ 0 & 2 & 2 \\ 1 & 1 & 1 \end{bmatrix} = \begin{bmatrix} 0 & 2 & -2 \\ -1 & 2 & -1 \\ -2 & -2 & 2 \end{bmatrix}.$$

3. Show that $\det(B^{-1}AB) = \det A$.

4. Find cof cof A, when

$$A = \begin{bmatrix} -1 & 2 & 1 & 1 \\ 0 & 1 & 0 & 0 \\ 0 & 1 & 1 & -1 \\ 2 & 0 & 0 & 3 \end{bmatrix}.$$ **[Ans.** 49A**]**

Find the inverse of the following (5 – 7):

5. $\begin{bmatrix} 1 & 1 & 0 \\ -1 & 2 & 1 \end{bmatrix}$

[Ans. No A_L, $A_R = \begin{bmatrix} 1 & 0 \\ 0 & 0 \\ 1 & 1 \end{bmatrix}$ with free variables each equal to 1 and

$$A_R = \begin{bmatrix} 0 & -1 \\ 1 & 1 \\ -2 & -2 \end{bmatrix} \text{ with free variables each equal to } -2]$$

6. $A = \begin{bmatrix} 1 & 1 \\ 0 & -1 \\ 1 & 4 \end{bmatrix}$ [**Ans.** Has no $A_R.A_L = \begin{bmatrix} 1 & 1 & 0 \\ -1 & 2 & 1 \end{bmatrix}, \begin{bmatrix} 0 & 4 & 1 \\ 0 & -1 & 0 \end{bmatrix}$]

7. (i) $\begin{bmatrix} 1 & 0 \\ 0 & 2 \\ 3 & 1 \end{bmatrix}$ [**Ans.** No A_R, $A_L = \begin{bmatrix} 7 & 1 & -2 \\ 0 & \frac{1}{2} & 0 \end{bmatrix}$]

(ii) $\begin{bmatrix} 3 & 1 \\ 5 & 2 \end{bmatrix}$ [**Ans.** $\begin{bmatrix} 2 & -1 \\ -5 & 3 \end{bmatrix}$]

8. Show that A_L and A_R of

$$A = \begin{bmatrix} 1 & 2 \\ 0 & 1 \end{bmatrix} \text{ are equal and the common value is } A^{-1} = \begin{bmatrix} 1 & -2 \\ 0 & 1 \end{bmatrix}.$$

9. Find the inverse of an 1×1 matrix.

[**Hints:** Let $A = [a_{11}]$, then $|A| = a_{11}$, cof $A = [1]$, adj $A = (\text{cof } A)^T = [1]$.

$$\therefore A^{-1} = \frac{\text{adj } A}{|A|} = [1/a_{11}],\ a_{11} \neq 0]$$

10. Show that the matrix

$B = \begin{bmatrix} 0 & 0 & 0 \\ 2 & -1 & 3 \\ 4 & 1 & 6 \end{bmatrix}$ does not have an inverse and that the matrix $\begin{bmatrix} 2 & 0 & 0 \\ 3 & 1 & 0 \\ 5 & 0 & 1 \end{bmatrix}$ is a right identity while the matrix

$$\begin{bmatrix} -5 & 3 & 3 \\ 0 & 1 & 0 \\ 4 & -2 & -1 \end{bmatrix}$$

is a left identity for B. Find one more right and left identity for B not of the form I_3 and verify the results.

11. Given that

$$A = \begin{bmatrix} 1 & 1 & 1 & 1 \\ 2 & -1 & 3 & 0 \\ 0 & 2 & 0 & 3 \\ -1 & 0 & 2 & 1 \end{bmatrix} \text{ and } A^{-1} = \frac{1}{14}\begin{bmatrix} 0 & 4 & 2 & 6 \\ 21 & -9 & -8 & 3 \\ 7 & -1 & -4 & 5 \\ -14 & 6 & 10 & -2 \end{bmatrix}.$$

Prove that

$$\begin{bmatrix} 0 & 4 & 2 & 6 \\ 21 & -9 & -8 & 3 \\ 7 & -1 & -4 & 5 \\ -14 & 6 & 10 & -2 \end{bmatrix}^{-1} = \frac{1}{14}\begin{bmatrix} 1 & 1 & 1 & 1 \\ 2 & -1 & 3 & 0 \\ 0 & 2 & 0 & 3 \\ -1 & 0 & 2 & 1 \end{bmatrix}.$$

[**Hints :** $\left(cA^{-1}\right) = \frac{1}{c}A$, c is a scalar]

12. If

$$A = \begin{bmatrix} A_1 & B_1 \\ A_2 & B_2 \end{bmatrix}$$

is a partitioned matrix such that A_1 and B_2 are nonsingular, show that

$$\begin{bmatrix} A_1 & B_1 \\ A_2 & B_2 \end{bmatrix}^{-1} = \begin{bmatrix} (A_1 - B_1B_2^{-1}A_2)^{-1} & (B_1B_2^{-1}A_2 - A_1)^{-1}B_1B_2^{-1} \\ (A_2A_1^{-1}B_1 - B_2)^{-1}A_2A_1^{-1} & (B_2 - A_2A_1^{-1}B_1)^{-1} \end{bmatrix}.$$

and that if A_2 and B_1 are nonsingular, this becomes

$$\begin{bmatrix} (A_1 - B_1B_2^{-1}A_2)^{-1} & (A_2 - B_1B_2^{-1}A_1)^{-1} \\ (B_1 - A_1A_2^{-1}B_2)^{-1} & (B_2 - A_2A_1^{-1}B_1)^{-1} \end{bmatrix}.$$

13. If the 3×3 partition matrix

$$\begin{bmatrix} A & B & C \\ B^T & D & 0 \\ C^T & 0 & E \end{bmatrix}$$

is symmetric and nonsingular, its inverse is given by

$$\begin{bmatrix} Q^{-1} & -Q^{-1}BD^{-1} & -Q^{-1}CE^{-1} \\ -D^{-1}B^TQ^{-1} & D^{-1} + D^{-1}B^TQ^{-1}BD^{-1} & D^{-1}B^TQ^{-1}CE^{-1} \\ -E^{-1}C^TQ^{-1} & E^{-1}C^TQ^{-1}BD^{-1} & E^{-1} + E^{-1}C^TQ^{-1}CE^{-1} \end{bmatrix},$$

where $Q = A - BD^{-1}B^T - CE^{-1}C^T$.

14. If

$$A^{-1} = \begin{bmatrix} 1 & 1 & 1 \\ 1 & 2 & 3 \\ 1 & 4 & 9 \end{bmatrix}, \text{ find A.}$$

[**Ans.**$(A^{-1})^{-1} = A$, then $A = \begin{bmatrix} 3 & -\frac{5}{2} & \frac{1}{2} \\ -3 & 4 & -1 \\ 1 & -\frac{3}{2} & \frac{1}{2} \end{bmatrix}$]

15. If

$$\left(A^T\right)^{-1} = \begin{bmatrix} 1 & 1 & 1 \\ 3 & 4 & 5 \\ 1 & 2 & 1 \end{bmatrix}, \text{ find A.}$$

[**Ans.** Since $\left(\left(A^T\right)^{-1}\right)^{-1} = A^T$, so $A = \begin{bmatrix} 3 & -1 & -1 \\ -\frac{1}{2} & 0 & \frac{1}{2} \\ -\frac{1}{2} & 1 & -\frac{1}{2} \end{bmatrix}$]

16. If

$A = \begin{bmatrix} 2 & 1 \\ 3 & 2 \end{bmatrix}$, find A^{-1}. Show by a direct computation that $(A^{-1})^3 = (A^3)^{-1}$.

[**Ans.** $A^{-3} = \begin{bmatrix} 26 & -15 \\ -45 & 26 \end{bmatrix}$]

17. If

$A = \begin{bmatrix} 6 & 7 \\ 5 & 6 \end{bmatrix}$, compute $\left(A^T\right)^{-2}$. Compare with $\left(A^{-2}\right)^T$.

[**Ans.** $\begin{bmatrix} 71 & -60 \\ -84 & 71 \end{bmatrix}$]

18. Show that $\left(A^3\right)^T$ is the inverse $\left(A^T\right)^{-3}$.

19. Prove that $\left(ABA^{-1}\right)^2 = AB^2A^{-1}$, and hence for positive integer *n*, $\left(ABA^{-1}\right)^n = AB^nA^{-1}$ if the respective matrix multiplication is coformable.

20. Prove that if A and B are symmetric and commute, then if A^{-1} and B^{-1} exist, $A^{-1}B$, AB^{-1} and $A^{-1}B^{-1}$ are symmetric.

21. Simplify $\left(\left(\left(A^{-1}\right)^{-1}\right)^{-1}\right)^{-1}$. [**Ans.** A]

22. If A^{-1} is

$\begin{bmatrix} 1 & 1 & 1 \\ 1 & 2 & 3 \\ 1 & 4 & 9 \end{bmatrix}$, find A. [**Hints:** $\left(A^{-1}\right)^{-1} = A = \begin{bmatrix} 1 & 1 & 1 \\ 1 & 2 & 3 \\ 1 & 4 & 9 \end{bmatrix}^{-1} = \begin{bmatrix} 3 & -\frac{5}{2} & \frac{1}{2} \\ -3 & 4 & -1 \\ 1 & -\frac{3}{2} & \frac{1}{2} \end{bmatrix}$]

23. If

$\left(A^T\right)^{-1} = \begin{bmatrix} 1 & 1 & 1 \\ 3 & 4 & 5 \\ 1 & 2 & 1 \end{bmatrix}$, find A.

[**Hints :** $A^T = \begin{bmatrix} 1 & 1 & 1 \\ 3 & 4 & 5 \\ 1 & 2 & 1 \end{bmatrix}^{-1} = \begin{bmatrix} 3 & -\frac{1}{2} & -\frac{1}{2} \\ -1 & 0 & 1 \\ -1 & \frac{1}{2} & -\frac{1}{2} \end{bmatrix}$. $\therefore A = \begin{bmatrix} 3 & -1 & -1 \\ -\frac{1}{2} & 0 & \frac{1}{2} \\ -\frac{1}{2} & 1 & -\frac{1}{2} \end{bmatrix}$]

24. If

$A=\begin{bmatrix}6 & 7\\ 5 & 6\end{bmatrix}$, compute $\left(A^T\right)^{-2}$. Compare with $\left(A^{-2}\right)^T$. [**Ans.** $\begin{bmatrix}71 & -84\\ 60 & 71\end{bmatrix}$]

25. If $Ax + B = C$, prove that $x = A^{-1}C - A^{-1}B$. What assumption has been made about A?

[**Ans.** A must be nonsingular]

26. For what values of x does A fail to have an inverse if

$$A=\begin{bmatrix}1 & 2 & x-2\\ x & 4 & 0\\ 2 & x & -3\end{bmatrix}?$$

[**Ans.** $x = 2,\ \pm\sqrt{2}$]

27. Find $(A \otimes B)^{-1}$ and $(AB)^{-1}$ each in two ways where

(i) $A=\begin{bmatrix}3 & 1\\ 1 & 3\end{bmatrix}$ and $B=\begin{bmatrix}1 & 3\\ 3 & 1\end{bmatrix}$; (ii) $A=\begin{bmatrix}1 & 2\\ -1 & 4\end{bmatrix}$ and $B=\begin{bmatrix}2 & 1\\ 1 & 2\end{bmatrix}$.

[**Ans.** (i) $\frac{1}{64}\begin{bmatrix}-3 & 9 & 1 & -3\\ 9 & -3 & -3 & 1\\ 1 & -3 & -3 & 9\\ -3 & 1 & 9 & -3\end{bmatrix}$ and $-\frac{1}{32}\begin{bmatrix}3 & -5\\ -5 & 3\end{bmatrix}$,

(ii) $\frac{1}{18}\begin{bmatrix}8 & -4 & -4 & 2\\ -4 & 8 & 2 & -4\\ 2 & -1 & 2 & -1\\ -1 & 2 & -1 & 2\end{bmatrix}$ and $\frac{1}{18}\begin{bmatrix}7 & -5\\ -2 & 4\end{bmatrix}$]

28. Find

$$\frac{4+3i}{1+2i}$$ using matrices.

[**Hints:**

$$\frac{4+3i}{1+2i} = (4+3i)(1+2i)^{-1}$$

$$=\begin{bmatrix}4 & 3\\ -3 & 4\end{bmatrix}\begin{bmatrix}1 & 2\\ -1 & 2\end{bmatrix}^{-1} = \begin{bmatrix}4 & 3\\ -3 & 4\end{bmatrix}\begin{bmatrix}\frac{1}{5} & -\frac{2}{5}\\ \frac{2}{5} & \frac{1}{5}\end{bmatrix} = \begin{bmatrix}2 & -1\\ 1 & 2\end{bmatrix} = 2+i\,]$$

29. Prove that adj (adj A) $= |A|^m$, $m = (n-1)^2$ and $A = [a_{ij}]_{n\times n}$.

30. If A is symmetric, how do cof A and adj A compare ?

31. For a square matrix A of order $n > 1$, does $(\text{cof } A)^T = \text{cof }(A^T)$? Show that the adjoint of a symmetric (Hermitian) matrix is itself symmetric (Hermitian).

[**Ans.** They are equal]

32. Find the inverse of

$$\begin{bmatrix}3 & 5 & 0\\ 1 & 2 & 0\\ 0 & 0 & 6\end{bmatrix}.$$

using the direct sum of blocks.

[**Hints :** $A^{-1} = \begin{bmatrix} 3 & 5 & 0 \\ 1 & 2 & 0 \\ 0 & 0 & 6 \end{bmatrix}^{-1} = \left[\begin{array}{cc|c} 3 & 5 & 0 \\ 1 & 2 & 0 \\ \hline 0 & 0 & 6 \end{array}\right]^{-1} = \left(\begin{bmatrix} 3 & 5 \\ 1 & 2 \end{bmatrix} \oplus [6]\right)^{-1} = \begin{bmatrix} 3 & 5 \\ 1 & 2 \end{bmatrix}^{-1} \oplus [6]^{-1}$

$= \begin{bmatrix} 2 & -5 \\ -1 & 3 \end{bmatrix} \oplus \left[\frac{1}{6}\right] = \begin{bmatrix} 2 & -5 & 0 \\ -1 & 3 & 0 \\ 0 & 0 & \frac{1}{6} \end{bmatrix}$]

33. Inverse of diagonal supermatrix

$$\begin{bmatrix} A_{11} & & \bigcirc \\ & \ddots & \\ \bigcirc & & A_{nn} \end{bmatrix}^{-1} = \begin{bmatrix} A_{11}^{-1} & & \bigcirc \\ & \ddots & \\ \bigcirc & & A_{nn}^{-1} \end{bmatrix}.$$

34. Find the inverse of the direct sum of

$$A = \begin{bmatrix} 1 & 2 \\ 3 & 5 \end{bmatrix} \quad \text{and} \quad B = \begin{bmatrix} 1 & 2 \\ 0 & 1 \end{bmatrix}.$$

35. Detemine the inverse of the following, given

$$\begin{bmatrix} 1 & 1 \\ 3 & 4 \end{bmatrix}^{-1} = \begin{bmatrix} 4 & -1 \\ -3 & 1 \end{bmatrix} \quad :$$

$A = \left[\begin{array}{cc|c} 1 & 1 & 0 \\ 3 & 4 & 0 \\ \hline 0 & 0 & 4 \end{array}\right]$, [**Ans.** $\left[\begin{array}{cc|c} 4 & -1 & 0 \\ -3 & 1 & 0 \\ \hline 0 & 0 & \frac{1}{4} \end{array}\right]$]

$B = \left[\begin{array}{cc|c} 1 & 1 & 2 \\ 3 & 4 & 1 \\ \hline 0 & 0 & 3 \end{array}\right]$, [**Ans.** $\left[\begin{array}{cc|c} 4 & -1 & \frac{7}{3} \\ -3 & 1 & \frac{5}{3} \\ \hline 0 & 0 & \frac{1}{3} \end{array}\right]$]

$C = \left[\begin{array}{cc|cc|c|c} 1 & 1 & 0 & 0 & 0 & 0 \\ 3 & 4 & 0 & 0 & 0 & 0 \\ \hline 0 & 0 & 1 & 1 & 0 & 0 \\ 1 & 1 & 3 & 4 & 0 & 0 \\ \hline 0 & 0 & 0 & 0 & 6 & 0 \\ \hline 0 & 0 & 0 & 0 & 0 & 9 \end{array}\right]$, [**Ans.** $\left[\begin{array}{cc|cc|c|c} 4 & -1 & 0 & 0 & 0 & 0 \\ -3 & 1 & 0 & 0 & 0 & 0 \\ \hline 0 & 0 & 4 & -1 & 0 & 0 \\ 0 & 0 & -3 & 1 & 0 & 0 \\ \hline 0 & 0 & 0 & 0 & \frac{1}{6} & 0 \\ \hline 0 & 0 & 0 & 0 & 0 & \frac{1}{9} \end{array}\right]$]

$D = \left[\begin{array}{ccc|cc} 3 & 0 & 0 & 1 & 1 \\ 0 & 2 & 0 & 1 & 1 \\ 0 & 0 & 1 & 1 & 1 \\ \hline 0 & 0 & 0 & 1 & 1 \\ 0 & 0 & 0 & 3 & 4 \end{array}\right]$. [**Ans.** $\left[\begin{array}{ccc|cc} \frac{1}{3} & 0 & 0 & -\frac{1}{3} & 0 \\ 0 & \frac{1}{2} & 0 & -\frac{1}{2} & 0 \\ 0 & 0 & 1 & -1 & 0 \\ \hline 0 & 0 & 0 & 4 & -1 \\ 0 & 0 & 0 & -3 & 1 \end{array}\right]$]

36. Obtain the inverse of the following matrix using given partitions

$$\left[\begin{array}{ccc:cc} 1 & 0 & 1 & 0 & 1 \\ 0 & 1 & 0 & 1 & 0 \\ 1 & 0 & 0 & 0 & 1 \\ \hdashline 0 & 1 & 0 & 0 & 1 \\ 1 & 0 & 1 & 1 & 1 \end{array}\right].$$

$$\text{[Ans.} \left[\begin{array}{ccc:cc} 0 & 0 & 1 & 0 & 0 \\ 1 & 1 & 0 & 0 & -1 \\ 2 & 1 & -1 & -1 & -1 \\ \hdashline -1 & 0 & 0 & 0 & 1 \\ -1 & -1 & 0 & 1 & 1 \end{array}\right]\text{]}$$

37. If A, B, C and D are nonsingular, square matrices of order n, show that

$$\left[\begin{array}{c:c} A & B \\ \hdashline C & D \end{array}\right]^{-1} = \left[\begin{array}{c:c} (A-BD^{-1}C) & (C-DB^{-1}A)^{-1} \\ \hdashline (B-AC^{-1}D)^{-1} & (D-CA^{-1}B)^{-1} \end{array}\right]$$

assuming that the large matrix is also nonsingular.

38. Given that,

$$\begin{bmatrix} 1 & -2 & 0 \\ 2 & 1 & 0 \\ 0 & 0 & 1 \end{bmatrix} A \begin{bmatrix} 1 & 2 & 0 \\ -2 & 1 & 0 \\ 0 & 0 & 1 \end{bmatrix} = 5I \text{, what is det A ?} \qquad \text{[Ans. } \tfrac{1}{5}\text{]}$$

39. Given that D is square, of the same order as A, and nonsingular, use the fact that

$$\det\begin{bmatrix} A & B \\ C & D \end{bmatrix} = \det\begin{bmatrix} A & B \\ C & D \end{bmatrix}\begin{bmatrix} I & 0 \\ -D^{-1}C & I \end{bmatrix}$$ to prove that, so that, if D^{-1} exists and C and D commute, $\det\begin{bmatrix} A & B \\ C & D \end{bmatrix} = \det[AD - BC]$.

40. Let

$$\left[\begin{array}{c:c} A_{11} & A_{12} \\ \hdashline 0 & A_{22} \end{array}\right]$$ where all four blocks are $n \times n$ matrices.

(i) If A_{11} and A_{22} are nonsingular, show that A must also be nonsingular and that

A^{-1} must be of the form $\left[\begin{array}{c:c} A_{11}^{-1} & C \\ \hdashline 0 & A_{22}^{-1} \end{array}\right]$.

(ii) Determine C. [**Hints:** Proceed as in sec. 8.4., Ans. $C = -A_{11}^{-1}A_{12}A_{22}^{-1}$]

41. Let A be an invertible matrix whose inverse is

$$\begin{bmatrix} 3 & 4 \\ 5 & 6 \end{bmatrix}.$$

Find the matrix A.

[**Hints:** $AA^{-1} = I$, **Ans.** $A = \begin{bmatrix} 3 & 4 \\ 5 & 6 \end{bmatrix}^{-1} = \begin{bmatrix} -3 & 2 \\ \frac{5}{2} & -\frac{3}{2} \end{bmatrix}$]

42. Find the matrix A such that the inverse of 7A is

$$\begin{bmatrix} -1 & 2 \\ 4 & -7 \end{bmatrix}.$$

[**Ans** . $\begin{bmatrix} 1 & \frac{2}{7} \\ \frac{4}{7} & \frac{1}{7} \end{bmatrix}$]

43. Let A be a square matrix such that

(i) $A^2 - 3A + I = 0$. Show that $A^{-1} = 3I - A$.

[**Hints:** $A^2 - 3A = -I$, $A - 3I = -A^{-1}$ post multiply by A^{-1}, $\therefore A^{-1} = 3I - A$]

(ii) Let

$$\begin{bmatrix} 2 & 0 & 1 \\ 3 & 1 & 2 \\ 0 & 1 & 1 \end{bmatrix}.$$

Show that $A^3 - 4A^2 + 3A = I$ and hence find A^{-1} in terms of powers of A.

[**Ans.** $A^{-1} = 3I - 4A + A^2$]

44. If

$$A = \begin{bmatrix} 2 & 2 & 1 \\ -1 & 0 & 2 \\ 4 & -5 & 2 \end{bmatrix},$$

find AA^T and hence find A^{-1}.

[**Hints:** $\because (A^T)^{-1} = (A^{-1})^T$, $\therefore A^{-1} = \left((A^T)^{-1}\right)^T = \begin{bmatrix} \frac{2}{9} & -\frac{1}{5} & \frac{4}{45} \\ \frac{2}{9} & 0 & -\frac{1}{9} \\ \frac{1}{9} & \frac{2}{5} & \frac{2}{45} \end{bmatrix}$]

45. If

$$A = \begin{bmatrix} 1 & 1 & 1 \\ a & b & c \\ a^2 & b^2 & c^2 \end{bmatrix}, \text{ then}$$

$$A^{-1} = \begin{bmatrix} bc^2 - cb^2 & b^2 - c^2 & c - b \\ ca^2 - ac^2 & c^2 - a^2 & a - c \\ ab^2 - ba^2 & a^2 - b^2 & b - a \end{bmatrix} \div (b - c)(c - a)(a - b)$$

or

$$A^{-1} = \begin{bmatrix} b - c & 0 & 0 \\ 0 & c - a & 0 \\ 0 & 0 & a - b \end{bmatrix} \begin{bmatrix} -bc & b + c & -1 \\ -ca & c + a & -1 \\ -ab & a + b & -1 \end{bmatrix} \div (b - c)(c - a)(a - b).$$

46. Determine integers x and y such that

$\begin{bmatrix} x & y \\ 3 & 5 \end{bmatrix}$ has an inverse whose elements are all integers.

[**Ans.** $x = -\frac{3}{2}k_1, y = \frac{5}{2}k_2, k_1, k_2$ are even integers]

47. Prove that $(A^2)^{-1} = (A^{-1})^2$ and in general $(A^k)^{-1} = (A^{-1})^k = A^{-k}$ here k is an integer and A^{-1} exists.

48. Solve the linear system $A\mathbf{x} = \mathbf{b}$, where

$$\text{(i) } A^{-1} = \begin{bmatrix} -1 & \frac{1}{2} & 0 \\ \frac{1}{2} & -1 & \frac{1}{2} \\ 0 & \frac{1}{2} & -1 \end{bmatrix}, \mathbf{b} = \begin{bmatrix} -3 \\ 1 \\ 2 \end{bmatrix};$$

$$\text{(ii) } A^{-1} = \begin{bmatrix} 1 & -1 & 3 & 1 \\ 0 & 1 & 0 & 2 \\ -2 & -1 & -3 & -1 \\ 1 & -1 & 0 & 0 \end{bmatrix}, \mathbf{b} = \begin{bmatrix} 1 \\ 2 \\ 3 \\ 4 \end{bmatrix}.$$

[**Ans.**(i) $[\frac{7}{2} \quad -\frac{3}{2} \quad -\frac{3}{2}]^T$, (ii)$[12 \quad 10 \quad -17 \quad -1]^T$

49. Solve the following four systems of equation using the method in sec. 8.7.1

$$\text{(i)} \left.\begin{aligned} x+4y+z+3t &= 0 \\ -y+3z-t &= 5 \\ 3x+y \quad\quad +2t &= -1 \\ x-2y+5z+t &= 2 \end{aligned}\right\}, \quad \text{(ii)} \left.\begin{aligned} x+4y+z+3t &= -1 \\ -y+3z-t &= 2 \\ 3x+y \quad\quad +2t &= 3 \\ x-2y+5z+t &= 0 \end{aligned}\right\},$$

$$\text{(iii)} \left.\begin{aligned} x+4y+z+3t &= 1 \\ -y+3z-t &= 2 \\ 3x+y \quad\quad +2t &= -2 \\ x-2y+5z+t &= 3 \end{aligned}\right\}, \quad \text{(iv)} \left.\begin{aligned} x+4y+z+3t &= \tfrac{1}{2} \\ -y+3z-t &= 3 \\ 3x+y \quad\quad +2t &= -3 \\ x-2y+5z+t &= 4 \end{aligned}\right\}.$$

[**Ans.**

$$\begin{cases} (i)\ x = \frac{20}{22}, y = \frac{30}{22}, z = \frac{13}{11}, t = -\frac{28}{11} \\ (ii)\ x = \frac{23}{11}, y = \frac{7}{11}, z = \frac{5}{22}, t = -\frac{43}{22} \\ (iii)\ x = -\frac{37}{44}, y = \frac{5}{44}, z = \frac{17}{22}, t = \frac{9}{44} \\ (iv)\ x = -\frac{93}{88}, y = \frac{9}{88}, z = \frac{23}{22}, t = \frac{3}{88} \end{cases}$$]

50. Find the inverse of the Vandermode matrix

$$V = \begin{bmatrix} 1 & 1 & 1 \\ 2 & 3 & 4 \\ 2^2 & 3^2 & 4^2 \end{bmatrix}$$

by Kaufman's algorithm.

[**Hints:** $x_1 = 2, x_2 = 3, a_1 = -(x_1 + x_2 + x_3) = -9, a_2 = x_1x_2 + x_2x_3 + x_3x_1 = 26,$

$$Q=\begin{bmatrix}12 & -7 & 1\\ 8 & -6 & 1\\ 6 & -5 & 1\end{bmatrix}, D=QV=\begin{bmatrix}12 & -7 & 1\\ 8 & -6 & 1\\ 6 & -5 & 1\end{bmatrix}\begin{bmatrix}1 & 1 & 1\\ 2 & 3 & 4\\ 2^2 & 3^2 & 4^2\end{bmatrix}=\begin{bmatrix}2 & 0 & 0\\ 0 & -1 & 0\\ 0 & 0 & 2\end{bmatrix},$$

$$V^{-1}=D^{-1}Q=\begin{bmatrix}\frac{1}{2} & 0 & 0\\ 0 & -1 & 0\\ 0 & 0 & \frac{1}{2}\end{bmatrix}\begin{bmatrix}12 & -7 & 1\\ 8 & -6 & 1\\ 6 & -5 & 1\end{bmatrix}=\begin{bmatrix}6 & -\frac{7}{2} & \frac{1}{2}\\ -8 & 6 & -1\\ 3 & -\frac{5}{2} & \frac{1}{2}\end{bmatrix}$$]

51. Solve the following matrix equations

(i) $$\begin{bmatrix}1 & -1\\ 2 & 3\end{bmatrix}\begin{bmatrix}x_{11} & x_{12} & x_{13}\\ x_{21} & x_{22} & x_{23}\end{bmatrix}=\begin{bmatrix}3 & 4 & 6\\ -1 & 2 & 3\end{bmatrix},$$ [**Ans.** $\frac{1}{5}\begin{bmatrix}8 & 14 & 11\\ -7 & -6 & -9\end{bmatrix}$]

(ii) $$\begin{bmatrix}1 & 2 & 2\\ 1 & 3 & 1\\ 2 & 5 & 4\end{bmatrix}\begin{bmatrix}x_{11} & x_{12} & x_{13} & x_{14}\\ x_{21} & x_{22} & x_{23} & x_{24}\\ x_{31} & x_{32} & x_{33} & x_{34}\end{bmatrix}=\begin{bmatrix}1 & 2 & 2 & 3\\ 1 & 4 & 2 & 5\\ 4 & -2 & 6 & 3\end{bmatrix}$$

[**Hints:** Find $A^{-1}=\begin{bmatrix}1 & 2 & 2\\ 1 & 3 & 1\\ 2 & 5 & 4\end{bmatrix}$.

Premultiply the equation by A^{-1}. **Ans.** $\begin{bmatrix}-7 & 22 & -6 & 11\\ 2 & -4 & 2 & -1\\ 2 & -6 & 2 & -3\end{bmatrix}$]

52. Verify that $|\text{cof } A| = |A|^2$, where

$$A=\begin{bmatrix}2 & 3 & -1\\ 1 & -2 & 4\\ 5 & -3 & 6\end{bmatrix}.$$

53. Verify cof AB = cof A cof B for the matrices

$$A=\begin{bmatrix}1 & 2 & -1\\ 0 & -1 & 4\\ 3 & 7 & -3\end{bmatrix} \text{ and } B=\begin{bmatrix}2 & 1 & -1\\ -1 & 0 & 4\\ 7 & -3 & 3\end{bmatrix}.$$

CHAPTER NINE

ELEMENTARY OPERATIONS AND MATRICES

Chapter summary: *Definitions & properties of elementary matrices, echelon matrix, reduced row (column) echelon matrix, are introduced. Linear dependence/ independence, proof of* $|AB| = |A|\,|B|$, *normal form, computation of* A^{-1} *by Gauss-Jordan elimination, evaluation of determinant by ERT, relaxation method for solving simultaneous equations & bordering method in the inversion of matrix with singular block(s) are the subject matters of this chapter. The chapter ends with congruent matrices and similarity matrices.*

9.1 Definitions and properties

Three special square matrices obtained from unit matrix, I_n by any one of the following three types of operations, called *elementary row operations* (ERO) [or *elementary column operations* (ECO)] are together termed as *elementary matrices*.

I. Interchanges of rows, $\mathbf{R}_j$ and $\mathbf{R}_k$ [or columns, $\mathbf{C}_j$ and $\mathbf{C}_k$], denoted by R_{jk} (or C_{jk}), $\mathbf{R}_j \leftrightarrow \mathbf{R}_k$ (or $\mathbf{C}_j \leftrightarrow \mathbf{C}_k$), $\mathbf{O}(j,k)$ [or $\mathbf{O}'(j,k)$].

II. Multiplication of a row, $\mathbf{R}_j$ (or column, $\mathbf{C}_j$) by a scalar λ, denoted by λR_j (or λC_j), $\mathbf{R}_j(\lambda)$ [or $\mathbf{C}_j(\lambda)$], $\mathbf{O}(\lambda j)$[or $\mathbf{O}'(\lambda j)$].

III. Addition of λ times row R_j to row R_k (or column C_j to column C_k) denoted by $R_{jk}(\lambda)$ [or $C_{jk}(\lambda)$], $\lambda R_j + R_k$ (or $\lambda C_j + C_k$), $O(\lambda j + k)$ [or $O'(\lambda j + k)$].

In I, II, and III above R stands for row, C for column and O & O' for operation. Among the notations *first* one in every case will be used in this book. Note that the elementary operations on a system of equations (sec. 7.2) are identical to ERO. The definitions and properties of elementary matrices are discussed below:

Any matrix obtained by any one of the above three elementary operations is called an *elementary matrix*. The elementary matrices are classified into three kinds on the basis of the elementary operations.

9.1(a) Elementary matrix of the first kind (EMFK)

It is a diagonal matrix of order n with unity along the main diagonal except one nonzero scalar α in the ith row. It is also called the *row multiplier matrix*. It is denoted by

$$D_i^n(\alpha) = \text{diag}(1,\ 1,\ \cdots,\ \underset{i\text{th position}}{\alpha},\ 1, \cdots, 1) \tag{9.1}$$

where superscript n stands for the order and subscript, i denotes position of α. This is also denoted by $D_i(\alpha)$ for any suitable square order, $\alpha \neq 0$.

Construction : The EMFK can be constructed from unit matrix, I_n using elementary operation II on the ith row (or column) and nonzero scalar α as multiplier.

As the unit matrix, I_n has been changed or transformed to a new matrix by the single elementary operation II, it is sometimes termed as the *elementary row* (or *column*) *transformation matrix*, abbreviated as ERT (or ECT) matrix or simply *elementary matrices* (EM) and denoted either by $E_i^n(\alpha)$ or by $R(\alpha i)$ [or either $F_i^n(\alpha)$ and $C(\alpha i)$]

Also s.u. vectors $\mathbf{e}_i$ or $\mathbf{e}_i^T$ can be used to denote I_n as

$$I_n = \begin{bmatrix} \mathbf{e}_1 & \mathbf{e}_2 & \cdots & \mathbf{e}_n \end{bmatrix} = \begin{bmatrix} \mathbf{e}_1^T \\ \mathbf{e}_2^T \\ \vdots \\ \mathbf{e}_n^T \end{bmatrix}.$$

Then the ERT matrices $E_i^n(\alpha)$, $R(\alpha i)$ can be represented in terms of s.u. vectors as

$$\begin{matrix} \\ \\ \\ i\text{th position} \rightarrow \\ \\ \\ \end{matrix}\begin{bmatrix} \mathbf{e}_1^T \\ \mathbf{e}_2^T \\ \vdots \\ \alpha\mathbf{e}_i^T \\ \vdots \\ \mathbf{e}_n^T \end{bmatrix} = [\mathbf{e}_1^T \quad \mathbf{e}_2^T \quad \cdots \quad \underset{\substack{\downarrow \\ i\text{th position}}}{\alpha\mathbf{e}_i^T} \quad \cdots \quad \mathbf{e}_n^T]$$

and ECT matrices can be similarly written in terms of s.u. vectors and

$$\begin{bmatrix} \mathbf{e}_1^T & \mathbf{e}_2^T & \cdots & \alpha\mathbf{e}_i^T & \cdots & \mathbf{e}_n^T \end{bmatrix}.$$

Don't confuse $E_i^n(\alpha)$ with the matrix units E_{ij}. Virtually,

$$\begin{aligned} D_i^n(\alpha) = E_i^n(\alpha) = F_i^n(\alpha) = R(\alpha i) = C(\alpha i) \\ = \begin{bmatrix} \mathbf{e}_1^T & \mathbf{e}_2^T & \cdots & \alpha\mathbf{e}_i^T & \cdots & \mathbf{e}_n^T \end{bmatrix}^T \\ = \begin{bmatrix} \mathbf{e}_1 & \mathbf{e}_2 & \cdots & \alpha\mathbf{e}_i & \cdots & \mathbf{e}_n \end{bmatrix} \end{aligned} \tag{9.2}$$

Properties :

(i) The matrix, $D_i^n(\alpha)$ is *non-singular..* Because,

$\det(D_i^n(\alpha)) = \alpha \det(I_n) = \alpha.1 = \alpha \neq 0$, by hypothesis.

(ii) $$\left[D_i^n(\alpha)\right]^{-1} = D_i^n(\alpha^{-1}) \tag{9.3}$$

i.e. the inverse of an elementary matrix of the first kind is an elementary matrix of the same kind with the *reciprocal* scalar.

[**Notations –** Let $I_{i.}$ and $I_{.j}$ denote the ith row and jth column of I_n respectively. The ith row of any matrix A is denoted by the row vector $\mathbf{A}_{i.}$ and jth column of A is denoted by the column vector $\mathbf{A}_{.j}$

Proof . The ith row of the matrix product

$$\left[D_i^n(\alpha).D_i^n\left(\frac{1}{\alpha}\right)\right] \text{ is}$$

$$\left[D_i^n(\alpha).D_i^n\left(\frac{1}{\alpha}\right)\right]_{i.} = \left[D_i^n(\alpha)\right]_{i.} D_i^n\left(\frac{1}{\alpha}\right) \quad \because (AB)_{i.} = A_i B$$

$$= \alpha\, I_{i.}\, D_i^n\left(\frac{1}{\alpha}\right) = \alpha\, I_{i.} \frac{1}{\alpha} I_{.i},$$

[$\because$ other columns are zero columns]

$$= \left(\alpha \frac{1}{\alpha}\right)(\mathbf{II})_{i.} = \mathbf{I}_{i.}$$

The kth row, $k \neq i$, of the matrix product is

$$\left[D_i^n(\alpha).D_i^n\left(\frac{1}{\alpha}\right)\right]_{(k\neq i).} = \left[D_i^n(\alpha)\right]_{(k\neq i).} D_i^n\left(\frac{1}{\alpha}\right)$$

$$= I_{k.}\, D_i^n\left(\frac{1}{\alpha}\right) = \left(I\, D_i^n\left(\frac{1}{\alpha}\right)\right)_{(k\neq i).}$$

$$= \left(D_i^n\left(\frac{1}{\alpha}\right)\right)_{(k\neq i).} = I_{k.}, \text{ the rows } k \neq i \text{ have no scalar } \alpha.$$

$\therefore$ the product $D_i^n(\alpha).D_i^n\left(\frac{1}{\alpha}\right) = I_n$.

Hence $D_i^n\left(\frac{1}{\alpha}\right) = I_n\left[D_i^n(\alpha)\right]^{-1} = \left[D_i^n(\alpha)\right]^{-1}$

i.e. $\left[D_i^n(\alpha)\right]^{-1} = D_i^n(\alpha^{-1})$

(iii) $\left[D_i^n(\alpha)\right]^T = D_i^n(\alpha)$. Truth can be easily verified.

(iv) Effect of multiplying an $m\times n$ matrix A by EMFK : The effect of pre (or post) multiplying an $m\times n$ matrix A by $m\times m$ (or $n\times n$) elementary matrix $D_i(\alpha)$ i.e. the effect of conformable product $D_i^m(\alpha)A$ [or $AD_i^n(\alpha)$] is the multiplication of the ith row (or column) of A by α only, keeping other rows (or columns) unchanged.

Proof . The effect of premultiplication is proved below. Others follow immediately. The ith row of the product $D_i^m(\alpha)A$ is given by

$$\left[D_i^m(\alpha)A\right]_{i.} = \left[D_i^n(\alpha)\right]_{i.} A = \alpha(IA)_{i.} = \alpha A_{i.}$$

$\therefore$ the ith row of A gets multiplied by α due to premultiplication of A by $D_i^m(\alpha)$.

Again, the other k rows, $(k \neq i)$, of this product are given by

$$\left[D_i^m(\alpha)A\right]_{k.,\,k\neq i} = \left[D_i^m(\alpha)\right]_{k.,\,k\neq i} A = I_{k.}A \qquad [\because I_{k.} \text{ have no } \alpha.]$$

$$= (IA)_{k.} = A_{k.},\ k \neq i.$$

$\therefore$ the other rows except ith row remain unchanged. Hence the assertion.

(v) Relationship of EMFK with ERO (or ECO) on A: The pre (or post) multiplication of an arbitrary $m \times n$ matrix A by conformable $D_i(\alpha)$ is in effect equivalent to the multiplication of the ith row (or column) of A by the scalar α which is denoted by $R_i(\alpha)$ [or $C_i(\alpha)$]. Thus

$$D_i^m(\alpha)A_{m\times m} = E_i^m(\alpha)A = (R_i(\alpha) \text{ of A}) \qquad (9.4)$$

which is elementary row operation (ERO) of multiplying ith column of A by α and

$$A_{m\times n}D_i^n(\alpha) = AF_i^n(\alpha) = (C_i(\alpha) \text{ of A}), \qquad (9.5)$$

which is elementary column operation (ECO) of multiplying ith column of A by α.

These are the relations of EMFK with ERO and ECO on A. Thus it is found here and will be found in the following two cases that elementary row and column operations of an arbitrary matrix are achieved by conformable elementary matrix multiplication.

Ex. Let

$$A = \begin{bmatrix} 1 & 2 & -1 \\ 3 & -4 & 5 \end{bmatrix}, \alpha = 3, i = 1.$$

Then $$D_1^2(3)A = E_1^2(3)A = \begin{bmatrix} 3 & 0 \\ 0 & 1 \end{bmatrix} \begin{bmatrix} 1 & 2 & -1 \\ 3 & -4 & 5 \end{bmatrix} = \begin{bmatrix} 3 & 6 & -3 \\ 3 & -4 & 5 \end{bmatrix}$$

which shows that the first row of A only gets multiplied by 3.

Now using $R_1(3)$ of A, i.e. multiplying first row of A by 3, we have,

$$B = \begin{bmatrix} 3 & 6 & -3 \\ 3 & -4 & 5 \end{bmatrix} = D_1^2(3)A = E_1^2(3)A.$$

Also $$AD_1^2(3) = AF_1^2(3) = \begin{bmatrix} 1 & 2 & -1 \\ 3 & -4 & 5 \end{bmatrix} \begin{bmatrix} 3 & 0 & 0 \\ 0 & 1 & 0 \\ 0 & 0 & 1 \end{bmatrix} = \begin{bmatrix} 3 & 2 & -1 \\ 9 & -4 & 5 \end{bmatrix}$$

which shows that first column of A gets only multiplied by 3.

Using $C_1(3)$ of A i.e. multiplying the first column of A by 3, we have,

$$B = \begin{bmatrix} 3 & 2 & -1 \\ 9 & -4 & 5 \end{bmatrix} = AD_1^3(3) = AF_1^3(3)$$

9.1(b) Elementary matrix of the second kind (EMSK)

It is an $n \times n$ matrix such that

(i) the diagonal elements are unity except two zeros in the kth and qth rows and

(ii) off diagonal elements are zero except two 1's in the kqth and qkth positions.

It is also called the *row interchange matrix*. It is denoted as

$$
P^n_{kq} = \begin{array}{c} \quad\; k \qquad\quad q \\ \begin{bmatrix} 1 & \cdots & 0 & \cdots & 0 & \cdots & 0 \\ \vdots & & \vdots & & \vdots & & \vdots \\ 0 & \cdots & 0 & \cdots & 1 & \cdots & 0 \\ \vdots & & \vdots & & \vdots & & \vdots \\ 0 & \cdots & 1 & \cdots & 0 & \cdots & 0 \\ \vdots & & \vdots & & \vdots & & \vdots \\ 0 & \cdots & 0 & \cdots & 0 & \cdots & 1 \end{bmatrix}_{m\times n} \end{array} \begin{array}{c} \\ \\ k \\ \\ q \\ \\ \\ \end{array} \tag{9.6}
$$

where k and q above and beside the matrix stand for corresponding columns and rows. The superscript n stands for the square order.

Construction : The matrix P^n_{kq} is a special permutation matrix, obtained from unit matrix by the single elementary operation I in which qth and kth rows (or columns) are interchanged keeping other rows (or columns) unchanged. This type of permutation matrix is called ERT (or ECT) matrix and is denoted either by E^n_{kq} or by $R(k,q)$ [or either F^n_{kq} or $C(k,q)$].

For example,

$$
P^4_{23} = \begin{array}{c} \;\; 2 \;\; 3 \\ \begin{bmatrix} 1 & 0 & 0 & 0 \\ 0 & 0 & 1 & 0 \\ 0 & 1 & 0 & 0 \\ 0 & 0 & 0 & 1 \end{bmatrix} \end{array} \begin{array}{c} \\ \\ 2 \\ 3 \\ \\ \end{array} = E^4_{23} = F^4_{23} = \begin{bmatrix} \mathbf{e}_1 & \mathbf{e}_2 & \mathbf{e}_3 & \mathbf{e}_4 \end{bmatrix}
$$

Thus $\quad P^n_{kq} = E^n_{kq} = F^n_{kq}$ in general. (9.7)

Don't confuse E^4_{23} with matrix units E_{23}.

Properties :

(i) The elementary matrix P^n_{kq} is nonsingular. Because

$$\det(P^n_{kq}) = -\det I_n = -1 \neq 0 .$$

(ii) $\qquad (P^n_{kq})^{-1} = P^n_{kq} = (P^n_{kq})^T$ (9.8)

i.e. the inverse and the transpose of the matrix P^n_{kq} is the matrix itself.

Proof . The kth row of matrix product $\left[P^n_{kq}.P^n_{kq}\right]$ is

$$\left[P^n_{kq}.P^n_{kq}\right]_{k\cdot} = \left[P^n_{kq}\right]_{k\cdot} P^n_{kq} = I_{q\cdot} P^n_{kq}$$

$$= \left[IP^n_{kq}\right]_{q\cdot} = \left[P^n_{kq}\right]_{q\cdot} = I_{k\cdot}$$

Similarly, the qth row of $\left[P^n_{kq}.P^n_{kq}\right]$ is $\left[P^n_{kq}.P^n_{kq}\right]_{q\cdot} = I_{q\cdot}$

and ith row , $i \neq k, q$ of the same product is

$$\left[P^n_{kq}.P^n_{kq}\right]_{i\cdot} = \left[P^n_{kq}\right]_{i\cdot} P^n_{kq} = I_{i\cdot} P^n_{kq} = \left(IP^n_{kq}\right)_{i\cdot} = I_{i\cdot} \; .$$

$\therefore$ combining these results, the product $P^n_{kq}.P^n_{kq} = I_n$.

$\therefore \qquad \mathrm{P}_{k\,q}^{n} = \mathrm{I}_n\left(\mathrm{P}_{k\,q}^{n}\right)^{-1} = \left(\mathrm{P}_{k\,q}^{n}\right)^{-1}.$

Also $\qquad \left[\mathrm{P}_{k\,q}^{n}\right]^{\mathrm{T}} = \mathrm{P}_{k\,q}^{n} \qquad \because \mathrm{I}_{.i} = \mathrm{I}_{i.}$ for all i.

Hence the assertion.

(iii) $\qquad \left[\mathrm{P}_{k\,q}^{n}\right]^{T} = \mathrm{P}_{k\,q}^{n}$ already shown in (ii).

(iv) Effect of multiplying a matrix A **by EMSK :** The effect of pre (or post) multiplying an arbitrary $m\times n$ matrix A by $\mathrm{P}_{k\,q}^{m}$ (or $\mathrm{P}_{k\,q}^{n}$) is just to interchange rows k and q (or columns k and q) of the matrix A keeping other rows (or columns) unchanged.

Proof . The effect of premultiplication is proved below. Other follows immediately.

The kth row of the product $\mathrm{P}_{k\,q}^{m}$ A is

$$\left[\mathrm{P}_{k\,q}^{m}\ \mathrm{A}\right]_{k.} = \left[\mathrm{P}_{k\,q}^{m}\right]_{k.}\mathrm{A} = \mathrm{I}_{q.}\mathrm{A} = [\mathrm{IA}]_{q.} = \mathrm{A}_{q.} \tag{9.9}$$

Similarly, the qth row of the product P_{kq}^{m} A is

$$\left[\mathrm{P}_{kq}^{m}\ \mathrm{A}\right]_{q.} = \left[\mathrm{P}_{kq}^{m}\right]_{q.}\mathrm{A} = \mathrm{I}_{k.}\mathrm{A} = [\mathrm{IA}]_{k.} = \mathrm{A} \tag{9.10}$$

$\therefore$ (9.9) and (9.10) show the interchange of kth and qth rows.

But the ith row , $i \neq k\,,\,q$ of the product given by

$$\left[\mathrm{P}_{kq}^{m}\ \mathrm{A}\right]_{i.} = \left[\mathrm{P}_{kq}^{m}\right]_{i.}\mathrm{A} = \mathrm{I}_{i.}\mathrm{A} = [\mathrm{IA}]_{i.} = \mathrm{A}_{i.} \tag{9.11}$$

remains unchanged.

$\therefore$ combining (9.9), (9.10), (9.11) the assertion is proved for premultiplication.

(v) Relationship of EMSK with ERO (or ECO) on A **:** The pre (or post) multiplication of an arbitrary $m\times n$ matrix A by P_{kq}^{m} (or P_{kq}^{n}) is in effect *equivalent* to interchanging rows R_k and R_q (or columns C_k and C_q) of the matrix A which is denoted by $\mathrm{R}_{k\,q}$ (or $\mathrm{C}_{k\,q}$) . Thus

$$\mathrm{P}_{kq}^{m}\mathrm{A}_{m\times n} = \mathrm{E}_{kq}^{m}\mathrm{A}_{m\times n} = (\mathrm{R}_{kq} \text{ of A}) \tag{9.12}$$

which is elementary row operation (ERO) of interchanging kth row and qth row of A.

And $\qquad \mathrm{A}_{m\times n}\mathrm{P}_{kq}^{n} = \mathrm{A}_{m\times n}\mathrm{F}_{kq}^{n} = (\mathrm{C}_{kq} \text{ of A}) \qquad (9.13)$

which is elementary column operation (ECO) of interchanging kth and qth column of A.

These are the relations of EMSK with ERO and ECO on A.

Ex. Let

$$\mathrm{A} = \begin{bmatrix} 1 & 2 & -1 \\ 3 & -4 & 5 \\ -2 & 7 & 8 \\ -7 & 0 & -8 \end{bmatrix}. \text{ Then}$$

$$P_{23}^{4}A=\begin{bmatrix}1&0&0&0\\0&0&1&0\\0&1&0&0\\0&0&0&1\end{bmatrix}\begin{bmatrix}1&2&-1\\3&-4&5\\-2&7&8\\-7&0&-8\end{bmatrix}=\begin{bmatrix}1&2&-1\\3&-4&5\\-2&7&8\\-7&0&-8\end{bmatrix}$$

$= E_{23}^{4}A = (R_{23}$ of A)

= (ERO of interchanging 2nd and 3rd rows)

Also $$AP_{23}^{4}=\begin{bmatrix}1&2&-1\\3&-4&5\\-2&7&8\\-7&0&-8\end{bmatrix}\begin{bmatrix}1&0&0\\0&0&1\\0&1&0\end{bmatrix}=\begin{bmatrix}1&2&-1\\3&-4&5\\-2&7&8\\-7&0&-8\end{bmatrix}$$

$= AF_{23}^{4} = (C_{23}$ of A) ,

(ECO of interchanging 2nd and 3rd columns of A)

9.1(c) Elementary matrix of the third kind (EMTK)

It is an $n \times n$ matrix such that the diagonal elements are unity and off diagonal elements are zero except one scalar α in the kqth position. It is also called the *row addition* matrix. It is denoted as

$$S_{kq}^{n}(\alpha)=\begin{bmatrix}1&\cdots&0&\cdots&0&\cdots&0\\\vdots&&\vdots&&\vdots&&\vdots\\0&\cdots&1&\cdots&\alpha&\cdots&0\\\vdots&&\vdots&&\vdots&&\vdots\\0&\cdots&0&\cdots&1&\cdots&0\\\vdots&&\vdots&&\vdots&&\vdots\\0&\cdots&0&\cdots&0&\cdots&1\end{bmatrix}_{n\times n}\quad \begin{matrix}\leftarrow k\text{th row}\\ \leftarrow q\text{th row}\end{matrix} \qquad (9.14)$$

(with kth Coln. and qth Coln. marked by arrows above the matrix)

where subscripts k, q indicate the row and the column position of α and the superscript n denotes the square order of the matrix.

Construction : The matrix $S_{kq}^{n}(\alpha)$ can be constructed from unit matrix I_n by the single elementary operation III in which α times of row, R_q is added to row R_k (or column C_k is added to column C_q). This operation is denoted either by $E_{kq}(\alpha)$ or by $R(k + \alpha q)$ [or either $F_{kq}(\alpha)$ or $C(k + \alpha q)$] and is another ERT (or ECT) matrix. In terms of s.u. vectors, these ERT's are identical to

$$\left[\mathbf{e}_1 \quad \mathbf{e}_2 \quad \cdots \quad (\mathbf{e}_k + \alpha\mathbf{e}_q) \quad \cdots \quad \mathbf{e}_n\right]$$

For example,

$$S_{32}^{4}(-5)=\begin{bmatrix}1&0&0&0\\0&1&0&0\\0&-5&1&0\\0&0&0&1\end{bmatrix}=E_{32}^{4}(-5)$$

$$=F_{32}^{4}(-5)=\begin{bmatrix}\mathbf{e}_1 & (\mathbf{e}_2+(-5)\mathbf{e}_3) & \mathbf{e}_3 & \mathbf{e}_4\end{bmatrix}$$

Thus, in general, $S_{kq}^{n}(\alpha)=E_{kq}^{n}(\alpha)=F_{kq}^{n}(\alpha)$. (9.15)

Properties :

(i) The matrix $S_{kq}^{n}(\alpha)$ is nonsingular. Because,

$\det(S_{kq}^{n}(\alpha))$ = [product of diagonal elements of the triangular determinant of the matrix $S_{kq}^{n}(\alpha)$]

$= (1.1.\ \cdots .n \text{ times}) = 1 \neq 0$

(ii) $$\left(S_{kq}^{n}(\alpha)\right)^{-1}=\left(S_{kq}^{n}(\alpha)\right)$$

i.e. the inverse of $S_{kq}^{n}(\alpha)$ is another matrix of the same kind with α replaced by $(-\alpha)$.

Proof . The kth row of the matrix product $[\, S_{kq}^{n}(\alpha).\, S_{kq}^{n}(-\alpha)\,]$ is

$[S_{kq}^{n}(\alpha).\, S_{kq}^{n}(-\alpha)\,]_{k.}$

$=[S_{kq}^{n}(\alpha).]_{k.}\; S_{kq}^{n}(-\alpha)$

$=[\, I_{k.} + \alpha I_{q.}]\, S_{kq}^{n}(-\alpha)$, (simplification is shown in the next page)

$= I_{k.} S_{kq}^{n}(-\alpha) + \alpha I_{q.} S_{kq}^{n}(-\alpha)$

$=[I\, S_{kq}^{n}(-\alpha)]_{k.} + \alpha[I\, S_{kq}^{n}(-\alpha)]_{q.}$

$= [S_{kq}^{n}(-\alpha)]_{k.} + \alpha[S_{kq}^{n}(-\alpha)]_{q.}$

$$\left\{\begin{aligned} \because [S^n_{kq}(\alpha)]_{k.} &= [0 \;\cdots\; \underset{\substack{\uparrow\\ k\text{th}\\ \text{position}}}{1} \;\cdots\; \underset{\substack{\uparrow\\ kq\text{th}\\ \text{position}}}{\alpha} \;\cdots\; 0] \\ &= [0 \;\cdots\; \underset{\substack{\uparrow\\ k\text{th}\\ \text{position}}}{1} \;\cdots\; \underset{\substack{\uparrow\\ q\text{th}\\ \text{position}}}{0} \;\cdots\; 0] \\ &\quad + \alpha[0 \;\cdots\; \underset{\substack{\uparrow\\ k\text{th}\\ \text{position}}}{0} \;\cdots\; \underset{\substack{\uparrow\\ q\text{th}\\ \text{position}}}{1} \;\cdots\; 0] \\ &= I_{k.} + \alpha I_{q.} \end{aligned}\right\}$$

$= I_{k.} - \alpha I_{k.} + \alpha I_{q.}$ $\quad\because$ q th row is devoid of $(-\alpha)$

$= I_{k.}$

Also the ith row $i \neq k$ of the same product is

$$[S^n_{kq}(\alpha).S^n_{kq}(-\alpha)]_{i.} = [S^n_{kq}(\alpha)]_{i.}\, S^n_{kq}(-\alpha) = I_{i.}\, S^n_{kq}(-\alpha)$$

$$= [S^n_{kq}(-\alpha)]_{i.} = [S^n_{kq}(-\alpha)]_{i.} = I_{i.}\,.$$

$\because$ ith row are devoid of both α & $(-\alpha)$

$\therefore$ combining two results, we have ,

$$S^n_{kq}(\alpha).S^n_{kq}(-\alpha) = I_n\,.$$

$$\therefore \qquad S^n_{kq}(-\alpha) = I_n\left[S^n_{kq}(\alpha)\right]^{-1} = \left[S^n_{kq}(\alpha)\right]^{-1}$$

Hence proved the requirement.

(iii) $\left[S^n_{kq}(\alpha)\right]^T = S^n_{qk}(\alpha)$ i.e. the transpose of EMTK is a matrix of the same kind but rows are transformed into columns in the usual sense of transpose.

Proof is left as an exercise.

(iv) Effect of multiplying a matrix A by EMTK : The effect of pre (or post) multiplication of an arbitrary $m \times n$ matrix A by $S^m_{kq}(\alpha)$ [or $S^n_{kq}(\alpha)$] is the addition of α times the qth row (or kth column) to the kth row (or qth column) keeping other rows (or columns) unchanged.

Proof. The effect of premultiplication is proved below. Other follows immediately.

The kth row of the product $S^m_{kq}(\alpha)A$ is

$$\left[S^m_{kq}(\alpha)A\right]_{k.} = \left[S^m_{kq}(\alpha)\right]_{k.} A$$

$$= \left[I_{k.} + \alpha I_{q.}\right]A \text{ , from (ii) above}$$

$$= I_{k.}A + \alpha I_{q.}A = [IA]_{k.} + \alpha[IA]_{q.}$$

$$= A_{k.} + \alpha A_{q.}$$

$$= [(\alpha \text{ times of } q\text{th row of A}) \text{ adds to } (k\text{th row of A})]$$

The ith rows $i \neq k$ of the product $S^m_{kq}(\alpha)A$ is

$$\left[S^m_{kq}(\alpha)A\right]_{i.} = \left[S^m_{kq}(\alpha)\right]_{i.}A$$
$$= I_i A\ ,\ [\because\ i\text{th rows are devoid of } \alpha]$$
$$= [IA]_{i.} = A_{i.}$$

i.e. *i*th rows remain unchanged by the multiplication. Hence proved.

(v) Relationship of EMTK with ERO (or ECO) on A : The pre (or post) multiplication of an arbitrary $m \times n$ matrix A by $S^m_{kq}(\alpha)$ [or $S^n_{kq}(\alpha)$] is in effect equivalent to adding α times of *q*th row (or column) to *k*th row (or column) of A and is denoted by $R_{kq}(\alpha)$ or $[C_{kq}(\alpha)]$. Thus,

$$S^m_{kq}(\alpha)A_{m\times n} = E^m_{kq}(\alpha)A_{m\times n} = (R_{kq}(\alpha)\text{of } A), \tag{9.16}$$

which is ERO of adding α times the *k*th column to *q*th column of A.

And $$A_{m\times n}S^n_{kq}(\alpha) = A_{m\times n}F^n_{qk}(\alpha) = (C_{qk}(\alpha) \text{ of } A) \tag{9.17}$$

which is ECO of adding α times the *k*th column to *q*th column of A. These show the relations of EMTK with ERO and ECO on A. For example, let

$$A = \begin{bmatrix} 1 & 2 & -1 \\ 3 & -4 & 5 \\ -2 & 7 & 8 \\ -7 & 0 & -8 \end{bmatrix}.$$

Then

$$S^4_{32}(4)A = \begin{bmatrix} 1 & 0 & 0 & 0 \\ 0 & 1 & 0 & 0 \\ 0 & 4 & 1 & 0 \\ 0 & 0 & 0 & 1 \end{bmatrix}\begin{bmatrix} 1 & 2 & -1 \\ 3 & -4 & 5 \\ -2 & 7 & 8 \\ -7 & 0 & -8 \end{bmatrix} = \begin{bmatrix} 1 & 2 & -1 \\ 3 & -4 & 5 \\ 10 & -9 & 28 \\ -7 & 0 & -8 \end{bmatrix}$$
$$= E^4_{32}(4)A$$
$$= (R_{32}(4) \text{ of } A),$$

ERO of adding 4 times the 2nd row to the 3rd row of A.

Also
$$AS^3_{32}(4) = \begin{bmatrix} 1 & 2 & -1 \\ 3 & -4 & 5 \\ -2 & 7 & 8 \\ -7 & 0 & -8 \end{bmatrix}\begin{bmatrix} 1 & 0 & 0 \\ 0 & 1 & 0 \\ 0 & 4 & 1 \end{bmatrix} = \begin{bmatrix} 1 & -2 & -1 \\ 3 & 16 & 5 \\ -2 & 39 & 8 \\ -7 & -32 & -8 \end{bmatrix}$$
$$= AF^3_{23}(4)$$
$$= (C_{23}(4) \text{ of } A),$$

ECO of adding 4 times the 3rd column to the 2nd column of A

Ex.1 Find, by inspection , the inverses of the following matrices:

$$\text{(i)}\begin{bmatrix} 1 & 0 & 0 & 0 \\ 0 & 0 & 1 & 0 \\ 0 & 1 & 0 & 0 \\ 0 & 0 & 0 & 1 \end{bmatrix}, \text{(ii)}\begin{bmatrix} 1 & 0 & 0 & 0 \\ 0 & \frac{1}{3} & 0 & 0 \\ 0 & 0 & 1 & 0 \\ 0 & 0 & 0 & 1 \end{bmatrix}, \text{(iii)}\begin{bmatrix} 1 & 0 & 0 & 4 \\ 0 & 1 & 0 & 0 \\ 0 & 0 & 1 & 0 \\ 0 & 0 & 0 & 1 \end{bmatrix}.$$

(i) It is an elementary matrix of the second kind, P^4_{23}. Therefore, by eqn.(9.7),

$$(P^4_{23})^{-1} = P^4_{23} \text{ i.e. the matrix itself.}$$

(ii) It is an elementary matrix of the first kind, $D^4_2(\frac{1}{3})$. Therefore, by eqn.(9.3),

$$(D^4_2(\tfrac{1}{3}))^{-1} = D^4_2(3) = \begin{bmatrix} 1 & 0 & 0 & 0 \\ 0 & 3 & 0 & 0 \\ 0 & 0 & 1 & 0 \\ 0 & 0 & 0 & 1 \end{bmatrix}.$$

(iii) It is an elementary matrix of third kind, $S^4_{41}(4))$. So, by eqn. (9.16), its inverse is given by

$$(S^4_{41}(4))^{-1} = S^4_{41}(-4) = \begin{bmatrix} 1 & 0 & 0 & -4 \\ 0 & 1 & 0 & 0 \\ 0 & 0 & 1 & 0 \\ 0 & 0 & 0 & 1 \end{bmatrix}.$$

9.2 Echelon matrices

A matrix can be either transferred to a *row echelon* or a *column echelon* matrix by elementary row or column transformation. If the transformed $m \times n$ matrix satisfies the following *four properties* then the matrix is called a *row echelon matrix* (RE M):

(i) The first (reading from top to bottom) k, $(1 \le k \le m)$ rows must be nonzero (i.e. *at least* one element in the row is *not* zero) and elements in other $(m-k)$ rows are all zeros i.e. all zero rows lie below the nonzero rows;

(ii) The first (reading from left to right) nonzero element in each of the first k nonzero rows must be 1; this entry is called the *leading entry* of row;

(iii) If the 1 in row i occurs in column C_i, i = (any one of 1, 2, $\cdots$, k) then $C_1 < C_2 < \cdots < C_k$, i.e. 1 in each of k nonzero rows occurs to the right of the preceding column and the 1's constitute a staircase:

$$\begin{array}{c|c|c|c} 1 & 0 & 0 & \cdots \\ \hline 0 & 1 & 0 & \cdots \\ \hline 0 & 0 & 1 & \cdots \\ \hline 0 & 0 & 0 & \cdots \end{array}$$

(iv) All elements of column C_i. in (iii) appearing in rows $i+1$, $i+2$, $\cdots$, m are zero, i.e. all elements below 1's are zeros.

Similarly, an $m \times n$ matrix is called a *column echelon matrix* with similar properties obtained by the following changes of wordings:

row $\rightleftarrows$ column , top $\rightleftarrows$ left, bottom $\rightleftarrows$ right,

$m \to n$, below $\rightleftarrows$ to the right of , $C_i \to R_i$.

9.2.1 Reduced row (column) echelon matrix (form) : An $m \times n$ row (column) echelon matrix (REM or CEM) is called a *reduced row (column) echelon matrix (form)* [RREM (or RREF) or RCEM (or RCEF)] if the first nonzero element 1 in each of the first k nonzero rows (columns) is the only nonzero element in the column (row) also. The terms RREM or RREF (RCEM or RCEF) are used interchangeably.

Echelon matrices are *not unique* while reduced echelon matrices are *unique*. All the intermediate transformed matrices between the echelon to the reduced echelon are echelon matrices as they all satisfy the four criterions for the echelon matrix.

Ex.1 Find some row echelon matrices for

$$\begin{bmatrix} 1 & 4 & 3 & 2 \\ 1 & 8 & 0 & 2 \\ 2 & 0 & 4 & 2 \end{bmatrix}.$$

Using different elementary row transformations shown & "≈" as an equivalent sign, we have,

$$\underset{A}{\begin{bmatrix} 1 & 4 & 3 & 2 \\ 1 & 8 & 0 & 2 \\ 2 & 0 & 4 & 2 \end{bmatrix}} \underset{R_{13}(-2)}{\overset{R_{12}(-1)}{\approx}} \underset{B}{\begin{bmatrix} 1 & 4 & 3 & 2 \\ 0 & 4 & -3 & 0 \\ 0 & -8 & -2 & -2 \end{bmatrix}} \overset{R_{23}(2)}{\approx} \underset{C}{\begin{bmatrix} 1 & 4 & 3 & 2 \\ 0 & 4 & -3 & 0 \\ 0 & 0 & -8 & -2 \end{bmatrix}} \underset{R_3(-\frac{1}{8})}{\overset{R_2(\frac{1}{4})}{\approx}}$$

$$\underset{D}{\begin{bmatrix} 1 & 4 & 3 & 2 \\ 0 & 1 & -\frac{3}{4} & 0 \\ 0 & 0 & 1 & \frac{1}{4} \end{bmatrix}} \overset{R_{21}(-4)}{\approx} \underset{E}{\begin{bmatrix} 1 & 0 & 6 & 2 \\ 0 & 1 & -\frac{3}{4} & 0 \\ 0 & 0 & 1 & \frac{1}{4} \end{bmatrix}} \overset{R_{31}(-6)}{\approx}$$

$$\underset{F}{\begin{bmatrix} 1 & 0 & 0 & \frac{1}{2} \\ 0 & 1 & -\frac{3}{4} & 0 \\ 0 & 0 & 1 & \frac{1}{4} \end{bmatrix}} \overset{R_{32}(\frac{3}{4})}{\approx} \underset{G}{\begin{bmatrix} 1 & 0 & 0 & \frac{1}{2} \\ 0 & 1 & 0 & \frac{3}{16} \\ 0 & 0 & 1 & \frac{1}{4} \end{bmatrix}}.$$

If the order of elementary transformation $R_{31}(-6)$, $R_{32}(\frac{3}{4})$ are reversed i.e. if ET $R_{32}(\frac{3}{4})$ is used first and then $R_{31}(-6)$ is used, another row echelon matrix

$$F' = \begin{bmatrix} 1 & 0 & 6 & 2 \\ 0 & 1 & 0 & \frac{3}{16} \\ 0 & 0 & 1 & \frac{1}{4} \end{bmatrix} \text{ results.}$$

Note that D, E, F, F′, G are the 5 row echelon matrices (REMs) of A and G is the only row reduced echelon matrix of A. Thus REM is *not unique* while RREM is *unique*. The stairs in D, E, F, F′, G are constructed from first nonzero element 1 in each row. These D, E, F, F′ are *not* RREM because leading element 1 in rows are *not* leading elements of columns.

Note that number of nonzero rows = k = 3 in each REMs and also in RREM. Let i = 2, i.e. leading element 1 occurs in R_2 and C_2 then $C_2 = 2 < C_3 = 3$ in each REMs and also in RREM. This verifies property (iii). Let us transform the above matrix to some *column echelon matrices* by elementary column transformation.

$$\underset{A}{\begin{bmatrix}1 & 4 & 3 & 2\\1 & 8 & 0 & 2\\2 & 0 & 4 & 2\end{bmatrix}} \underset{C_{13}(-3)\ C_{14}(-2)}{\overset{C_{12}(-4)}{\approx}} \underset{B}{\begin{bmatrix}1 & 0 & 0 & 0\\1 & 4 & -3 & 0\\2 & -8 & -2 & -2\end{bmatrix}} \overset{C_{23}(\frac{3}{4})}{\approx} \underset{C}{\begin{bmatrix}1 & 0 & 0 & 0\\1 & 4 & 0 & 0\\2 & -8 & -8 & -2\end{bmatrix}} \overset{C_{34}(-\frac{1}{4})}{\approx}$$

$$\underset{D}{\begin{bmatrix}1 & 0 & 0 & 0\\1 & 4 & 0 & 0\\2 & -8 & -8 & 0\end{bmatrix}} \underset{C_3(-\frac{1}{8})}{\overset{C_2(\frac{1}{4})}{\approx}} \underset{E}{\begin{bmatrix}1 & 0 & 0 & 0\\1 & 1 & 0 & 0\\2 & -2 & 1 & 0\end{bmatrix}} \overset{C_{21}(-1)}{\approx} \underset{F}{\begin{bmatrix}1 & 0 & 0 & 0\\0 & 1 & 0 & 0\\4 & -2 & 1 & 0\end{bmatrix}} \underset{C_{32}(2)}{\overset{C_{31}(-4)}{\approx}}$$

$$\underset{G}{\begin{bmatrix}1 & 0 & 0 & 0\\0 & 1 & 0 & 0\\0 & 0 & 1 & 0\end{bmatrix}}.$$

The matrices E, F, G are column echelon matrices (CEM) of A while G is a reduced column echelon matrix. The elementary column transformation $C_{31}(-2)$ on E, $C_{31}(-4)$ on F only and $C_{32}(2)$ on F only result in 3 more column echelon matrices.

Ex.2 Identify the following with reasons as echelon and reduced echelon matrices:

(i) I_n, (ii) Null matrix, (iii) $\begin{bmatrix}1 & 0 & 0 & 0\\0 & 0 & 0 & 0\\0 & 0 & 0 & 1\end{bmatrix}$, (iv) $\begin{bmatrix}1 & 0 & 0 & 0\\0 & 0 & 1 & 0\\0 & 0 & 0 & 1\end{bmatrix}$,

(v) $\begin{bmatrix}0 & 1 & 4 & 6\\0 & 0 & 1 & 0\\0 & 0 & 0 & 0\end{bmatrix}$, (vi) $\begin{bmatrix}1 & 4 & 3 & 0\\0 & 1 & 2 & 0\\0 & 1 & 3 & 0\\0 & 0 & 1 & 9\end{bmatrix}$, (vii) $\begin{bmatrix}1 & 0 & 0 & 0 & 0\\0 & 0 & 1 & 0 & 0\\-\frac{1}{3} & 1 & \frac{5}{3} & 0 & 0\end{bmatrix}$.

(i) I_n is an unit matrix. It is both RREM and RCEM. Because each row (column) is nonzero, the first nonzero element in each row (column) is unity which is also the nonzero element in its columns (rows). Each of these unities occurs *to the right of* (*below*) column (row) in which first unity of the preceding column (row) appeared.

(ii) Null matrix is neither an echelon nor a reduced echelon matrix; because each row (column) is zero row (column).

(iii) The matrix is neither echelon nor reduced echelon matrix. Because second row (second and third columns) is (are) zero row (columns) not at the bottom (left) of the nonzero row (columns).

(iv) The matrix is both REM and RREM. Because all the three rows are nonzero rows, the first nonzero element in each of the three rows is 1 which is also the nonzero element in its column. Each 1 occurs to the right of the column in which 1 appears in the preceding row *but* not RCM because of the presence of zero 2nd column between two nonzero 1st and 3rd columns.

(v) The matrix is only a REM. Because the first two rows are nonzero rows, the first nonzero element in those two rows are 1 and that 1 in second row belongs to the third column to the right of the second column in the first row. It is not RREM because of the presence of the nonzero element 4 in the second column.

(vi) The matrix is neither a REM nor a RREM because the first nonzero element 1 in the second and third rows belong to the same second column. The 1 in third row second column violates preceding condition.
(vii) The matrix is not CEM due to second column in which the first nonzero element 1 does not belong to the right in lower i.e. third row.

9.2.2 RREF (RCEF) of a matrix deleting some of its columns (rows) : Let an $m\times n$ matrix A be reduced to an $m\times n$ RREF, R. Let some columns of A be deleted to form $A_1, A_2, \cdots$ of lower order say $m\times(n-1)$, $m\times(n-2), \cdots$. The RREF of these A_i's , $i = 1, 2, \cdots$ can be easily read off from R by deletion of corresponding columns of R without further reduction if a row does not appear with leading element other than 1 or a zero row; otherwise few reduction may be required. This statement is obvious from the criteria of RREF because removal of a column in A removes the corresponding column in R. From section 9.2.1 note the RREF of

$$A = \begin{bmatrix} 1 & 4 & 3 & 2 \\ 1 & 8 & 0 & 2 \\ 2 & 0 & 4 & 2 \end{bmatrix} \text{ is } R = G = \begin{bmatrix} 1 & 0 & 0 & \frac{1}{2} \\ 0 & 1 & 0 & \frac{3}{16} \\ 0 & 0 & 1 & \frac{1}{4} \end{bmatrix}$$

Now if A_3 is obtained by deleting 4th column of A then it can be easily verified that corresponding RREF of A_3 is

$$R_3 = \begin{bmatrix} 1 & 0 & 0 \\ 0 & 1 & 0 \\ 0 & 0 & 1 \end{bmatrix}$$

obtained from R by deleting 4th column of R. But if 3rd column of A is deleted to obtain A_3, then its echelon form

$$R_3 = \begin{bmatrix} 1 & 0 & \frac{1}{2} \\ 0 & 1 & \frac{3}{16} \\ 0 & 0 & \frac{1}{4} \end{bmatrix}$$

is not in RREF. It requires further row operations $R_3(4)$, $R_{13}(-\frac{1}{2})$, $R_{23}(-\frac{3}{16})$ to make it RREF, i.e. RREF of R_3 is

$$\begin{bmatrix} 1 & 0 & 0 \\ 0 & 1 & 0 \\ 0 & 0 & 1 \end{bmatrix}.$$

This is because 3rd column of A corresponds to s.u. vector of R while 4th column of R is not a s.u. vector.

Similar conclusions may be drawn for reduced column echelon form (RCEF).

9.3 Factorization and inverse of nonsingular matrices

The elementary matrices help to *factor* a matrix into elementary matrices and also to find the *inverse* of a nonsingular matrix. The following theorem ensures these properties.

Theorem. A nonsingular matrix can be reduced to a unit matrix by elementary row operations. Symbolically, if $E_1, E_2, \cdots, E_p$ be a sequence of elementary row operations on a nonsingular matrix A, then $E_p E_{p-1} \cdots E_2 E_1 A = I$.

Conversely, such a reduced matrix has a factorization of A into elementary matrices as $A = E_1^{-1} E_2^{-1}, \cdots E_p^{-1}$ as well an inverse $A^{-1} = E_p E_{p-1} \cdots E_2 E_1$.

Proof. (i) Let A be an $n \times n$ square matrix. Since A is non-singular, i.e. det $A \neq 0$, a sequence of elementary row operations $E_1, E_2, \cdots, E_m$ reduces the matrix A to a upper triangular matrix T, i.e.

$$T = E_m \, E_{m-1} \cdots E_2 E_1 A . \tag{9.18}$$

Let
$$T = \begin{bmatrix} t_{11} & t_{12} & \cdots & \cdots & \cdots & t_{1n} \\ 0 & t_{22} & \cdots & \cdots & \cdots & t_{2n} \\ \vdots & \vdots & \vdots & \vdots & \vdots & \vdots \\ 0 & 0 & \cdots & \cdots & \cdots & t_{nn} \end{bmatrix} \tag{9.19}$$

This T is again nonsingular, because of the nonsingularity of A and no $t_{ii} = 0$.

(ii) The matrix T can be reduced to a diagonal matrix D by the elementary row operations $E_{m+1}, E_{m+2}, \cdots, E_r$. The ERO E_{m+1} of last row operates on the $(n-1)$ top rows to make all, $t_{in} = 0,\ i = 1, 2, \cdots, (n-1)$. Similarly, ERO E_{m+2} of $(n-1)$th row operates on the $(n-1)$ top rows to make all $t_{i(n-1)} = 0,\ i = 1, 2, \cdots, (n-2)$. Proceeding in this way, the ERO E_r of 2nd row operates on the first row to make $t_{ii} = 0$. Thus

$$\begin{aligned} D &= E_r E_{r-1} \cdots E_{m+1} T \\ &= E_r E_{r-1} \cdots E_{m+1} (E_m E_{m-1} \cdots E_2 E_1 A) \quad \text{by (9.18)} \\ &= E_r E_{r-1} \cdots E_2 E_1 A \end{aligned} \tag{9.20}$$

(iii) Next the ERO $E_{r+1} E_{r+2} \cdots E_{p-1} E_p$ operates on D to reduce it to I. Thus

$$\begin{aligned} I &= E_p E_{p-1} \cdots E_{r+1} D \\ &= E_p E_{p-1} \cdots E_{r+1} (E_r \cdots E_2 E_1 A) \\ &= E_p E_{p-1} \cdots E_2 E_1 A \end{aligned} \tag{9.21}$$

This proves the first part of the theorem.

Converse, from (9.21)

$$\begin{aligned} A &= I(E_p \, E_{p-1} \cdots E_2 \, E_1)^{-1} \\ &= (E_p \, E_{p-1} \cdots E_2 \, E_1)^{-1} \\ &= E_1^{-1} \, E_2^{-1} \cdots E_p^{-1} \end{aligned} \tag{9.22}$$

which is the factorization of the matrix into elementary matrices as the inverses of elementary matrices are again elementary matrices. In other words, an n square matrix is expressed as the product of elementary matrices.

Again from (9.22),

$$\begin{aligned} A^{-1} &= (E_1^{-1} \, E_2^{-1} \cdots E_p^{-1})^{-1} \\ &= E_p \, E_{p-1} \cdots E_2 \, E_1 \end{aligned} \tag{9.23}$$

$\therefore$ the inverse of the nonsingular square matrix A is the product of the elementary matrices in order to reduce A to I.

Cor. (a) **Algorithm for computing** A^{-1}: The equation (9.23) provides with an algorithm for computing A^{-1}. Thus ERO's on A are to be continued until I is obtained. A convenient way of performing this is to write the partitioned matrix [A : I]. Then

$$\left(E_p\, E_{p-1} \cdots E_2\, E_1\right)[A:I]$$
$$= \left[E_p\, E_{p-1} \cdots E_2\, E_1 A \vdots E_p\, E_{p-1} \cdots E_2\, E_1\right]$$
$$= \left[I \vdots A^{-1}\right] \text{ by } (9.21) \text{ and } (9.23)$$

Cor. (b) If t_{ii} in T are converted to 1's during first set of ERO's and then elementary column operations $F_1, F_2, \cdots, F_q$ are performed sequentially to reduce the resulting upper triangular matrix to I then the alternative form of (9.21) is

$$E_m\, E_{m-1} \cdots E_2\, E_1 A F_1\, F_2 \cdots F_q = I$$

Hence
$$A = \left(E_m\, E_{m-1} \cdots E_2\, E_1\right)^{-1} I \left(F_1\, F_2 \cdots F_q\right)^{-1}$$
$$= \left(E_1^{-1}\, E_2^{-1} \cdots E_m^{-1}\right)\left(F_q^{-1}\, F_{q-1}^{-1} \cdots F_2^{-1}\, F_1^{-1}\right)$$

Thus A is factored into both elementary row and column matrices. Therefore,

$$A^{-1} = \left[\left(E_1^{-1}\, E_2^{-1} \cdots E_m^{-1}\right)\left(F_q^{-1}\, F_{q-1}^{-1} \cdots F_2^{-1}\, F_1^{-1}\right)\right]^{-1}$$
$$= F_1\, F_2 \cdots F_q\, E_m\, E_{m-1} \cdots E_2\, E_1$$

So, the inverse of A is the product of elementary column matrices in order and then row matrices in reverse order.

Cor. (c) If at the first instance instead of ERO's a sequence of ECO's $F_1, F_2, \cdots, F_q$ are performed, then A will be reduced to a lower triangular matrix

$$L = \begin{bmatrix} l_{11} & 0 & 0 & \cdots & \cdots & 0 \\ l_{21} & l_{22} & 0 & \cdots & \cdots & 0 \\ l_{31} & l_{32} & l_{33} & \cdots & \cdots & 0 \\ \vdots & \vdots & \vdots & \ddots & \vdots & \vdots \\ \vdots & \vdots & \vdots & \vdots & \ddots & \vdots \\ l_{n1} & l_{n2} & l_{n3} & \cdots & \cdots & l_{nn} \end{bmatrix}$$

and $A F_1 F_2 \cdots F_q = L$ which can be reduced to I either by another set of elementary ECO's or ERO's. Then

$$\text{either} \quad A F_1\, F_2 \cdots F_q F_{q+1} F_{q+2} \cdot \cdots F_m = I \quad \text{or} \quad E_m E_{m-1} \cdots E_1\; A F_1, F_2 \cdots F_q = I.$$

From which A and A^{-1} can be obtained in terms of EC matrices only or both EC and ER matrices.

Cor. (d) Since det T or det L is the product of diagonal elements, so if any one of the diagonal elements becomes zero during reduction to I, the original determinant of the matrix A is zero and hence the given matrix is *singular*.

Cor. (e) Appearance of a zero element along the principal diagonal during the process of reduction to I identifies that the matrix is *singular* and no separate test is required for the existence of inverse.

9.4 Equivalent matrices

Definition: If an $m \times n$ matrix A is transformed to another $m \times n$ matrix B by one or more elementary row (column) operations, then A is said to be *row* (*column*) *equivalent* to B and vice-versa. This binary relation (relation between two sets or objects) is denoted as

$$A \overset{\text{row(coln.)}}{\approx} B,\quad A \overset{\text{RE(CE)}}{=} B,\quad A \underset{\text{CE}}{\overset{\text{RE}}{\rightarrow}} B$$

Definition : If an $m \times n$ matrix A is transformed to another $m \times n$ matrix B by a finite number of elementary operations — row or column, or both, then A is said to *equivalent* to B and vice-versa. This binary relation is denoted by $A \approx B$.

Definition : If an equivalent matrix takes *any one* of the forms

$$\begin{bmatrix} I_r & \vdots & 0 \\ \cdots & & \cdots \\ 0 & \vdots & 0 \end{bmatrix},\ [I_r \ \vdots\ 0],\ \begin{bmatrix} I_r \\ 0 \end{bmatrix} \text{ and } I_r,\ r \le m \text{ or } n$$

where I_r is an unit matrix of order r then each of the form is known as *normal form* or *canonical form* (cf. more in chapter 14) for equivalence.

The first form in which null matrices border the unit matrix I_r, is sometimes called the *rank normal form*. The normal form of a zero matrix is a zero matrix.

Ex. Reduce the matrix

$$\begin{bmatrix} -1 & 0 & -1 & 3 & 2 \\ -1 & 4 & 0 & 0 & 1 \\ 0 & 0 & 0 & 4 & 2 \\ 1 & 0 & 1 & 1 & 0 \\ 0 & 4 & 1 & 1 & 1 \end{bmatrix}$$

to normal form.

Let

$$A = \overset{A}{\begin{bmatrix} -1 & 0 & -1 & 3 & 2 \\ -1 & 4 & 0 & 0 & 1 \\ 0 & 0 & 0 & 4 & 2 \\ 1 & 0 & 1 & 1 & 0 \\ 0 & 4 & 1 & 1 & 1 \end{bmatrix}} \begin{matrix} R_{12}(-1) \\ \approx \\ R_{14}(1) \end{matrix} \overset{B}{\begin{bmatrix} -1 & 0 & -1 & 3 & 2 \\ 0 & 4 & 1 & -3 & -1 \\ 0 & 0 & 0 & 4 & 2 \\ 0 & 0 & 0 & 4 & 2 \\ 0 & 4 & 1 & 1 & 1 \end{bmatrix}} \begin{matrix} R_1(-1) \\ \approx \\ R_{24}(-1) \end{matrix}$$

$$\underset{C}{\begin{bmatrix} 1 & 0 & 1 & -3 & -2 \\ 0 & 4 & 1 & -3 & -1 \\ 0 & 0 & 0 & 4 & 2 \\ 0 & 0 & 0 & 4 & 2 \\ 0 & 0 & 0 & 4 & 2 \end{bmatrix}} \underset{R_{35}(-1)}{\overset{R_{34}(-1)}{\approx}} \underset{D}{\begin{bmatrix} 1 & 0 & 1 & -3 & -2 \\ 0 & 4 & 1 & -3 & -1 \\ 0 & 0 & 0 & 4 & 2 \\ 0 & 0 & 0 & 0 & 0 \\ 0 & 0 & 0 & 0 & 0 \end{bmatrix}} \overset{R_3(\frac{1}{2})}{\approx}$$

$$\underset{E}{\begin{bmatrix} 1 & 0 & 1 & -3 & -2 \\ 0 & 4 & 1 & -3 & -1 \\ 0 & 0 & 0 & 2 & 1 \\ 0 & 0 & 0 & 0 & 0 \\ 0 & 0 & 0 & 0 & 0 \end{bmatrix}} \underset{\substack{C_{14}(3) \\ C_{15}(2)}}{\overset{C_{13}(-1)}{\approx}} \underset{F}{\begin{bmatrix} 1 & 0 & 0 & 0 & 0 \\ 0 & 4 & 1 & -3 & -1 \\ 0 & 0 & 0 & 2 & 1 \\ 0 & 0 & 0 & 0 & 0 \\ 0 & 0 & 0 & 0 & 0 \end{bmatrix}}$$

$$\underset{\substack{C_{34}(3) \\ C_{35}(1)}}{\overset{C_{32}(-4)}{\approx}} \underset{G}{\begin{bmatrix} 1 & 0 & 0 & 0 & 0 \\ 0 & 0 & 1 & 0 & 0 \\ 0 & 0 & 0 & 2 & 1 \\ 0 & 0 & 0 & 0 & 0 \\ 0 & 0 & 0 & 0 & 0 \end{bmatrix}} \overset{C_{54}(-2)}{\approx} \underset{H}{\begin{bmatrix} 1 & 0 & 0 & 0 & 0 \\ 0 & 0 & 1 & 0 & 0 \\ 0 & 0 & 0 & 0 & 1 \\ 0 & 0 & 0 & 0 & 0 \\ 0 & 0 & 0 & 0 & 0 \end{bmatrix}} \overset{C_{23}}{\approx}$$

$$\underset{I}{\begin{bmatrix} 1 & 0 & 0 & 0 & 0 \\ 0 & 1 & 0 & 0 & 0 \\ 0 & 0 & 0 & 0 & 1 \\ 0 & 0 & 0 & 0 & 0 \\ 0 & 0 & 0 & 0 & 0 \end{bmatrix}} \overset{C_{35}}{\approx} \underset{J}{\left[\begin{array}{ccc|cc} 1 & 0 & 0 & 0 & 0 \\ 0 & 1 & 0 & 0 & 0 \\ 0 & 0 & 1 & 0 & 0 \\ \hline 0 & 0 & 0 & 0 & 0 \\ 0 & 0 & 0 & 0 & 0 \end{array}\right]} = \left[\begin{array}{c|c} I_3 & 0 \\ \hline 0 & 0 \end{array}\right].$$

N.B. Note that $A \overset{row}{\approx} B, A \overset{row}{\approx} C, A \overset{row}{\approx} D, A \overset{row}{\approx} E; B \overset{row}{\approx} C, B \overset{row}{\approx} D, B \overset{row}{\approx} E; C \overset{row}{\approx} D, C \overset{row}{\approx} E; D \overset{row}{\approx} E$. Also $D \overset{col}{\approx} F$, $D \overset{col}{\approx} G$, $D \overset{col}{\approx} H$, $D \overset{col}{\approx} I$, $D \overset{col}{\approx} J$; $E \overset{col}{\approx} F$, $E \overset{col}{\approx} G$, $E \overset{col}{\approx} H$, $E \overset{col}{\approx} I$, $E \overset{col}{\approx} J$; $F \overset{col}{\approx} G$, $F \overset{col}{\approx} H$, $F \overset{col}{\approx} J$; $G \overset{col}{\approx} H$, $G \overset{col}{\approx} I$, $G \overset{col}{\approx} J$; $H \overset{col}{\approx} I$, $H \overset{col}{\approx} J$; $I \overset{col}{\approx} J$.

9.4.1 Some theorems: The subsection deals with some theorems concerning the properties of ERO (ECO) and equivalent matrices.

Theorem 1. If an ERO (ECO) of type I, II, or III on an $m \times n$ matrix A yields the matrix B and the same ERO (ECO) on $I_m (I_n)$ yields the elementary matrix E then $B = EA$ (or AE).

Proof . The theorem can be established in three stages depending on the three types of ERO (ECO) I, II, III. The theorem is proved only for the ERO of type I, II, and III.

Stage 1. Using ERO of type I, let $A = \left[a_{ij}\right]_{m \times n}$. Since $A_{i.}$'s denotes the rows,

$i = 1,2, \cdots, m$, then $A = \begin{bmatrix} A_{1.} \\ A_{2.} \\ \vdots \\ A_{q.} \\ \vdots \\ A_{k.} \\ \vdots \\ A_{m.} \end{bmatrix}$ changes to $B = \begin{bmatrix} A_{1.} \\ A_{2.} \\ \vdots \\ A_{k.} \\ \vdots \\ A_{q.} \\ \vdots \\ A_{m.} \end{bmatrix}$ in which qth and kth row of A are interchanged by ERO of type I . Now

$$I_m = \begin{bmatrix} 1 & \cdots & 0 & \cdots & 0 & \cdots & \cdots & 0 \\ \vdots & & \vdots & & \vdots & & & \vdots \\ 0 & \cdots & 1 & \cdots & 0 & \cdots & \cdots & 0 \\ \vdots & & \vdots & & \vdots & & & \vdots \\ 0 & \cdots & 0 & \cdots & 1 & \cdots & \cdots & 0 \\ \vdots & & \vdots & & \vdots & & & \vdots \\ 0 & \cdots & 0 & \cdots & 0 & \cdots & \cdots & 1 \end{bmatrix} \begin{matrix} \\ \\ \leftarrow k\text{th row} \\ \\ \leftarrow q\text{th row} \\ \\ \\ \end{matrix}$$

(the 1's in the kth and qth rows standing in the qth coln. and kth coln. respectively)

$$\therefore \quad E = P_{kq}^{m} = \begin{bmatrix} 1 & \cdots & 0 & \cdots & 0 & \cdots & \cdots & 0 \\ \vdots & & \vdots & & \vdots & & & \vdots \\ 0 & \cdots & 0 & \cdots & 1 & \cdots & \cdots & 0 \\ \vdots & & \vdots & & \vdots & & & \vdots \\ 0 & \cdots & 1 & \cdots & 0 & \cdots & \cdots & 0 \\ \vdots & & \vdots & & \vdots & & & \vdots \\ 0 & \cdots & 0 & \cdots & 0 & \cdots & \cdots & 1 \end{bmatrix} \begin{matrix} \\ \\ \leftarrow k\text{th row} \\ \\ \leftarrow q\text{th row} \\ \\ \\ \end{matrix}$$

(columns marked qth coln. and kth coln.)

obtained from I_m by ERO of type I where kth and qth rows are interchanged. Now $EA = P_{kq}^{m} A = B$ by prop (iv) in sec. 9.1(b) .

Stage II. Using ERO of type II. Here multiplying ith row of A and I_m by α using ERO of type II, we have,

$$B = \begin{bmatrix} \mathbf{A}_{1.} \\ \mathbf{A}_{2.} \\ \vdots \\ \alpha\mathbf{A}_{i.} \\ \vdots \\ \mathbf{A}_{m.} \end{bmatrix} \text{ and } E = \begin{bmatrix} 1 & 0 & \cdots & 0 & \cdots & 0 \\ 0 & 1 & \cdots & 0 & \cdots & 0 \\ \vdots & \vdots & \ddots & \vdots & \vdots & \vdots \\ 0 & 0 & \cdots & \alpha & \cdots & 0 \\ \vdots & \vdots & \vdots & \vdots & \ddots & \vdots \\ 0 & 0 & \cdots & 0 & \cdots & 1 \end{bmatrix}_{m\times n} \leftarrow i\text{th row}$$

$\therefore$ EA = $D_i^m(\alpha)A = B$ by prop (iv) in sec 9.1(a).

Stage III. Using ERO of type III,

$$A = \begin{bmatrix} \mathbf{A}_{1.} \\ \mathbf{A}_{2.} \\ \vdots \\ \mathbf{A}_{k.} \\ \vdots \\ \mathbf{A}_{q.} \\ \vdots \\ \mathbf{A}_{m.} \end{bmatrix} \begin{matrix} \\ \\ \\ \leftarrow k\text{th row} \\ \\ \leftarrow q\text{th row} \\ \\ \\ \end{matrix} \quad \text{becomes} \quad B = \begin{bmatrix} \mathbf{A}_{1.} \\ \mathbf{A}_{2.} \\ \vdots \\ \mathbf{A}_{k.} + \alpha\mathbf{A}_{q.} \\ \vdots \\ \mathbf{A}_{q.} \\ \vdots \\ \mathbf{A}_{m.} \end{bmatrix} \begin{matrix} \\ \\ \\ \leftarrow k\text{th row} \\ \\ \leftarrow q\text{th row} \\ \\ \\ \end{matrix}.$$

$$\text{Also E} = S_{kq}^m(\alpha) = \begin{bmatrix} 1 & \cdots & 0 & \cdots & 0 & \cdots & \cdots & 0 \\ \vdots & & \vdots & & \vdots & & & \vdots \\ 0 & \cdots & 1 & \cdots & 0 & \cdots & \cdots & 0 \\ \vdots & & \vdots & & \vdots & & & \vdots \\ 0 & \cdots & 0 & \cdots & 1 & \cdots & \cdots & 0 \\ \vdots & & \vdots & & \vdots & & & \vdots \\ 0 & \cdots & 0 & \cdots & 0 & \cdots & \cdots & 1 \end{bmatrix} \begin{matrix} \\ \\ \leftarrow k\text{th row} \\ \\ \leftarrow q\text{th row} \\ \\ \\ \end{matrix}$$

(with the 3rd and 5th columns marked "qth coln." and "kth coln." respectively)

$\therefore$ E $= S_{kq}^m(\alpha) = B$ by prop. (iv) in sec. 9.1(c)

Hence the theorem is established.

Theorem 2. The $m\times n$ matrices A and B are row (column) equivalent i.e. $A \overset{\text{row(col)}}{\approx} B$ if and only if

$$B = E_r E_{r-1} \cdots E_2 E_1 A \quad (\text{or } B = AF_1F_2 \cdots F_{r-1}F_r)$$

where $E_1, E_2, \cdots, E_{r-1}, E_r$ are elementary matrices.

In other words if $m\times n$ matrices A and B are row (column) equivalent then B = PA (or B =AQ) where P or Q is a product of r elementary row (column) matrices i.e. P (or Q) is nonsingular.

Proof . The theorem is proved only for row equivalence. Let $A \overset{\text{row(col)}}{\approx} B$ then B results by above theorem 1 by a sequence of ERO's. This means that there exist elementary matrices $E_1, E_2, \cdots, E_{r-1}, E_r$ such that

$$B = E_{r-1}E_r \cdots E_2E_1A = PA \text{ where } B = E_{r-1}E_r \cdots E_2E_1A = PA,$$

as E_i's are nonsingular, their product is nonsingular, so P is nonsingular.

Conversely, since $B = E_rE_{r-1} \cdots E_2E_1A$ where E_i, $i = 1, 2, \cdots, r$, are elementary matrices, so B is obtained from A by a sequence of ERO's. Then, by definition, $A \overset{\text{row}}{\approx} B$.

Theorem 3. The $m \times n$ matrices A and B are equivalent matrices, i.e. $A \approx B$ if and only if B = PAQ where P and Q are $m \times m$ and $n \times n$ nonsingular matrices respectively.

Proof. Let $A \approx B$. That is, B is obtained from A by finite sequence of ERO's and ECO's. So by above theorem 2, B = PAQ where P is a product of appropriate ERT matrices and Q is a product of appropriate ECT matrices. ERT matrices by prop (i) sec. 9.1 of (a), (b), (c) are nonsingular, so their products are nonsingular. Hence P and Q are nonsingular.

N.B. Theorem 3 provides with a *method of construction* of an equivalent matrix from a given matrix and also the two nonsingular matrices needed for the purpose. Sometimes this is used as a definition of equivalent matrix. Note the example followed.

Ex. Obtain an equivalent matrix

$$B = \begin{bmatrix} 1 & 0 & 0 & 0 \\ 0 & 1 & 0 & 1 \\ 0 & 0 & 0 & -\frac{1}{2} \end{bmatrix} \text{ from the matrix } A = \begin{bmatrix} 1 & 3 & -1 & 3 \\ 3 & 4 & -3 & 4 \\ 1 & 3 & 1 & 2 \end{bmatrix}.$$

and hence write the nonsingular matrices P and Q such that B = PAQ.

Let us begin the problem with known nonsingular matrices I_3 and I_4 such that $A = I_3AI_4$. Now apply the elementary transformations so that L.H. matrix A becomes B while I_3 and I_4 transform to P and Q by row & column transforms, respectively.

Performing $R_{12}(-3)$, $R_{13}(-1)$, we have,

$$\begin{bmatrix} 1 & 3 & -1 & 3 \\ 0 & -5 & 0 & -5 \\ 0 & 0 & 2 & -1 \end{bmatrix} = \begin{bmatrix} 1 & 0 & 0 \\ -3 & 1 & 0 \\ -1 & 0 & 1 \end{bmatrix} A \begin{bmatrix} 1 & 0 & 0 & 0 \\ 0 & 1 & 0 & 0 \\ 0 & 0 & 1 & 0 \\ 0 & 0 & 0 & 1 \end{bmatrix}$$

Performing $C_{12}(-3)$, $C_{13}(1)$, $C_{14}(-3)$, we obtain,

$$\begin{bmatrix} 1 & 0 & 0 & 0 \\ 0 & -5 & 0 & -5 \\ 0 & 0 & 2 & -1 \end{bmatrix} = \begin{bmatrix} 1 & 0 & 0 \\ -3 & 1 & 0 \\ -1 & 0 & 1 \end{bmatrix} A \begin{bmatrix} 1 & -3 & 1 & -3 \\ 0 & 1 & 0 & 0 \\ 0 & 0 & 1 & 0 \\ 0 & 0 & 0 & 1 \end{bmatrix}$$

Performing $R_2(-\frac{1}{5})$, $R_3(\frac{1}{2})$, we have,

$$\begin{bmatrix} 1 & 0 & 0 & 0 \\ 0 & 1 & 0 & 1 \\ 0 & 0 & 1 & -\frac{1}{2} \end{bmatrix} = \begin{bmatrix} 1 & 0 & 0 \\ \frac{3}{5} & -\frac{1}{5} & 0 \\ -\frac{1}{2} & 0 & \frac{1}{2} \end{bmatrix} A \begin{bmatrix} 1 & -3 & 1 & -3 \\ 0 & 1 & 0 & 0 \\ 0 & 0 & 1 & 0 \\ 0 & 0 & 0 & 1 \end{bmatrix}$$

The L. H. matrix is the required matrix B and

$$P=\begin{bmatrix}1 & 0 & 0\\ \frac{3}{5} & -\frac{1}{5} & 0\\ -\frac{1}{2} & 0 & \frac{1}{2}\end{bmatrix} \text{ and } Q=\begin{bmatrix}1 & -3 & 1 & -3\\ 0 & 1 & 0 & 0\\ 0 & 0 & 1 & 0\\ 0 & 0 & 0 & 1\end{bmatrix}.$$

Alternatively, the solution may be proceeded as follows:

$$A=\begin{bmatrix}1 & 3 & -1 & 3\\ 3 & 4 & -3 & 4\\ 1 & 3 & 1 & 2\end{bmatrix} \underset{R_{13}(-1)}{\overset{R_{12}(-3)}{\approx}} \begin{bmatrix}1 & 3 & -1 & 3\\ 0 & -5 & 0 & -5\\ 0 & 0 & 2 & -1\end{bmatrix} \underset{C_{13}(1),\ C_{14}(-3)}{\overset{C_{12}(-3)}{\approx}}$$

$$\begin{bmatrix}1 & 0 & 0 & 0\\ 0 & -5 & 0 & -5\\ 0 & 0 & 2 & -1\end{bmatrix} \underset{R_3(\frac{1}{2})}{\overset{R_2(-\frac{1}{5})}{\approx}} \begin{bmatrix}1 & 0 & 0 & 0\\ 0 & 1 & 0 & 1\\ 0 & 0 & 1 & -\frac{1}{2}\end{bmatrix} = B \text{ (given)}.$$

Now $P = R_3(\frac{1}{2})\ R_2(-\frac{1}{5})\ R_{13}(-1)\ R_{12}(-3)$, in reverse order by theorem 2 in sec.9.4.1

$$=\begin{bmatrix}1 & 0 & 0\\ 0 & 1 & 0\\ 0 & 0 & \frac{1}{2}\end{bmatrix}\begin{bmatrix}1 & 0 & 0\\ 0 & -\frac{1}{5} & 0\\ 0 & 0 & 1\end{bmatrix}\begin{bmatrix}1 & 0 & 0\\ 0 & 1 & 0\\ -1 & 0 & 1\end{bmatrix}\begin{bmatrix}1 & 0 & 0\\ -3 & 1 & 0\\ 0 & 0 & 1\end{bmatrix} = \begin{bmatrix}1 & 0 & 0\\ \frac{3}{5} & -\frac{1}{5} & 0\\ -\frac{1}{2} & 0 & \frac{1}{2}\end{bmatrix}$$

and $Q = C_{12}(-3)\ C_{13}(1)\ C_{14}(-3)$ in order

$$=\begin{bmatrix}1 & -3 & 0 & 0\\ 0 & 1 & 0 & 0\\ 0 & 0 & 1 & 0\\ 0 & 0 & 0 & 1\end{bmatrix}\begin{bmatrix}1 & 0 & 1 & 0\\ 0 & 1 & 0 & 0\\ 0 & 0 & 1 & 0\\ 0 & 0 & 0 & 1\end{bmatrix}\begin{bmatrix}1 & 0 & 0 & -3\\ 0 & 1 & 0 & 0\\ 0 & 0 & 1 & 0\\ 0 & 0 & 0 & 1\end{bmatrix} = \begin{bmatrix}1 & -3 & 1 & -3\\ 0 & 1 & 0 & 0\\ 0 & 0 & 1 & 0\\ 0 & 0 & 0 & 1\end{bmatrix}.$$

Theorem 4. Equivalences of matrices are reflexive, symmetric and transitive.

Proof . Let A be an $m \times n$ matrix.

Reflexive means we have to prove that A is equivalent to itself, i.e. $A \approx A$.

By theorem 3, $A = I_m A I_n$ where both I_m and I_n are nonsingular.

$\therefore\ A \approx A$. Hence equivalence is reflexive.

Symmetric means if $A \approx B$ then $B \approx A$.

By theorem 3, B= PAQ where P, Q, P^{-1}, Q^{-1} are nonsingular.

$$\therefore \quad P^{-1}BQ^{-1} = P^{-1}(PAQ)Q^{-1}$$
$$= (P^{-1}P)A(QQ^{-1}), \text{ by associative law for multiplication}$$
$$= I_m A I_m$$

$\therefore$ $B \approx A$. Hence the proof.

Transitive means if $A \approx B$, $B \approx C$ then $A \approx C$.

Since $A \approx B$, so by theorem 3, B = PAQ, P & Q are nonsingular and $B \approx C$, C = SBT, S & T are nonsingular.

$= S(PAQ)T = (SP)A(QT)$ where SP and QT are nonsingular

$\therefore$ $C \approx A$. Hence the proof.

$\therefore$ the theorem is established for all three cases.

Ex. Find the inverse of

$$A = \begin{bmatrix} 1 & 1 & 2 & 1 \\ 2 & 3 & 4 & 1 \\ 3 & 3 & 3 & 1 \\ 1 & 2 & 3 & 1 \end{bmatrix}$$

using cor. (a) of theorem of sec. 9.3 . Write both A and A^{-1} as the product of elementary matrices.

Let us form the partition matrix $[A_4 \vdots I_4]$ where A_4 is the given matrix.

Let us perform ERO's that transforms $[A_4 \vdots I_4]$ to $[I_4 \vdots A_4^{-1}]$. Let us consider $[A_4 \vdots I_4]$ as a 4×8 matrix and whatever we do to a row of A_4 we also do the same on the corresponding row of I_4. Let us proceed with the following ERO's.

$$\left[\begin{array}{cccc|cccc} 1 & 1 & 2 & 1 & 1 & 0 & 0 & 0 \\ 2 & 3 & 4 & 1 & 0 & 1 & 0 & 0 \\ 3 & 3 & 3 & 1 & 0 & 0 & 1 & 0 \\ 1 & 2 & 3 & 1 & 0 & 0 & 0 & 1 \end{array}\right] \underset{\substack{R_3 - 3R_1 \\ R_4 - R_1}}{\overset{R_2 - 2R_1}{\approx}} \left[\begin{array}{cccc|cccc} 1 & 1 & 2 & 1 & 1 & 0 & 0 & 0 \\ 0 & 1 & 0 & -1 & -2 & 1 & 0 & 0 \\ 0 & 0 & -3 & -2 & -3 & 0 & 1 & 0 \\ 0 & 1 & 1 & 0 & -1 & 0 & 0 & 1 \end{array}\right]$$

$$\overset{R'_4 - R'_2}{\approx} \left[\begin{array}{cccc|cccc} 1 & 1 & 2 & 1 & 1 & 0 & 0 & 0 \\ 0 & 1 & 0 & -1 & -2 & 1 & 0 & 0 \\ 0 & 0 & -3 & -2 & -3 & 0 & 1 & 0 \\ 0 & 0 & 1 & 1 & 1 & -1 & 0 & 1 \end{array}\right] \overset{-\frac{1}{3}R^{ii}}{\approx}$$

$$\left[\begin{array}{cccc|cccc} 1 & 1 & 2 & 1 & 1 & 0 & 0 & 0 \\ 0 & 1 & 0 & -1 & -2 & 1 & 0 & 0 \\ 0 & 0 & 1 & \frac{2}{3} & 1 & 0 & -\frac{1}{3} & 0 \\ 0 & 0 & 1 & 1 & 1 & -1 & 0 & 1 \end{array}\right] \overset{R_4^{iii} - R_3^{iii}}{\approx}$$

$$\left[\begin{array}{cccc|cccc} 1 & 1 & 2 & 1 & 1 & 0 & 0 & 0 \\ 0 & 1 & 0 & -1 & -2 & 1 & 0 & 0 \\ 0 & 0 & 1 & \frac{2}{3} & 1 & 0 & -\frac{1}{3} & 0 \\ 0 & 0 & 0 & \frac{1}{3} & 0 & -1 & \frac{1}{3} & 1 \end{array}\right] \overset{R_1^{iv} - R_2^{iv}}{\approx}$$

$$\left[\begin{array}{cccc|cccc} 1 & 0 & 2 & 2 & 3 & -1 & 0 & 0 \\ 0 & 1 & 0 & -1 & -2 & 1 & 0 & 0 \\ 0 & 0 & 1 & \frac{2}{3} & 1 & 0 & -\frac{1}{3} & 0 \\ 0 & 0 & 0 & \frac{1}{3} & 0 & -1 & \frac{1}{3} & 1 \end{array}\right] \overset{R_1^{v} - 2R_3^{v}}{\approx}$$

$$\begin{bmatrix} 1 & 0 & 0 & \frac{2}{3} & \vdots & 1 & -1 & \frac{2}{3} & 0 \\ 0 & 1 & 0 & -1 & \vdots & -2 & 1 & 0 & 0 \\ 0 & 0 & 1 & \frac{2}{3} & \vdots & 1 & 0 & -\frac{1}{3} & 0 \\ 0 & 0 & 0 & \frac{1}{3} & \vdots & 0 & -1 & \frac{1}{3} & 1 \end{bmatrix} \underset{\approx}{3R_4^{vi}} \begin{bmatrix} 1 & 0 & 0 & \frac{2}{3} & \vdots & 1 & -1 & \frac{2}{3} & 0 \\ 0 & 1 & 0 & -1 & \vdots & -2 & 1 & 0 & 0 \\ 0 & 0 & 1 & \frac{2}{3} & \vdots & 1 & 0 & -\frac{1}{3} & 0 \\ 0 & 0 & 0 & 1 & \vdots & 0 & -3 & 1 & 3 \end{bmatrix}$$

$$\begin{array}{c} R_1^{vii} - \frac{2}{3}R_4^{vii} \\ \approx \\ R_2^{vii} + R_4^{vii} \\ R_3^{vii} - \frac{2}{3}R_4^{vii} \end{array} \begin{bmatrix} 1 & 0 & 0 & 0 & \vdots & 1 & 1 & 0 & -2 \\ 0 & 1 & 0 & 0 & \vdots & -2 & -2 & 1 & 3 \\ 0 & 0 & 1 & 0 & \vdots & 1 & 2 & -1 & -2 \\ 0 & 0 & 0 & 1 & \vdots & 0 & -3 & 1 & 3 \end{bmatrix} = \left[I_4 \vdots A_4^{-1}\right].$$

It is easy to verify that $AA^{-1} = A^{-1}A = I_4$.

Second part

Let us express each of the ERO's in terms of elementary row matrices. Thus

$$R_2 - 2R_1 \equiv E_{12}^4(-2) = \begin{bmatrix} 1 & 0 & 0 & 0 \\ -2 & 1 & 0 & 0 \\ 0 & 0 & 1 & 0 \\ 0 & 0 & 0 & 1 \end{bmatrix};\ R_3 - 3R_1 \equiv E_{13}^4(-3) = \begin{bmatrix} 1 & 0 & 0 & 0 \\ 0 & 1 & 0 & 0 \\ -3 & 0 & 1 & 0 \\ 0 & 0 & 0 & 1 \end{bmatrix};$$

$$R_4 - R_1 \equiv E_{14}^4(-1) = \begin{bmatrix} 1 & 0 & 0 & 0 \\ 0 & 1 & 0 & 0 \\ 0 & 0 & 1 & 0 \\ -1 & 0 & 0 & 1 \end{bmatrix};\ R_4' - R_2' \equiv E_{24}^4(-1) = \begin{bmatrix} 1 & 0 & 0 & 0 \\ 0 & 1 & 0 & 0 \\ 0 & 0 & 1 & 0 \\ 0 & -1 & 0 & 1 \end{bmatrix};$$

$$-\frac{1}{3}R_3'' \equiv E_3^4(-\tfrac{1}{3}) = \begin{bmatrix} 1 & 0 & 0 & 0 \\ 0 & 1 & 0 & 0 \\ 0 & 0 & -\frac{1}{3} & 0 \\ 0 & 0 & 0 & 1 \end{bmatrix};\ R_4''' - R_3''' \equiv E_{34}^4(-1) = \begin{bmatrix} 1 & 0 & 0 & 0 \\ 0 & 1 & 0 & 0 \\ 0 & 0 & 1 & 0 \\ 0 & 0 & -1 & 1 \end{bmatrix};$$

$$R_1^{iv} - R_2^{iv} \equiv E_{21}^4(-1) = \begin{bmatrix} 1 & -1 & 0 & 0 \\ 0 & 1 & 0 & 0 \\ 0 & 0 & 1 & 0 \\ 0 & 0 & 0 & 1 \end{bmatrix};\ R_1^{v} - 2R_3^{v} \equiv E_{31}^4(-2) = \begin{bmatrix} 1 & 0 & -2 & 0 \\ 0 & 1 & 0 & 0 \\ 0 & 0 & 1 & 0 \\ 0 & 0 & 0 & 1 \end{bmatrix};$$

$$3R_4^{vi} \equiv E_4^4(3) = \begin{bmatrix} 1 & 0 & 0 & 0 \\ 0 & 1 & 0 & 0 \\ 0 & 0 & 1 & 0 \\ 0 & 0 & 0 & 3 \end{bmatrix};\ R_1^{vii} - \frac{2}{3}R_4^{vii} \equiv E_{41}^4(-\tfrac{2}{3}) = \begin{bmatrix} 1 & 0 & 0 & -\frac{2}{3} \\ 0 & 1 & 0 & 0 \\ 0 & 0 & 1 & 0 \\ 0 & 0 & 0 & 1 \end{bmatrix};$$

$$R_2^{vii} + R_4^{vii} \equiv E_{42}^4(1) = \begin{bmatrix} 1 & 0 & 0 & 0 \\ 0 & 1 & 0 & 1 \\ 0 & 0 & 1 & 0 \\ 0 & 0 & 0 & 1 \end{bmatrix};$$

$$R_3^{vii} - \frac{2}{3}R_4^{vii} \equiv E_{43}^4(-\tfrac{2}{3}) = \begin{bmatrix} 1 & 0 & 0 & 0 \\ 0 & 1 & 0 & 0 \\ 0 & 0 & 1 & -\frac{2}{3} \\ 0 & 0 & 0 & 1 \end{bmatrix}$$

$$\therefore\ A^{-1} = E_{43}^4(-\tfrac{2}{3})\, E_{42}^4(1)\, E_{41}^4(-\tfrac{2}{3})\, E_4^4(3)\, E_{31}^4(-2)\, E_{21}^4(-1)\, E_{34}^4(-1)$$
$$E_3^4(-\tfrac{1}{3})\, E_{24}^4(-1)\, E_{14}^4(-1)\, E_{13}^4(-3)\, E_{12}^4(-2)$$

in reverse order and therefore, taking inverse of both sides,

$$A = E_{12}^4(-\tfrac{1}{2})\, E_{13}^4(-\tfrac{1}{3})\, E_{14}^4(-1)\, E_{24}^4(-1)\, E_3^4(-3)\, E_{34}^4(-1)\, E_{21}^4(-1)$$
$$E_{31}^4(-\tfrac{1}{2})E_4^4(\tfrac{1}{3})E_{41}^4(-\tfrac{3}{2})E_{42}^4(1)E_{43}^4(-\tfrac{3}{2})$$

Theorem 5. The following statements are equivalent:

(1) The matrix A is nonsingular.

(2) The homogeneous equations $\mathbf{Ax} = 0$ has only the trivial solution.

(3) The matrix A is row equivalent to I.

Proof . The sequence of the proof is

(a) (1) implies (2);

(b) (2) implies (3);

(c) (3) implies (1) .

Now if A is non-singular, i.e. A^{-1} exists and $\mathbf{x}$ is A solution to $\mathbf{Ax} = 0$ then

$$\mathbf{x} = \mathbf{Ix} = (A^{-1} A)\mathbf{x} = A^{-1} (\mathbf{Ax}) = A^{-1} (0) = 0.$$

$\therefore$ $\mathbf{Ax} = 0$ has only the trivial solution. Thus proof of (a) is complete.

Using ERO on $\mathbf{Ax} = 0$, we have the equivalent row echelon from $(rE)\mathbf{x} = 0$, (rE) is in row echelon form. If one of the diagonal element of (rE) were zero, the last row of (rE) would consists entirely of 0's. Then $\mathbf{Ax} = 0$ would be equivalent to an over determined system i.e. with more equations than unknowns and the system would have nontrivial solution. Thus (rE) must be a triangular matrix with diagonal elements all equal to 1. If follows that I is the reduced row echelon form of A. Hence $A \overset{\text{row}}{\approx} I$. Thus proof of (b) is complete.

Finally, if $A \overset{\text{row}}{\approx} I$ then there exists elementary matrices $E_1, E_2, \cdots, E_p$ such that

$$A = E_p E_{p-1} \cdots E_2 E_1 I = E_p E_{p-1} \cdots E_2 E_1$$

Hence A is nonsingular as each E_i is nonsingular.

9.5 Linear dependence / independence

Row reduced echelon form of a matrix A consisting of vectors as columns of A and the following definition together provides a method for testing the linear dependence / independence of the vectors and simultaneously determines the coefficient of dependence. For supporting theorems see sec.10.10.

Definition . Elementary columns of a matrix are of s.u. vector.

Testing : Let $\mathbf{u}_1, \mathbf{u}_2, \cdots, \mathbf{u}_n$ be n vectors each with m components. Now construct an $m \times n$ matrix A with jth column $A_{.j}$ by the vectors $\mathbf{u}_j$, $i=1, 2, \cdots, n$. The vectors $\mathbf{u}_1, \mathbf{u}_2, \cdots, \mathbf{u}_n$ are linearly independent if row reduced echelon form of A contains *only distinct* elementary columns. If any one or more columns of the reduced row echelon form is *not* an elementary column, then the vectors are linearly dependent. The vector (s) corresponding to the *nonelementary* column is (are) dependent on the others and the numbers along the nonelementary columns are the *coefficients* of dependence of the preceding vectors in order. Thus elementary operations on matrix provide a powerful and easy computational method for linear dependence and coefficients of dependence.

Ex. Find the linearly independent rows of the following matrix

$$\begin{bmatrix} 1 & 2 & -1 & 0 \\ 0 & 1 & 1 & 2 \\ 2 & 3 & -1 & -2 \\ 1 & 4 & 1 & 4 \end{bmatrix}$$

and write other rows as a linear combination of the independent rows.

Let $$A = \begin{bmatrix} 1 & 2 & -1 & 0 \\ 0 & 1 & 1 & 2 \\ 2 & 3 & -1 & -2 \\ 1 & 4 & 1 & 4 \end{bmatrix}.$$

To use above testing procedure first obtain the row reduced echelon form of A^T, i.e.

$$A^T = \begin{bmatrix} 1 & 0 & 2 & 1 \\ 2 & 1 & 3 & 4 \\ -1 & 1 & -1 & 1 \\ 0 & 2 & -2 & 4 \end{bmatrix} \underset{R_3 + R_1}{\overset{R_2 - 2R_1}{\approx}} \begin{bmatrix} 1 & 0 & 2 & 1 \\ 0 & 1 & -1 & 2 \\ 0 & 1 & 1 & 2 \\ 0 & 2 & -2 & 4 \end{bmatrix} \underset{R_4 - 2R_2}{\overset{R_3 - R_2}{\approx}} \begin{bmatrix} 1 & 0 & 2 & 1 \\ 0 & 1 & -1 & 2 \\ 0 & 0 & 2 & 0 \\ 0 & 0 & 0 & 0 \end{bmatrix}$$

$$\overset{\frac{1}{2}R_3}{\approx} \begin{bmatrix} 1 & 0 & 2 & 1 \\ 0 & 1 & -1 & 2 \\ 0 & 0 & 1 & 0 \\ 0 & 0 & 0 & 0 \end{bmatrix} \underset{R_2 + R_3}{\overset{R_1 - 2R}{\approx}} \begin{bmatrix} 1 & 0 & 0 & 1 \\ 0 & 1 & 0 & 2 \\ 0 & 0 & 1 & 0 \\ 0 & 0 & 0 & 0 \end{bmatrix}.$$

$\therefore$ 1st, 2nd, 3rd columns are elementary columns while 4th is not. So fourth column of A^T i.e. 4th row of A is *dependent* while 1st, 2nd, 3rd rows of A are independent. Since nonelementary column of A^T has numbers 1, 2, 0 along first, 2nd and 3rd row, so these are coefficient of dependence of 1st, 2nd and 3rd columns of A^T i.e. 1st, 2nd, and 3rd rows of A. Hence

$$[1 \ \ 4 \ \ 1 \ \ 4] = (1)[1 \ \ 2 \ \ -1 \ \ 0] + 2[0 \ \ 1 \ \ 1 \ \ 2] + (0)[2 \ \ 3 \ \ -1 \ \ -2].$$

Alter. (without A^T) Use column transformation as follows :

$$\begin{bmatrix} 1 & 2 & -1 & 0 \\ 0 & 1 & 1 & 2 \\ 2 & 3 & -1 & -2 \\ 1 & 4 & 1 & 4 \end{bmatrix} \underset{C_3+C_1}{\overset{C_2-2C_1}{\approx}} \begin{bmatrix} 1 & 0 & 0 & 0 \\ 0 & 1 & 1 & 2 \\ 2 & -1 & 1 & -2 \\ 1 & 2 & 2 & 4 \end{bmatrix} \underset{C_4-2C_2}{\overset{C_3-C_2}{\approx}} \begin{bmatrix} 1 & 0 & 0 & 0 \\ 0 & 1 & 0 & 0 \\ 2 & -1 & 2 & 0 \\ 1 & 2 & 0 & 0 \end{bmatrix} \overset{\frac{1}{2}C_3}{\approx}$$

$$\begin{bmatrix} 1 & 0 & 0 & 0 \\ 0 & 1 & 0 & 0 \\ 2 & -1 & 1 & 0 \\ 1 & 2 & 0 & 0 \end{bmatrix} \underset{C_2+C_3}{\overset{C_1-2C_3}{\approx}} \begin{bmatrix} 1 & 0 & 0 & 0 \\ 0 & 1 & 0 & 0 \\ 0 & 0 & 1 & 0 \\ 1 & 2 & 0 & 0 \end{bmatrix}.$$

The last equivalent matrix is column echelon form in which 1st, 2nd & 3rd rows are elementary rows while 4th one is not. So, the matrix representing row vectors is *linearly dependent* and 4th row depends on the first three rows. Let

R_i , $i = 1, 2, 3, 4$ be the rows. Then

$$R_4 = 1.R_1 + 2.R_2 \quad \text{i.e.} \quad [1 \ \ 4 \ \ 1 \ \ 4] = (1)[1 \ \ 2 \ \ -1 \ \ 0] + 2[0 \ \ 1 \ \ 1 \ \ 2].$$

Ex. Find a linearly independent set of equations with same solution set for the system of equations

$$\left.\begin{aligned} 2x_1 \qquad\quad + 6x_3 + x_4 + 2x_5 + 4x_6 &= 0 \\ 2x_1 + 2x_2 + 10x_3 \qquad + x_5 + 3x_6 &= 0 \\ x_1 + x_2 + 5x_3 + x_4 + 3x_5 + x_6 &= 0 \\ 3x_1 \qquad\quad + 9x_3 + x_4 + x_5 + 7x_6 &= 0 \\ 5x_1 + 3x_2 + 21x_3 + 5x_4 + 2x_5 + 18x_6 &= 0 \end{aligned}\right\}$$

This system of homogeneous equations is equivalent to testing the vectors

$$\begin{bmatrix} 2 \\ 2 \\ 1 \\ 3 \\ 5 \end{bmatrix}, \begin{bmatrix} 0 \\ 2 \\ 1 \\ 0 \\ 3 \end{bmatrix}, \begin{bmatrix} 6 \\ 10 \\ 5 \\ 9 \\ 21 \end{bmatrix}, \begin{bmatrix} 1 \\ 0 \\ 1 \\ 1 \\ 5 \end{bmatrix}, \begin{bmatrix} 2 \\ 1 \\ 3 \\ 1 \\ 2 \end{bmatrix} \& \begin{bmatrix} 4 \\ 3 \\ 1 \\ 7 \\ 18 \end{bmatrix}$$

for linear independence/dependence and x_i , i=1, 2, 3, 4, 5, are the scalars required for linear combinations of six vectors. So, the test for linear independence/dependence of a system of homogeneous equations is the same as the test for linear indepencence/dependence of columns of the coefficient matrix of the system by row transformations.

The coefficient matrix is

$$A = \begin{bmatrix} 2 & 0 & 6 & 1 & 2 & 4 \\ 2 & 2 & 10 & 0 & 1 & 3 \\ 1 & 1 & 5 & 1 & 3 & 1 \\ 3 & 0 & 9 & 1 & 1 & 7 \\ 5 & 3 & 21 & 5 & 2 & 18 \end{bmatrix}$$

$$\therefore\ A \approx \begin{bmatrix} 2 & 0 & 6 & 1 & 2 & 4 \\ 0 & 2 & 4 & -1 & -1 & -1 \\ 0 & 1 & 2 & \frac{1}{2} & 2 & -1 \\ 0 & 0 & 0 & -\frac{1}{2} & -2 & 1 \\ 0 & 3 & 6 & \frac{5}{2} & -3 & 8 \end{bmatrix} \approx \begin{bmatrix} 2 & 0 & 6 & 1 & 2 & 4 \\ 0 & 2 & 4 & -1 & -1 & -1 \\ 0 & 0 & 0 & 1 & \frac{5}{2} & -\frac{1}{2} \\ 0 & 0 & 0 & -\frac{1}{2} & -2 & 1 \\ 0 & 0 & 0 & 4 & -\frac{3}{2} & \frac{19}{2} \end{bmatrix}$$

$$\approx \begin{bmatrix} 2 & 0 & 6 & 1 & 2 & 4 \\ 0 & 1 & 2 & -\frac{1}{2} & -\frac{1}{2} & -\frac{1}{2} \\ 0 & 0 & 0 & 1 & \frac{5}{2} & -\frac{1}{2} \\ 0 & 0 & 0 & -\frac{1}{2} & -2 & 1 \\ 0 & 0 & 0 & 4 & -\frac{3}{2} & \frac{19}{2} \end{bmatrix} \approx \begin{bmatrix} 2 & 0 & 6 & 1 & 2 & 4 \\ 0 & 1 & 2 & -\frac{1}{2} & -\frac{1}{2} & -\frac{1}{2} \\ 0 & 0 & 0 & 1 & \frac{5}{2} & -\frac{1}{2} \\ 0 & 0 & 0 & 0 & -\frac{3}{4} & \frac{3}{4} \\ 0 & 0 & 0 & 0 & -\frac{23}{2} & \frac{23}{2} \end{bmatrix}$$

$$\approx \begin{bmatrix} 2 & 0 & 6 & 1 & 2 & 4 \\ 0 & 1 & 2 & -\frac{1}{2} & -\frac{1}{2} & -\frac{1}{2} \\ 0 & 0 & 0 & 1 & \frac{5}{2} & -\frac{1}{2} \\ 0 & 0 & 0 & 0 & 1 & -1 \\ 0 & 0 & 0 & 0 & -1 & 1 \end{bmatrix} \approx \begin{bmatrix} 2 & 0 & 6 & 1 & 2 & 4 \\ 0 & 1 & 2 & -\frac{1}{2} & -\frac{1}{2} & -\frac{1}{2} \\ 0 & 0 & 0 & 1 & \frac{5}{2} & -\frac{1}{2} \\ 0 & 0 & 0 & 0 & 1 & -1 \\ 0 & 0 & 0 & 0 & 0 & 0 \end{bmatrix}$$

$$\approx \begin{bmatrix} 2 & 0 & 6 & 1 & 0 & 6 \\ 0 & 1 & 2 & -\frac{1}{2} & 0 & -1 \\ 0 & 0 & 0 & 1 & 0 & 2 \\ 0 & 0 & 0 & 0 & 1 & -1 \\ 0 & 0 & 0 & 0 & 0 & 0 \end{bmatrix} \approx \begin{bmatrix} 1 & 0 & 3 & 0 & 0 & 2 \\ 0 & 1 & 2 & 0 & 0 & 0 \\ 0 & 0 & 0 & 1 & 0 & 2 \\ 0 & 0 & 0 & 0 & 1 & -1 \\ 0 & 0 & 0 & 0 & 0 & 0 \\ & & \uparrow & & & \uparrow \end{bmatrix}$$

which is in row reduced echelon form. In this form 3rd & 5th columns (marked by vertical arrows) are not elementary columns. So, the column vectors are linearly dependent i.e. the given system is linearly dependent.

∴ the linearly independent set of equations is given by

$$\begin{bmatrix} 1 & 0 & 3 & 0 & 0 & 2 \\ 0 & 1 & 2 & 0 & 0 & 0 \\ 0 & 0 & 0 & 1 & 0 & 2 \\ 0 & 0 & 0 & 0 & 1 & -1 \\ 0 & 0 & 0 & 0 & 0 & 0 \end{bmatrix} \begin{bmatrix} x_1 \\ x_2 \\ x_3 \\ x_4 \\ x_5 \\ x_6 \end{bmatrix} = 0 \quad \text{i.e.} \quad \left.\begin{aligned} x_1 \quad + 3x_3 \quad + 2x_6 &= 0 \\ x_2 + 2x_3 \qquad &= 0 \\ x_4 + 2x_6 &= 0 \\ x_4 - \ x_6 &= 0 \end{aligned}\right\}$$

9.6 Alternative proof of |AB|= |A||B| and a related theorem

Theorem 1. If A and B are *n*-square matrices then $|AB| = |A||B|$.

Proof . The theorem is proved again with the help of elementary matrices. There arises four cases.

Case (i) Let A be an EMFK. Then from sec. 9.1 (a), $A = D_i^k(\alpha)$

$\therefore\ |A| = \left|D_i^n(\alpha)\right| = \alpha \neq 0$ by prop (i) sec.9.1 (a).

$$\text{Also } AB = D_i^n(\alpha)B = \begin{bmatrix} 1 & 0 & \cdots & 0 & \cdots & 0 \\ 0 & 1 & \cdots & 0 & \cdots & 0 \\ \vdots & \vdots & \ddots & \vdots & & \vdots \\ 0 & 0 & \cdots & \alpha & \cdots & 0 \\ \vdots & \vdots & & \vdots & \ddots & \vdots \\ 0 & 0 & \cdots & 0 & \cdots & 1 \end{bmatrix} \begin{bmatrix} b_{11} & b_{12} & \cdots & b_{1i} & \cdots & b_{1n} \\ b_{21} & b_{22} & \cdots & b_{2i} & \cdots & b_{2n} \\ \vdots & \vdots & \ddots & \vdots & \vdots & \vdots \\ b_{i1} & b_{i2} & \cdots & b_{ii} & \cdots & b_{in} \\ \vdots & \vdots & \vdots & \vdots & \ddots & \vdots \\ b_{n1} & b_{n2} & \cdots & b_{ni} & \cdots & b_{nn} \end{bmatrix} \quad (i\text{th row} \rightarrow)$$

$$= \begin{bmatrix} b_{11} & b_{12} & \cdots & b_{1i} & \cdots & b_{1n} \\ b_{21} & b_{22} & \cdots & b_{2i} & \cdots & b_{2n} \\ \vdots & \vdots & \ddots & \vdots & \vdots & \vdots \\ \alpha b_{i1} & \alpha b_{i2} & \cdots & \alpha b_{ii} & \cdots & \alpha b_{in} \\ \vdots & \vdots & \vdots & \vdots & \ddots & \vdots \\ b_{n1} & b_{n2} & \cdots & b_{ni} & \cdots & b_{nn} \end{bmatrix}$$

For **alternative** proof of this see prop. (iii) sec.9.1 (a).

$\therefore |AB| = \alpha|B| = |A||B|$. Since $A = D_i^k(\alpha)$., so the theorem is proved when A is an EMFK.

Case (ii) Let A be an EMSK. Then from sec. 9.1 (b), let $A = P_{kq}^n$.

Therefore, $|A| = \left|P_{kq}^n\right|$, by prop (i) sec. 9.1 (b).

Now $A = P_{kq}^n \cdot$ = (interchanges k and q rows of B)

$\therefore\ |AB| = -|B|$, $\because$ two rows of B are interchanged

$=(-1)\ |B| = |A|\ |B|$.

As $P_{kq}^n = E_{kq}^n = F_{kq}^n$, so the theorem is again proved when A is an EMSK.

Case(iii) Let A be an EMTK. Then from sec. 9.1 (c), let $A = S_{kq}^n(\alpha)$,

$\therefore\quad |A| = \left|S_{kq}^n(\alpha)\right| = 1,$ by prop (i) sec. 9.1(c).

Now AB= $S_{kq}^n(\alpha)$ B

= (B with the addition of α times the qth row to the kth row)

$\therefore\quad |AB| = |B|$, $\because$ theorem 6 sec. 5.4

$= (1)\ |B| = |A||B|$.

As $S_{kq}^n(\alpha) = E_{kq}^n(\alpha) = F_{kq}^n(\alpha)$, so theorem is proved when A is an EMTK.

Finally, $|AB| = |A||B|$ when either A or B is an elementary matrix.

Case (iv) Proof of $|AB| = |A||B|$ for case when $|A| = 0$. Since $|A| = 0$ then L.H.S $= |AB| = 0$ by theorem above and R.H.S= $|A||B| = 0$ by obvious reason. Hence $|AB| = |A|\ |B|$. Let $|A| \neq 0$ i.e. A is invertible. Hence A can be expressed as a product of elementary matrices.

$$\therefore\quad |AB| = |E_1 \cdots E_p B|$$

$$= |E_1|\ |E_2| \cdots |E_p|\ |B|$$

$$= \left|E_1 E_2 \cdots E_p\right| |B|$$

$$= |A|\,|B|$$

$\because$ by case (i) to (iii) $|E_1| = \alpha, -1$, or 1 can be bring outside as a common factor. The repeated application gives the next step. The back application produces the third step.

Theorem 2. If A and B are n square matrices and if $|A| = 0$ then $|AB| = 0$.

Proof. Case 1. If A is a null matrix, then AB = 0.

$\therefore$ $|AB| = |0| = 0$

Case 2. If A is not a null matrix Then A can be reduced to normal form N. So there exists nonsingular or invertible matrices P and Q such that

$$A = PNQ$$

or $AB = PNQB$, post multiplication by B

or $P^{-1}(AB) = NQB$, premultiplication by P^{-1}

But $\left|P^{-1}(AB)\right| = \left|(E_1 E_2 \cdots E_p)(AB)\right| \quad \because P^{-1} = E_1 E_2 \cdots E_p$

$= \left|E_1 E_2 \cdots E_p\right| |AB|$, arguing as in case (iv) of theorem above.

$= \left|P^{-1}\right| |AB|.$

$\therefore$ $\left|P^{-1}\right| |AB| = |NQB|$.

Since $|AB| = 0$, and $A \approx N$, so $N \neq I_n$.

Hence the last row of N is zero row and automatically last row of NQ and that of NQB are zero rows. Therefore, $|NQB|$ must be equal to zero. Hence $\left|P^{-1}\right||AB| = 0$.

or $|AB| = 0$ $\quad \because \left|P^{-1}\right| \neq 0$ by hypothesis.

Ex.1 Let

$$A = \begin{bmatrix} 3 & 1 \\ 9 & 3 \end{bmatrix} \text{ and } B = \begin{bmatrix} 1 & 3 \\ 2 & 7 \end{bmatrix}. \text{ Verify the theorems.}$$

Note that $|A| = 0$ and $|B| = 1 \neq 0$. Therefore,

$$|AB| = \det\left\{\begin{bmatrix} 3 & 1 \\ 9 & 3 \end{bmatrix}\begin{bmatrix} 1 & 3 \\ 2 & 7 \end{bmatrix}\right\} = \det\begin{bmatrix} 5 & 16 \\ 15 & 48 \end{bmatrix} = 240 - 240 = 0$$

Also $|A||B| = 0$.

$\therefore$ $|AB| = 0 = |A|\,|B|$ which verifies both first and second theorems above.

Ex.2 Let

$$A = \begin{bmatrix} 3 & 1 \\ 9 & 2 \end{bmatrix} \text{ and } B = \begin{bmatrix} 1 & 3 \\ 2 & 7 \end{bmatrix}. \text{ Verify the theorems.}$$

Note that $|A| = 6 - 9 = -3 \neq 0$ and $|B| = 1 \neq 0$. Therefore,

Now $|AB| = \det\begin{bmatrix} 3 & 1 \\ 9 & 2 \end{bmatrix}\begin{bmatrix} 1 & 3 \\ 2 & 7 \end{bmatrix} = \det\begin{bmatrix} 5 & 16 \\ 13 & 41 \end{bmatrix}$

$$\therefore \qquad |AB| = \det\begin{bmatrix} 5 & 16 \\ 13 & 41 \end{bmatrix} = 205 - 208 = -3$$

$$\text{And} \qquad |A||B| = (-3)(1) = -3$$

$$\therefore \qquad |AB| = |A|\,|B|$$

verifying the first theorem when none of A & B is singular .

9.7 Normal form of any matrix

The following theorem assures that any matrix is always equivalent to a unique normal form .

Theorem . A nonzero $m \times n$ matrix A can be reduced to normal form by elementary operations.

Proof . Let $A = [a_{ij}]_{m \times n}$ be a nonsingular matrix. Also let $a_{11} \neq 0$. If $a_{11} = 0$ then there is an element $a_{ij} \neq 0$ as A is nonsingular. Now move a_{ij} to (1 , 1) position by elementary operation I. So , without loss of generality, we can assume $a_{11} \neq 0$. Then by ERO'S

$$(\frac{1}{a_{11}} R_1),\ (-a_{21}R_1 + R_2),\ \cdots,(-a_{m1}R_1 + R_m) \ \text{ and}$$

ECO 's $\quad (-a_{12}C_1 + C_2),\ \cdots,(-a_{1n}C_1 + C_n)$ the given matrix A

transforms to an equivalent matrix of the form

$$\left[\begin{array}{c|c} 1 & 0 \\ \hline 0 & B \end{array}\right] \quad \text{if } m>1,\ n>1 \text{ and B is } (m-1)\times(n-1) \ .$$

Now if either $m = 1$ or $n = 1$ or both $m = n = 1$ then A is equivalent to the normal forms

$$[1 \ \ 0 \ \ 0 \ \cdots \ \ 0] \text{ or } [1 \ \ 0 \ \ 0 \ \cdots \ \ 0]^T \ \text{ or } I_1 \text{ respectively.}$$

Hence proof is complete.

Other possibilities of B

(i) If B is null then of course

$$\left[\begin{array}{c|c} 1 & 0 \\ \hline 0 & 0 \end{array}\right]$$

is normal form and the proof is complete.

(ii) If B is nonsingular then repeat the previous procedure and for $m > 2$, $n > 2$

$$A \approx \left[\begin{array}{c|c} I & 0 \\ \hline 0 & C \end{array}\right] \quad \text{where C is } (m-2)\times(n-2) .$$

If either m=2 or n =2 or $m = n$ =2 then A is equivalent to the normal forms

$$[I_2 \mid 0] \text{ or } \left[\begin{array}{c} I_2 \\ \hline 0 \end{array}\right] \text{ or } I_2 \text{ respectively.}$$

Hence the proof is complete.

Proceeding in this way, until a null matrix appears in the lower right hand corner or until all of the rows or columns are exhausted, finally A is reduced to equivalent normal form. This proves the assertion of the theorem.

9.8 Computation of A^{-1} by Gauss-Jordan elimination i.e. $[A \mid I_n] \to [I_n \mid A^{-1}]$

Computation of A^{-1} from adjoint and determinant is time consuming and inefficient for large matrices. Let us first describe the procedure and then describe the motivation of the procedure.

Procedure for finding A^{-1} : Let A be an n square matrix and let I_n be an nth order identity matrix. First form the augmented matrix $[A \mid I_n]$. Now transform $[A \mid I_n]$ to row reduced echelon form $[C \mid D]$. Then

(i) if C is the identity matrix I_n , then $D = A^{-1}$;

(ii) if C is *not* the identity matrix , then A is not invertible i.e. A^{-1} does not exist.

Motivation of the procedure :

The motivation of this procedure comes from the fundamental definition of inverse of a matrix. Since A^{-1} of a square matrix A exists if

$$AB = BA = I.$$

This definition tells that inverse of an n square matrix A is the solution of the matrix equation

$$AX = I_n$$

An illustration : To make the discussion lucid and palatable, let us consider a numerical example. Let

$$A = \begin{bmatrix} 0 & 4 & 2 & -6 \\ 21 & -9 & -8 & 3 \\ 7 & -1 & -4 & 5 \\ -14 & 6 & 10 & -2 \end{bmatrix}.$$

Then if we can have a 4×4 matrix X such that $AX = I_4$. That is, $\begin{bmatrix} 0 & 4 & 2 & -6 \\ 21 & -9 & -8 & 3 \\ 7 & -1 & -4 & 5 \\ -14 & 6 & 10 & -2 \end{bmatrix}$

$$\begin{bmatrix} x_{11} & x_{12} & x_{13} & x_{14} \\ x_{21} & x_{22} & x_{23} & x_{24} \\ x_{31} & x_{32} & x_{33} & x_{34} \\ x_{41} & x_{42} & x_{43} & x_{44} \end{bmatrix} = \begin{bmatrix} 1 & 0 & 0 & 0 \\ 0 & 1 & 0 & 0 \\ 0 & 0 & 1 & 0 \\ 0 & 0 & 0 & 1 \end{bmatrix}.$$

Now by matrix multiplication, respective columns of the product in the left equals the corresponding columns of I_n. Symbolically, let

$$(\mathbf{AX})_{.j} = I_{.j} , \ j = 1, 2, 3, 4.$$

Therefore, using matrix multiplication,

$$(\mathbf{AX})_{.1} = \begin{bmatrix} \mathbf{A}_{1.}\mathbf{X}_{.1} \\ \mathbf{A}_{2.}\mathbf{X}_{.1} \\ \mathbf{A}_{3.}\mathbf{X}_{.1} \\ \mathbf{A}_{4.}\mathbf{X}_{.1} \end{bmatrix} = \begin{bmatrix} 0x_{11} + 4x_{21} + 2x_{31} - 6x_{41} \\ 21x_{11} - 9x_{21} - 8x_{31} + 3x_{41} \\ 7x_{11} - x_{21} - 4x_{31} + 5x_{41} \\ -14x_{11} + 6x_{21} + 10x_{31} - 2x_{41} \end{bmatrix}$$

$$\text{But } \mathbf{AX}_{.1} = \begin{bmatrix} 0 & 4 & 2 & -6 \\ 21 & -9 & -8 & 3 \\ 7 & -1 & -4 & 5 \\ -14 & 6 & 10 & -2 \end{bmatrix} \begin{bmatrix} x_{11} \\ x_{21} \\ x_{31} \\ x_{41} \end{bmatrix} \begin{bmatrix} 0x_{11} + 4x_{21} + 2x_{31} - 6x_{41} \\ 21x_{11} - 9x_{21} - 8x_{31} + 3x_{41} \\ 7x_{11} - x_{21} - 4x_{31} + 5x_{41} \\ -14x_{11} + 6x_{21} + 10x_{31} - 2x_{41} \end{bmatrix}$$

$$\therefore \quad (\mathbf{AX})_{.1} = \mathbf{AX}_{.1} = \mathbf{I}_{.1} \qquad \text{(i)}$$

Similarly,

$$\mathbf{AX}_{.2} = \mathbf{I}_{.2} \qquad \text{(ii)}$$

$$\mathbf{AX}_{.3} = \mathbf{I}_{.3} \qquad \text{(iii)}$$

and $\quad \mathbf{AX}_{.4} = \mathbf{I}_{.4} \qquad$ (iv)

Now X can be obtained by solving the system of matrix equations (i) through (iv). This can be done by Gauss-Jordan elimination forming the augmented matrices

$$\left[\begin{array}{cccc|c} 0 & 4 & 2 & -6 & 1 \\ 21 & -9 & -8 & 3 & 0 \\ 7 & -1 & -4 & 5 & 0 \\ -14 & 6 & 10 & -2 & 0 \end{array}\right], \left[\begin{array}{cccc|c} 0 & 4 & 2 & -6 & 0 \\ 21 & -9 & -8 & 3 & 1 \\ 7 & -1 & -4 & 5 & 0 \\ -14 & 6 & 10 & -2 & 0 \end{array}\right], \left[\begin{array}{cccc|c} 0 & 4 & 2 & -6 & 0 \\ 21 & -9 & -8 & 3 & 0 \\ 7 & -1 & -4 & 5 & 1 \\ -14 & 6 & 10 & -2 & 0 \end{array}\right],$$

$$\left[\begin{array}{cccc|c} 0 & 4 & 2 & -6 & 0 \\ 21 & -9 & -8 & 3 & 0 \\ 7 & -1 & -4 & 5 & 0 \\ -14 & 6 & 10 & -2 & 1 \end{array}\right]$$

and transforming each to its row reduced echelon form

$$\left[\begin{array}{cccc|c} 1 & 0 & 0 & 0 & \frac{1}{14} \\ 0 & 1 & 0 & 0 & \frac{1}{7} \\ 0 & 0 & 1 & 0 & 0 \\ 0 & 0 & 0 & 1 & -\frac{1}{14} \end{array}\right], \left[\begin{array}{cccc|c} 1 & 0 & 0 & 0 & \frac{1}{14} \\ 0 & 1 & 0 & 0 & -\frac{1}{14} \\ 0 & 0 & 1 & 0 & \frac{1}{7} \\ 0 & 0 & 0 & 1 & 0 \end{array}\right], \left[\begin{array}{cccc|c} 1 & 0 & 0 & 0 & -\frac{1}{14} \\ 0 & 1 & 0 & 0 & \frac{3}{14} \\ 0 & 0 & 1 & 0 & 0 \\ 0 & 0 & 0 & 1 & \frac{1}{7} \end{array}\right],$$

$$\left[\begin{array}{cccc|c} 1 & 0 & 0 & 0 & \frac{1}{14} \\ 0 & 1 & 0 & 0 & 0 \\ 0 & 0 & 1 & 0 & \frac{3}{14} \\ 0 & 0 & 0 & 1 & \frac{1}{14} \end{array}\right]$$

$$\therefore \quad \mathbf{X}_{.1}^{\mathrm{T}} = \begin{bmatrix} \frac{1}{14} & \frac{1}{7} & 0 & -\frac{1}{14} \end{bmatrix}^{\mathrm{T}}; \qquad \mathbf{X}_{.2}^{\mathrm{T}} = \begin{bmatrix} \frac{1}{14} & -\frac{1}{14} & \frac{1}{7} & 0 \end{bmatrix}^{\mathrm{T}}$$

$$\mathbf{X}_{.3}^{\mathrm{T}} = \begin{bmatrix} -\frac{1}{14} & \frac{3}{14} & 0 & \frac{1}{7} \end{bmatrix}^{\mathrm{T}}; \qquad \mathbf{X}_{.4}^{\mathrm{T}} = \begin{bmatrix} \frac{1}{14} & 0 & \frac{3}{14} & \frac{1}{14} \end{bmatrix}^{\mathrm{T}}$$

$$\therefore \qquad A^{-1} = \begin{bmatrix} \frac{1}{14} & \frac{1}{14} & \frac{1}{14} & \frac{1}{14} \\ \frac{1}{7} & -\frac{1}{14} & \frac{3}{14} & 0 \\ 0 & \frac{1}{7} & 0 & \frac{3}{14} \\ -\frac{1}{14} & 0 & \frac{1}{7} & \frac{1}{14} \end{bmatrix} = \frac{1}{14}\begin{bmatrix} 1 & 1 & 1 & 1 \\ 2 & -1 & 3 & 0 \\ 0 & 2 & 0 & 3 \\ -1 & 0 & 2 & 1 \end{bmatrix}.$$

Discussion on the solution procedure :

The above procedure has a gross inefficiency. An efficient procedure is discussed below after constructing an augmented matrix. Since the coefficient matrices of all the four matrix equations (i) through (iv) are same, all four augmented matrices can be transformed using same row operations and hence lot of work and time may be saved by doing all the row reductions at the same time. To do this, a 4×8 augmented matrix is formed from the columns of A in the first four places and those of I_4 in the next four places. Thus

$$\left[\begin{array}{cccc|cccc} 0 & 4 & 2 & -6 & 1 & 0 & 0 & 0 \\ 21 & -9 & -8 & 3 & 0 & 1 & 0 & 0 \\ 7 & -1 & -4 & 5 & 0 & 0 & 1 & 0 \\ -14 & 6 & 10 & -2 & 0 & 0 & 0 & 1 \end{array}\right] \begin{array}{c} R_1 \leftrightarrow R_3 \\ \approx \end{array}$$

$$\left[\begin{array}{cccc|cccc} 7 & -1 & -4 & 5 & 0 & 0 & 1 & 0 \\ 21 & -9 & -8 & 3 & 0 & 1 & 0 & 0 \\ 0 & 4 & 2 & -6 & 1 & 0 & 0 & 0 \\ -14 & 6 & 10 & -2 & 0 & 0 & 0 & 1 \end{array}\right] \begin{array}{c} R_2 - 3R_1 \\ \approx \\ R_4 + 2R_1 \end{array}$$

$$\left[\begin{array}{cccc|cccc} 7 & -1 & -4 & 5 & 0 & 0 & 1 & 0 \\ 0 & -6 & 4 & -12 & 0 & 1 & -3 & 0 \\ 0 & 4 & 2 & -6 & 1 & 0 & 0 & 0 \\ 0 & 4 & 2 & 8 & 0 & 0 & 2 & 1 \end{array}\right] \qquad \begin{array}{l} \text{two steps: first, } -\frac{1}{6}R_2 \\ \text{then } R_3 - 4R_2 \text{ and} \\ \approx \\ R_4 - 4R_2 \end{array}$$

$$\left[\begin{array}{cccc|cccc} 7 & -1 & -4 & 5 & 0 & 0 & 1 & 0 \\ 0 & 1 & -\frac{2}{3} & 2 & 0 & -\frac{1}{6} & \frac{1}{2} & 0 \\ 0 & 0 & \frac{14}{3} & -14 & 1 & \frac{2}{3} & -2 & 0 \\ 0 & 0 & \frac{14}{3} & 0 & 0 & \frac{2}{3} & 0 & 1 \end{array}\right] \begin{array}{l} \frac{3}{14}R_3 \text{, then } R_1 + R_2 \\ \approx \\ \frac{3}{14}R_4 \text{, then } R_2 - R_3 \end{array}$$

$$\left[\begin{array}{cccc|cccc} 7 & 0 & -\frac{14}{3} & 7 & 0 & -\frac{1}{6} & \frac{3}{2} & 0 \\ 0 & 1 & 0 & 0 & \frac{1}{7} & -\frac{1}{14} & \frac{3}{14} & 0 \\ 0 & 0 & 1 & -3 & \frac{3}{14} & \frac{1}{7} & -\frac{3}{7} & 0 \\ 0 & 0 & 1 & 0 & 0 & \frac{1}{7} & 0 & \frac{3}{14} \end{array}\right] \begin{array}{c} R_4 - R_3 \\ \approx \\ R_1 + \frac{3}{14}R_3 \end{array} \qquad \begin{array}{l} \text{next } \frac{1}{7}R_1 \\ R_3 + R_4 \text{ etc.} \end{array}$$

$$\left[\begin{array}{cccc|cccc} 1 & 0 & 0 & 0 & \frac{1}{14} & \frac{1}{14} & \frac{1}{14} & \frac{1}{14} \\ 0 & 1 & 0 & 0 & \frac{1}{7} & -\frac{1}{14} & \frac{3}{14} & 0 \\ 0 & 0 & 1 & 0 & 0 & \frac{1}{7} & 0 & \frac{3}{14} \\ 0 & 0 & 0 & 1 & -\frac{1}{14} & 0 & \frac{1}{7} & \frac{1}{14} \end{array}\right] = \left[I_n \mid A^{-1}\right]$$

$$\therefore \qquad A^{-1} = \frac{1}{14}\begin{bmatrix} 1 & 1 & 1 & 1 \\ 2 & -1 & 3 & 0 \\ 0 & 2 & 0 & 3 \\ -1 & 0 & 2 & 1 \end{bmatrix}.$$

Ex. Find the inverse of

$$A = \begin{bmatrix} 1 & -1 & 2 \\ 2 & -1 & 1 \\ 0 & -1 & 3 \end{bmatrix} \text{ if it exists.}$$

Let us reduce the augmented matrix

$$\left[\begin{array}{ccc|ccc} 1 & -1 & 2 & 1 & 0 & 0 \\ 2 & -1 & 1 & 0 & 1 & 0 \\ 0 & -1 & 3 & 0 & 0 & 1 \end{array}\right] \underset{\approx}{R_2 - 2R_1} \left[\begin{array}{ccc|ccc} 1 & -1 & 2 & 1 & 0 & 0 \\ 0 & 1 & -3 & -2 & 1 & 0 \\ 0 & -1 & 3 & 0 & 0 & 1 \end{array}\right] \underset{\approx}{R_3 + R_2}$$

$$\left[\begin{array}{ccc|ccc} 1 & -1 & 2 & 1 & 0 & 0 \\ 0 & 1 & -3 & -2 & 1 & 0 \\ 0 & 0 & 0 & -2 & 1 & 1 \end{array}\right].$$

Since the third row of A becomes zero row, so further reduction is not necessary. It can be concluded that A will not reduce to I_3. So, A does not have any inverse, i.e. A is a singular matrix.

Alternatively,

$$|A| = \begin{vmatrix} 1 & -1 & 2 \\ 2 & -1 & 1 \\ 0 & -1 & 3 \end{vmatrix} = 0.$$

So A is a singular matrix.

9.9 Evaluation of determinant by ERT

The elementary row operations I, II and III in sec. 9.1 are respectively identical with properties of determinants governed by theorems 2, 5, 6 of secs.6.4 or 6.5.3. Thus, the determinant is affected by ERT's in the following ways:

(i) Interchange of two rows of a determinant A by ERT I changes the sign to the value of the original determinant;

(ii) Multiplying a row by a scalar k by ERT II multiplies the value of the original determinant by k. Alternatively, a common factor to a row can be taken outside the determinant.

(iii) Adding kth times of the elements of one row to those of another by ERT III keeps the value of the determinant unchanged.

Therefore, noting the above effects of ERT'S, a determinant can be reduced to an upper triangular form or row echelon form and hence its value is the product of the diagonal elements (cf. sec.6.12). The following examples illustrate the above fact.

Ex 1. Let

$$A = \begin{vmatrix} 1 & 2 & 3 \\ 0 & 1 & 4 \\ 1 & 2 & 1 \end{vmatrix} = -2$$

Let

$$A_1 = \begin{vmatrix} 0 & 1 & 4 \\ 1 & 2 & 3 \\ 1 & 2 & 1 \end{vmatrix}$$ obtained from A by interchanging the first and second rows.

$= 2 = -\text{A}$ illustrating the effect of ERT I.

Let

$$A_2 = \begin{vmatrix} 1 & 2 & 3 \\ 0 & 1 & 4 \\ 3 & 6 & 3 \end{vmatrix}$$ obtained from A by multiplying third row by 3.

$= -6 = 3\,(-2) = 3A$ illustrating the effect of ERT II .

Let

$$A_3 = \begin{vmatrix} 1 & 2 & 3 \\ 2 & 5 & 6 \\ 1 & 2 & 1 \end{vmatrix}$$ obtained from A by adding 2 times of the elements of row 3 to those of row 2.

$= -2 = A$ illustrates the truth of ERT III.

Ex 2 . Evaluate the determinant

$$\begin{vmatrix} 3 & 6 & 9 & 3 \\ -1 & 0 & 1 & 0 \\ 1 & 3 & 2 & -1 \\ -1 & -2 & -2 & 1 \end{vmatrix}.$$

Let

$$A = \begin{vmatrix} 3 & 6 & 9 & 3 \\ -1 & 0 & 1 & 0 \\ 1 & 3 & 2 & -1 \\ -1 & -2 & -2 & 1 \end{vmatrix} = 3\begin{vmatrix} 1 & 2 & 3 & 1 \\ -1 & 0 & 1 & 0 \\ 1 & 3 & 2 & -1 \\ -1 & -2 & -2 & 1 \end{vmatrix}.$$

Alternate of ERT II common factor 3 in row 1 has taken outside the determinant. Then

$$A = 3\begin{vmatrix} 1 & 2 & 3 & 1 \\ 0 & 2 & 4 & 1 \\ 0 & 1 & -1 & -2 \\ 0 & 0 & 1 & 2 \end{vmatrix} \begin{matrix} R_2 + R_1 \\ R_3 - R_1 \\ R_4 + R_1 \end{matrix} \qquad (1)$$

$$= \frac{3}{2}\begin{vmatrix} 1 & 2 & 3 & 1 \\ 0 & 2 & 4 & 1 \\ 0 & 0 & -6 & -5 \\ 0 & 0 & 1 & 2 \end{vmatrix} \qquad (2)$$

$2R_3 - R_2$ $\because R_3$ is multiplied by 2. So, to balance the value,division by 2 is essential.

$$=\frac{3}{26}\begin{vmatrix} \mathbf{1} & 2 & 3 & 1 \\ 0 & \mathbf{2} & 4 & 1 \\ 0 & 0 & \mathbf{-6} & -5 \\ 0 & 0 & 0 & 7 \end{vmatrix} \qquad (3)$$

$6R_4 + R_3$, $\because$ R_4 is multiplied by 6. So, to balance the value, division by 6 is needed.

$$= \frac{1}{4}.(1)(2)(-6)(7) = -21$$

Note: Alternatively, if the *bold* elements, also called the *pivot element* in (1), (2), (3) were already unity as in (1), then no division as in (2), (3) is required.

9.10 Method of relaxation for solving simultaneous equations

Before starting the description of the method, a definition is needed. It is the *residual vector.* It is a vector $R = \mathbf{b} - A\mathbf{x}_i$ characterzing the exactness of the approximate solution $\mathbf{x}_i$ of the system $A\mathbf{x} = \mathbf{b}$. The magnitude of R is called *residual* only. The smallness of R measures the correctness of the solution $\mathbf{x}_i$. The method of relaxation is a method which starts with an initial guess and each step of relaxation reduces the maximum residual to zero and finally gives the solution of all the unknowns simultaneously which is the basic difference with elimination procedures. The reduction of maximum residual to zero requires to determine which variable has the greatest influence of that residual and how much the identified variable is to be adjusted. This can be performed in two ways —

(i) unit relaxation which means a unit change in one variable keeping others unchanged and this unit change has greatest influence on particular residual;

(ii) group relaxation means a combination of more than one unit relaxation. This latter relaxation is more efficient than the previous one.

In this method, the given system of simultaneous equations

$$\sum_{j=1}^{n} a_{ij}x_j = b_j, \quad i = 1,2,\cdots,n$$

or the corresponding matrix equation $A\mathbf{x} = \mathbf{b}$ is written as

$$\sum_{j=1}^{n} a_{ij}x_j - b_j = R_i, \quad i = 1, 2,\cdots,n,$$

or

$$A\mathbf{x} - \mathbf{b} = R$$

This R_i or R is the residual for some assumed value of x's and this residual approach zero for some improvement of the assumed x value. The residuals are shown as function of x's.

Let the method of unit relaxation be explained with the help of an example.

Ex. Solve the following system of equations by relaxation method.

$$\left.\begin{aligned} 3x_1 - 3x_2 + 5x_3 &= 14 \\ x_1 + 2x_2 - 6x_3 &= 3 \\ 2x_1 - x_2 + 3x_3 &= 1 \end{aligned}\right\} \qquad (1)$$

The unit relaxation table for the concerned problem is

Δx_1	Δx_2	Δx_3	ΔR_1	ΔR_2	ΔR_3
1	0	0	$\underline{3}$	1	2
0	1	0	−3	$\underline{2}$	−1
0	0	1	5	−6	$\underline{3}$

(2)

This table consists of two parts. (i) L.H.S shows the unit changes of x_1, x_2 and x_3 along the diagonal. The off diagonal elements '0' indicates that while one changes others are kept unchanged. (ii) R.H.S. is the transpose of the coefficient matrix of the system (1) and the underlined elements along the diagonal shows these residuals will approach zero due to the changes of x values. To start the method, (1) is written as

$$\left.\begin{aligned} 3x_1 - 3x_2 + 5x_3 - 14 &= R_1 \\ x_1 + 2x_2 - 6x_3 - 3 &= R_2 \\ 2x_1 - x_2 + 3x_3 - 1 &= R_3 \end{aligned}\right\} \quad (3)$$

by transferring the constants to the left and equating the corresponding expressions to the components R_1, R_2, R_3 of the residual vector $\mathbf{R}$.

First relaxation :

Let the initial guess to the solution be

$$x_1 = x_2 = x_3 = 0 \quad \text{(any other values may be assumed)}$$

∴ the components of the residual $R_1 = -14,\ R_2 = -3,\ R_3 = -1$.

∴ the maximum residual is $|R_1| = 14$ and according to the table of unit relaxation if x_1 changes from 0 to $\frac{14}{3} = 4.67$ (approximate) $\left[\frac{|R_1|}{(coeff.\text{ of } x_1)} = \frac{14}{3}\right]$ keeping x_2, x_3 unchanged then R_1 reduces to zero. Therefore, corresponding to new set of x values $x_1 = 4.67,\ x_2 = 0,\ x_3 = 0$ the new components of the residual vector are

$$\left.\begin{aligned} R_1 &= 3(4.67) - 3(0) + 5(0) - 14 = 0 \\ R_2 &= 4.67 \quad + 2(0) - 6(0) - 3 = 1.67 \\ R_3 &= 2(4.67) - 2(0) + 3(0) - 1 = 8.34 \end{aligned}\right\}.$$

Second relaxation :

Since $|R_3| = 8.34$ is the maximum component of the residual vector, so a change in x_3 will reduce R_3 to zero. This change in x_3 can be obtained by dividing R_3 by the coefficient of x_3 in (3) or (1). Therefore, if x_3 changes from 0 to $-\frac{8.34}{3} = -2.78$. Then R_3 reduces to zero. Thus, the corresponding second set of new x values $x_1 = 4.67,\ x_2 = 0,\ x_3 = -2.78$ and the new components of the residual vector are

$$\left.\begin{aligned} R_1 &= 3(4.67) - 3(0) + 5(-2.78) - 14 = -13.91 \\ R_2 &= \quad 4.67 + 2(0) - 6(-2.78) - 3 = -18.35 \\ R_3 &= 2(4.67) - \quad (0) + 3(-2.78) - 1 = 0 \end{aligned}\right\}.$$

Third relaxation :

Since $|R_2| = 18.35$ is the maximum, so x_2 must be changed. Now if R_2 changes from 0 to $-\frac{18.35}{2} = -9.18$ then R_2 reduces to zero. Thus corresponding third set of new x values $x_1 = 4.67,\ x_2 = -9.18,\ x_3 = -2.78$. The new R's are

$$\left.\begin{aligned} R_1 &= 3(4.67) - 3(-9.18) + 5(-2.78) - 14 = 13.65 \\ R_2 &= 4.67 + 2(-9.18) - 6(-2.78) - 3 = 0 \\ R_3 &= 2(4.67) - (-9.18) + 3(-2.78) - 1 = 9.18 \end{aligned}\right\}.$$

Fourth relaxation :

For each of the relaxation from 1st to 17 th, the resultant x values are considered to calculate the residual components. Thus $R_1 = 13.65$ is maximum in third relaxation. So new x_1 changes from previous 4.67 to $\left(4.67 - \frac{13.65}{3}\right) = 4.67 - 4.55 = .12$ and hence the residual components are

$$\left.\begin{aligned} R_1 &= 3(.12) - 3(-9.18) + 5(-2.78) - 14 = 0 \\ R_2 &= .12 + 2(-9.18) - 6(-2.78) - 3 = -4.56 \\ R_3 &= 2(.12) - (-9.18) + 3(-2.78) - 1 = .08 \end{aligned}\right\}.$$

The following table summarises the solution of the system of equations using this method. In all cases, the residual being reduced to zero is underlined.

Relaxation	$x_1 = 0$	$x_2 = 0$	$x_3 = 0$	$R_1 = -14$	$R_2 = -3$	$R_3 = -1$
1	4.67	0	0	0	1.67	<u>8.34</u>
2	4.67	0	−2.78	−13.91	<u>−18.35</u>	0
3	4.67	−9.18	−2.78	<u>13.65</u>	0	9.18
4	0.12	−9.18	−2.78	0	<u>− 4.56</u>	0.08
5	0.12	−6.9	"	<u>− 6.84</u>	0	−2.2
6	2.4	"	"	0	<u>2.28</u>	2.04
7	"	−8.04	"	3.42	0	<u>3.5</u>
8	"	"	−3.95	− 2.43	<u>7.02</u>	0
9	"	−11.55	"	<u>8.1</u>	0	3.5
10	−0.3	"	"	0	<u>− 2.7</u>	−1.9
11	"	−10.2	"	<u>−4.05</u>	0	−3.25
12	1.05	"	"	0	<u>1.35</u>	−0.55
13	"	−10.88	"	<u>2.03</u>	0	0.13
14	0.38	"	"	0	− 0.68	<u>−1.23</u>
15	"	"	−3.54	2.05	<u>− 3.14</u>	0
16	"	−9.31	"	<u>−2.65</u>	0	−1.56
17	1.13	"	"	−0.01	0.88	0.21
Solutions	1.13	−9.31	−3.54	−0.01	0.88	0.21

Group relaxation :

By elementary row operation, the unit relaxation table becomes

$$\begin{array}{ccc|ccc} 1 & 0 & 0 & 3 & 1 & 2 \\ 0 & 1 & 0 & -3 & 2 & -1 \\ 0 & 3 & 1 & 4 & 0 & 0 \end{array}.$$

9.11 Bordering method in the inversion of matrix with singular block(s)

The bordering method is also applicable to find the inverse of a nonsingular matrix having one or more singular block or blocks. Let A_i be a singular block in A. Then the simplest procedure is to transfer the ith row of A in the last place, continue the usual method by the bordering process. This transformation is performed by premultiplication of elementary row transformation P_{ni}^{n}. The inverse matrix so obtained is then post multiplied by the same elementary row transformation to get the required inverse matrix. The following example illustrates the procedure.

Ex. Find the inverse of

$$A = \begin{bmatrix} 1 & 1 & 2 & 1 \\ 1 & 2 & 4 & 3 \\ 1 & 0 & 0 & 2 \\ 3 & -1 & 1 & 5 \end{bmatrix}$$

using the bordering method.

Let the first bordering be

$$A = A_4 = \left[\begin{array}{ccc|c} 1 & 1 & 2 & 1 \\ 1 & 2 & 4 & 3 \\ 1 & 0 & 0 & 2 \\ \hline 3 & -1 & 1 & 5 \end{array}\right] = \left[\begin{array}{c|c} A_3 & \mathbf{u}_4 \\ \hline \mathbf{v}_4 & u_{44} \end{array}\right]$$

where $A_3 = \begin{bmatrix} 1 & 1 & 2 \\ 1 & 2 & 4 \\ 1 & 0 & 0 \end{bmatrix}$ which is singular i.e.

$|A_3| = 0$. So, premultiplying A by P_{34}^{4}, i.e.

$$P_{34}^{4}A = \begin{bmatrix} 1 & 0 & 0 & 0 \\ 0 & 1 & 0 & 0 \\ 0 & 0 & 0 & 1 \\ 0 & 0 & 1 & 0 \end{bmatrix}\begin{bmatrix} 1 & 1 & 2 & 1 \\ 1 & 2 & 4 & 3 \\ 1 & 0 & 0 & 2 \\ 3 & -1 & 1 & 5 \end{bmatrix} = \begin{bmatrix} 1 & 1 & 2 & 1 \\ 1 & 2 & 4 & 3 \\ 3 & -1 & 1 & 5 \\ 1 & 0 & 0 & 2 \end{bmatrix}$$

and let it be denoted A^{new}. Now let the first bordering be

$$A^{new} = \left[\begin{array}{ccc|c} 1 & 1 & 2 & 1 \\ 1 & 2 & 4 & 3 \\ 3 & -1 & 1 & 5 \\ \hline 1 & 0 & 0 & 2 \end{array}\right] = \left[\begin{array}{c|c} A_3^{new} & \mathbf{u}_4^{new} \\ \hline \mathbf{v}_4^{new} & a_{44}^{new} \end{array}\right]$$

where

$$B = A_3^{new} = \begin{bmatrix} 1 & 1 & 2 \\ 1 & 2 & 4 \\ 3 & -1 & 1 \end{bmatrix}$$

is nonsingular, $u_4^{new} = [1 \;\; 2 \;\; 3]^T$, $v_4^{new} = [1 \;\; 0 \;\; 0]$ and $a_4^{new} = 2$. Using the second bordering, let

$$B = A_3^{new} = \left[\begin{array}{cc|c} 1 & 1 & 2 \\ 1 & 2 & 4 \\ \hline 3 & -1 & 1 \end{array}\right] = \left[\begin{array}{c|c} A_2^{new} & u_3^{new} \\ \hline v_4^{new} & a_{33}^{new} \end{array}\right] \text{ where}$$

$$A_2^{new} = \begin{bmatrix} 1 & 1 \\ 1 & 2 \end{bmatrix},$$

is nonsingular $u_3^{new} = [2 \;\; 4]^T$, $v_3^{new} = [3 \;\; -1]$ and $a_{33}^{new} = 1$. Let

$$B^{-1} = \left(A_3^{new}\right)^{-1} = \left[\begin{array}{c|c} P_2 & r_3 \\ \hline q_3 & \frac{1}{\alpha_3} \end{array}\right]$$

Here $\left|A_2^{new}\right| = 1$ and $\left(A_2^{new}\right)^{-1} = \begin{bmatrix} 2 & -1 \\ -1 & 1 \end{bmatrix}$

Also from (8.42), $\alpha_3 = a_{33}^{new} - v_3^{new}\left(A_2^{new}\right)^{-1} u_3^{new}$

$$= 1 - [3 \;\; -1]\begin{bmatrix} 2 & -1 \\ -1 & 1 \end{bmatrix}\begin{bmatrix} 2 \\ 4 \end{bmatrix} = 1 - [-2] = 3$$

From (8.41), $r_3 = -\frac{1}{\alpha_3}\left(A_2^{new}\right)^{-1} u_3^{new}$

$$= \tfrac{1}{3}\begin{bmatrix} 2 & -1 \\ -1 & 1 \end{bmatrix}\begin{bmatrix} 2 \\ 4 \end{bmatrix} = \tfrac{1}{3}\begin{bmatrix} 0 \\ -2 \end{bmatrix}$$

From (8.44), $q_3 = -\frac{1}{\alpha_3} v_3^{new}\left(A_2^{new}\right)^{-1} = -\frac{1}{3}[3 \;\; -1]\begin{bmatrix} 2 & -1 \\ -1 & 1 \end{bmatrix} = \frac{1}{3}[-7 \;\; 4]$

From (8.45), $P_2 = \left(A_2^{new}\right)^{-1} - \left(A_2^{new}\right)^{-1} u_3^{new} q_3$

$$= \begin{bmatrix} 2 & -1 \\ -1 & 1 \end{bmatrix} - \begin{bmatrix} 2 & -1 \\ -1 & 1 \end{bmatrix}\begin{bmatrix} 2 \\ 4 \end{bmatrix}\tfrac{1}{3}[-7 \;\; 4]$$

$$= \tfrac{1}{3}\begin{bmatrix} 6 & -3 \\ 11 & -5 \end{bmatrix}$$

$$\therefore \quad B^{-1} = \left(A_3^{new}\right)^{-1} = \left[\begin{array}{cc|c} 2 & -1 & 0 \\ \frac{11}{3} & -\frac{5}{3} & -\frac{2}{3} \\ \hline -\frac{7}{3} & \frac{4}{3} & \frac{1}{3} \end{array}\right] = \frac{1}{3}\left[\begin{array}{cc|c} 6 & -3 & 0 \\ 11 & -5 & -2 \\ \hline -7 & 4 & 1 \end{array}\right]$$

Now to find $\left(A_3^{new}\right)^{-1}$ with the help of $\left(A_4^{new}\right)^{-1}$ let

$$\left(A_3^{new}\right)^{-1} = (A_4^{new})^{-1} = \left[\begin{array}{c|c} P_3 & \mathbf{r}_4 \\ \hline \mathbf{q}_4 & \frac{1}{\alpha_4} \end{array}\right]$$

Proceeding with (8.42), (8.41), (8.44), (8.45), we have

$$\alpha_4 = a_{44}^{new} - \mathbf{v}_4^{new}\left(A_3^{new}\right)^{-1}\mathbf{u}_4^{new}$$

$$= 2 - \begin{bmatrix} 1 & 0 & 0 \end{bmatrix}\tfrac{1}{3}\begin{bmatrix} 6 & -3 & 0 \\ 11 & -5 & -2 \\ -7 & 4 & 1 \end{bmatrix}\begin{bmatrix} 1 \\ 3 \\ 5 \end{bmatrix} = 3.$$

$$\mathbf{r}_4 = -\frac{\left(A_3^{new}\right)^{-1}\mathbf{u}_4^{new}}{\alpha_4} = -\tfrac{1}{3}\cdot\tfrac{1}{3}\begin{bmatrix} 6 & -3 & 0 \\ 11 & -5 & -2 \\ -7 & 4 & 1 \end{bmatrix}\begin{bmatrix} 1 \\ 3 \\ 5 \end{bmatrix} = -\tfrac{1}{9}\begin{bmatrix} -3 \\ -14 \\ 10 \end{bmatrix}.$$

$$\mathbf{q}_4 = -\frac{\mathbf{v}_4^{new}\left(A_3^{new}\right)^{-1}}{\alpha_4} = -\tfrac{1}{3}[1 \quad 0 \quad 0]\ \tfrac{1}{3}\begin{bmatrix} 6 & -3 & 0 \\ 11 & -5 & -2 \\ -7 & 4 & 1 \end{bmatrix} = \tfrac{1}{9}\begin{bmatrix} -6 & 3 & 0 \end{bmatrix}$$

$$P_3 = \left(A_3^{new}\right)^{-1} - \left(A_3^{new}\right)^{-1}\mathbf{u}_4^{new}\mathbf{q}_4$$

$$= \tfrac{1}{3}\begin{bmatrix} 6 & -3 & 0 \\ 11 & -5 & -2 \\ -7 & 4 & 1 \end{bmatrix} - \tfrac{1}{3}\begin{bmatrix} 6 & -3 & 0 \\ 11 & -5 & -2 \\ -7 & 4 & 1 \end{bmatrix}\begin{bmatrix} 1 \\ 3 \\ 5 \end{bmatrix}\tfrac{1}{9}\begin{bmatrix} -6 & 3 & 0 \end{bmatrix}$$

$$= \tfrac{1}{9}\begin{bmatrix} 12 & -6 & 0 \\ 5 & -1 & -6 \\ 1 & 2 & 3 \end{bmatrix}$$

$$\therefore \quad \left(A^{new}\right)^{-1} = \left[\begin{array}{ccc|c} \frac{4}{3} & -\frac{2}{3} & 0 & \frac{1}{3} \\ \frac{5}{9} & -\frac{1}{9} & -\frac{2}{3} & \frac{14}{9} \\ -\frac{1}{9} & \frac{2}{9} & \frac{1}{3} & -\frac{10}{9} \\ \hline -\frac{2}{3} & \frac{1}{3} & 0 & \frac{1}{3} \end{array}\right]$$

$$\therefore \qquad A^{-1} = \left(A^{new}\right)^{-1}P_{34}^4$$

$$\therefore \qquad A^{-1} = \tfrac{1}{14}\begin{bmatrix} 1 & 1 & 1 & 1 \\ 2 & -1 & 3 & 0 \\ 0 & 2 & 0 & 3 \\ -1 & 0 & 2 & 1 \end{bmatrix}.$$

Ex. Find the inverse of

$$A = \begin{bmatrix} 1 & -1 & 2 \\ 2 & -1 & 1 \\ 0 & -1 & 3 \end{bmatrix} \text{ if it exists.}$$

$$= \begin{bmatrix} \frac{4}{3} & -\frac{2}{3} & 0 & \frac{1}{3} \\ \frac{5}{9} & -\frac{1}{9} & -\frac{2}{3} & \frac{14}{9} \\ -\frac{1}{9} & \frac{2}{9} & \frac{1}{3} & -\frac{10}{9} \\ -\frac{2}{3} & \frac{1}{3} & 0 & \frac{1}{3} \end{bmatrix} \begin{bmatrix} 1 & 0 & 0 & 0 \\ 0 & 1 & 0 & 0 \\ 0 & 0 & 0 & 1 \\ 0 & 0 & 1 & 0 \end{bmatrix}$$

$$= \frac{1}{9} \begin{bmatrix} 12 & -6 & 3 & 0 \\ 5 & -1 & 14 & -6 \\ -1 & 2 & -10 & 3 \\ -6 & 3 & 3 & 0 \end{bmatrix}$$

9.12 Congruent matrices

This is a class of matrices which play an important role in the study of matrix theory. It arises in linear transformation, quadratic forms etc. This section defines and assembles the properties of congruent matrices in the form of theorems.

Definition. Two n square matrices A and B are said to be congruent iff $B = P^T AP$ or $A = PBP^T = (P^T)B(P^T)$ for nonsingular matrix P. The symbol used for the purpose is $B \overset{c}{=} A$.

Note the fundamental difference with equivalent matrices A & B where A = PBQ or B = PAQ with nonsingular matrices P & Q obtained from elementary row & column operations, respectively. Equivalent matrices A & B *may* be rectangular as well as square but congruent matrices A & B *must* be square.

Properties & construction :

(i) Congruent matrices are reflexive, symmetric & transitive.

(ii) If A and B are n-square matrices, then $B \overset{c}{=} A$ iff B is obtainable by a finite sequence of elementary cogredient operation i.e. ERO followed by ECO. Thus if $P^T = E_1 E_2 \cdots E_k$ where E_i are ERT then

$$B = E_1 E_2 \cdots E_k A E_k^T \cdots E_2^T E_1^T = P^T AP$$

which shows that B is obtainable from A by a sequence of k elementary cogredient operations.

In literature, this is also known as the *congruent transformation. Explicitly*, this transformation consists of

I. Interchange of ith and jth row followed by that of ith and jth column or conversely, i.e. $R_{ij} AC_{ij}$ or $C_{ij} AR_{ij}$.

II. Multiplication of the ith row followed by that if ith column or conversely, i.e.

$$R_i(k)AC_i(k) \text{ or } C_i(k)AR_i(k).$$

III. Addition of k times the jth row to the ith row followed by that of jth column to the ith column or conversely

$$\text{i.e. } R_{ij}(k)AC_{ij}(k) \text{ or } C_{ij}(k)AR_{ij}(k).$$

In practice, the following algorithms may be adapted:

Algorithm 1

Step 1. Write A = I A I

Step 2. Perform row operations on A in L.H.S. and on I in the premultiplicant of R.H.S.

Step 3. Perform column operation indentical to step 2 on A in L.H.S and on I in the post multiplicant of R.H.S.

Step 4. Repeat step 2 and step 3 until the desired congruent form on the left is reached.

Step 5. On the right premultiplicant is P^T and post multiplicant is P. Note that no operation is performed on A in the right.

Algorithm 2

This is identical to algorithm 1. In this case write A as AI or (A:I). Perform row operations on A and column operations on I until A reduces to the desired form. Then I will be transformed to P.

Algorithm 3

Step 1. Perform the congruent transforms on A only until the desired congruent form is reached.

Step 2. Write $P^T = E_1 E_2 \cdots E_k$

and $P = E_k^T A E_{k-1}^T \cdots E_2^T E_1^T$

N.B. 1.In case of Hermitian matrices A & B, B= $P^* AP$ is called the *Hermitian congruent* or *conjuctive.* The congruent transformation III is to be changed by taking *conjugate.* Thus $R_{ij}(k)AC_{ij}(k)$ or $C_{ij}(k)AR_{ij}(k)$. is to be changed to $R_{ij}(a+ib)AC_{ij}(a-ib)$.

2. In reducing diagonal element, *k* to 1 in (*i*, *i*) position, perform

$$R_i\left(\frac{1}{\sqrt{k}}\right) \text{ and } C_i\left(\frac{1}{\sqrt{k}}\right).$$

(iii) If $B \overset{c}{=} A$, then B is symmetric or Hermitian (skew symmetric or skew Hermitian) iff A is symmetric or Hermetian (skew symmetric or skew Hermitian).

(iv) If

$$A = \begin{bmatrix} A_1 & 0 \\ 0 & A_2 \end{bmatrix} \text{ and } B = \begin{bmatrix} B_1 & 0 \\ 0 & B_2 \end{bmatrix}, \quad A_1 \overset{c}{=} B_1 \text{ and } A_2 \overset{c}{=} B_2 \text{ ,then } A \overset{c}{=} B .$$

(v) Congruent matrices (symmetric/Hermitian) have the same rank and the same index (cf. chapter 10).

Ex 1. Prove the property (iii).

Let A and B be both *n* square congruent matrices and A be symmetric. Then by the first condition of definition,

$$B = P^T AP \qquad (1)$$

Taking transpose of both sides,

$$\begin{aligned} B^T &= (P^T AP)^T = P^T A^T (P^T)^T = P^T A^T P \\ &= P^T AP \quad \because \text{ A is symmetric.} \\ &= B \qquad \text{by (1)} \end{aligned}$$

$\therefore$ B is symmetric. Hence proved.

Ex 2. Find a nonsingular matrix P such that B= $P^* AP$ where

$$A=\begin{bmatrix} 1 & 1+i & 2-3i \\ 1-i & 2 & -5i \\ 2+3i & 5i & 0 \end{bmatrix} \text{ and } B=\begin{bmatrix} 1 & 0 & 0 \\ 0 & -1 & 0 \\ 0 & 0 & 1 \end{bmatrix}.$$

Hence verify that A and B are Hermitely congruent.

The matrix A is Hermitian. Follow any one of the above 3 algorithms and perform the following congruent transformations

$$R_{21}(-1+i),\ C_{21}(-1-i);\ R_{31}(-2-3i),\ C_{31}(-2+3i);$$

$$R_{23},\ C_{23}\ ;\ R_2(\tfrac{1}{\sqrt{13}})\ ,\ C_2(\tfrac{1}{\sqrt{13}})\ ;$$

$$R_{32}(\tfrac{1}{\sqrt{13}})\ ,\ C_2(\tfrac{1}{\sqrt{13}})\ ;\ \text{and } R_3(\sqrt{13}),\ C_3(\sqrt{13}).$$

Then

$$P=\begin{bmatrix} 1 & \frac{-15-10i}{\sqrt{13}} & \frac{-2+3i}{\sqrt{13}} \\ 0 & \sqrt{13} & 0 \\ 0 & \frac{1}{\sqrt{13}} & \frac{1}{\sqrt{13}} \end{bmatrix}.$$

Readers are requested to complete the task remains.

9.13 Similar matrices

Definition: Two $n\times n$ matrices A and B are said to ***similar*** if there exists a nonsingular matrix P such that $B = P^{-1}AP$ or $A = PBP^{-1}$. P is not unique and so B is not fixed for a fixed A.

9.13.1 Properties: Some of the properties of similar matrices are

(i) similarity of matrices is an equivalence relation;

(ii) similar matrices have identical traces;

(iii) similar matrices have equal determinants ;

(iv) similar matrices are either both singular or nonsingular;

(v) similar matrices of same integral powers are similar.

(vi) inverse of similar matrices are similar.

These matrices play an important role in applications. More on properties and uses of similar matrices will be discussed in sec.13.39.

Proofs.

(i) This means that (a) A is reflexive; (b) A is symmetric; (c) A is transitive.

(a) The unit matrix I is nonsingular and $I = I^{-1}$. Since $A = I^{-1}AI$, so A is similar to A.

(b) Let $A = P^{-1}BP$. Now pre and post multiplication by P and P^{-1} result into $PAP^{-1} = B$. But $P = P^{-1}(P^{-1})^{-1}$. Therefore $B = (P^{-1})^{-1}AP^{-1}$) and thus B is similar to A.

(c) Let $A = P^{-1}BP$ and $B = Q^{-1}CQ$ for nonsingular matrices P and Q. Now

$$\begin{aligned} A &= P^{-1}(Q^{-1}CQ)P \\ &= (P^{-1}Q^{-1})C(QP) \\ &= (QP)^{-1}C(QP) \end{aligned}$$

As the product of nonsingular matrices is again nonsingular, so A is similar to C.

(ii) If B is similar to A, then tr B = tr A.

Since $\quad tr(AB) = tr\,BA$ and $B = P^{-1}AP$, then

$$\text{tr B} = \text{tr}(P^{-1}(AP)) = \text{tr}((AP)\,P^{-1}) = \text{tr}(A(PP^{-1}))$$
$$= \text{tr}(AI) = \text{tr A}.$$

[Note that tr (AB) = sum of diagonal terms of AB

$$= (a_{11}b_{11} + \cdots + a_{1n}b_{n1}) + (a_{21}b_{21} + \cdots + a_{2n}b_{n2}) + (a_{n1}b_{1n} + \cdots + a_{nn}b_{nn})$$
$$= (b_{11}a_{11} + \cdots + b_{1n}a_{n1}) + (b_{21}a_{21} + \cdots + b_{2n}a_{n2}) + (b_{n1}a_{1n} + \cdots + b_{nn}a_{nn})$$

= sum of diagonal terms of BA = tr (BA)].

(iii) Let $B = P^{-1}AP$, P is nonsingular

$\therefore$ $\det B = \det P^{-1}AP = \det P^{-1} \det A \det P$

$= \det A \det P^{-1} \det P = \det A \det P^{-1} P$

$= \det A \det I = \det A \quad \because \det I = 1.$

i.e. similar matrices have equal determinants.

(iv) Let A and B be similar matrices. Then, either both A and B are singular matrices or both nonsingular matrices.

Let $B = P^{-1}AP$. As det B= det A, so det B = 0 if det A= 0 and det B$\neq$ 0 if det A$\neq$ 0. Hence the conclusion.

(v) If $B = P^{-1}AP$. Then $B^k = P^{-1}A^kP$

i.e. if B is similar to A. Then B^k is similar to A^k.

Now $B^2 = (P^{-1}AP)(P^{-1}AP) = P^{-1}A(PP^{-1})AP = P^{-1}A^2P$

$B^3 = B.B^2 = (P^{-1}AP)(P^{-1}A^2P) = P^{-1}A(PP^{-1})A^2P = P^{-1}A^3P$

Proceeding in this way,

$B^k = P^{-1}A^kP$ for integer k.

(vi) If B is similar to A then B^{-1} is similar to A^{-1}.

Since $B = P^{-1}AP$

So, $B^{-1} = (P^{-1}AP)^{-1} = P^{-1}A^{-1}(P^{-1})^{-1} = P^{-1}A^{-1}P$

$\therefore$ B^{-1} is similar to A^{-1}.

9.13.2 Test for similar matrices by definition

Two matrices will be similar if they hold the properties. Another general procedure follows from definition $B = P^{-1}AP$. This definition is equivalent to PB = AP. If the solution of this matrix equation is a nonsingular matrix P, then it can be concluded that the matrix A is similar to the matrix B.

Ex. Test whether the matrices

$$A = \begin{bmatrix} 6 & 2 \\ -2 & 1 \end{bmatrix} \text{ and } B = \begin{bmatrix} 8 & 6 \\ -3 & -1 \end{bmatrix}$$

are similar or not.

From properties, (*i*) note that tr (A) = 6+1=7 = 8+(−1) = tr (B)

$\therefore$ A is similar to B.

(ii) det A = 6+4 = 10, det B = −8+18 = 10

$\therefore$ det A = det B. Hence A is similar to B.

(iii) Since det A ≠ 0 & det B ≠ 0, so both A & B are nonsingular and they are similar matrices.

(iv) A general procedure is to solve AP = PB for P.

Let $P = \begin{bmatrix} p_{11} & p_{12} \\ p_{21} & p_{22} \end{bmatrix}$

Therefore,

$$\begin{bmatrix} 6 & 2 \\ -2 & 1 \end{bmatrix}\begin{bmatrix} p_{11} & p_{12} \\ p_{21} & p_{22} \end{bmatrix} = \begin{bmatrix} p_{11} & p_{12} \\ p_{21} & p_{22} \end{bmatrix}\begin{bmatrix} 8 & 6 \\ -3 & -1 \end{bmatrix}$$

$$\text{i.e. } \left.\begin{aligned} 6p_{11} + 2p_{21} &= 8p_{11} - 3p_{12} \\ 6p_{12} + 2p_{22} &= 6p_{11} - p_{12} \\ -2p_{11} + p_{21} &= 8p_{21} - 3p_{22} \\ -2p_{12} + p_{22} &= 6p_{21} - p_{22} \end{aligned}\right\}$$

$$\text{i.e. } \left.\begin{aligned} 2p_{11} - 3p_{12} - 2p_{21} \qquad\qquad &= 0 \\ 6p_{11} - 7p_{12} \qquad\qquad - 2p_{22} &= 0 \\ 2p_{11} \qquad\qquad + 7p_{21} - 3p_{22} &= 0 \\ 2p_{12} + 6p_{21} - 2p_{22} &= 0 \end{aligned}\right\}$$

which is a system of 4 equations with 4 unknowns.

The echelon form of the system is

$$\left.\begin{aligned} 2p_{11} - 3p_{12} - 2p_{21} \qquad &= 0 \\ p_{12} + 3p_{21} - p_{22} &= 0 \end{aligned}\right\}$$

in which p_{21} & p_{22} are free variables.

Let $p_{21} = 2s$, $p_{22} = 2t$. Then $p_{11} = 3t - 7s$, $p_{22} = 2t - 6s$.

$$\begin{bmatrix} 3t - 7s & 2t - 6s \\ 2s & 2t \end{bmatrix}, \text{ provided det P} \neq 0.$$

i.e. provided $6t - 18st + 12s \neq 0$ or $(6t - 12s)(t - s) \neq 0$ or $t \neq 2s$ and $t \neq s$.

Hence A and B are similar.

To be specific, let s = 0, $t = 1$, then

$$P = \begin{bmatrix} 3 & 2 \\ 0 & 2 \end{bmatrix}, \; P^{-1} = \begin{bmatrix} \frac{1}{3} & -\frac{1}{3} \\ 0 & \frac{1}{2} \end{bmatrix}.$$

$$\text{and } P^{-1}AP = \begin{bmatrix} \frac{1}{3} & -\frac{1}{3} \\ 0 & \frac{1}{2} \end{bmatrix}\begin{bmatrix} 6 & 2 \\ -2 & 1 \end{bmatrix}\begin{bmatrix} 3 & 2 \\ 0 & 2 \end{bmatrix} = \begin{bmatrix} 8 & 6 \\ -3 & -1 \end{bmatrix} = B.$$

N.B. The procedure (iv) may be extended to higher orders. But as the order increases labor also increases.

9.13.3 Construction of similar matrix by elementary transformation : It is to be noted that elementary matrices E^n_{kq}, $E^n_i(\alpha)$, $E^n_{kq}(\alpha)$ obtained by elementary row operations I, II & III respectively and F^n_{kq}, $F^n_i(\alpha)$, $F^n_{kq}(\alpha)$ obtained by elementary column operations I, II & III respectively are connected by the following relations :

$$E^n_{kq}\, F^n_{kq} = I$$

$$E^n_i(\alpha)\; F^n_i(\tfrac{1}{\alpha}) = I$$

$$E^n_{kq}(\alpha)\; F^n_{kq}(-\alpha) = I$$

which shows that one is the inverse of the other. Readers can easily establish the truth of the above statements. Similar matrix B can be constructed from A by the pairwise elementary transformation R_{kq}, C_{kq} ; $R_i(\alpha)$, $C_i\left(\frac{1}{\alpha}\right)$; and $R_{ij}(\alpha)$, $C_{ji}(-\alpha)$ — one followed by the other of the pairs.

These transformations may be called elementary *similarity transformations.* Then P^{-1} is the product of elementary matrices corresponding to elementary row (column) operations taken in reverse order while P is the products of elementary matrices corresponding to elementary column (row) operations taken in order. This procedure is not suitable for obtaining a matrix similar to a diagonal matrix.

9.13.4 Comparison of equivalent, congruent & similar matrices: Four sets of comparison are mentioned below:

A . The definitions are identical in form as

(i) A is equivalent to B if

$A = PBQ$, P and Q are nonsingular.

Here A may be square and rectangular.

(ii) A is congruent to B if

$A = P^T BP$, P is nonsingular and A is a square matrix.

(iii) A is similar to B if

$A = P^{-1} BP$, P is nonsingular and A is a square matrix.

Existence of the nonsingular matrix is the primary condition for all the three types of matrices. If a given matrix B is pre and post multiplied by two nonsingular matrices P and Q, an equivalent matrix A results. Similarly, for congruence and similar matrices, transpose P^T and inverse P^{-1} of a nonsingular matrix P respectively premultiplies the given matrix B while postmultiplies by P to obtain A.

B. Congruent and similar matrices are confined to square matrices only while the equivalent matrix is not.

C. No restriction or rule in elementary row & column operations for equivalent matrix but in congruent and similar matrices successive row & column elementary operations are needed.

D. In congruent matrix row operations is the transpose of column operation while in similar matrix one is the inverse of the other. The equivalent matrix is free of such considerations.

Ex.1 Find a matrix similar to

$$A = \begin{bmatrix} 1 & 0 & 2 \\ -1 & 3 & 4 \\ 0 & 1 & 3 \end{bmatrix}$$

and the corresponding nonsingular matrix P using three pairs elementary matrices at least once. Verify that $B = P^{-1}AP$.

$$A = \begin{bmatrix} 1 & 0 & 2 \\ -1 & 3 & 4 \\ 0 & 1 & 3 \end{bmatrix} \underset{\approx}{R_1(10)} \begin{bmatrix} 10 & 0 & 20 \\ -1 & 3 & 4 \\ 0 & 1 & 3 \end{bmatrix} \underset{\approx}{C_1(\tfrac{1}{10})} \begin{bmatrix} 1 & 0 & 20 \\ -\frac{1}{10} & 3 & 4 \\ 0 & 1 & 3 \end{bmatrix}$$

$$\underset{\approx}{R_{12}} \begin{bmatrix} -\frac{1}{10} & 3 & 4 \\ 1 & 0 & 20 \\ 0 & 1 & 3 \end{bmatrix} \underset{\approx}{C_{12}} \begin{bmatrix} 3 & -\frac{1}{10} & 4 \\ 0 & 1 & 20 \\ 1 & 0 & 3 \end{bmatrix} \underset{\approx}{R_{32}(-1)} \begin{bmatrix} 3 & -\frac{1}{10} & 4 \\ -1 & 1 & 17 \\ 1 & 0 & 3 \end{bmatrix}$$

$$\underset{\approx}{C_{23}(1)} \begin{bmatrix} 3 & -\frac{1}{10} & \frac{39}{10} \\ -1 & 1 & 18 \\ 1 & 0 & 3 \end{bmatrix} = B.$$

Note that tr A = 1 + 3 + 3 = 7 which remains invariant at the end of each pair of elementary operations. This may be used as a *check point* of the procedure. Also note that det A = det B.

$$P^{-1} = R_{23}(-1)\ R_{12}\ R_1(10)$$

$$= \begin{bmatrix} 1 & 0 & 0 \\ 0 & 1 & -1 \\ 0 & 0 & 1 \end{bmatrix} \begin{bmatrix} 0 & 1 & 0 \\ 1 & 0 & 0 \\ 0 & 0 & 1 \end{bmatrix} \begin{bmatrix} 10 & 0 & 0 \\ 0 & 1 & 0 \\ 0 & 0 & 1 \end{bmatrix} = \begin{bmatrix} 0 & 1 & 0 \\ 10 & 0 & -1 \\ 0 & 0 & 1 \end{bmatrix}$$

$$P = C_1(\tfrac{1}{10})\ C_{12}\ C_{32}(1)$$

$$= \begin{bmatrix} \frac{1}{10} & 0 & 0 \\ 0 & 1 & 0 \\ 0 & 0 & 1 \end{bmatrix} \begin{bmatrix} 0 & 1 & 0 \\ 1 & 0 & 0 \\ 0 & 0 & 1 \end{bmatrix} \begin{bmatrix} 1 & 0 & 0 \\ 0 & 1 & 1 \\ 0 & 0 & 1 \end{bmatrix} = \begin{bmatrix} 0 & \frac{1}{10} & \frac{1}{10} \\ 1 & 0 & 0 \\ 0 & 0 & 1 \end{bmatrix}$$

It can be easily verified that $P^{-1}P = I$.

$$\text{Now } P^{-1}AP = \begin{bmatrix} 0 & 1 & 0 \\ 10 & 0 & -1 \\ 0 & 0 & 1 \end{bmatrix} \begin{bmatrix} 1 & 0 & 2 \\ -1 & 3 & 4 \\ 0 & 1 & 3 \end{bmatrix} \begin{bmatrix} 0 & \frac{1}{10} & \frac{1}{10} \\ 1 & 0 & 0 \\ 0 & 0 & 1 \end{bmatrix} = \begin{bmatrix} 3 & -\frac{1}{10} & \frac{39}{10} \\ -1 & 1 & 18 \\ 1 & 0 & 3 \end{bmatrix} \text{ which is B.}$$

Ex.2 Transform A into $P^{-1}AP$ using the given nonsingular matrix P where

$$A = \begin{bmatrix} 1 & 0 & 2 \\ -1 & 3 & 4 \\ 0 & 1 & 3 \end{bmatrix}, \quad P = \begin{bmatrix} 0 & 3 & 3 \\ 1 & 2 & 3 \\ 1 & 4 & 6 \end{bmatrix}.$$

OR Find a similar matrix B such that $B = PAP^{-1}$ where A & P are as above.

Here $|P| = -3 \neq 0$. So P is nonsingular and

$$P^{-1} = \begin{bmatrix} 0 & 2 & -1 \\ 1 & 1 & -1 \\ -\frac{2}{3} & -1 & 1 \end{bmatrix}.$$

$$\therefore\ P^{-1}AP = \begin{bmatrix} 0 & 2 & -1 \\ 1 & 1 & -1 \\ -\frac{2}{3} & -1 & 1 \end{bmatrix} \begin{bmatrix} 1 & 0 & 2 \\ -1 & 3 & 4 \\ 0 & 1 & 3 \end{bmatrix} \begin{bmatrix} 0 & 3 & 3 \\ 1 & 2 & 3 \\ 1 & 4 & 6 \end{bmatrix}. = \begin{bmatrix} 10 & 24 & 39 \\ 5 & 16 & 24 \\ -\frac{13}{3} & -\frac{37}{3} & -19 \end{bmatrix}$$

which is the required similarity transformation of A into $P^{-1}AP$ under a given nonsingular matrix P.

OR the required similar matrix B is

$$\begin{bmatrix} 10 & 24 & 39 \\ 5 & 16 & 24 \\ -\frac{13}{3} & -\frac{37}{3} & -19 \end{bmatrix}.$$

Ex.3 Verify the properties of similar matrices for A and B in the example 1 above.

Prop. (ii) It can be easily verified as

tr A = 1 + 3 + 3=7 and tr B= 10 + 16 – 19 = 7.

$\therefore$ tr A = tr B for similar matrices A & B. Similarly, readers can verify the remaining properties.

Exercises Nine

1. The following pairs of matrices satisfy matrix equations of the type EA=B or AE=B where E is an elementary matrix. Find E and the equation.

(i) $A = \begin{bmatrix} 1 & 2 & 3 \\ 1 & -1 & 4 \\ 1 & 1 & 2 \end{bmatrix},\ B = \begin{bmatrix} 1 & 5 & 3 \\ 1 & 3 & 4 \\ 1 & 3 & 2 \end{bmatrix}$; (ii) $A = \begin{bmatrix} 1 & 2 \\ -3 & 5 \\ 4 & 1 \end{bmatrix},\ B = \begin{bmatrix} 1 & 2 \\ 4 & 1 \\ -3 & 5 \end{bmatrix}$;

(iii) $A = \begin{bmatrix} 1 & -1 & 4 \\ 3 & 2 & 1 \end{bmatrix},\ B = \begin{bmatrix} 2 & -1 & 4 \\ 6 & 2 & 1 \end{bmatrix}$;

(iv) $A = \begin{bmatrix} 1 & 3 & 2 & 1 \\ -1 & 2 & 3 & 4 \\ 3 & 0 & 1 & 2 \end{bmatrix},\ B = \begin{bmatrix} -5 & 3 & 0 & -3 \\ -1 & 2 & 3 & 4 \\ 3 & 0 & 1 & 2 \end{bmatrix}$.

[**Ans.** (i) $E = S_{32}^{3}(1)$, AE = B; (ii) $E = P_{23}^{2}$, EA = B;
(iii) $E = D_{1}^{3}(2)$, AE = B; (iv) $E = S_{23}(-2)$, EA = B]

2. Which of the following are echelon, reduced echelon matrices or neither. Identify each of these with explanation. State the value of *k* and verify property (iii) for each.

(i) $\begin{bmatrix} 0 & 1 & 4 & 6 & 3 & 7 \\ 0 & 0 & 1 & 7 & 8 & 9 \\ 0 & 0 & 0 & 0 & 1 & 4 \\ 0 & 0 & 0 & 0 & 0 & 0 \end{bmatrix}$; (ii) $\begin{bmatrix} 1 & -8 & 0 & 0 & 7 & 8 \\ 0 & 0 & 1 & 0 & 9 & -7 \\ 0 & 0 & 0 & 1 & 4 & 4 \\ 0 & 0 & 0 & 0 & 0 & 0 \end{bmatrix}$;

(iii) Any null matrix; (iv) I_6 ; (v) $\begin{bmatrix} 1 & 2 \\ 0 & 2 \end{bmatrix}$; (vi) $\begin{bmatrix} 1 & 0 & 0 \\ 0 & 0 & 0 \\ 0 & 1 & 0 \end{bmatrix}$; (vii) $\begin{bmatrix} 0 & 1 \\ 0 & 0 \\ 0 & 0 \end{bmatrix}$;

(viii) $\begin{bmatrix} 1 & 2 \\ 0 & 1 \\ 0 & 0 \end{bmatrix}$; (ix) $\begin{bmatrix} 1 & 2 & 0 \\ 0 & 0 & 1 \\ 0 & 0 & 0 \end{bmatrix}$.

[**Ans.** (i) Row echelon, not row reduced echelon because 3rd and 4th columns contain nonzero elements above 1; $k = 3, c_1 = 2, c_2 = 3, c_3 = 4$ and $c_1 < c_2 < c_3$.

(ii) Row reduced echelon; $k = 3, c_1 = 1, c_2 = 3, c_3 = 4$ and $c_1 < c_2 < c_3$.

(iii) None, see example.

(iv) Both, see example.

(v) None as first nonzero element in second row is not 1.

(vi) neither, 2nd row is zero row .

(vii) Both, $k = 1, c_1 = 2$.

(viii) Row echelon, $k = 2$ $c_1 = 1$, $c_2 = 2$.

(ix) Both, $k = 2, c_1 = 1, c_2 = 3$]

3. Find (a) column echelon and (b) reduced column echelon matrices corresponding to matrices

(i) $\begin{bmatrix} 0 & 0 & -1 & 2 & 3 \\ 0 & 2 & 3 & 4 & 5 \\ 0 & 1 & 3 & -1 & 2 \\ 0 & 3 & 2 & 4 & 1 \end{bmatrix}$; (ii) $\begin{bmatrix} 1 & 2 & 3 & 4 & 5 \\ 2 & 1 & 3 & -1 & 2 \\ 3 & 1 & 2 & 4 & 1 \end{bmatrix}$.

[**Ans.** (i) (a) Some possible answers are

$$\begin{bmatrix} 1 & 0 & 0 & 0 & 0 \\ -3 & 1 & 0 & 0 & 0 \\ -3 & \frac{1}{2} & 1 & 0 & 0 \\ -2 & \frac{3}{2} & -\frac{7}{2} & 1 & 0 \end{bmatrix}, \begin{bmatrix} 1 & 0 & 0 & 0 & 0 \\ -3 & 1 & 0 & 0 & 0 \\ -3 & \frac{1}{2} & 1 & 0 & 0 \\ -2 & \frac{3}{2} & -\frac{7}{2} & 1 & 0 \end{bmatrix};$$

(b) $\begin{bmatrix} 1 & 0 & 0 & 0 & 0 \\ 0 & 1 & 0 & 0 & 0 \\ 0 & 0 & 1 & 0 & 0 \\ 0 & 0 & 0 & 1 & 0 \end{bmatrix}$

(ii) (a) Some possible answers are

$$\begin{bmatrix}1&0&0&0&0\\2&1&0&0&0\\3&\frac{5}{3}&1&0&0\end{bmatrix},\begin{bmatrix}1&0&0&0&0\\2&1&0&0&0\\3&\frac{5}{3}&1&0&0\end{bmatrix},\begin{bmatrix}1&0&0&0&0\\0&1&0&0&0\\3&0&1&0&0\end{bmatrix}$$

$$\text{(b)}\begin{bmatrix}1&0&0&0&0\\0&1&0&0&0\\0&0&1&0&0\\0&0&0&1&0\end{bmatrix}]$$

4. Write each of the following matrices as the product of the elementary matrices

$$\text{(i)}\begin{bmatrix}1&-1&2\\3&1&4\\4&0&1\end{bmatrix},\ \text{(ii)}\begin{bmatrix}1&-1&-2\\2&1&1\\1&5&3\end{bmatrix}.$$

[**Ans.** (i) $\begin{bmatrix}1&0&0\\3&1&0\\4&0&1\end{bmatrix}\begin{bmatrix}1&0&0\\0&1&0\\0&1&1\end{bmatrix}\begin{bmatrix}1&0&0\\0&4&0\\0&0&-5\end{bmatrix}\begin{bmatrix}1&0&2\\0&1&0\\0&0&1\end{bmatrix}\begin{bmatrix}1&0&0\\0&1&-\frac{1}{2}\\0&0&1\end{bmatrix}\begin{bmatrix}1&-1&0\\0&1&0\\0&0&1\end{bmatrix}$

(ii) $\begin{bmatrix}1&0&0\\2&1&0\\1&0&1\end{bmatrix}\begin{bmatrix}1&0&0\\0&1&0\\0&2&1\end{bmatrix}\begin{bmatrix}1&0&0\\0&3&0\\0&0&-5\end{bmatrix}\begin{bmatrix}1&0&-2\\0&1&\frac{5}{3}\\0&0&1\end{bmatrix}\begin{bmatrix}1&-1&0\\0&1&0\\0&0&1\end{bmatrix}$]

5. Find the inverse of each of the follwing matrices by inspection.

$$\text{(i)}\begin{bmatrix}1&0&0&0\\0&1&0&0\\0&0&1&0\\0&-3&0&1\end{bmatrix},\ \text{(ii)}\begin{bmatrix}1&0&0&0\\0&5&0&0\\0&0&1&0\\0&0&0&1\end{bmatrix},\ \text{(iii)}\begin{bmatrix}0&0&0&1\\0&1&0&0\\0&0&1&0\\1&0&0&0\end{bmatrix},\ \text{(iv)}\begin{bmatrix}1&0&0&2\\0&1&0&0\\0&0&1&0\\0&0&0&1\end{bmatrix}.$$ [**Ans.**

$$\text{(i)}\begin{bmatrix}1&0&0&0\\0&1&0&0\\0&0&1&0\\0&3&0&1\end{bmatrix},\ \text{(ii)}\begin{bmatrix}1&0&0&0\\0&\frac{1}{5}&0&0\\0&0&1&0\\0&0&0&1\end{bmatrix},\ \text{(iii)}\begin{bmatrix}0&0&0&1\\0&1&0&0\\0&0&1&0\\1&0&0&0\end{bmatrix},\ \text{(iv)}\begin{bmatrix}1&0&0&-2\\0&1&0&0\\0&0&1&0\\0&0&0&1\end{bmatrix}$$]

6. Prove that (i) $E_{kq}^{n}(\alpha)=F_{qk}^{n}(\alpha)$,

(ii) $E_{kq}=F_{kq}$,

(iii) $E_k(\alpha)=F_k(\alpha)$.

7. Let

$$A=\begin{bmatrix}2&3&-1\\1&0&3\\0&2&-3\\-2&1&3\end{bmatrix}.$$

Find the elementary matrix which as a post multiplier of A performs the following column operations on A:

(i) Multiplies the third column of A by 4.

(ii) Interchanges first and third columns of A.

(iii) Adds 4 times the first column of A to the second column of A.

[**Ans.** (i) $\begin{bmatrix} 1 & 0 & 0 \\ 0 & 1 & 0 \\ 0 & 0 & 4 \end{bmatrix}$, (ii) $\begin{bmatrix} 0 & 0 & 1 \\ 0 & 1 & 0 \\ 1 & 0 & 0 \end{bmatrix}$, (iii) $\begin{bmatrix} 1 & 4 & 0 \\ 0 & 1 & 0 \\ 0 & 0 & 1 \end{bmatrix}$]

8. Factor

$$\begin{bmatrix} 1 & -1 & 0 \\ -1 & 2 & 0 \\ 0 & 0 & 1 \end{bmatrix}$$

into a product of ERT matrices.

[**Hints:** $A = \begin{bmatrix} 1 & -1 & 0 \\ -1 & 2 & 0 \\ 0 & 0 & 1 \end{bmatrix} \overset{R_{12}(1)}{\approx} \begin{bmatrix} 1 & -1 & 0 \\ 0 & 1 & 0 \\ 0 & 0 & 1 \end{bmatrix} \overset{R_{21}(1)}{\approx} \begin{bmatrix} 1 & 0 & 0 \\ 0 & 1 & 0 \\ 0 & 0 & 1 \end{bmatrix}$

$$\therefore \begin{bmatrix} 1 & 1 & 0 \\ 0 & 1 & 0 \\ 0 & 0 & 1 \end{bmatrix}\begin{bmatrix} 1 & 0 & 0 \\ 1 & 1 & 0 \\ 0 & 0 & 1 \end{bmatrix} A = I$$

$$\text{i.e. } A = \begin{bmatrix} 1 & 0 & 0 \\ 1 & 1 & 0 \\ 0 & 0 & 1 \end{bmatrix}^{-1}\begin{bmatrix} 1 & 1 & 0 \\ 0 & 1 & 0 \\ 0 & 0 & 1 \end{bmatrix}^{-1}$$

$$= \begin{bmatrix} 1 & 0 & 0 \\ -1 & 1 & 0 \\ 0 & 0 & 1 \end{bmatrix}\begin{bmatrix} 1 & -1 & 0 \\ 0 & 1 & 0 \\ 0 & 0 & 1 \end{bmatrix}$$]

9. Prove that

$$A = \begin{bmatrix} 1 & 2 & 3 \\ 0 & 1 & 2 \\ 1 & 0 & 3 \end{bmatrix}$$

is nonsingular without using determinant and write it as a product of elementary matrices.

[**Ans.** $\begin{bmatrix} 1 & 0 & 0 \\ 0 & 1 & 0 \\ 1 & 0 & 1 \end{bmatrix}\begin{bmatrix} 1 & 0 & 0 \\ 0 & 1 & 0 \\ 0 & -2 & 1 \end{bmatrix}\begin{bmatrix} 1 & 0 & 0 \\ 0 & 1 & 0 \\ 0 & 0 & 4 \end{bmatrix}\begin{bmatrix} 1 & 0 & 0 \\ 0 & 1 & 2 \\ 0 & 0 & 1 \end{bmatrix}\begin{bmatrix} 1 & 0 & 3 \\ 0 & 1 & 0 \\ 0 & 0 & 1 \end{bmatrix}\begin{bmatrix} 1 & 2 & 0 \\ 0 & 1 & 0 \\ 0 & 0 & 1 \end{bmatrix}$]

10. Find a possible column echelon matrix of the matrix

$$\begin{bmatrix} 1 & 2 & 3 & 4 & 5 \\ 2 & 1 & 3 & -1 & 2 \\ 3 & 1 & 2 & 4 & 1 \end{bmatrix}.$$

Hence obtain the corresponding reduced column echelon matrix.

[**Ans.** A possible column echelon matrix and the reduced column echelon matrix are

$$\begin{bmatrix} 1 & 0 & 0 & 0 & 0 \\ 2 & 1 & 0 & 0 & 0 \\ 3 & \frac{5}{3} & 1 & 0 & 0 \end{bmatrix} \text{ and } \begin{bmatrix} 1 & 0 & 0 & 0 & 0 \\ 0 & 1 & 0 & 0 & 0 \\ 0 & 0 & 1 & 0 & 0 \end{bmatrix} \text{ respectively]}$$

11. Obtain the ECT's of matrices in exercises 4 and 8 and verify the results.

12. Is

$$\begin{bmatrix} 2 & 1 & 4 \\ 0 & 1 & 3 \\ 2 & 0 & 1 \end{bmatrix}$$

row equivalent of I_3 ? Why ?

$$\text{[\textbf{Ans.} No, because } \begin{bmatrix} 2 & 1 & 4 \\ 0 & 1 & 3 \\ 2 & 0 & 1 \end{bmatrix} \overset{\text{row}}{\approx} \begin{bmatrix} 1 & 0 & 1 \\ 0 & 1 & 3 \\ 0 & 0 & 0 \end{bmatrix} \text{]}$$

13. Which of the following matrices are row equivalent?

$$A = \begin{bmatrix} 2 & 1 \\ 4 & 3 \end{bmatrix}; \; B = \begin{bmatrix} 2 & 1 & 1 \\ 4 & 0 & 1 \end{bmatrix}; \; C = \begin{bmatrix} -2 & 1 & 0 \\ 4 & 3 & 2 \\ 0 & 5 & 2 \end{bmatrix}; \; D = \begin{bmatrix} 1 & 0 & 0 \\ 0 & 4 & 0 \\ 2 & 0 & 1 \end{bmatrix};$$

$$E = \begin{bmatrix} 1 & 3 & 0 \\ 2 & 1 & 4 \end{bmatrix}; \; F = \begin{bmatrix} 4 & 2 \\ 3 & 2 \\ 0 & 1 \end{bmatrix}; \; G = \begin{bmatrix} 0 & 1 \\ 4 & 3 \end{bmatrix}.$$

[**Ans.** $A \overset{\text{row}}{\approx} G$, $B \overset{\text{row}}{\approx} E$.]

14. Find two different echelon forms for the matrix

$$\begin{bmatrix} 1 & 4 & 3 & 2 \\ 1 & 8 & 0 & 2 \\ 2 & 0 & 4 & 2 \end{bmatrix}.$$

$$\text{[\textbf{Ans.} } \begin{bmatrix} 1 & 4 & 3 & 2 \\ 0 & 1 & -\frac{3}{4} & 0 \\ 0 & 0 & 1 & \frac{1}{4} \end{bmatrix}, \begin{bmatrix} 1 & 0 & 6 & 2 \\ 0 & 1 & -\frac{3}{4} & 0 \\ 0 & 0 & 1 & \frac{1}{4} \end{bmatrix} \text{]}$$

15. Premultiply by suitable elementary matrices to show that the matrices

$$\text{(i)} \begin{bmatrix} -1 & 0 & 0 & 1 \\ -1 & 1 & 0 & 0 \\ -1 & 0 & 1 & 0 \end{bmatrix} \text{ and (ii)} \begin{bmatrix} 5 & 3 & 8 & 9 \\ 1 & -1 & 2 & 3 \\ 1 & 0 & 1 & 0 \\ 3 & 4 & 5 & 6 \end{bmatrix}$$

transform to echelon forms

$$\begin{bmatrix} 1 & 0 & 0 & -1 \\ 0 & 1 & 0 & -1 \\ 0 & 0 & 1 & -1 \end{bmatrix} \text{ and } \begin{bmatrix} 1 & \frac{3}{5} & \frac{8}{5} & \frac{9}{5} \\ 0 & 1 & -\frac{1}{4} & -\frac{3}{4} \\ 0 & 0 & 1 & 3 \\ 0 & 0 & 0 & 0 \end{bmatrix} \text{ respectively.}$$

16. Determine whether each of the following sets of vectors is linearly independent or linearly dependent. Express the dependent vector(s) as linear combination of the independent vectors.

$$\text{(i)} \left\{ \alpha_1 = \begin{bmatrix} 0 \\ 1 \\ -1 \\ 2 \end{bmatrix}, \alpha_2 = \begin{bmatrix} 1 \\ 1 \\ 1 \\ 0 \end{bmatrix}, \alpha_3 = \begin{bmatrix} 5 \\ 1 \\ 9 \\ -8 \end{bmatrix} \right\}, \quad \text{(ii)} \left\{ \alpha_1 = \begin{bmatrix} 1 \\ 0 \\ 3 \end{bmatrix}, \alpha_2 = \begin{bmatrix} 0 \\ 1 \\ 0 \end{bmatrix}, \alpha_3 = \begin{bmatrix} 1 \\ 1 \\ 1 \end{bmatrix} \right\}$$

$$\text{(iii)} \left\{ \alpha_1 = \begin{bmatrix} 1 \\ 0 \\ 3 \end{bmatrix}, \alpha_2 = \begin{bmatrix} 0 \\ 1 \\ 0 \end{bmatrix}, \alpha_3 = \begin{bmatrix} 1 \\ 1 \\ 1 \end{bmatrix}, \alpha_4 = \begin{bmatrix} 7 \\ -5 \\ 4 \end{bmatrix}, \alpha_5 = \begin{bmatrix} 5 \\ 0 \\ 4 \end{bmatrix} \right\}$$

$$\text{(iv)} \left\{ \alpha_1 = \begin{bmatrix} 1 \\ -1 \\ 2 \\ 0 \end{bmatrix}, \alpha_2 = \begin{bmatrix} 0 \\ 1 \\ 4 \\ 1 \end{bmatrix}, \alpha_3 = \begin{bmatrix} 0 \\ 1 \\ 1 \\ 0 \end{bmatrix} \right\}.$$

[**Ans.** (i) Dependent, $\alpha_3 = -4\alpha_1 + 5\alpha_2$; (ii) Independent ; (iii) Dependent, $\alpha_4 = -\frac{3}{2}\alpha_1 - \frac{27}{2}\alpha_2 + \frac{17}{2}\alpha_3$, $\alpha_5 = -\frac{1}{2}\alpha_1 - \frac{11}{2}\alpha_2 + \frac{11}{2}\alpha_3$; (iv) Independent]

17. For each of the following systems of equations, find a linearly independent set of equations with same solution set.

$$\text{(i)} \left.\begin{aligned} 2x_1 - 3x_2 - 8x_3 &= 0 \\ x_1 + 2x_2 + 5x_3 &= 0 \\ 3x_1 + 4x_2 + 5x_3 &= 0 \end{aligned}\right\} \qquad \text{(ii)} \left.\begin{aligned} 4x_1 - 8x_2 + 16x_3 &= 0 \\ x_1 - 3x_2 + 6x_3 &= 0 \\ 2x_1 + x_2 - 2x_3 &= 0 \end{aligned}\right\}$$

(iii) $\left.\begin{aligned} x_1 - x_2 + 2x_3 + x_4 + x_5 + 4x_6 &= 0 \\ -x_1 + x_2 - x_3 + x_4 - x_5 - 2x_6 &= 0 \\ 2x_1 - x_2 + 4x_3 + 3x_4 + 2x_5 + 8x_6 &= 0 \\ x_1 + 3x_3 + 4x_4 + x_5 + 6x_6 &= 0 \\ 3x_1 - 2x_2 + 6x_3 + 4x_4 + 4x_5 + 11x_6 &= 0 \end{aligned}\right\}$

[**Ans.** (i) $\left.\begin{aligned} x_1 - x_3 &= 0 \\ x_2 + 2x_3 &= 0, \end{aligned}\right\}$ (ii) $\left.\begin{aligned} x_1 &= 0 \\ x_2 - 2x_3 &= 0 \end{aligned}\right\}$

(iii) $\left.\begin{aligned} x_1 - 2x_4 + x_6 &= 0 \\ x_2 + x_4 &= 0 \\ x_3 + 2x_4 + 2x_6 &= 0 \\ x_5 - x_6 &= 0 \end{aligned}\right\}$]

18. Find the inverse of the following matrices, if exist, by elementary operations.

(i) $\begin{bmatrix} 1 & 1 & 1 & 1 \\ 2 & -1 & 3 & 0 \\ 0 & 2 & 0 & 3 \\ -1 & 0 & 2 & 1 \end{bmatrix}$, (ii) $\begin{bmatrix} 1 & -1 & 2 & 3 \\ 2 & 1 & -2 & 3 \\ 1 & -1 & 2 & 5 \\ 1 & -1 & 3 & 3 \end{bmatrix}$, (iii) $\begin{bmatrix} 2 & 0 & -1 & 1 \\ 1 & 1 & -2 & 0 \\ 0 & 3 & 1 & -2 \\ 0 & 6 & 4 & -1 \end{bmatrix}$

[**Ans.**

(I) $\begin{bmatrix} 0 & \frac{2}{7} & \frac{1}{7} & -\frac{3}{7} \\ \frac{3}{2} & -\frac{9}{14} & -\frac{4}{7} & \frac{3}{14} \\ \frac{1}{2} & -\frac{1}{14} & -\frac{2}{7} & \frac{5}{14} \\ -1 & \frac{3}{7} & \frac{5}{7} & -\frac{1}{7} \end{bmatrix}$, (ii) $\frac{1}{6}\begin{bmatrix} 8 & 2 & -6 & 0 \\ -19 & 2 & 3 & 12 \\ -6 & 0 & 0 & 6 \\ -3 & 0 & 3 & 0 \end{bmatrix}$, (iii) $\frac{1}{35}\begin{bmatrix} 25 & -15 & 15 & -5 \\ -7 & 14 & -7 & 7 \\ 9 & -18 & 4 & 1 \\ -6 & 12 & -26 & 11 \end{bmatrix}$]

19. (i) Let

$$A = \begin{bmatrix} 3 & 1 & 0 \\ -1 & 2 & 4 \\ 5 & 6 & -2 \end{bmatrix} \text{ and } B = \begin{bmatrix} 1 & -3 & 2 \\ -8 & 0 & 9 \\ 4 & 5 & -7 \end{bmatrix}.$$

Find det(AB).

(ii) Let

$$A = \begin{bmatrix} 0 & 0 & 3 \\ 0 & 4 & 1 \\ 2 & 3 & 1 \end{bmatrix} \;\&\; B = \begin{bmatrix} 1 & 1 & 1 & 3 \\ 0 & 3 & 1 & 1 \\ 0 & 0 & 2 & 2 \\ -1 & -1 & -1 & 2 \end{bmatrix}.$$

Find $|A|$ & $|B|$ by inspection and hence find det(AB).

[**Ans.** $|A| = -24$, $|B| = 30$ & det(AB) = –720]

20. Let

$$A = \begin{bmatrix} 1 & 0 & 1 \\ 3 & 3 & 4 \\ 2 & 2 & 3 \end{bmatrix}.$$

(a) Verify that $A^{-1} = \begin{bmatrix} 1 & 2 & -3 \\ -1 & 1 & -1 \\ 0 & -2 & 3 \end{bmatrix}$.

(b) Use A^{-1} to solve $A\mathbf{x} = \mathbf{b}$ for the follwing choices of $\mathbf{b}$:

(i) $\mathbf{b} = (1, 1, 1)^T$, (ii) $\mathbf{b} = (1, 2, 3)^T$, (iii) $\mathbf{b} = (-2, 1, 0)^T$.

[**Ans.** (b) (i) $(0, -1, 1)^T$, (ii) $(-4, -2, 5)^T$, (iii) $(0, 3, -2)^T$]

21. Let

$$A = \begin{bmatrix} 0 & 1 & 2 & 3 \\ 1 & 1 & 1 & 1 \\ -2 & -2 & 3 & 3 \\ 1 & 2 & -2 & -3 \end{bmatrix}$$

(a) Use the elemination method to evaluate $|A|$.

(b) Use the value of $|A|$ to evaluate

$$\begin{bmatrix} 0 & 1 & 2 & 3 \\ -2 & -2 & 3 & 3 \\ 1 & 2 & -2 & -3 \\ 1 & 1 & 1 & 1 \end{bmatrix} + \begin{bmatrix} 0 & 1 & 2 & 3 \\ 1 & 1 & 1 & 1 \\ -1 & -1 & 4 & 4 \\ 2 & 3 & -1 & -2 \end{bmatrix}.$$

[**Ans.** (a) 10 (b) 20. **Hints:** Add 2nd & 3rd rows to get 3rd row and add 2nd & 4th rows to get 4th row of 2nd determinant. Interchange rows to obtain first determinant]

22. Reduce the following matrices to normal form.

(i) $\begin{bmatrix} 3 & 2 & -1 \\ 7 & 8 & 0 \\ 4 & 6 & 1 \end{bmatrix}$, (ii) $\begin{bmatrix} 3 & 1 & 4 & 6 & 2 \\ 4 & 1 & i & 9 & 6 \end{bmatrix}$, (iii) $\begin{bmatrix} 4 & 9 & -3 & 1 \\ 6 & 9 & -4 & 0 \\ 2 & 9 & -2 & 2 \\ -2 & 0 & 1 & 1 \end{bmatrix}$

(iv) $\begin{bmatrix} 1 & 2 & -1 & 4 \\ 5 & 1 & 2 & -3 \\ 2 & 1 & 4 & 3 \\ 2 & 0 & 1 & 2 \\ 5 & 1 & 2 & 3 \end{bmatrix}$, (v) $\begin{bmatrix} 1 & 3 & -1 & 2 \\ 2 & -4 & -2 & 1 \\ 3 & 1 & 2 & -3 \\ 7 & 3 & -2 & 5 \end{bmatrix}$

[**Ans.** (i) $\left[\begin{array}{c|c} I_2 & 0 \\ \hline 0 & 0 \end{array}\right]$, (ii) $[I_2 \mid 0]$, (iii) $\left[\begin{array}{c|c} I_2 & 0 \\ \hline 0 & 0 \end{array}\right]$, (iv) $\left[\begin{array}{c} I_4 \\ \hline 0 \end{array}\right]$, (v) I_4]

23. Assume that

$$B = \begin{bmatrix} 3 & 2 \\ 1 & 4 \\ 6 & 0 \end{bmatrix}$$

is obtained from a matrix A by adding 2 times the second column of A to the first column of A. This elementary operation can be expressed as

$$AE = B \text{ or } AEE^{-1} = BE^{-1} \text{ or } A = BE^{-1}.$$

Find A by first finding E, then E^{-1} and then multiply as shown above. Check by calculating AE.

[**Ans.** Possible answers are

$$\text{(a)} \begin{bmatrix} 1 & -2 & 3 & 0 \\ 0 & -1 & 4 & 3 \\ 0 & 2 & -5 & -2 \end{bmatrix}, \text{(b)} \begin{bmatrix} 1 & 0 \\ 0 & 0 \end{bmatrix}, \text{(c)} \begin{bmatrix} 1 & 0 & 0 & 0 & 0 \\ 0 & 1 & -2 & 0 & 2 \\ 0 & 5 & 5 & 4 & 4 \end{bmatrix}]$$

24. Find nonsingular matrices P and Q such that B=PAQ is in normal form for the matrices in Exercise 23.

[(a) **Hints:** Write $A = I_3 A_{3\times4} I_4$. Perform row operations in $I_3 A_{3\times4}$ keeping I_4 unchanged and succesively perform column operation on $A_{3\times4} I_4$ keeping I_3 unchanged. Proceed on until A is in normal form and consequently I_3 & I_4 will be converted to desired P and Q respectively.

$$\textbf{Ans.}\text{(a) } B = [I_3 \,\vdots\, 0],\ P = \begin{bmatrix} 1 & 0 & 0 \\ 0 & -1 & 0 \\ -\frac{1}{3} & \frac{2}{3} & \frac{1}{3} \end{bmatrix},\ Q = \begin{bmatrix} 1 & 2 & 5 & -\frac{5}{3} \\ 0 & 1 & 4 & -\frac{7}{3} \\ 0 & 0 & 1 & -\frac{4}{3} \\ 0 & 0 & 0 & 1 \end{bmatrix}$$

$$\text{(b) } B = \begin{bmatrix} 1 & 0 \\ 0 & 0 \end{bmatrix},\ P = \begin{bmatrix} 1 & 0 \\ -2 & 1 \end{bmatrix},\ Q = \begin{bmatrix} 1 & -3 \\ 0 & 1 \end{bmatrix}$$

$$\text{(c) } B = \begin{bmatrix} 1 & 0 & 0 & 0 & 0 \\ 0 & 1 & 0 & 0 & 0 \\ 0 & 0 & 1 & 0 & 0 \end{bmatrix},\ P = \begin{bmatrix} 1 & 0 & 0 \\ 0 & 1 & 0 \\ \frac{1}{15} & -\frac{1}{3} & \frac{1}{15} \end{bmatrix},\ Q = \begin{bmatrix} 1 & -2 & -7 & -\frac{32}{15} & -\frac{9}{5} \\ 0 & 1 & 2 & -\frac{8}{15} & -\frac{6}{5} \\ 0 & 0 & 1 & -\frac{4}{15} & \frac{2}{5} \\ 0 & 0 & 0 & 1 & 0 \\ 0 & 0 & 0 & 0 & 1 \end{bmatrix}]$$

25. Let

$$A = \begin{bmatrix} 1 & 2 & 3 \\ 0 & 1 & 2 \\ 1 & 0 & -2 \end{bmatrix}.$$

Find a matrix B in the normal form and equivalent to A. Also find nonsingular matrices P and Q such that B = P A Q.

[**Ans.** $B = I_3$, $P = \begin{bmatrix} 1 & 0 & 0 \\ 0 & 1 & 0 \\ 1 & -2 & -1 \end{bmatrix}$ and $Q = \begin{bmatrix} 1 & -2 & 1 \\ 0 & 1 & -2 \\ 0 & 0 & 1 \end{bmatrix}$]

26. Repeat Exercise 17 by Gauss - Jordan elimination.

27. Solve the following systems of equations by relaxation method with residuals:

(i) $\begin{cases} 4x_1 + 2x_2 = 2 \\ 2x_1 + 10x_2 + 4x_3 = 6 \\ 4x_2 + 5x_3 = 5 \end{cases}$ (two decimal places), (ii) $\begin{cases} 1.6x_1 + 2.0x_2 + 0.8x_3 - 1.2x_4 = 37.2 \\ 1.8x_1 + 0.7x_2 - 1.6x_3 - 0.8x_4 = 42.0 \\ 2.0x_1 - 1.6x_2 + 0.4x_3 - 3.0x_4 = 13.8 \\ 1.2x_1 - 1.4x_2 + 1.4x_3 + 2.0x_4 = -30.4 \end{cases}$ (four decimal places)

(iii) $\begin{cases} 10x_1 - 2x_2 - 2x_3 = 6 \\ -x_1 + 10x_2 - 2x_3 = 7 \\ -x_1 - x_2 + 10x_3 = 5 \end{cases}$ (two decimal places)

[**Ans.** (i) $x_1 = 0.41$, $x_2 = 0.17$, $x_3 = 0.86$; $R_1 = -0.02$, $R_2 = -0.04$, $R_3 = -0.02$
(ii) $x_1 = 8.4734$, $x_2 = 11.5090$, $x_3 = -8.5665$, $x_4 = -6.2313$;
$R_1 = -0.0020$, $R_2 = -0.0014$, $R_3 = -0.0010$, $R_4 = -0.0022$
(iii) exact solutions are $x_1 = 1.00$, $x_2 = 1.00$, $x_3 = 1.00$]

28. Use the boarding method to obtain the inverse of

$$\begin{bmatrix} 1 & 4 & 1 & 3 \\ 0 & -1 & 3 & -1 \\ 3 & 1 & 0 & 2 \\ 1 & -2 & 5 & 1 \end{bmatrix}.$$

[**Ans.** $\frac{1}{44}\begin{bmatrix} -5 & 15 & 19 & -8 \\ 9 & 17 & 1 & -12 \\ 4 & 10 & -2 & 2 \\ 3 & -31 & -7 & 18 \end{bmatrix}$]

29. Find matrices A & B such that

$$A\begin{bmatrix} 2 & 2 & -6 \\ 1 & -2 & -2 \end{bmatrix}B$$

is a canonical matrix.

[**Ans.** $A = \begin{bmatrix} \frac{1}{2} & 0 \\ \frac{1}{6} & -\frac{1}{3} \end{bmatrix}$, $B = \begin{bmatrix} 1 & -1 & \frac{8}{3} \\ 0 & 1 & \frac{1}{3} \\ 0 & 0 & 1 \end{bmatrix}$]

30. Find a nonsingular matrix P such that each of the matrices

$$\text{(i) } A = \begin{bmatrix} 1 & -2 & 1 & -2 \\ -2 & 0 & 3 & -5 \\ 1 & 3 & 2 & 2 \\ 2 & -5 & 2 & -9 \end{bmatrix} \text{ and (ii) } A = \begin{bmatrix} 3 & 0 & 1 & -1 \\ 0 & 0 & -4 & 2 \\ 1 & -4 & 5 & -2 \\ -1 & 2 & -2 & 1 \end{bmatrix}$$

is congruent to the diagonal matrices

$$\text{(i) } D = \begin{bmatrix} 1 & 0 & 0 & 0 \\ 0 & 1 & 0 & 0 \\ 0 & 0 & -29 & 0 \\ 0 & 0 & 0 & 0 \end{bmatrix} \text{ and (ii) } D = \begin{bmatrix} 3 & 0 & 0 & 0 \\ 0 & \frac{1}{2} & 0 & 0 \\ 0 & 0 & \frac{2}{3} & 0 \\ 0 & 0 & 0 & -8 \end{bmatrix} \text{ respectively.}$$

[**Hints :** for (ii): Use prop.(ii) of sec. 9.12.

$$A = \begin{bmatrix} 3 & 0 & 1 & -1 \\ 0 & 0 & -4 & 2 \\ 1 & -4 & 5 & -2 \\ -1 & 2 & -2 & 1 \end{bmatrix} \underset{\approx}{R_3 - \tfrac{1}{3}R_1} \begin{bmatrix} 3 & 0 & 1 & -1 \\ 0 & 0 & -4 & 2 \\ 1 & -4 & \frac{14}{3} & -\frac{5}{3} \\ -1 & 2 & -2 & 1 \end{bmatrix} \overset{(1)}{\underset{\approx}{C_3 - \tfrac{1}{3}C_1}}$$

$$\underset{\approx}{R_4 + \tfrac{1}{3}R_1} \begin{bmatrix} 3 & 0 & 0 & -1 \\ 0 & 0 & -4 & 2 \\ 1 & -4 & \frac{14}{3} & -\frac{5}{3} \\ -1 & 2 & -\frac{5}{3} & 1 \end{bmatrix} \overset{(2)}{\underset{\approx}{C_4 + \tfrac{1}{3}C_1}} \begin{bmatrix} 3 & 0 & 0 & -1 \\ 0 & 0 & -4 & 2 \\ 1 & -4 & \frac{14}{3} & -\frac{5}{3} \\ 0 & 2 & -\frac{5}{3} & \frac{2}{3} \end{bmatrix} \underset{\approx}{R_3 + 2R_4}$$

$$\begin{bmatrix} 3 & 0 & 0 & 0 \\ 0 & 0 & -4 & 2 \\ 0 & 0 & \frac{4}{3} & -\frac{1}{3} \\ 0 & 2 & -\frac{5}{3} & \frac{2}{3} \end{bmatrix} \overset{(3)}{\underset{\approx}{C_3 + 2C_4}} \begin{bmatrix} 3 & 0 & 0 & 0 \\ 0 & 0 & 0 & 2 \\ 0 & 0 & \frac{2}{3} & -\frac{1}{3} \\ 0 & 2 & -\frac{1}{3} & \frac{2}{3} \end{bmatrix} \underset{\approx}{R_4 + \tfrac{1}{2}R_3} \begin{bmatrix} 3 & 0 & 0 & 0 \\ 0 & 0 & 0 & 2 \\ 0 & 0 & \frac{2}{3} & -\frac{1}{3} \\ 0 & 2 & 0 & \frac{1}{2} \end{bmatrix}$$

$$\overset{(4)}{\underset{\approx}{C_4 + \tfrac{1}{2}C_3}} \begin{bmatrix} 3 & 0 & 0 & 0 \\ 0 & 0 & 0 & 2 \\ 0 & 0 & \frac{2}{3} & 0 \\ 0 & 2 & 0 & \frac{1}{2} \end{bmatrix} \underset{\approx}{R_4 \leftrightarrow R_2} \begin{bmatrix} 3 & 0 & 0 & 0 \\ 0 & 2 & 0 & \frac{1}{2} \\ 0 & 0 & \frac{2}{3} & 0 \\ 0 & 0 & 0 & 2 \end{bmatrix} \overset{(5)}{\underset{\approx}{C_2 \leftrightarrow C_4}} \begin{bmatrix} 3 & 0 & 0 & 0 \\ 0 & \frac{1}{2} & 0 & 2 \\ 0 & 0 & \frac{2}{3} & 0 \\ 0 & 2 & 0 & 0 \end{bmatrix}$$

$$\underset{\approx}{R_4 - 4R_2} \begin{bmatrix} 3 & 0 & 0 & 0 \\ 0 & \frac{1}{2} & 0 & 2 \\ 0 & 0 & \frac{2}{3} & 0 \\ 0 & 0 & 0 & -4 \end{bmatrix} \overset{(6)}{\underset{\approx}{C_4 - 4C_2}} \begin{bmatrix} 3 & 0 & 0 & 0 \\ 0 & \frac{1}{2} & 0 & 0 \\ 0 & 0 & \frac{2}{3} & 0 \\ 0 & 0 & 0 & -8 \end{bmatrix}. \text{ Then}$$

$$P = \begin{bmatrix} 1 & 0 & -\frac{1}{3} & 0 \\ 0 & 1 & 0 & 0 \\ 0 & 0 & 1 & 0 \\ 0 & 0 & 0 & 1 \end{bmatrix} \begin{bmatrix} 1 & 0 & 0 & \frac{1}{3} \\ 0 & 1 & 0 & 0 \\ 0 & 0 & 1 & 0 \\ 0 & 0 & 0 & 1 \end{bmatrix} \begin{bmatrix} 1 & 0 & 0 & 0 \\ 0 & 1 & 0 & 0 \\ 0 & 0 & 1 & 0 \\ 0 & 0 & 2 & 1 \end{bmatrix} \begin{bmatrix} 1 & 0 & 0 & 0 \\ 0 & 1 & 0 & 0 \\ 0 & 0 & 1 & \frac{1}{2} \\ 0 & 0 & 0 & 1 \end{bmatrix} \begin{bmatrix} 1 & 0 & 0 & 0 \\ 0 & 0 & 0 & 1 \\ 0 & 0 & 1 & 0 \\ 0 & 1 & 0 & 0 \end{bmatrix}$$

$$\begin{bmatrix} 1 & 0 & 0 & 0 \\ 0 & 1 & 0 & -4 \\ 0 & 0 & 1 & 0 \\ 0 & 0 & 0 & 1 \end{bmatrix} = \begin{bmatrix} 1 & \frac{1}{2} & \frac{1}{3} & -2 \\ 0 & 0 & 0 & 1 \\ 0 & \frac{1}{2} & 1 & -2 \\ 0 & 2 & 2 & -8 \end{bmatrix}. \text{ (i) } P = \begin{bmatrix} 1 & -1 & 7 & -1 \\ 0 & 0 & 1 & -1 \\ 0 & 1 & -5 & 1 \\ 0 & 0 & 0 & 1 \end{bmatrix}]$$

31. Find a nonsingular matrix P such that the matrix (ii) in Exercise (Prob.32 above) is congruent to the matrix

$$B = \begin{bmatrix} 3 & 0 & 0 & 0 \\ 0 & 0 & -4 & 2 \\ 0 & -4 & \frac{14}{3} & -\frac{5}{3} \\ 0 & 2 & -\frac{5}{3} & \frac{2}{3} \end{bmatrix}. \qquad \textbf{[Ans. } P = \begin{bmatrix} 1 & 0 & -\frac{1}{3} & \frac{1}{3} \\ 0 & 1 & 0 & 0 \\ 0 & 0 & 1 & 0 \\ 0 & 0 & 0 & 1 \end{bmatrix}]$$

32. Find by inspection, the product of six matrices

$$\begin{bmatrix} 1 & 0 & -\frac{1}{3} & 0 \\ 0 & 1 & 0 & 0 \\ 0 & 0 & 1 & 0 \\ 0 & 0 & 0 & 1 \end{bmatrix} \begin{bmatrix} 1 & 0 & 0 & \frac{1}{3} \\ 0 & 1 & 0 & 0 \\ 0 & 0 & 1 & 0 \\ 0 & 0 & 0 & 1 \end{bmatrix} \begin{bmatrix} 1 & 0 & 0 & 0 \\ 0 & 1 & 0 & 0 \\ 0 & 0 & 1 & 0 \\ 0 & 0 & 2 & 1 \end{bmatrix} \begin{bmatrix} 1 & 0 & 0 & 0 \\ 0 & 1 & 0 & 0 \\ 0 & 0 & 1 & \frac{1}{2} \\ 0 & 0 & 0 & 1 \end{bmatrix}$$

$$\begin{bmatrix} 1 & 0 & 0 & 0 \\ 0 & 0 & 0 & 1 \\ 0 & 0 & 1 & 0 \\ 0 & 1 & 0 & 0 \end{bmatrix} \begin{bmatrix} 1 & 0 & 0 & 0 \\ 0 & 1 & 0 & -4 \\ 0 & 0 & 1 & 0 \\ 0 & 0 & 0 & 1 \end{bmatrix}. \qquad \textbf{[Ans.} \begin{bmatrix} 1 & \frac{1}{2} & \frac{1}{3} & -2 \\ 0 & 0 & 0 & 1 \\ 0 & \frac{1}{2} & 1 & -2 \\ 0 & 2 & 2 & -8 \end{bmatrix}]$$

33. Find a nonsingular matrix P such that $B = P^T AP$, where

$$A = \begin{bmatrix} 1 & i & 1-i \\ i & 0 & 2+i \\ 1-i & 2+i & 1-2i \end{bmatrix} \text{ and } B = \begin{bmatrix} 1 & 0 & 0 \\ 0 & 1 & 0 \\ 0 & 0 & 0 \end{bmatrix}.$$

34. Show that the following matrices could be obtained from each other by elementary operations:

$$\begin{bmatrix} 1 & 2 & 1 & -1 \\ 0 & 0 & 1 & -4 \\ 0 & 0 & 0 & 0 \end{bmatrix}, \begin{bmatrix} 1 & 0 & 0 & 0 \\ 0 & 1 & 0 & 0 \\ 0 & 0 & 0 & 0 \end{bmatrix}.$$

35. For each of the following matrices A, find a nonsingular matrix P such that $P^T AP$ is diagonal:

$$\text{(i) } A = \begin{bmatrix} 2 & 3 \\ 3 & 4 \end{bmatrix}, \text{ (ii) } A = \begin{bmatrix} 1 & -2 & 3 \\ -2 & 6 & -9 \\ 3 & -9 & 4 \end{bmatrix}, \text{ (iii) } A = \begin{bmatrix} 1 & 1 & -2 & -3 \\ 1 & 2 & -5 & -1 \\ -2 & -5 & 6 & 9 \\ -2 & -1 & 9 & 11 \end{bmatrix}.$$

[**Ans.** (i) $P=\begin{bmatrix}1 & -3\\ 0 & 2\end{bmatrix}$, $P^TAP=\begin{bmatrix}2 & 0\\ 0 & -2\end{bmatrix}$;

(ii) $P=\begin{bmatrix}1 & 2 & 0\\ 0 & 1 & 3\\ 0 & 0 & 2\end{bmatrix}$, $P^TAP=\begin{bmatrix}1 & 0 & 0\\ 0 & 2 & 0\\ 0 & 0 & -38\end{bmatrix}$,

(iii) $P=\begin{bmatrix}1 & -1 & -1 & 26\\ 0 & 1 & 3 & 13\\ 0 & 0 & 1 & 9\\ 0 & 0 & 0 & 7\end{bmatrix}$, $p^TAP=\begin{bmatrix}1 & 0 & 0 & 0\\ 0 & 1 & 0 & 0\\ 0 & 0 & -7 & 0\\ 0 & 0 & 0 & 469\end{bmatrix}$]

36. For each of the following Hermitian matrices, find nonsingular matrix P such that $P^TH\overline{P}$ is diagonal:

(i) $H=\begin{bmatrix}1 & i\\ -i & 2\end{bmatrix}$, (ii) $H=\begin{bmatrix}1 & i & 2+i\\ -i & 2 & 1-i\\ 2-i & 1+i & 2\end{bmatrix}$. [**Hints:** P is conjugate of P. Follow prop.(ii) of sec.9.12 **Ans.** (i) $P=\begin{bmatrix}1 & i\\ 0 & 1\end{bmatrix}$, $P^TH\overline{P}=\begin{bmatrix}1 & 0\\ 0 & 1\end{bmatrix}$;

(ii) $P=\begin{bmatrix}1 & i & -3+i\\ 0 & 1 & i\\ 0 & 0 & 1\end{bmatrix}$, $P^TH\overline{P}=\begin{bmatrix}1 & 0 & 0\\ 0 & 1 & 0\\ 0 & 0 & -4\end{bmatrix}$]

37. Let RREF of A be R where

$$A=\begin{bmatrix}0 & 1 & 3 & -2 & -1 & 2\\ 0 & 2 & 6 & -4 & -2 & 4\\ 0 & 1 & 3 & -2 & 1 & 4\\ 0 & 2 & 6 & 1 & -1 & 0\end{bmatrix} \text{ and } R=\begin{bmatrix}0 & 1 & 3 & 0 & 0 & 1\\ 0 & 0 & 0 & 1 & 0 & -1\\ 0 & 0 & 0 & 0 & 1 & 1\\ 0 & 0 & 0 & 0 & 0 & 0\end{bmatrix}.$$

Find the RREF of

$$A_1=\begin{bmatrix}1 & 3 & -2 & -1 & 2\\ 2 & 6 & -4 & -2 & 4\\ 1 & 3 & -1 & 1 & 4\\ 2 & 6 & 1 & -1 & 0\end{bmatrix},\ A_2=\begin{bmatrix}1 & 3 & -2 & -1 & 2\\ 0 & 6 & -4 & -2 & 4\\ 0 & 3 & -1 & 1 & 4\\ 0 & 6 & 1 & -1 & 0\end{bmatrix}$$

from R directly or with some row operations, if necessary.

[**Ans.** $R_1=\begin{bmatrix}1 & 3 & 0 & 0 & 1\\ 0 & 0 & 1 & 0 & -1\\ 0 & 0 & 0 & 1 & 1\\ 0 & 0 & 0 & 0 & 0\end{bmatrix}$, $R_2=\begin{bmatrix}1 & 1 & 0 & 0 & \frac{1}{3}\\ 0 & 0 & 1 & 0 & -1\\ 0 & 0 & 0 & 1 & 1\\ 0 & 0 & 0 & 0 & 0\end{bmatrix}$]

38. The matrix

$$A=\begin{bmatrix}1&4&0&-6&5&1\\2&1&1&1&3&0\\-1&2&1&-7&1&-1\\1&1&2&-2&2&1\\3&0&3&3&3&2\end{bmatrix}\text{ reduces to }R=\begin{bmatrix}1&0&0&2&1&0\\0&1&0&-2&1&0\\0&0&1&-1&0&0\\0&0&0&0&0&1\\0&0&0&0&0&0\end{bmatrix}.$$

Each of the following is obtained by deleting some columns of A. Determine their RREF without using any row operations.

$$\text{(a)}\begin{bmatrix}1&4&0&1\\2&1&1&0\\-1&2&1&-1\\1&1&2&1\\3&0&3&2\end{bmatrix},\ \text{(b)}\begin{bmatrix}1&4&5&1\\2&1&3&0\\-1&2&1&-1\\1&1&2&1\\3&0&3&2\end{bmatrix},\ \text{(c)}\begin{bmatrix}1&4&0&1&-6\\2&1&1&0&1\\-1&2&1&-1&-7\\1&1&2&1&-2\\3&0&3&2&3\end{bmatrix}.$$

$$\textbf{[Ans.}\text{ (a)}\begin{bmatrix}1&0&0&0\\0&1&0&0\\0&0&1&0\\0&0&0&1\\0&0&0&0\end{bmatrix},\ \text{(b)}\begin{bmatrix}1&0&1&0\\0&1&1&0\\0&0&0&1\\0&0&0&0\\0&0&0&0\end{bmatrix},\ \text{(c)}\begin{bmatrix}1&0&0&0&2\\0&1&0&0&-2\\0&0&1&0&-1\\0&0&0&1&0\\0&0&0&0&0\end{bmatrix}]$$

39. Let A, B, C be obtained from A and R by deleting second & 4th columns of each . Suppose R is RREF.

(i) Give an example to show that C may not be a RREF.

(ii) What conditions on R will guarantee that C is a RREF?

[**Ans.** (ii) If column two is a multiple of column one & column four is a linear combination of columns one & two. Other answers as possible]

40. Test whether the following pairs of matrices are similar or not, by definition:

$$\text{(i)}\begin{bmatrix}-2&-1\\0&11\end{bmatrix}\ \&\ \begin{bmatrix}0&1\\2&3\end{bmatrix};\ \text{(ii)}\begin{bmatrix}2&2\\1&3\end{bmatrix}\ \&\ \begin{bmatrix}2&1\\0&2\end{bmatrix}.$$

[**Ans.** (i) not similar, (ii) not similar]

41. List all kinds of equivalences encountered in this chapter.

[**Ans.** (1) Row equivalence, (2) column equivalence, (3) Row & column combined, equivalent, (4) congruent, (5) Hermitian equivalence or conjunctivity, (6) similarity]

42. Find two more similar matrices similar to one in the Ex.1, 2 in sec.9.13.4 verify the results so obtained.

43. Find a similar matrix B such that $B = P^{-1}AP$,

(i) $A = \begin{bmatrix} 0 & 2 & 1 \\ 1 & -4 & 0 \\ 3 & 0 & 0 \end{bmatrix}$ and $P = \begin{bmatrix} 1 & 1 & 1 \\ 1 & 1 & 0 \\ 1 & 0 & 0 \end{bmatrix}$. [**Ans.** $B = \begin{bmatrix} 3 & 3 & 3 \\ -6 & -6 & -2 \\ 6 & 5 & -1 \end{bmatrix}$]

(ii) $A = \begin{bmatrix} 0 & 2 \\ 3 & -1 \end{bmatrix}$ and $P = \begin{bmatrix} 1 & 2 \\ 3 & 5 \end{bmatrix}$. [**Ans.** $B = \begin{bmatrix} -30 & -48 \\ 18 & 29 \end{bmatrix}$]

(iii) $A = \begin{bmatrix} 1 & 1 & 0 \\ 0 & 0 & 1 \\ 2 & -1 & 0 \end{bmatrix}$ & $P = \begin{bmatrix} 2 & 1 & -1 \\ 1 & 0 & 1 \\ 3 & 1 & 1 \end{bmatrix}$. [**Ans.** $B = \begin{bmatrix} 6 & 1 & 5 \\ -12 & -1 & -14 \\ -3 & 0 & -4 \end{bmatrix}$]

(iv) $A = \begin{bmatrix} 2 & 1 & 3 \\ 1 & -2 & 1 \\ 0 & 2 & 1 \end{bmatrix}$ & $P = \begin{bmatrix} 1 & 0 & 1 \\ 2 & 1 & 2 \\ 0 & 2 & 1 \end{bmatrix}$. [**Ans.** $B = \begin{bmatrix} -22 & -25 & -30 \\ -11 & -14 & -16 \\ 26 & 32 & 37 \end{bmatrix}$]

44. If A is similar to B, prove that det A = det B.

45. Let A & B square matrices of the same order with det A = det B. Show that it is not necessarily true that A is similar to B.

46. If A & B are similar,

(a) Show that the squares, cubes, $\cdots$, of A and B are also similar.

(b) Show that A^T and B^T are similar.

(c) Show that A^{-1} and B^{-1} are similar if A is nonsingular.

CHAPTER TEN

RANK AND NULLITY OF A MATRIX

Chapter summary: *In the beginning, the chapter defines and illustrates rank and nullity of matrix. Theorems on ranks, sum and product of ranks, rank of Kronecker product and related theorems are discussed. Gerstein algorithm, echelon form are incorporated for determining rank. Simultaneous equations, linear dependence and independence are revisited in the light of rank. Index and signature are defined, illustrated and used for definiteness test of matrices.*

10.1 Introduction

The concept of rank is of fundamental importance and is introduced by G.F. Fröbenius in 1879 in his studies of determinant. It appears in many occasions such as in linear independence of vectors, vector space (cf. chapter 11), linear transformation (cf. chapter 12) and in other mathematical disciplines and matrices too. If S is a set of vectors of which r is the number of linearly independent vectors of S then r is called the *rank* of S and is written as rank S = r. This implies that r is the size of the largest linearly independent subset of vectors contained in S.

This chapter deals with the concept of the rank of a matrix. This is an extension of the idea described in the previous paragraph. Because matrix is nothing but a collection of row (column) vectors (cf. chapter 1). Hence rank of a matrix obviously determines the size of the largest linearly independent subset of row (column) vectors contained in a matrix.

10.2 Definitions and illustrations

In this section the definition is based on determinant and so it is also called the *determinantal rank.*

10.2.1 Rank : Rank of a rectangular or square matrix is the greatest order r of a nonzero minor of a matrix. In other words, a matrix A has rank r if

(1) there exists at least one nonzero minor of order r;

(2) all minors of order $(r+1)$ and higher are equal to zero .

If A denote the matrix, the rank is denoted either by rank (A) or rank A or $\rho(A)$.

10.2.2 Determination of rank : The following rules are useful while determining rank of a matrix:

(1) Begin with minors of smaller to larger orders (even with minors of order 1 i.e. with individual elements).

(2) Suppose nonzero minor D of order, r is found then compute minors of order $(r+1)$ bordering the minor D. If all these bordered minors of order $(r+1)$ are zero, then the required rank is r. But if one of the minors of order $(r+1)$ is found nonzero, then this procedure is to be repeated and the required rank is definitely greater than r.

10.2.3 Maximum possible rank : The smaller of m and n of an $m \times n$ matrix is called the *maximum possible rank* while for an n square nonsingular matrix, it is n and for singular matrix it is less than n. Thus the maximum possible rank of 4×6 matrix is 4.

10.2.4 Full rank : Full rank of a matrix is the maximum possible rank.

10.2.5 Nullity : Complement of the rank with respect to the order of the matrix A is called the *nullity* of the matrix and is denoted by $\nu(A)$. Thus if r is the rank of an n square matrix then nullity, $\nu(A)$ is $(n - r)$ = (order, n) –(rank, r).

The rectangular $m \times n$ matrix possesses two nullities—

(i) row nullity and (ii) column nullity.

Thus if r is the rank of a rectangular $m \times n$ matrix with $r < m$ or n then

row nullity = $(m-r)$ and column nullity = $(n-r)$.

In other words,

(rank A + row nullity) = m, number of rows of A and

(rank A + column nullity) = n, number of columns of A.

Nullity is also defined as the dimension of null space. Null space consists of the solution vectors of the homogeneous system of linear equations $Ax = 0$. This nullity is actually the column nullity of $A_{m \times n}$. More on nullity will be obtained under transformation. Readers are advised to relate among the statements in different context.

Illustrations : From sec. 10.2 & its subsections following can be easily seen.

(i) An $m \times n$ null matrix is of rank 0, row nullity = m and column nullity = n. Thus only null matrix has rank zero and any nonzero matrix has at least rank one.

(ii) A nonzero row or column vector which is of order $m \times 1$ or $1 \times n$ has rank 1 and it is the maximum possible rank and is of full rank. The nullity of such vector is either $(m - 1)$ or $(n - 1)$.

(iii) The unit matrix, I_n has rank n and nullity 0.

(iv) The $m \times n$ matrix whose every element is unity has rank 1 and

row nullity = $(m-1)$ and column nullity = $(n-1)$.

Ex. Find the rank and the nullities of the 3×4 matrix

$$\begin{bmatrix} 1 & 2 & 1 & -1 \\ 0 & -1 & -2 & 3 \\ 2 & 3 & 0 & 1 \end{bmatrix}.$$

Let A denote the matrix. Let nonzero 1×1 minor be 1 in the (1,1) position. Bordering this 1 in (1,1) position the 2 square minors are

$$\begin{vmatrix} 1 & 2 \\ 0 & -1 \end{vmatrix} = -1 \neq 0, \qquad \begin{vmatrix} 2 & 1 \\ -1 & -2 \end{vmatrix} \neq 0 .$$

Other possible second order minors can be easily checked to be nonzero. Now bordering the nonzero 2×2 minor

$\begin{vmatrix} 1 & 2 \\ 0 & -1 \end{vmatrix}$, the 3×3 minor is $\begin{vmatrix} 1 & 2 & 1 \\ 0 & -1 & -2 \\ 2 & 3 & 0 \end{vmatrix} = 0$. Similarly, bordering nonzero 2×2 minor

$\begin{vmatrix} 2 & 1 \\ -1 & -2 \end{vmatrix}$, the 3×3 minors are $\begin{vmatrix} 1 & 2 & 1 \\ 0 & -1 & -2 \\ 2 & 3 & 0 \end{vmatrix} = 0$ and $\begin{vmatrix} 1 & 2 & 1 \\ 0 & -1 & -2 \\ 2 & 3 & 0 \end{vmatrix} = 0$.

∴ no more possible 3×3 minor exists, hence ,
rank (A) = 2. Then row nullity (A) = 3 – 2 = 1 and column nullity = 4 – 2 = 2.

10.3 Rank characterizations, its inequalities & equalities

The objective of this section is to enlist the characteristics of rank and its properties in terms of inequalities and equalities.

10.3.1. Rank characterizations : The following statements about a given $m \times n$ matrix A are equivalent and each can be useful in different context.

(i) Rank A = r;

(ii) There exist r and no more than r rows (columns) of A that constitute a linearly independent set;

(iii) There is a $r \times r$ nonzero minor of A and all $(r+1)\times(r+1)$ zero minors of A;

(iv) The dimension of range of A is r ;

(v) There is a set of r but no more than r linearly independent vectors **b** such that the linear system $A\mathbf{x} = \mathbf{b}$ is consistent;

(vi) r = (n, number of columns of A) – (nullity , the dimension of null space of A) i.e. for any matrix A, column rank + nullity = number of columns.

10.3.2 Rank inequalities & equalities : Some fundamental inequalities and equalities about rank are collected together. Some of them are restated from above sections, some from following sections and rest are just stated for their importance.

Inequalities:

(i) Rank A $\le \min\{m,n\}$ for an $m \times n$ matrix A.

(ii) Rank of a partitioned block cannot exceed the rank of the original matrix.

(iii) If A and B are $m \times n$ and $n \times p$ matrices then

$$(\text{rank A} + \text{rank B} - n) \le \text{rank AB} \le \min\{\text{rank A}, \text{rank B}\}.$$

It provides a bound for the rank of a product.

(iv) Fröbenius extended (iii) as rank AB + rank BC ≤ rank B + rank ABC for matrices A, B, C of orders $m \times n$, $n \times p$, $p \times q$ respectively .

(v) If A and B are $m \times n$ matrices, not necessarily of full rank, then

(a) rank (A + B) ≤ rank A + rank B,

(b) rank(A– B) ≥ |rank A – rank B|.

(vi) For n square matrix A, rank $(A^2) \le$ rank (A).

Equalities :

(i) Rank $A^* = \text{rank } A^T = \text{rank } \overline{A} = \text{rank } A$ for $m \times n$ C - matrix A.

(ii) If B is an $m \times n$ matrix while A and C are nonsingular m and n square matrices, then

$$\text{rank AB} = \text{rank B} = \text{rank BC} = \text{rank ABC}$$

i.e. rank is unchanged upon post and pre as well as both multiplication by a non-singular matrix (cf. converse theorem 5 of sec.10.5).

(iii) If A and B are both $m \times n$ matrices, then rank A = rank B iff there exist nonsingular m square matrix X and nonsingular n square matrix Y such that

$$B = XAY.$$

(iv) If A is an $m \times n$ C-matrix, then rank (A^*A) = rank A .

(v) If an $m \times n$ matrix A has rank r, then A = XBY where B is an n square nonsingular matrix and X and Y are $m \times n$ and $n \times p$ respectively.

10.4 Theorems on rank

Five theorems on rank are stated and proved in this section.

Theorem 1. The order and the rank of a matrix remain unchanged by an elementary transformation.

Proof. Since elementary transformations do not decrease or increase the number of rows and columns of a matrix. So, the order of a matrix remains unchanged by elementary transformations. Proof for rank is considered below in three steps according to three different types of elementary transforms.

Step 1. Some elementary transformation of type I on a determinant changes sign only, so, all the minors may at best change the sign or are left unchanged due to interchange of rows (columns). Hence all nonzero minors remain nonzero and zero
minors remain zero. As no new zero minors are created, so the rank remains unchanged.

Step 2. For the elementary transformation of type II, the minors of the matrix are either left unchanged by multiplication of a row (column) by $c \neq 0$ or are multiplied by c. Again, all nonzero minors remain nonzero and zero minors remain zero. As no new zero minors are created, so the rank remains unchanged.

Step 3. Since elementary transformation of type III does not change the value of the determinants, so arguing as the above two steps, it can be easily concluded that in this case also, the rank remains unchanged.

Hence the proof is complete.

Ex. Using elementary transformation, find the rank of the matrix

$$A = \begin{bmatrix} 1 & 2 & 1 & 4 \\ 0 & 1 & 1 & -1 \\ 1 & -1 & 0 & 2 \end{bmatrix}.$$

Applying elementary transformations one at a time we get

$$A = \begin{bmatrix} 1 & 2 & 1 & 4 \\ 0 & 1 & 1 & -1 \\ 1 & -1 & 0 & 2 \end{bmatrix} \underset{R_3 - R_1}{\approx} \begin{bmatrix} 1 & 2 & 1 & 4 \\ 0 & 1 & 1 & -1 \\ 0 & -3 & -1 & -2 \end{bmatrix} \underset{R_3 + 2R_2}{\approx} \begin{bmatrix} 1 & 2 & 1 & 4 \\ 0 & 1 & 1 & -1 \\ 0 & 0 & 2 & -5 \end{bmatrix}$$

$$\underset{\approx}{R_1 - 2R_2} \begin{bmatrix} 1 & 0 & -1 & 6 \\ 0 & 1 & 1 & -1 \\ 0 & 0 & 2 & -5 \end{bmatrix} \underset{\approx}{C_3 + C_1} \begin{bmatrix} 1 & 0 & 0 & 6 \\ 0 & 1 & 1 & -1 \\ 0 & 0 & 2 & -5 \end{bmatrix} \underset{\approx}{C_4 - 6C_1} \begin{bmatrix} 1 & 0 & 0 & 0 \\ 0 & 1 & 1 & -1 \\ 0 & 0 & 2 & -5 \end{bmatrix}$$

$$\underset{\approx}{C_3-C_2}\begin{bmatrix}1&0&0&0\\0&1&0&-1\\0&0&2&-5\end{bmatrix}\underset{\approx}{C_4+C_2}\begin{bmatrix}1&0&0&0\\0&1&0&0\\0&0&2&-5\end{bmatrix}\underset{\frac{1}{2}R_3}{\approx}\begin{bmatrix}1&0&0&0\\0&1&0&0\\0&0&1&-\frac{5}{2}\end{bmatrix}.$$

Here $(-\frac{5}{2})$ is 1×1 nonzero minor. Its bordering 2×2 minor is $\begin{vmatrix}0&0\\1&-\frac{5}{2}\end{vmatrix}=0$.

Again 1 in (1,1) position is 1×1 nonzero minor. Its bordering 2×2 minor is

$\begin{bmatrix}1&0\\0&1\end{bmatrix}\neq 0$. So 3×3 minor bordering $\begin{vmatrix}1&0\\0&1\end{vmatrix}$ is $\begin{vmatrix}1&0&0\\0&1&0\\0&0&1\end{vmatrix}=1\neq 0$ and no other possible 3×3 bordering minor exists. So, rank (A) = 3.

N.B. Only elementary row (column) operation will give rank from the number of nonzero rows (columns).

Theorem 2. The order and the rank of a matrix A remain unchanged by pre or post multiplying A by elementary matrices .

Proof. Since pre or post multiplication of a matrix by an elementary matrix is nothing but performing elementary transformation of the matrix A. So, this theorem is a direct consequence of theorem 1 already established.

Theorem 3. Let A be a matrix of any size and its rank be $r \leq$ (full rank of A) . Let P and Q be nonsingular matrices such that both are conformable for various multiplication with A. Then

(i) rank (PA) = rank A,

(ii) rank (AQ) = rank A and

(iii) rank (PAQ) = rank A.

Proof . Since A is of any size and is conformable for various multiplication with P and Q, so premultiply A by the nonsingular matrix P which is a product of elementary matrices (cf. theorem 2 of sec.9.4.1), does not alter the rank of A by theorem 2 above. Hence rank (PA) = rank A.

Similarly, post multiplication of A by Q does not alter the rank of A by theorem 2 again. Hence rank (AQ) = rank A.

Now post multiplying PA by Q and using (ii) we have rank (PAQ) = rank A .

Cor. Combining (i), (ii) , (iii) we can conclude that

rank (PA) = rank (AQ) = rank (PAQ) for any A with rank $r \leq$ (full rank of A.).

Theorem 4. If A is a $m\times n$ matrix with rank (A) = r and P and Q are nonsingular matrices of orders $m\times m$ & $n\times n$ respectively then

$$PAQ=\begin{bmatrix}I_r & 0_{r,n-r}\\ 0_{m-r,r} & 0_{m-r,n-r}\end{bmatrix},\ r<m,\ r<n.$$

where I_r is the rth order unit matrix and $0_{k\times l}$ is a null matrix with k rows and l columns.

Proof. Since rank (PAQ) = rank A and the matrix A can be reduced to the normal form

$$\begin{bmatrix}I_r & 0_{r,n-r}\\ 0_{m-r,r} & 0_{m-r,n-r}\end{bmatrix}$$

by elementary transformation. This normal form shows that rank A = r. Also as P and Q are non-singular, i.e. each of them is a product of elementary matrices, so by the theorem 2 and 3 of sec .9.4.1, PAQ is reduced to the same normal with rank (PAQ) = r.

Hence $$PAQ = \begin{bmatrix} I_r & O_{r,n-r} \\ O_{m-r,r} & O_{m-r,n-r} \end{bmatrix}.$$

Cor. (i) If $r = m$, the second row of the partitioned matrix on the right, is deleted. Similarly, second column is omitted when $r = n$. Thus for $m = n = r$, the normal form reduces to I_n. In that case, $P = A^{-1}$, $Q = I_n$.

Cor. (ii) The matrices P and Q are not uniquely determined.

Theorem 5 (on rank of direct sum): If the block matrices A and B are of rank r_1 and r_2 respectively, then

$$\text{rank}\left(\begin{bmatrix} A & O \\ 0 & B \end{bmatrix}\right) = \text{rank A} + \text{rank B}.$$

or $$\text{rank } (A \oplus B) = \text{rank A} + \text{rank B}$$

i.e. rank of a direct sum is the sum of individual ranks .

Proof. Since rank A = r_1 then A has r_1 independent rows or r_1 nonzero rows. Similarly, B has r_2 nonzero rows. Also the elementary transformation does not alter the rank, so by appropriate elementary transformation the given matrix can be put in the normal form

$$R_{r_1+r_2} = \begin{bmatrix} I_{r_1+r_2} & 0 \\ 0 & 0 \end{bmatrix}$$

$$\therefore \quad \text{rank}(R_{r_1+r_2}) = r_1 + r_2 = \text{rank A} + \text{rank B}.$$

Hence $$\text{rank}\left(\begin{bmatrix} A & 0 \\ 0 & B \end{bmatrix}\right) = \text{rank A} + \text{rank B}$$

i.e. $$\text{rank } (A \oplus B) = \text{rank A} + \text{rank B}.$$

10.4.1 Equality of row & column rank : The following theorem establishes that the row rank and the column rank of any matrix are equal.

Theorem. The row rank and the column rank of a matrix are equal.

Proof. Since the zero matrix has zero rank and it becomes obvious that both row and column ranks are equal. Let A be any nonzero matrix. Then there exist nonsingular matrices P and Q (theorem 2 & 3 in 9.4.1) such that PAQ reduces to the normal form

$$PAQ = \begin{bmatrix} I_r & 0 \\ 0 & 0 \end{bmatrix} \tag{10.1}$$

This shows that there are r independent rows as well as columns. Hence,

row rank (PAQ) = r = column rank (PAQ).

Again from (10.1), $$PA \equiv \begin{bmatrix} I_r & 0 \\ 0 & 0 \end{bmatrix} Q^{-1}$$

$$= \begin{bmatrix} I_r & 0 \\ 0 & 0 \end{bmatrix} \begin{bmatrix} Q_r \\ \hline Q_{n-r} \end{bmatrix}$$

where $Q^{-1} \equiv \begin{bmatrix} Q_r \\ \hline Q_{n-r} \end{bmatrix}$ and Q_1 contains first r rows of Q^{-1}

i.e. $$PA \equiv \begin{bmatrix} Q_r \\ 0 \end{bmatrix}$$

$\therefore$ row rank (PA) = row rank A.

$\therefore$ row rank (A) = r.

Also row rank A = row rank (A^T) = column rank A.

$\therefore$ row rank A = r = column rank A.

Ex. Show that the row rank and the column rank of

$$A = \begin{bmatrix} 1 & 2 & 2 & 0 \\ 0 & 1 & 2 & 1 \\ -2 & -1 & 2 & 3 \end{bmatrix}$$

are equal. Hence obtain rank A and nullities of A.

By elementary row operations, A yields,

$$A \underset{\approx}{R_3+2R_1} \begin{bmatrix} 1 & 2 & 2 & 0 \\ 0 & 1 & 2 & 1 \\ 0 & 3 & 6 & 3 \end{bmatrix} \begin{matrix} R_1-2R_2 \\ \approx \\ R_3-3R_2 \end{matrix} \left[\begin{array}{cc|cc} 1 & 0 & -2 & -2 \\ 0 & 1 & 2 & 1 \\ \hline 0 & 0 & 0 & 0 \end{array}\right]$$

which shows that 1st and 2nd rows are nonzero, so row rank A = 2. The 3rd and 4th columns are linear combinations of 1st and 2nd columns i.e.

3rd column = (–2)×1st col.+ 2×2nd col.

and 4th column = (–2) ×1st. column.+ 2nd column.

Therefore, there are two linearly independent columns or two nonzero columns and further elementary column operations give

$$A \approx \left[\begin{array}{cc|cc} 1 & 0 & 0 & 0 \\ 0 & 1 & 0 & 0 \\ \hline 0 & 0 & 0 & 0 \end{array}\right].$$

Hence, column rank (A) = 2

$\therefore$ row rank (A) = 2 = column rank(A).

$\therefore$ row nullity A = 3–2 = 1 and column nullity A = 4 –2 = 2.

10.4.2 Determination of P and Q of theorem 4 in sec. 10.40.1 : Unfortunately, in most cases, it is laborious and time consuming to obtain a pair of nonsingular matrices P and Q such that PAQ reduces to normal form. A procedure is being described here which is more amenable to routine calculations.

Let A be an $m \times n$ matrix of rank r which means that r rows of A are linearly independent. If these r rows were not together in the top, then premultiplying A by elementary matrices P_1 i.e. P_1 A will collect all r rows together in the top. These r rows now form an $r \times n$ matrix of rank r and this $r \times n$ matrix must have r independent columns as the rank is r. If these r columns were not together in the left, by post multiplication of P_1 A by elementary matrices Q_1, i.e. P1AQ will collect all r columns together in the left. Hence

$$B = PAQ = \begin{bmatrix} K_{r\times r} & L_{r,n-r} \\ M_{(m-r)r} & N_{(m-r)(n-r)} \end{bmatrix} \tag{10.2}$$

where $K_{r\times r}$ is nonsingular. Let us define

$$P_2 = \begin{bmatrix} I_r & O \\ -MK^{-1} & I_{m-r} \end{bmatrix} \text{ and } Q_2 = \begin{bmatrix} K^{-1} & -K^{-1}L \\ O_{m-r,r} & I_{m-r} \end{bmatrix}$$

$$\therefore \quad P_2BQ_2 = \begin{bmatrix} I_r & O \\ -MK^{-1} & I_{m-r} \end{bmatrix}\begin{bmatrix} K_{r\times r} & L_{r,n-r} \\ M_{(m-r)r} & N_{(m-r)(n-r)} \end{bmatrix}\begin{bmatrix} K^{-1} & -K^{-1}L \\ O_{m-r,r} & I_{m-r} \end{bmatrix}$$

$$= \begin{bmatrix} K & L \\ O & N-MK^{-1}L \end{bmatrix}\begin{bmatrix} K^{-1} & -K^{-1}L \\ O_{m-r,r} & I_{m-r} \end{bmatrix}$$

$$= \begin{bmatrix} I & O \\ O & N-MK^{-1}L \end{bmatrix} \tag{10.3}$$

But by theorem 3,

$$\text{rank } P_2BQ_2 = \text{rank B} = \text{rank A} = r \tag{10.4}$$

Since $\quad \text{rank}\begin{bmatrix} X & 0 \\ 0 & Y \end{bmatrix} = \text{rank X} + \text{rank Y},$

so from (10.3) it is easy to infer that

$$N - MK^{-1}L = 0$$

otherwise (10.4) cannot hold true. Therefore, (10.3) becomes

$$P_2BQ_2 = \begin{bmatrix} I & O \\ O & O \end{bmatrix}.$$

or $\quad P_2P_1AQ_1Q_2 = \begin{bmatrix} I & O \\ O & O \end{bmatrix}$, using $B = P_1AQ_1$ from (10.2)

or $\quad PAQ = \begin{bmatrix} I & 0 \\ 0 & 0 \end{bmatrix}$, where $P = P_2P_1$ and $Q = Q_1Q_2$

Ex. Find nonsingular matrices P and Q so that PAQ is in normal form where

$$A = \begin{bmatrix} 1 & 0 & 2 & 3 & -1 \\ 2 & 1 & 0 & 2 & 0 \\ -4 & -3 & 4 & 0 & -2 \\ 1 & 1 & -2 & -1 & 1 \end{bmatrix}.$$

$$\text{Now } A = \begin{bmatrix} 1 & 0 & 2 & 3 & -1 \\ 2 & 1 & 0 & 2 & 0 \\ -4 & -3 & 4 & 0 & -2 \\ 1 & 1 & -2 & -1 & 1 \end{bmatrix} \underset{\substack{R_3+4R_1 \\ R_4-R_1}}{\overset{R_2-2R_1}{\approx}} \begin{bmatrix} 1 & 0 & 2 & 3 & -1 \\ 0 & 1 & -4 & -4 & 2 \\ 0 & -3 & 12 & 12 & -6 \\ 0 & 1 & -4 & -4 & 2 \end{bmatrix} \underset{R_4-R_2}{\overset{R_3+3R_2}{\approx}}$$

$$\begin{bmatrix} 1 & 0 & 2 & 3 & -1 \\ 0 & 1 & -4 & -4 & 2 \\ 0 & 0 & 0 & 0 & 0 \\ 0 & 0 & 0 & 0 & 0 \end{bmatrix}.$$

$\therefore$ rank(A) = 2 as $\begin{vmatrix} 1 & 0 \\ 0 & 1 \end{vmatrix} = 1 \neq 0$ and higher order minors are all zero.

$\therefore$ $m = 4$, $n = 5$, $r = 2$. So, let us partition the matrix A into four blocks in such a way that block in left top corner is nonsingular of rank = r = 2.

$$\text{Let } A = \left[\begin{array}{cc|ccc} 1 & 0 & 2 & 3 & -1 \\ 2 & 1 & 0 & 2 & 0 \\ \hline -4 & -3 & 4 & 0 & -2 \\ 1 & 1 & -2 & -1 & 1 \end{array}\right] = \left[\begin{array}{c|c} K & L \\ \hline M & N \end{array}\right] \text{ where } K_{2\times 2} = \begin{bmatrix} 1 & 0 \\ 2 & 1 \end{bmatrix} \text{ is nonsingular.}$$

$$\text{and } K^{-1} = \begin{bmatrix} 1 & 0 \\ -2 & 1 \end{bmatrix}, \ MK^{-1} = \begin{bmatrix} 2 & -3 \\ -1 & 1 \end{bmatrix} \ \& \ K^{-1}L = \begin{bmatrix} 2 & 3 & -1 \\ -4 & -4 & 2 \end{bmatrix}.$$

Since no interchange is required, so $P_2 = P$ and $Q_2 = Q$.

$$\therefore \quad P_2 = \begin{bmatrix} I_2 & O \\ -MK^{-1} & I_{4-2=2} \end{bmatrix} = \left[\begin{array}{cc|cc} 1 & 0 & 0 & 0 \\ 0 & 1 & 0 & 0 \\ \hline -2 & 3 & 1 & 0 \\ 1 & -1 & 0 & 1 \end{array}\right]$$

and

$$Q_2 = \begin{bmatrix} K_2 & -K^{-1}L \\ 0_{5-2,2=3,2} & I_3 \end{bmatrix} = \left[\begin{array}{cc|ccc} 1 & 0 & -2 & -3 & 1 \\ -2 & 1 & 4 & 4 & -2 \\ \hline 0 & 0 & 1 & 0 & 0 \\ 0 & 0 & 0 & 1 & 0 \\ 0 & 0 & 0 & 0 & 1 \end{array}\right]$$

Now it can be verified that

$$P_2BQ_2 = \begin{bmatrix} I & O \\ O & O \end{bmatrix}.$$

10.4.3 Rank condition for equivalent matrices: The theorem below describes how an equivalent matrix B can be obtained from a given matrix A and finally gives the rank condition for the equivalence of two matrices.

Theorem. The $m\times n$ matrix A and B are equivalent if B can be obtained by pre and post multiplying A by a finite number of elementary matrices.

Proof . (This is an alternate proof of Theorem 3 of sec .9.4.1).Let A and B are equivalent and both have rank r.

Then by theorem 4 of sec.10.4, there exist nonsingular matrices C, D ,E and F such that

$$\text{CAD} = \begin{bmatrix} I_r & O \\ O & O \end{bmatrix} \quad \text{and} \quad \text{EBF} = \begin{bmatrix} I_r & 0 \\ 0 & 0 \end{bmatrix}.$$

$\therefore$ CAD = EBF or $E^{-1}CADF^{-1} = B$.

Since by hypothesis C, D, E, F are nonsigular, so each of them is a product of elementary matrices. The matrices $E^{-1}C$ and DF^{-1} are also products of elementary matrices. Therefore , B is obtained from A by pre and post multiplying A by finite set of elementary matrices.

Conversely, let there exist nonsingular matrices P and Q such that PAQ = B. Then the rank of A remains unchanged by pre and post multiplication of A by P and Q respectively.

That is, rank (PAQ) = rank A = rank B.

Thus, A and B are equivalent.

Cor. Two $m \times n$ matrices A and B are equivalent iff they have the same rank.

10.5 Rank of sum and product of matrices

The ranks of the sum and the product of two conformable matrices are determined by theorems in subsections 10.5.1 and 10.5.2 respectively.

10.5.1 Rank of a sum : The rank of a sum of matrices is estimated by following theorem.

Theorem 1. The rank of a sum of matrices cannot exceed the sum of ranks of matrices.

Proof . Let A and B be two $m \times n$ matrices. Note that

$$\begin{bmatrix} I_m & I_m \\ O_m & O_m \end{bmatrix} \begin{bmatrix} A \\ B \end{bmatrix} = \begin{bmatrix} A + B \\ O \end{bmatrix} = C, \text{ (say)}.$$

$\therefore$ by theorem 3(i) of sec10.4 and considering row ranks, it follows that

$$\text{rank } (A + B) = \text{rank } C \le \text{rank} \begin{bmatrix} A \\ B \end{bmatrix} \le \text{rank } (A) + \text{rank } (B).$$

i.e. $\quad \text{rank } (A + B) \le \text{rank } (A) + \text{rank } (B).$

Cor. Replacing A by (A –B) in the theorem,

$$\text{rank } ((A - B) + B) \le \text{rank } (A - B) + \text{rank } (B)$$

or $\quad \text{rank } (A) \le \text{rank } (A - B) + \text{rank } (B)$

or $\quad \text{rank } (A - B) \ge \text{rank } (A) - \text{rank } (B).$

10.5.2 Rank of a product : Following theorems estimate the rank of product of matrices .

Theorem 1. Let A be an $m \times n$ matrix with rank r and B be an $n \times p$ matrix. Then the rank of the product AB cannot exceed the rank of either A or B, namely

$$\text{rank } AB \le \text{rank} A \tag{10.5}$$

$$\text{rank } AB \le \text{rank } B \tag{10.6}$$

Proof . Since A is of rank r, then there exist nonsingular matrices P and Q such that

$$PAQ = \begin{bmatrix} I_r & O \\ O & O \end{bmatrix} = R_r \quad \text{(say)} \tag{10.7}$$

$$A = P^{-1} \begin{bmatrix} I_r & O \\ O & O \end{bmatrix} Q^{-1} = P^{-1} R_r Q^{-1} \tag{10.8}$$

$$\therefore \quad AB = P^{-1}R_rQ^{-1}B = P^{-1}(R_rQ^{-1}B). \tag{10.9}$$

Since P is nonsingular, so P^{-1} is nonsingular. Hence by theorem 3(i) of sec. 10.4

$$\text{Rank (AB)} = \text{Rank } (R_rQ^{-1}B)$$
$$= \text{Rank } \{R_r(Q^{-1}B)\} \tag{10.10}$$

Now the last $(m-r)$ rows of R_r are zero rows, so the last $(m-r)$ rows of the product $R_r(Q^{-1}B)$ must be zero. Hence rank (AB) cannot exceed r, the rank of A

i.e. $\quad$ rank (AB) $\leq$ rank A .

Similarly, it can be proved that rank (AB) $\leq$ rank B.

Hence the theorem is proved.

Cor. (i) The theorem gives an estimate of the rank of the product

(ii) Eqn. (10.10) gives a procedure for finding rank (AB).

(iii) rank (AB) = min {rank A , rank B}.

Ex.1 Find the rank (AB) and verify the theorem where

$$A = \begin{bmatrix} 1 & 0 & 2 & 3 & -1 \\ 2 & 1 & 0 & 2 & 0 \\ -4 & -3 & 4 & 0 & -2 \\ 1 & 1 & -2 & -1 & 1 \end{bmatrix}_{4\times5} \quad \text{and} \quad B = \begin{bmatrix} 1 & 2 & 3 & 4 \\ 0 & 1 & -1 & 0 \\ 1 & 0 & 1 & 2 \\ 1 & 1 & 0 & 2 \\ 2 & 2 & 0 & 4 \end{bmatrix}_{5\times4}$$

The rank (A) =2 and rand (B) =3. (cf. w.o ex. under sec .10.4.2 & 10. 6). Now

$$AB = \begin{bmatrix} 4 & 3 & 5 & 10 \\ 4 & 7 & 5 & 12 \\ -4 & -15 & -5 & -16 \\ 0 & 4 & 0 & 2 \end{bmatrix} \approx \begin{bmatrix} 4 & 3 & 5 & 10 \\ 0 & 4 & 0 & 2 \\ 0 & 0 & 0 & 0 \\ 0 & 0 & 0 & 0 \end{bmatrix}$$

$\therefore$ rank (AB) =2 = rank (A) and < rank (B)

$\therefore$ rank (AB) does not exceed the rank of either A or B . Hence verified.

Also note that rank (AB) = min {rank (A) , rank (B) } = min { 2,3} = 2 which verifies cor. (iii).

Theorem 2. If A is an $m \times n$ matrix of rank r and B is an $n \times p$ matrix such that AB = 0 then

(i) rank B $\leq (n-r)$ i.e. rank cannot exceed $(n-r)$ and

(ii) there exists a matrix B of rank $(n-r)$ such that AB = 0.

Proof. Since rank (A) = r, so by theorem 4 0f sec 10.4, there exist nonsingular matrices P and Q such that

$$PAQ = \begin{bmatrix} I_r & 0_{r,n-r} \\ 0_{m-r,r} & 0_{m-r,n-r} \end{bmatrix} = R_r \text{ (say)} \tag{10.11}$$

(i) To prove the first case, let AB = 0 . Then obviously

$$A\,I\,B = 0 \text{ or } AQQ^{-1}B = 0\,.$$

Hence $\quad PAQQ^{-1}B = 0,\quad$ premultiplying by P.

or $\quad (R_rQ^{-1}B) = 0$

Since Q^{-1} is nonsingular, so rank $Q^{-1}B$ = rank B by theorem 3(i) of sec 10.4.

Let $Q^{-1}B = Y.$

Then $R_r Y = 0.$

or $$\begin{bmatrix} I_r & O \\ O & O \end{bmatrix}\begin{bmatrix} Y_{r,k} & Y_{r,p-k} \\ Y_{n-r,k} & Y_{n-r,p-k} \end{bmatrix} = 0, \quad 0 \le k \le p \text{ by (10.11)} \qquad (10.12)$$

Since R.H.S is a null matrix. $Y_{r,k} = 0$ and $Y_{r,p-k} = 0$ and $Y_{n-r,k}, Y_{n-r,p-k}$ are arbitrary blocks with $(n - r)$ rows. Thus rank $Y \le (n - r)$. Hence by theorem 3, rank B $\le (n-r)$, i.e. rank B cannot exceed $(n-r)$, which proves the first part (i) of the theorem.

(ii) Let us construct an $n \times p$ matrix Y with the partitioned blocks as shown in eqn. (10.12) such that $Y_{r,k}$, $Y_{r,p-k}$ are zero blocks while $Y_{n-r,k}$ is of full rank $(n-r)$ and $Y_{n-r,p-k}$ is arbitrary. Thus, rank Y= $(n-r)$, by construction.

Then $$R_r Y = \begin{bmatrix} I_r & O \\ O & O \end{bmatrix}\begin{bmatrix} O & O \\ Y_{n-r,k} & Y_{n-r,p-k} \end{bmatrix} = 0$$

or PAQY= 0 by (10.5)

or AQY= 0 premultiplying by P^{-1}

or AB = 0, where B = QY and Y= $Q^{-1}B$

i.e. rank Y= rank $Q^{-1}B$ = rank B

$\therefore$ B has been constructed with rank $(n-r)$ such that AB = 0.

Ex.2 Show that A is singular if

$$A = \begin{bmatrix} 2 & 2 & 4 \\ 1 & 2 & 3 \\ 3 & -3 & 0 \end{bmatrix}.$$

Determine a nonzero singular matrix B such that AB = 0.

First part :

$$A = \begin{bmatrix} 2 & 2 & 4 \\ 1 & 2 & 3 \\ 3 & -3 & 0 \end{bmatrix} \underset{\approx}{R_2 - \tfrac{1}{2}R_1} \begin{bmatrix} 2 & 2 & 4 \\ 0 & 1 & 1 \\ 3 & -3 & 0 \end{bmatrix} \underset{\approx}{C_2 - C_1} \begin{bmatrix} 2 & 0 & 4 \\ 0 & 1 & 1 \\ 3 & -6 & 0 \end{bmatrix}$$

$$\underset{\approx}{R_3 - \tfrac{3}{2}R_1} \begin{bmatrix} 2 & 0 & 4 \\ 0 & 1 & 1 \\ 0 & -6 & -6 \end{bmatrix} \underset{\approx}{C_3 - 2C_1} \begin{bmatrix} 2 & 0 & 0 \\ 0 & 1 & 1 \\ 0 & -6 & -6 \end{bmatrix} \underset{\approx}{R_3 - 6R_2} \begin{bmatrix} 2 & 0 & 0 \\ 0 & 1 & 1 \\ 0 & 0 & 0 \end{bmatrix}.$$

$$\underset{\approx}{C_3 - C_2} \begin{bmatrix} 2 & 0 & 0 \\ 0 & 1 & 0 \\ 0 & 0 & 0 \end{bmatrix} \underset{\approx}{\tfrac{1}{2}R_1} \begin{bmatrix} 1 & 0 & 0 \\ 0 & 1 & 0 \\ 0 & 0 & 0 \end{bmatrix} \approx \begin{bmatrix} I_2 & 0 \\ 0 & 0 \end{bmatrix}.$$

Since the equivalent matrix of A has one zero row as well as one zero column, so det A = 0. Hence A is a singular matrix. Also, det $I_2 = 1 \ne 0$. Therefore, rank A = 2.

Second part : By second part theorem 3(i) of sec. 10.4 ,

$$\text{rank } B \le n - r = 3 - 2 = 1.$$

Again B is a nonzero matrix, so rank B must be one i.e. rank (B) =1.
Also by theorem 3 (i) of sec. 10.4, B = QY where Q is elementary column transforrmation. From the first part,

$$Q=\begin{bmatrix}1 & -1 & 0\\ 0 & 1 & 0\\ 0 & 0 & 1\end{bmatrix}\begin{bmatrix}1 & 0 & -2\\ 0 & 1 & 0\\ 0 & 0 & 1\end{bmatrix}\begin{bmatrix}1 & 0 & 0\\ 0 & 1 & -1\\ 0 & 0 & 1\end{bmatrix}=\begin{bmatrix}1 & -1 & 1\\ 0 & 1 & -1\\ 0 & 0 & 1\end{bmatrix}$$

$$\therefore \quad B=\begin{bmatrix}1 & -1 & 1\\ 0 & 1 & -1\\ 0 & 0 & 1\end{bmatrix}\begin{bmatrix}O_{2\times 1} & O_{2\times 2}\\ \hline Y_{3-2,1} & Y_{3-2,3-1}\end{bmatrix}_{3\times 3}$$

$\because$ B is singular, so Y must be 3×3.

$$=\left[\begin{array}{cc|c}1 & -1 & 1\\ 0 & 1 & -1\\ \hline 0 & 0 & 1\end{array}\right]\left[\begin{array}{c|cc}0 & 0 & 0\\ 0 & 0 & 0\\ \hline y_{11} & y_{12} & y_{13}\end{array}\right]$$

which shows conformal partitions of Q.

$$\therefore B=\left[\begin{array}{c|c}\begin{bmatrix}1 & -1\\ 0 & 1\end{bmatrix}\begin{bmatrix}0\\0\end{bmatrix}+\begin{bmatrix}1\\-1\end{bmatrix}[y_{11}] & \begin{bmatrix}1 & -1\\ 0 & 1\end{bmatrix}\begin{bmatrix}0 & 0\\ 0 & 0\end{bmatrix}+\begin{bmatrix}1\\-1\end{bmatrix}[y_{12} \;\; y_{13}]\\ \hline [0 \;\; 0]\begin{bmatrix}0\\0\end{bmatrix}+[1]\,[y_{11}] & [0 \;\; 0]\begin{bmatrix}0 & 0\\ 0 & 0\end{bmatrix}+[1]\,[y_{12} \;\; y_{13}]\end{array}\right]$$

$$=\left[\begin{array}{c|cc}y_{11} & y_{12} & y_{13}\\ -y_{11} & -y_{12} & -y_{13}\\ \hline y_{11} & y_{12} & y_{13}\end{array}\right] \quad \text{which is nonsingular.}$$

Here any two columns must be a multiple of the remaining for rank B = 1.

Theorm 3. If A and B are two n square matrices, then

$$\text{rank (AB)} \geq \text{rank (A)} + \text{rank (B)} - n.$$

Proof. Let rank (A) = r. Then there exist nonsingular matrices P and Q such that

$$PAQ=\begin{bmatrix}I_r & O\\ O & O\end{bmatrix}$$

$$\therefore \qquad A=P^{-1}\begin{bmatrix}I_r & O\\ O & O\end{bmatrix}Q^{-1} \tag{10.13}$$

Let
$$C=P^{-1}\begin{bmatrix}0 & 0\\ 0 & I_{n-r}\end{bmatrix}Q^{-1} \tag{10.14}$$

$$\therefore \text{ adding (10.13) \& (10.14)}, \quad A+C = P^{-1}\left\{\begin{bmatrix}I_r & 0\\ 0 & 0\end{bmatrix}+\begin{bmatrix}0 & 0\\ 0 & I_{n-r}\end{bmatrix}\right\}Q^{-1}$$

$$= P^{-1}\left\{\begin{bmatrix}I_r & 0\\ 0 & 0\end{bmatrix}+\begin{bmatrix}0 & 0\\ 0 & I_{n-r}\end{bmatrix}\right\}Q^{-1}$$

$$= P^{-1}I_nQ^{-1} = P^{-1}Q^{-1}$$

$\therefore$ A + C is nonsingular and C is equivalent to

$$\begin{bmatrix} 0 & 0 \\ 0 & I_{n-r} \end{bmatrix} Q^{-1} . \qquad (10.15)$$

Hence rank C = $n - r$.

Since (A + C) is nonsingular, so by theorem 3 (i) of sec 10.4,

$$\begin{aligned} \text{rank B} &= \text{rank ((A+C) B)} \\ &= \text{rank (AB+ CB)} \\ &\leq \text{rank (AB)+rank (CB)}, \text{ by rank of a sum} \\ &\leq \text{rank (AB)+rank (C)} \quad \text{by sec.10.4} \\ &= \text{rank (AB)} + (n - r) \quad \text{by (10.15)} \end{aligned}$$

or $\qquad \text{rank (B)} - n + r \leq \text{rank (AB)}$

or $\qquad \text{rank (AB)} \geq \text{rank (A)} + \text{rank(B)} - n$

Hence proved.

10.5.3 Rank of Kronecker product : The rank of the Kronecker product can be obtained by the theorem below.

Theorem. If A is an $m \times n$ matrix of rank r and B is a $p \times q$ matrix of rank s then

$$\text{rank } (A \otimes B) = rs = \text{rankA rank B}.$$

Proof. Since rank (A) = r, so there are at least r nonzero rows. Then the simplest from of the reduced matrix A is diag (1, 1, $\cdots$, r times, 0, 0, $\cdots$, $(m - r)$ times).

Similarly, the simplest form of B matrix to have rank (B) = s is

diag (1, 1, $\cdots$, s times, 0, 0, $\cdots$, $(p - s)$ times).

Therefore, by definition of the Kronecker product,

A$\otimes$B = diag (1,1, $\cdots$,r times, 0,0, $\cdots$,$(m - r)$ times)$\otimes$diag (1,1, $\cdots$,s times 0, 0, $\cdots$ $(p - s)$ times).

= diag (1, 1, $\cdots$, rs times, 0, 0, $\cdots$, $(mp - rs)$ times) .

$\therefore$ rank (A$\otimes$B) = rs = rank A rank B.

To be specific, let $A_{3\times3} \otimes B_{3\times3}$ = diag (1, 1, 0) $\otimes$ diag (1,1, 1)

$$= \begin{bmatrix} 1 & 0 & 0 \\ 0 & 1 & 0 \\ 0 & 0 & 0 \end{bmatrix} \otimes \begin{bmatrix} 1 & 0 & 0 \\ 0 & 1 & 0 \\ 0 & 0 & 1 \end{bmatrix}$$

$$= \left[\begin{array}{ccc|ccc|ccc} 1 & 0 & 0 & 0 & 0 & 0 & 0 & 0 & 0 \\ 0 & 1 & 0 & 0 & 0 & 0 & 0 & 0 & 0 \\ 0 & 0 & 1 & 0 & 0 & 0 & 0 & 0 & 0 \\ \hline 0 & 0 & 0 & 1 & 0 & 0 & 0 & 0 & 0 \\ 0 & 0 & 0 & 0 & 1 & 0 & 0 & 0 & 0 \\ 0 & 0 & 0 & 0 & 0 & 1 & 0 & 0 & 0 \\ \hline 0 & 0 & 0 & 0 & 0 & 0 & 0 & 0 & 0 \\ 0 & 0 & 0 & 0 & 0 & 0 & 0 & 0 & 0 \\ 0 & 0 & 0 & 0 & 0 & 0 & 0 & 0 & 0 \end{array}\right]_{9\times 9}$$

$\therefore$ rank $(A\otimes B) = 6$

But rank (A) = 2 and rank (B) = 3 .

So, rank $(A\otimes B) = 6 = 2\times 3$ = rank A rank B.

10.6 Sylvester's law of nullity

J.J. Sylvester in 1884 developed the following laws for the nullity of a product of matrices.

Law 1. The nullity of the product of two square matrices is at least as great as the nullity of either factor, i.e. $\nu(AB) \geq \max.(\nu(A), \nu(B))$ (10.16)

Law 2. The nullity of the prduct of two matrices $A_{m\times n}$ and $B_{n\times p}$ is not greater than the sum of their nullities i.e. $\nu(AB) \leq \nu(A) + \nu(B)$. (10.17)

For the square matrices A & B, one can combine (10.16) and (10.17) & the Sylvester's law stands as

$$\max.(\nu(A), \nu(B)) \leq \nu(AB) \leq \nu(A) + \nu(B). \tag{10.18}$$

In words, Sylvester's law in (10.18) states that the nullity of the product of two square matrices is at least as great as the nullity of either factors, and at most as great as the sum of the nullities of the two factors.

Ex.1 Discuss different nullities for the matrix

$$A = \begin{bmatrix} 1 & 0 & 2 & 3 & -1 \\ 2 & 1 & 0 & 2 & 0 \\ -4 & -3 & 4 & 0 & -2 \\ 1 & 1 & -2 & -1 & 1 \end{bmatrix}_{4\times 5}$$

Check that rank (A) = 2, column nullity = 5 – 2 = 3 , row nullity = 4 – 2 = 2.

$\therefore$ dimension of solution space generated by $A_{4\times 5}\, x_{5\times 1} = 0$ is the column nullity of A and it is 3.

Ex. 2 Find the nullity of AB such that

$$A = \begin{bmatrix} 1 & 0 & 2 & 3 & -1 \\ 2 & 1 & 0 & 2 & 0 \\ -4 & -3 & 4 & 0 & -2 \\ 1 & 1 & -2 & -1 & 1 \end{bmatrix}_{4\times 5} \quad \text{and } B = \begin{bmatrix} 1 & 2 & 3 & 4 \\ 0 & 1 & -1 & 0 \\ 1 & 0 & 1 & 2 \\ 1 & 1 & 0 & 2 \\ 2 & 2 & 0 & 4 \end{bmatrix}_{5\times 4}$$

and verify Sylvester's law of nullity.

Now rank (A) = 2. Therefore,

row $\nu(A) = 4 - 2 = 2$ and column $\nu(A) = 5 - 2 = 3$.

Also rank (B) = 3. Therefore, row $\nu(B) = 5 - 3 = 2$ and column $\nu(B) = 4 - 2 = 2$.

$$\text{Now } AB = \begin{bmatrix} 4 & 3 & 5 & 10 \\ 4 & 7 & 5 & 12 \\ -4 & -15 & -5 & -16 \\ 0 & 4 & 0 & 2 \end{bmatrix}_{4\times4}. \text{ Hence rank (AB)} = 2.$$

Therefore, $\nu(AB) = 4 - 2 = 2$.

Hence $\nu(AB) = 2 <$ row $\nu(A)$ + row $\nu(B) = 2 + 2 = 4$.

Also $\nu(AB) = 2 <$ col. $\nu(A)$ + col. $\nu(B) = 3 + 2 = 5$.

Ex. 3 Find $\nu(AB)$, $\nu(BA)$ and verify Sylvester's law of nullity where

$$A = \begin{bmatrix} 1 & 2 & 3 \\ 2 & 2 & 0 \\ 3 & 0 & -2 \end{bmatrix} \text{ and } B = \begin{bmatrix} 2 & 2 & 4 \\ 1 & 2 & 3 \\ 3 & -3 & 0 \end{bmatrix}.$$

Part 1. Rank (A) = 3, rank B = 2. Therefore $\nu(A) = 3 - 3 = 0$, $\nu(B) = 3 - 2 = 1$ and $\nu(A) + \nu(B) = 1$.

Also $$AB = \begin{bmatrix} 13 & -3 & 10 \\ 6 & 8 & 14 \\ 0 & 12 & 12 \end{bmatrix}.$$

$\therefore$ rank (AB) = 3. So, $\nu(AB) = 3 - 3 = 0$.

$\therefore$ $\nu(AB) = 0 < \nu(A) + \nu(B) = 0 + 1 = 1$.

Also as both A & B are square matrices, so max.$\{\nu(A), \nu(B)\}$ = max $\{0, 1\} = 1$.

Part 2.

$$BA = \begin{bmatrix} 2 & 2 & 4 \\ 1 & 2 & 3 \\ 3 & -3 & 0 \end{bmatrix}\begin{bmatrix} 1 & 2 & 3 \\ 2 & 2 & 0 \\ 3 & 0 & -2 \end{bmatrix} = \begin{bmatrix} 18 & 8 & -2 \\ 14 & 6 & -3 \\ -3 & 0 & 9 \end{bmatrix} \underset{\approx}{R_2 - \tfrac{7}{9}R_1} \begin{bmatrix} 18 & 8 & -2 \\ 0 & -\frac{2}{9} & \frac{13}{9} \\ -3 & 0 & 9 \end{bmatrix}$$

$$\underset{\approx}{C_2 - \tfrac{4}{9}C_1} \begin{bmatrix} 18 & 0 & -2 \\ 0 & -\frac{2}{9} & \frac{13}{9} \\ -3 & \frac{4}{3} & 9 \end{bmatrix} \underset{\approx}{R_3 + \tfrac{1}{6}R_1} \begin{bmatrix} 18 & 0 & -2 \\ 0 & -\frac{2}{9} & \frac{13}{9} \\ 0 & \frac{4}{3} & \frac{26}{3} \end{bmatrix} \underset{\approx}{C_3 + \tfrac{1}{9}C_1}$$

$$\begin{bmatrix} 18 & 0 & 0 \\ 0 & -\frac{2}{9} & \frac{13}{9} \\ 0 & \frac{4}{3} & \frac{26}{3} \end{bmatrix} \underset{\approx}{R_3 + 6R_2} \begin{bmatrix} 18 & 0 & 0 \\ 0 & -\frac{2}{9} & \frac{13}{9} \\ 0 & 0 & \frac{52}{3} \end{bmatrix} \underset{\approx}{C_3 + \tfrac{13}{2}C_1} \begin{bmatrix} 18 & 0 & 0 \\ 0 & -\frac{2}{9} & 0 \\ 0 & 0 & \frac{52}{3} \end{bmatrix}.$$

$\therefore$ rank (AB) = 3. So $\nu(AB) = 3 - 3 = 0$.

$\therefore$ $\nu(AB) = 0 < \nu(A) + \nu(B) = 0 + 1 = 1$.

Ex.4 Find the nullity of AB to verify Sylvester's law of nullity such that

$$A=\begin{bmatrix}1 & -1 & 0\\ 2 & -2 & 1\\ 3 & 0 & 1\end{bmatrix} \text{ and } B=\begin{bmatrix}2 & 2 & 4\\ 1 & 2 & 3\\ 3 & 3 & 0\end{bmatrix}.$$

Now $$A=\begin{bmatrix}1 & -1 & 0\\ 2 & -2 & 1\\ 3 & 0 & 1\end{bmatrix}\approx\begin{bmatrix}1 & -1 & 0\\ 0 & 0 & -1\\ 3 & 0 & 1\end{bmatrix}\approx\begin{bmatrix}1 & 0 & 0\\ 0 & 3 & 0\\ 0 & 0 & -1\end{bmatrix}\approx\begin{bmatrix}1 & 0 & 0\\ 0 & 0 & -1\\ 0 & 3 & 1\end{bmatrix}$$

$$\approx\begin{bmatrix}1 & 0 & 0\\ 0 & 3 & 1\\ 0 & 0 & -1\end{bmatrix}\approx\begin{bmatrix}1 & 0 & 0\\ 0 & 3 & 0\\ 0 & 0 & -1\end{bmatrix}$$

$\therefore$ rank (A) = 3 , ν(A) = 0.

Note that

$$\text{rank (B)} = \text{rank}\begin{bmatrix}2 & 2 & 4\\ 1 & 2 & 3\\ 3 & -3 & 0\end{bmatrix}. = 2, \nu(B) = 1.$$

Also $$AB=\begin{bmatrix}1 & -1 & 0\\ 2 & -2 & 1\\ 3 & 0 & 1\end{bmatrix}\begin{bmatrix}2 & 2 & 4\\ 1 & 2 & 3\\ 3 & -3 & 0\end{bmatrix}=\begin{bmatrix}1 & 0 & 1\\ 5 & -3 & 2\\ 9 & 3 & 12\end{bmatrix}\approx\begin{bmatrix}1 & 0 & 1\\ 0 & -3 & -3\\ 0 & 3 & 3\end{bmatrix}$$

$$\approx\begin{bmatrix}1 & 0 & 0\\ 0 & -3 & -3\\ 0 & 3 & 3\end{bmatrix}\approx\begin{bmatrix}1 & 0 & 0\\ 0 & -3 & -3\\ 0 & 0 & 0\end{bmatrix}\approx\begin{bmatrix}1 & 0 & 0\\ 0 & -3 & 0\\ 0 & 0 & 0\end{bmatrix}$$

$\therefore$ rank (AB) = 2, ν(AB) = 1 and ν(AB) = max (ν(A), ν(B)) = max (0, 1) = 1.

Ex.5 Verify Sylvester's law of nullity for the symmetric matrices

$$A=\begin{bmatrix}1 & 2 & 3\\ 2 & 2 & 0\\ 3 & 0 & -2\end{bmatrix} \text{ and } B=\begin{bmatrix}2 & -1 & 1\\ -1 & 0 & 2\\ 1 & 2 & 3\end{bmatrix}.$$

Now

$$A=\begin{bmatrix}1 & 2 & 3\\ 2 & 2 & 0\\ 3 & 0 & -2\end{bmatrix} \text{ gives } \nu(A)=0 \text{. Also}$$

$$B=\begin{bmatrix}2 & -1 & 1\\ -1 & 0 & 2\\ 1 & 2 & 3\end{bmatrix}\xrightarrow{R_2+\frac{1}{2}R_1}\begin{bmatrix}2 & -1 & 1\\ 0 & -\frac{1}{2} & \frac{5}{2}\\ 1 & 2 & 3\end{bmatrix}\xrightarrow{C_2+\frac{1}{2}C_1}$$

$$\begin{bmatrix} 2 & 0 & 1 \\ 0 & -\frac{1}{2} & \frac{5}{2} \\ 1 & \frac{5}{2} & 3 \end{bmatrix} \xrightarrow{R_2-\frac{1}{2}R_1} \begin{bmatrix} 2 & 0 & 1 \\ 0 & -\frac{1}{2} & \frac{5}{2} \\ 0 & \frac{5}{2} & \frac{5}{2} \end{bmatrix} \xrightarrow{C_3-\frac{1}{2}C_1} \begin{bmatrix} 2 & 0 & 0 \\ 0 & -\frac{1}{2} & \frac{5}{2} \\ 0 & \frac{5}{2} & \frac{5}{2} \end{bmatrix}$$

$$\xrightarrow{R_3+5R_2} \begin{bmatrix} 2 & 0 & 0 \\ 0 & -\frac{1}{2} & \frac{5}{2} \\ 0 & 0 & 15 \end{bmatrix} \xrightarrow{C_3+5C_2} \begin{bmatrix} 2 & 0 & 0 \\ 0 & -\frac{1}{2} & 0 \\ 0 & 0 & 15 \end{bmatrix}$$

$\therefore$ rank(B) = 3 , ν(B) = 0 and max.{ ν(A) , ν(B)} = max.{0,0 } = 0 .

$$AB = \begin{bmatrix} 1 & 2 & 3 \\ 2 & 2 & 0 \\ 3 & 0 & -2 \end{bmatrix} \begin{bmatrix} 2 & -1 & 1 \\ -1 & 0 & 2 \\ 1 & 2 & 3 \end{bmatrix} = \begin{bmatrix} 3 & 5 & 14 \\ 2 & -2 & 6 \\ 4 & -7 & -3 \end{bmatrix} \xrightarrow{R_2-\frac{2}{3}R_1}$$

$$\begin{bmatrix} 3 & 5 & 14 \\ 0 & -\frac{16}{3} & -\frac{10}{3} \\ 4 & -7 & -3 \end{bmatrix} \xrightarrow{C_3-\frac{5}{3}C_1} \begin{bmatrix} 3 & 0 & 14 \\ 0 & -\frac{16}{3} & -\frac{10}{3} \\ 4 & -7-\frac{20}{3} & -3 \end{bmatrix} \xrightarrow{R_2-\frac{4}{3}R_1}$$

$$\begin{bmatrix} 3 & 0 & 14 \\ 0 & -\frac{16}{3} & -\frac{10}{3} \\ 4 & -\frac{41}{3} & -\frac{65}{3} \end{bmatrix} \to \begin{bmatrix} 3 & 0 & 0 \\ 0 & -\frac{16}{3} & -\frac{10}{3} \\ 0 & -\frac{41}{3} & -\frac{65}{3} \end{bmatrix}$$

$$\xrightarrow{R_3-\frac{41}{16}R_2} \begin{bmatrix} 3 & 0 & 0 \\ 0 & -\frac{16}{3} & -\frac{10}{3} \\ 0 & 0 & -\frac{65}{3}+\frac{41}{16}\cdot\frac{10}{3} \end{bmatrix} \to \begin{bmatrix} 3 & 0 & 0 \\ 0 & -\frac{16}{3} & 0 \\ 0 & 0 & -\frac{105}{8} \end{bmatrix}.$$

$\therefore$ rank (AB) = 3 , ν(AB) = 0.

$\therefore$ max.{ ν(A), ν(B)} = 0 = ν(AB).

10.7 Gerstein algorithm for computing the rank of a matrix

L.J. Gerstein developed a new algorithm for computing the rank of a matrix & appeared in American Mathematical monthly in December, 1988. This is based on Chio's pivotal condensation for the expansion of determinant. This was not mentioned by Gerstein. Yet he used the sub determinant.

$$d_{ij} = \begin{vmatrix} a_{11} & a_{1j} \\ a_{i1} & a_{ij} \end{vmatrix} = a_{11}a_{ij} - a_{i1}a_{1j},\ 2 \le i \le m,\ 2 \le j \le n. \tag{10.19}$$

which is due to Chio (cf. sec. 6.7.4). The procedure is based on his recursive formula stated in the theorem below:

Theorem. Let A = $[a_{ij}]$ and $a_{11} \neq 0$. Then

$$\text{rank A} = 1 + \text{rank} \begin{bmatrix} d_{22} & d_{23} & \cdots & d_{2n} \\ \vdots & \vdots & \vdots & \vdots \\ d_{m2} & d_{m3} & \cdots & d_{mn} \end{bmatrix} \tag{10.20}$$

where d_{ij} is given by (10.19) .

This theorem is applied recursively until a matrix with only one row or column is obtained.

Proof . Since A = $[a_{ij}]$ and $a_{11} \neq 0$. Then multiplying rows 2 to m by a_{11}, we get,

$$A \approx \begin{bmatrix} a_{11} & a_{12} & \cdots & \cdots & a_{1m} \\ a_{11}a_{21} & a_{11}a_{22} & \cdots & \cdots & a_{11}a_{2n} \\ \vdots & \vdots & \vdots & \vdots & \vdots \\ a_{11}a_{m1} & a_{11}a_{m2} & \cdots & \cdots & a_{11}a_{mn} \end{bmatrix}$$

$$\approx \begin{bmatrix} a_{11} & a_{12} & \cdots & \cdots & a_{1n} \\ 0 & d_{22} & \cdots & \cdots & d_{2n} \\ \vdots & \vdots & \vdots & \vdots & \vdots \\ 0 & d_{m2} & \cdots & \cdots & d_{mn} \end{bmatrix}$$

by subtracting a_{ij} times of 1st row from each row i, $2 \le i \le m$,

$$\approx \left[\begin{array}{c|cccc} a_{11} & a_{12} & \cdots & \cdots & a_{1m} \\ \hline 0 & d_{22} & \cdots & \cdots & d_{2m} \\ \vdots & \vdots & \vdots & \vdots & \vdots \\ 0 & d_{m2} & \cdots & \cdots & d_{mn} \end{array}\right],$$

dividing into four blocks by partitioning after the first row and the first column. So

$$A \approx \left[\begin{array}{c|c} a_{11} & B \\ \hline 0 & D \end{array}\right].$$

Now rank A = rank (a_{11}) + rank D. Since rank (a_{11}) = rank B as both blocks belong to the same first row. So,

$$\text{rank } A = 1 + \text{rank} \begin{bmatrix} d_{22} & d_{23} & \cdots & d_{2n} \\ \vdots & \vdots & \vdots & \vdots \\ d_{m2} & d_{m3} & \cdots & d_{mn} \end{bmatrix}.$$

Ex. Find the rank of the matrix

$$\begin{bmatrix} 1 & 2 & 3 & 4 \\ 0 & 1 & -1 & 0 \\ 1 & 0 & 1 & 2 \\ 1 & 1 & 0 & 2 \\ 2 & 2 & 0 & 4 \end{bmatrix}$$

using the Gerstein method.

Since $a_{11} = 1 \neq 0$. Let us first calculate d_{ij}' s, $2 \le i \le 5$, $2 \le j \le 4$.

$$d_{22} = \begin{vmatrix} a_{11} & a_{12} \\ a_{21} & a_{22} \end{vmatrix} = \begin{vmatrix} 1 & 2 \\ 0 & 1 \end{vmatrix} = 1; \quad d_{23} = \begin{vmatrix} a_{11} & a_{13} \\ a_{21} & a_{23} \end{vmatrix} = \begin{vmatrix} 1 & 3 \\ 0 & -1 \end{vmatrix} = -1;$$

$$d_{24} = \begin{vmatrix} a_{11} & a_{14} \\ a_{21} & a_{24} \end{vmatrix} = \begin{vmatrix} 1 & 4 \\ 0 & 0 \end{vmatrix} = 0;$$

$$d_{32} = \begin{vmatrix} a_{11} & a_{13} \\ a_{31} & a_{32} \end{vmatrix} = \begin{vmatrix} 1 & 2 \\ 1 & 0 \end{vmatrix} = -2; \quad d_{33} = \begin{vmatrix} a_{11} & a_{13} \\ a_{31} & a_{33} \end{vmatrix} = \begin{vmatrix} 1 & 3 \\ 1 & 1 \end{vmatrix} = -2;$$

$$d_{34} = \begin{vmatrix} a_{11} & a_{14} \\ a_{31} & a_{34} \end{vmatrix} = \begin{vmatrix} 1 & 4 \\ 1 & 2 \end{vmatrix} = -2$$

$$d_{42}=\begin{vmatrix} a_{11} & a_{12} \\ a_{41} & a_{42}\end{vmatrix}=\begin{vmatrix} 1 & 2 \\ 1 & 1\end{vmatrix}=-1\,;\; d_{43}=\begin{vmatrix} a_{11} & a_{13} \\ a_{41} & a_{43}\end{vmatrix}=\begin{vmatrix} 1 & 3 \\ 1 & 0\end{vmatrix}=-3\,;$$

$$d_{44}=\begin{vmatrix} a_{11} & a_{14} \\ a_{41} & a_{44}\end{vmatrix}=\begin{vmatrix} 1 & 4 \\ 1 & 2\end{vmatrix}=-2\,;$$

$$d_{52}=\begin{vmatrix} a_{11} & a_{12} \\ a_{51} & a_{52}\end{vmatrix}=\begin{vmatrix} 1 & 2 \\ 2 & 2\end{vmatrix}=-2\,;\; d_{53}=\begin{vmatrix} a_{11} & a_{13} \\ a_{51} & a_{53}\end{vmatrix}=\begin{vmatrix} 1 & 3 \\ 2 & 0\end{vmatrix}=-6\,;$$

$$d_{54}=\begin{vmatrix} a_{11} & a_{14} \\ a_{51} & a_{54}\end{vmatrix}=\begin{vmatrix} 1 & 4 \\ 2 & 4\end{vmatrix}=-4\,;$$

$$\therefore \text{rank A}=1+\text{rank}\begin{bmatrix} d_{22} & d_{23} & d_{24} \\ d_{32} & d_{33} & d_{34} \\ d_{42} & d_{43} & d_{44} \\ d_{52} & d_{53} & d_{54}\end{bmatrix}=1+\text{rank}\begin{bmatrix} 1 & -1 & 0 \\ -2 & -2 & -2 \\ -1 & -3 & -2 \\ -2 & -6 & -4\end{bmatrix}$$

$$=1+1+\text{rank}\begin{bmatrix} 0 & -2 \\ -2 & -2 \\ -8 & -4\end{bmatrix}=2+\text{rank}\begin{bmatrix} -4 \\ -16\end{bmatrix}=2+1=3.$$

N.B. Procedure is simple. It needs no division as in Chio's pivotal condensation. All the time evaluate a second order determinant.

10.8 Rank from echelon form

Rank is equal to the number of nonzero rows (columns) in echelon form of a matrix. This is established by the following theorem.

Theorem. The rank of an $m\times n$ matrix A is the number of nonzero rows (columns) in row (column) echelon form.

Proof. Let $A=[a_{ij}]_{m\times n}$, $a_{11}\neq 0$. Let the $m\times n$ matrix be transformed to row echelon form B by elementary row operations. Let

$$B=\begin{bmatrix} b_{11} & b_{12} & b_{13} & \cdots & b_{1j} & \cdots & b_{1n} \\ 0 & b_{22} & b_{23} & \cdots & b_{2j} & \cdots & b_{2n} \\ \vdots & \vdots & \vdots & \vdots & \vdots & \vdots & \vdots \\ 0 & 0 & 0 & \cdots & b_{kj} & \cdots & b_{kn} \\ \vdots & \vdots & \vdots & \vdots & \vdots & \vdots & \vdots \\ 0 & 0 & 0 & \cdots & \cdots & \cdots & 0\end{bmatrix}_{m\times n} \longleftarrow \text{zero rows}$$

Now rank B = k, number of nonzero rows if it is possible to show that these rows are linearly independent. Let $R_1, R_2, \cdots, R_k$ be the nonzero rows. Now let

$$c_1R_1+c_2R_2+\cdots+c_kR_k=0$$

or $c_1\begin{bmatrix} b_{11} & b_{12} & b_{13} & \cdots & b_{1j} & \cdots & b_{1n}\end{bmatrix}+c_2\begin{bmatrix} 0 & b_{22} & b_{23} & \cdots & b_{2j} & \cdots & b_{2n}\end{bmatrix}+$

$$\cdots+c_k\begin{bmatrix} 0 & 0 & 0 & \cdots & b_{kj} & \cdots & b_{kn}\end{bmatrix}=0$$

which implies $b_{11}c_1 = 0,\ \ b_{12}c_1 + b_{22}c_2 = 0,\ \cdots,\ \ c_1b_{1j} + c_2b_{2j} + \cdots + c_kb_{kj} = 0$.

$\therefore\ c_1 = 0$ as $b_{11} \neq 0$. Similarly, $c_2 = 0, \cdots, c_k = 0$.

Hence rows $R_1, R_2, \cdots, R_k$ are linearly independent. Therefore, rank B = k = number of nonzero rows.

10.8.1 Rank of a matrix deleting some of its columns: From sec. 9.2.2 and 10.8, it is obvious that deletion of some columns of A= $[a_{ij}]_{m\times n}$ deletes those corresponding columns from its RREF B and RREF B_R of the new matrix A_R obtained after deletion can be easily read off from B with and without further row operations. So, evidently rank of B_R can be easily read off from B_R. Rank (A) and rank (A_R) remain the same if no new zero row results otherwise rank, may be different. Thus from examples in sec. 9.2.2, it appears that

$$\text{rank(A)} = 3 = \text{rank}(R_2) = \text{rank}(R_3)$$

while different ranks result in problem 2 in exercises given below.

10.9 Rank condition for solving linear system of equation

This section establishes different rank condition for the solution of linear system of equations. Since the linear system of equation arises as both homogeneous and nonhomogeneous, the discussion about the solution begins with homogeneous and then with nonhomogeneous.

Homogeneous : To introduce rank conditions for the solution of homogeneous linear system of equations, let us first define the rank of such system.

The *rank* of a linear system of homogeneous equations is the rank of its coefficient matrix. On the basis of the rank of a system, the types of solutions are discussed below.

If the rank of a homogeneous linear system equals the number, *n*, of variables then the system has *unique* solution i.e. trivial or zero solution. In other words, if the row reduced coefficient matrix has *n* nonzero rows then the system has unique solution.

If the rank of a linear system of homogeneous equations is *less than* the number, *n* of variables then the system has *more than one* solution. In other words, if the row reduced coefficient matrix has (n–1) nonzero rows then the system has a solution in which at least one variable is not zero. In this case there remains *free variable*(s) corresponding not to leading entry of reduced matrix.

The rank of a homogeneous system with *more unknowns than equations* is less than the number of unknowns and it always has at least one solution in which at least one variable is not zero. This is also a case of free variables.

No. of free variables = (no. of variables) – (rank of system).

Nonhomogeneous : The following theorems and sec. 10.9.1 together collect the rank condition for solving the linear system of nonhomogeneous equations.

Theorem 1. The linear system A**x** = **b** has at least one solution iff the rank of the coefficient matrix A is equal to the rank of the augmented matrix i.e.

$$\text{rank (A)} = \text{rank } [\text{A} \vdots \mathbf{b}].$$

Proof. Let A be an $m\times n$ matrix and let rank (A) = rank $[\text{A} \vdots \mathbf{b}]$. Then row reduced echelon form of $[\text{A} \vdots \mathbf{b}]$ must have at least one row in the form $\begin{bmatrix}0 & 0 & 0 & \cdots & 0 & \vdots & 1\end{bmatrix}$ which means $x_n 0 = 1$, i.e. $0 = 1$, an absurdity.

Therefore, rank (A) must necessarily be equal to rank $[A \vdots \mathbf{b}]$ to have at least one solution.

Conversely, let rank A = rank $[A \vdots \mathbf{b}]$. Let the row reduced echelon form of $[A \vdots \mathbf{b}]$ be $[C \vdots \mathbf{d}]$. Let us discard zero rows in $[C \vdots \mathbf{d}]$ and assume that first $r \times r$ diagonal block contains the leading variables of C. Let the equivalent system be

$$\left.\begin{array}{l} x_1 + \cdots\cdots + c_{1,r+1}x_{r+1} + \cdots\cdots\cdots + c_{1n}x_n = d_1 \\ x_2 + \cdots + c_{2,r+1}x_{r+1} + \cdots\cdots\cdots + c_{2n}x_n = d_2 \\ \vdots \qquad \vdots \qquad \vdots \qquad \vdots \\ x_r + c_{r,r+1}x_{r+1} + \cdots\cdots\cdots + c_{rn}x_n = d_r \end{array}\right\} \tag{10.21}$$

Setting the free variables $x_{r+1} = x_{r+2} = \cdots = x_n = 0$ the one particular solution is obtained as $(d_1, d_2, \cdots, d_r, 0, 0, \cdots, 0)$. Hence the theorem is proved.

Theorem 2. Let A be an $m \times n$ matrix. Then the system $\mathbf{Ax} = \mathbf{b}$ has a unique solution iff rank of coefficient and augmented matrices are both equal to n, the number of unknowns i.e.

$$\text{rank } A = \text{rank } [A \vdots \mathbf{b}] = n.$$

Proof. Let $\mathbf{Ax} = \mathbf{b}$ have a unique solution. Then by theorem above

$$\text{rank } A = \text{rank } [A \vdots \mathbf{b}].$$

Hence it remains to show that rank = n. Since there exists a "1" in each nonzero row of the row reduced n matrix of A and there are at most n such rows. Hence rank A $\leq n$. Suppose rank A $\leq n$. Then the equivalent system is of the form (10.21) in theorem 1 above and there would be infinite solutions. [For example set $x_{r+1} = t, x_{r+2} = \cdots = x_n = 0$ to get solution

$$(d_1 - c_{1,r+1}t, \cdots, d_r - c_{r,r+1}t, t, 0, \cdots, 0)].$$

Thus a contradiction appears. Hence for unique solution rank A must be n.

Conversely, let rank A = rank$[A \vdots \mathbf{b}] = n$. After discarding zero rows of row reduced echelon form, the equivalent system is obtained as

$$\left.\begin{array}{lll} x_1 & & = d_1 \\ & x_2 & = d_2 \\ \cdots & \cdots & \cdots \\ & & x_n = d_n \end{array}\right\}$$

which has unique solution $(d_1, d_2, \cdots, d_n)$. Hence the theorem is proved.

Cor. 1 If the rank (A) = rank $[A \vdots \mathbf{b}]$ but $< n$ then the system $\mathbf{Ax} = \mathbf{b}$, where A is an $m \times n$ matrix, has infinitely many solutions.

Cor. 2 A linear system with more equations than unknowns either has no solution or infinitely many solutions.

Cor. 3 If rank (A) = n then homogeneous system $\mathbf{Ax} = 0$ with A, an $m \times n$ matrix, has a unique solution (zero or trivial one).

Cor. 4 If rank (A) $< n$ then the homogeneous system $\mathbf{Ax} = 0$ with $m \times n$ matrix, has a nontrivial solution (i.e. infinitely many).

Cor. 5 A homogeneous linear system with more unknowns than equations has infinitely many solutions, i.e. if $m < n$ then $\mathbf{Ax} = 0$ always possess nontrivial solutions.

Cor. 6 If A is singular then $\mathbf{Ax} = 0$ possesses nontrivial solutions.

10.9.1 Synopsis of solutions of Ax = **b**: Table of synopsis of the conditions of Ax = **b**, m equations & n unknowns.

Relationship of m & n	Rank (A) = r Rank [A⋮**b**] = p		Solutions to Ax = **b**
$m < n$	$r < m < n$	$r \neq p$	No solution
	$r < m < n$	$r = p$	Infinite solutions
	$r = m$	p must = r	Infinite solutions
$m = n$	$r < m$	$r \neq p$	No solution
	$r < m$	$r = p$	Infinite solutions
	$r = m$	p must = r	Unique solutions
$m > n$	$r \leq n < m$	$r \neq p$	No solution
	$r < n < m$	$r = p$	Infinite solutions
	$r = n$	$r = p$	Unique solutions

10.9.2 Particular and complementary solutions of Ax = **b** : This section identifies two solutions of nonhomogeneous systems of linear equations as in the case of nonhomogeneous linear ordinary differential equations (ODE) with constant coefficients. The procedure for the purpose is completely identical with that of ODE. The following theorem provides with the procedure.

Theorem. Let Ax = **b** be the matrix form of the system of m linear equations with n unknowns. Let rank (A) = $r < n$. Let y be a particular solution Ax = **b** and $x_1, x_2, \cdots, x_{n-r}$ be $(n-r)$ linearly independent complementary solutions of the associated homogeneous equations Ax = 0 . Then the general solution of Ax = **b** is

$$\mathbf{x} = \mathbf{y} + c_1 x_1 + c_2 x_2 + \cdots + c_{n-r} x_{n-r} \tag{10.22}$$

Conversely every vector of the form (10.22) is a solution of Ax = **b**.

Proof. Let x be a solution of Ax = **b**. Since **y** is a given particular solution, then both **x** and **y** must satisfy the equation. So,

$$\mathrm{A}\mathbf{x} = \mathbf{b} \text{ and } \mathrm{A}\mathbf{y} = \mathbf{b}.$$

Therefore, $\mathrm{A}\mathbf{x} - \mathrm{A}\mathbf{y} = \mathbf{b} - \mathbf{b} = 0$

or $\mathrm{A}(\mathbf{x} - \mathbf{y}) = 0.$

Hence, (**x** – **y**) is a solution of Ax = 0. So, by cor.4 of theorem 2 of sec. 10.8,

$$\mathbf{x} - \mathbf{y} = c_1 \mathbf{x}_1 + c_2 \mathbf{x}_2 + \cdots + c_{n-r} \mathbf{x}_{n-r}$$

or $$\mathbf{x} = \mathbf{y} + c_1 \mathbf{x}_1 + c_2 \mathbf{x}_2 + \cdots + c_{n-r} \mathbf{x}_{n-r}$$

= (particular solution, **y**) + (complementary solution, a linear combination $(n-r)$ linearly independent solutions $\mathbf{x}_1, \mathbf{x}_2, \cdots, \mathbf{x}_{n-r}$)

Conversely, let $\mathbf{x} = \mathbf{y} + c_1 \mathbf{x}_1 + c_2 \mathbf{x}_2 + \cdots + c_{n-r} \mathbf{x}_{n-r}$

Using this **x** in L.H.S. of Ax = **b** , gives

$$\text{L.H.S.} = \mathrm{A}\mathbf{x} = \mathrm{A}(\mathbf{y} + c_1 \mathbf{x}_1 + c_2 \mathbf{x}_2 + \cdots + c_{n-r} \mathbf{x}_{n-r})$$
$$= \mathrm{A}\mathbf{y} + c_1 \mathrm{A}\mathbf{x}_1 + c_2 \mathrm{A}\mathbf{x}_2 + \cdots + c_{n-r} \mathrm{A}\mathbf{x}_{n-r}$$

$$= \mathbf{b} + 0 + 0 + \cdots + 0 = \mathbf{b}$$

$\because$ $\mathbf{x}_i$ are solutions of $A\mathbf{x} = 0$ and $\mathbf{y}$ is a solution of $A\mathbf{y} = \mathbf{b}$

$\therefore$ $\mathbf{x}$ is a general solution of $A\mathbf{x} = \mathbf{b}$.

Working procedure : The following steps may be used for working out the problems $A\mathbf{x} = \mathbf{b}$ under sec. 10.8.1.

Step 1. Write augmented matrix $[A \vdots \mathbf{b}]$.

Step 2. Use elementary row operations only to $[A \vdots \mathbf{b}]$ to the form

$$\left[\begin{array}{cc:c} \underline{I}_r & 0 & \underline{C} \\ 0 & 0 & 0 \end{array}\right]$$

Step 3. If rank A = rank $[A \vdots \mathbf{b}]$ then conclude that there exists unique solution if no free variable exists or infinitely many solutions if free variables exist.

Step 4. Give the free variables some particular values specially, zero and obtain the particular solution from the reduced nonhomogeneous equations.

Step 5. Solve the reduced homogeneous equations for some convenient values of the free variable and obtain the complementary solution.

Step 6. In case of unique solution, complementary solution is the trivial one and particular solution, is the unique solution itself.

Ex 1. Find the general solution of

$$\left.\begin{aligned} x + y - z &= 1 \\ x + 2y + 3z &= 6 \\ 2x + 3y + 2z &= 7 \\ 4x + 5y \quad &= 9 \\ x \quad - 5z &= -4 \end{aligned}\right\}$$

$$[A \vdots \mathbf{b}] = \left[\begin{array}{ccc:c} 1 & 1 & -1 & 1 \\ 1 & 2 & 3 & 6 \\ 2 & 3 & 2 & 7 \\ 4 & 5 & 0 & 9 \\ 1 & 0 & -5 & -4 \end{array}\right] \underset{\substack{R_4 - 4R_1 \\ R_5 - R_1}}{\overset{\substack{R_2 - R_1 \\ R_3 - 2R_1}}{\approx}} \left[\begin{array}{ccc:c} 1 & 1 & -1 & 1 \\ 0 & 1 & 4 & 5 \\ 0 & 1 & 4 & 5 \\ 0 & 1 & 4 & 5 \\ 0 & -1 & -4 & -5 \end{array}\right] \underset{\substack{R_4 - R_2 \\ R_5 + R_2}}{\overset{R_3 - R_2}{\approx}}$$

$$\left[\begin{array}{ccc:c} 1 & 1 & -1 & 1 \\ 0 & 1 & 4 & 5 \\ 0 & 0 & 0 & 0 \\ 0 & 0 & 0 & 0 \\ 0 & 0 & 0 & 0 \end{array}\right] \overset{R_1 - R_2}{\approx} \left[\begin{array}{ccc:c} 1 & 0 & -5 & -4 \\ 0 & 1 & 4 & 5 \\ 0 & 0 & 0 & 0 \\ 0 & 0 & 0 & 0 \\ 0 & 0 & 0 & 0 \end{array}\right].$$

$\therefore$ Rank(A) = 2 = Rank $(A \vdots \mathbf{b})$.

Hence the system is consistent. As rank $2 < n = 3$, so the system has infinitely many solutions. The complementary solution is given by the reduced associated homogeneous equations

$$\left.\begin{aligned} x \quad -5z &= 0 \\ y+4z &= 0 \end{aligned}\right\} \quad \text{or} \quad \left.\begin{aligned} x &= 5z \\ y &= -4z \end{aligned}\right\}.$$

Letting $z=1$, the required complementary solution is $(x, y, z) = c\,(5, -4, 1)$ for some constant c. The particular solution is given by

$$\left.\begin{aligned} x - \quad 5z &= -4 \\ y+4z &= \ \ 5 \end{aligned}\right\}$$

Letting $z = 0$, $(x, y, z) = (-4, 5, 0)$.

$\therefore$ the general solution is

$$\begin{aligned}(x, y, z) &= (\text{complementary solution}) + (\text{particular solution}) \\ &= (-4, 5, 0) + c\,(5, -4, 1).\end{aligned}$$

Ex 2. Find the general solution

$$\left.\begin{aligned} x-2y-z &= 0 \\ x+y \ \ +z &= 4 \\ x+2y+2z &= 6 \end{aligned}\right\},$$

as a sum of two solutions.

The problem is being solved below in two procedures side by side. Left side of the table shows the Gauss-Jordan reduction to simultaneous algebraic equations whereas right side shows equivalent matrix form reducing the matrix to RREF.

<table>
<tr><th>Gauss-Jordan reduction to simultaneous equations.</th><th>Equivalent matrix form (RREF)</th></tr>
<tr><td>$$\left.\begin{aligned} x-2y-z &= 0 \\ x+y \ \ +z &= 4 \\ x+2y+2z &= 6 \end{aligned}\right\}$$</td><td>$$\left[\begin{array}{ccc|c} 1 & -2 & -1 & 0 \\ 1 & 1 & 1 & 4 \\ 1 & 2 & 2 & 6 \end{array}\right] \underset{R_3-R_1}{\overset{R_2-R_1}{\approx}}$$</td></tr>
<tr><td>$$\text{or} \quad \left.\begin{aligned} x-2y-z &= 0 \\ 3y+2z &= 4 \\ 4y+3z &= 6 \end{aligned}\right\},$$</td><td>$$\left[\begin{array}{ccc|c} 1 & -2 & -1 & 0 \\ 0 & 3 & 2 & 4 \\ 0 & 4 & 3 & 6 \end{array}\right] \overset{R_3-\frac{4}{3}R_2}{\approx}$$</td></tr>
<tr><td>$$\text{or} \quad \left.\begin{aligned} x-2y-z &= 0 \\ 3y+2z &= 4 \\ \tfrac{1}{3}z &= \tfrac{2}{3} \end{aligned}\right\},$$</td><td>$$\left[\begin{array}{ccc|c} 1 & -2 & -1 & 0 \\ 0 & 3 & 2 & 4 \\ 0 & 0 & \frac{1}{3} & \frac{2}{3} \end{array}\right] \overset{3R_3}{\approx}$$</td></tr>
<tr><td>$$\text{or} \quad \left.\begin{aligned} x-2y-z &= 0 \\ 3y+2z &= 4 \\ z &= 2 \end{aligned}\right\},$$</td><td>$$\left[\begin{array}{ccc|c} 1 & -2 & -1 & 0 \\ 0 & 3 & 2 & 4 \\ 0 & 0 & 1 & 2 \end{array}\right] \overset{R_1+\frac{2}{3}R_2}{\approx}$$</td></tr>
</table>

$$\left.\begin{aligned} x \quad + \tfrac{1}{3}z &= \tfrac{8}{3} \\ 3y + 2z &= 4 \\ z &= 2 \end{aligned}\right\}, \quad \text{or} \qquad \left[\begin{array}{ccc|c} 1 & -2 & \frac{1}{3} & \frac{8}{3} \\ 0 & 3 & 2 & 4 \\ 0 & 0 & 1 & 2 \end{array}\right] \begin{array}{c} R_1 - \frac{1}{3}R_3 \\ \approx \\ R_2 - 2R_3 \end{array}$$

$$\left.\begin{aligned} x \quad &= 2 \\ 3y \quad &= 0 \\ z &= 2 \end{aligned}\right\}, \quad \text{or} \qquad \left[\begin{array}{ccc|c} 1 & 0 & 0 & 2 \\ 0 & 3 & 0 & 0 \\ 0 & 0 & 1 & 2 \end{array}\right] \begin{array}{c} \frac{1}{3}R_2 \\ \approx \end{array}$$

$$\left.\begin{aligned} x \quad &= 2 \\ y \quad &= 0 \\ z &= 2 \end{aligned}\right\}, \quad \text{or} \qquad \left[\begin{array}{ccc|c} 1 & 0 & 0 & 2 \\ 0 & 1 & 0 & 0 \\ 0 & 0 & 1 & 2 \end{array}\right]$$

rank (A) = 3 = rank $[A \vdots \mathbf{b}]$. So there exists unique solution and the solutions are (2, 0, 2). The complementary solutions are the trivial solutions (0, 0, 0).

$\therefore$ the general solution is

$$\mathbf{x} = \begin{bmatrix} 2 \\ 0 \\ 2 \end{bmatrix} + c_1 \begin{bmatrix} 0 \\ 0 \\ 0 \end{bmatrix}.$$

10.10 Linear dependence and independence

The following theorems provide tests of linear dependence and independence of a set of vectors and coefficients of linear dependence too (stated and used in sec.7.9).

Theorem 1. Let S = $\{\mathbf{u}_1, \mathbf{u}_2, \cdots, \mathbf{u}_n\}$ be a set of m-vectors (each of n vectors $\mathbf{u}_i$ has m components). Let A be the $m \times n$ matrix with column vectors $A_{.j} = \mathbf{u}_i$. Let the row reduced echelon form of A be B. Then the set S is linearly independent if and only if the columns of the matrix B consists of elementary vectors $\mathbf{e}_1, \mathbf{e}_2, \cdots, \mathbf{e}_n$.

Proof. The given set of vectors is linearly independent if and only if the vector equation

$$c_1\mathbf{u}_1 + c_2\mathbf{u}_2 + \cdots + c_n\mathbf{u}_n = 0$$

is only satisfied for the scalars $c_1 = c_2 = \cdots = c_n = 0$ (10.23)

The matrix eqn. of (10.23) is Ac = 0 (10.24)

and (10.24) is true only when rank A = n by cor.3 of theorem 2 of sec. 10.9, i.e. only when the number of nonzero rows of B, the row reduced echelon form of A, is n. Since in the row reduced echelon form each nonzero row has leading 1 and each column with leading 1 has all other entries equal to 0, the rank of A is n iff the columns of B are s.u. m-vectors $\mathbf{e}_1, \mathbf{e}_2, \cdots, \mathbf{e}_n$.

Theorem 2. Let S = $\{\mathbf{u}_1, \mathbf{u}_2, \cdots, \mathbf{u}_n\}$ be a set of m-vectors. Let A be the $m \times n$ matrix with column vectors $A_{.j} = \mathbf{u}_i$, $i, j = 1, 2, \cdots, n$ and let the row reduced echelon form of A be B. Let one of the columns of B, say $B_{.j}$ is not a s.u. vector or is duplication of a preceding s.u. vector. The S is linearly dependent and $u_j = b_{1j}\mathbf{w}_1 + b_{2j}\mathbf{w}_2 + \cdots + b_{kj}\mathbf{w}_k$ where $\mathbf{w}_1, \mathbf{w}_2, \cdots, \mathbf{w}_k$

are original vectors in S corresponding to new s.u. vectors $\mathbf{e}_1, \mathbf{e}_2, \cdots, \mathbf{e}_k$ the distinct s.u. columns of B that precede the jth column.

Proof. Since all the columns of B, the row reduced echelon form of A, are not distinct elementary vectors due to presence of $B_{\cdot j}$. So by the previous theorem, the given set S is not linearly independent i.e. S is linearly dependent. Now the vector equation

$$c_i \mathbf{u}_i = 0 \quad i = 1, 2, \quad \cdots, n. \tag{10.25}$$

is equivalent to the matrix equation $\mathbf{Ac} = 0$ which is virtually equivalent of $\mathbf{Bc} = 0$ i.e. in explicit form , it is

$$\left.\begin{array}{l} c_1 + \cdots\cdots \quad + b_{1j}c_j + \cdots\cdots = 0 \\ \quad c_2 + \cdots\cdots \quad + b_{2j}c_j + \cdots\cdots = 0 \\ \quad \vdots \qquad\qquad \vdots \qquad\qquad \vdots \\ \quad c_k + \cdots\cdots + b_{kj}c_j + \cdots\cdots = 0 \end{array}\right\} \tag{10.26}$$

Here $c_{k+1}, \quad \cdots, c_n$ are free variables. So, letting $c_j = 1$ and $c_i = 0$ for variables other than basic variables $c_1, c_2, \quad \cdots, c_k$ we have the solutions

$$c_1 = -b_{1j}, c_2 = -b_{2j}, \quad \cdots, c_k = -b_{kj}, c_j = 1 \;\&\; c_i = 0 \text{ for } i > k \;\&\; i \neq j. \tag{10.27}$$

But c_i are the coefficients of the vector equation (10.25).

$\therefore$ using (10.27) in (10.25), we get

$-b_{1j}\mathbf{u}_1 - b_{2j}\mathbf{u}_2 - \quad \cdots - b_{kj}\mathbf{u}_k + 0\mathbf{u}_{k+1} + \quad \cdots + 0\,\mathbf{u}_{j-1} + 1\mathbf{u}_j + 0\mathbf{u}_{j+1} + \quad \cdots + 0\mathbf{u}_n = 0$

or solving for $\mathbf{u}_j$,

$$\mathbf{u}_j = b_{1j}\mathbf{u}_1 + b_{2j}\mathbf{u}_2 + \quad \cdots + b_{kj}\mathbf{u}_k$$

Hence the proof.

Ex. Determine whether the set of vectors

$$\begin{bmatrix} 1 \\ 0 \\ 3 \end{bmatrix}, \begin{bmatrix} 0 \\ 1 \\ 0 \end{bmatrix}, \begin{bmatrix} 1 \\ 1 \\ 1 \end{bmatrix}, \begin{bmatrix} 7 \\ -5 \\ 4 \end{bmatrix} \;\&\; \begin{bmatrix} 5 \\ 0 \\ 4 \end{bmatrix}$$

is linearly independent or dependent. If dependent, express the dependent one as a linear combination of the preceding vectors.

Let the matrix with the vectors as columns be

$$A = \begin{bmatrix} 1 & 0 & 1 & 7 & 5 \\ 0 & 1 & 1 & -5 & 0 \\ 3 & 0 & 1 & 4 & 4 \end{bmatrix}.$$

Using elementary row operations

$$A \approx \begin{bmatrix} 1 & 0 & 1 & 7 & 5 \\ 0 & 1 & 1 & -5 & 0 \\ 0 & 0 & -2 & -17 & -11 \end{bmatrix} \approx \begin{bmatrix} 1 & 0 & 1 & 7 & 5 \\ 0 & 1 & 1 & -5 & 0 \\ 0 & 0 & 1 & \frac{17}{2} & \frac{11}{2} \end{bmatrix} \approx \begin{bmatrix} 1 & 0 & 0 & -\frac{3}{2} & \frac{1}{2} \\ 0 & 1 & 0 & -\frac{27}{2} & -\frac{11}{2} \\ 0 & 0 & 1 & \frac{17}{2} & \frac{11}{2} \end{bmatrix}$$

The first three columns are s.u. vectors $\mathbf{e}_1, \mathbf{e}_2, \mathbf{e}_3$ but the last two are not. So, the given vectors are linearly dependent. The fourth and fifth vectors are linear combinations of the preceding original vectors which reduce to s.u. vectors as

$$\begin{bmatrix}7\\-5\\4\end{bmatrix}=-\tfrac{3}{2}\begin{bmatrix}1\\0\\3\end{bmatrix}-\tfrac{27}{2}\begin{bmatrix}0\\1\\0\end{bmatrix}+\tfrac{17}{2}\begin{bmatrix}1\\1\\1\end{bmatrix} \text{ and } \begin{bmatrix}5\\0\\4\end{bmatrix}=\tfrac{1}{2}\begin{bmatrix}1\\0\\3\end{bmatrix}-\tfrac{11}{2}\begin{bmatrix}0\\1\\0\end{bmatrix}+\tfrac{11}{2}\begin{bmatrix}1\\1\\1\end{bmatrix}.$$

10.11 Index and signature of a matrix

Index and signature are two more ways of describing the status of real symmetric and complex Hermitian matrices as the rank does. These two properties are next found useful in the study of forms (chapter 15)

Index. It is the number, p of positive 1's in the diagonal form

$$A=\begin{bmatrix}I_p & & \bigcirc\\ \bigcirc & -I_{r-p} & \bigcirc\\ & & \end{bmatrix}=\text{diag}(I_p,-I_{r-p},\ \bigcirc\ )\text{ where }\ \bigcirc\text{stands for set of}$$

zero elements, of a real symmetric or complex Hermitian matrix A of rank r. The plural of the index is indices.

Signature. It is defined as the number of differences of positive 1's and (−1)'s of the above diagonal form of a real symmetric or complex Hermitian matrix A. It is denoted by Sig (A) or s. Thus, by definition,

$$\text{Sig(A) or } s = p-(r-p) \quad \because p \text{ is number of +ve 1's and } (r-p) \text{ is the no. of −ve 1's}$$
$$= 2p - r.$$

Note that index and signature go hand in hand and

$$\text{rank (A)} = r = 2p-s = \text{(twice the index)} - \text{(signature)}.$$

Ex. Find the index and the signature of the matrix

$$A=\begin{bmatrix}1 & -2 & 1 & -2\\ -2 & 0 & 3 & -5\\ 1 & 3 & 2 & 2\\ -2 & -5 & 2 & -9\end{bmatrix}.$$

This can be reduced to diagonal form

$$A\approx\begin{bmatrix}1 & 0 & 0 & 0\\ 0 & 1 & 0 & 0\\ 0 & 0 & -1 & 0\\ 0 & 0 & 0 & 0\end{bmatrix}$$

by elementary row transformation followed by elementary column transformation. Then the (number of + ve 1's) = 2 and (that of (−1)'s) =1. Therefore, index $p = 2$ and sig., $s = 2-1= 1$. Note that rank (A) = 3.

10.11.1 Definite, semidefinite and indefinite matrices: Discussion on definiteness of matrices begins in sec.1.3. This subsection uses rank and index on one hand and then congruence on the other for testing definiteness of matrices.

(a) Using rank and index

Let A be a real symmetric $n\times n$ matrix of rank r and index p. Then A is said to be

(i) positive definite iff $r = n,\ p = n$;

(ii) negative definite iff $r = n,\ p = 0$;

(iii) positive semidefinite iff $r < n,\ p = r$;

(iv) negative semidefinite iff $r < n,\ p = 0$;

(v) indefinite in every other cases.

(b) Using congruence

Congruence can be needed for the definiteness test of symmetric R-matrices as follows:

If A is a nonzero $n\times n$ symmetric matrix having rank r, then A is

(i) positive definite iff A is congruent to I_n,

(ii) positive semidefinite iff A is congruent to

$$\begin{bmatrix} I_r & O \\ O & O \end{bmatrix}_{n\times n}, n \neq r,$$

(iii) negative definite iff A is congruent to $(-I_n)$.

(iv) negative semidefinite iff A is congruent to

$$\begin{bmatrix} -I_r & O \\ O & O \end{bmatrix}_{n\times n}, n \neq r$$

(v) indefinite in every other cases.

Ex. Test for definiteness using (a) rank, index and (b) congruence:

$$\text{(i)} \begin{bmatrix} 2 & -1 & 1 \\ -1 & 2 & -1 \\ 1 & -1 & 2 \end{bmatrix} \ \& \ \text{(ii)} \begin{bmatrix} 1 & -2 & 0 \\ -2 & 4 & 0 \\ 0 & 0 & 3 \end{bmatrix}.$$

(i) (a) Let $A = \begin{bmatrix} 2 & -1 & 1 \\ -1 & 2 & -1 \\ 1 & -1 & 2 \end{bmatrix} \underset{\approx}{2R_2 + R_1} \begin{bmatrix} 2 & -1 & 1 \\ 0 & 3 & -1 \\ 1 & -1 & 2 \end{bmatrix} \underset{\approx}{2C_2 + C_1} \begin{bmatrix} 2 & 0 & 1 \\ 0 & 6 & -1 \\ 1 & -1 & 2 \end{bmatrix}$

$$\underset{\approx}{2R_3 - R_1} \begin{bmatrix} 2 & 0 & 1 \\ 0 & 6 & -1 \\ 0 & -2 & 3 \end{bmatrix} \underset{\approx}{2C_3 - C_1} \begin{bmatrix} 2 & 0 & 0 \\ 0 & 6 & -2 \\ 0 & -2 & 6 \end{bmatrix} \underset{\approx}{R_3 + \tfrac{1}{3}R_2}$$

$$\begin{bmatrix} 2 & 0 & 0 \\ 0 & 6 & -2 \\ 0 & 0 & \frac{16}{3} \end{bmatrix} \underset{\approx}{C_3 + \tfrac{1}{3}C_2} \begin{bmatrix} 2 & 0 & 0 \\ 0 & 6 & 0 \\ 0 & 0 & \frac{16}{3} \end{bmatrix}$$

$\therefore\ r = p = n = 3$. $\therefore$ A is positive definite.

(b) $A = \begin{bmatrix} 2 & -1 & 1 \\ -1 & 2 & -1 \\ 1 & -1 & 2 \end{bmatrix} \underset{\approx}{2R_2 + R_1} \begin{bmatrix} 2 & -1 & 1 \\ 0 & 3 & -1 \\ 1 & -1 & 2 \end{bmatrix} \underset{\approx}{2C_2 + C_1} \begin{bmatrix} 2 & 0 & 1 \\ 0 & 6 & -1 \\ 1 & -1 & 2 \end{bmatrix}$

$$\underset{\approx}{2R_3 - R_1}\begin{bmatrix}2 & 0 & 1\\0 & 6 & -1\\0 & -2 & 3\end{bmatrix}\underset{\approx}{2C_3 - C_1}\begin{bmatrix}2 & 0 & 0\\0 & 6 & -2\\0 & -2 & 6\end{bmatrix}\underset{\approx}{R_3 + \frac{1}{3}R_2}\begin{bmatrix}2 & 0 & 0\\0 & 6 & -2\\0 & 0 & \frac{16}{3}\end{bmatrix}$$

$$\underset{\approx}{C_3 + \frac{1}{3}C_2}\begin{bmatrix}2 & 0 & 0\\0 & 6 & 0\\0 & 0 & \frac{16}{3}\end{bmatrix}\begin{matrix}R_1(\frac{1}{\sqrt{2}})\\ \approx \\ R_2(\frac{1}{\sqrt{6}})\\ R_3(\frac{\sqrt{3}}{4})\end{matrix}\begin{bmatrix}\sqrt{2} & 0 & 0\\0 & \sqrt{6} & 0\\0 & 0 & \frac{4}{\sqrt{3}}\end{bmatrix}\begin{matrix}C_1(\frac{1}{\sqrt{2}})\\ \approx \\ C_2(\frac{1}{\sqrt{6}})\\ C_3(\frac{\sqrt{3}}{4})\end{matrix}\begin{bmatrix}1 & 0 & 0\\0 & 1 & 0\\0 & 0 & 1\end{bmatrix} = I_3$$

$\therefore \quad A_3 \approx I_3$.

So, A_3 is positive definite.

(ii) (a) Let $A = \begin{bmatrix}1 & -2 & 0\\-2 & 4 & 0\\0 & 0 & 3\end{bmatrix}\underset{\approx}{R_2 + 2R_1}\begin{bmatrix}1 & -2 & 0\\0 & 0 & 0\\0 & 0 & 3\end{bmatrix}\underset{\approx}{C_2 + 2C_1}$

$$\begin{bmatrix}1 & 0 & 0\\0 & 0 & 0\\0 & 0 & 3\end{bmatrix}.$$

$\therefore \quad r = 2, p = 2, n = 3.$

$\therefore \quad r < n$ and $p = r$.

$\therefore \quad$ A is positive semidefinite.

(b) $A = \begin{bmatrix}2 & -1 & 1\\-1 & 2 & -1\\1 & -1 & 2\end{bmatrix}\underset{\approx}{R_2 + 2R_1}\begin{bmatrix}1 & -2 & 0\\0 & 0 & 0\\0 & 0 & 3\end{bmatrix}\underset{\approx}{C_2 + 2C_1}\begin{bmatrix}1 & 0 & 0\\0 & 0 & 0\\0 & 0 & 3\end{bmatrix}\underset{\approx}{R_{23}}\begin{bmatrix}1 & 0 & 0\\0 & 0 & 3\\0 & 0 & 0\end{bmatrix}$

$$\underset{\approx}{C_{23}}\begin{bmatrix}1 & 0 & 0\\0 & 3 & 0\\0 & 0 & 0\end{bmatrix}\underset{\approx}{R_2(\frac{1}{\sqrt{3}})}\begin{bmatrix}1 & 0 & 0\\0 & \sqrt{3} & 0\\0 & 0 & 0\end{bmatrix}\underset{\approx}{C_2(\frac{1}{\sqrt{3}})}\begin{bmatrix}1 & 0 & 0\\0 & 1 & 0\\0 & 0 & 0\end{bmatrix} = \left[\begin{array}{c|c}I_2 & 0\\ \hline 0 & 0\end{array}\right]$$

$\therefore$ A is positive semidefinite as A_2 is congruent to

$$\left[\begin{array}{c|c}I_2 & 0\\ \hline 0 & 0\end{array}\right].$$

10.11.2 Congruence by signature: Since signature measures some kind of status of real symmetric and complex Hermitian matrices, congruence of these types of matrices can be determined by signature. Thus, two nonsingular symmetric (Hermitian) matrices A and B are defined to be congruent iff they have the same signature.

Ex . Determine whether or not the following two matrices

$$A = \begin{bmatrix}1 & 2 & 3\\2 & 1 & 0\\3 & 0 & 1\end{bmatrix} \text{ and } B = \begin{bmatrix}1 & 3 & 3\\3 & 1 & 0\\2 & 0 & 1\end{bmatrix}$$

are congrurent (use signature test) .

Now,

$$A=\begin{bmatrix}1&2&3\\2&1&0\\3&0&1\end{bmatrix}\underset{\approx}{R_2-2R_1}\begin{bmatrix}1&2&3\\0&-5&-6\\3&0&1\end{bmatrix}\underset{\approx}{C_2-C_1}$$

$$\begin{bmatrix}1&0&3\\0&-5&-6\\3&-6&1\end{bmatrix}\underset{\approx}{R_3-3R_1}\begin{bmatrix}1&0&3\\0&-5&-6\\0&-6&-8\end{bmatrix}\underset{\approx}{C_3-3C_1}$$

$$\begin{bmatrix}1&0&0\\0&-5&-6\\0&-6&-8\end{bmatrix}\underset{\approx}{R_3-\tfrac{6}{5}R_2}\begin{bmatrix}1&0&0\\0&-5&-6\\0&0&-\tfrac{4}{5}\end{bmatrix}\underset{\approx}{C_3-\tfrac{6}{5}C_2}$$

$$\begin{bmatrix}1&0&0\\0&-5&0\\0&0&-\tfrac{4}{5}\end{bmatrix}\begin{matrix}R_2(\tfrac{1}{\sqrt5})\\ \approx\\ R_3(\tfrac{2}{\sqrt5})\end{matrix}\begin{bmatrix}1&0&0\\0&-\sqrt5&0\\0&0&-\tfrac{2}{\sqrt5}\end{bmatrix}$$

$$\begin{matrix}C_2(\tfrac{1}{\sqrt5})\\ \approx\\ C_3(\tfrac{2}{\sqrt5})\end{matrix}\begin{bmatrix}1&0&0\\0&-1&0\\0&0&-1\end{bmatrix}$$

$\therefore$ sig (A) =1−2 = −1 .

Similarly, sig (B) =2−1=1.

$\therefore$ the signatures of the given matrices are not same.

Hence the given matrices are not congruent.

Exercises Ten

1. Show that A, A^*, A^*A, AA^* all have the same rank.

2. Let

$$M=\begin{bmatrix}A_1&0&0\\0&A_2&0\\0&0&A_3\end{bmatrix}$$

where A_1, A_2, A_3 are square matrices. Show that

$$\text{rank }(M)=\text{rank }(A_1)+\text{rank}(A_2)+\text{rank}(A_3)\ .$$

3. Find the ranks of A, $A^2 A^3 A^4$ where

$$A=\begin{bmatrix}0&1&0&0\\0&0&1&0\\0&0&0&1\\0&0&0&0\end{bmatrix}.$$

[**Ans.** rank (A) = rank $(A^2)=2$, rank$(A^3)=1$, rank$(A^4)=0$]

4. Show that a matrix of rank *r* can be expressed as the sum of *r* matrices each of rank 1. Verify the statement from the above exercise.

5. Determine the ranks of A, B, A+B, A–B, AB & BA if

$$A=\begin{bmatrix}2&1\\-3&3\end{bmatrix} \text{ and } B=\begin{bmatrix}-2&5\\3&-3\end{bmatrix}.$$

[**Ans.** rank (A) =2, rank (B)=2, rank (A + B)=1, rank(A–B)=1, rank (AB)=2, rank (BA)=2]

6. Prove that the rank of A+ B is less than or equal to the rank of A plus rank of B, i.e. rank (A+B) ≤ rank A + rank B.

7. Determine the rank and the nullity of the matrix,

$$A=\begin{bmatrix}2&1&4&4&0\\3&1&2&3&1\\5&2&6&7&1\\1&0&-2&-1&1\end{bmatrix}.$$

[**Ans.** rank (A) =2, row nullity (A)= 4 –2 =2, column nullity (A) =5–2=3]

8. Find the general solutions as a sum of two solutions of the following:

(i) $\left.\begin{aligned}x+y+z&=6\\-x+y-2z&=-5\\2y-z&=1\end{aligned}\right\}$; (ii) $\left.\begin{aligned}x+2y\quad -v&=2\\x+3y-z+v&=1\\x\quad -2z+3v&=0\\2y+z-2v&=2\end{aligned}\right\}$; (iii) $\left.\begin{aligned}-3x_1+x_2+x_3+5x_4&=4\\2x_2-3x_3+5x_4&=4\\-x_3+5x_4&=4\end{aligned}\right\}$;

(iv) $x_1+x_2+x_3+x_4=1$ |

[**Ans.** (i) $(x,y,z)=(4,1,1)+c(-3,1,2)$; (ii) $(x,y,z,v)=(\frac{7}{6},\frac{1}{3},1,-\frac{1}{6})+c(0,0,0,0)$;
(iii) $(x_1,x_2,x_3,x_4)=(-4,-4,-4,0)+c(5,5,5,1)$;
(iv) $(x_1,x_2,x_3,x_4)=(1,0,0,0)+c_1(0,1,0,0)+c_2(0,0,1,0)+c_3(0,0,0,1)$]

9. For what value of *x* is the rank of the matrix 3 ? Is it possible for the rank to be 1 ? why?

$$\begin{bmatrix}2&4&2\\2&1&2\\1&0&x\end{bmatrix}.$$

[**Ans.** $x\neq 1$; no, because there exists a 2×2 nonzero minor]

10. Test for equivalence the following matrices:

$$\text{(i)} \begin{bmatrix} 1 & -1 & 2 & 3 & 1 \\ 1 & 1 & 1 & -1 & 2 \\ 1 & -2 & 1 & 3 & 1 \\ 1 & 0 & 0 & -1 & 2 \end{bmatrix}, \text{(ii)} \begin{bmatrix} 2 & 1 & 3 & 4 & -1 \\ 1 & -3 & 1 & 2 & -1 \\ 3 & 4 & -1 & 2 & 1 \\ 0 & 6 & 5 & 4 & 3 \end{bmatrix},$$

$$\text{(iii)} \begin{bmatrix} 1 & 2 & -1 & 3 & 4 \\ 1 & -1 & 2 & 1 & 0 \\ 2 & 1 & 1 & 4 & 4 \\ 1 & -4 & 5 & -1 & -4 \end{bmatrix}$$

[**Ans.** (i) and (ii) are equivalent. **Hints:** show equal ranks]

11. Find the rank & the nullity of

$$\text{(i) } A = \begin{bmatrix} 0 & 1 & 0 & 1 & 2 & 0 & 3 \\ 0 & 2 & 0 & 2 & 4 & 0 & 6 \\ 0 & 1 & 0 & 2 & 3 & -1 & 4 \end{bmatrix} ; \text{ (ii) } A = \begin{bmatrix} 2 & 1 & 3 & 1 \\ 1 & 0 & 1 & 1 \\ 2 & 1 & 3 & 1 \\ -1 & 0 & -1 & -1 \end{bmatrix}.$$

[**Ans.** (i) Rank A=2, row nullity=3−2=1, column nullity =7−2= 5,
(ii) rank A= 2, row nullity = column nullity = 4−2 =2]

12. Prove that rank A = rank A^T.

[**Hints:** Blocks of A^T are the transpose of the blocks of A. Hence the result]

13. Are the following statements true or false ? Give your reasons.

(i) If $\underline{A} = [a_{ij}]$ has rank 3, then the determinant of every 4th order block is zero & conversely, if every such 4th order determinant is zero, the rank of the matrix is 3.

(ii) Rank of A is not equal to rank of A^T.

(iii) If $A = [a_{ij}]_{3\times 3}$ is skew symmetric, its rank is less than 3.

[**Ans.** (i) False. Since converge is not true, rank may be less than 2.
(ii) False. Block of A^T is transpose of block of A and det A=det A^T.
(iii) True; Since $A = -A^T$, so $|A| = A^T$ if $|A| = 0$ then rank is less than 3]

14. Determine the values of x, if any, that will make G(x) of rank 1; of rank 2; of rank 3 ; if

$$G(x) = \begin{bmatrix} x & 2 & x \\ x & x & 2 \\ 2 & x & x \end{bmatrix}.$$

[**Hints:** $P^TGP = D = \begin{bmatrix} x & 0 & 0 \\ 0 & x-2 & 0 \\ 0 & 0 & \frac{2}{x}(x-2)(x+1) \end{bmatrix}$

Ans. rank G =1 if x =2, rank G =2 if $x = -1$, rank G =3 if $x \neq 2, x \neq 0, x \neq -1$]

15. Explain why the rank of A is 3 but the rank of C is only 2 after A is premultiplied by B in

$$BA = \begin{bmatrix} 2 & 1 & 0 \\ 1 & 0 & 4 \\ 3 & 2 & -4 \end{bmatrix} \begin{bmatrix} 1 & 3 & 0 & 1 \\ 4 & 1 & 2 & 5 \\ 3 & 2 & 2 & 4 \end{bmatrix} = \begin{bmatrix} 6 & 7 & 2 & 7 \\ 13 & 11 & 8 & 17 \\ -1 & 3 & -4 & -3 \end{bmatrix} = C$$

[**Ans.** B is singular, theorem 3 fails to satisfy]

16. Find a matrix B of the minimum order and maximum rank such that AB = 0 where

$$A = \begin{bmatrix} 1 & 4 & -1 & 2 \\ -1 & 2 & 0 & -1 \\ 3 & 0 & -1 & 4 \\ -1 & 8 & -1 & 0 \end{bmatrix}$$

[**Hints :** rank B $= n - r = 4-2 = 2$.

$$B = QY = Q\begin{bmatrix} O & O \\ Y_{2,k} & Y_{2,p-k} \end{bmatrix} = Q\begin{bmatrix} O_{2,2} & O_{2,p-2} \\ Y_{2,2} & Y_{2,p-2} \end{bmatrix}$$

for the minimum order as $0 \le k \le p$. Since rank Y = rank B, so p must be equal to 2. Now using procedure of w.o. Ex 2 of sec.10.5.2

$$Q = Q_2 = \begin{bmatrix} K^{-1} & K^{-1}L \\ O & I_{m-r} \end{bmatrix} = \begin{bmatrix} \frac{1}{3} & -\frac{2}{3} & \frac{1}{3} & -\frac{4}{3} \\ \frac{1}{6} & \frac{1}{6} & \frac{1}{6} & -\frac{1}{6} \\ 0 & 0 & 1 & 0 \\ 0 & 0 & 0 & 1 \end{bmatrix}$$

$$\text{where } K = \begin{bmatrix} 1 & 4 \\ -1 & 2 \end{bmatrix} \text{ and } K^{-1} = \begin{bmatrix} \frac{1}{3} & -\frac{2}{3} \\ \frac{1}{6} & \frac{1}{6} \end{bmatrix}$$

$$\therefore B = QY = \begin{bmatrix} \frac{1}{3} & -\frac{2}{3} & \frac{1}{3} & -\frac{4}{3} \\ \frac{1}{6} & \frac{1}{6} & \frac{1}{6} & -\frac{1}{6} \\ 0 & 0 & 1 & 0 \\ 0 & 0 & 0 & 1 \end{bmatrix}\begin{bmatrix} 0 & 0 \\ 0 & 0 \\ y_{31} & y_{32} \\ y_{41} & y_{42} \end{bmatrix} = \frac{1}{6}\begin{bmatrix} 2(a-4b) & 2(c-4d) \\ (a-b) & (c-d) \\ 6a & 6c \\ 6b & 6d \end{bmatrix},$$

where $a = y_{31}, b = y_{41}, c = y_{32}, d = y_{42}$ and $a,b,c,d \neq 0$]

17. Find nonsingular matrices P and Q such that PAQ is in normal from where

$$A = \begin{bmatrix} 1 & 4 & -1 & 2 \\ -1 & 2 & 0 & -1 \\ 3 & 0 & -1 & 4 \\ -1 & 8 & -1 & 0 \end{bmatrix}.$$

[**Ans.** $P = \begin{bmatrix} 1 & 0 & 0 & 0 \\ 0 & 1 & 0 & 0 \\ -1 & 2 & 1 & 0 \\ -1 & -2 & 0 & 1 \end{bmatrix}$ and $Q = \frac{1}{6}\begin{bmatrix} 2 & -4 & 2 & -8 \\ 1 & 1 & 1 & -1 \\ 0 & 0 & 6 & 0 \\ 0 & 0 & 0 & 6 \end{bmatrix}$]

18. Find a matrix X of rank 2 such that AX = 0 if

$$A = \begin{bmatrix} 1 & 2 & 3 \\ 2 & 4 & 6 \\ 2 & 4 & 6 \end{bmatrix}$$

[**Ans.** $X = QY = \left[\begin{array}{c|cc} 1 & -2 & -3 \\ \hline 0 & 1 & 0 \\ 0 & 0 & 1 \end{array}\right]_{3\times 3} \left[\begin{array}{c|c} O_{1,k} & O_{1,p-k} \\ \hline Y_{3-1,k} & Y_{3-1,p-k} \end{array}\right]_{3\times p}$

$$= \begin{bmatrix} (-2y_{11} - 3y_{21}) & \cdots & (-2y_{1k} - 3y_{2k}) & (-2y_{1,k+1} - 3y_{2,k+1}) & \cdots & (-2y_{1,p-k} - 3y_{2,p-k}) \\ y_{11} & \cdots & y_{1k} & y_{1,k+1} & \cdots & y_{1,p-k} \\ y_{21} & \cdots & y_{2k} & y_{2,k+1} & \cdots & y_{2,p-k} \end{bmatrix}$$

where any $(p-2)$ columns are linear combinations of the remaining two for rank B =2]

19. By inspection, determine the possible number of solutions for each of the following systems.

(a) $\left.\begin{array}{l} 2x_1 + x_2 - x_3 = 0 \\ 3x_1 - 2x_2 + x_3 = 0 \\ x_1 - x_2 + x_3 = 0 \end{array}\right\}$, (b) $\left.\begin{array}{l} 2x + 3y - z = 1 \\ y + 2z - 3w = 0 \\ 2x - 4z + w = 0 \end{array}\right\}$,

(c) $\left.\begin{array}{l} 2x + 3y = 4 \\ -x + 4y = 0 \\ 4x + 3y = 0 \end{array}\right\}$, (d) $\left.\begin{array}{l} 2x + 3y - z = 0 \\ y + 2z - 3w = 0 \\ 2x - 4z + w = 0 \end{array}\right\}$.

[**Hints: Ans.** (a): The system is homogeneous. The number of equations 3 equals the number of unknowns. The rank of the coefficient matrix A is given by

$$A = \begin{bmatrix} 2 & 1 & -1 \\ 3 & -2 & 1 \\ 1 & -1 & 1 \end{bmatrix} \approx \begin{bmatrix} 1 & -1 & 1 \\ 3 & -2 & 1 \\ 2 & 1 & -1 \end{bmatrix} \approx \begin{bmatrix} 1 & -1 & 1 \\ 0 & 1 & -2 \\ 0 & 3 & -3 \end{bmatrix} \approx \begin{bmatrix} 1 & -1 & 1 \\ 0 & 1 & -2 \\ 0 & 0 & 3 \end{bmatrix}$$

∴ rank A=3 .

∴ 3 unique solutions i.e. 3 trivial or zero solutions by cor 3.

(b). The system is nonhomogeneous with 3 equations and 4 unknowns. So, system has infinitely many solutions by cor. 5.

(c). Nonhomogeneous system with more equations than unknowns and rank of coefficient matrix is less than that of augmented matrix. So no solutions exist by cor. 2.

(d). Homogeneous system with more unknowns than equations. So, system has infinitely many solutions by cor. 5]

20. By inspection of the following systems, determine the number of possible solutions:

a. $x_1 + 2x_2 - x_3 = 0$

$2x_1 - x_2 + 3x_3 = 0$

[**Ans.**Infinitely many solutions]

b. $2x_1 + x_2 + x_3 = 1$ [**Ans.** No solution or infinitely many solutions]

$x_1 - x_2 + x_3 = 0$

[**Ans.**rank (A) = rank([A ⋮ b]) but $<n=4$.

Infinitely many solutions]

c. $x_1 \quad +2x_4 = \ \ 1$

$\quad -x_2 \quad -x_4 = -1$

$\quad\quad x_3 + x_4 = \ \ 0$

21. Let the linear system be $A\mathbf{x} = \mathbf{b}$ where

$$A = \begin{vmatrix} a & b \\ c & d \end{vmatrix}, \ \mathbf{b} = \begin{bmatrix} e \\ d \end{bmatrix} \text{ and } |A| \neq 0.$$

(i) Determine conditions on the constant a, b, c, d, e & f so that

(a) the rank of A is 2,

(b) the rank of A is 1 but the rank of $[A \vdots \mathbf{b}]$ is 2,

(c) the rank of A and the rank of $[A \vdots \mathbf{b}]$ are each 1.

(ii) Using the results of exercise 21(i), give conditions on a, b, c, d, e & f so that the system has

(a) no solutions,

(b) one solution,

(c) infinitely many solutions.

[**Ans.** (i) (a) $\begin{vmatrix} a & b \\ c & d \end{vmatrix} \neq 0$;(b) $ad = bc$ but $af \neq ce$;(c) $ad = bc$ & $af = ce$.

(ii) (a) $ad = bc$ and $e \neq kf$, k is a constant; (b) $ad \neq bc$ and $af \neq ce$

(c) $ad = bc$ and $af = ce$]

22. Find a homogeneous system of linear equations with a complementary solution,

$$\mathbf{x} = t_1 \begin{bmatrix} 1 \\ 1 \\ 0 \\ 0 \end{bmatrix} + t_2 \begin{bmatrix} 0 \\ 1 \\ 1 \\ 1 \end{bmatrix}.$$

[**Ans.** $\left.\begin{matrix} x_2 - x_1 - x_4 = 0 \\ x_3 - x_4 = 0 \end{matrix}\right\}$ in reduced form. Number of equations may be increased, dependent on those two equations]

23. Find a nonhomogeneous system of linear equations with the general solution

$$\mathbf{x} = \begin{bmatrix} 1 \\ 1 \\ 1 \\ 1 \end{bmatrix} + t_1 \begin{bmatrix} 1 \\ 1 \\ 0 \\ 0 \end{bmatrix} + t_2 \begin{bmatrix} 0 \\ 1 \\ 1 \\ 1 \end{bmatrix}.$$

[**Hints :** Solutions shows that the number of unknowns $n = 4$ and $n - r = 2$ as the solution contains two arbitrary constants. Then $r = 2$. Hence there are two linearly independent equations in the required system. From the complementary solution, it is easy to write $x_1 = t_1$, $x_2 = t_1 + t_2$, $x_3 = t_2$, $x_4 = t_2$. Now eliminate t_1 & t_2 so that the complementary part of solution is true. Thus, $x_2 = x_1 + x_4$ and $x_3 = x_4$. So the linearly independent homogeneous system is $\begin{cases} x_2 - x_1 - x_4 = 0 \\ x_3 - x_4 = 0 \end{cases}$.

From the particular solution, it is seen that the free variables $x_1 = 1$, $x_4 = 1$. Since $x_2 = 1$ and $x_3 1$ in a particular solution so $x_2 - x_1 - x_4 = -1$ and $x_3 - x_4 = 0$. Hence the linearly independent nonhomogenous system is $\begin{cases} x_2 - x_1 - x_4 = -1 \\ x_3 - x_4 = 0 \end{cases}$.]

24. Find a nonhomogenous system of linear equations with general solution

$$\mathbf{x} = \begin{bmatrix} 1 \\ 1 \\ 0 \end{bmatrix} + t \begin{bmatrix} -1 \\ 0 \\ 1 \end{bmatrix}.$$

[**Ans.** $x_1 + x_3 = 1$

$x_2 = 1$]

25. Determine whether the following set of vectors are linearly independent or not. Express the dependent vector, if any, as a linear combination of the preceding vectors.

(i) $\mathbf{a}_1 = \begin{bmatrix} 1 \\ 0 \\ -2 \\ 4 \end{bmatrix}, \mathbf{a}_2 = \begin{bmatrix} 1 \\ 1 \\ 1 \\ -2 \end{bmatrix}, \mathbf{a}_3 = \begin{bmatrix} 4 \\ 3 \\ 1 \\ -2 \end{bmatrix}$; (ii) $\mathbf{a}_1 = \begin{bmatrix} 1 \\ 2 \\ -1 \end{bmatrix}, \mathbf{a}_2 = \begin{bmatrix} 3 \\ 0 \\ 2 \end{bmatrix}, \mathbf{a}_3 = \begin{bmatrix} 5 \\ 4 \\ 0 \end{bmatrix}$;

(iii) $\mathbf{a}_1 = (2, -1, 0, 3)$, $\mathbf{a}_2 = (1, 2, 5, -1)$, $\mathbf{a}_3 = (7, -1, 5, 8)$;

(iv) $\mathbf{a}_1 = (1, 2, 1, -2)$, $\mathbf{a}_2 = (0, -2, -2, 0)$, $\mathbf{a}_3 = (0, 2, 3, 1)$, $\mathbf{a}_4 = (3, 0, -3, 6)$;

(v) $\mathbf{a}_1 = (3, 0, 4, 1)$, $\mathbf{a}_2 = (6, 2, -1, 2)$, $\mathbf{a}_3 = (-1, 3, 5, 1)$, $\mathbf{a}_4 = (-3, 7, 8, 3)$.

[**Ans.** (i) linearly dependent, $\mathbf{a}_3 = \mathbf{a}_1 + 3\mathbf{a}_2$;

(ii) linearly dependent, $\mathbf{a}_3 = 2\mathbf{a}_1 + \mathbf{a}_2$;

(iii) linearly dependent, $\mathbf{a}_3 = 3\mathbf{a}_1 + \mathbf{a}_2$;

(iv) independent;

(v) independent]

26. Express the vector

$$\begin{bmatrix} a_1 \\ a_2 \\ a_3 \end{bmatrix}$$

as a linear combination of the vectors

$$\begin{bmatrix} 0 \\ 1 \\ 1 \end{bmatrix}, \begin{bmatrix} 1 \\ 0 \\ 1 \end{bmatrix}, \begin{bmatrix} 1 \\ 1 \\ 0 \end{bmatrix}.$$

[**Ans.** $\begin{bmatrix} a_1 \\ a_2 \\ a_3 \end{bmatrix} = \frac{1}{2}(a_1 + a_2 - a_3) + \frac{1}{2}(a_1 + a_2 - a_3)\begin{bmatrix} 1 \\ 0 \\ 1 \end{bmatrix} + \frac{1}{2}(a_1 + a_2 - a_3)\begin{bmatrix} 1 \\ 1 \\ 0 \end{bmatrix}$]

27. Find the rank of A⊗B when

$$A = \begin{bmatrix} 1 & 2 & 3 \\ 4 & 5 & 6 \\ 2 & 4 & 6 \end{bmatrix} \text{ and } B = \begin{bmatrix} 0 & 0 & 0 & 1 \\ 1 & 2 & 0 & 2 \\ -1 & 2 & 5 & 2 \end{bmatrix}.$$

[**Ans.** 6]

28. Find the rank of the following matrix as a function of *h* and *k*:

$$\begin{bmatrix} 1 & -1 & 2 & 3 \\ 2 & 1 & h & 1 \\ 0 & -3 & k & 5 \\ 3 & 3 & 4 & -1 \end{bmatrix}.$$

[**Ans.** Rank is 3 unless h=3, k =1, in this case rank is 2]

29. Find index and signature of the following matrices

(i) $A = \begin{bmatrix} 1 & 2 & 3 \\ 2 & 2 & 0 \\ 3 & 0 & -2 \end{bmatrix}$, (ii) $A = \begin{bmatrix} 1 & 2 & -3 \\ 2 & 5 & 4 \\ -3 & -4 & 8 \end{bmatrix}$,

(iii) $A = \begin{bmatrix} 1 & -3 & 2 \\ -3 & 7 & -5 \\ 2 & -5 & 8 \end{bmatrix}$, (iv) $A = \begin{bmatrix} 1 & 1 & -2 & -3 \\ 1 & 2 & -5 & -1 \\ -2 & -5 & 6 & 9 \\ -3 & -1 & 9 & 11 \end{bmatrix}$,

(v) $A = \begin{bmatrix} 1 & i & 2+i \\ -i & 2 & 1-i \\ 2-i & 1+i & 2 \end{bmatrix}$, (vi) $A = \begin{bmatrix} 1 & -2 & 1 & -2 \\ -2 & 0 & 3 & -5 \\ 1 & 3 & 2 & 2 \\ -2 & -5 & 2 & -9 \end{bmatrix}$.

(vii) $A - \begin{bmatrix} 3 & 0 & 1 & -1 \\ 0 & 0 & -4 & 2 \\ 1 & -4 & 5 & -2 \\ -1 & 2 & -2 & 1 \end{bmatrix}$ (viii) $A = \begin{bmatrix} 0 & 0 & 1 \\ 0 & 0 & 0 \\ 1 & 0 & 0 \end{bmatrix}$.

[**Ans.** (i) p =2, r =3, s =2−1=1 ; (ii) p =2, r =3, s =1 ;
(iii) p =2, r =3, s =1; (iv) p =2, r =3, s =2 ;
(v) p =2, r =3, s =1; (vi) p =2, r =3, s =1 ;
(vii) r = 4 , p =2, s =0; (viii) p =2, r =3, s =2]

30. (i) For which real values of λ are the following matrices positive definite,

$$\begin{bmatrix} 1 & \lambda & \lambda \\ \lambda & 1 & \lambda \\ \lambda & \lambda & 1 \end{bmatrix}, \begin{bmatrix} 1 & 1 & \lambda \\ 1 & 1 & 1 \\ \lambda & 1 & 1 \end{bmatrix} ?$$

[**Hints:** $A = \begin{bmatrix} 1 & \lambda & \lambda \\ \lambda & 1 & \lambda \\ \lambda & \lambda & 1 \end{bmatrix} \underset{\approx}{R_2 - \lambda R_1} \begin{bmatrix} 1 & \lambda & \lambda \\ 0 & 1-\lambda^2 & \lambda-\lambda^2 \\ \lambda & \lambda & 1 \end{bmatrix} \quad \underset{\approx}{C_2 - \lambda C_1}$

$$\begin{bmatrix} 1 & 0 & \lambda \\ 0 & 1-\lambda^2 & \lambda-\lambda^2 \\ \lambda & \lambda-\lambda^2 & 1 \end{bmatrix} \underset{\approx}{R_3-\lambda R_1} \begin{bmatrix} 1 & 0 & \lambda \\ 0 & 1-\lambda^2 & \lambda(1-\lambda) \\ 0 & \lambda(1-\lambda) & 1-\lambda^2 \end{bmatrix} \underset{\approx}{C_3-\lambda C_1}$$

$$\begin{bmatrix} 1 & 0 & 0 \\ 0 & 1-\lambda^2 & \lambda(1-\lambda) \\ 0 & \lambda(1-\lambda) & 1-\lambda^2 \end{bmatrix} \underset{\approx}{R_3-\frac{\lambda}{1+\lambda}R_2} \begin{bmatrix} 1 & 0 & 0 \\ 0 & 1-\lambda^2 & \lambda(1-\lambda) \\ 0 & 0 & \frac{(1-\lambda)(1+2\lambda)}{1+\lambda} \end{bmatrix}$$

$$\underset{\approx}{C_3-\frac{\lambda}{1+\lambda}C_2} \begin{bmatrix} 1 & 0 & 0 \\ 0 & 1-\lambda^2 & 0 \\ 0 & 0 & \frac{(1-\lambda)(1+2\lambda)}{1+\lambda} \end{bmatrix}$$

$\therefore$ rank (A) = 3, if $\lambda \neq 1$, $\lambda \neq -\frac{1}{2}$, $\lambda \neq \pm 1$]

(ii) Test for definiteness

$$A=\begin{bmatrix} 2 & -1 & 0 & 0 \\ -1 & 2 & -1 & 0 \\ 0 & -1 & 2 & -1 \\ 0 & 0 & -1 & 2 \end{bmatrix}$$

[**Ans.** $n = r = p = 4$, $\therefore$ A is positive definite]

31. Find the particular solution and the complementary solution of the system

$$\begin{cases} 2x_1 + 3x_2 - x_3 - x_4 = -1 \\ 4x_1 + x_2 + x_3 - 2x_4 = 9 \end{cases}$$

[**Ans.** Particular solution is $(\frac{14}{5}, -\frac{11}{5}, 0, 0)$ and complementary solution is $c_1(-\frac{2}{5}, \frac{3}{5}, 1, 0) + c_2(\frac{1}{2}, 0, 0, 1)$]

32. (i) If a homogeneous system has rank 5 and at least two solutions, what is the minimum number of variables the system could have ? State reasons.

[**Hints :** 4th para of homogeneous under sec.10.9. rank = $n-1$. **Ans.** 6]

(ii) If a homogeneous system has five variables and exactly one solution, what are the minimum and the maximum possible values for its rank .

[**Hints :** 3rd para on homogeneous under sec.10.9. **Ans.** max .= 5 = min.]

33. Find the rank of the following homogeneous systems and determine number of possible solutions:

(i) $2x - y = 0$; (ii) $\left.\begin{array}{l} x + y + z = 0 \\ x + 2y + z = 0 \\ x + 4y + z = 0 \\ -x + y - z = 0 \end{array}\right\}$; (iii) $\left.\begin{array}{l} 3x - 2y + z = 0 \\ 2x - 4y + 2z = 0 \\ -x - y + 3z = 0 \end{array}\right\}$

[**Ans.**(i) rank =1<(no. of variables , infinite number of solutions),

(ii) rank = 2< no. of variables , infinite solutions ,

(iii) rank 3= no. of variables , unique solution i.e. trivial solution]

34. Determine whether or not the following pair of matrices are congruent (use signature test).

(i) $\begin{bmatrix} 1 & i & 2+i \\ -i & 2 & 1-i \\ 2-i & 1+i & 2 \end{bmatrix}$ & $\begin{bmatrix} 1 & 1+i & 2i \\ 1-i & 4 & 2-3i \\ -2i & 2+3i & 7 \end{bmatrix}$;

(ii) $\begin{bmatrix} 1 & -3 & 2 \\ -3 & 7 & -5 \\ 2 & -5 & 8 \end{bmatrix}$ & $\begin{bmatrix} 0 & 1 & 1 \\ 1 & -2 & 2 \\ 1 & 2 & -1 \end{bmatrix}$;

(iii) $\begin{bmatrix} 1 & i \\ -i & 2 \end{bmatrix}$ & $\begin{bmatrix} 1 & 2+3i \\ 2-3i & -1 \end{bmatrix}$;

(iv) from Ex.28 above (vi) & (vii);

(v) from Ex.28 above (ii) & (iii);

(vi) $\begin{bmatrix} 0 & 1 \\ 1 & 0 \end{bmatrix}$ & $\begin{bmatrix} 0 & 4 \\ 4 & 5 \end{bmatrix}$;

(vii) $\begin{bmatrix} 1 & 0 \\ 0 & -1 \end{bmatrix}$ & $\begin{bmatrix} 0 & i \\ -i & 0 \end{bmatrix}$; (viii) from Ex.28 above (iii) & (v).

[**Ans.** (i) Congruent, both sig =1;

(ii) not congruent, because sig of matrices are 1 & −1 respectively;

(iii) not congruent as sig. are 2 and 0 respectively;

(iv) not congruent as signatures are different;

(v) congruent as both signatures are equal;

(vi) congruent as both signatures are zero;

Hints : congruent transformations of 1st one are $R_{21}(i), C_{21}(i)$; $R_{12}(-\frac{1}{2}), C_{12}$; $R_1(\sqrt{2})$, $C_1(\sqrt{2})$; $R_2(\frac{1}{\sqrt{2}})$, $C_2(\frac{1}{\sqrt{2}})$; R_{12}, C_{12} ; similarly for 2nd one except first pair of operations ;

(vii) congruent as both signatures are zero. congruent operation for 2nd one are $R_2(i)$, $C_2(-i)$; $R_{12}(-\frac{1}{2}), C_{12}(-\frac{1}{2})$; $R_1(\sqrt{2}), C_1(\sqrt{2})$; $R_2(\frac{1}{\sqrt{2}}), C_2(\frac{1}{\sqrt{2}})$;

(viii) congruent , because both signatures are 1]

35. Find a nonsingular matrix P with rational elements such that $P^T AP$ is diagonal where

$$A=\begin{bmatrix} 1 & 2 \\ 2 & 4 \end{bmatrix}.$$

Verify yourself the statement.

[Hints : $A \overset{R_2-2R_1}{\approx} \begin{bmatrix} 1 & 2 \\ 0 & 0 \end{bmatrix} \overset{C_2-2C_1}{\approx} \begin{bmatrix} 1 & 0 \\ 0 & 0 \end{bmatrix}$. **Ans.** $P = \begin{bmatrix} 1 & -2 \\ 0 & 1 \end{bmatrix}$]

36. What are the rank and the indices of the matrices

$$A_1 = \begin{bmatrix} 0 & 1 & 2 \\ 1 & 0 & 3 \\ 2 & 3 & 0 \end{bmatrix} \text{ and } A_2 = \begin{bmatrix} 0 & 0 & 1 \\ 0 & 0 & 0 \\ 1 & 0 & 0 \end{bmatrix}$$

such that the diagonal contains at least one nonzero element.

[Ans. congruent matrices of A_1 & A_2 are

$$\begin{bmatrix} 1 & 0 & 0 \\ 0 & 1 & 0 \\ 0 & 0 & 1 \end{bmatrix} \text{ and } \begin{bmatrix} 1 & 0 & 0 \\ 0 & 1 & 0 \\ 0 & 0 & 0 \end{bmatrix}$$

respectively. Also rank $A_1 = 3$,

index p_1 of $A_1 = 3$ = number of +ve 1's in diagonal form

and rank $A_2 = 2$ and index $p_2 = 2$]

37. Show that rank (A), rank (A_1), and rank (A_2) of Ex.37 (a) of exercises nine, are equal and equal to 3.

38. Find the ranks of the new matrices (a), (b), (c) in Ex.40 of exercises nine. Conclude that

rank (b) ≠ rank (a).

[**Ans.** (a) rank = 4, (b) rank = 3, (c) rank = 4]

CHAPTER ELEVEN

VECTOR (LINEAR) SPACES

Chapter summary: *In the usual manner this chapter also begins with introduction. Then introduces the algebraic structures like set, group, ring & field. Definition and properties of vector space, its clarification of abstractness, physical interpretation; subspace & its identification criteria, algebra of subspaces — intersection, union, linear & direct sum; linear combinations and spans, equality of two linear spans, linear independence/dependence; basis & dimension, related theorems, dimension of linear & direct sums; rank of a matrix, system of equations; reducing a spanning set to a basis, extension of basis, change of basis.*

11.1 Introduction

This chapter concerns with the algebraic system — vector spaces (linear space). The concept of vector is essential in functional analysis, which has applications to differential equations , numerical analysis and other fields of practical interest to engineers. Vector subspace plays an important role in providing the geometrical basis of matrix theory. Algebraic system on the other hand needs some knowledge of set and its operations because an algebraic system or structure is defined as a nonempty set of elements S with one or more operations over S satisfying certain postulates or axioms. So, before going into some details of vector spaces, let the learners be introduced in a nutshell with sets and correlation of vector space with other algebraic system like group, ring and field.

11.1.1 Set and set operations : It is the basic object of mathematics. It is defined as a collection of objects called *elements* having similar characteristics. The elements are enclosed into a pair of curl braces "{" and "}" and usually denoted by capital letters like S, A, B, etc. Thus

$$S = \{0, 2, 5, 6\},\ S = \{\text{students of Electrical Engineering}\}\ \text{etc.}$$

The symbol "$\in$" and "$\notin$" are used to mean "belong to " and "does not belong to ". Thus $2 \in S$ while $4 \notin S$. The symbol $|S|$ is used to mean number of elements in S and $\phi = \{\ \}$ is used for null set i.e. set ϕ has no elements. Thus $|S| = 4$ and $|\phi| = 0$.

There are three operations on set. They are union, intersection and complement while 4th combined operation of intersection & complement is difference. They are defined and illustrated.

(i) Union : Union of two sets A and B is denoted by $A \cup B$ to mean the elements which belong to A or B or both. Thus $A = \{2, 4, 7\}$, $B = \{1, 2, 3\}$. Then $A \cup B = \{1, 2, 3, 4, 7\}$.

(ii) Intersection : Intersection of two sets A and B is denoted by $A \cap B$ to mean the elements common to both A and B. Thus $A \cap B = \{2\}$ of the sets in (i).

(iii) Complement or **negation :** Complement or negation of a set A is defined with respect to a bigger set U, called *universal set*. It is denoted by A' or A^c. Thus if $U = \{1, 0, -1, 2, 3\}$,

A = {0 , –1 , 2} and B = {1 , 2 , 3} then A^c = {1 , 3}, B^c = {0 , –1}. In this book A′ will *not be used* as it denotes the derivative in the calculus part.

(iv) Difference : Difference of two sets A and B denoted by (A – B) means elements belonging to A but not in B. So, using sets in (iii),

$$(A - B) = \{0, -1\} = A \cap B^c \text{ and } B - A = \{1, 3\} = B \cap A^c.$$

Thus, the difference operation can be expressed in terms of intersection and comlementation.

Equality of sets : As in the equality of matrices and vectors, two sets A & B are equal if both have exactly the same elements irrespective of their position and is written as A = B if A = {1 , 0 , 5} and B = {0 , 1 , 5}.

Subset : Set A is a subset of the set B written symbolically as $A \subseteq B$ if every element of A is also an element of B. Thus,

$$A \subseteq B,\ B \subseteq B \text{ and } A \subseteq A \text{ if } A = \{1, 2, 4\},\ B = \{0, -1, 7, 4, 2, 1\}.$$

Empty set is a subset of every set. Thus $\phi \subseteq A$, $\phi \subseteq B$, $\phi \subseteq \phi$.

Equality in terms of subsets : Set A = set B iff $A \subseteq B$ and $B \subseteq A$.

Proper subset : Set A is a proper subset of B written $A \subset B$ if $A \subseteq B$ but $A \neq B$ then B must contain at least one element not in A. Thus

$$A = \{1, 2, 3\} \subset B = \{0, 1, 2, 3\} \text{ and } \phi \subset \{5\}.$$

Disjoint set : Two sets A and B are disjoint if their intersection is an empty or null set. Thus if A={0 ,–1 , 2} and B = {5 , –2 , 7 , 8} the $A \cap B = \phi$ and A and B are disjoint sets.

11.1.2 Group : A *group* is an algebraic system consisting of

(i) a nonempty set of elements S and,

(ii) a binary operation "∘" over S satisfying the following 4 postulates ,

(1) closure, i.e. if $a \in S$ and $b \in S$ then $a \circ b \in S$

(2) "∘" is associative, i.e. $\forall a, b, c \in S$, $a \circ (b \circ c) = (a \circ b) \circ c$,

(3) S contains an identity element for "∘" i.e $a \circ e = a$, $\forall a, e \in S$ where e is an identity element ;

(4) each element in S has an inverse element in S with respect to "∘" i.e. $a \circ b = e$, $\forall a, b, e \in S$.

Commutative/abelian group : It is a group with an additional 5th postulate

(5) commutative : $a \circ b = b \circ a$, $\forall a, b \in S$.

The symbol "∀" stands " for all ".

The stated binary operation may be ordinary algebraic operations or an arbitrary defined operation.

Ex.1 Determine whether the S = {a , b} and the binary operation "∘" defined by

∘	a	b
a	a	b
b	b	a

form a group ?

(i) The given set S is evidently nonempty as it contains two elements a and b.

(ii) Let us test 4 postulates of (ii). (1) $a \circ b = b \in S$ which is satisfied.

(2) $(a \circ b) \circ a = b \circ a = b$ and $a \circ (b \circ a) = a \circ b = b$.

$\therefore$ $(a \circ b) \circ a = a \circ (b \circ a)$.

$\therefore$ (2) is satisfied.

(3) $a \circ b = b$.

∴ a is the identity element in S.

(4) $a \circ b = a$.

∴ b is the inverse element in S.

So, (3) & (4) of (ii) are satisfied.

∴ S forms a group.

Ex.2 Determine whether all 2×3 real matrices under matrix addition form a group or not.

Let $$M = \left\{ \begin{bmatrix} a_{11} & a_{12} & a_{13} \\ a_{21} & a_{22} & a_{23} \end{bmatrix} \right\}.$$

The set M is nonempty. So condition (i) is satisfied.

Closure: $$\begin{bmatrix} a_{11} & a_{12} & a_{13} \\ a_{21} & a_{22} & a_{23} \end{bmatrix} + \begin{bmatrix} b_{11} & b_{12} & b_{13} \\ b_{21} & b_{22} & b_{23} \end{bmatrix} = (\text{another } 2\times3 \text{ matrix}) \in M.$$

Associative: $A + (B + C) = (A + B) + C \quad \forall$ matrices $A, B, C \in M$.

Identity: $A + 0 = A$, where 0 is a 2×3 null matrix $\in M$.

Inverse: $A - A = 0$

∴ $(-A)$ is inverse & belongs to M.

∴ the set of all 2×3 matrices under matrix addition forms a group.

11.1.3 Ring : A *ring* is an algebraic system consisting of :

(i) a nonempty set of elements, S.

(ii) two binary operations " ® " and " © " over S subject to the 3 postulates

(1) S forms an abelian group w. r. t " ® ";

(2) " © " is associative i.e. a © (b © c) = (a © b) © c $\quad \forall\, a, b, c \in S$

(3) " © " is right and left distributive w. r. t " ® " i.e.

(A) (a ® b) © c = (a © c) ® (b © c) and

(B) a © (b ® c) = (a © b) ® (a © c).

The definition implies that set must be commutative group w. r. t. " ® " first and then it must satisfy three more postulates i.e. altogether eight postulates must be satisfied by S.

Ring with unity : A ring with postulate that S contains an identity element for " © ".

Commutative ring : Ring must be commutative w. r. t " © ".

Division ring : A ring with unity, with at least two elements and additional property : every element except identity for " ® " has an inverse w. r. t " © ".

Ex. Note the definition and example under the field.

11.1.4 Field : A commutative division ring is called a *field.* Thus a field is an algebraic system consisting of

(i) a set S containing at least two elements,

(ii) two binary operations " ® " and " © " over S, subject to the 3 postulates :

(1) S forms an abelian group w. r. t " ® " ;

(2) S, without the identity for " ® ", forms an abelian group w. r. t " © ";

(3) " © " is right and left distributive w. r. t " ® ".

In practice an arbitrary field is denoted by F instead of S. An *explicit* definition of the field is given below. A field is a nonempty set of elements $F = \{a, b, c, ...\}$ together with two binary operations or compositions in F, called *addition* " ® "and *multiplication* " © " for which the following axioms or postulates hold true : (For simplicity " ® " & " © " are replaced by usual addition & multiplication signs of algebra and note that signs " ® " & " © " are arbitrary)

(a) Addition possesses

(i) Closure : $a + b \in F \quad \forall\, a, b \in F$, where $\forall$ stands for all.

(ii) Commutativity : $a + b = b + a, \quad \forall\, a, b \in F$.

(iii) Associativity : $a + (b + c) = (a + b) + c \;, \forall\, a, b \;\&\; c \in F$.

(iv) Identity element : There is a unique identity element called the *zero element* such that

$$a + 0 = a \quad \forall\, a \in F.$$

(v) Inverse element : For each $a \in F$, there exists an element in F called the *negative* of a and designated by $(-a)$, such that $a + (-a) = 0 =$ identity element.

(b) Multiplication possesses

(i) Closure : $ab \in F \quad \forall\, a, b \in F$.

(ii) Commutativity : $ab = ba \quad \forall\, a, b \in F$.

(iii) Associativity : $a\,(bc) = (ab)\,c\,, \quad \forall\, a, b, c \in F$.

(iv) Identity element : There is a unique identity element called the *unit element*, 1(one) in F such that $a.1 = a, \quad \forall\, a \in F$.

(v) Inverse element : For each a in F , $a \neq 0$, there exists an element called the *inverse* of a and designted by a^{-1} such that $a.\, a^{-1} = 1$, multiplicative identity.

(vi) Distributivity over addition : $a\,(b + c) = ab + ac, \quad \forall\, a, b, c$ in F.

The axioms (a) constitute abelian group w.r.t. " ® " while first five axioms (i) to (v) of (b) constitute abelian group w. r. t. " © ". Hence a field, may be *compactly defined* as a nonempty set which forms an abelian group w. r. t. two operations " ® " & " © " and distribution over addition.

Ex. Show that all the 2×2 matrices of the form

$$\begin{bmatrix} a & -b \\ b & a \end{bmatrix}$$

where a and b are real numbers, form a field under usual matrix addition and multiplication.

Addition postulates : Let

$$M=\left\{\begin{bmatrix} a & -b \\ b & a \end{bmatrix}\right\}.$$

(i) Closure : Let

$$A=\begin{bmatrix} a & -b \\ b & a \end{bmatrix}, B=\begin{bmatrix} c & -d \\ d & c \end{bmatrix}$$

$$\therefore \quad A + B = \begin{bmatrix} a+c & -(b+d) \\ b+d & a+c \end{bmatrix}$$

which is of the same form as M and belongs 2×2 matrices.

(ii) Commutativity : A + B = B + A since real numbers $a + c = c + a$ etc.

(iii) Associativity : Let

$$C = \begin{bmatrix} e & -f \\ f & e \end{bmatrix}.$$

$\therefore$ A + (B + C) = (A + B) + C $\quad \because$ real numbers $a + (c + f) = (a + c) + f$, etc.

(iv) Identity :

$$0_{2\times2}=\begin{bmatrix}0 & -0\\ 0 & 0\end{bmatrix} \text{ exists such that } A+0=A.$$

(v) Inverse:

$$(-1)A=\begin{bmatrix}-a & b\\ -b & -a\end{bmatrix}. \quad \therefore A+(-A)=0.$$

$\therefore$ M forms an abelian group w. r. t matrix addition '+'.

Multiplication postulates :

(i) Closure :

$$AB=\begin{bmatrix}ac-bd & -(ad+bc)\\ bc+ad & (ac-db)\end{bmatrix}\in M$$

(ii) Commutivity :

$$BA=\begin{bmatrix}ac-bd & -(bc+ad)\\ ad+bc & (ac-bd)\end{bmatrix}. \quad \therefore AB=BA$$

(iii) Associativity:

$$(AB)C=A\,(B\,C)=\begin{bmatrix}e(ac-bd)-f(ad+bc) & -(e(ad+bc)+f(ac-bd))\\ e(ad+bc)+f(ac-bd) & e(ac-bd)-f(ad+bc)\end{bmatrix}.$$

(iv) Identity :

$$I=\begin{bmatrix}1 & 0\\ 0 & 1\end{bmatrix} \text{ exists such that } AI=I.$$

(v) Inverse : $|A|=a^2+b^2 \neq 0$ for any one of $a, b \neq 0$ so A^{-1} exists. So

$$AA^{-1}=A^{-1}A=I.$$

$\therefore$ M is an abelian group w. r. t matrix multiplication.

(vi) Distribution over addition:

$$A\,(B+C)=\begin{bmatrix}a & -b\\ b & a\end{bmatrix}\begin{bmatrix}c+e & -(d+f)\\ d+f & c+e\end{bmatrix}$$

$$=\begin{bmatrix}a(c-e)-(d+f) & -a(d+f)-b(c+e)\\ b(c+e)+a(d+f) & -b(d+f)+a(c+e)\end{bmatrix}$$

$$=\begin{bmatrix}(ac-bd)+(ae-bf) & -(ad+bc)-(af+be)\\ (bc+ad)+(be+af) & (ac-bd)+(ae-bf)\end{bmatrix}$$

$$=\begin{bmatrix}ac-bd & -(ad+bc)\\ bc+ad & ac-bd\end{bmatrix}+\begin{bmatrix}ae-bf & -(af+be)\\ be+af & ac-bf\end{bmatrix}=AB+AC.$$

Thus all the conditions are satisfied. So, M forms a field.

11.2 Vector (Linear) space

It is another algebraic system consisting of

(i) a nonempty set of elements called *vectors*,

(ii) an associated field and

(iii) two binary operations — addition and scalar multiplication.

A formal definition in definition 1 and explicit definition in definition 2 are given below.

11.2.1 Definition 1. A set V of elements, called *vectors*, forms a vector space (or linear space) over a field F if

(A) V forms an abelian group w. r. t a binary operation,

(B) a unique element of V can be obtained from the premultiplication of any element of V by an arbitrary element of the associated field F,

(C) the following postulates are valid :

(1) $(c_1 + c_2)\mathbf{u} = c_1\mathbf{u}$ ® $c_2\,\mathbf{u}$, $\forall$ $\mathbf{u} \in$ V and $\forall c_1 , c_2 \in$ F and " + " is a field operation while " ® " group operation

(2) $(c_1.c_2)\mathbf{u} = c_1c_2\mathbf{u}$, $\forall$ $\mathbf{u} \in$ V and $\forall$ $c_1 , c_2 \in$ F and ". " is a field operation

(3) c_1 (**u** ® **v**) = $c_1\mathbf{u}$ ® $c_1\mathbf{v}$, $\forall$ **u** , **v** $\in$ V and $\forall c_1 \in$ F and " ® " is group operation

(4) $1\mathbf{u} = \mathbf{u}$, $\forall \mathbf{u} \in$ V and 1 is the multiplicative identity of F.

N.B. The elements of V are called *vectors* while those of F are called *scalars*. The vector space over a field is denoted by V(F) and sometimes by V_n(F) or V^n(F) where n specifies the dimension of vector space, V.

Note that a field must contain an identity element, but a vector space does not.

Vectors of order *n* : A vector of order n is described as $\mathbf{x} = (x_1 , x_2 , \cdots , x_n)$, $x_k \in$ F, k =1, 2 $\cdots$, n and is expressed as n-tuple, n atributes or n-vectors. Some text described it as the n-dimensional vector. The exact meaning of dimension will be found in sec.11.12. It will be clear in that section that n-tuple vectors may constitute a vector space but that vector space may have dimension other than n.

11.2.2 Definition 2. A vector space (or linear space) V consists of the following: (To remove complexity of notations distinction between field and group operations are not made and ordinary addition and multiplication signs are used)

1. a field F of scalars ;
2. a nonempty set **v** of objects, called *vectors, elements* or *points* ;
3. a rule (or operation) of vector addition, which associates with each pair of vectors **u, v** in V, a vector **u** + **v** in V (closure for addition), called the *sum* of **u** and **v**, in such a way that addition satisfies 4 axioms ;

(a_1) commutativity : **u** + **v** = **v** + **u**;

(a_2) associativity : **u** + (**v** + **w**) = (**u** + **v**) + **w**;

(a_3) additive identity : There is a unique vector 0 in V, called the *zero vector,* such that **u** + 0 = **u** $\forall$ **u** in V;

(a_4) additive inverse : for each vector **u** in V there is a unique vector (– **u**) in V such that **u** + (– **u**) = 0;

4. a rule (or operation) of scalar multiplication, which associates with each scalar c in F and vector **u** in V, a vector c**u** in V (closure for scalar multiplication) called the product of c and **u** in such a way that scalar multiplication satisfies 4 other axioms;

(m_1) multiplicative identity : 1**u** = **u** for every **u** in V;

(m_2) associativity for scalar multiplication : $(c_1c_2)\mathbf{u} = c_1(c_2\,\mathbf{u})$;

(m_3) distributivity over vector addition : $c(\mathbf{u} + \mathbf{v}) = c\mathbf{u} + c\,\mathbf{v}$;

(m_4) distributivity over scalar addition : $(c_1 + c_2)\mathbf{u} = c_1\mathbf{u} + c_2\mathbf{u}$.

11.2.3 Examples of vector spaces: The following are some familiar examples of vector space. It can be easily shown that these vector spaces satisfy all the eight axioms of section 11.2.2. as well as the formal definition of section 11.2.1. Some of them are verified in Ex. 1 to 7 after sec. 11.3.

(1) The set of all tuples of real numbers, $R^n = \{(x_1, x_2, \cdots, x_n) : x_i \in R\}$. To be more specific, for $n = 1, 2, 3$ the vector spaces are real line, plane and 3-dimensional spaces respectively.

(2) Similarly, the set of n-tuples of complex numbers constitute a vector space (cf. Ex. 4).

(3) The set of all polynomials in single variable x with real (complex) coefficients such as

$\sum_{i=1}^{n} a_i x^i$ is a vector space over R (or C) [cf. Ex. 3].

(4) The set of all continuous (sectionally continuous) functions on an interval $[a, b]$ constitutes a vector space [cf. Ex. 5, 6].

(5) The set of all Riemann-integrable functions on an interval $[a , b]$ is a vector space.

(6) The set of $m \times n$ matrices with entries from R (or C) constitutes a vector space over R (or C) [cf. Ex. 7].

(7) The set of all polynomials of degree at most 5 constitutes a vector space.

(8) The set of all solutions of $\mathbf{Ax} = 0$, where A is an $m \times n$ matrix, $\mathbf{x}$ is an n-vector in R^n as well as the set of all solutions of the differential

$$\frac{d^2y}{dx^2} + 5\frac{dy}{dx} + 6y = 0$$

separately constitute two vector spaces called *solution spaces* or *null spaces*.

(9) The set of all infinite sequences $a_1, a_2, \cdots$ of real or complex numbers is a vector space.

(10) Zero vector space: Let V consists of a single object which is denoted by 0 and addition and scalar multiplication by a scalar c are defined as $0 + 0 = 0$ and $c\,0 = 0$, $\forall$ scalars c. This type of vector space is called the *zero vector space*. It is easy to check all the conditions of the vector space are satisfied by 0.

11.3 Clarification of abstractness and testing

This section makes an attempt to clarify the abstractness involved in vector space.

Clarification of abstractness : In all the four algebraic system defined above the binary addition and the binary multiplication used not only to mean the usual addition and multiplications, but also to mean arbitrarily defined operations such as

(i) binary addition of a & b may be defined as $(a \circledR b) = a + b - ab$ and by a table such as

®	a	b
a	a	b
b	b	a

etc.

(ii) binary multiplication of a & b may be defined as $a \cdot b = a + b + ab$ and by a table such as

·	a	b
a	a	b
b	b	a

.

Again, the word "vector" or "element" used in vector space has abstractness in the fact that it is devoid of any *geometrical* and *physical* attributes in contrast to 2 and 3 dimensional vectors of vector analysis. It may represent an n-tuple of real (complex) numbers, functions, matrices, polynomials etc. depending of the nature of the vector space. Thus, due to the arbitrariness of both operations and vectors, vector spaces are sometimes understood as abstract vector space. Finally, the algebra of vector spaces is called *linear algebra.* Now a days, matrix theory and linear algebra are two synonyms. The representation of vectors by bold faced letters is also unnecessary for n-tuples, $n > 3$.

Testing : To test whether a given set forms a vector space or not, verify all the 10 conditions i.e. 2 closure properties and 8 axioms of the vector space. To save time and space, it is sufficient to test the existence of two closure properties, additive inverse, multiplicative identity. Note the examples given below.

Ex.1 Let S denote the set of all ordered pairs of real numbers

$$\alpha = (a_1 , a_2) , \ \beta = (b_1 , b_2) , \gamma = (c_1 , c_2) , \ \cdots \text{ in } R^2.$$

Determine whether or not R^2 constitutes a vector space if addition and scalar multiplication are defined as

$$\alpha + \beta = (a_1 + b_1 , \ a_2 + b_2) \text{ and } c\alpha = (ca_1, ca_2)$$

Since closure properties are true as $\alpha + \beta \in R^2$ and $c\alpha \in R^2$

(a_1) $\alpha + \beta = (a_1 + b_1 , a_2 + b_2)$, $\beta + \alpha = (b_1 + a_1 , b_2 + a_2)$, so $\alpha + \beta = \beta + \alpha$

(a_2) $(\alpha + \beta) + \gamma = (a_1 + b_1 , a_2 + b_2) + (c_1 , c_2)$

$= ((a_1 + b_1) + c_1 , (a_2 + b_2) + c_2)$

$= (a_1 + b_1 + c_1 , a_2 + b_2 + c_2)$

and $\alpha + (\beta + \gamma) = (a_1 , a_2) + (b_1 + c_1 , b_2 + c_2)$

$= (a_1 + (b_1 + c_1) , a_2 + (b_2 + c_2))$

$= (a_1 + b_1 + c_1 , a_2 + b_2 + c_2)$

$\therefore$ $(\alpha + \beta) + \gamma = \alpha + (\beta + \gamma)$

(a_3) A null vector is $0 = (0 , 0)$. So, $\alpha + 0 = (a_1 + 0 , a_2 + 0) = (a_1 , a_2) = \alpha$

(a_4) Let $\alpha = (a_1 , a_2)$ then $(-\alpha) = -(a_1 , a_2) = (- a_1 , - a_2)$

$\therefore$ $\alpha + (- \alpha) = (a_1 , a_2) + (-a_1 , -a_2) = (a_1 - a_1) + (a_2 - a_2) = (0 , 0)$

$\therefore$ R^2 from a commutative group under the defined addition.

The set is closed under defined scalar multiplication as $c\alpha \in R^2$.

(m_1) $1\alpha = (1a_1 , 1a_2) = (a_1 , a_2) = \alpha$

(m_2) $c_1(c_2 \alpha) = c_1 (c_2 (a_1 , a_2)) = c_1(c_2 a_1 , c_2 a_2)$

$= (c_1c_2a_1 , c_1c_2 a_2) = (c_1c_2) (a_1 , a_2) = c_1c_2\alpha$

(m_3) (i) $c(\alpha + \beta) = c (a_1 + b_1 , a_2 + b_2) = (c(a_1 + b_1) , c(a_2 + b_2))$

and $c\alpha + c\beta = c(a_1 , a_2) + c(b_1 , b_2) = (ca_1 , ca_2) + (cb_1 , cb_2)$

$= ((ca_1 + cb_1), (ca_2 + cb_2)) = (c (a_1 + b_1) , c(a_2 + b_2))$

So , $c(\alpha + \beta) = c\alpha + c\beta$.

(ii) $(c_1 + c_2) \alpha = (c_1 + c_2) (a_1, a_2)$

$= ((c_1 + c_2)a_1 , (c_1 + c_2)a_2) = (c_1a_1 + c_2a_1 , c_1a_2 + c_2a_2)$

and $c_1\alpha + c_2\alpha = (c_1a_1 , c_1a_2) + (c_2a_1 , c_2a_2) = (c_1a_1 + c_2a_1 , c_1a_2 + c_2a_2)$

So , $(c_1 + c_2) \alpha = c_1\alpha + c_2\alpha$.

$\therefore$ R^2 forms a vector space over the field R of real numbers.

Ex.2 Do, Ex. 1, if the addition and scalar multiplication are defined as

$$\alpha + \beta = ((a_1^3 + b_1^3)^{\frac{1}{3}} , (a_2^3 + b_2^3)^{\frac{1}{3}}) \text{ and } c\alpha = (ca_1, ca_2)$$

Two closure properties are satisfied as both $\alpha+\beta$ and $c\alpha \in R^2$

(a₁) $\alpha+\beta=(a_1, a_2)+(b_1,b_2)=((a_1^3+b_1^3)^{\frac{1}{3}},(a_2^3+b_2^3)^{\frac{1}{3}})$ and

$\beta+\alpha=(b_1, b_2)+(a_1, a_2)$

$=((b_1^3+a_1^3)^{\frac{1}{3}},(b_2^3+a_2^3)^{\frac{1}{3}})=((a_1^3+b_1^3)^{\frac{1}{3}},(a_2^3+b_2^3)^{\frac{1}{3}})$

So, $\alpha+\beta=\beta+\alpha$.

(a_2) $(\alpha+\beta)+\gamma$

$=((a_1^3+b_1^3)^{\frac{1}{3}},(a_2^3+b_2^3)^{\frac{1}{3}})+(c_1,c_2) = (a_1^3+b_1^3+c_1^3)^{\frac{1}{3}},(a_2^3+b_2^3+c_2^3)^{\frac{1}{3}})$

and $\alpha+(\beta+\gamma)$

$= (a_1,a_2)+((b_1^3+c_1^3)^{\frac{1}{3}},(b_2^3+c_2^3)^{\frac{1}{3}}) =((a_1^3+b_1^3+c_1^3)^{\frac{1}{3}},(a_2^3+b_2^3+c_2^3)^{\frac{1}{3}})$

So, $(\alpha+\beta)+\gamma=\alpha+(\beta+\gamma)$.

(a_3) $\alpha+0=(a_1, a_2)+(0, 0)= ((a_1^3+0^3)^{\frac{1}{3}},(a_2^3+0^3)^{\frac{1}{3}})=\alpha$

(a_4) $\alpha+\mathbf{x}=(a_1, a_2)+(x_1, x_2)=((a_1^3+x_1^3)^{\frac{1}{3}},(a_2^3+x_2^3)^{\frac{1}{3}})$

Now $\alpha+\mathbf{x}=0 \Rightarrow (a_1^3+x_1^3)^{\frac{1}{3}}=0$ and $(a_2^3+x_2^3)^{\frac{1}{3}}=0$

i.e. $a_1^3+x_1^3=0$ and $a_2^3+x_2^3=0$. Here x_1, and x_2 are unique for real values only. Because $x_1=-a_1, \frac{1}{2}(1\pm i\sqrt{3})a_1$ and $x_2=-a_2, \frac{1}{2}(1\pm i\sqrt{3})a_2$.

$\therefore$ $\mathbf{x}=(-a_1,-a_2)=-\alpha$, the negative of α.

$\therefore$ R^2 forms a group under defined addition.

(m_1) $1\alpha=1(a_1, a_2)=(a_1, a_2) = \alpha$.

(m_2) $c_1(c_2\alpha) = c_1(c_2a_1, c_2a_2) = (c_1c_2a_1, c_1c_2a_2) = (c_1c_2)(a_1, a_2) = (c_1c_2)\alpha$ (m_3) $(c_1+c_2)\alpha = ((c_1+c_2)a_1, (c_1+c_2)a_2)$ and

$c_1\alpha+c_2\alpha=(c_1a_1,c_1a_2)+(c_2a_1,c_2a_2) = ((c_1^3+c_2^3)^{\frac{1}{3}}a_1,(c_1^3+c_2^3)^{\frac{1}{3}}a_2)$.

So, $(c_1+c_2)\alpha \neq c_1\alpha+c_2\alpha$.

Hence the given set under the given operation of addition and c-multiple do not constitute a vector space.

Ex.3 (Polynomial space) Prove that P is a real vector space if it denotes the set of all polynomials in x with real coefficients.

Let $P(x)=a_0+a_1x+a_2x^2+\cdots+a_nx^n$ where $a_0, a_1, \cdots, a_n$ are real numbers and addition and scalar multiplication are defined as in high school algebra. Let

$$P_1(x)=a_{01}+a_{11}x+a_{21}x^2+\cdots+a_{n1}x^n,$$
$$P_2(x)=a_{02}+a_{12}x+a_{22}x^2+\cdots+a_{n2}x^n,$$
$$P_3(x)=a_{03}+a_{13}x+a_{23}x^2+\cdots+a_{n3}x^n,$$

be the set of polynomials. Two closure properties hold true.

(i) $P_1(x)+P_2(x) = (a_{01}+a_{02})+(a_{11}+a_{12})x+\cdots+(a_{n1}+a_{n2})x^n$

$= (a_{02}+a_{01})+(a_{12}+a_{11})x+\cdots+(a_{n2}+a_{n1})x^n$

$=(a_{02}+a_{12}x+a_{22}x^2+\cdots+a_{n2}x^n)+$

$(a_{01}+a_{11}x+a_{21}x^2+\cdots+a_{n1}x^n)$,

$= P_2(x) + P_1(x)$.

(ii) Similarly, $P_1(x) + (P_2(x) + P_3(x)) = (P_1(x) + P_2(x)) + P_3(x)$

(iii) $P + 0 = P$ follows easily from ordinary algebra

(iv) $P + (-P) = 0$ also follows from ordinary algebra.

(v) $\alpha(P_1(x) + P_2(x)) = (\alpha a_{01} + \alpha a_{02}) + (\alpha a_{11} + \alpha a_{12})x + \cdots + (\alpha a_{n1} + \alpha a_{n2})x^n$.

$= \alpha [a_{01} + a_{11}x + \cdots + a_{n1}x^n] + \alpha [a_{02} + a_{12}x + \cdots + a_{n2}x^n]$

$= \alpha P_1(x) + \alpha P_2(x)$.

(vi) $(\alpha + \beta)P(x) = (\alpha + \beta)(a_0 + a_1x + a_2x^2 + \cdots + a_nx^n)$

$= \alpha [a_0 + a_1x + a_2x^2 + \cdots + a_nx^n] + \beta [a_0 + a_1x + a_2x^2 + \cdots + a_nx^n]$

$= \alpha P(x) + P(x)$.

(vii) $(\alpha\beta)P(x) = \alpha(\beta P(x))$ can be shown easily

(viii) $1.P(x) = P(x)$.

∴ $P(x)$ constitute a vector space under ordinary algebraic definition of addition and scalar multiplication.

Ex.4 Let C denote the set of all complex numbers, i.e. all numbers of the form $a + bi$, a and b real, & $i = \sqrt{-1}$. With addition of complex numbers defined by $(a + bi) + (c + di) = (a + c) + (b + d)i$,

and scalar multiplication defined by

$$\alpha(a + bi) = \alpha a + \alpha bi,$$

prove that C is a vector space. Compare this space with R^2.

Two closure properties are true from given definitions.

(a_1) Commutative : $(a + bi) + (c + di) = (c + a) + (d + b)i = (c + di) + (a + bi)$

(a_2) Associative: $\{(a + bi) + (c + di)\} + (e + fi) = \{(a+bi)+(c+ di)\}+(e+fi)$

$= (a+c+e)+(b+d+f)i$

$= \{a + (c + e)\}+\{b + (d + f)i\}$

$= (a + bi) + \{(c + e) + (d + f)i\}$

$= (a + bi) + \{(c + di) + (e + fi)\}$

(a_3) $(a + bi) + (0 + i0) = (a + 0) + (b+0)i = (a + bi)$

(a_4) $(a + bi) + (-(a + bi)) = (a - a) + (b - b)i = 0 + 0i = 0$

(m_1) $1(a + bi) = a + bi$

(m_2) $(\alpha\beta)(a + bi) = \alpha\beta a + \alpha\beta bi = \alpha(\beta(a + bi))$

(m_3) $\alpha[(a + bi) + (c + di)] = \alpha(a + bi) + \alpha(c + di)$

(m_4) $(\alpha + \beta)(a + bi) = (\alpha+\beta)a + (\alpha+\beta)bi = \alpha(a + bi) + \beta(a + bi)$

∴ C forms a vector space.

Ex.5 (Function space) With addition and scalar multiplication defined as in C[a, b] determine whether all functions which are continuous every where on [a, b] except at a finite number of points (or the set of functions having a finite number of discontinuities), is a real vector space or not.

The graph of such function will be like the ***figure*** 11.1. This type of functions is nothing but sectionally continuous functions. Within the subintervals the set of functions are continuous. So, they can be added, there exits 0 function and reflection about the x-axis. That is,

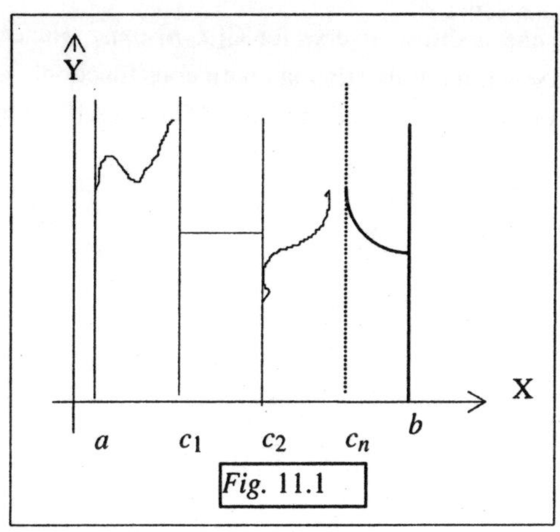

Fig. 11.1

$$f + g = g + f$$
$$(f + g) + h = f + (g + h)$$
$$f + 0 = f$$
$$(f) + (-f) = 0$$
$$\alpha(f + g) = \alpha f + \alpha g$$
$$(\alpha + \beta)f = \alpha f + \beta f$$
$$1f = f$$

exist. Hence the given set of functions constitutes a real vector space.

Ex.6 (From calculus) The set C[a, b] consists of all real valued continuous functions defined on a closed interval [a, b] of the real line. Addition of f and g in C[a, b] and scalar multiplication by c are defined by

(i) $(f + g)(x) = f(x) + g(x)$ for all real numbers x,

(ii) $(cf)(x) = c\,(f(x))$.

Prove that C[a, b] forms a vector space.

It is important to note that since the sum of two continuous functions is continuous, this definition is meaningful in the sense that (f + g) is again a vector in C[a, b] (*figure* 11.2). Therefore, f is closed under addition then it follows from definition that

$$f + (g + h) = (f + g) + h$$

and $$f + g = g + f$$

∴ (a_2) and (a_3) are satisfied.

If 0 denotes the function whose value is zero at each point of [a, b], then $f + 0 = f$ for every f in C[a, b]. Finally if (–f) is the function whose value at x is $(-f(x))$, i.e. (–f) is the reflection of f across 0 then f + (–f) has the value zero at each point of [a, b] and we have

$$f + (-f) = 0 \text{ and } f = f + 0.$$

Thus additive identity and additive inverse for C[*a*, *b*] exist. Hence C[*a*, *b*] forms a group under usual addition & scalar multiplication of continuous function.

Next, if f is any vector in C[*a*, *b*] and α is an arbitrary real number, the closure for scalar multiplication is satisfied.

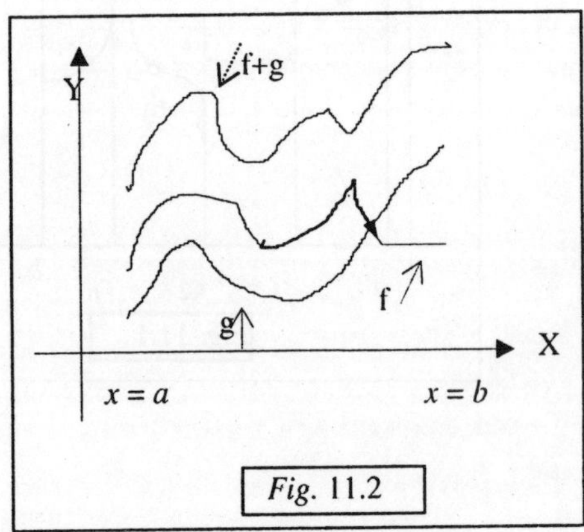

Fig. 11.2

Therefore, from the analogy with R^2, we have

$$\alpha(f + g) = \alpha f + \alpha g$$
$$(\alpha + \beta)f = \alpha f + \beta f$$
$$\alpha\beta f = \alpha(\beta f)$$
$$1f = f$$

∴ [*a*, *b*] forms a vector space under the defined addition & scalar multiplication.

Ex.7 (Vector space of $m\times n$ matrices). Prove that $m\times n$ matrices whose elements are real, form a vector space under ordinary matrix addition and scalar multiplication.

Let $A = [a_{ij}]$, $B = [b_{ij}]$ and $C = [c_{ij}]$ be all $m\times n$ matrices.

Then $A + B = [a_{ij} + b_{ij}]$ is another $m\times n$ matrix and $cA = [ca_{ij}]$ is also an $m\times n$ matrix. Thus the two closure properties are satisfied. Now from matrix algebra, it is well known that matrix addition is commutative as well as associative. There exists $m\times n$ null matrix which when adds to an $m\times n$ matrix keeps the later matrix unchanged. Also by scalar multiplication if $c = -1$ it multiplies all the mn elements of the $m\times n$ matrix. So there exists a negative matrix

$$(-A) = [-a_{ij}]$$

which when adds to A makes the result a null matrix. Therefore, there exists an additive inverse and hence $m\times n$ matrices form a commutative group under the usual matrix addition.

The multiplicative properties hold easily. Hence $m\times n$ matrices form a vector space under usual matrix addition and scalar multiplication.

N.B. Step by step procedure is avoided to save time & space.

11.4 Properties of a vector space

The properties of a vector space are embodied in the following theorem.

Theorem. In a vector space V(F), let 0 be the zero vector in V, then (symbol "∀" means "for all ", "∈" means "belong to " and " ⇒ "means " implies")

(i) $\alpha.\mathbf{0} = \mathbf{0}\ \ \forall\alpha\in F$ and $\mathbf{0}\in V$

(ii) $0.\mathbf{x} = \mathbf{0}$ $\forall$ $\mathbf{x} \in V$ and 0 being the zero of F, result is a zero vector

(iii) $(-\alpha)\mathbf{x} = \alpha.(-\mathbf{x})$ $\forall \alpha \in F$ and $\forall \mathbf{x} \in V$.

(iv) $\alpha(\mathbf{x}-\mathbf{y}) = \alpha.\mathbf{x} - \alpha.\mathbf{y}$ $\forall \alpha \in F$ and $\forall \mathbf{x}, \mathbf{y} \in V$.

(v) $\alpha\mathbf{x} = 0 \Rightarrow$ either $\alpha = 0$ or $\mathbf{x} = \mathbf{0}$ where $\alpha \in F$ and $\mathbf{x} \in V$.

(vi) for α, $\beta \in F$ and nonzero vector $\mathbf{x}$ in V, $\alpha.\mathbf{x} = \beta.\mathbf{x} \Rightarrow \alpha = \beta$;

(vii) for $\mathbf{x}, \mathbf{y} \in V$ and nonzero scalar α in F, $\alpha.\mathbf{x} = \alpha.\mathbf{y} \Rightarrow \mathbf{x} = \mathbf{y}$.

Proof . The above results directly follow from the definition of a vector space and may be derived as under :

(i) $\alpha.\mathbf{0} = \alpha.(\mathbf{0} + \mathbf{0})$ $\because$ $\mathbf{0} + \mathbf{0} = \mathbf{0}$ additive property so that

$\alpha.\mathbf{0} = \alpha.\mathbf{0} + \alpha.\mathbf{0}$ [distributivity of vector addition]. Adding to both sides the additive inverse of $\alpha.\mathbf{0}$, we have,

$$-(\alpha.\mathbf{0}) + (\alpha.\mathbf{0}) = -(\alpha.\mathbf{0}) + \{\alpha.\mathbf{0} + \alpha.\mathbf{0}\}$$
$$= \{-(\alpha.\mathbf{0}) + (\alpha.\mathbf{0})\} + (\alpha.\mathbf{0}), \text{ associativity}$$
$$\therefore \quad \mathbf{0} = \mathbf{0} + \alpha.\mathbf{0}, \text{ additive identity}$$
$$\text{or} \quad \mathbf{0} = \alpha.\mathbf{0}$$
$$\text{i.e.} \quad \alpha.\mathbf{0} = \mathbf{0}.$$

(ii) $0.\mathbf{x} = (0 + 0).\mathbf{x} = 0.\mathbf{x} + 0.\mathbf{x}$ distributivity of scalar addition.

Adding to both sides the additive inverse of $0.\mathbf{x}$, we have,

$$-(0.\mathbf{x}) + 0.\mathbf{x} = -(0.\mathbf{x}) + \{0.\mathbf{x} + 0.\mathbf{x}\}$$
$$= \{-(0.\mathbf{x}) + (0.\mathbf{x})\} + 0.\mathbf{x}$$
$$= 0 + 0.\mathbf{x}$$
$$\text{or} \quad 0 = 0.\mathbf{x}$$
$$\text{i.e.} \quad 0.\mathbf{x} = 0$$

(iii) We may write, $0 = 0.\mathbf{x}$

$$= [\alpha + (-\alpha)].\mathbf{x} \text{ for any scalar}$$
$$= \alpha.\mathbf{x} + (-\alpha).\mathbf{x}$$

Thus $(-\alpha).\mathbf{x}$ is the additive inverse of $\alpha.\mathbf{x}$, i.e. $(-\alpha).\mathbf{x} = -(\alpha.\mathbf{x})$

Also $-0 = 0.(-\mathbf{x})$

$$=[\alpha + (-\alpha)](-\mathbf{x}) = \alpha(-\mathbf{x}) + (-\alpha).(-\mathbf{x}) = \alpha.(-\mathbf{x}) + \alpha.\mathbf{x}$$

Thus $\alpha(-\mathbf{x})$ is the additive inverse of $\alpha.\mathbf{x}$ i.e. $\alpha(-\mathbf{x}) = -(\alpha.\mathbf{x})$

$\therefore$ $(-\alpha)\mathbf{x} = \alpha.(-\mathbf{x})$

(iv) $\alpha.(\mathbf{x}-\mathbf{y}) = \alpha[\mathbf{x} + (-\mathbf{y})] = \alpha.\mathbf{x} + \alpha.(-\mathbf{y}) = \alpha.\mathbf{x} + \{(-\alpha.\mathbf{y})\}$ by (iii)

$$= \alpha.\mathbf{x} - \alpha.\mathbf{y}$$

(v) Let $\alpha.\mathbf{x} = 0$ and $\alpha \neq 0$, then α^{-1} exists in F & therefore,

$$\alpha.\mathbf{x} = \mathbf{0} \Rightarrow \alpha^{-1}(\alpha.\mathbf{x}) = \alpha^{-1}.\mathbf{0}$$
$$\Rightarrow (\alpha^{-1}\alpha).\mathbf{x} = \alpha^{-1}.\mathbf{0}$$
$$\Rightarrow 1.\mathbf{x} = \mathbf{0} \quad \because \alpha^{-1}.\mathbf{0} = \mathbf{0} \;\&\; \alpha^{-1}\alpha = 1$$
$$\Rightarrow \mathbf{x} = \mathbf{0}.$$

Again, let $\alpha.\mathbf{x} = \mathbf{0}$ and $\mathbf{x} \neq \mathbf{0}$ & let if possible $\alpha \neq 0$, then α^{-1} exists & therefore,

$$\alpha.\mathbf{x} = \mathbf{0} \Rightarrow \alpha^{-1}(\alpha.\mathbf{x}) = \alpha^{-1}.\mathbf{0}$$
$$\Rightarrow (\alpha^{-1}\alpha).\mathbf{x} = \mathbf{0}$$
$$\Rightarrow 1.\mathbf{x} = \mathbf{0}$$
$$\Rightarrow \mathbf{x} = \mathbf{0} \text{ which is a contradiction.}$$

So, whenever $\alpha.\mathbf{x} = \mathbf{0}$ and $\mathbf{x} \neq 0$ we have $\alpha = 0$.

Hence $\alpha.\mathbf{x} = \mathbf{0} \Rightarrow$ either $\alpha = 0$ or $\mathbf{x} = \mathbf{0}$.

(vi) Let α, β be any two scalars and **x** be a nonzero vector such that $\alpha.\mathbf{x} = \beta.\mathbf{x}$, then

$$\begin{aligned}\alpha.\mathbf{x} = \beta.\mathbf{x} \Rightarrow \alpha.\mathbf{x}-\beta.\mathbf{x} = \mathbf{0} &\Rightarrow (\alpha-\beta).\mathbf{x} = \mathbf{0}\\ &\Rightarrow \text{either } (\alpha-\beta) = 0 \text{ or } \mathbf{x} = \mathbf{0} \text{ from (v)}\\ &\Rightarrow \alpha-\beta = 0 \because \mathbf{x} \neq \mathbf{0}\\ &\Rightarrow \alpha = \beta\end{aligned}$$

Hence $\alpha.\mathbf{x} = \beta.\mathbf{x}$ and $\mathbf{x} \neq \mathbf{0} \Rightarrow \alpha = \beta$.

(vii) Let α be a nonzero scalar and **x**,**y** be any two vectors such that $\alpha.\mathbf{x}=\beta.\mathbf{y}$, then

$$\begin{aligned}\alpha.\mathbf{x} = \beta.\mathbf{y} &\Rightarrow \alpha.\mathbf{x}-\beta.\mathbf{y} = \mathbf{0}\\ &\Rightarrow \alpha.(\mathbf{x}-\mathbf{y}) = \mathbf{0}\\ &\Rightarrow \text{either } \alpha = 0 \text{ or } (\mathbf{x}-\mathbf{y}) = \mathbf{0}\\ &\Rightarrow \mathbf{x}-\mathbf{y} = \mathbf{0} \because \alpha \neq 0\\ &\Rightarrow \mathbf{x} = \mathbf{y}\end{aligned}$$

Hence $\alpha.\mathbf{x} = \beta.\mathbf{y}, \alpha \neq 0 \Rightarrow \mathbf{x} = \mathbf{y}$.

11. 5 Physical interpretation and kinds of vector spaces

The word "space" in English literature is a common word and is used in different context in our daily life. For example space may mean "room" in a bench, "a wide open region" "interval" between any two or more objects, lines, words etc. Vector space is a generalization of the same thing in mathematics and sciences. Vector is a general name of the objects. We live in 3-dimensional objects like houses whose dimensions are length, breadth and height. The formal mathematical definition of the dimension will be encountered next in sec. 11.1.2. Thus every thing around us with length, breadth and height are 3-dimensional vector spaces including human beings and living objects too.

Matrix is a cut short media for representing those vector things and their problems as it is met in secs.1.19 etc. So, vector space is a common geometrical platform for the study of matrix theory. It is well known that analytic geometry provides with a means to investigate the properties of geometric object algebraically and to interpret algebraic problems geometrically. This circumstance makes the introduction of geometric terminology into linear algebra and to treat linear algebra geometrically, requires the concept of vector. This is affected by the vector of linear space.

Vector space may be of different *kinds* depending on the nature or type of vector consisting of the space. Following are some vector spaces:

(i) Real vector space denoted by $R^n = \{(x_1, x_2, \cdots, x_n) : x_i \in R$, the real number$\}$ is a n-dimensional real vector space of real numbers i.e. the set of real numbers are vectors which comprise the space. When $n = 1$, $R = \{x_1: x_1 \in R\}$ is the plane, when $n = 3$, $R^3 = \{(x_1, x_2, x_3) : x_i \in R\}$ is the ordinary 3-dimensional space in which we live in. For larger values of n, R^n can be visualize only and cannot be pictured.

(ii) Complex vector space denoted by C^n = {all ordered n-tuple of complex numbers} is a n-dimensional complex vector space of complex numbers, as the complex numbers are vectors which comprise the space (cf. Ex. 4 after sec. 11.3), multiplication is defined for complex numbers, field is complex.

(iii) Function space denoted by $C[a, b]$ = {all continuous functions in the closed interval between a and b} is a vector function space or function space only as the vectors involved are functions (cf. Ex. 5, 6 after sec. 11.3).

(iv) ***n*th differentiable function space** denoted by $C^n[a, b]$ = {all continuous functions with *n*th derivative in the closed interval between *a* and *b*} is also the vector space of functions with *n*th derivative as vectors involved are continuous functions with *n*th derivatives.
(v) **Polynomial function space** denoted by P_n = {all polynomials of degree *n*} is a vector space of *n*th degree polynomial functions as the vectors involved are *n*th degree polynomials. (cf. Ex. 3 after sec. 11.3).
(vi) **Spaces of matrices** denoted by $M_{m\times n}$ = {real or complex $m\times n$ matrices} is vector space of matrices of dimension *mn* as vectors comprising the space are matrices (cf. Ex. 7 after sec. 11.3).

There are other vector spaces in practice like affine space (linear manifold), Euclidean vector space , unitary vector space, arithmetic vector space, metric vector space, Minkowski space in special theory of relativity, Banach space, Sobolev space, Hilbert space, etc. which require the additional property of the inner product and the norm. So, they are just introduced in chapter14 next.

11.6 Subspaces

Definition. Let V be a vector space over the field F.A subspace of V is a nonempty subset W of V i.e. $W \subseteq V$ and is itself a vector space over F with the operations of vector addition and scalar multiplication on V.

11.6.1. Examples of subspaces : Following are examples of subspaces.
(1) Let V be any vector space. Then the set {0} consisting of the zero vector alone, and also the entire space V are subspaces of V.
(2) Let V be the vector space R^3 of real numbers. Then the set W consisting of those vectors whose third component is zero, i.e. W = {*a*, *b*, 0 : *a*, *b* ∈ R}, is a subspace of V.
(3) Any particular solution of the system of homogeneous equations in *n* unknowns may be viewed as a point in R^n. The set W of all solutions of the homogeneous system is a subspace of R^n called the *solution space*.
(4) Every vector space V has at least two subspaces —

(i) V itself is a subspace and

(ii) the set {0} consisting of just the zero vector in V is a subspace called *zero subspace*.

(5) Lines through origin are subspaces of R^2.
(6) In R^3, the proper subspaces are lines and planes through origin.
(7) Lines and planes which do not pass through origin cannot correspond to subspaces of R^3 because 0^3 is not in these set as every vector space of 3-components must contain 0^3.
(8) Rows and columns of a matrix separately constitute subspaces of the vector space specified by the matrix under usual addition and scalar multiplication, and known as row space and column space respectively.

11.6.2 Closure for addition and scalar multiplications : Two necessary definitions are :
Definition 1. Closure under addition : Closure under addition of a subset S of a vector space V over the field F means that $\alpha + \beta$ is in S for all α, β in S.
Definition 2. Closure under scalar multiplication : Closure under scalar multiplication of a subset S of a vector space V over the field F means that $k\alpha$ is in S for all α in S and all *k* in F.

11.6.3 Criteria for identifying a subspace : Let us now prove the theorems which give necessary and sufficient conditions for a subset W of V to be a subspace of V.

Theorem 1. A nonempty subset W of a vector space V(F) is a subspace of V, if and only if,

(i) $\alpha, \beta \in W$ implies ($\Rightarrow$) $\alpha + \beta \in W$ i.e. W is closed under addition

(ii) $a \in F$, $\alpha \in W$ implies ($\Rightarrow$) $a\alpha \in W$ i.e.W is closed under scalar multiplication.

Proof. Necessary conditions . Let W be a subspace of V(F), then by definition W is a vector space in its own right. Consequently, for all $\alpha, \beta \in W$ there must be $\alpha + \beta \in W$, i.e. W must be closed under vector addition. Also for all $\alpha \in W$ and all $a \in F$ there must be $a\alpha \in W$, i.e. W must be closed under scalar multiplication. Hence the conditions are necessary.

Sufficient conditions . Here it needs to show that W possesses the additive identity and inverse and other conditions of vector space.

Let W be a nonempty subset of V(F) and satisfies the given condition. Now if α be an arbitrary element of W and 1 be the unit element of F, then $(-1) \in F$ and therefore by given condition (ii),

$$(-1) \in F,\ \alpha \in W \Rightarrow (-1)\ \alpha \in W$$
$$\Rightarrow (-1.\alpha) \in W \Rightarrow (-\alpha) \in W.$$

Thus every element in W has an additive inverse in W.

Also, $\alpha \in W$, $(-\alpha) \in W \Rightarrow \alpha + (-\alpha) \in W \Rightarrow 0 \in W$ by (i).This shows that the additive identity, zero vector of V is also the zero vector of W.

Moreover, all the elements of W being the elements of V, the addition of vectors being commutative as well as associative in V, so it is in W. Hence additive properties of vector space are satisfied by W. Further $W \subseteq V$ i.e. every element of W is also an element of V, therefore, all the multiplicative properties of vector space will hold in W since they hold in V.

Hence W is a vector space and accordingly W is a subspace of V.

Theorem 2. A nonempty subset W of vector space V(F) is a subspace of V, if and only if α, $\beta \in W$ and $a, b \in F$ implies $a\alpha + b\beta \in W$ i.e. iff, W is closed under addition and scalar multiplication on V.

Proof. The condition is necessary. Let W be a subspace of V, then we need to show the existence of $a\alpha + b\beta \in W$ for all $\alpha, \beta \in W$ and $a, b \in F$.

As W is a vector space over F for the defined vector addition and scalar multiplication in V, it must be closed for these two operations of V i.e. by scalar multiplication in W, we have $a\alpha \in W$, for all $a \in F$, $\alpha \in W$ and $b\beta \in W$ for all $b \in F$, $\beta \in W$. Hence by vector addition in W, we have $a\alpha + b\beta \in W$ for all $a\alpha$, $b\beta \in W$. Hence the condition is necessary.

The condition is sufficient. Let W be a non-empty subset of V such that $a\alpha + b\beta$ belongs to W for all $\alpha, \beta \in W$ and $a, b \in F$, i.e. W is closed under vector addition and scalar multiplication in V. Now we have to show that W is a vector space of V(F).

Let $\alpha \in W$ and $1 \in F$, i.e. $(-1) \in F$, $(-1) \in F$, $\alpha \in W \Rightarrow (-1)\ \alpha \in W \Rightarrow -(1.\alpha) \in W, \Rightarrow (-\alpha) \in W$. [$\because$ $\alpha \in W \Rightarrow \alpha \in V$ and in V, $1\alpha = \alpha$].

Thus the additive inverse of each element of W is also an element of W. Therefore, we have, $\alpha \in W$, $(-\alpha) \in W \Rightarrow \alpha + (-\alpha) \in W$ [$\because$ W is closed for addition] $\Rightarrow 0 \in W$. The elements of W are also the elements of V so that zero vector of V is also the zero vector of W. Therefore, the additive properties of vector space are verified.

Also as given W is closed under scalar multiplication and hence all the remaining multiplicative properties of vector space will hold in W since they hold in V of which W is a subset. Hence W is a vector space and accordingly a subspace of V(F).

11.6.4 An easy technique for identifying a vector space : The above theorems 1 and 2 provide separately easy techniques for the identification of a vector space if one can identify a subspace of a known vector space. This procedure needs only two operations for verification using theorem 1 and one operation by theorem 2. Note the following examples.

Ex.1 Let P_n be the set of all polynomials of degree less than equal to n together with the usual polynomial arithmetic. Show that P_n is a vector space for each positive integer n.

Since it has already been proved in Ex.3 after sec.11.3 that

$$P(x) = a_0 + a_1x + a_2x^2 + \cdots + a_nx^n$$

is a vector space. Now if it is possible to show that $P_n(x)$ is a subspace of P then $P_n(x)$ is proved a vector space.

Since adding two polynomials does not increase the degree higher than any one of those two polynomials. This establishes the closure for addition. Also, multiplication by a nonzero scalar leaves the degree of the polynomial unchanged. Thus closure for scalar multiplication is established.

Therefore, P_n is a subspace of the known vector space P. Hence P_n is a vector space by its own right.

Ex.2 Consider the following subsets of the 3-tuple space $R^3(R)$:—

$$W_1 = \{(x_1,\ x_2,\ 0) :\ x_1,\ x_2 \in R\}.$$

By testing the validity of the condition required for a subspace of vector space, show that the above subset is a subspace of $R^3(R)$.

Thus in order to show that, W_1 is a subspace of $R^3(R)$, let $\mathbf{x} = (x_1, x_2, 0)$ and $\mathbf{y} = (y_1, y_2, 0)$ be any two arbitrary points of W_1, then $\forall a, b \in R$ such that

$$\begin{aligned} a\mathbf{x}+b\mathbf{y} &= a(x_1, x_2, 0) + b\,(y_1, y_2, 0) \\ &= (ax_1, ax_2, a0) + (by_1, by_2, b0) \\ &= (ax_1, ax_2, 0) + (by_1, by_2, 0) \\ &= (ax_1 + by_1, ax_2 + by_2, 0) \in W_1 \end{aligned}$$

Hence W_1 is a subspace of $R^3(R)$.

Ex.3 Show that $V_2(R)$ is not a subspace of vector space $V_3(R)$.

The elements of $V_3(R)$ being 3-tuple vectors are of the form (a_1, a_2, a_3) and the elements of $V_2(R)$ being 2-tuple vectors are of the form (a_1, a_2).

Since the elements of $V_3(R)$ and $V_2(R)$ are distinct elements and therefore $V_2(R)$ is not a subset of $V_3(R)$. Hence $V_2(R)$ is not a subspace of vector space $V_3(R)$.

Ex.4 Solution space W is a subspace of the vector space $V_n(F)$ — prove.

Let a homogeneous system of linear equations in n unknowns $x_1, x_2, \cdots, x_n$ over a field F be

$$\left.\begin{array}{llllll} a_{11}x_1 + & a_{12}x_2 + & \cdots & + & a_{1n}x_n & = 0 \\ a_{21}x_1 + & a_{22}x_2 + & \cdots & + & a_{1n}x_n & = 0 \\ \vdots & \vdots & \vdots & & \vdots & \vdots \\ a_{m1}x_1 + & a_{m2}x_2 + & \cdots & + & a_{mn}x_n & = 0 \end{array}\right\}$$

Here the trivial solution $0 = (0, 0, 0, \cdots, 0) \in W$ since clearly

$$a_{i1}.0 + a_{i2}.0 + \cdots + a_{in}.0 = 0 \text{ for } i = 1, 2, \cdots, m.$$

Let $\alpha = (u_1, u_2, \cdots, u_n)$ and $\beta = (v_1, v_2, \cdots, v_n) \in W$
i.e. for $i = 1, 2, \cdots, m$,

$$a_{i1}u_1 + a_{i2}u_2 + \cdots + a_{in}u_n = 0$$

and
$$a_{i1}v_1 + a_{i2}v_2 + \cdots + a_{in}v_n = 0$$

Let p, q be scalars in the field F.
Then $p\alpha + q\beta = (pu_1 + qv_1, \cdots, pu_n + qv_n)$
$\therefore$ for $i = 1, 2, \cdots, m$,

$$a_{i1}(pu_1 + qv_1) + a_{i2}(pu_2 + qv_2) + \cdots + a_{i1}(pu_n + qv_n)$$
$$= p(a_{i1}u_1 + a_{i2}u_2 + \cdots + a_{in}u_n) + q(a_{i1}v_1 + a_{i2}v_2 + \cdots + a_{in}v_n)$$
$$= p.0 + q.0 = 0 \in W$$

$\therefore p\alpha + q\beta$ is solution of the system & belongs to W.
$\therefore$ W, the solution space, is a subspace of $V_n(R)$.

Ex.5 Which of the following sets of vectors
$\alpha = (a_1, a_2, \cdots, a_n)$ in R_n are subspaces of R_n $(n \geq 3)$,
(a) all $\alpha : a_1 \geq 0$; (b) all $\alpha : a_1 + 3a_2 = a_3$; (c) all $\alpha : a_2 = a_1^2$;
(d) all $\alpha : a_1a_2 = 0$; (e) all $\alpha : a_2$ is rational.

(a) Let $W = \{\alpha : \alpha \in R_n \text{ and } a_1 \geq 0\}$
Let $\alpha = (a_1, a_2, \cdots, a_n)$, $\beta = (b_1, b_2, \cdots, b_n)$
Evidently $a_1 \geq 0, b_1 \geq 0$ by the given condition.
Now $a, b \in R \Rightarrow a\alpha + b\beta = (aa_1 + bb_1, aa_2 + bb_2, \cdots, aa_n + bb_n)$.
But $aa_1 + bb_1$ is not necessarily ≥ 0 although $a_1 \geq 0, b_1 \geq 0$; e.g. if we take $a_1 = 5, b_1 = 5, a = -3, b = -3$ then $aa_1 + bb_1 = -15 - 15 = -30 < 0$.
Hence $a, b \in R, \alpha, \beta \in W \Rightarrow a\alpha + b\beta \notin W$
i.e. W is not a vector subspace of R_n.
(b) In this case

$$a\alpha + b\beta = (aa_1 + bb_1, aa_2 + bb_2, \cdots, aa_n + bb_n).$$
$$= (p_1, p_2, \cdots, p_n) \text{ where } p_i = aa_i + bb_i,\ i = 1, 2, 3, \cdots, n.$$

Now $p_1 + 3p_2 = (aa_1 + bb_1 + 3(aa_2 + bb_2))$
$$= a(a_1 + 3a_2) + b(b_1 + 3b_2)$$
$$= aa_3 + bb_3 = p_3$$

$\therefore a\alpha + b\beta \in W$.
$\therefore$ W is a vector subspace of R_n.
(c) In this case

$$a\alpha + b\beta = (aa_1 + bb_1, aa_2 + bb_2, \cdots, aa_n + bb_n).$$
$$= (p_1, p_2, \cdots, p_n) \text{ where } p_i = aa_i + bb_i,\ i = 1, 2, 3, \cdots, n$$

Now $p_2 = aa_2 + bb_2 = aa_1^2 + bb_1^2 \neq (aa_1 + bb_1)^2 \neq p_1^2$
i.e. $a\alpha + b\beta \notin W$.
$\therefore$ W is not a vector subspace of R_n.
(d) In this case

$$a\alpha + b\beta = (aa_1 + bb_1, aa_2 + bb_2, \cdots, aa_n + bb_n).$$
$$= (p_1, p_2, \cdots, p_n) \text{ where } p_i = aa_i + bb_i,\ i = 1, 2, 3, \cdots, n$$

Now $p_1p_2 = (aa_1 + bb_1)(aa_2 + bb_2)$

= not necessarily equal to zero

i.e. $a\alpha + b\beta \notin$ W

i.e. W is not a vector subspace of R_n.

(e) In this case

$a\alpha + b\beta = (aa_1 + bb_1, aa_2 + bb_2, \cdots, aa_n + bb_n).$

$= (p_1, p_2, \cdots, p_n)$ where $p_i = aa_i + bb_i, i = 1, 2, 3, \cdots, n$

Now $p_2 = aa_2 + bb_2$ = Not necessarily rational

e.g if $a_2 = 3, b_2 = 7$ & $a = \sqrt{5}, b =$ then

$p_2 = \sqrt{5}.3 + 7\sqrt{7}$ = not rational

$\therefore$ $a\alpha + b\beta \notin$ W, i.e. W is not a vector subspace of R_n.

Ex.6 Let A be a $m \times n$ matrix. Let N(A) denote the set of all solutions to the homogeneous system Ax = 0. Thus

$$N(A) = \left\{\mathbf{x} \in R^n \middle| A\mathbf{x} = 0\right\}.$$

Show that N(A) is subspace of R^n.

If $\mathbf{x} \in N(A)$ and α is a scalar, then

$A(\alpha\mathbf{x}) = \alpha A\mathbf{x} = \alpha 0 = 0 \quad \because A\mathbf{x} = 0.$

$\therefore$ $\alpha\mathbf{x} \in N(A)$.

Again if **x** and **y** are two vectors of N(A), then

$A(\mathbf{x} + \mathbf{y}) = A\mathbf{x} + A\mathbf{y} = 0 + 0 = 0$

$\therefore$ $\mathbf{x} + \mathbf{y} \in N(A)$.

$\therefore$ N(A) is a subspace of R^n.

N.B. The set of all solutions to the homogeneous system Ax = 0 forms a subspace of R^n. The subspace N(A) is called the *null space* of A (cf.sec.11.2.3).

Ex.7 Let $C^n[a, b]$ be the set of all functions that have a continuous *n*th derivative on [*a, b*].

Prove that

$$S = \{f \in C^2[a,b] : f''(x) + f(x) = 0 \quad \forall x \in [a,b]\}$$

forms a subspace.

Since $f \in S$ and for any scalar α then for any $x \in [a,b]$,

$$(\alpha f)''(x) + (\alpha f)(x) = \alpha f''(x) + \alpha f(x)$$
$$= \alpha[f''(x) + f(x)] = \alpha.0 = 0$$

$\therefore$ $\alpha f \in S$.

Again let f and g both $\in S$, then

$$(f+g)''(x) + (f+g)(x) = f''(x) + g''(x) + f(x) + g(x)$$
$$= [f''(x) + f(x)] + [g''(x) + g(x)] = 0 + 0 = 0.$$

$\therefore$ the set of all solutions f on [*a, b*] of the differential eqn. $f''(x) + f(x) = 0$ forms a subspace of $C^2[a, b]$. If we note that f(*x*) = sin*x* and g(*x*) = cos*x* are both in S, it follows that any function of the form $c_1\sin x + c_2\cos x$ must also belong to S. It is easy to verify that the solution of

$$y'' + y = 0 \quad \text{is} \quad y = c_1\sin x + c_2\cos x.$$

Ex.8 Show that the $D_{2\times2}$ of all 2×2 diagonal matrices is a subspace of the vector space $M_{2\times2}$, a set of all 2×2 matrices.

Let $$A=\begin{bmatrix} a_{11} & 0 \\ 0 & a_{22}\end{bmatrix},\ B=\begin{bmatrix} b_{11} & 0 \\ 0 & b_{22}\end{bmatrix}$$

be two diagonal matrices of the set $D_{2\times2}$.

Now $$A+B=\begin{bmatrix} a_{11}+b_{11} & 0 \\ 0 & a_{22}+b_{22}\end{bmatrix}$$

which is still a diagonal matrix. Therefore, $A+B\in D_{2\times2}$, i.e., $D_{2\times2}$ is closed under matrix addition.

Also, for nonzero scalar c,

$$cA=\begin{bmatrix} ca_{11} & 0 \\ 0 & ca_{22}\end{bmatrix}$$

which is also a diagonal matrix and $cA\in D_{2\times2}$. So $D_{2\times2}$ is closed under scalar multiplication. Hence $D_{2\times2}$ is a subspace of $M_{2\times2}$.

Ex.9 Show that P_3' the set of all polynomials of degree exactly equal to three together with 0, is not a subspace of P_3, the set of all polynomials of degree less than or equal to three.

To show this it only needs to show that either of two conditions in theorem 1 in sec.11.6.2 is not satisfied for specific choice of polynomials and/or scalars.

Let $p(x)=-2x^3+2x-1$ and $q(x)=2x^3+x^2+3$, then

$(p+q)(x)=p(x)+q(x)=x^2+2x+2$, a polynomial of degree 2

$\therefore$ $(p+q)(x)\notin P_3'$.

So, P_3' is not a subspace of P_3.

N.B. Once one test fails, then no need to proceed further.

Ex.10 Verify whether the following sets are subspaces of $V_3(\mathrm{IR})$:—

(i) $\{(x, 2y, 5): x, y\in \mathrm{IR}\}$ i.e. points the plane $z=5$

(ii) $\{(x, x+y, 3z): x, y, z\in \mathrm{IR}\}$.

(i) Let $=\{(x, 2y, 5): x, y \in \mathrm{IR}\}$.

Let $\mathbf{a}=(x_1, 2y_1, 5)$ and $\mathbf{b}=(x_2, 2y_2, 5)$ be any two points/vectors of W, then for all scalars p and $q\in \mathrm{IR}$.

$$p\mathbf{a}+q\mathbf{b}=p(x_1, 2y_1, 5)+q(x_2, 2y_2, 5)$$
$$=(px_1+qx_2, 2(py_1+qy_2), 5(p+q))\notin W$$

[$\because$ third element is a fixed number 5]

Hence W is not a subspace.

(ii) Let $W=\{(x, x+y, 3z): x, y, z\in \mathrm{IR}\}$

Let $\mathbf{a}=(x_1, x_1+y_1, 3z_1)$ and $\mathbf{b}=(x_2, x_2+y_2, 3z_2)$ be any two points/vectors of W, then for all scalars $p, q\in \mathrm{IR}$.

$$p\alpha+q\beta=p(x_1, x_1+y_1, 3z_1)+q(x_2, x_2+y_2, 3z_2)$$
$$=(px_1+qx_2, p(x_1+y_1)+q(x_2+y_2), 3(z_1p+z_2q))\in W$$

as $px_1 + qx_2, p(x_1+y_1)+q(x_2+y_2), z_1p+z_2q \in \text{IR}$

$\therefore$ W is a subspace of V_3(IR).

Ex.11 Show that the set $W = \{(a,b,c) : a,b,c \in \text{IR}$ and $2a - b + c = 0\}$ is a subspace of V_3(IR).

Here $W = \{(a,b,c) : a,b,c \in \text{IR}$ and $2a - b + c = 0\}$ is clearly a nonempty subset of V_3(IR).

Now for $\mathbf{a} = (a_1,b_1,c_1)$, $\mathbf{b} = (a_2,b_2,c_2) \in W$ and $p, q \in R$

We have $p\mathbf{a} + q\mathbf{b} = p(a_1,b_1,c_1) + q(a_2,b_2,c_2)$

$= p(a_1, 2a_1 + c_1, c_1) + q(a_2, 2a_2 + c_2, c_2)$

$\because 2a_1 - b_1 + c_1 = 0$ & $2a_2 - b_2 + c_2 = 0$ for set W

$= (pa_1 + qa_2, 2(a_1p + qa_2) + pc_1 + qc_2, pc_1 + qc_2) \in W$

$\because (pa_1 + qa_2, 2(a_1p + qa_2) + pc_1 + qc_2, pc_1 + qc_2c_2)$ satisfies $2a - b + c = 0$

$\therefore p, q \in R$ & $\alpha, \beta \in W \Rightarrow p\alpha + q\alpha \in W$.

Ex.12 Prove that the x - axis & the y - axis are subspace of the space V_2(IR), where IR is the field of real number.

Let W_1 be any point on the x - axis.

$\therefore$ $W_1 = \{(x,0) : x \in \text{IR}\}$

which is a nonempty subset of $V_2(\text{IR}) = \{(x,y) | x, y \in \text{IR}\}$

Now for $\mathbf{a} = (x_1, 0)$, $\mathbf{b} = (x_2, 0) \in W_1$ & $p, q \in R$

We have $p\mathbf{a} + q\mathbf{b} = p(x_1,0) + q(x_2,0) = (px_1 + qx_2, 0) \in W_1$

$\therefore$ W_1 is a subspace of V_2(IR)

Similarly, let W_2 be any point on the y-axis.

$\therefore W_2 = \{(0, y) : y \in \text{IR}\}$ which is a nonempty subset of V_2 (IR).

Now for $\mathbf{a} = (0, y_1)$, $\mathbf{b} = (0, y_2) \subset W_2$ and $p, q \subset R$.

We have $p\mathbf{a} + q\mathbf{b} = p(0, y_1) + q(0, y_2) = (0, py_1 + qy_2) \in W_2$

$\therefore$ W_2 is a subspace of V_2 (IR).

Ex.13 Let W consists of the zero function and all real polynomial function

$$\phi(x) = a_0 + a_1x + \cdots + a_nx^n \quad (1)$$

with degree $\leq n$, positive integer, and $a_0, a_1, \cdots, a_n$ as real numbers. Prove that W is a subspace of the vector space of all real valued functions.

Let P and Q be the polynomials

$$P = p(x) = a_0 + a_1x + \cdots + a_nx^n$$

$$Q = q(x) = b_0 + b_1x + \cdots + b_nx^n$$

Then $(P + Q)x = p(x) + q(x) = (a_0 + b_0) + (a_1 + b_1)x + \cdots + (a_n + b_n)x^n$ (2)

and $(kP)x = kp(x) = (ka_0) + (ka_1)x + (ka_2)x^2 + \cdots + (ka_n)x^n$ (3)

R.H.S. of (2) and (3) are of the same form as (1).

$\therefore$ P + Q and kP lie in W. So, W is a subspace. It may be denoted by P_n.

11.7 Algebra of subspaces

Algebra of submatrices is concerned with the intersection, union, disjoint, linear sum and direct sum of subspaces and the related theorems. The concerned definitions and theorems are discussed below one after another.

11.7.1 Intersection : If W_1 and W_2 are two subspace of V(F) then the intersection of the subspaces W_1 and W_2 is the set of all vectors common to both W_1 and W_2. Symbolically, $W_1 \cap W_2 = \{\alpha : \alpha \in W_1 \text{ and } \alpha \in W_2\}$

In other words, $W_1 \cap W_2 \subseteq W_1$ and $W_1 \cap W_2 \subseteq W_2$

Ex. If W_1 and W_2 are two distinct planes throuh the origin in R^3 then $W_1 \cap W_2$ is the set of all points in a line as two distinct planes interset in a line. Similarly, if W_1 is plane and W_2 is a line not lying in W_1 then $W_1 \cap W_2$ is a single point, origin, the zero vector.

Theorem 1. The intersection of any two subspaces of a vector space V(F) is a subspace of V(F).

Proof. Let W_1 and W_2 be any two subspaces of V(F). Then it is to show that $W_1 \cap W_2$ is also a subspace of V(F).

Now W_1 and W_2 are subspaces. Then 0(zero) is an element of both W_1 and W_2. Hence, $W_1 \cap W_2$ is not empty. This satisfies the 1st criteria of subspace.

Let $\mathbf{a}, \mathbf{b} \in W_1 \cap W_2$ and $p, q \in F$, then

$$\mathbf{a} \in W_1 \cap W_2 \Rightarrow \mathbf{a} \in W_1 \text{ and } \mathbf{a} \in W_2$$

and $$\mathbf{b} \in W_1 \cap W_2 \Rightarrow \mathbf{b} \in W_1 \text{ and } \mathbf{b} \in W_2$$

Again since W_1 is a subspace of V then $p, q \in F$ and $\mathbf{a}, \mathbf{b} \in W_1 \Rightarrow p\mathbf{a} + q\mathbf{b} \in W_1$.

Similarly, W_2 is subspace of V, then

$$p, q \in F \text{ and } \mathbf{a}, \mathbf{b} \in W_2 \Rightarrow p\mathbf{a} + q\mathbf{b} \in W_2.$$

Now as $p\mathbf{a} + q\mathbf{b} \in W_1$ and $p\mathbf{a} + q\mathbf{b} \in W_2$ then $p\mathbf{a} + q\mathbf{b} \in W_1 \cap W_2$.

Thus $p, q \in F$ and $\mathbf{a}, \mathbf{b} \in W_1 \cap W_2 \Rightarrow p\mathbf{a} + q\mathbf{b} \in W_1 \cap W_2$.

Hence $W_1 \cap W_2$ is a subspace of V(F), i.e., intersection of two subspaces of a vector space is again a subspace.

Ex. If $V_3R) = \{(a, b, c) : a, b, c \in R\}$ is a vector space, then that $W_1 \cap W_2$ is a subspace where

$$W_1 = \{(0, b, c) : b, c \in R\} \ \& \ W_2 = \{(a, 0, c) : \in R\}$$

As $W_1 = \{(0, b, c) : b, c \in R\} \ \& \ W_2 = \{(a, 0, c) : a, c \in R\}$

then $W_1 \cap W_2 = \{(0, 0, c) : c \in R\}$.

Let $\mathbf{x} = (0, 0, c_1)$, $\mathbf{y} = (0, 0, c_2) \in W_1 \cap W_2$ and $p, q \in R$.

Therefore, $p\mathbf{x} + q\mathbf{y} = p(0, 0, c_1) + q(0, 0, c_2)$

$= (0, 0, pc_1 + qc_2) \in W_1 \cap W_2$

Hence $W_1 \cap W_2 = \{(0, 0, c) : c \in R\}$ is a subspace of $V_3(R)$ which is the set of all vectors along z-axis.

Theorem 2. The intersection of an arbitrary collection of subspaces of a vector space is also a subspace of the same.

Proof. Let $\{W_k : k \in \Lambda\}$ be an arbitrary collection of subspaces of a vector space V(F), then zero vector of V is surely contained in each W_k and therefore, this zero vector is also contained in $\bigcap\{W_k : k \in \Lambda\}$. Consequently, $\bigcap\{W_k : k \in \Lambda\} \neq \phi$.

Let **a** and **b** be any two arbitrary elements of $\bigcap\{w_k : k \in \Lambda\}$. Then

$$\mathbf{a}, \mathbf{b} \in \bigcap\{W_k : k \in \Lambda\} \Rightarrow \mathbf{a}, \mathbf{b} \in \text{each } W_k .$$

$\therefore$ for all p, q in F and **a, b**$\in \bigcap\{W_k : k \in \Lambda\}$, we have

$$p\mathbf{a} + q\mathbf{b} \in \bigcap\{W_k : k \in \Lambda\} .$$

Hence $\bigcap\{W_k : k \in \Lambda\}$ is a subspace.

11.7.2. Union : The union of two subspaces W_1 and W_2 is the collection of all vectors common and uncommon to both subspaces W_1 and W_2, i.e. all vectors either in W_1 or in W_2 or both. Symbolically,

$$W_1 \cup W_2 = \{\alpha : \alpha \in W_1 \text{ or } \alpha \in W_2\} .$$

Ex. Let $W_1 = \{(1, 0, 3), (2, 1, 0)\}$ & $W_2 = \{(1, 0, 3), (1, 1, 1)\}$ then $W_1 \cup W_2 = \{(1, 0, 3), (2, 1, 0), (1, 1, 1)\}$.

Theorem 1. The union of two subspaces of a vector space is not necessarily a subspace.

Proof. The theorem is proved by giving a counter example as follows:
Let $V_3(R) = \{(a,b,c) : a,b,c \in R\}$ be a vector space and
$W_1 = \{(a,0,0) : a \in R\}$ & $W_2 = \{(0,b,0) : a \in R\}$ be its two subspaces. Then
$W_1 \cup W_2 = \{\alpha : \alpha = (a, 0, 0) \text{ or } (0, b, 0)\}$.
Let $\alpha = (a, 0, 0)$, $\beta = (0, b, 0) \in W_1 \cup W_2$ and $p, q \in R$.
Then $p\alpha + q\beta = p(a,0,0) + q(0,b,0)$

$= (pa, qb, 0)$ is not necessarily in $W_1 \cup W_2$

To be ,more specific, let $\alpha = (1, 0, 0)$, $\beta = (0, 2, 0)$ in $W_1 \cup W_2$
and $p = 2$, $q = 3$, then
$p\alpha + q\beta = 2(1,0,0) + 3(0,2,0) = (2,6,0) \notin W_1 \cup W_2$.
Hence $W_1 \cup W_2$ is not a subspace of $V_3(R)$.

Theorem 2. The union of two subspaces is a subspace, if and only if, one is contained in the other.

Proof. Let W_1 and W_2 be any two subspaces of a vector space V.
Also, if $W_1 \subseteq W_2$ or $W_2 \subseteq W_1$ then $W_1 \cup W_2 = W_2$ or W_1
i.e. $W_1 \cup W_2$ is also a subspace of V.

Conversely, let $W_1 \cup W_2$ be a subspace, then we have to show that

$$W_1 \subseteq W_2 \text{ or } W_2 \subseteq W_1 .$$

Let W_1 and W_2 be independent subsets of V, i.e., let
W_1 is not a $\subseteq W_2$ and W_2 is not a $\subseteq W_1$.

Now W_1 is not a $\subseteq W_2 \Rightarrow \exists\ \alpha \in W_1$ and $\alpha \notin W_2$ (11.1)

and W_2 is not a $\subseteq W_1 \Rightarrow \exists\ \beta \in W_2$ and $\beta \notin W_2$ (11.2)

From (11.1) and (11.2), we have

$$\alpha \in W_1 \cup W_2 \text{ and } \beta \in W_1 \cup W_2 .$$

Now as $W_1 \cup W_2$ is a subspace, therefore,

$$\alpha + \beta \text{ is also an element of } W_1 \cup W_2 .$$

But $\alpha+\beta \in W_1 \cup W_2 \Rightarrow \alpha+\beta \in W_1$ or W_2.

Let $\alpha+\beta \in W_1$. Now as W_1 is a subspace and $\alpha \in W_1$, therefore, $(\alpha+\beta)+(-\alpha)=\beta$ is an element of W_1. But from (11.2), we have $\beta \notin W_1$, i.e. a contradiction.

Hence either $W_1 \subseteq W_2$ or $W_2 \subseteq W_1$.

11.7.3 Linear Sum (Join) : If W_1 and W_2 are any two subspaces of a vector space V(F), then the linear sum of these two subspaces denoted by $W_1 + W_2$ is a set which contains all the elements obtained by adding each element of W_1 to each element of W_2.

Thus if $\alpha \in W_1$ & $\beta \in W_2$, then $W_1 + W_2$ is the set of all sums $\alpha+\beta$, i.e.

$W_1+W_2=\{\alpha+\beta : \alpha \in W_1, \beta \in W_2\}$.

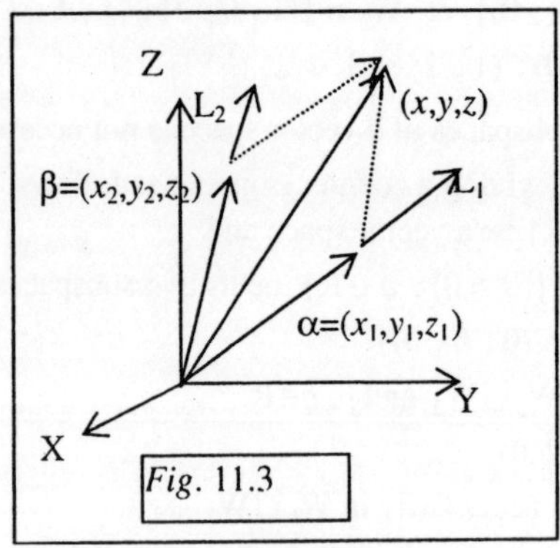

Fig. 11.3

It is evident from above that an arbitrary element α of W_1 may be written as α of W_1 where $\alpha \in W_1$ and $0 \in W_2$.

So, $\alpha \in W_1 \Rightarrow \alpha \in W_1 + W_2$

$\therefore W_1 \subseteq W_1 + W_2$

Similarly, if $0 \in W_1$ and $\beta \in W_2$ then $\beta \in W_2 \Rightarrow 0+\beta=\beta \in W_1+W_2$.

$\therefore \quad W_2 \subseteq W_1 + W_2$

Let us consider a familiar example from 3-space $V_3(R)$. Two distinct lines L_1 and L_2 through origin are two subspaces, and the intersection of two lines determines a plane. Hence L_1 and L_2 must lie in a plane, say S. That is plane S joins lines L_1 and L_2 (*figure* 11.3). Therefore, every vector or point (x, y, z) on S is a sum $\alpha = (x_1, y_1, z_1)$ on L_1 and $\beta = (x_2, y_2, z_2)$ on L_2. Thus the linear sum of two subspaces L_1 and L_2 i.e. $L_1 + L_2$ is a subspace S. This is generalized in the theorem below.

Theorem. If W_1 and W_2 are two subspaces of V(F), then $W_1 + W_2$ is a subspace of V(F). [Or "The linear sum of two subspaces of a vector space is also a vector space"].

Proof. Since W_1 & W_2 are two subspaces of V(F), then each of them is nonempty, so their linear sum, $W_1 + W_2$ is nonempty.

Let $\alpha_1, \alpha_2 \in W_1$ and $\beta_1, \beta_2 \in W_2$. Then $\alpha = \alpha_1 + \beta_1 \in W_1 + W_2$ and $\beta = \alpha_2 + \beta_2 \in W_1 + W_2$ by definition. Now to prove that $W_1 + W_2$ is a subspace, it is required to show that

$$a\alpha + b\beta \in W_1 + W_2 \text{ for } a, b \in F.$$

$$\alpha_1, \alpha_2 \in W_1 \Rightarrow a\alpha_1 + b\alpha_2 \in W_1 \text{ for } a, b \in F, \ [\because W_1 \text{ is a subspace}]$$

$$\beta_1, \beta_2 \in W_2 \Rightarrow a\beta_1 + b\beta_2 \in W_2 \text{ for } a, b \in F, \ [\because W_2 \text{ is a subspace}]$$

Thus by definition of linear sum

$$(a\alpha_1 + b\alpha_2) + (a\beta_1 + b\beta_2) \in W_1 + W_2$$

i.e. $$a(\alpha_1 + \beta_1) + b(\alpha_2 + \beta_2) \in W_1 + W_2$$

Hence, $W_1 + W_2$ is a subspace of V(F) i.e. the linear sum of two subspaces of a vector space is also a subspace.

11.7.4 Direct sum: Let W_1 and W_2 be two subspaces of a vector space V(F). Then V is said to be the *direct sum* of W_1 and W_2 if every vector (element) γ of V can be written in one and only way as the sum of an element of W_1 and an element of W_2

i.e. as $\gamma = \alpha + \beta$ for unique elements $\alpha = W_1$ and $\beta = W_2$.

Symbolically, the direct sum is expressed as $V = W_1 \oplus W_2$.

Ex. Vector space R^3 is the direct sum of its two subspaces

$$R^1 \ \& \ R^2 \text{ i.e., } R^3 = R^1 \oplus R^2.$$

Complementary subspace : If $V = W_1 \oplus W_2$ then the two subspaces W_1 and W_2 are said to be *complementary subspaces.*

Disjoint : If V(F) is a vector-space and W_1, W_2 are its two subspaces, then W_1 and W_2 are said to be *disjoint* if their intersection is the zero subspace. i.e if $W_1 \cap W_2 = \{0\}$ then W_1 & W_2 are disjoint.

Extension of the definition of direct sum : Let $W_1, W_2, \cdots, W_m$ be subspaces of V(F). Then V is said to be the *direct sum* of $W_1, W_2, \cdots, W_m$; if & only if every $\gamma \in V$ is uniquely expressed as

$$\gamma = \alpha_1 + \alpha_2 + \cdots + \alpha_m \text{ where}$$

$$\alpha_1 \in W_1, \ \alpha_2 \in W_2, \ \cdots, \ \alpha_m \in W_m$$

then $$V = W_1 \oplus W_2 \oplus W_3 \oplus \cdots \oplus W_m.$$

Theorem. The necessary and sufficient conditions for a vector space V to be the direct sum of two if its subspaces W_1 and W_2 are

$$\text{(i) } V = W_1 + W_2, \quad \text{(ii) } W_1 \cap W_2 = \{0\}.$$

Proof. The proof consists of two parts —proving of necessary condition which explicitly states "If V is a direct sum then prove that (i) & (ii) above" while sufficient proves the converse i.e " If two subspaces W_1 and W_2 of V satisfy (i) & (ii) above then prove that V is a direct sum of W_1 and W_2".

Proof of necessary conditions:

Let $V = W_1 \oplus W_2$ i.e. V is a direct sum of two subspaces W_1 and W_2 then we have to prove that

$$V = W_1 + W_2 \qquad \text{(i)}$$
and $W_1 \cap W_2 = \{0\}$. (ii)

Since each element of V is expressible as sum of an element of W_1 and an element of W_2. We have,

$V = W_1 + W_2$, [(i) is verified]

Let, if possible, $0 \neq \alpha \in W_1 \cap W_2$ is a nonzero vector common to both W_1 and W_2 and therefore, we may write,

$\alpha = 0 + \alpha$ for $0 \in W_1$, $\alpha \in W_2$

and $\alpha = \alpha + 0$ for $\alpha \in W_1$, $0 \in W_2$.

Thus $\alpha \in V$ is expressible in two ways as sum of an element of W_1 and an element of W_2 and accordingly expressible as sum of an element of W_1 and an element of W_2 for $V = W_1 \oplus W_2$ is contradicted. It is possible only when $\alpha = 0$.

Hence $W_1 \cap W_2 = \{0\}$. $\therefore$ [ii] is verified.

Conditions are sufficient :

Let (i) $V = W_1 + W_2$, (ii) $W_1 \cap W_2 = \{0\}$, then it is sufficient to show that $V = W_1 \oplus W_2$.

Since $V = W_1 + W_2$, each element of V, say γ can be expressed as sum of an element of W_1 and an element of W_2 i.e.,

$$\gamma = \alpha + \beta \text{ for } \alpha \in W_1,\ \beta \in W_2 \qquad (11.3)$$

Now it is required only to show that the above representation is unique.

Let, if possible, there may be two representations for γ, i.e.,

$\gamma = \alpha_1 + \beta_1$; $\gamma_1 \in W_1$, $\beta_1 \in W_2$ and

$\gamma = \alpha_2 + \beta_2$; $\gamma_2 \in W_1$, $\beta_2 \in W_2$.

These imply that, $\alpha_1 + \beta_1 = \alpha_2 + \beta_2 \Rightarrow \alpha_1 - \alpha_2 = \beta_2 - \beta_1$

where $\alpha_1 - \alpha_2 \in W_1$ & $\beta_2 - \beta_1 \in W_2$.

Thus $\alpha_1 - \alpha_2 = \beta_2 - \beta_1 \in W_1 \cap W_2$

But as $W_1 \cap W_2 = \{0\}$, therefore, $\alpha_1 - \alpha_2 = 0 = \beta_2 - \beta_1$

i.e. $\alpha_1 - \alpha_2 = 0 \Rightarrow \alpha_1 = \alpha_2$ and $\beta_2 - \beta_1 = 0 \Rightarrow \beta_1 = \beta_2$.

So that each element γ of V is uniquely expressible as the sum of an element of W_1 and an element of W_2. Hence, $V = W_1 \oplus W_2$.

Ex.1 If $W_1 = \{(a, 0, 0) : a \in R\}$ and $W_2 = \{(b, 2b, 3b) : b \in R\}$ are two subspaces of R^3, then find the linear sum $W_1 + W_2$ and give the geometrical interpretation of the subspaces W_1, W_2 and $W_1 + W_2$.

Let $\alpha = (a, 0, 0) \in W_1$ and $\beta = (b, 2b, 3b) \in W_2$.

Now $\alpha + \beta = (a + b, 2b, 3b)$.

Therefore, the linear sum $W_1 + W_2$ consists of all vectors of the form

$$\alpha + \beta = (a + b, 2b, 3b).$$

$\therefore$ $W_1 + W_2 = \{(a + b, 2b, 3b) : a, b \in R\}$.

The subspace W_1 corresponds to the subspace of vectors which have tails at the origin & heads on the *x*-axis and subspace W_2 corresponds to the subspace of vectors with heads on the line through (0, 0, 0) and the point (1, 2, 3). Therefore, the set of all sums of each vectors clearly consists of all vectors with heads on the plane containing these two lines & $W_1 + W_2$ denotes coordinates of all points in this plane.

Ex.2 If $V_3(R)$ is a vector space and

$$W_1 = \{(a, b, 0): a, b \in R\}$$

$$W_2 = \{(0, 0, c): c \in R\}$$

are two subspaces of $V_3(R)$. Then prove that $V_3(R) = W_1 \oplus W_2$.

Let $V_3(R)$. Then $(a, b, c) = (a, b, 0) + (0, 0, c)$

i.e. $\gamma = \alpha + \beta$ (1)

where $\gamma \in V$; $\alpha \in W_1$ & $\beta \in W_2$

$\therefore$ $V_3(R) = W_1 + W_2$.

It is evident that $W_1 \cap W_2 = \{0\}$.

Also there cannot be any other way of expressing $(a,b,c) \in V_3(R)$ in terms of elements of W_1 and W_2, i.e., representation (1) is unique and

$$V_3(R) = W_1 \oplus W_2$$

i.e. $V_3(R)$ is the direct sum of subspaces W_1 & W_2.

Ex.3 If $V_3(R)$ is a vector space and

$$W_1 = \{(a, 0, c): a, c \in R\}$$

$$W_2 = \{(0, b, c): b, c \in R\}$$

are two subspaces of $V_3(R)$. Then show that

$$V_3(R) = W_1 + W_2 \text{ and } V \neq W_1 \oplus W_2.$$

Let $\gamma(a,b,c)$ be any element of $V_3(R)$.

Then $(a, b, c) = (a, 0, \tfrac{1}{2}c) + (0, b, \tfrac{1}{2}c)$,

also $(a, b, c) = (a, 0, \tfrac{1}{4}c) + (0, b, \tfrac{3}{4}c)$

i.e. $\gamma = \alpha_1 + \beta_1 = \alpha_2 + \beta_2$ for $\alpha_1, \alpha_2 \in W_1$; $\beta_1, \beta_2 \in W_2$.

This shows that $\gamma \in V$ can be expressed in two ways in terms of the elements of subspaces W_1 and W_2, i.e. representation $\gamma = \alpha + \beta$ where

$\gamma \in V$, $\alpha \in W_1$, $\beta \in W_2$, is not unique in this case.

Also $W_1 \cap W_2 = \{c\} \neq \{0\}$.

Hence V is a linear sum of two subspaces & not the direct sum i.e.

$$V_3(R) = W_1 + W_2 \text{ and } V \neq W_1 \oplus W_2.$$

11.8 Linear combinations and linear span

Definitions and theorems on linear combination and linear spans will be introduced in this section.

Linear combination : Let V(F) be a vector space and $\mathbf{x}_1, \mathbf{x}_2, \cdots, \mathbf{x}_n$ be *n* vectors of V. Then for some ordered set $\{\lambda_1, \lambda_2, \cdots, \lambda_n\}$ of scalars in the field F, a vector of the form $\lambda_1\mathbf{x}_1 + \lambda_2\mathbf{x}_2 + \cdots + \lambda_n\mathbf{x}_n$ is called a *linear combination* (L.C) of the vectors $\mathbf{x}_1, \mathbf{x}_2, \cdots, \mathbf{x}_n$.

The linear combination of an infinite set of vectors is nothing but the linear combination of a finite subset of this infinite set.

Examples of linear combinations of the vectors $\mathbf{x}_1, \mathbf{x}_2, \mathbf{x}_3$ are

$$-\mathbf{x}_1 + 3\mathbf{x}_2 + 7\mathbf{x}_3 \ , \ \ \mathbf{x}_3 = 0.\mathbf{x}_1 + 0.\mathbf{x}_2 + 1.\mathbf{x}_3 \text{ etc.}$$

Also the solution set of homogeneous equation is a linear combination of vectors shown below:

Let $\qquad x_1 + 2x_2 + 3x_3 + 4x_4 = 0 \qquad (1)$

be a single homogeneous equation. This is in echelon form with free variables x_2 , x_3 , x_4 . Therefore, $x_1 = -2x_2 - 3x_3 - 4x_4$. So, the solution set

$$(x_1, x_2, x_3, x_4) = (-2x_2 - 3x_3 - 4x_4, x_2, x_3, x_4)$$
$$= (-2x_2, x_2, 0, 0) + (-3x_3, 0, x_3, 0) + (-4x_4, 0, 0, x_4)$$
$$= x_2(-2, 1, 0, 0) + x_3(-3, 0, 1, 0) + x_4(-4, 0, 0, 1) \qquad (2)$$

$\therefore$ the solution set (x_1, x_2, x_3, x_4) of equation (1) is a linear combination of the set of vectors

$$\{(-2, 1, 0, 0), (-3, 0, 1, 0) + (-4, 0, 0, 1)\}.$$

Let λ denote (x_1, x_2, x_3, x_4) and $\lambda_1 = (-2, 1, 0, 0)$, $\lambda_2 = (-3, 0, 1, 0)$, $\lambda_3 = (-4, 0, 0, 1)$. So $\{\lambda_1, \lambda_2, \lambda_3\}$ generates/spans the solution set. This idea can be extended to system of homogeneous equations.

Linear span : If S is a nonempty subset of a vector space V(F) then the set of all linear combinations of finite number of elements of S is called the *linear span* of S and is denoted by L(S) or $< S > = \{ a_1\lambda_1 + a_2\lambda_2 + \cdots + a_n\lambda_n \ ; \lambda_i \in S, a_i \in F, i = 1, 2, \cdots, n \}$.

Such a set of vectors in a sense defines the vector space since every vector in the space can be obtained from this set. So, this is a method of constructing a vector space from a given set of vectors.

For example :

(i) The subspace spanned by $\mathbf{u} = (1, 4)$ consists of all points $(x, 4x)$ and hence is the line $y = 4x$ which passes through the origin and the point (1, 4).

(ii) The subspace of $V_3(R)$ spanned by $\mathbf{u} = (1, 3, 2)$ consists of all points $(x, 3x, 2x)$ and hence is the locus of the equations $y = 3x$, $z = 2x$. This is the line segment of the line of intersection of the planes $y = 3x$ and $z = 2x$ connecting the origin and the point (1, 3, 2).

(iii) The subspace spanned by $\mathbf{u} = (1, 0, 0)$ and $\mathbf{u} = (0, 1, 0)$ consists of all points $(x, y, 0)$ and hence is the *XY* plane. This subspace can also be spanned by $\frac{1}{2}\mathbf{u} = (\frac{1}{2}, 0, 0)$, $\frac{1}{3}\mathbf{u} = (0, \frac{1}{3}, 0)$ and every vector

$$(x, y, 0) = 2x(\tfrac{1}{2}\mathbf{u}) + 3y(\tfrac{1}{3}\mathbf{u}).$$

This example shows that spanning set for a subspace is by no means unique.

Ex. Write the vector $\lambda = (1, -2, 5)$ as a linear combination of vectors $\lambda_1 = (1, 1, 1)$, $\lambda_2 = (1, 2, 3)$, $\lambda_3 = (2, -1, 1)$ in the vector space $V_3(R)$.

Let $\lambda = x\lambda_1 + y\lambda_2 + z\lambda_3$ where $x, y, z \in R$.

$$\therefore (1, -2, 5) = x(1, 1, 1) + y(1, 2, 3) + z(2, -1, 1)$$
$$= (x + y + 2z, x + 2y - z, x + 3y + z).$$

Equating the components, we have the following system of equations

$$\left.\begin{array}{l} x+y+2z = 1 \\ x+2y-z = -2 \\ x+3y+z = 5 \end{array}\right\} \qquad (1)$$

Let us reduce the system to the echelon form

$$\left.\begin{array}{r} x+y+2z = 1 \\ y-3z = -3 \\ 2y-z = 4 \end{array}\right\} \Rightarrow \left.\begin{array}{r} x+y+2z = 1 \\ y-3z = -3 \\ 5z = 10 \end{array}\right\}.$$

The last set is in echelon form and the system is uniquely consistent. So λ is a inear combination of $\lambda_1, \lambda_2, \lambda_3$. Solving for x, y, z, we have $x = -6, y = 3, z = 2$.
$\therefore$ the required linear combination is $\lambda = -\lambda_1 + 3\lambda_2 + 2\lambda_3$.

N.B. Procedure for testing linear combination : Equivalent system of nonhomogeneous equations in echelon f m must be consistent for a linear combination.

11.8.1 Equality of two linear spans: The procedure for testing equality of two linear spans are described and illustrated in this subsection.

Procedure :

(1) Express each vector of one set as a linear combination of the vectors of another set ;

(2) Reduce the resulting equivalent system of nonhomogeneous equations in procedure (1) to echelon form ;

(3) If all the equivalent nonhomogeneous systems in procedure (1) are consistent then the L.C. of two sets are equal otherwise not.

Ex. 1 Are the following subspaces identical

$$V = L\{(1,-1,2,0),(2,1,1,1),(3,-1,2,-1)\}$$
$$W = L\{(3,0,3,1),(1,2,-1,1),(4,-1,5,1)\} \ ?$$

By procedure (1) in sec.11.8.1, let the first vector of W as a L.C. of the vectors in V be $(3,0,3,1) = x(1,-1,2,0) + y(2,1,1,1) + z(3,-1,2,-1)$ which implies

$$\left.\begin{array}{r} x+2y+3z = 3 \\ -x+y-z = 0 \\ 2x+y+2z = 3 \\ y-z = 1 \end{array}\right\} \Rightarrow \left.\begin{array}{r} x+2y+3z = 3 \\ 3y+2z = 3 \\ -3y-4z = -3 \\ y-z = 1 \end{array}\right\} \Rightarrow \left.\begin{array}{r} x+2y+3z = 3 \\ 3y+2z = 3 \\ -2z = 0 \\ 5z = 0 \end{array}\right\} \Rightarrow$$

$$\left.\begin{array}{r} x+2y+3z = 3 \\ 3y+2z = 3 \\ z = 0 \end{array}\right\}$$

$\therefore$ consistent, $z = 0$, $y = 1$, $x = 1$.

$\therefore \quad (3,0,3,1) = 1.(1,-1,2,0) + 1.(2,1,1,1) + 0.(3,-1,2,-1)$

By procedure (1) in sec.11.8.1 , let the second vector of W as a L.C. of the vectors in V be $(1, 2, -1, 1) = x(1,-1,2,0) + y(2,1,1,1) + z(3,-1,2,-1)$ which implies

$$\left.\begin{array}{r} x+2y+3z=1 \\ -x+y-z=2 \\ 2x+y+2z=-1 \\ y-z=1 \end{array}\right\} \Rightarrow \left.\begin{array}{r} x+2y+3z=1 \\ 3y+2z=3 \\ -3y-4z=-3 \\ y-z=1 \end{array}\right\} \Rightarrow \left.\begin{array}{r} x+2y+3z=1 \\ 3y+2z=3 \\ -2z=0 \\ 5z=0 \end{array}\right\} \Rightarrow \left.\begin{array}{r} x+2y+3z=1 \\ 3y+2z=3 \\ z=0 \\ 0=0 \end{array}\right\}$$

$\therefore$ consistent giving $z=0, y=1, x=-1$ and

$$(1, 2, -1, 1) = -(1, -1, 2, 0) + (2, 1, 1, 1).$$

Similarly, the third vector of W can be represented as a L.C. of the vectors of V as

$$(4, -1, 5, 1) = x(1,-1, 2, 0) + y(2, 1, 1, 1) + z(3, -1, 2, -1) \text{ giving}$$

$$\left.\begin{array}{r} x+2y+3z=4 \\ -x+y-z=-1 \\ 2x+y+2z=5 \\ y-z=1 \end{array}\right\} \Rightarrow \left.\begin{array}{r} x+2y+3z=4 \\ 3y+2z=3 \\ -3y-4z=-3 \\ y-z=1 \end{array}\right\} \Rightarrow$$

$$\left.\begin{array}{r} x+2y+3z=4 \\ 3y+2z=3 \\ -2z=0 \\ 5z=0 \end{array}\right\} \Rightarrow \left.\begin{array}{r} x+2y+3z=4 \\ 3y+2z=3 \\ z=0 \end{array}\right\}$$

$\therefore$ consistent giving $z=0, y=1, x=2$

and $(4, -1, 5, 1) = 2(1,-1, 2, 0) + 1(2, 1, 1, 1)$

As all the equivalent systems in procedure (3) in sec.11.8.1are consistent, so subspaces are identical.

$\therefore$ $\langle V \rangle = \langle W \rangle$.

Ex.2 In IR^3, let $S = \{(1, 2, 1), (3, 5, 0)\}$ and

$$T = \{(1, 2, 1), (3, 5, 0), (2, 3, -1)\}.$$

Examine whether $\langle S \rangle = \langle T \rangle$.

The set of vectors which is a linear combination of $\alpha_1 = (1, 2, 1)$, $\alpha_2 = (3, 5, 0)$, $\alpha_3 = (2, 3, -1)$ is a linear span of T and hence

$$\langle T \rangle = \{x\alpha_1 + y\alpha_2 + z\alpha_3 \; : x, y, z \in R\}$$

$$= \{x(1, 2, 1) + y(3, 5, 0) + z(2, 3, -1)\} \tag{1}$$

Also let $(2, 3, -1) = \{b_1(1, 2, 1) + b_2(3, 5, 0) \; : b_1, b_2 \in R\}$

$$= (b_1+3b_2,\; 2b_1+5b_2,\; b_1) \tag{2}$$

Consequently,

$$\left.\begin{array}{r} b_1+3b_2=2 \\ 2b_1+5b_2=3 \\ b_1=-1 \end{array}\right\}$$

$\therefore$ solving $b_1 = -1,\ b_2 = 1.$

$\therefore$ $(2, 3, -1) = -(1, 2, 1) + (3, 5, 0)$

or $z(2, 3, -1) = -z(1, 2, 1) + z(3, 5, 0)$ (3)

Using (3) in (1), we have

$$< T > = \{ x(1, 2, 1) + y(3, 5, 0) - z(1, 2, 1) + z(3, 5, 0)\}$$
$$= \{(x - z)(1, 2, 1) + (y + z)(3, 5, 0)\}$$
$$= \{x_1(1, 2, 1) + y_1(3, 5, 0) : x_1, y_1 \in R\} = < S >$$

and the answer to the question is "yes".

Alternatively

Let $(a, b, c) = L < S > = x(1, 2, 1) + y(3, 5, 0)$ for $x, y \in R$.

$$\Rightarrow \left.\begin{array}{l} x + 3y = a \\ 2x + 5y = b \\ x \quad\quad = c \end{array}\right\} \rightarrow \left.\begin{array}{l} x + 3y = a \\ -y = b - 2a \\ -3y = c - a \end{array}\right\} \rightarrow \begin{array}{l} x + y = a \\ y = 2a - b \\ 0 = 5a - 3b + c \end{array}$$

$\therefore$ the system is consistent if $5a - 3b + c = 0$ i.e. $c = 3b - 5a$.

$y = 2a - b,\ x = a - 3(2a - b) = -5a + 3b.$

$\therefore$ $L< S > = (-5a + 3b)(1, 2, 1) + (2a - b)(3, 5, 0)$

$= (-5a + 3b + 6a - 3b,\ -10a + 6b + 10a - 5b,\ -5a + 3b)$

$= (a,\ b,\ -5a + 3b)$ (1)

Also, let $(a, b, c) = L< T > = x(1, 2, 1) + y(3, 5, 0) + z(2, 3, -1)$

$$\Rightarrow \left.\begin{array}{l} x + 3y + 2z = a \\ 2x + 5y + 3z = b \\ x \quad\quad - z = c \end{array}\right\} \rightarrow \left.\begin{array}{l} x + 3y + 2z = a \\ -y - z = b - 2a \\ -3y - 3z = c - a \end{array}\right\} \rightarrow \begin{array}{l} x + 3y + 2z = a \\ y + z = 2a - b \\ 0 = 5a - 3b + c \end{array}$$

$\therefore$ the system is consistent if $5a - 3b + c = 0$.

Hence $x = -5a + 3b + z,\ y = 2a - b - z.$

Hence $L< T > = (-5a + 3b + z)(1, 2, 1) + (2a - b - z)(3, 5, 0) + z(2, 3, -1)$

$= (-5a + 3b + z + 6a - 3b - 3z + 2z,\ -10a + 6b + 2z + 10a - 5b - 5z + 3z,\ -5a + 3b + z + 0 - z)$

$= (a, b, -5a + 3b)$ (2)

$\therefore$ from (1) and (2), $L< S > = L< T >$.

Ex. 3 If $\alpha_1 = (1, 2, -1)$, $\alpha_2 = (2, -3, 2)$, $\alpha_3 = (4, 1, 3)$, $\alpha_4 = (-3, 1, 2)$ are the vectors in $V_3(R)$, show that $L(\{\alpha_1, \alpha_2\}) = L(\{\alpha_3, \alpha_4\})$.

Let $L(\{\alpha_1, \alpha_2\}) = L(\{\alpha_3, \alpha_4\})$, then there exists $a_1, a_2 \in R$ for arbitrary $a_3, a_4 \in R$ such that

$$a_1\alpha_1 + a_2\alpha_2 = a_3\alpha_3 + a_4\alpha_4$$

or $\quad a_1(1,2,-1)+a_2(2,-3,2)=a_3(4,1,3)+a_4((-3,1,2)$

or $\quad (a_1+2a_2-4a_3+3a_4,\ 2a_1-3a_2-a_3-a_4,\ -a_1+2a_2-3a_3-2a_4)=0$

$$\Rightarrow \left.\begin{array}{l} a_1+2a_2-4a_3+3a_4=0 \\ 2a_1-3a_2-a_3-a_4=0 \\ -a_1+2a_2-3a_3-2a_4=0 \end{array}\right\} \to \left.\begin{array}{l} a_1+2a_2-4a_3+3a_4=0 \\ -7a_2+7a_3-7a_4=0 \\ 4a_2-7a_3+a_4=0 \end{array}\right\} \to$$

$$\left.\begin{array}{r} a_1+2a_2-4a_3+3a_4=0 \\ a_2-a_3+a_4=0 \\ a_3+a_4=0 \end{array}\right\}$$

$\therefore\ a_3=-a_4,\ a_2=-2a_4,\ a_1=-3a_4.$

N.B. In echelon form a_4 is a free variable.

$\therefore \quad L(\{\alpha_1,\alpha_2\})=a_1(1,2,-1)+a_2(2,-3,2)$
$\qquad = -3a_4(1,2,-1)-2a_4(2,-3,2)=(-7a_4,0,-a_4)$

Also, $\quad L(\{\alpha_3,\alpha_4\})=a_3(4,1,3)+a_4(-3,1,2)$
$\qquad = -a_4(4,1,3)+a_4(-3,1,2)=(-7a_4,0,-a_4)$

$\therefore \quad L(\{\alpha_1,\alpha_2\})=L(\{\alpha_3,\alpha_4\}).$

Hence the problem is established.

Ex.4 Let $\mathbf{u}_1=(1,1,-1)$, $\quad \mathbf{u}_2=(2,3,-1)$, $\quad \mathbf{u}_3=(3,1,-5)$;
$\mathbf{v}_1=(1,-1,-3)$, $\quad \mathbf{v}_2=(3,-2,-8)$, $\mathbf{v}_3=(2,1,-3)$.

Show that the subspace of R^3 generated by $\mathbf{u}_i$ is the same as the subspace generated by $\mathbf{v}_i$.

We need to show that an arbitrary vector (a, b, c) is a linear combination of $\mathbf{u}_i$ & $\mathbf{v}_i$ respectively. So that $\mathbf{u}_i$ & $\mathbf{v}_i$ will generate or span a subspace of R^3. Lastly if the linear combination of $\mathbf{u}_i$ is the same as the linear combination of $\mathbf{v}_i$ then the subspaces are same.

Let $\quad (a,b,c)=x\mathbf{u}_1+y\mathbf{u}_2+z\mathbf{u}_3$ for $x, y, z\in R$
$\qquad = x(1,1,-1)+y(2,3,-1)+z(3,1,-5)$
$\qquad = (x+2y+3z,\ x+3y+z,\ -x-y-5z)$

$$\Rightarrow \left.\begin{array}{rrrl} x & +2y & +3z & =a \\ x & +3y & +z & =b \\ -x & -y & -5z & =c \end{array}\right\} \to \left.\begin{array}{rrrl} x & +2y & +3z & =a \\ & y & -2z & =b-a \\ & y & -2z & =a+c \end{array}\right\} \to$$

$$\left.\begin{array}{rrrl} x & +2y & +3z & =a \\ & y & -2z & =b-a \\ & & 0 & =2a-b+c \end{array}\right\}$$

This is consistent if $2a-b-c=0$.

$\therefore\ \mathbf{u}_i$ generates a subspace if $2a-b-c=0$.

Now $y=b-a+2z$ and $x=3a-2b-7z$.

$\therefore\ L(\{\mathbf{u}_1,\mathbf{u}_2,\mathbf{u}_3\})=(x\mathbf{u}_1+y\mathbf{u}_2+z\mathbf{u}_3)$
$\qquad = (3a-2b-7z)(1,1,-1)+(b-a+2z)(2,3,-1)$

$+z(3, 1, -5))$

$$= (3a - 2b - 7z + 2(b-a+2z) + 3z,\ 3a - 2b - 7z + 3(b-a+2z) + z,\ -(3a - 2b - 7z) - (b-a+2z) - 5z)$$

$$= (a, b, -2a+b) \qquad \text{(i)}$$

Again let $(a, b, c) = x\mathbf{v}_1 + y\mathbf{v}_2 + z\mathbf{v}_3$ for $x, y, z \in \mathrm{R}$

$$= x(1, -1, -3) + y(3, -2, -8) + z(2, 1, -3)$$

$$= (x + 3y + 2z, -x - 2y + z, -3x - 8y - 3z)$$

$$\Rightarrow \left.\begin{aligned} x + 3y + 2z &= a \\ -x - 2y + z &= b \\ -3x - 8y - 3z &= c \end{aligned}\right\} \to \left.\begin{aligned} x + 3y + 2z &= a \\ y + 3z &= a+b \\ y + 3z &= c+3a \end{aligned}\right\} \to$$

$$\left.\begin{aligned} x + 3y + 2z &= a \\ y + 3z &= a+b \\ 0 &= 2a-b+c \end{aligned}\right\}.$$

This system will be consistent if $2a - b + c = 0$ i.e. if $c = b - 2a$.

$\therefore$ $\mathbf{v}_i$ generates a subspace if $2a - b + c = 0$.

$\therefore$ $y = b + a - 3z$

$x = a - 3y - 2z = a - 3(b + a - 3z) - 2z = -2a - 3b + 7z.$

$\therefore$ $\mathrm{L}(\{\mathbf{v}_1, \mathbf{v}_2, \mathbf{v}_3\}) = (x\mathbf{v}_1 + y\mathbf{v}_2 + z\mathbf{v}_3)$

$$= \{(-2a - 3b + 7z)(1, -1, -3) + (b + a - 3z)(3, -2, -8) + z(2, 1, -3)\}$$

$$= (-2a - 3b + 7z + 3b + 3a - 9z + 2z,\ 2a + 3b - 7z - 2b - 2a + 6z + z,\ 6a + 9b - 21z - 8b - 8a + 24z - 3z)$$

$$= (a, b, -2a + b) \qquad \text{(ii)}$$

$\therefore$ from (i) and (ii), $\mathrm{L}(\{\mathbf{u}_1, \mathbf{u}_2, \mathbf{u}_3\}) = \mathrm{L}(\{\mathbf{v}_1, \mathbf{v}_2, \mathbf{v}_3\})$.

Ex.5 Is the vector $(3, -1, 0, -1)$ in the subspace of R^4 spanned by the vectors $(2, -1, 3, 2)$, $(-1, 1, 1, -3)$ & $(1, 1, 9, -5)$?

If the vector $(3, -1, 0, -1)$ is a linear combination of

$$(2, -1, 3, 2), (-1, 1, 1, -3) \ \& \ (1, 1, 9, -5)$$

then the first vector in the subspace of R^4 are spanned by the last 3 vectors. To do that let

$$(3, -1, 0, -1) = x(2, -1, 3, 2) + y(-1, 1, 1, -3) + z(1, 1, 9, -5)$$

for $x, y, z \in \mathrm{R}$.

$$\Rightarrow \left.\begin{aligned} 2x - y + z &= 3 \\ -x + y + z &= -1 \\ 3x + y + 9z &= 0 \\ 2x - 3y - 5z &= -1 \end{aligned}\right\} \to \left.\begin{aligned} 2x - y + z &= 3 \\ y + 3z &= 1 \\ 5y + 15z &= -9 \\ -2y - 6z &= -4 \end{aligned}\right\} \to \left.\begin{aligned} 2x - y + z &= 3 \\ y + 3z &= 1 \\ 0 &= -\tfrac{14}{5} \\ 0 &= -2 \end{aligned}\right\}.$$

which is a inconsistent system of equations in echelon form.

$\therefore$ 1st vector is not a linear combination of the last 3 vectors & so it is not spanned by them.

Ex.6 Show that the vectors $\mathbf{u} = (1, 2, 3)$, $\mathbf{v} = (0, 1, 2)$ and $\mathbf{w} = (0, 0, 1)$ generates R^3.

We need to show that an arbitrary vector $(a, b, c) \in \mathrm{R}^3$ is a linear combination of **u, v** & **w**.

Let $(a, b, c) = x\mathbf{u} + y\mathbf{v} + z\mathbf{w}$

or $(a, b, c) = x(1, 2, 3) + y(0, 1, 2) + z(0, 0, 1)$

$$\Rightarrow \left.\begin{aligned} x &= a \\ 2x + y &= b \\ 3x + 2y + z &= c \end{aligned}\right\} \to \left.\begin{aligned} x &= a \\ y &= b - 2a \\ y + z &= 3c - a \end{aligned}\right\} \to \left.\begin{aligned} x &= a \\ y &= b - 2a \\ z &= a - 2b + c \end{aligned}\right\}.$$

which is in echelon form & is consistent.

[In fact, $x = a$, $y = b - 2a$, $z = a - 2b + c$]

$\therefore$ **u, v, w** generates R^3.

Ex.7 Is the vector $\alpha = (2, -5, 3) \in V_3(R)$ a linear combination of vectors $\alpha_1 = (1, -3, 2)$, $\alpha_2 = (2, -4, -1)$, $\alpha_3 = (1, -5, 7)$.

Let $\alpha = x\alpha_1 + y\alpha_2 + z\alpha_3$.

$\therefore$ $(2, -5, 3) = x(1, -3, 2) + y(2, -4, -1) + z(1, -5, 7)$

$= (x + 2y + z, -3x - 4y - 5z, 2x - y + 7z)$

Equating components,

$$\Rightarrow \left.\begin{aligned} x + 2y + z &= 2 \\ -3x - 4y - 5z &= -5 \\ 2x - y + 7z &= 3 \end{aligned}\right\} \Rightarrow \left.\begin{aligned} x + 2y + z &= 2 \\ -2y + 2z &= -1 \\ -5y + 5z &= -1 \end{aligned}\right\} \Rightarrow \left.\begin{aligned} x + 2y + z &= 2 \\ -2y + 2z &= -1 \\ 0 &= 3 \end{aligned}\right\}$$

which is inconsistent.

$\therefore$ α cannot be expressed as linear combination α_1, α_2, α_3.

Ex. 8 Show that the span of $\{\alpha, \beta\}$ is the plane $5x - 4y + 2z = 0$ where

$$\alpha = \begin{bmatrix} 2 \\ 3 \\ 1 \end{bmatrix} \text{ and. } \beta = \begin{bmatrix} 0 \\ 1 \\ 2 \end{bmatrix}.$$

The span of $\{\alpha, \beta\}$ is the linear combination of α and β i.e.

$$c_1\alpha + c_2\beta = c_1 \begin{bmatrix} 2 \\ 3 \\ 1 \end{bmatrix} + c_2 \begin{bmatrix} 0 \\ 1 \\ 2 \end{bmatrix} = \begin{bmatrix} 2c_1 \\ 3c_1 + c_2 \\ c_1 + 2c_2 \end{bmatrix} = \begin{bmatrix} x \\ y \\ z \end{bmatrix}, \text{ (say)}$$

Then

$$x = 2c_1$$
$$y = 3c_1 + c_2$$
$$z = c_1 + 2c_2$$

Eliminating c_1 and c_2 from these equations, we have the equation of the plane as $5x - 4y + 2z = 0$.

11.8.2 Theorems of linear span : The following theorems establish that linear span is a subspace.

Theorem. 1 The linear span L(S) of a nonempty subset S of a vector space V(F) is the smallest subspace of V containing S.

Alternatively, let S be nonempty subset of V. The set of all linear combinations of vectors in S, denoted by L(S), is a subspace of V containing S. Furthermore, if W is any other subspace of V containing S. Then L(S) $\subset$ W.

Proof. The proof consists of two parts — **(i)** L(S) is a subspace of V(F) and **(ii)** L(S) is the smallest subspace of V(F).

(i) Let $\mathbf{x} = \alpha_1 x_1 + \alpha_2 x_2 + \cdots + \alpha_m x_m$,

$$x_i \in S \ \& \ \alpha_i \in F, \ i = 1, 2, 3, \cdots, m.$$

and $\mathbf{y} = \beta_1 y_1 + \beta_2 y_2 + \cdots + \beta_m y_m$,

$$y_i \in S \ \& \ \beta_i \in F, \ i = 1, 2, \cdots, m,$$

be two arbitray elements of L(S). Then $\forall\ \alpha, \beta \in F$,

$$\alpha\mathbf{x} + \beta\mathbf{y} = \alpha(\alpha_1 x_1 + \alpha_2 x_2 + \cdots + \alpha_m x_m) + \beta(\beta_1 y_1 + \beta_2 y_2 + \cdots + \beta_m y_m)$$
$$= (\alpha\alpha_1 x_1 + \alpha\alpha_2 x_2 + \cdots + \alpha\alpha_m x_m) + (\beta\beta_1 y_1 + \beta\beta_2 y_2 + \cdots + \beta\beta_m y_m)$$

which is a linear combination of finite number of elements of S and therefore, belongs to L(S).
Thus $\mathbf{x}, \mathbf{y} \in L(S) \Rightarrow \alpha\mathbf{x} + \beta\mathbf{y} \in L(S) \ \forall\ \alpha, \beta \in F$

$\therefore$ L(S) is a subspace of V.

(ii) Next let W be any subspace of V containing S and let $\mathbf{x}_1, \mathbf{x}_2, \cdots, \mathbf{x}_n \in S \subset W$, then W being a subspace,

$$\alpha_1 x_1 + \alpha_2 x_2 + \cdots + \alpha_m x_m, \ x_i \in W \text{ for all scalar } \alpha_i .$$

This shows that W contains all linear combinations of finite number of vectors in S. Consequently, L(S)$\subset$ W. Hence L(S) is the smallest subspace of V containing S.

Theorem 2. The linear sum of two subspaces W_1 and W_2 of a vector space V(F) is generated (spanned) by their union i.e.

$$W_1 + W_2 = L(W_1 \cup W_2) .$$

Proof. We know that the linear sum of two subspaces of a vector space is a subspace and the linear span of every nonempty subset of the vector space is a subspace of the same vector space. So, $W_1 + W_2$, as well as $L(W_1 \cup W_2)$ is a subspace of V(F).

Now in order to establish their equality, let **x** be any arbitrary element of $W_1 + W_2$. Then

$$\mathbf{x} = \mathbf{x}_1 + \mathbf{x}_2 \text{ for some } \mathbf{x}_1 \in W_1 \text{ and } \mathbf{x}_2 \in W_2 .$$

Also we may write $\quad \mathbf{x} = 1.\mathbf{x}_1 + 1.\mathbf{x}_2$

which shows that **x** is expressible as a linear combination of finite number of elements, namely $\mathbf{x}_1$ and $\mathbf{x}_2$ of $L(W_1 \cup W_2)$ and so $\mathbf{x} \in L(W_1 \cup W_2)$.

Thus $\mathbf{x} \in W_1 + W_2 \Rightarrow \mathbf{x} \in L(W_1 \cup W_2)$.

Consequently , $W_1 + W_2 \subseteq L(W_1 \cup W_2)$. (11.4)

Again $\quad W_1 \subseteq W_1 + W_2 \quad \& \quad W_2 \subseteq W_1 + W_2$.

$\therefore \quad W_1 \cup W_2 \subseteq W_1 + W_2$ (11.5)

$\therefore$ from (11.4) and (11.5), we have, (cf. sec. 11.1.1 equality of subsets)

$\therefore \quad W_1 \cup W_2 \subseteq W_1 + W_2 \subseteq L(W_1 \cup W_2)$.

But $L(W_1 \cup W_2)$ being the smallest subspace containing $(W_1 \cup W_2)$,

so $\quad L(W_1 \cup W_2) \subseteq W_1 + W_2$.

Hence, $\quad W_1 + W_2 = L(W_1 \cup W_2)$.

11.9 Linear independence and dependence revisited

The definitions and concerned theorem on linear dependence/ independence of vectors are the theme of this section.

Definition. Let V be a vector space over a field F. Then the vectors

$$\alpha_1, \alpha_2, \cdots, \alpha_n \in V$$

are said to be *linearly dependent* over F, if there exists scalars $a_1, a_2, \cdots, a_n \in F$, called *weights*, not all of them zero, such that

$$a_1\alpha_1 + a_2\alpha_2 + \cdots + a_n\alpha_n = 0.$$

If, however, all the scalars $a_1, a_2, \cdots, a_n$ are zero

i.e. $$\sum_{i=1}^{n} a_i\alpha_i = 0 \Rightarrow a_i = 0 \quad \text{for } i = 1, 2, \cdots, n,$$

then the vectors $\alpha_1, \alpha_2, \cdots, \alpha_n \in V$ are said to be *linearly independent* over F.

i.e. the set $\{\alpha_1, \alpha_2, \cdots, \alpha_n\}$ of which $\{\alpha_1, \alpha_2, \cdots, \alpha_k\}$ is a subset, must be linearly independent.

Conversely. Let S be linearly dependent. Then $\{\alpha_1\}$ is obviously linearly independent as $\alpha_1 \neq 0$. Let us consider the set $\{\alpha_1, \alpha_2\}$. If $\{\alpha_1, \alpha_2\}$ is linearly dependent, then there exists a_1, $a_2 \in$ F (both not zero), such that

$$a_1\alpha_1 + a_2\alpha_2 = 0 \tag{11.6}$$

Here $a_2 \neq 0$ otherwise one must have $a_1\alpha_1 = 0$ which implies $a_1 = 0 \quad \because \alpha_1 \neq 0$. In other words, both a_1 and a_2 are zero which is not possible.

$\therefore$ (11.6) $\Rightarrow \alpha_2 = -a_1a_2^{-1}\alpha_1 \qquad \because a_2 \neq 0, a_2^{-1} \in F$

i.e. $-a_1a_2^{-1}\alpha_1 \in F$ with the help of which α_2 may be expressed as a linear combination of α_1.

Further, if $\{\alpha_1, \alpha_2\}$ is linearly independent, we consider $\{\alpha_1, \alpha_2, \alpha_3\}$ and if it is linearly dependent, then it can be easily shown that α_3 can be expressed as a linear combination of α_1 and α_2 whereas on the other hand if $\{\alpha_1, \alpha_2, \alpha_3\}$ is linearly independent we proceed further & repeat the same steps.

Proceeding in this way, let k is an integer such that $2 \leq k \leq n$ & $\{\alpha_1, \alpha_2, \cdots, \alpha_{k-1}\}$ is linearly independent.

Now let $\{\alpha_1, \alpha_2, \cdots, \alpha_{k-1}, \alpha_k\}$ be linearly dependent,

then one must have $$\sum_{i=1}^{k} b_i\alpha_i = 0 \tag{11.7}$$

where at least one $b_i \neq 0$ (say $b_k \neq 0$).

This is because if $b_k = 0$ the set $\{\alpha_1, \alpha_2, \cdots, \alpha_{k-1}\}$ becomes linearly dependent from (11.7) and this is contrary to the choice of k. Thus, from (11.7) one gets

$$\sum_{i \neq k} b_i\alpha_i + b_k\alpha_k = 0$$

$$\therefore \qquad \alpha_k = \sum_{i=1}^{k-1}(-b_ib_k^{-1})\alpha_i \quad \because -b_ib_k^{-1} \in F$$

Thus α_k is a linear combination of its preceding vectors.

Therefore, a subset S of V is said to be *linearly dependent* or *linearly independent* respectively as the vectors in S are linearly dependent or linearly independent.

In the same way, any infinite set of vectors of V is said to be *linearly independent* if its every finite subset is linearly independent, otherwise it is *linearly dependent.*

Note. The vectors corresponding to nonzero rows of a matrix in echelon form are *linearly independent.*

Theorem.1 If V(F) is a vector space and S = $\{\alpha_1, \alpha_2, \cdots, \alpha_n\}$ is a subset of some nonzero elements of V, then S is linearly dependent, if and only if some element of S is a linear combination of the others.

Proof. Let S = $\{\alpha_1, \alpha_2, \cdots, \alpha_n\}$ be linearly dependent and let $a_1, a_2, \cdots, a_n \in$ F not all zero, such that

$$a_1\alpha_1 + a_2\alpha_2 + \cdots + a_n\alpha_n = 0$$

or $$\sum a_i\alpha_i = 0$$

or $$\sum_{i \neq k} a_i\alpha_i + a_k\alpha_k = 0$$

or $$a_k\alpha_k = -\sum_{i \neq k} a_i\alpha_i$$

or $$\alpha_n = \sum_{i \neq k}\left(\frac{-a_i}{a_k}\right)\alpha_i \quad \because a_k \neq 0 \Rightarrow a_k^{-1} \in \text{F}.$$

$\therefore\ \alpha_k \in$ S can be expressed as a linear combination of the others.

Conversely. Let α_j be any element of S. It can be expressed as a linear combination of the others

i.e. $$\alpha_j = \sum_{i \neq j} b_i\alpha_i \text{ for } b_i \text{ is not all zero belongs to F}$$

or $$\sum_{i \neq j} b_i\alpha_i + (-1)\alpha_j = 0.$$

Here, $-1 \neq 0 \Rightarrow$ S is linearly dependent.

Theorem 2. If V(F) be a vector space and S = $\{\alpha_1, \alpha_2, \cdots, \alpha_n\}$ be a subset of some nonzero elements of V, then S is linearly dependent if and only if some one of the elements of S (say α_k) $2 \leq k \leq n$, can be expressed as a linear combination of the preceding one's.

Proof. If $\alpha_k \in$ S ($2 \leq k \leq n$) is expressed as a linear combination of $\alpha_1, \alpha_2, \cdots, \alpha_{k-1}$, then it needs to prove that the set S is linearly dependent

$$\alpha_k = a_1\alpha_1 + a_2\alpha_2 + \cdots + a_{k-1}\alpha_{k-1} \text{ for } a_i \in \text{F}, i = 1, 2, \cdots, (k-1)$$

or $$a_1\alpha_1 + a_2\alpha_2 + \cdots + a_{k-1}\alpha_{k-1} + (-1)\alpha_k = 0$$

$\therefore$ the set $\{\alpha_1, \alpha_2, \cdots, \alpha_{k-1}, \alpha_k\}$ is linearly dependent.

Hence the nonzero vectors $\{\alpha_1, \alpha_2, \cdots, \alpha_n\}$ are linearly dependent if & only if one of them, say α_k, is a linear combination of the preceding vectors, i.e.

$$\alpha_k = a_1\alpha_1 + a_2\alpha_2 + \cdots + a_{k-1}\alpha_{k-1}.$$

Thus, the theorem is established in all respects.

Ex.1 Show that the set of row vectors

$$\{(2, 3, 1, 0), (0, 1, 1, -2), (1, 0, -1, 3), (5, 3, -2, 9)\}$$

is linearly dependent and extract a linearly independent subset which generates the same subspace of V_4(R). Write each remaining vector as a linear combination of the members of this subset.

The matrix of the given row vectors is

$$\begin{bmatrix} 2 & 3 & 1 & 0 \\ 0 & 1 & 1 & -2 \\ 1 & 0 & -1 & 3 \\ 5 & 3 & -2 & 9 \end{bmatrix} \xrightarrow[2R_4 - 5R_1]{2R_3 - R_1} \begin{bmatrix} 2 & 3 & 1 & 0 \\ 0 & 1 & 1 & -2 \\ 0 & -3 & -3 & 6 \\ 0 & -9 & -9 & 18 \end{bmatrix} \xrightarrow[R_4 + 9R_2]{R_3 + 3R_2}$$

$$\begin{bmatrix} 2 & 3 & 1 & 0 \\ 0 & 1 & 1 & -2 \\ 0 & 0 & 0 & 0 \\ 0 & 0 & 0 & 0 \end{bmatrix}$$

$\because$ the echelon matrix contains two zero rows, therefore, the row vectors are linearly dependent. Of which the 1st 2 row vectors are linearly independent which generates $V_4(R)$. Let the linear combination of the remaining vector (5, 3, –2, 9) be

$$(5, 3, -2, 9) = x(2, 3, 1, 0) + y(0, 1, 1, -2) \quad \text{where } x, \ y, \ z \text{ are scalars in F.}$$

This implies

$$\left.\begin{aligned} 2x + 0.y &= 5 \\ 3x + \ y &= 3 \\ x + \ y &= -2 \\ 0.x - 2y &= 9 \end{aligned}\right\} \rightarrow \left.\begin{aligned} 2x \quad &= 5 \\ 2y &= -9 \\ 2y &= -9 \\ -2y &= 9 \end{aligned}\right\} \rightarrow \left.\begin{aligned} 2x \quad &= 5 \\ 2y &= -9 \\ 0 &= 0 \\ 0 &= 0 \end{aligned}\right\}$$

$\therefore \quad x = \frac{5}{2}, \ y = -\frac{9}{2}$

$(5, 3, -2, 9) = L < (2, 3, 1, 0), (0, 1, 1, -2) >$

$\qquad = \frac{5}{2}(2, 3, 1, 0) - \frac{9}{2}(0, 1, 1, -2).$

Similarly, $(1, 0, -1, 3) = L < (2, 3, 1, 0), (0, 1, 1, -2) >$

$\qquad = \frac{1}{2}(2, 3, 1, 0) - \frac{3}{2}(0, 1, 1, -2).$

Ex.2 Find the linearly independent subset A of the set $S = \{\alpha_1, \alpha_2, \alpha_3, \alpha_4\}$ in $V_3(R)$, where

$$\alpha_1 = (1, 2, -1), \ \alpha_2 = (-3, -6, 3), \ \alpha_3 = (2, 1, 3), \alpha_4 = (8, 7, 7)$$

which spans the same space as S.

Let us form the matrix whose rows are the given vectors and reduce it to the echelon matrix using elementary row operations as follows:

Let us form the matrix whose rows are the given vectors & reduce it to the echelon matrix using elementary row operations as follows:

$$\begin{bmatrix} 1 & 2 & -1 \\ -3 & -6 & 3 \\ 2 & 1 & 3 \\ 8 & 7 & 7 \end{bmatrix} \xrightarrow[R_4 - 8R_1]{\substack{R_2 + 3R_1 \\ R_3 - R_1}} \begin{bmatrix} 1 & 2 & -1 \\ 0 & 0 & 0 \\ 0 & -3 & 5 \\ 0 & -9 & 15 \end{bmatrix} \xrightarrow{R_2 \leftrightarrow R_3} \begin{bmatrix} 1 & 2 & -1 \\ 0 & -3 & 5 \\ 0 & 0 & 0 \\ 0 & -9 & 15 \end{bmatrix}$$

$$\xrightarrow{R_3 \leftrightarrow R_4} \begin{bmatrix} 1 & 2 & -1 \\ 0 & -3 & 5 \\ 0 & -9 & 15 \\ 0 & 0 & 0 \end{bmatrix} \xrightarrow{R_3 - 3R_2} \begin{bmatrix} 1 & 2 & -1 \\ 0 & -3 & 5 \\ 0 & 0 & 0 \\ 0 & 0 & 0 \end{bmatrix}.$$

$\because$ we have two zero rows. Therefore, the given vectors are linearly dependent. Here 1st two nonzero rows evidently arise from α_1 & α_3. Hence the subset $A = \{\alpha_1, \alpha_3\}$ is linearly independent.

Ex.3 Let $\alpha_1, \alpha_2, \cdots, \alpha_n$ be linearly independent vectors and suppose α is a linear combination of α_i's , $1 \le i \le n$, (say) $\alpha = a_1\alpha_1 + a_2\alpha_2 + \cdots + a_n\alpha_n$ where α_i's are scalars. Show that above representation of α is unique.

If the given linear combination is not unique, let another linear combination be

$$\alpha = b_1\alpha_1 + b_2\alpha_2 + \cdots + b_n\alpha_n,$$

where b_i's are scalars. Subtracting this from the given linear combination in terms of α_i's , we get,

$$0 = \alpha - \alpha = (a_1 - b_1)\alpha_1 + (a_2 - b_2)\alpha_2 + \cdots + (a_n - b_n)\alpha_n.$$

Now as α_i' s are linearly independent, therefore, the coefficients in the above relation are each equal to 0. Thus

$$(a_1 - b_1) = 0, (a_2 - b_2) = 0, \cdots, (a_n - b_n) = 0$$

$$\therefore \quad a_1 = b_1,\ a_2 = b_2,\ \cdots, a_n = b_n$$

Hence the above representation of α as a linear combination of $\alpha_1, \alpha_2, \cdots, \alpha_n$ is unique.

Ex. 4 Let U be the subspace of $V_3(R)$ which consists of all vectors of the form $(a, b, 0)$ and W, the subspace spanned by the vectors (1, 1, 1) and (3, 4, 5). Determine the subspaces $U \cap W$ and U + W.

1st part : Since the subspace W is spanned by the vectors (1, 1, 1) and (3, 4, 5), so all vectors of W must be L. C. of these two vectors, i.e. the form of these vectors is

$$p(1, 1, 1) + q(3, 4, 5) \tag{1}$$

for scalars p and q.

Again, subspace U and W intersects, so the vectors of W must be of the form $(a, b, 0) \in U$. Hence third component of the vectors in (1) must be zero. Hence $p + 5q = 0$ or $p = -5q$. Therefore, the vectors in $U \cap W$ have the form

$-5q(1, 1, 1) + q(3, 4, 5)$, i.e. $U \cap W = q(3, 4, 5) - 5q(1, 1, 1) = q(-2, -1, 0)$.

Second part : U + W consists of all vectors of the form

$$p(1, 1, 1) + q(3, 4, 5) + (a, b, 0) = p(1, 1, 1) + q(3, 4, 5) + a(1, 0, 0) + b(0, 1, 0).$$

If three of the four vectors (1, 1, 1), (3, 4, 5), (1, 0, 0) and (0, 1, 0) are linearly independent then U + W = $V_3(R)$.

To check linear independence one should reduce the matrix whose rows are the above four vectors

$$\begin{bmatrix} 1 & 1 & 1 \\ 3 & 4 & 5 \\ 1 & 0 & 0 \\ 0 & 1 & 0 \end{bmatrix} \approx \begin{bmatrix} 1 & 1 & 1 \\ 0 & 1 & 2 \\ 0 & -1 & -1 \\ 0 & 1 & 0 \end{bmatrix} \approx \begin{bmatrix} 1 & 1 & 1 \\ 0 & 1 & 2 \\ 0 & 0 & 1 \\ 0 & 0 & -2 \end{bmatrix} \approx \begin{bmatrix} 1 & 1 & 1 \\ 0 & 1 & 2 \\ 0 & 0 & 1 \\ 0 & 0 & 0 \end{bmatrix}$$

which shows that first three vectors are linearly independent. Therefore,

$$U + W = p(1, 1, 1) + q(3, 4, 5) + a(1, 0, 0) = (p + 3q + a,\ p + 4q,\ p + 5q).$$

Ex.5 If $U = L\{(1, 0, 3), (2, -1, 6)\}$ find two other subspaces W such that $V_3 = U \oplus W$.

Since vectors spanning U are linearly independent then it is required to find a third independent vector α such that $W = L\{\alpha\}$. Let $\alpha = (a, b, c)$. There is an infinity of choices for α in the form (a, b, c). Now to ensure the independence of (1, 0, 3), (2, −1, 6) and (a, b, c), we have,

$$\begin{vmatrix} 1 & 0 & 3 \\ 2 & -1 & 6 \\ a & b & c \end{vmatrix} \neq 0 \tag{1}$$

i.e., we have reduced the matrix

$$\begin{bmatrix} 1 & 0 & 3 \\ 2 & -1 & 6 \\ a & b & c \end{bmatrix} \approx \begin{bmatrix} 1 & 0 & 3 \\ 0 & -1 & 0 \\ 0 & b & c-3a \end{bmatrix} \approx \begin{bmatrix} 1 & 0 & 3 \\ 0 & -1 & 0 \\ 0 & 0 & c-3a \end{bmatrix}.$$

For (1) to be true $c \neq 3a$. Therefore, two choices of

(a, b, c) such that $c \neq 3a$ are (1, 0, 0) and (0, 0, 1).

$\therefore W = L\{(1, 0, 0)\}$ or $L\{(0, 0, 1)\}$ are two required subspaces.

11.10 Basis of a vector space

It has been seen in previous sections that a vector space or a subspace is generated or spanned by a set of vectors. But nothing has been said about the minimum number of the vectors required for the generation of the vector space. For example construction of a house or building in a rectangular 3-dimensional space we live in requires some specified length, breadth and height. These are measured in some units and the set of units is a "basis" or "building blocks" of measurements. The set of units needed to construct a 3-dimensional conical tent is not identical to that of a house. So different bases (pural of basis) are necessary for the construction of 3-dimensional objects around us. These concepts are extended for construction of different arbitrary spaces with the suporting theorems in this section. Let us begin with the formal definition of basis.

11.10.1 Basis: A subset S of a vector space V(F) is called its *basis* if

(i) S is linearly independent and (ii) S generates or spans V.

The word basis is synonymous with geometric terms "frame of reference" or "system of coordinate axes". The basis subset provides with a common distinct features among different spaces. For example, a set of two vectors $\mathbf{v}_1, \mathbf{v}_2$ with the common initial point spans an entire plane.

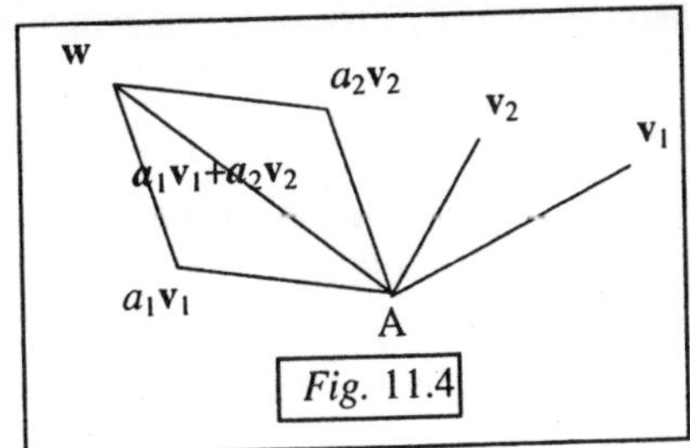

Fig. 11.4

Let this two vectors originate (*figure* 11.4) from A and scalar multiplication by a_1 and a_2 produce vectors $a_1\mathbf{v}_1$ and $a_2\mathbf{v}_2$, respectively. By the parallelogram law of addition the vector $\mathbf{w}=a_1\mathbf{v}_1+a_2\mathbf{v}_2$ is obtained. So for different choices of a_1 and a_2 the whole plane is created or generated or spanned by the two distinct vectors $\mathbf{v}_1$ and $\mathbf{v}_2$. So, $\mathbf{v}_1$ and $\mathbf{v}_2$ are bases of the plane. But from the knowledge of geometry the standard bases of a plane in orthogonal Cartesian coordinates are (1 , 0) and (0, 1) which are linearly independent. Any vector can be written in terms of these standard bases as $(a, b) = a(1, 0) + b(0, 1)$. A vector space may have different bases but with same number of vectors (cf. theorem in sec.11.11.1) satisfying the definition. Thus, the sets {(2, 0), (1, –3)} and {(1, 1), (1,–1)} are also bases for the plane as any vector (a, b) can be generated as

$$(a\ ,\ b\ ) = (\tfrac{1}{2}a+\tfrac{1}{6}b)\ (2\ ,\ 0) + (-\tfrac{1}{3}b)\ (1\ ,\ -3)$$

and $$(a\ ,\ b) = \tfrac{1}{2}(a+b)(1\ ,\ 1)\ +\tfrac{1}{2}(a-b)(1\ ,-1) \text{ respectively.}$$

Note that in all the three cases individual sets {(1 , 0) , (0 , 1)} , {(2 , 0) , (1, –3)} and {(1, 1) , (1, –1)} are separately linearly independent set of two vectors. But for a chosen set the generation of a vector is unique (theorem.1). Thus the distinguishing feature of 2-dimensional space is that there exists a set of two vectors and no set lies along a line in the same direction i.e. one cannot be expressible in terms of others. Similarly, the distinguishing feature of 3-dimensional space is that there exists a set of three vectors, none is expressible in terms of others.

Lot of examples may be cited for different kinds of spaces. But their common features are (i) a linearly independent set and (ii) set generates the space.

Infinite bases : An infinite set S is a basis for V(F) if

(i) S is linearly independent i.e. no element of S is a linear combination of other elements of S,

(ii) S generates or spans V(F) .

Thus one basis for the polynomial vector space is the set S of powers of

$$x : \mathrm{S} = \{1\ ,\ x\ ,x^2,\ \cdots\}$$

Standard (natural) bases for some vector spaces :

The standard (canonical) basis set for

(i) R^n is $\{\mathbf{e}_1, \mathbf{e}_2,\ \cdots, \mathbf{e}_n\}$ where $\mathbf{e}_i$ are s.u. vectors;

(ii) nth degree polynomial is $\{1\ , t\ ,\ \cdots\ , t^n\}$;

(iii) $m\times n$ matrix is $\{\mathrm{E}_{ij}\ ,\ i=1\ ,2\ ,\ \cdots,\ m\ ,\ j=1\ ,2\ ,\ \cdots,\ n\}$

Theorem 1. If $\alpha_1, \alpha_2 \cdots \alpha_n$ form a basis for the vector space $\mathrm{V}_n(\mathrm{F})$ then every vector $\alpha \in \mathrm{V}$ is expressible uniquely as a linear combination of $\alpha_1, \alpha_2, \cdots, \alpha_n$.

Proof. Since α_i , $i = 1, 2, \cdots, n$ form a basis of $V_n(F)$, they are linearly independent and span $V_n(F)$ by definition of basis. So, every vector α is expressible as a linear combination of α_i , $i = 1, 2, \cdots, n$. That is,

$$\alpha = b_1\alpha_1 + b_2\alpha_2 + \cdots + b_n\alpha_n, a_i \in F, i = 1, 2, \cdots, n, \tag{11.8}$$

Let this representation is *not* unique. Let a second representation be

$$\alpha = b_1\alpha_1 + b_2\alpha_2 + \cdots + b_n\alpha_n, \; b_i \in F, \; i = 1, 2, \cdots, n, \tag{11.9}$$

$\therefore$ R.H.S of (11.8) and (11.9) must be equal, i.e.

$$\alpha = a_1\alpha_1 + a_2\alpha_2 + \cdots + a_n\alpha_n = b_1\alpha_1 + b_2\alpha_2 + \cdots + b_n\alpha_n$$

or $\quad (a_i - b_i)\alpha_i = 0$.

$\therefore \quad a_i - b_i = 0$ for linear independence of α_i. i.e. $a_i = b_i$`

Hence two representations (11.8) and (11.9) are identical and hence the theorem.

11.10.2 Distinction between spanning set and basis set : Both basis set and spanning set characterize a vector space. But a basis set characterizes the vector space more exactly and uniquely (theorem 1 in sec.11.10) than a general spanning set does. All basis sets are spanning sets but all spanning sets are not basis sets. Because a vector may be represented by spanning set in many different ways. This is expressed by the following theorem.

Theorem. If $\{\alpha_1, \alpha_2, \cdots, \alpha_n\}$ is a spanning set which is not a basis set then each vector in V can be expressed in more than one way as a linear combination of the set of vectors.

Proof. Since given spanning set is not a basis, so the spanning is linearly dependent. Therefore, there exist constants c_i, $i = 1, 2, \cdots, n$, not all zero, such that

$$\alpha_1c_1 + \alpha_2c_2 + \cdots + \alpha_nc_n = 0 \tag{11.10}$$

Again an arbitrary vector α can be expressed in terms of the vectors of the spanning set i.e.

$$\alpha = \alpha_1d_1 + \alpha_2d_2 + \cdots + \alpha_nd_n. \tag{11.11}$$

Now $\quad \alpha = \alpha_1d_1 + \alpha_2d_2 + \cdots + \alpha_nd_n + \mathbf{0}$

$\quad = \alpha_1d_1 + \alpha_2d_2 + \cdots + \alpha_nd_n + \alpha_1c_1 + \alpha_2c_2 + \cdots + \alpha_nc_n$ by (11.10)

$\quad = \alpha_1(d_1 + c_1) + \alpha_2(d_2 + c_2) + \cdots + \alpha_n(d_n + c_n)$

which is another distinct expression for α in terms of $\alpha_1, \alpha_2, \cdots, \alpha_n$.

Ex.1 Show that the set S = $\{M_1, M_2, M_3, M_4\}$ is a basis for the vector space $M_{2\times2}$ of 2×2 matrices where

$$M_1 = \begin{bmatrix} 1 & 0 \\ 0 & 0 \end{bmatrix}, M_2 = \begin{bmatrix} 0 & 1 \\ 0 & 0 \end{bmatrix}, M_3 = \begin{bmatrix} 0 & 0 \\ 1 & 0 \end{bmatrix}, M_4 = \begin{bmatrix} 0 & 0 \\ 0 & 1 \end{bmatrix}.$$

To see that S spans $M_{2\times2}$ consider a typical vector (matrix)

$$\begin{bmatrix} a & b \\ c & d \end{bmatrix}$$

which can be written as

$$\begin{bmatrix} a & b \\ c & d \end{bmatrix} = c_1\begin{bmatrix} 1 & 0 \\ 0 & 0 \end{bmatrix} + c_2\begin{bmatrix} 0 & 1 \\ 0 & 0 \end{bmatrix} + c_3\begin{bmatrix} 0 & 0 \\ 1 & 0 \end{bmatrix} + c_4\begin{bmatrix} 0 & 0 \\ 0 & 1 \end{bmatrix}.$$

It can be readily seen that $c_1 = a$, $c_2 = b$, $c_3 = c$ and $c_4 = d$. Therefore,

$$\begin{bmatrix} a & b \\ c & d \end{bmatrix} = aM_1 + bM_2 + cM_3 + dM_4$$

$$= L\{M_1, M_2, M_3, M_4\}$$

Hence S spans $M_{2\times 2}$.

Now to see that S is linearly independent, let

$$c_1\begin{bmatrix}1 & 0\\0 & 0\end{bmatrix} + c_2\begin{bmatrix}0 & 1\\0 & 0\end{bmatrix} + c_3\begin{bmatrix}0 & 0\\1 & 0\end{bmatrix} + c_4\begin{bmatrix}0 & 0\\0 & 1\end{bmatrix} = \begin{bmatrix}0 & 0\\0 & 0\end{bmatrix}$$

From which it can seen easily that $c_1 = c_2 = c_3 = c_4 = 0$.

$\therefore$ S is linearly independent.

Hence S forms a basis for the vector space $M_{2\times 2}$.

11.10.3 Basis — a minimal spanning set : The following theorem shows that basis is a minimal spanning set.

Theorem. A basis is a minimal spanning set.

Alternatively, if a vector space V over F is spanned by a set S and one of the vectors of S is a linear combination of others, the deletion of that vector from S still spans V.

Proof. Let $S = \{\alpha_1, \alpha_2, \cdots, \alpha_n\}$. Since one of the vectors in S, say α_2 is a linear combination of $\alpha_1, \alpha_3, \cdots, \alpha_n$, so $\alpha_2 = c_1\alpha_1 + c_3\alpha_3 + \cdots + c_n\alpha_n$, for $c_1, c_3, \cdots, c_n \in F$,

(11.12)

Again S spans V, then any vector α is a linear combination of the vectors in S i.e. $\alpha = a_1\alpha_1 + a_2\alpha_2 + a_3\alpha_3 + \cdots + a_n\alpha_n$, $\forall a_i \in F$,

$$= a_1\alpha_1 + a_2(c_1\alpha_1 + c_3\alpha_3 + \cdots + c_n\alpha_n) + a_3\alpha_3 + \cdots + a_n\alpha_n, \text{ by (11.12)}$$

$$= (a_1 + a_2c_1)\alpha_1 + (a_2c_3 + a_3)\alpha_3 + \cdots + (a_2c_n + a_n)\alpha_n$$

which shows that α is a linear combination of $S' = \{\alpha_1, \alpha_3, \cdots, \alpha_n\}$ which is obtained from S by deleting α_2.

$\therefore$ V is spanned by S'. Thus the theorem is established.

N.B. The theorem is useful in solving linear programming problems suing simplex technique.

Ex. Let W be the subspace of R^3 spanned by the set S={(1, −1, 1), (−2, 1, 3), (−4, 1, 11)}. Show that the subset S′={(1, −1, 1), (−2, 1, 3)} of S spans W.

Here (−4, 1, 11) = 2(1, −1, 1) + 3(−2, 1, 3), a linear combination of (1, −1, 1) and (−2, 1, 3). This linear combination can be obtained in different ways — (i) solving nonhomogeous equations, or (ii) elementary transformation of matrices (cf. examples in secs.11.8 and 11.9). Hence by the above theorem S' spans V.

11.10.4 Reducing a spanning set to a basis : The following theorem with a procedure describes a method for reducing a spanning set to a basis of a vector space V.

Theorem. If the set $S = \{\mathbf{v}_1, \mathbf{v}_2, \cdots, \mathbf{v}_n\}$ spans the vector space V then the basis of V is a subset of S. That is, spanning set may be reduced to a basis.

Proof. A procedure for the purpose of proving the theorem has been developed.

Procedure for reducing a spanning set to a basis : Let an $m \times n$ matrix A be constructed with columns $A_{\cdot j} = \mathbf{v}_j$, $j = 1, 2, \cdots, n$ assuming that each vector $\mathbf{v}_j$ consists of m components.

Let the row reduced echelon form of A be B such that some columns of B are elementary vectors $\mathbf{e}_i$, $i = 1, 2, \cdots, n$. Let the original vectors $\mathbf{v}_s$ which are transformed to $\mathbf{e}_i$'s be denoted by the set $S_0 = \{\mathbf{w}_1, \mathbf{w}_2, \cdots, \mathbf{w}_s\}$ for distinctness. Then the subset S_0 of S is the basis of V. This completes the procedure.

Now the only thing needs to be proved is that the subset S_0 is basis of V i.e. S_0 is linearly independent and S_0 spans V.

Since $\mathbf{e}_i$ is linearly independent and derived from vectors of S_0, so S_0 must be linearly independent and by theorem in sec.11.10.2 other vectors in S are linear combinations of the vectors in S_0. Finally, by theorem in sec.11.10.3, S_0 spans V. Hence S_0 forms a basis of V.

Cor. The equivalent theorem & procedure for subspace is the replacement of vector space by subspace everywhere in the theorem and procedure.

Ex. Find a basis for the vector space V generated by the set

$$S = \{(-1, 1, 2, -1), (1, 0, -1, 1), (-1, 2, 3, -1), (1, 1, 0, 1)\}$$

To use the procedure in sec.11.10.4, let us construct a 4×4 matrix A with columns $A_{.j} = \mathbf{v}_j$ i.e.

$$A = \begin{bmatrix} -1 & 1 & -1 & 1 \\ 1 & 0 & 2 & 1 \\ 2 & -1 & 3 & 0 \\ -1 & 1 & -1 & 1 \end{bmatrix}.$$

(columns headed $\mathbf{v}_1$, $\mathbf{v}_2$, $\mathbf{v}_3$, $\mathbf{v}_4$)

Let us reduce A to the row reduced echelon form as follows:

$$A = \begin{bmatrix} -1 & 1 & -1 & 1 \\ 1 & 0 & 2 & 1 \\ 2 & -1 & 3 & 0 \\ -1 & 1 & -1 & 1 \end{bmatrix} \underset{\substack{R_3+2R_1 \\ R_4-R_1}}{\overset{R_2+R_1}{\approx}} \begin{bmatrix} -1 & 1 & -1 & 1 \\ 0 & 1 & 1 & 2 \\ 0 & 1 & 1 & 2 \\ 0 & 0 & 0 & 0 \end{bmatrix} \overset{R_3-R_2}{\approx}$$

$$\begin{bmatrix} -1 & 1 & -1 & 1 \\ 0 & 1 & 1 & 2 \\ 0 & 0 & 0 & 0 \\ 0 & 0 & 0 & 0 \end{bmatrix} \overset{R_1-R_2}{\approx} \begin{bmatrix} -1 & 0 & -2 & -1 \\ 0 & 1 & 0 & 2 \\ 0 & 0 & 0 & 0 \\ 0 & 0 & 0 & 0 \end{bmatrix} \overset{(-1)R_1}{\approx}$$

$$\begin{bmatrix} 1 & 0 & 2 & 1 \\ 0 & 1 & 0 & 2 \\ 0 & 0 & 0 & 0 \\ 0 & 0 & 0 & 0 \end{bmatrix} = B \text{ (say)}$$

(each matrix with columns headed $\mathbf{v}_1$, $\mathbf{v}_2$, $\mathbf{v}_3$, $\mathbf{v}_4$)

which shows that $\mathbf{v}_1, \mathbf{v}_2$ have been transformed to s.u vectors $\mathbf{e}_1, \mathbf{e}_2$ respectively and $\mathbf{v}_1, \mathbf{v}_2$ correspond to $\mathbf{w}_1, \mathbf{w}_2$ of the procedure. Hence the basis of V is the subset $S_0 = \{\mathbf{w}_1, \mathbf{w}_2\} = \{\mathbf{v}_1, \mathbf{v}_2\} = \{1, 1, 2, -1), (1, 0, -1, 1)\}$.

11.10.5 Extension of basis : The previous theorem in sec.11.10.4 discusses a governing theorem for reducing a spanning set to a basis with procedure and example. Thus naturally raises the question of any possibility of extending a linearly independent set to a basis. The answer to the question is positive. The next theorem and accompanying procedure describe the process.

Theorem. If the set $S = \{v_1, v_2, \cdots, v_n\}$ spans the vector space V and $S_1 = \{u_1, u_2, \cdots, u_s\}$ is linearly independent subset of V, then the basis of V is the union of S_1 and S_2 i.e. $S_1 \cup S_2$ where S_2 is a subset S. That is, a linearly independent set may be *extended* to a basis.

Proof. A procedure for the purpose of proving the theorem has been developed.

Procedure for extending an independent set to a basis :

Let an $m\times(s+n)$ matrix A be constructed with columns $A._j = u_j$, $j = 1,2,\cdots,n$ and columns $A_{.(s+i)} = v_i$ where i=1, 2, $\cdots$, n assuming that each vector u_j & v_i consists of m components. In other words, the first s columns of A are the vectors in S_1 and the last n columns are the vectors in S. Let the row reduced echelon form of A be B such that some columns of B are s.u vectors e_i , $i = 1, 2, \cdots, s, s+1, \cdots, t$. Let the original vectors of S which are transformed to e's be denoted by the set $S_2 = \{w_{s+1}, w_{s+2}, \cdots, w_t, \}$ for distinctness. Then the subset $S_1 \cup S_2$ becomes the basis of V. This completes the procedure.

Since $S_1 \cup S_2$ consists of s.u vectors e_i , $i = 1, 2, \cdots, s, s+1, \cdots, t$ which are linearly independent. Of these s.u vectors the first s vectors correspond to S_1 which were given linearly independent and the rest $s + 1$, $\cdots$, t vectors correspond to a subset S_2 of S which by theorem in sec.11.10.3 spans V as S spans V. The vectors in S_2 must be linearly independent too. So, subset $S_1 \cup S_2$ spans V and is linearly independent. Hence the subset $S_1 \cup S_2$ becomes the extended basis of V. Hence the proof.

Ex. Extend the linearly independent set $\{(-2, -1, 1, 0, 0), (-1, 0, 0, 1, 1)\}$ to a basis of R^5.

Since spanning set of the vector space R^5 is not given let us take the standard spanning set $S = \{e_1, e_2, e_3, e_4, e_5\}$ of R^5.

Let

$$S_1 = \{u_1, u_2\} = \{(-2, -1, 1, 0, 0), (-1, 0, 0, 1, 1)\}$$

linearly independent subset of R^5. To extend S_1 to a basis of R^5, let us construct a $5\times(2+5)$ matrix A with first two columns by S_1 and next 5 columns by S as

$$A = \begin{array}{c} \begin{array}{ccccccc} u_1 & u_2 & e_1 & e_2 & e_3 & e_4 & e_5 \end{array} \\ \left[\begin{array}{cc|ccccc} -2 & -1 & 1 & 0 & 0 & 0 & 0 \\ -1 & 0 & 0 & 1 & 0 & 0 & 0 \\ 1 & 0 & 0 & 0 & 1 & 0 & 0 \\ 0 & 1 & 0 & 0 & 0 & 1 & 0 \\ 0 & 1 & 0 & 0 & 0 & 0 & 1 \end{array}\right] \end{array}.$$

Now reduce A to row reduced echelon form B as follows.

$$A \approx \left[\begin{array}{rr|rrrrr} -2 & -1 & 1 & 0 & 0 & 0 & 0 \\ 0 & \frac{1}{2} & -\frac{1}{2} & 1 & 0 & 0 & 0 \\ 0 & -\frac{1}{2} & \frac{1}{2} & 0 & 1 & 0 & 0 \\ 0 & 1 & 0 & 0 & 0 & 1 & 0 \\ 0 & 1 & 0 & 0 & 0 & 0 & 1 \end{array}\right] \approx \left[\begin{array}{rr|rrrrr} -2 & -1 & 1 & 0 & 0 & 0 & 0 \\ 0 & 1 & -1 & 1 & 0 & 0 & 0 \\ 0 & 0 & 0 & 1 & 2 & 0 & 0 \\ 0 & 0 & 1 & -1 & 0 & 1 & 0 \\ 0 & 0 & 1 & -1 & 0 & 0 & 1 \end{array}\right]$$

$$\approx \left[\begin{array}{rr|rrrrr} -2 & -1 & 1 & 1 & 0 & 0 & 0 \\ 0 & 1 & -1 & 0 & 0 & 0 & 0 \\ 0 & 0 & 0 & 1 & 2 & 0 & 0 \\ 0 & 0 & 1 & -1 & 0 & 0 & 0 \\ 0 & 0 & 1 & -1 & 0 & 1 & 1 \end{array}\right] \approx \left[\begin{array}{rr|rrrrr} -2 & 0 & 0 & 0 & -2 & 0 & 0 \\ 0 & 1 & -1 & 0 & -2 & 0 & 0 \\ 0 & 0 & 0 & 1 & 2 & 0 & 0 \\ 0 & 0 & 1 & 0 & 2 & 1 & 0 \\ 0 & 0 & 1 & 0 & 2 & 0 & 1 \end{array}\right]$$

$$\approx \left[\begin{array}{rr|rrrrr} -2 & 0 & 0 & 0 & -2 & 0 & 0 \\ 0 & 1 & 0 & 0 & 0 & 1 & 0 \\ 0 & 0 & 0 & 1 & 2 & 0 & 0 \\ 0 & 0 & 1 & 0 & 2 & 1 & 0 \\ 0 & 0 & 0 & 0 & 0 & 1 & 1 \end{array}\right] \approx \begin{array}{c} \begin{array}{ccccccc} \mathbf{u}_1 & \mathbf{u}_2 & \mathbf{e}_1 & \mathbf{e}_2 & \mathbf{e}_3 & \mathbf{e}_4 & \mathbf{e}_5 \end{array} \\ \left[\begin{array}{rr|rrrrr} 1 & 0 & 0 & 0 & 1 & 0 & 0 \\ 0 & 1 & 0 & 0 & 0 & 0 & -1 \\ 0 & 0 & 0 & 1 & 2 & 0 & 0 \\ 0 & 0 & 1 & 0 & 2 & 0 & -1 \\ 0 & 0 & 0 & 0 & 0 & 1 & 1 \end{array}\right] \\ \begin{array}{ccccccc} \uparrow & \uparrow & \uparrow & \uparrow & & \uparrow & \end{array} \end{array} = \text{B}.$$

The upward arrows indicate elementary vectors in the row reduced echelon form B in which $S_2 = \{\mathbf{e}_1, \mathbf{e}_2, \mathbf{e}_4\}$, a subset of S.

$\therefore$ the required extended basis is $S_1 \cup S_2 = \{\mathbf{u}_1, \mathbf{u}_2, \mathbf{e}_1, \mathbf{e}_2, \mathbf{e}_4\}$.

11.11 Coordinates and change of basis

Basis and coordinate system are related to each other. This section will explore their relationships and some important results about the change of the basis for vector space. The change of basis is necessary to understand the matrix representation of linear transformation in chapter 12 and the later provides with one of the major interpretations of the matrices.

Normally the coordinates (a, b) of a point P in the plane are indicated with respect to a pair of orthogonal axes what is called *coordinate axes*. But coordinates can also be introduced with the help of vectors without any reference to coordinate axes. This concept generalizes the notion of coordinates in a vector space of more than three dimensions. The extra requirement for the purpose is a basis of the vector space and no particular coordinate axes.

Relationship of coordinates and bases :

Let us begin with the idea from our conventional two dimensional Euclidean analytic geometry of Descartes (*figure* 11.5).

Let P(a, b) be a point in the xy plane with respect to the orthogonal coordinate axes OX and OY. Now let $\hat{\mathbf{v}}_1$ and $\hat{\mathbf{v}}_2$ be two unit vectors orthogonal to each other & with the same initial point O. These vectors form a basis of R^2. From a point P in the plane draw perpendiculars PM and PN on the extended lines of actions of the unit vectors $\hat{\mathbf{v}}_1$ and $\hat{\mathbf{v}}_2$ such that $\overrightarrow{OM} = a\,\hat{\mathbf{v}}_1$ and $\overrightarrow{ON} = b\,\hat{\mathbf{v}}_2$ where 'a' & 'b' are two scalars. Evidently OMPN is a rectangle and by parallelogram law of vector addition

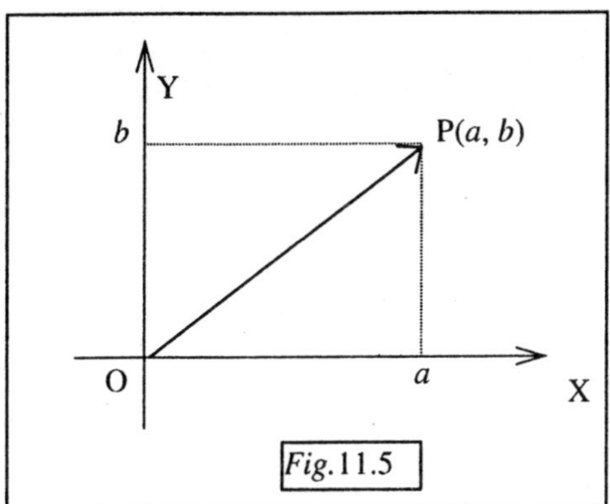

Fig. 11.5

$\overrightarrow{OP} = \overrightarrow{OM} + \overrightarrow{ON} = a\hat{\mathbf{v}}_1 + b\hat{\mathbf{v}}_2$.

Clearly $\overrightarrow{OP}$ is a linear combination of two unit vectors $\hat{\mathbf{v}}_1$ and $\hat{\mathbf{v}}_2$ and the

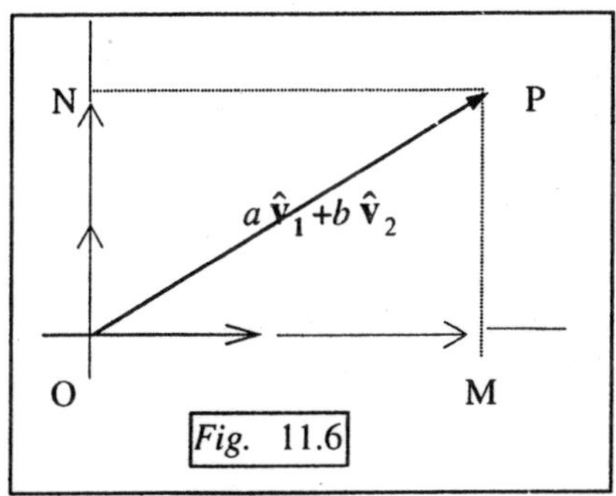

Fig. 11.6

coefficients a and b in order of $\hat{\mathbf{v}}_1$ and $\hat{\mathbf{v}}_2$ are the same as the coordinates of P (*figure* 11.6). This is one particular instance that coordinates are coefficients of basis vectors in order of a linear combination. The above idea can be extended by removing the restrictions that the basis vectors are perpendicular and of unit length. Let the basis vectors $\mathbf{v}_1$ and $\mathbf{v}_2$ are inclined to each other at angle $\theta \neq \frac{\pi}{2}$ at O. Constructing vectors $\mathbf{v}_1$ and $\mathbf{v}_2$ for scalars a and b and by parallelogram law of addition (*figure* 11.7) of vectors, the resultant vector $\overrightarrow{OP} = a\mathbf{v}_1 + b\mathbf{v}_2$ can be easily obtained. Then we can regard (a , b) as the coordinates of P relative to the basis set $\{ \mathbf{v}_1, \mathbf{v}_2 \}$.

11.11.1 Definition of coordinates relative to a basis : In the beginning, distinction must be made between order and disorder basis. If $S = \{\mathbf{v}_1, \mathbf{v}_2, \cdots, \mathbf{v}_n\}$ be an ordered basis then $S = \{\mathbf{v}_2, \mathbf{v}_1, \mathbf{v}_3, \cdots, \mathbf{v}_n\}$ is a disorder basis.

Let $S = \{\mathbf{v}_1, \mathbf{v}_2, \cdots, \mathbf{v}_n\}$ be an ordered basis for a finite dimensional vector space V. Since S spans V and any vector $\mathbf{v} \in V$ can be uniquely expressed as a linear combination of the basis vectors in S, so

$$\mathbf{v} = c_1\mathbf{v}_1 + c_2\mathbf{v}_2 + \cdots + c_n\mathbf{v}_n \text{ for scalars } c_1, c_2, \cdots, c_n.$$

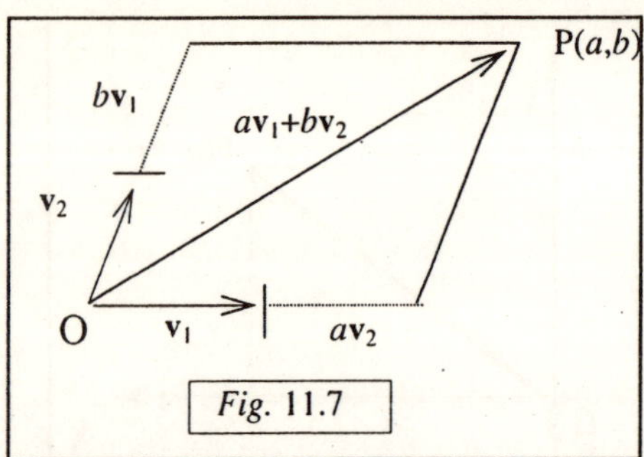

Fig. 11.7

These scalars $c_1, c_2, \cdots, c_n$ are called the *coordinates* of **v** relative to the ordered basis S. The coordinate vector of **v** (coordinate of the terminal point of vector **v** or coordinate of a vector **v**) relative to S is denoted by $(\mathbf{v})_s$ and is the vector in R^n defined by

$$(\mathbf{v})_s = (c_1, c_2, \cdots, c_n)$$

The coordinate matrix of **v** relative to the ordered basis S is denoted by $[\mathbf{v}]_s$ and the $n \times 1$ column matrix is defined by

$$[\mathbf{v}]_s = \begin{bmatrix} c_1 \\ c_2 \\ \vdots \\ c_n \end{bmatrix} \text{ or by } [c_1 \quad c_2 \quad \cdots \quad c_n]^T \text{ to save space.}$$

N.B. (i) Coordinate vectors and coordinate matrices depend on the order in which basis vectors are written. A change in the order of basis vectors results in a corresponding change of order of entries (components) in coordinate vectors and coordinate matrices.
(ii) Some texts do not differentiate between the coordinate vector and the coordinate matrix. In such cases, only the coordinate vector is used with the notation

$$[\mathbf{v}]_s = (c_1, c_2, \cdots, c_n)$$

without making any mention of the coordinate matrix and vice-versa. Beginners should be *cautious* in the notations.

11.11.2 Change of basis : If the basis of a vector space V has been changed from old basis B to some new basis B′ (say), then there arise a problem about the relationship between the old coordinate matrix $[\mathbf{v}]_B$ and the new coordinate matrix $[\mathbf{v}]_{B'}$ for a vector **v** in V.

Let us proceed to solve the problem first for a two dimensional space V. Let

$$B = \{\mathbf{u}_1, \mathbf{u}_2\} \text{ and } B' = \{\mathbf{u}_1', \mathbf{u}_2'\} \qquad (11.13)$$

be old and new basis sets, respectively. Let the coordinate matrices for the old basis vectors relative to the new basis be

$$[\mathbf{u}_1]_{B'} = \begin{bmatrix} a \\ b \end{bmatrix} \text{ and } [\mathbf{u}_2]_{B'} = \begin{bmatrix} c \\ d \end{bmatrix} \qquad (11.14)$$

$$\therefore \quad \left.\begin{aligned} \mathbf{u}_1 &= a\mathbf{u}_1' + b\mathbf{u}_2' \\ \mathbf{u}_2 &= c\mathbf{u}_1' + d\mathbf{u}_2' \end{aligned}\right\} \tag{11.15}$$

Let **v** be any vector in V and let the coordinate matrix for **v** relative to old basis B be

$$[\mathbf{v}]_B = \begin{bmatrix} k_1 \\ k_2 \end{bmatrix}. \tag{11.16}$$

$$\therefore \quad \mathbf{v} = k_1\mathbf{u}_1 + k_2\mathbf{u}_2 \tag{11.17}$$

$$= k_1(a\mathbf{u}_1' + b\mathbf{u}_2') + k_2(c\mathbf{u}_1' + d\mathbf{u}_2') \text{ by (11.15)}$$

$$= (k_1a + k_2c)\,\mathbf{u}_1' + (k_1b + k_2d)\,\mathbf{u}_2'$$

which gives the coordinate matrix relative to the new basis as

$$[\mathbf{v}]_{B'} = \begin{bmatrix} k_1a + k_2c \\ k_1b + k_2d \end{bmatrix} = \begin{bmatrix} a & c \\ b & d \end{bmatrix} \begin{bmatrix} k_1 \\ k_2 \end{bmatrix}$$

$$= \left[[\mathbf{u}_1]_{B'} \vdots [\mathbf{u}_2]_{B'}\right] [\mathbf{v}]_B = P[\mathbf{v}]_B \tag{11.18}$$

where the partition matrix P= $\left[[\mathbf{u}_1]_{B'} \vdots [\mathbf{u}_2]_{B'}\right]$ is obtained from coordinate matrices for old basis vectors relative to the new basis vectors and is called the *transition matrix* from B-basis to B′-basis. This may be denoted by $P\Big]_B^{B'} = P\Big]_{\text{old basis}}^{\text{new basis}}$ keeping conformity with definite integral of calculus.

This procedure can be extended to n-dimensional vector space. If the basis of a vector space V has been changed from old basis $B = \{\mathbf{u}_1, \mathbf{u}_2, \cdots, \mathbf{u}_n\}$ to some new basis $B' = \{\mathbf{u}_1', \mathbf{u}_2', \cdots, \mathbf{u}_n'\}$ then the relationship between the old coordinate matrix $[\mathbf{v}]_B$ and the new coordinate matrix $[\mathbf{v}]_{B'}$ for a vector **v** in V is

$$[\mathbf{v}]_{B'} = P[\mathbf{v}]_B \tag{11.19}$$

where the transition matrix

$$P\Big]_B^{B'} = P\Big]_{\text{old basis}}^{\text{new basis}} \left[[\mathbf{u}_1]_{B'} \vdots [\mathbf{u}_2]_{B'} \vdots \cdots \vdots [\mathbf{u}_n]_{B'}\right]. \tag{11.20}$$

Cor. One can write (11.14) as

$$[\mathbf{u}_1']_B = \begin{bmatrix} a_1 \\ b_1 \end{bmatrix} \text{ and } [\mathbf{u}_2']_B = \begin{bmatrix} c_1 \\ d_1 \end{bmatrix}$$

i.e. one can express coordinate matrices of new basis vectors relative to the old basis.

$$\therefore \quad \left.\begin{aligned} \mathbf{u}_1' &= a_1\mathbf{u}_1 + b_1\mathbf{u}_2 \\ \mathbf{u}_2' &= c_1\mathbf{u}_1 + d_1\mathbf{u}_2 \end{aligned}\right\}.$$

Considering the coordinate matrix of any vector $\mathbf{v} \in V$ relative to the new basis as

$$[\mathbf{v}]_{B'} = \begin{bmatrix} k_1 \\ k_2 \end{bmatrix} \text{ i.e.}$$

$$\mathbf{v} = k_1\mathbf{u}_1' + k_2\mathbf{u}_2' = k_1(a\mathbf{u}_1 + b\mathbf{u}_2) + k_2(c\mathbf{u}_1 + d\mathbf{u}_2)$$

$$= (k_1a_1 + k_2c_1)\mathbf{u}_1 + (k_1b_1 + k_2d_1)\mathbf{u}_2$$

$$\therefore\ [\mathbf{v}]_B = \begin{bmatrix} k_1a_1 + k_2c_1 \\ k_1b_1 + k_2d_1 \end{bmatrix} = \begin{bmatrix} a_1 & c_1 \\ b_1 & d_1 \end{bmatrix} \begin{bmatrix} k_1 \\ k_2 \end{bmatrix}$$

$$= \left[[\mathbf{u}_1]_B \ \vdots\ [\mathbf{u}_2]_B \right] [\mathbf{v}]_{B'} \qquad (11.21)$$

N.B. Care should be taken using either relation (11.19) or (11.21).

11.11.3 Algorithm for obtaining transition matrix and coordinate matrix : Algorithm is based on solving several systems of equations with common coefficient matrix. The equations (11.15) correspond to several systems in which common coefficient matrix results from column vectors $\mathbf{u}_i'$, $i = 1, 2, \cdots, n$ in the new basis set B'. Therefore, following procedure in sec.8.7.1, the transition matrix $P]_B^{B'}$ can be obtained using the following algorithm.

Step 1. Write the augmented matrix, the columns of which consist of the new basis vectors in order then the old basis vectors in order with a partition separating the two sets of bases.

Step 2. Reduce the matrix obtained in step 1 to row reduced echelon form such that the matrix formed by new basis vectors i.e. the common coefficient matrix reduces to unit matrix.

Step 3. Reduced matrix corresponding to old basis vectors is the required transition matrix $P]_B^{B'}$.

To obtain $P]_{B'}^{B}$ proceed as above replacing the words "new" to "old" and "old" to "new" althrough.

The algorithm is equally applicable if the basis vectors are matrices, polynomials or functions. In case of matrices, replace matrices A_i by $\text{vec}A_i$, coefficients of fundamental bases of polynomials and functions are considered as vectors.

Note that $P]_B^{B'} \cdot P]_{B'}^{B} = I$ i.e. one is the inverse of the other. Also *note* that columns of $P]_B^{B'}$ are respectively the coordinate matrix of each old basis vector with respect to the new basis set B'. Thus coordinate matrix $[\mathbf{v}]_{B'}$ of any vector $\mathbf{v}$ can be easily obtained directly as above where coefficient matrix consisting of the columns of set B', is augmented with $\mathbf{v}$. All these are illustrated by the following examples.

Ex.1 Find the coordinate vector and the coordinate matrix for $\mathbf{v} \in V^3$ relative to $S = \{\mathbf{v}_1, \mathbf{v}_2, \mathbf{v}_3\}$ where

$$\mathbf{v} = (2, -1, 3),\ \mathbf{v}_1 = (1, 0, 0),\ \mathbf{v}_2 = (2, 2, 0),\ \mathbf{v}_3 = (3, 3, 3).$$

Method 1. Since $\mathbf{v} \in V^3$ and S is the basis of V, so $\mathbf{v}$ can be uniquely expressed as a linear combination of the basis vectors in S for some scalars $c_1, c_2, c_3 \in F$.

Let $\mathbf{v} = c_1\mathbf{v}_1 + c_2\mathbf{v}_2 + c_3\mathbf{v}_3$

or $(2, -1, 3) = c_1(1, 0, 0) + c_2(2, 2, 0) + c_3(3, 3, 3)$

$$\therefore \quad \left.\begin{aligned} c_1 + 2c_2 + 3c_3 &= 2 \\ 2c_2 + 3c_3 &= -1 \\ 3c_3 &= 3 \end{aligned}\right\}$$

which is in echelon form and $c_1 = 3$, $c_2 = -2$, $c_3 = 1$.

$\therefore$ coordinate vector of $\mathbf{v}$ relative to S is

$$(\mathbf{v})_s = (c_1 , c_2 , c_3) = (3 ,-2 , 1)$$

and the coordinate matrix of $\mathbf{v}$ relative to S is $[\mathbf{v}]_s = \begin{bmatrix} 3 \\ -2 \\ 1 \end{bmatrix}$.

Method 2 : This is simply a new terminology for the old technique of producing coefficients to show that $\mathbf{v}$ is in the span of S . We write an augmented matrix with columns given by vectors of S augmented with $\mathbf{v}$ as the last column. This is nothing but the augmented matrix corresponding the nonhomogeneous system of equations in the method. Let us reduce the resulting matrix to row reduced echelon form such that the coefficient matrix corresponding to S reduce to unit matrix.This is shown below:

$$\left[\begin{array}{ccc|c} 1 & 2 & 3 & 2 \\ 0 & 2 & 3 & -1 \\ 0 & 0 & 3 & 3 \end{array}\right] \approx \left[\begin{array}{ccc|c} 1 & 2 & 3 & 2 \\ 0 & 2 & 3 & -1 \\ 0 & 0 & 1 & 1 \end{array}\right] \approx \left[\begin{array}{ccc|c} 1 & 2 & 0 & -1 \\ 0 & 2 & 0 & -4 \\ 0 & 0 & 1 & 1 \end{array}\right] \approx \left[\begin{array}{ccc|c} 1 & 0 & 0 & 3 \\ 0 & 1 & 0 & -2 \\ 0 & 0 & 1 & 1 \end{array}\right]$$

The last column is the coefficient of linear combination of $\mathbf{v}_1 , \mathbf{v}_2 , \mathbf{v}_3$ to form $\mathbf{v}$ with respect to the basis vectors in S. So the coordinate vector and the coordinate matrix of $\mathbf{v}$

relative to the basis set S are $(\mathbf{v})_s = (3 ,-2 , 1)$ and $[\mathbf{v}]_s = \begin{bmatrix} 3 \\ -2 \\ 1 \end{bmatrix}$ respectively.

Ex.2 Consider the bases $B = \{\mathbf{u}_1 , \mathbf{u}_2 , \mathbf{u}_3\}$ and $B' = \{\mathbf{v}_1 , \mathbf{v}_2 , \mathbf{v}_3\}$ for R^3

where $\mathbf{u}_1 = \begin{bmatrix} -3 \\ 0 \\ -3 \end{bmatrix}$, $\mathbf{u}_2 = \begin{bmatrix} -3 \\ 2 \\ -1 \end{bmatrix}$, $\mathbf{u}_3 = \begin{bmatrix} 1 \\ 6 \\ -1 \end{bmatrix}$, and $\mathbf{v}_1 = \begin{bmatrix} -6 \\ -6 \\ 0 \end{bmatrix}$, $\mathbf{v}_2 = \begin{bmatrix} -2 \\ -6 \\ 4 \end{bmatrix}$, $\mathbf{v}_3 = \begin{bmatrix} -2 \\ -3 \\ 7 \end{bmatrix}$

(a) Find the transition matrix from B-basis to B' -basis.

(b) Compute the coordinate matrix $[\mathbf{w}]_B$ where $\mathbf{w} = \begin{bmatrix} -5 \\ 8 \\ -5 \end{bmatrix}$ and hence compute $[\mathbf{w}]_{B'}$.

(c) Check the result in (b) by computing $[\mathbf{w}]_{B'}$, directly.

(d) Find the transition matrix from B'-basis to B-basis and hence obtain $[\mathbf{w}]_B$ using the result from (b) or (c).

(e) Verify that $P]_B^{B'} . P]_{B'}^{B} = I$.

(a) Using the algorithm in subsec.11.11.1, let us form a 3×6 matrix whose first 3-columns are formed by $\mathbf{v}_1$, $\mathbf{v}_2$, $\mathbf{v}_3$ and last 3 columns by $\mathbf{u}_1$, $\mathbf{u}_2$, $\mathbf{u}_3$ separated by a partition and reduce it to the row reduced echelon form as

$$\begin{array}{c} \begin{matrix} \mathbf{v}_1 & \mathbf{v}_2 & \mathbf{v}_3 & \mathbf{u}_1 & \mathbf{u}_2 & \mathbf{u}_3 \end{matrix} \\ \left[\begin{array}{ccc:ccc} -6 & -2 & -2 & -3 & -3 & 1 \\ -6 & -6 & -3 & 0 & 2 & 6 \\ 0 & 4 & 7 & -3 & -1 & -1 \end{array}\right] \end{array} \approx \left[\begin{array}{ccc:ccc} -6 & -2 & -2 & -3 & -3 & 1 \\ 0 & -4 & -1 & 3 & 5 & 5 \\ 0 & 4 & 7 & -3 & -1 & -1 \end{array}\right]$$

$$\approx \left[\begin{array}{ccc:ccc} -6 & -2 & -2 & -3 & -3 & 1 \\ 0 & -4 & -1 & 3 & 5 & 5 \\ 0 & 0 & 6 & 0 & 4 & 4 \end{array}\right] \approx \left[\begin{array}{ccc:ccc} -6 & -2 & -2 & -3 & -3 & 1 \\ 0 & -4 & -1 & 3 & 5 & 5 \\ 0 & 0 & 1 & 0 & \frac{2}{3} & \frac{2}{3} \end{array}\right]$$

$$\approx \left[\begin{array}{ccc:ccc} -6 & -2 & 0 & -3 & -\frac{5}{3} & \frac{7}{3} \\ 0 & -4 & 0 & 3 & \frac{17}{3} & \frac{17}{3} \\ 0 & 0 & 1 & 0 & \frac{2}{3} & \frac{2}{3} \end{array}\right] \approx \left[\begin{array}{ccc:ccc} -6 & -2 & 0 & -3 & -\frac{5}{3} & \frac{7}{3} \\ 0 & 1 & 0 & -\frac{3}{4} & -\frac{17}{12} & -\frac{17}{12} \\ 0 & 0 & 1 & 0 & \frac{2}{3} & \frac{2}{3} \end{array}\right]$$

$$\approx \left[\begin{array}{ccc:ccc} -6 & 0 & 0 & -\frac{9}{2} & -\frac{9}{2} & -\frac{1}{2} \\ 0 & 1 & 0 & -\frac{3}{4} & -\frac{17}{12} & -\frac{17}{12} \\ 0 & 0 & 1 & 0 & \frac{2}{3} & \frac{2}{3} \end{array}\right] \approx \left[\begin{array}{ccc:ccc} 1 & 0 & 0 & \frac{3}{4} & \frac{3}{4} & \frac{1}{12} \\ 0 & 1 & 0 & -\frac{3}{4} & -\frac{17}{12} & -\frac{17}{12} \\ 0 & 0 & 1 & 0 & \frac{2}{3} & \frac{2}{3} \end{array}\right]$$

$\therefore$ the transition matrix from B-basis to B′-basis is

$$P]_B^{B'} = \left[[\mathbf{u}_1]_{B'} \vdots [\mathbf{u}_2]_{B'} \vdots [\mathbf{u}_3]_{B'}\right] = \begin{bmatrix} \frac{3}{4} & \frac{3}{4} & \frac{1}{12} \\ -\frac{3}{4} & -\frac{17}{12} & -\frac{17}{12} \\ 0 & \frac{2}{3} & \frac{2}{3} \end{bmatrix}$$

(b)

$$\begin{array}{c} \begin{matrix} \mathbf{u}_1 & \mathbf{u}_2 & \mathbf{u}_3 & \mathbf{w} \end{matrix} \\ \left[\begin{array}{ccc:c} -3 & -3 & 1 & -5 \\ 0 & 2 & 6 & 8 \\ -3 & -1 & -1 & -5 \end{array}\right] \end{array} \approx \left[\begin{array}{ccc:c} -3 & -3 & 1 & -5 \\ 0 & 2 & 6 & 8 \\ 0 & 2 & -2 & 0 \end{array}\right]$$

$$\approx \left[\begin{array}{ccc:c} -3 & -3 & 1 & -5 \\ 0 & 2 & 6 & 8 \\ 0 & 0 & -8 & -8 \end{array}\right] \approx \left[\begin{array}{ccc:c} -3 & -3 & 1 & -5 \\ 0 & 1 & 3 & 4 \\ 0 & 0 & 1 & 1 \end{array}\right] \approx \left[\begin{array}{ccc:c} -3 & 0 & 10 & 7 \\ 0 & 1 & 0 & 1 \\ 0 & 0 & 1 & 1 \end{array}\right]$$

$$\approx \left[\begin{array}{ccc:c} -3 & 0 & 0 & -3 \\ 0 & 1 & 0 & 1 \\ 0 & 0 & 1 & 1 \end{array}\right] \approx \left[\begin{array}{ccc:c} 1 & 0 & 0 & 1 \\ 0 & 1 & 0 & 1 \\ 0 & 0 & 1 & 1 \end{array}\right].$$

$$\therefore \quad [\mathbf{w}]_B = \begin{bmatrix} 1 \\ 1 \\ 1 \end{bmatrix}, \quad [\mathbf{w}]_{B'} = P[\mathbf{w}]_B = \begin{bmatrix} \frac{3}{4} & \frac{3}{4} & \frac{1}{12} \\ -\frac{3}{4} & -\frac{17}{12} & -\frac{17}{12} \\ 0 & \frac{2}{3} & \frac{2}{3} \end{bmatrix} \begin{bmatrix} 1 \\ 1 \\ 1 \end{bmatrix} = \begin{bmatrix} \frac{19}{12} \\ -\frac{43}{12} \\ \frac{4}{3} \end{bmatrix}, \text{ by (a).}$$

(c) To obtain $[\mathbf{w}]_{B'}$ directly, let us form the augmented matrix with the new basis set B′ and the given vector **w** separated by a partition, then use the procedure of subsec. 11.11.1 as follows:

$$\begin{array}{c}\mathbf{v}_1 \quad \mathbf{v}_2 \quad \mathbf{v}_3 \quad \mathbf{w}\\ \left[\begin{array}{ccc|c} -6 & -2 & -2 & -5\\ -6 & -6 & -3 & 8\\ 0 & 4 & 7 & -5 \end{array}\right]\end{array} \approx \left[\begin{array}{ccc|c} -6 & -2 & -2 & -5\\ 0 & -4 & -1 & 13\\ 0 & 4 & 7 & -5 \end{array}\right] \approx \left[\begin{array}{ccc|c} -6 & -2 & -2 & -5\\ 0 & -4 & -1 & 13\\ 0 & 0 & 6 & 8 \end{array}\right]$$

$$\approx \left[\begin{array}{ccc|c} -6 & -2 & -2 & -5\\ 0 & -4 & -1 & 13\\ 0 & 0 & 1 & \frac{4}{3} \end{array}\right] \approx \left[\begin{array}{ccc|c} -6 & -2 & 0 & -\frac{7}{3}\\ 0 & -4 & 0 & \frac{43}{3}\\ 0 & 0 & 1 & \frac{4}{3} \end{array}\right] \approx \left[\begin{array}{ccc|c} -6 & -2 & 0 & -\frac{7}{3}\\ 0 & 1 & 0 & -\frac{43}{12}\\ 0 & 0 & 1 & \frac{4}{3} \end{array}\right] \approx$$

$$\left[\begin{array}{ccc|c} -6 & 0 & 0 & -\frac{57}{6}\\ 0 & 1 & 0 & -\frac{43}{12}\\ 0 & 0 & 1 & \frac{4}{3} \end{array}\right] \approx \left[\begin{array}{ccc|c} 1 & 0 & 0 & \frac{19}{12}\\ 0 & 1 & 0 & -\frac{43}{12}\\ 0 & 0 & 1 & \frac{4}{3} \end{array}\right]. \qquad \therefore [\mathbf{w}]_{B'} = \begin{bmatrix} \frac{19}{12}\\ -\frac{43}{12}\\ \frac{4}{3} \end{bmatrix}$$

which checks the result of $[\mathbf{w}]_{B'}$ obtained in (b).

(d) To obtain $[\mathbf{w}]_B$, the augmented matrix is formed first with columns of the basis set B and then with those B' and follow the procedure adopted in (a).

$$\begin{array}{c}\mathbf{u}_1 \quad \mathbf{u}_2 \quad \mathbf{u}_3 \quad \mathbf{v}_1 \quad \mathbf{v}_2 \quad \mathbf{v}_3\\ \left[\begin{array}{ccc|ccc} -3 & -3 & 1 & -6 & -2 & -2\\ 0 & 2 & 6 & -6 & -6 & -3\\ -3 & -1 & -1 & 0 & 4 & 7 \end{array}\right]\end{array} \approx \left[\begin{array}{ccc|ccc} -3 & -3 & 1 & -6 & -2 & -2\\ 0 & 2 & 6 & -6 & -6 & -3\\ 0 & 2 & -2 & 6 & 6 & 9 \end{array}\right]$$

$$\left[\begin{array}{ccc|ccc} -3 & -3 & 1 & -6 & -2 & -2\\ 0 & 2 & 6 & -6 & -6 & -3\\ 0 & 0 & -8 & 12 & 12 & 12 \end{array}\right] \approx \left[\begin{array}{ccc|ccc} -3 & -3 & 1 & -6 & -2 & -2\\ 0 & 2 & 6 & -6 & -6 & -3\\ 0 & 0 & 1 & -\frac{3}{2} & -\frac{3}{2} & -\frac{3}{2} \end{array}\right]$$

$$\approx \left[\begin{array}{ccc|ccc} -3 & -3 & 0 & -\frac{9}{2} & -\frac{1}{2} & -\frac{1}{2}\\ 0 & 2 & 0 & 3 & 3 & 6\\ 0 & 0 & 1 & -\frac{3}{2} & -\frac{3}{2} & -\frac{3}{2} \end{array}\right] \approx \left[\begin{array}{ccc|ccc} -3 & -3 & 0 & -\frac{9}{2} & -\frac{1}{2} & -\frac{1}{2}\\ 0 & 1 & 0 & \frac{3}{2} & \frac{3}{2} & 3\\ 0 & 0 & 1 & -\frac{3}{2} & -\frac{3}{2} & -\frac{3}{2} \end{array}\right]$$

$$\approx \left[\begin{array}{ccc|ccc} -3 & 0 & 0 & 0 & 4 & \frac{17}{2}\\ 0 & 1 & 0 & -\frac{3}{2} & -\frac{3}{2} & 3\\ 0 & 0 & 1 & -\frac{3}{2} & -\frac{3}{2} & -\frac{3}{2} \end{array}\right] \approx \left[\begin{array}{ccc|ccc} 1 & 0 & 0 & 0 & -\frac{4}{3} & -\frac{17}{6}\\ 0 & 1 & 0 & \frac{3}{2} & \frac{3}{2} & 3\\ 0 & 0 & 1 & -\frac{3}{2} & -\frac{3}{2} & -\frac{3}{2} \end{array}\right].$$

Transition matrix R from B' - basis to B - basis is

$$P\Big]_{B'\text{-basis}}^{B\text{-basis}} = \begin{bmatrix} 0 & -\frac{4}{3} & -\frac{17}{6}\\ \frac{3}{2} & \frac{3}{2} & 3\\ -\frac{3}{2} & -\frac{3}{2} & -\frac{3}{2} \end{bmatrix}.$$

$$\therefore [\mathbf{w}]_B = P[\mathbf{w}]_{B'} = \begin{bmatrix} 0 & -\frac{4}{3} & -\frac{17}{6} \\ \frac{3}{2} & \frac{3}{2} & 3 \\ -\frac{3}{2} & -\frac{3}{2} & -\frac{3}{2} \end{bmatrix} \begin{bmatrix} \frac{19}{12} \\ -\frac{43}{12} \\ \frac{4}{3} \end{bmatrix}$$

$$= \begin{bmatrix} \frac{4.43}{36} - \frac{68}{18} = \frac{172-136}{36} = \frac{36}{36} = 1 \\ \frac{57}{24} - \frac{129}{24} + 4 = \frac{57-129+96}{24} = \frac{153-127}{24} = 1 \\ \frac{-57+129-48}{24} = 1 \end{bmatrix} = \begin{bmatrix} 1 \\ 1 \\ 1 \end{bmatrix}.$$

(e) Now $P]_B^{B'} . P]_{B'}^B = \begin{bmatrix} \frac{3}{4} & \frac{3}{4} & \frac{1}{12} \\ -\frac{3}{4} & -\frac{17}{12} & -\frac{17}{12} \\ 0 & \frac{2}{3} & \frac{2}{3} \end{bmatrix} \begin{bmatrix} 0 & \frac{-3}{4} & \frac{-17}{6} \\ \frac{3}{2} & \frac{3}{2} & 3 \\ -\frac{3}{2} & -\frac{3}{2} & -\frac{3}{2} \end{bmatrix} = \begin{bmatrix} 1 & 0 & 0 \\ 0 & 1 & 0 \\ 0 & 0 & 1 \end{bmatrix} = I$. Hence verified.

11.12 Dimension

The number of vectors or elements contained in the basis of a finitely generated vector space V(F) is called its *dimension* and is written as dim.V.

The dimension does not, in general, mean the number of components of the basis vector. Let the vector space V(F) has n basis vectors. Then dim.V= n. The vector space consisting of the zero vector only, i.e. *zero vector space* has dimension zero. Thus the dimension is a positive integer. The dimension may be fractional in fractal geometry in which dimension 1.7 of a set C in the plane means the properties of set "between" those of the 1-dimensional curve and an open 2-dimensional region. A vector space may be finite and infinite dimensional.

Finite dimensional vector space : A vector space is called *finite dimensional* if the least numbers of vectors $\mathbf{x}_1, \mathbf{x}_2, \cdots, \mathbf{x}_n$ are such that every vector in space may be represented in the form

$$c_1\mathbf{x}_1 + c_2\mathbf{x}_2 + \cdots + c_n\mathbf{x}_n$$

(i.e. every vector is a linear combination of $\mathbf{x}_1, \mathbf{x}_2, \cdots\cdots, \mathbf{x}_n$) in addition to the usual axioms of vector space.

Infinite dimensional vector space : If the additional axiom for the finite dimensional vector space is not satisfied then the space is called *the infinite dimensional* vector space. The study of this type of space with more limitations is the subject matter of functional analysis.

Examples. (i) Zero is the dimension of zero vector space consisting of only the zero vector. It has no basis as it has no linearly independent sets.

(ii) n is the dimension of R^n ; the basis of R^n is $\{\mathbf{e}_1, \mathbf{e}_2, \cdots, \mathbf{e}_n\}$.

(iii) $(n+1)$ is the dimension of all polynomials of the degree n or less and the basis is

$$\{1, t, \cdots, t_n\}.$$

(iv) mn is the dimension of the vector space of $m \times n$ matrices

(v) R^n , C^n , P^n and zero vector spaces $M_{2\times 2}$ are finite dimensional vector spaces

(vi) Set of all polynomials of all degrees with real or complex coefficients is a real or complex infinite dimensional vector space.

Ex. Find a basis and the dimension of the solution space V of the ODE

$$\frac{d^2u}{dx^2} - k^2u = 0, \quad k > 0 \tag{1}$$

Let $y = e^{mx}$ be the trial solution of the given second order homogeneous ODE. The auxilliary equation is

$$m^2 - k^2 = 0 \text{ , } k > 0 \text{ i.e. } m = \pm k \text{ , } k > 0.$$

$\therefore$ $y_1 = e^{kx}$ and $y_2 = e^{-kx}$ are the solutions.

Let c_1 and c_2 be two scalars such that

$c_1 y_1 + c_2 y_2 = 0$ for linear independence / dependence.

i.e $c_1 e^{kx} + c_2 e^{-kx} = 0$ (2)

To find c_1 and c_2, let us obtain another equation by differentiating (2)

$$kc_1 e^{kx} - kc_2 e^{-kx} = 0$$

$$c_1 e^{kx} - c_2 e^{-kx} = 0 \tag{3}$$

Solving c_1 and c_2 from (2) & (3), we have, $c_1 = c_2 = 0$. Therefore y_1 and y_2 are two independent solutions of (1).

It can be verified that there is no other independent solution of (1). If possible, let y_3 be another solution. Therefore, for linear independence/ dependence let

$$c_1 y_1 + c_2 y_2 + c_3 y_3 = 0 \tag{4}$$

Differentiating twice successively,

$$c_1 \frac{dy_1}{dx} + c_2 \frac{dy_2}{dx} + c_3 \frac{dy_3}{dx} = 0 \tag{5}$$

$$c_1 \frac{d^2 y_1}{dx^2} + c_2 \frac{d^2 y_2}{dx^2} + c_3 \frac{d^2 y_3}{dx^2} = 0 \tag{6}$$

Denoting derivatives by y_1', y_1'' etc, zero solutions of c_1, c_2, c_3 the determinant or the coefficient matrix must not vanish, i.e.

$$D = \begin{vmatrix} y_1 & y_2 & y_3 \\ y_1' & y_2' & y_3' \\ y_1'' & y_2'' & y_3'' \end{vmatrix} \neq 0 \tag{7}$$

$\because$ y_1, y_2, y_3 are assumed as a solutions of (1), so $\frac{d^2 y_3}{dx^2} - k^2 y_3 = 0$ etc. , i.e. $y_3'' = k^2 y_3$ etc.

$$\therefore \quad D = k^2 \begin{vmatrix} y_1 & y_2 & y_3 \\ y_1' & y_2' & y_3' \\ y_1 & y_2 & y_3 \end{vmatrix} \text{ which is a contradiction to (7).}$$

$\therefore$ there can be no independent other solutions than y_1, y_2.

Hence the dimension of he solution space V is 2 and the basis is $\{e^{kx}, e^{-kx}\}$. Any other vector in V is a linear combination of y_1 and y_2. This is why the general solution is written as

$$y = c_1 y_1 + c_2 y_2 = c_1 e^{kx} + c_2 e^{-kx},$$

a linear combination of $y_1 = e^{kx}$ and $y_2 = e^{-kx}$.

11.12.1 Theorems on basis and dimensions: The theorems describe the important aspects about the invariance of basis and dimensions of vector spaces and subspaces.

Theorem 1. Every basis of a finite dimensional vector space has the same number of elements (vectors).

ALTER. Let V(F) be a finite dimensional vector space then any two bases of V have the same number of elements.

ALTER. The number of elements (vectors) in the bases of a finite dimensional vector space is invariant.

Proof. Let V(F) be a finite dimensional vector space and $B_1 = \{\alpha_1, \alpha_2, \cdots, \alpha_n\}$ and $B_2 = \{\beta_1, \beta_2, \cdots, \beta_m\}$ be any two bases of V. Then we have to show that $m = n$ i.e. the number of elements or vectors in B_1 is equal to that in B_2.

Now B_1 being the basis of V, each element of V is uniquely expressible as a linear combination of elements of B_1. In particular β_1 is expressible uniquely as a linear combination of elements of B_1 and therefore, the set $S = \{\beta_1, \alpha_1, \alpha_2, \cdots, \alpha_n\}$ obtained by adjoining β_1 in B_1 is linearly dependent. Consequently, $\exists$ a vector α_k other than β_1 in S which is expressible as a L.C. of preceeding vectors $\beta_1, \alpha_1, \alpha_2, \cdots, \alpha_{k-1}$.

Now each element of V can be expressed as a L.C. of $S = \{\beta_1, \alpha_1, \alpha_2, \cdots, \alpha_n\}$ and α_k is expressible as a L.C. of $\beta_1, \alpha_1, \alpha_2, \cdots, \alpha_{k-1}$ which means that each vector in V is expressible as a L.C. of $\beta_1, \alpha_1, \alpha_2, \cdots, \alpha_{k-1}$.

In other words the set $S_1 = \{\beta_1, \alpha_1, \alpha_2, \cdots, \alpha_{k-1}, \alpha_{k+1}, \cdots, \alpha_n\}$ obtained by deleting α_k from S generates V.

Now S_1 generates V and $\beta_2 \in V$, so β_2 is a L.C. of elements of S_1 and therefore the set obtained by adjoining β_2 to S_1 is linearly dependent. As argued before, the set

$$S_2 = \{\beta_1, \beta_2, \alpha_1, \alpha_2, \cdots, \alpha_i, \alpha_{i+1}, \alpha_{k-1}, \alpha_{k+1}, \cdots, \alpha_n\}$$

obtained by adjoining β_2 to S_1 and deleting some vector α_l $(l > k)$ other than β_1, β_2 from S_1 generates V.

We continue to proceed in the above manner. Each step consists of the inclusion of one β and the exclusion of one α and the resulting set generates V. Now all the α's can be exhausted before the β's because it would mean that a proper subset of B_2 generates V and thus contradicting the linear independence of B_2. Hence $m \not< n$.

Similarly, by changing the roles of the basis, it follows that $n \not< m$.

Hence $m = n$.

Theorem 2. (Linear dependence) Each set of $(n+1)$ or more vectors of a finite dimensional vector space V(F) of dimension n is linearly dependent.

Proof. If a set of $(n+1)$ or more vectors of a finite dimensional vector space V(F) of dimension n is linearly independent then either it itself is a basis set or it can be extended to form a basis of

V(F), i.e. in either case the basis set of V(F) will consists of $(n+1)$ or more vectors and this contradicts that the dimension of V(F) is n.

Hence the set of $(n+1)$ or more vectors of a finite dimensional vector space V(F) of dimension n is linearly dependent.

Theorem.3 (Basis) In an n-dimensional vector space V(F) any linearly independent set of n-vectors is a basis set of V(F).

Proof. Each linearly independent set can be extended to form a basis and as the given vector space is n-dimensional its basis set will contain n linearly independent elements.

Hence the given linearly independent set of n vectors cannot be further extended but it itself forms a basis set of V(F).

Cor. The finite dimensional vector space of dimension n cannot be generated by fewer than n-vectors.

11.12.2 Dimension of subspaces

Since subspaces by definition are also vector spaces. So, the concepts of linear combination, span, basis and dimension are equally defined and extended to subspaces.

Theorem.1 Every nonzero subspace W of $V_n(F)$ has a basis. All bases of W have the same number of vectors.

Proof. Since W is a nonzero subspace of $V_n(F)$, so W contains a nonzero linearly independent vector. Also as $V_n(F)$ has n linearly independent vectors, so W cannot contain more than n linearly independent vectors. Thus, let W contains $V_n(F)$ linearly independent vectors $\alpha_1, \alpha_2, \cdots, \alpha_r$. Now any vector $\alpha \in$ w must be written as a linear combination of $\alpha_1, \alpha_2, \cdots, \alpha_r$.

i.e. $\alpha = c_1\alpha_1 + c_2\alpha_2 + \cdots + c_r\alpha_r$ for $c_i \in F$, $i = 1, 2, \cdots, r$.

i.e. $W = L\{\alpha_1, \alpha_2, \cdots, \alpha_r\}$.

As α_i, $i = 1, 2, \cdots, r$ are linearly independent in W and span W, hence α_i, $i = 1, 2, \cdots, r$, form a basis of W.

Now if there is another set of basis of W, then the statement that their total number is also r can be easily proved as the theorem 1 in sec.11.10.3.

Hence the theorem.

Cor.1 If subspace W has dimension r, any set of $(r+1)$ vectors in W are linearly dependent.

Cor.2 Every linearly independent set of vectors of subspace W can be extended to a basis of the subspace W as well as the whole space $V_n(F)$.

Theorem 2. (Dimension of subspace and vector space) If V(F) is a finite dimensional vector space of dimension n and if W is any subspace of V, then W is a finite dimensional space having the dimension at most n i.e. dim.W $\leq$ dim.V. Also V= W, iff dim.V= dim.W.

Proof. Since V is of dimension n, any $(n+1)$ or more vectors in V are linearly dependent. Also W is a vector subspace of V. Hence if it consists of more than n vectors then it will be linearly dependent. Hence W may contain at most n linearly independent vectors.

Now let the largest set of linearly independent vectors in W be

$$S = \{\alpha_1, \alpha_2, \cdots, \alpha_m\} \quad \text{for } m \leq n.$$

To claim that S is a basis of W, we have to prove that (i) S is linearly independent subset of W and (ii) $L(S) = W$.

Let $\alpha \in W$, then the $(m+1)$ vectors $\alpha, \alpha_1, \alpha_2, \cdots, \alpha_m$ of W are linearly dependent. (because the largest linearly independent set in W contains only m elements). Hence an element of this set can be expressed as a linear combination of the preceding ones. Now $\alpha_1, \alpha_2, \cdots, \alpha_m$ being L.I, this element can be α itself. Therefore α can be expressed as a L.C. of $\alpha_1, \alpha_2, \cdots, \alpha_m$. i.e., W= L(S) and hence S is a basis of W

i.e. $\dim. W = m$ where $m \le n$.

i.e. $\dim. W \le \dim. V$.

In particular, if $\alpha_1, \alpha_2, \cdots, \alpha_n$ is a basis of W then since it is an independent set of n elements, it is also a basis of V.

Thus W = V when dim.W= n, i.e. iff dim.W = dim.V.

Ex.1 In $V_n(F)$ consider the subspace W of all vectors $(\alpha_1, \alpha_2, \alpha_3, \alpha_4)$ sastisfying $(\alpha_1 + 2\alpha_2 = \alpha_3 + 2\alpha_4)$. Show that

$$\mathbf{x}_1 = (1, 0, 1, 0) \text{ and } \mathbf{x}_2 = (0, 1, 0, 1)$$

are linearly independent and lie in W. Then extend this set of two vectors to a basis of W.

Part 1. For linear independency, there may be a relation

$c_1\mathbf{x}_1 + c_2\mathbf{x}_2 = 0$ i.e. $c_1(1, 0, 1, 0) + c_2(0, 1, 0, 1) = 0 \Rightarrow c_1 = 0,\ c_2 = 0$

$\therefore$ $\mathbf{x}_1$ and $\mathbf{x}_2$ are linearly independent.

Alter. Let the 4×2 matrix with columns $A_{.j} = x_j,\ j = 1, 2$ be

$$\begin{bmatrix} 1 & 0 \\ 0 & 1 \\ 1 & 0 \\ 0 & 1 \end{bmatrix} \approx \begin{bmatrix} 1 & 0 \\ 0 & 1 \\ 0 & 0 \\ 0 & 1 \end{bmatrix} \approx \begin{bmatrix} 1 & 0 \\ 0 & 1 \\ 0 & 0 \\ 0 & 0 \end{bmatrix}$$

in row reduced echelon form and it has two nonzero rows, hence the given two vectors are linearly independent.

Part 2. Any vector $(a, b, c, d) \in W$ can be written as

$$(a, b, c, d) = c_1(1, 0, 1, 0) + c_2(0, 1, 0, 1)$$

Also $a + 2b = c + 2d$ (given)

or $a - c = 2(d - b)$

i.e. $c_1 = a,\ c_2 = b,\ c_3 = c,\ c_4 = d$

$\therefore$ $c_1 = a = c$ and $c_2 = b = d$

$\therefore$ any vector $(a, b, a, b) = a(1, 0, 1, 0) + b(0, 1, 0, 1) = a\,\mathbf{x}_1 + b\,\mathbf{x}_2$, a linear combination of the given vectors. Hence $\mathbf{x}_1, \mathbf{x}_2$ lie in W.

Part 3. Constructing a 4×6 matrix A with the first two columns by $\mathbf{x}_1$ and $\mathbf{x}_2$ and the last 4 columns by standard bases $\mathbf{e}_1, \mathbf{e}_2, \mathbf{e}_3, \mathbf{e}_4$ of $V_4(R)$ and reducing to the row reduced echelon form as follows, we get,

$$A = \begin{bmatrix} \mathbf{x}_1 & \mathbf{x}_2 & \mathbf{e}_1 & \mathbf{e}_2 & \mathbf{e}_3 & \mathbf{e}_4 \\ 1 & 0 & 1 & 0 & 0 & 0 \\ 0 & 1 & 0 & 1 & 0 & 0 \\ 1 & 0 & 0 & 0 & 1 & 0 \\ 0 & 1 & 0 & 0 & 0 & 1 \end{bmatrix} \approx \begin{bmatrix} 1 & 0 & 1 & 0 & 0 & 0 \\ 0 & 1 & 0 & 1 & 0 & 0 \\ 0 & 0 & -1 & 0 & 1 & 0 \\ 0 & 1 & 0 & 0 & 0 & 1 \end{bmatrix} \approx \begin{bmatrix} 1 & 0 & 1 & 0 & 0 & 0 \\ 0 & 1 & 0 & 1 & 0 & 0 \\ 0 & 0 & -1 & 0 & 1 & 0 \\ 0 & 0 & 0 & -1 & 0 & 1 \end{bmatrix}$$

$$\approx \begin{bmatrix} 1 & 0 & 1 & 0 & 0 & 0 \\ 0 & 1 & 0 & 1 & 0 & 0 \\ 0 & 0 & 1 & 0 & -1 & 0 \\ 0 & 0 & 0 & 1 & 0 & -1 \end{bmatrix} \approx \begin{bmatrix} 1 & 0 & 0 & 0 & 1 & 0 \\ 0 & 1 & 0 & 0 & 0 & 1 \\ 0 & 0 & 1 & 0 & -1 & 0 \\ 0 & 0 & 0 & 1 & 0 & -1 \\ \uparrow & \uparrow & \uparrow & \uparrow & & \end{bmatrix}$$

$\therefore$ the required extended basis of the subspace W are $\{\mathbf{x}_1, \mathbf{x}_2, \mathbf{e}_1, \mathbf{e}_2\}$, shown by vertical arrows while $\mathbf{e}_3$ and $\mathbf{e}_4$ can be expressed as linear combination of four basis vectors as follows:

$$\mathbf{e}_3 = 1\mathbf{x}_1 + 0.\mathbf{x}_2 - \mathbf{e}_1 + 0\ \ \mathbf{e}_2 = (0, 0, 1, 0)$$
$$\mathbf{e}_4 = 0\mathbf{x}_1 + 1.\mathbf{x}_2 + 0\mathbf{e}_1 - \mathbf{e}_2 \quad = (0, 0, 0, 1)$$

Therefore, dim.W = 4.

Ex. 2 Prove that the set S = {(1, 2, 1), (2, 1, 0), (1, –1, 2)} forms a basis of $V_3(R)$.

By definition S must be (i) L. I. & (ii) S generates V(F) i. e. V(F) = L(S).

(i) To show that S is a L.I. system, let $a_1, a_2, a_3 \in R$ such that

$$a_1(1, 2, 1) + a_2(2, 1, 0) + a_3(1, -1, 2) = 0 = (0, 0, 0)$$

$$\Rightarrow \left.\begin{aligned} a_1 + 2a_2 + a_3 &= 0 \\ 2a_1 + a_2 - a_3 &= 0 \\ a_1 + 2a_3 &= 0 \end{aligned}\right\} \to \left.\begin{aligned} a_1 + 2a_2 + a_3 &= 0 \\ -3a_2 - 3a_3 &= 0 \\ -2a_2 + a_3 &= 0 \end{aligned}\right\} \to$$

$$\left.\begin{aligned} a_1 + 2a_2 + a_3 &= 0 \\ a_2 + a_3 &= 0 \\ -2a_2 + a_3 &= 0 \end{aligned}\right\} \to \left.\begin{aligned} a_1 + 2a_2 + a_3 &= 0 \\ a_2 + a_3 &= 0 \\ a_3 &= 0 \end{aligned}\right\}$$

which is in echelon form and the homogeneous system has a zero solution.

$\therefore$ S is L.I.

(ii) Let (*a, b, c*) be any vector in $V_3(R)$. Let

$$(a,b,c) = a_1(1, 2, 1) + a_2(2, 1, 0) + a_3(1, 01,2) \text{ for } a_1, a_2, a_3 \in R$$

$$\Rightarrow \left.\begin{aligned} a_1 + 2a_2 + a_3 &= a \\ 2a_1 + a_2 - a_3 &= b \\ a_1 + 2a_3 &= c \end{aligned}\right\} \to \left.\begin{aligned} a_1 + 2a_2 + a_3 &= a \\ -3a_2 - 3a_3 &= b - 2a \\ -2a_2 + a_3 &= c - a \end{aligned}\right\} \to$$

$$\left.\begin{aligned} a_1 + 2a_2 + a_3 &= a \\ a_2 + a_3 &= \tfrac{1}{3}(2a - b) \\ a_3 &= \tfrac{2}{9}(a - 2b + 3c) \end{aligned}\right\}$$

which is in the echelon form and the nonhomogeneous system is consistent.

$\therefore$ any vector $(a,b,c) \in R$ can be expressed as L.C. of elements of the set S i.e. V(F)=L(S).

$\therefore$ S forms a basis of $V_3(R)$.

Ex.3 Show that the set S = {(1, 0, 0), (1, 1, 0), (1, 1, 1), (0, 1, 0)} spans the vector space $V_3(R)$ but is not a basis set. Express (4, 5, 6) in many different ways in terms of the vectors in S.

First part: Let $a_1, a_2, a_3, a_4 \in R$.

(i) For linear independence or dependence,

$$a_1(1, 0, 0) + a_2(1, 1, 0) + a_3(1, 1, 1) + a_4(0, 1, 0) = 0$$

$$\Rightarrow \left.\begin{array}{l} a_1 + a_2 + a_3 = 0 \\ a_2 + a_3 + a_4 = 0 \\ a_3 = 0 \end{array}\right\}$$

which is in the echelon form and the system has nonzero solution. The systems of the homogeneous equation are L.D.

$\therefore$ S is L.D.

(ii) Let (a, b, c) be any vector in $V_3(R)$.

Let $(a, b, c) = a_1(1, 0, 0) + a_2(1, 1, 0) + a_3(1, 1, 1) + a_4(0, 1, 0)$

$$\Rightarrow \left.\begin{array}{l} a_1 + a_2 + a_3 = a \\ a_2 + a_3 + a_4 = b \\ a_3 = c \end{array}\right\}$$

which is in the echelon form & the nonhomogeneous system is consistent.

$\therefore$ any vector $(a, b, c) \in V_3(R)$ can be expressed as a L.C. of the set S, i.e. V(F) = L(S).

$\because$ S satisfies one of the two conditions of basis, so the given S spans $V_3(R)$ but not a basis of $V_3(R)$.

Second part: It can be easily shown that

$$(4, 5, 6) = -(a_4 + 1)(1, 0, 0) + (a_4 - 1)(1, 1, 0) + 6(1, 1, 0) + a_4(0, 1, 0).$$

Thus for different values of a_4, the vector (4, 5, 6) can be represented differently by the spanning vectors.

Ex.4 Find a basis & dimension for the subspace S of $V_5(R)$ generated by the set of row vectors {(–3, 3, 1, –3, 2), (0, 7, 1, –2, 3), (5, –11, 2, 4, –5), (2, –1, 4, –1, 0)}.

We reduce the matrix

$$\begin{bmatrix} -3 & 3 & 1 & -3 & 2 \\ 0 & 7 & 1 & -2 & 3 \\ 5 & -11 & 2 & 4 & -5 \\ 2 & -1 & 4 & -1 & 0 \end{bmatrix}$$

to the echelon form & note the numbers of nonzero rows which form a basis for S.

$$\xrightarrow[2R_1+3R_4]{5R_1+3R_3}\begin{bmatrix}-3 & 3 & 1 & -3 & 2\\ 0 & 7 & 1 & -2 & 3\\ 0 & -18 & 11 & -3 & -5\\ 0 & 3 & 14 & -9 & 4\end{bmatrix}\xrightarrow[7R_4-3R_2]{18R_2+7R_3}$$

$$\begin{bmatrix}-3 & 3 & 1 & -3 & 2\\ 0 & 7 & 1 & -2 & 3\\ 0 & 0 & 95 & -57 & 19\\ 0 & 0 & 95 & -57 & 19\end{bmatrix}\xrightarrow{R_4-R_3}\begin{bmatrix}-3 & 3 & 1 & -3 & 2\\ 0 & 7 & 1 & -2 & 3\\ 0 & 0 & 95 & -57 & 19\\ 0 & 0 & 0 & 0 & 0\end{bmatrix}$$

which is an echelon matrix. It has 3 nonzero rows. Therefore, the required basis and dimension are

B= {(–3, 3, 1, –3, 2), (0, 7, 1, –2, 3), (5, –11, 2, 4, –5) } and the dimension = 3.

11.12.3 Dimension of a direct sum of subspaces : The following two theorems are concerned about direct sum of subspaces.

Theorem. 1 If a finite dimensional vector space V(F) be direct sum of its two subspaces W_1 and W_2 i.e.

$$V = W_1 \oplus W_2 \text{ then } \dim.V = \dim.W_1 + \dim.W_2.$$

Proof. If W_1 and W_2 are subspaces of V(F), a finite dimensional vector space, therefore, W_1 and W_2 are also finite dimensional.

Let $\quad \dim. W_1 = l$ & $\dim.W_2 = m$.

Also as given $V = W_1 \oplus W_2$ i.e.

(i) $\quad \gamma = \alpha + \beta, \quad \forall\, \gamma \in V,\ \alpha \in W_1,\ \beta \in W_2$

(ii) $\quad W_1 \cap W_2 = \{0\}$.

Also let $\quad S_1 = \{\alpha_1, \alpha_2, \cdots, \alpha_l\}$ be a basis for W_1, i.e. $\dim.W_1 = l$

and $\quad S_2 = \{\beta_1, \beta_2, \cdots, \beta_l\}$ be a basis for W_2, i.e. $\dim.W_2 = m$.

Now consider the set

$$S = \{\alpha_1, \alpha_2, \cdots, \alpha_l, \beta_1, \beta_2, \cdots, \beta_m\}.$$

We have to prove that this set S of dim.= $(l + m)$ is a basis of V. Hence we have to show that

(A) S is L.I

(B) L(S) = V.

(A) S is liearly independent : To show that S is L.I. consider the relation

$$a_1\alpha_1 + a_2\alpha_2 + \cdots + a_l\alpha_l + (b_1\beta_1 + b_2\beta_2 + \cdots + b_m\beta_m) = 0 \text{ for } \alpha_i\text{'s},\ \beta_i\text{'s} \in F$$

or $\quad a_1\alpha_1 + a_2\alpha_2 + \cdots + a_l\alpha_l = -(b_1\beta_1 + b_2\beta_2 + \cdots + b_m\beta_m)$

Hence $(a_1\alpha_1 + a_2\alpha_2 + \cdots + a_l\alpha_l)$ and $(b_1\beta_1 + b_2\beta_2 + \cdots + b_m\beta_m)$ are both elements of $W_1 \cap W_2$ but we are given that $V = W_1 \oplus W_2$. Hence from (ii), 0 is the only vector common to W_1 and W_2.

Therefore, we have,

$$a_1\alpha_1 + a_2\alpha_2 + \cdots + a_l\alpha_l = 0 \qquad (11.22)$$

$$b_1\beta_1 + b_2\beta_2 + \cdots + b_m\beta_m = 0 \qquad (11.23)$$

Now as $\{\alpha_1+\alpha_2+\cdots+\alpha_m\}$ and $\{\beta_1+\beta_2+\cdots+\beta_m\}$ are L. I. sets, so we must have from (11.22) & (11.23)

$$a_1=0,\ a_2=0,\cdots,\ a_l=0 \text{ and}$$
$$b_1=0,\ b_2=0,\cdots,\ b_m=0.$$

Hence S is linearly independent.

(B) L(S) = V

For $\alpha\in W_1,\ \beta\in W_2$, we have any element γ of V given by

$$\gamma=\alpha+\beta=\sum a_i\alpha_i+\sum b_j\beta_j$$

[because $\alpha\in W_1$, S_1 is a basis set of $W_1\Rightarrow\alpha=\sum a_i\alpha_i$ for $1\le i\le l$ and S_2 is a basis set of $W_2\Rightarrow\beta=\sum b_j\beta_j$ for $1\le j\le m$ i.e.

$$\gamma=a_1\alpha_1+a_2\alpha_2+\cdots+a_l\alpha_l+b_1\beta_1+b_2\beta_2+\cdots+b_m\beta_m,$$

showing that the set S generates V].

Hence S is linearly independent.

Thus S is a basis set for V containing $(l+m)$ vectors such that

$$\dim.V=(l+m)=\dim.W_1+\dim.W_2.$$

Theorem 2. If V is a finite dimensional vector space and W_1, W_2 are subspaces of V such that $V=W_1+W_2$ and $\dim.V=\dim.W_1+\dim.W_2$ then $V=W_1\oplus W_2$.

Proof. This theorem is converse of theorem 1. Here if $\dim.W_1=l$ and $\dim.W_2=m$ then

$$\dim.V=l+m.$$

Also let $S_1=\{\alpha_1,\alpha_2,\cdots,\alpha_l\}$ be a basis of W_1

and $S_2=\{\beta_1,\beta_2,\cdots,\beta_m\}$ be a basis of W_2

Firstly, we show that $S_1\cup S_2$ is a basis of V. Le $\gamma\in V$, then as $V=W_1+W_2$, we have $\gamma=\alpha+\beta$ for $a\in W_1$ and $\beta\in W_2$.

Now $\alpha=\sum a_i\alpha_i$, a_i 's$\in$ F, α_i 's$\in S_1$ and $\beta=\sum b_j\beta_j$ for b_j 's $\in$ F, β_i 's$\in S_2$

Therefore $\gamma=\alpha+\beta=\sum a_i\alpha_i+\sum b_j\beta_j$

i.e. can be expressed as a L.C. of elements of $S_1\cup S_2$. Hence, we have

$$L(S_1\cup S_2)=V.$$

Also as $\dim.V=(l+m)$ and $V=L(S_1\cup S_2)$, therefore $S_1\cup S_2$, cannot have less than $(l+m)$ distinct elements i.e. it can have only $(l+m)$ distinct elements. Thus $S_1\cup S_2$ is a basis set of V. Hence the set

$$S=\{\alpha_1,\alpha_2,\cdots,\alpha_l,\beta_1,\beta_2,\cdots,\beta_m\} \text{ is L.I.}$$

Let us now show that $W_1\cap W_2=\{0\}$. We have

$$\alpha\in W_1\cap W_2\Rightarrow\alpha\in W_1 \text{ and } \alpha\in W_2$$
$$\alpha\in W_1\Rightarrow\alpha=a_1\alpha_1+a_2\alpha_2+\cdots+a_l\alpha_l$$

and $\alpha\in W_2\Rightarrow\alpha=b_1\beta_1+b_2\beta_2+\cdots+b_m\beta_m$

$$\therefore\quad \alpha=a_1\alpha_1+a_2\alpha_2+\cdots+a_l\alpha_l=b_1\beta_1+b_2\beta_2+\cdots+b_m\beta_m$$

or
$$a_1\alpha_1+a_2\alpha_2+\cdots+a_l\alpha_l+(-b_1)\beta_1+(-b_2)\beta_2+\cdots+(-b_m)\beta_m=0$$

$\Rightarrow a_1=0,\ a_2=0,\ \cdots,a_l=0,\ b_1=0,\ b_2=0,\ \cdots,b_m=0,\because$ S is L.I.

$\Rightarrow\ \alpha = 0$, i.e $\quad W_1 \cap W_2 = \{0\} \quad \Rightarrow\ V = W_1 + W_2$

and $\quad W_1 \cap W_2 = \{0\} \Rightarrow V = W_1 \oplus W_2$.

11.12.4 Dimension of a linear sum (Join) : Following is an important theorem which determines the dimension of a linear sum.

Theorem. If W_1, W_2 are two subspaces of a finite dimensional vector space V(F), then

$$\dim (W_1 + W_2) = \dim. W_1 + \dim. W_2 - \dim. W_1 \cap W_2$$

where dim. W_1 and dim. W_2 denote the dimensions of the subspaces W_1 and W_2 respectively.

Proof. V(F) being a finite dimensional vector space, all its subspaces W_1, W_2, $W_1 \cap W_2$ and $W_1 + W_2$ are finite dimensional.

Let $B_2 = \{\gamma_1, \gamma_2, \cdots, \gamma_r, \alpha_1, \alpha_2, \cdots, \alpha_{m-r}\}$ be a basis set of W_1 of dimension, m

and $B_3 = \{\gamma_1, \gamma_2, \cdots, \gamma_r, \beta_1, \beta_2, \cdots, \beta_{n-r}\}$ be a basis set of W_2 of dimension, n.

Now, consider the set $S = B_2 \cup B_3 = \{\gamma_1, \gamma_2, \cdots, \gamma_r, \alpha_1, \alpha_2, \cdots, \alpha_{m-r}, \beta_1, \beta_2, \cdots, \beta_{n-r}\}$.

We shall now show that the set S is a basis of $W_1 + W_2$.

Firstly, we will show that S is L. I., if possible

(i) $$\sum_{i=1}^{r} a_i \gamma_i + \sum_{j=1}^{m-r} b_j \gamma_j + \sum_{k=1}^{n-r} c_k \beta_k = 0 \qquad (11.24)$$

where a_i's, b_j's, c_k's $\in$ F

Therefore, we have $\displaystyle \sum_{j=1}^{m-r} b_j \gamma_j + \sum_{i=1}^{r} a_i \gamma_i = -\sum_{k=1}^{n-r} c_k \beta_k$.

This $\Rightarrow \sum c_k \beta_k$ may be expressed as a L.C. of elements of basis set B_2

i.e. it is an element of W_1.

Also $\quad \sum c_k \beta_k = \sum c_k \beta_k + \sum 0 \gamma_i$ (0 being additive identity of F)

$\Rightarrow \quad \sum c_k \beta_k$ may be expressed as a L.C. of elements of basis set B_2 i.e. it is an element of W_2.

$\Rightarrow \quad \sum c_k \beta_k \in W_1 \cap W_2$

i.e. $\quad \displaystyle \sum c_k \beta_k = \sum_{i=1}^{r} d_i \gamma_i \quad (\because \sum_{i=1}^{r} d_i \gamma_i \in W_1 \cap W_2)$

$$\sum c_k \beta_k - \sum d_i \gamma_i = 0$$

$$\Rightarrow c_1 = 0,\ c_2 = 0,\ \cdots,\ c_{n-r} = 0;\ d_1 = 0,\ d_2 = 0,\ \cdots,\ d_n = 0 \qquad (11.25)$$

($\because$ B_3 being basis set of W_2 is L.I.)

Hence by (11.24), we have

$$\sum_{i=1}^{r} a_i \gamma_i + \sum_{j=1}^{m-r} b_j \alpha_j = -\sum_{k=1}^{n-r} 0 \beta_k = 0$$

Now B_2 being a basis set of W_1' i.e. L.I., we have ,

$a_1 = 0, a_2 = 0, \cdots, a_r = 0; b_1 = 0, b_2 = 0, \cdots, b_{m-r} = 0 \qquad (11.26)$

Thus, from (11.25) and (11.26), we conclude that a_i's , b_i's , c_k's $\in$F used to express (11.24), are all equal to zero. Hence, the set S is L.I.

(ii) To show that $L(S) = W_1 + W_2$, let $\gamma \in W_1 + W_2 \Rightarrow \gamma = \alpha + \beta$, $\alpha \in W_1$, $\beta \in W_2$.

Also $\alpha = \left(\sum a_i \gamma_i + \sum b_j \alpha_j\right)$, B_2 being basis set of W_1

and $\beta = \left(\sum d_i \gamma_i + \sum c_k \beta_k\right)$, B_3 being basis set of W_2.

$$\gamma = \left(\sum a_i \gamma_i + \sum b_j \alpha_j\right) + \left(\sum d_i \gamma_i + \sum c_k \beta_k\right)$$
$$= \sum (a_i + d_i)\gamma_i + \sum b_j \alpha_j + \sum c_k \beta_k$$
$$= \text{L.C. of } \gamma_i\text{'s},\ \alpha_j\text{'s},\ \beta_k\text{'s of S}.$$

$\Rightarrow$ $\gamma \in L(S)$.

Thus $W_1 + W_2 \subseteq L(S)$. (11.27)

Again, let $\gamma \in L(S)$ and proceeding as above in reverse order, we can easily prove that $\gamma \in W_1 + W_2$ i.e. $L(S) \subseteq W_1 + W_2$. (11.28)

From (11.27) and (11.28), we get, $L(S) \subseteq W_1 + W_2$.

Hence S is a basis set of $W_1 + W_2$, such that,

$$\dim.(W_1 + W_2) = r + (m - r) + (n - r) = m + n - r$$
$$= \dim. W_1 + \dim. W_2 - \dim.(W_1 \cap W_2)$$

This may also be written as

$$\dim. W_1 + \dim. W_2 = \dim.(W_1 + W_2) + \dim.(W_1 \cap W_2)$$

Cor. As a particular case if $W_1 \cap W_2 = \{0\}$ i.e. $\dim.(W_1 \cap W_2) = 0$ then $W_1 + W_2$ and $W_1 \oplus W_2$ becomes the same & then we have

$$\dim.(W_1 + W_2) = \dim.(W_1 \oplus W_2).$$

Ex.1 Find a basis and the dimension of the solution space W of the homogeneous system :

$$\begin{cases} x + 3y + 2z = 0 \\ x + 5y + z = 0 \\ x + 5y + 8z = 0 \end{cases}$$

Let us reduce the system to echelon form :

$$\left.\begin{aligned} x + 3y + 2z &= 0 \\ 2y - z &= 0 \\ -4y + 2z &= 0 \end{aligned}\right\} \rightarrow \left.\begin{aligned} x + 3y + 2z &= 0 \\ 2y - z &= 0 \end{aligned}\right\}$$

The system in echelon form has 2 equations in 3 unknowns. Hence dimension of the solution space $= 3 - 2 = 1$.

The free variable is z. Therefore, the solution is $y = \frac{1}{2}z$, $x = -3y - 2z = -\frac{7}{2}z$.

$\therefore$ the basis of the solution space is a particular solution of the above general solution. Let $z = -2$. So, the basis set is $\{(7, -1, -2)\}$.

Ex.2 Find the solution space of the homogeneous system

$$\left.\begin{aligned} -2x_2+2x_3&=0\\ 4x_1-\ x_2+5x_3&=0\\ 2x_1+\ x_2+\ x_3&=0 \end{aligned}\right\}$$

and obtain a basis of the solution space.

Let us reduce the system to echelon form:

$$\left.\begin{aligned} 4x_1-x_2+5x_3&=0\\ -2x_2+2x_3&=0\\ 2x_1+x_2+\ x_3&=0 \end{aligned}\right\}\to \left.\begin{aligned} 4x_1-x_2+5x_3&=0\\ -2x_2+2x_3&=0\\ x_2-3x_3&=0 \end{aligned}\right\}\to \left.\begin{aligned} 4x_1-x_2+5x_3&=0\\ -2x_2+2x_3&=0\\ -4x_3&=0 \end{aligned}\right\}\to$$

$$\left.\begin{aligned} 4x_1-x_2+5x_2&=0\\ x_2-x_2\ &=0\\ x_2\ &=0 \end{aligned}\right\}$$

The system in echelon form has 3 equations in 3 unknowns

$\therefore$ the solution space is $\{0, 0, 0\}$. So the dimension =3–3 = 0 and basis = 0.

Ex.3 Find a homogeneous system whose solution set W is generated by

$$\{(1,-2,0,3,-1),(2,-3,2,5,-3),(1,-2,1,2,-2)\}.$$

We know that $\mathbf{v}=(x,\ y,\ z,\ s,\ t)\in\text{W}$ iff $\mathbf{v}$ is a L.C. of the generators of W. Then $(x,\ y, z, s, t)=a_1(1,-2,0,3,-1)+a_2(2,-3,2,5,-3)+a_3(1,-2,1,2,-2)$

$$\Rightarrow\quad \left.\begin{aligned} a_1+2a_2+a_3&=x\\ -2a_1-3a_2-2a_3&=y\\ 2a_2+\ a_3&=z\\ 3a_1+5a_2+2a_3&=s\\ -a_1-3a_2-2a_3&=t \end{aligned}\right\}\to \left.\begin{aligned} a_1+2a_2+a_3&=x\\ a_2\qquad&=2x+y\\ 2a_2+a_3&=z\\ -a_2-a_3&=s-3x\\ -a_2-a_3&=x+t \end{aligned}\right\}\to$$

$$\left.\begin{aligned} a_1+2a_2+a_3&=x\\ a_2\qquad&=2x+y\\ a_3&=z-4x-2y\\ -a_3&=-x+y+s\\ -a_3&=3x+y+t \end{aligned}\right\}\to \left.\begin{aligned} a_1+2a_2+a_3&=x\\ a_2\qquad&=2x+y\\ a_3&=z-4x-2y\\ 0&=-5x-y+z+s\\ 0&=-x-y+z+t \end{aligned}\right\}$$

Thus $\mathbf{v}\in\text{W}$ iff the above system is consistent if

$$\left.\begin{aligned} 0&=-5x-y+z+s\\ 0&=-\ x-y+z+t \end{aligned}\right\}$$ which is the required homogeneous system of equations.

Ex.4 Find the solution space of the system of equations:

$$\left.\begin{array}{r} 6x_1 - 2x_2 + 5x_3 + x_4 = 0 \\ 3x_1 + x_2 + x_3 - 2x_4 = 0 \\ x_1 + 3x_2 - x_3 + 4x_4 = 0 \end{array}\right\}$$

and give a basis of this solution space.

Let us reduce the system to the echelon form:

$$\left.\begin{array}{r} 6x_1 - 2x_2 + 5x_3 + x_4 = 0 \\ 4x_2 - 3x_3 - 5x_4 = 0 \\ 4x_3 + 48x_4 = 0 \end{array}\right\} \to \left.\begin{array}{r} 6x_1 - 2x_2 + 5x_3 + x_4 = 0 \\ 4x_2 - 3x_3 - 5x_4 = 0 \\ x_3 + 12x_4 = 0 \end{array}\right\}$$

which is in echelon form and has 3 eqns. in 4 unknowns with one free variable x_4. $\therefore$

$x_3 = -12x_4$, $x_2 = -\frac{31}{4}x_4$, $x_1 = \frac{87}{12}x_4$.

$\therefore$ the solution space is the solution set $\{(\frac{87}{12}x_4, -\frac{31}{4}x_4, -12x_4, x_4)\}$

$\therefore$ dimension of the solution space = 4 −3 = 1 and basis $=\{(87, -93, -144, 12)\}$ when $x_4 = 12$.

Ex.5 Let V and W be the following subspaces of $V_4(R)$

V=$\{(a, b, c, d) : b - 2c + d = 0\}$, W = $\{(a, b, c, d) : a = d, b = 2c\}$

Find a basis & the dimension of (i) V, (ii) W.

We seek a basis of the set of solutions (a, b, c, d).

(i) the equation $b - 2c + d = 0$ which is in echelon form and has 1 equation with 4 unknowns.

$\therefore$ dim.V = 4 −1= 3. i.e $0.a + 1.b - 2.c + 1.d = 0$. The free variables are a, c and d. Let (1) $a=1, c=0, d=0$; (2) $a=0, c=1, d=0$; (3) $a=0, c=0, d=1$

to obtain the respective solutions

$$\mathbf{v}_1 = (1, 0, 0, 0), \ \mathbf{v}_2 = (0, 2, 1, 0), \ \mathbf{v}_3 = (0, -1, 0, 1)$$

$\therefore$ the set $\{\mathbf{v}_1, \mathbf{v}_2, \mathbf{v}_3\}$ is a basis & dim. V = 3.

(ii) Here the equations

$$\left.\begin{array}{r} a = d \\ b = 2c \end{array}\right\} \text{ or } \left.\begin{array}{r} a - d = 0 \\ b - 2c = 0 \end{array}\right\}$$

which is in the echelon form with 2 equns. & 4 unknowns. So dim.W=4 −2=2.The free variables are d & c. Giving particular values to d & c, we have particular solution sets when $d = 1, c = 0$; $d = 0$, $c = 1$and these are $\mathbf{w}_1 = \{1, 0, 0, 1\}$ & $\mathbf{w}_2 = \{0, 2, 1, 0\}$. Therefore the set $\{\mathbf{w}_1, \mathbf{w}_2\}$ is a basis & dim.W = 2.

Ex.6 Find the dimension of the vector space spanned by

$$(1, -2, 3, -1) \ \& \ (1, 1, -2, 3).$$

Two nonzero vectors span a space W of dimension 2. If they are independent & of dimension 2 if they are dependent. Recall that two vectors are dependent iff one is a multiple of the other.

Hence the two vectors are independent or dependent if

$$a_1(1, -2, 3, -1) + a_2(1, 1, -2, 3) = 0$$

$$\Rightarrow \left.\begin{array}{rcrl} a_1 & + & a_2 & =0 \\ -2a_1 & + & a_2 & =0 \\ 3a_1 & - & 2a_2 & =0 \\ -a_1 & + & 3a_2 & =0 \end{array}\right\} \to \Rightarrow \left.\begin{array}{rcrl} a_1 & + & a_2 & =0 \\ & & 3a_2 & =0 \\ & - & 5a_2 & =0 \\ & & 3a_2 & =0 \end{array}\right\} \to \left.\begin{array}{r} a_1+a_2=0 \\ a_2=0 \\ 0=0 \\ 0=0 \end{array}\right\}.$$

The reduced equations have zero solution. So the given vectors are independent.

$\therefore$ the dimension of the space spanned by the given vectors is 2.

Ex.7 Let W_1 be the subspace of $V_4(R)$ generated by the set of vectors

$$S_1 = \{(1, 1, 0, -1), (1, 2, 3, 0), (2, 3, 3, -1)\}$$

and W_2 is the subspace of $V_4(R)$ generated by the set of vectors

$$S_2 = \{(1, 2, 2, -2), (2, 3, 2, -3), (1, 3, 4, -3)\} .$$

Find the dimension of

(i) $W_1 + W_2$; (ii) $W_1 \cap W_2$.

(i) We known that $L(W_1 \cup W_2) = W_1 + W_2$, therefore, $W_1 + W_2$ is a subspace generated by the set of vectors

$$S_1 \cup S_2 = \{(1, 1, 0, -1), (1, 2, 3, 0), (2, 3, 3, -1), (1, 2, 2, -2), (2, 3, 2, -3), (1, 3, 4, -3)\}.$$

Hence forming the matrix whose rows are the given six vectors and then reducing the rows to the echelon form as follows:

$$\begin{bmatrix} 1 & 1 & 0 & -1 \\ 1 & 2 & 3 & 0 \\ 2 & 3 & 3 & -1 \\ 1 & 2 & 2 & -2 \\ 2 & 3 & 2 & -3 \\ 1 & 3 & 4 & -3 \end{bmatrix} \xrightarrow[R_4-R_1,\ R_5-2R_1,\ R_6-R_1]{R_2-R_1,\ R_3-2R_1} \begin{bmatrix} 1 & 1 & 0 & -1 \\ 0 & 1 & 3 & 1 \\ 0 & 1 & 3 & 1 \\ 0 & 1 & 2 & -1 \\ 0 & 1 & 2 & -1 \\ 0 & 2 & 4 & -2 \end{bmatrix} \xrightarrow[R_5-R_2,\ R_6-2R_2]{R_3-R_2,\ R_4-R_2}$$

$$\begin{bmatrix} 1 & 1 & 0 & -1 \\ 0 & 1 & 3 & 1 \\ 0 & 0 & 0 & 0 \\ 0 & 0 & -1 & -2 \\ 0 & 0 & -1 & -2 \\ 0 & 0 & -2 & -4 \end{bmatrix} \xrightarrow{R_3 \leftrightarrow R_2} \begin{bmatrix} 1 & 1 & 0 & -1 \\ 0 & 1 & 3 & 1 \\ 0 & 0 & -1 & -2 \\ 0 & 0 & 0 & 0 \\ 0 & 0 & -1 & -2 \\ 0 & 0 & -2 & -4 \end{bmatrix} \xrightarrow[R_6-2R_3]{R_5-R_3} \begin{bmatrix} 1 & 1 & 0 & -1 \\ 0 & 1 & 3 & 1 \\ 0 & 0 & -1 & -2 \\ 0 & 0 & 0 & 0 \\ 0 & 0 & 0 & 0 \\ 0 & 0 & 0 & 0 \end{bmatrix}$$

which is in echelon matrix with 3 nonzero rows.

$\therefore \quad \dim(W_1 + W_2) = 3.$

(ii) First let us find the dimensions of subspaces W_1 & W_2. So, we form the matrix with rows as the generators of W_1 & then reduce the rows to echelon form i.e.

$$\begin{bmatrix} 1 & 1 & 0 & -1 \\ 1 & 2 & 3 & 0 \\ 2 & 3 & 3 & -1 \end{bmatrix} \xrightarrow[R_2-2R_1]{R_2-R_1} \begin{bmatrix} 1 & 1 & 0 & -1 \\ 0 & 1 & 3 & 1 \\ 0 & 1 & 3 & 1 \end{bmatrix} \xrightarrow{R_3-R_2} \begin{bmatrix} 1 & 1 & 0 & -1 \\ 0 & 1 & 3 & 1 \\ 0 & 0 & 0 & 0 \end{bmatrix}.$$

and, similarly, we form the matrix with rows as the generators of W_2, then reduce the rows to echelon form, i.e.

$$\begin{bmatrix} 1 & 2 & 2 & -2 \\ 2 & 3 & 2 & -3 \\ 1 & 3 & 4 & -3 \end{bmatrix} \xrightarrow[R_3 - R_1]{R_2 - 2R_1} \begin{bmatrix} 1 & 2 & 2 & -2 \\ 0 & -1 & -2 & 1 \\ 0 & 1 & 2 & -1 \end{bmatrix} \xrightarrow{R_3 + R_2} \begin{bmatrix} 1 & 2 & 2 & -2 \\ 0 & -1 & -2 & 1 \\ 0 & 0 & 0 & 0 \end{bmatrix}.$$

As both the echelon matrices above have two nonzero rows. So dim. W_1= dim.W_2. But we know that

$$\text{dim.}(W_1 + W_2) = \text{dim.}\, W_1 + \text{dim.}\, W_2 - \text{dim.}(W_1 \cap W_2)$$

or $$3 = 2 + 2 - \text{dim.}(W_1 \cap W_2)$$

$\therefore$ $$\text{dim}(W_1 \cap W_2) = 2 + 2 - 3 = 1.$$

Ex.8 Let U be the space of R^5 generated by

$$S_1 = \{(1, 3, -3, -1, -4), (1, 4, -1, -2, -2), (2, 9, 0, -5, -2)\}$$

and W be the subspace generated by

$$S_2 = \{(1, 6, 2, -2, 3), (2, 8, -1, -6, -5), (1, 3, -1, -5, -6)\}$$

Find a basis and the dimension of (i) $U + W$, (ii) $U \cap W$.

(i) We know that $L(W_1 \cap W_2) = W_1 + W_2$, therefore, $W_1 + W_2$ is a subspace generated by the set of vectors

$$S_1 \cup S_2 = \{(1, 3, -3, -1, -4), (1, 4, -1, -2, -2), (2, 9, 0, -5, -2), (1, 6, 2, -2, 3), (2, 8, -1, -6, -5), (1, 3, -1, -5, -6)\}.$$

Hence forming the matrix whose rows are the given six vectors and then reducing the rows to echelon form i.e.

$$\begin{bmatrix} 1 & 3 & -3 & -1 & -4 \\ 1 & 4 & -1 & -2 & -2 \\ 2 & 9 & 0 & -5 & -2 \\ 1 & 6 & 2 & -2 & 3 \\ 2 & 8 & -1 & -6 & -5 \\ 1 & 3 & -1 & -5 & -6 \end{bmatrix} \xrightarrow[R_4 - R_1,\; R_5 - 2R_1,\; R_6 - R_1]{R_2 - R_1,\; R_3 - 2R_1} \begin{bmatrix} 1 & 3 & -3 & -1 & -4 \\ 0 & 1 & 2 & -1 & 2 \\ 0 & 3 & 6 & -3 & 6 \\ 0 & 3 & 5 & -1 & 7 \\ 0 & 2 & 5 & -4 & 3 \\ 0 & 0 & 2 & -4 & -2 \end{bmatrix} \xrightarrow[R_5 - 2R_2]{R_3 - 3R_2,\; R_4 - 3R_2}$$

$$\begin{bmatrix} 1 & 3 & -3 & -1 & -4 \\ 0 & 1 & 2 & -1 & 2 \\ 0 & 0 & 0 & 0 & 0 \\ 0 & 0 & -1 & 2 & 1 \\ 0 & 0 & 1 & -2 & -1 \\ 0 & 0 & 2 & -4 & -2 \end{bmatrix} \xrightarrow{R_4 \leftrightarrow R_3} \begin{bmatrix} 1 & 3 & -3 & -1 & -4 \\ 0 & 1 & 2 & -1 & 2 \\ 0 & 0 & -1 & 2 & 1 \\ 0 & 0 & 0 & 0 & 0 \\ 0 & 0 & 1 & -2 & -1 \\ 0 & 0 & 2 & -4 & -2 \end{bmatrix} \xrightarrow[R_6 + 2R_3]{R_5 + R_3}$$

$$\begin{bmatrix} 1 & 3 & -3 & -1 & -4 \\ 0 & 1 & 2 & -1 & 2 \\ 0 & 0 & -1 & 2 & 1 \\ 0 & 0 & 0 & 0 & 0 \\ 0 & 0 & 0 & 0 & 0 \\ 0 & 0 & 0 & 0 & 0 \end{bmatrix}$$

which is an echelon matrix with 3 nonzero rows.
Hence $\dim(U+W) = 3$ and
the basis is $\{(1, 3, -3, -1, -4), (0, 1, 2, -1, 2)(0, 0, -1, 2, 1)\}$

(ii) First find homogeneous systems whose solution sets are U and W respectively. Form the matrix whose 4 first rows are the generators of U and whose last row is (x, y, z, s, t) and then the row reduced to the echelon form is given by

$$\begin{bmatrix} 1 & 3 & -3 & -1 & -4 \\ 1 & 4 & -1 & -2 & -2 \\ 2 & 9 & 0 & -5 & -2 \\ x & y & z & s & t \end{bmatrix} \xrightarrow[\substack{R_3 - 2R_1 \\ R_4 - xR_1}]{R_2 - R_1} \begin{bmatrix} 1 & 3 & -3 & -1 & -4 \\ 0 & 1 & 2 & -1 & 2 \\ 0 & 3 & 6 & -3 & 6 \\ 0 & y-3x & z+3x & s+x & t+4x \end{bmatrix}$$

$$\xrightarrow[R_4 - (y-3x)R_3]{R_3 - 3R_2} \begin{bmatrix} 1 & 3 & -3 & -1 & -4 \\ 0 & 1 & 2 & -1 & 2 \\ 0 & 0 & 0 & 0 & 0 \\ 0 & 0 & \begin{Bmatrix} z+3x- \\ 2(y-3x) \end{Bmatrix} & \begin{Bmatrix} s+x+ \\ (y-3x) \end{Bmatrix} & \begin{Bmatrix} t+4x- \\ 2(y-3x) \end{Bmatrix} \end{bmatrix} \xrightarrow{R_4 \leftrightarrow R_3}$$

$$\begin{bmatrix} 1 & 3 & -3 & -1 & -4 \\ 0 & 2 & 2 & -1 & 2 \\ 0 & 0 & 9x-2y+z & -2x+y+s & 10x-2y+t \\ 0 & 0 & 0 & 0 & 0 \end{bmatrix}$$

Set the entries of the 3rd row equal to 0 to obtain the homogeneous system whose solution set is

$$U: 9x-2y+z=0,\ 2x-y-s=0,\ 10x-2y+t=0.$$

Now form the matrix whose first rows are the generators of W and whose last row is (x, y, z, s, t) and then reduce to echelon form :

$$\begin{bmatrix} 1 & 6 & 2 & -2 & 3 \\ 2 & 8 & -1 & -6 & -5 \\ 1 & 3 & -1 & -5 & -6 \\ x & y & z & s & t \end{bmatrix} \xrightarrow[\substack{R_3 - R_1 \\ R_4 - xR_1}]{R_2 - R_1} \begin{bmatrix} 1 & 6 & 2 & -2 & 3 \\ 0 & -4 & -5 & -2 & -11 \\ 0 & -3 & -3 & -3 & -9 \\ 0 & y-6x & z-2x & s+2x & t-3x \end{bmatrix}$$

$$\xrightarrow[4R_4 + (y-6x)R_2]{4R_3 - 3R_2} \begin{bmatrix} 1 & 6 & 2 & -2 & 3 \\ 0 & -4 & -5 & -2 & -11 \\ 0 & 0 & 3 & -6 & -3 \\ 0 & 0 & \begin{Bmatrix} 4(z-2x)- \\ 5(y-6x) \end{Bmatrix} & \begin{Bmatrix} 4(s+2x)- \\ 2(y-6x) \end{Bmatrix} & \begin{Bmatrix} 4(t-3x)- \\ 11(y-6x) \end{Bmatrix} \end{bmatrix} \rightarrow$$

$$\begin{bmatrix} 1 & 6 & 2 & -2 & 3 \\ 0 & -4 & -5 & -2 & -11 \\ 0 & 0 & 3 & -6 & -3 \\ 0 & 0 & (22x-5y+4z) & (20x-2y+4s) & (54x-11y+4t) \end{bmatrix} \xrightarrow{\substack{3R_4-(22x\\ -5y+4z)R_3}}$$

$$\begin{bmatrix} 1 & 6 & 2 & -2 & 3 \\ 0 & -4 & -5 & -2 & -11 \\ 0 & 0 & 3 & -6 & -3 \\ 0 & 0 & 0 & \begin{cases}3(20x-2y+4s)\\ +6(22x-5y+4z)\end{cases} & \begin{cases}3(54x-11y+4t)\\ +3(22x-5y+4z)\end{cases} \end{bmatrix} \rightarrow$$

$$\begin{bmatrix} 1 & 6 & 2 & -2 & 3 \\ 0 & -4 & -5 & -2 & -11 \\ 0 & 0 & 3 & -6 & -3 \\ 0 & 0 & 0 & \begin{cases}(192x-36y+\\ 12s+24z)\end{cases} & \begin{cases}(218x-48y+\\ 12z+12t)\end{cases} \end{bmatrix}.$$

Set the entries of the fourth row equal to 0 to obtain the homogeneous system whose solution set is W, namely,

$$12(16x - 3y + 2z + s) = 0 \quad \text{and} \quad 218x - 48y + 12z + 12t = 0.$$

Combining both systems, we obtain below the homogeneous system whose solution set in $U \cap W$:

$$\left.\begin{aligned} 9x - 2y + z &= 0 \\ 2x - y - s &= 0 \\ 10x - 2y + t &= 0 \\ 6x + 3y - 2z - s &= 0 \\ 6x - 4y + z + t &= 0 \end{aligned}\right\} \rightarrow \left.\begin{aligned} 9x - 2y + z &= 0 \\ -5y - 2z - 9s &= 0 \\ 2y - 10z + 9t &= 0 \\ 13y - 8z - 3s &= 0 \\ -8y + z + 3t &= 0 \end{aligned}\right\} \rightarrow$$

$$\left.\begin{aligned} 9x - 2y + z &= 0 \\ 5y + 2z + 9s &= 0 \\ -54z - 18s + 45t &= 0 \\ -66z - 132s &= 0 \\ 21z + 72s + 15t &= 0 \end{aligned}\right\} \rightarrow$$

$$\left.\begin{aligned} 9x - 2y + z &= 0 \\ 5y + 2z + 9s &= 0 \\ 6z + 2s - 5t &= 0 \\ 2s + t &= 0 \\ 7z + 24s + 5t &= 0 \end{aligned}\right\} \rightarrow$$

$$\left.\begin{array}{rrrrrl} 9x- & 2y+ & z & & & =0\\ & 5y+ & 2z+ & 9s & & =0\\ & & 6z+ & 2s+ & 5t & =0\\ & & & 10s+ & 5t & =0\\ & & & 130s+ & 65t & =0 \end{array}\right\} \to$$

$$\left.\begin{array}{rrrrrl} 9x- & 2y+ & z & & & =0\\ & 5y+ & 2z+ & 9s & & =0\\ & & 6z+ & 2s+ & 5t & =0\\ & & & 2s+ & t & =0\\ & & & & 0 & =0 \end{array}\right\}$$

There is one free variable which is t. Hence dim. ($U \cap W$) = 5 – 4= 1. Setting $t = -90$, we obtain $x=7$, $y = 27$, $z = -90$, $s= 45$, $t = -90$. Thus {(7, 27, –90, 45, –90)} is a basis of $U \cap W$.

11.13 Vector space associated with matrix

This section deals with certain vector spaces associated with matrices — a simple procedure for finding bases. The procedure is nothing but the familiar row echelon form of a matrix. As the rows of a matrix A constitutes the row vectors and the columns, the column vectors, the subspace of a vector space spanned by row vectors is called the *row space* of A and is denoted by row (A) while that spanned by column vectors is called the *column space* of A and is denoted by col(A).

Thus if A= $[a_{ij}]_{m \times n}$ over V(F) and if $A_{i\cdot}$ and $A_{\cdot j}$ denote the rows and the columns respectively then by definition

$$\text{row (A)} = L\,(A_{1\cdot}, A_{2\cdot}, \cdots, A_{m\cdot})$$

and

$$\text{col (A)} = L\,(A_{\cdot 1}, A_{\cdot 2}, \cdots, A_{\cdot n}).$$

Therefore, $V_n(F)$ is the totality of $m \times 1$ or $n \times 1$ matrices over F.

These vector subspaces may differ vastly but they have a common property. It is the common dimension called rank (A) supported by theorem. The following theorem will help to find bases for a vector space.

Theorem 1. Elementary row (column) operations keep the row (column) space invariant. In other words, if A is row (column) equivalent to B then

row (A) = row (B) [or col(A) = col(B)].

Proof. (Row case only) Let the row vectors of a matrix A be $A_{1\cdot}, A_{2\cdot}, \cdots, A_{m\cdot}$ and let the row equivalent matrix of A be B. It is to be shown that every vector in the row space of B is also in the row space of A and vice-versa.

Case 1. Let interchange of rows be the row operation, then B and A have the same row vectors and consequently the same row space. Because no increase or decrease of the number of rows results by the interchange of rows.

Case 2 and case 3. If either a row is multiplied by a scalar or addition of a multiple of one row to another, then the new rows $B_{1\cdot}, B_{2\cdot}, \cdots, B_{m\cdot}$ of B are nothing but linear combinations of $A_{1\cdot}, A_{2\cdot}, \cdots, A_{m\cdot}$, thus they lie in the row space of A. Since a vector space is closed under addition and scalar multiplication, all linear combinations $B_{1\cdot}, B_{2\cdot}, \cdots, B_{m\cdot}$ lie in the row space of A. Therefore, each vector in the row space of B is in the row space of A.

Conversely, as B is obtained by elementary row operations from A, so A can be obtained from B by the inverse operations which have already performed on A. Arguing as above it can be shown that row space of A is contained in the row space of B.

Hence row space remains invariant under elementary row operations.

Similarly, either column operations or row operations on A^T keeps column space invariant.

Cor. The nonzero row vectors in a row echelon form of a matrix A (or A^T) form a *basis* for the row (A) [or col (A)].

Theorem 2. The dimension of the row (A) and that of col(A) is invariant.

Proof. Let $A = [a_{ij}]_{m \times n}$ and let its rows be $L(A_{.1}, A_{.2}, \cdots, A_{.n})$.

Let dim.(row (A)) = k and $S = \{\mathbf{b}_1, \mathbf{b}_2, \cdots, \mathbf{b}_m\}$ be a basis where $\mathbf{b}_i = (b_{i1}, b_{i2}, \cdots, b_{in})$. Since S is the basis for the row space, each row is expressible as a linear combination of $\mathbf{b}_1, \mathbf{b}_2, \cdots, \mathbf{b}_k$, thus

$$\left.\begin{aligned} A_{1\cdot} &= c_{11}\mathbf{b}_1 + c_{12}\mathbf{b}_2 + \cdots + c_{1k}\mathbf{b}_k \\ A_{2\cdot} &= c_{21}\mathbf{b}_1 + c_{22}\mathbf{b}_2 + \cdots + c_{2k}\mathbf{b}_k \\ &\vdots \\ A_{m\cdot} &= c_{m1}\mathbf{b}_1 + c_{m2}\mathbf{b}_2 + \cdots + c_{mk}\mathbf{b}_k \end{aligned}\right\} \quad (11.29)$$

By the definition of equality of two vectors, let us equate the jth component of both sides of (11.29), we get,

$$\left.\begin{aligned} a_{1j} &= c_{11}b_{1j} + c_{12}b_{2j} + \cdots + c_{1k}b_{kj} \\ a_{2j} &= c_{21}b_{1j} + c_{22}b_{2j} + \cdots + c_{2k}b_{kj} \\ &\vdots \\ a_{mj} &= c_{m1}b_{1j} + c_{m2}b_{2j} + \cdots + c_{mk}b_{kj} \end{aligned}\right\}$$

or equivalently

$$\begin{bmatrix} a_{1i} \\ a_{2i} \\ \vdots \\ a_{mi} \end{bmatrix} = b_{1i}\begin{bmatrix} c_{11} \\ c_{21} \\ \vdots \\ c_{m1} \end{bmatrix} + b_{2i}\begin{bmatrix} c_{12} \\ c_{22} \\ \vdots \\ c_{m2} \end{bmatrix} + \cdots + b_{ri}\begin{bmatrix} c_{1r} \\ c_{2r} \\ \vdots \\ c_{mr} \end{bmatrix} \quad (11.30)$$

The L.H.S of (11.30) is the jth column of A. So each column vector of A lies in the vector space spanned by k vectors in the R.H.S. of (11.30).

Therefore, dim.(col(A)) $\le k$. (11.31)

Since, k = dim.(row (A)) (11.32)

$\therefore$ from (11.31) and (11.32),

$$\text{dim.(col(A))} \le \text{dim.(row (A))} \quad (11.33)$$

i.e.

$$\text{dim.(col}\,(A^T)) \le \text{dim.(row}\,(A^T)) \quad (11.34)$$

$\because$ (A^T) transposes columns into rows of A.

Therefore, col (A^T) = row (A) and row(A^T) = col(A).

Then (11.34) can be written as

$$\text{dim.(row (A))} \leq \text{dim.(col(A))} \tag{11.35}$$

The inequalities (11.33) and (11.35) can be true only when

$$\text{dim.(row (A))} = \text{dim.(col(A))}$$

Hence the proof.

11.14 Rank of a matrix revisited

Rank of a matrix is discussed here in terms of dimension of space. Dimensions of the row space and the column space of an arbitrary matrix A are respectively called the *row rank* and the *column rank* of the matrix A. The row rank and the column rank of a matrix are equal and thus the rank of a matrix A, written rank (A) is either the row rank or the column rank of the matrix A.

Thus the rank of a matrix is the maximum number of linearly independent rows or columns. In other words, dim. (row (A)) = dim. (col(A)) = rank A.

Theorem 1. The row rank and the column rank of any matrix are equal.

Or dimensions of row space and column space of any matrix are equal.

This theorem is same as the theorem in sec.10.4.1 with different approaches of proofs.

Proof. Let A be an arbitrary $m \times n$ matrix over a field F.

Let
$$A = \begin{bmatrix} a_{11} & a_{12} & \cdots & \cdots & a_{1n} \\ a_{21} & a_{22} & \cdots & \cdots & a_{2n} \\ \vdots & \vdots & \vdots & \vdots & \vdots \\ a_{m1} & a_{m2} & \cdots & \cdots & a_{mn} \end{bmatrix}.$$

Let $R_{1.}, R_{2.}, \cdots, R_{m.}$ denote its rows :

$$R_{1.} = (a_{11}, a_{12}, \cdots, a_{1n}), \cdots, R_{m.} = (a_{m1}, a_{m2}, \cdots, a_{mn})$$

Let the row rank is r and the following r vectors form a basis for row space:

$$\begin{aligned} S_1 &= (b_{11}, b_{12}, \cdots, b_{1n}), \\ S_2 &= (b_{21}, b_{22}, \cdots, b_{2n}) \\ &\vdots \\ S_r &= (b_{r1}, b_{r2}, \cdots, b_{rn}). \end{aligned}$$

Then each of the row vectors is a L.C, of S_i:

$$\begin{aligned} R_1 &= k_{11}S_1 + k_{12}S_2 + \cdots + k_{1r}S_r \\ R_2 &= k_{21}S_1 + k_{22}S_2 + \cdots + k_{2r}S_r \\ &\vdots \\ R_m &= k_{m1}S_1 + k_{m2}S_2 + \cdots + k_{mr}S_r \end{aligned}$$

where the k_{ij} are scalars. Setting the ith components of each of the above vector equations equal to each other, we obtain the following system of equations, each valid for $i = 1, 2, \cdots, n$:

$$\begin{aligned} a_{1i} &= k_{11}b_{1i} + k_{12}b_{2i} + \cdots + k_{1r}b_{ri} \\ a_{2i} &= k_{21}b_{1i} + k_{22}b_{2i} + \cdots + k_{2r}b_{ri} \\ &\vdots \\ a_{mi} &= k_{m1}b_{1i} + k_{m2}b_{2i} + \cdots + k_{mr}b_{ri} \end{aligned}$$

Thus for i=1, 2, $\cdots$, n

$$\begin{bmatrix} a_{1i} \\ a_{2i} \\ \vdots \\ a_{mi} \end{bmatrix} = b_{1i}\begin{bmatrix} k_{11} \\ k_{21} \\ \vdots \\ k_{m1} \end{bmatrix} + b_{2i}\begin{bmatrix} k_{12} \\ k_{22} \\ \vdots \\ k_{m2} \end{bmatrix} + \cdots + b_{ri}\begin{bmatrix} k_{1r} \\ k_{2r} \\ \vdots \\ k_{mr} \end{bmatrix}$$

In other words, each of the columns of A is a linear combination of the vectors

$$\begin{bmatrix} k_{11} \\ k_{21} \\ \vdots \\ k_{m1} \end{bmatrix}, \begin{bmatrix} k_{12} \\ k_{22} \\ \vdots \\ k_{m2} \end{bmatrix}, \ldots, \begin{bmatrix} k_{1r} \\ k_{2r} \\ \vdots \\ k_{mr} \end{bmatrix}.$$

Thus the column space of A has dimension at most r

i.e. column rank $\leq r$.

Hence, column rank $\leq$ row rank.

Similarly or considering the transpose matrix A^T we obtain row rank $\leq$ column rank. Thus the row rank and the column rank are equal. In other words dimensions of row space and column space are equal.

11.14.1. Rank inequalites for sum and product of matrices : These inequalities of rank have already established in chapter 10 are again established using dimension.

Theorem 1. If A and B are both $m \times n$, then rank $(A + B) \leq$ rank A + rank B.

Proof. Let the solution spaces of $A\mathbf{x} = 0$ and $B\mathbf{x} = 0$ be S_1 and S_2 which are each a subspace of dimension (n – rank A) and (n – rank B) (cf. Theorem 1 in sec.11.14.3 next) respectively

i.e. $\dim.S_1 = n - \text{rank A}$ and $\dim.S_2 = n - \text{rank B}$.

But $\dim.(S_1 + S_2) + \dim.(S_1 \cap S_2) = \dim.S_1 + \dim.S_2$ (11.36)

Also $\dim.(S_1 + S_2) \leq n$, then (11.36) gives

$$\dim.(S_1 \cap S_2) \geq \dim.S_1 + \dim.S_2 - n$$
$$= (n - \text{rank A}) + (n - \text{rank B}) - n$$
$$= n - \text{rank A} - \text{rank B}$$

i.e. $\dim.(S_1 \cap S_2) \geq n - \text{rank A} - \text{rank B}$ (11.37)

Again $S_1 \cap S_2$ contains all vectors $\mathbf{x}$ which satisfy both $A\mathbf{x} = 0$ and $B\mathbf{x} = 0$. Such vectors also satisfy $(A + B)\mathbf{x} = 0$. The dimension of the solution space of

$$(A + B)\mathbf{x} = 0 \text{ is } n - \text{rank } (A + B).$$

Therefore, by theorem 2 in sec.11.12.2 and eqn.(11.37) above,

$$n - \text{rank}(A + B) \geq \dim.(S_1 \cap S_2) \geq n - \text{rankA} - \text{rankB}$$

or rank $(A + B) \leq$ rank A + rank B.

Theorem 2. Rank(AB) $\leq$ min.(rank A, rank B).

Proof. Since in matrix multiplication, column vectors of AB are linear combinations of the column vectors of A. Then the column space of AB is the column space of A. That is,

$$\text{rank (AB)} \leq \text{rank A}. \quad (11.38)$$

Again row vectors of AB are linear combinations of the row vectors of B. Hence the row space of AB is the row space of B. That is,

$$\text{rank (AB)} \leq \text{rank B}. \quad (11.39)$$

Since row rank = column rank, therefore, from (11.38) & (11.39).

$$\text{Rank}(AB) \leq \text{min.}(\text{rank } A, \text{rank } B).$$

11.14.2 Rank for inversion of rectangular matrices : Rank may be used for determining left and right inverses of a rectangular matrix. Let A be an $m \times n$ matrix.

Then

(i) A has a left inverse iff rank A = n = number of columns of A.

(ii) A has a right inverse iff rank A = m = number of rows of A.

Proof. (i) Let B be a left inverse of A, then BA = I_n. Evidently, rows of I_n lie in row space of A and are linearly independent. Then row rank A, $r = n$ at least, i.e. $r \geq n$. The column rank is also r. So, A must have r linearly independent columns. Since the matrix has n columns only, so $r \leq n$. Hence $r = n$. This proves the necessary condition that if A has a left inverse then rank A = n.

Conversely, let rank A = n. Then A has n linearly independent rows $A_{1.}, A_{2.}, \cdots, A_{n.}$ belong to the vector space defined by A. Therefore, s.u. vectors $\mathbf{e}_i$ are linear combinations of $A_{i.}$

i.e. $\mathbf{e}_i = b_{i1}A_{1.} + b_{i2}A_{2.} + \cdots + b_{in}A_{n.}, \; i = 1, 2, \cdots, n$

for suitable scalars b_{ij}. Hence the corresponding matrix equation is I_n = BA where B is $n \times m$ as A is $m \times n$ i.e. B is the left inverse of A.

Proof of (ii). This may be established as (i). A simpler procedure is taking transpose as rank (A) = rank (A^T) = m. So $B^T A^T = I_m$ by (i). Taking transpose, AB = I_m. Hence A has a right inverse if rank (A) = m.

Converse is the reverse of the above steps as (i). Hence second statement is established.

11.14.3 Dimension of solution space : The following theorem provides a formula for the dimension of solution space.

Theorem.1 The dimension of the solution space W of the homogeneous system of linear equns. $A\mathbf{x} = 0$ is $n - r$ where n is the number of unknowns and r is the rank of the coefficient matrix A.

OR. The number of linearly independent solutions of the equation $A\mathbf{x} = 0$, is $n - r$, where n is the number of unknowns and r is rank of $m \times n$ matrix A.

OR. If s denotes dim.(solution space of $A\mathbf{x} = 0$), c denotes dim.(col(A)) and n is the usual number of columns of an $m \times n$ matrix A, then $s = n - c$ i.e. $s + c = n$.

Proof. Let the system $A\mathbf{x} = 0$ be reduced to the echelon form and it has precisely $(n - r)$ free variables. Let these free variables be $x_{i_1}, x_{i_2}, \cdots, x_{i_{n-r}}$. Let $\mathbf{v}_j$ be the solution obtained by setting $x_{i_j} = 1$ and all other free variables equal to zero.

Let us show that these solutions $\mathbf{v}_1, \mathbf{v}_2, \cdots, \mathbf{v}_{n-r}$ are linearly independent and form a basis for the solution space. Let A be the matrix whose rows are $\mathbf{v}_j$. We interchange column 1 and column i_1, then column 2 and column i_2 and then column r & column i_r. and obtain the $r \times n$ matrix

$$B = (I, C) = \begin{bmatrix} 1 & 0 & \cdots & 0 & c_{1,r+1} & \cdots & c_{1n} \\ 0 & 1 & \cdots & 0 & c_{2,r+2} & \cdots & c_{2n} \\ \vdots & \vdots & \ddots & \vdots & \vdots & \vdots & \vdots \\ 0 & 0 & 0 & 1 & c_{r,r+1} & \cdots & c_{rn} \end{bmatrix}.$$

The above matrix B is in echelon form & so its rows are independent. Hence rank (B) = r. Since A and B are column equivalent, they have the same rank i.e. rank (A) = r. But A has r rows, hence these rows i.e. $\mathbf{v}_j$ are linearly independent and so form the basis of the solution space. Therefore, the dimension of the solution space is $(n - r)$.

i.e. $s = n - c$

or $s + c = n$ on replacing r by c, dim.(col(A)).

An alter proof. The matrix A with rank r has r L.I. columns without any loss of generality, we suppose that it is the first r columns to the left which are L.D. for it only amounts to renaming the components of **x**.

Let $$A = \left[\mathbf{C}_{\cdot 1}\ \mathbf{C}_{\cdot 2} \cdots \mathbf{C}_{\cdot r}\ \ \mathbf{C}_{\cdot (r+1)} \cdots \mathbf{C}_{\cdot n}\right].$$

So, the given equation is equivalent to

$$\left[\mathbf{C}_{\cdot 1}\ \mathbf{C}_{\cdot 2} \cdots \mathbf{C}_{\cdot r}\ \ \mathbf{C}_{\cdot (r+1)} \cdots \mathbf{C}_{\cdot n}\right] \begin{bmatrix} x_1 \\ x_2 \\ \vdots \\ x_r \\ x_{r+1} \\ \vdots \\ x_n \end{bmatrix} = 0$$

$$\text{i.e } x_1\mathbf{C}_{\cdot 1} + x_2\mathbf{C}_{\cdot 2} + \cdots + x_r\mathbf{C}_{\cdot r} + x_{r+1}\mathbf{C}_{\cdot (r+1)} + \cdots + x_n\ \mathbf{C}_{\cdot n} = 0 \tag{11.40}$$

As each of $\mathbf{C}_{\cdot (r+1)}, \cdots, \mathbf{C}_{\cdot n}$ is a L.C. of $\mathbf{C}_{\cdot 1}, \mathbf{C}_{\cdot 2}, \cdots, \mathbf{C}_{\cdot r}$, we have relations of the form,

$$\left.\begin{aligned} \mathbf{C}_{\cdot (r+1)} &= k_{11}\mathbf{C}_{\cdot 1} + k_{12}\mathbf{C}_{\cdot 2} + \cdots + k_{1r}\mathbf{C}_{\cdot r} \\ \mathbf{C}_{\cdot (r+2)} &= k_{21}\mathbf{C}_{\cdot 1} + k_{22}\mathbf{C}_{\cdot 2} + \cdots + k_{2r}\mathbf{C}_{\cdot r} \\ \vdots\quad &\quad \vdots \qquad\quad \vdots \qquad\qquad \vdots \\ \mathbf{C}_{\cdot n} &= k_{p1}\mathbf{C}_{\cdot 1} + k_{p2}\mathbf{C}_{\cdot 2} + \cdots + k_{pr}\mathbf{C}_{\cdot r} \end{aligned}\right\} \tag{11.41}$$

where $p = n - r$.

Comparing (11.40) and (11.41) we see that

$$\mathbf{x}_1 = \begin{bmatrix} k_{11} \\ k_{21} \\ \vdots \\ k_{1r} \\ -1 \\ 0 \\ \vdots \\ 0 \end{bmatrix}, \mathbf{x}_2 = \begin{bmatrix} k_{12} \\ k_{22} \\ \vdots \\ k_{2r} \\ 0 \\ -1 \\ \vdots \\ 0 \end{bmatrix}, \cdots, \mathbf{x}_{ri-r} = \begin{bmatrix} k_{p1} \\ k_{p2} \\ \vdots \\ k_{pr} \\ 0 \\ 0 \\ \vdots \\ -1 \end{bmatrix}$$

are $(n - r)$ solutions of the given equation.

We show that these $(n-r)$ solutions form a linearly independent system.

Let $$k_1\mathbf{x}_1 + k_2\mathbf{x}_2 + \cdots + k_{n-r}\mathbf{x}_{n-r} = 0.$$

Considering $(r+1)$th, $(r+2)$ th, $\cdots$ nth components of the vector on th left, we may see that
$$k_1 = 0, \; \cdots, \; k_{n-r} = 0.$$

It follows that $\mathbf{x}_1, \mathbf{x}_2, \cdots, \mathbf{x}_{n-r}$ form $(n-r)$ L. I. set of solutions.

11.14.4 Consistency of Ax = b: The following theorem provides with a condition about the consistent of the system of equations **Ax** = **b**.

Theorem. A system of linear equations **Ax** = **b** is consistent if **b** is in the column space of A.

Proof. Let the explicit form of the matrix equation **Ax** = **b** be

$$\begin{bmatrix} a_{11} & a_{12} & \cdots & \cdots & \cdots & \cdots & a_{1n} \\ a_{21} & a_{22} & \cdots & \cdots & \cdots & \cdots & a_{2n} \\ \vdots & \vdots & \vdots & \vdots & \vdots & \vdots & \vdots \\ a_{m1} & a_{m2} & \cdots & \cdots & \cdots & \cdots & a_{mn} \end{bmatrix} \begin{bmatrix} x_1 \\ x_2 \\ \vdots \\ x_n \end{bmatrix} = \begin{bmatrix} b_1 \\ b_2 \\ \vdots \\ b_m \end{bmatrix}$$

By matrix multiplication

$$\begin{bmatrix} a_{11}x_1 + & a_{12}x_2 + & \cdots & \cdots & \cdots & + & a_{1n}x_n \\ a_{21}x_1 + & a_{22}x_2 + & \cdots & \cdots & \cdots & + & a_{2n}x_n \\ \vdots & \vdots & \vdots & \vdots & \vdots & \vdots & \vdots \\ a_{m1}x_1 + & a_{m2}x_2 + & \cdots & \cdots & \cdots & + & a_{mn}x_n \end{bmatrix} = \begin{bmatrix} b_1 \\ b_2 \\ \vdots \\ b_m \end{bmatrix}$$

By converse of matrix addition and scalar multiplication, this can be written as

$$x_1\begin{bmatrix} a_{11} \\ a_{21} \\ \vdots \\ a_{m1} \end{bmatrix} + x_2\begin{bmatrix} a_{12} \\ a_{22} \\ \vdots \\ a_{m2} \end{bmatrix} + \cdots + x_n\begin{bmatrix} a_{1n} \\ a_{2n} \\ \vdots \\ a_{mn} \end{bmatrix} = \begin{bmatrix} b_1 \\ b_2 \\ \vdots \\ b_m \end{bmatrix}$$

Since L.H.S of this equation is a linear combination of the column vectors of A and is equal to **b** then it follows that the system **Ax** = **b** is consistent if **b** is the column space of A.

Ex.1 Find the rank of the matrix

$$A = \begin{bmatrix} 1 & 2 & -1 \\ -1 & 1 & 1 \\ 0 & 3 & 0 \end{bmatrix} \text{ by any method.}$$

$$A = \begin{bmatrix} 1 & 2 & -1 \\ -1 & 1 & 1 \\ 0 & 3 & 0 \end{bmatrix} \xrightarrow{R_2 + R_1} \begin{bmatrix} 1 & 2 & -1 \\ 0 & 3 & 0 \\ 0 & 3 & 0 \end{bmatrix} \xrightarrow{R_3 - R_2} \begin{bmatrix} 1 & 2 & -1 \\ 0 & 3 & 0 \\ 0 & 0 & 0 \end{bmatrix}$$

$\because$ the row space consists of 2 linearly independent rows. Therefore, dim. of the row space = 2 i.e. rank (A) = 2.

Ex.2 Determine the rank of the matrix

$$A = \begin{bmatrix} 1 & -1 & 2 & 0 \\ 1 & 1 & 1 & 1 \\ -1 & -3 & 0 & -2 \end{bmatrix}.$$

$$A = \begin{bmatrix} 1 & -1 & 2 & 0 \\ 1 & 1 & 1 & 1 \\ -1 & -3 & 0 & -2 \end{bmatrix} \xrightarrow[R_3+R_1]{R_2-R_1} \begin{bmatrix} 1 & -1 & 2 & 0 \\ 0 & 2 & -1 & 1 \\ 0 & -4 & 2 & -2 \end{bmatrix} \xrightarrow{R_3+2R_2}$$

$$\begin{bmatrix} 1 & -1 & 2 & 0 \\ 0 & 2 & -1 & 1 \\ 0 & 0 & 0 & 0 \end{bmatrix}$$

∵ the row space consists of 2 linearly independent rows. Therefore, dim. of the row space = 2 i.e. rank (A) = 2.

Ex.3 Determine which of the following matrices have the same row space:

$$A=\begin{bmatrix} 1 & -2 & -1 \\ 3 & -4 & 5 \end{bmatrix}, B = \begin{bmatrix} 1 & -1 & 2 \\ 2 & 3 & -1 \end{bmatrix}, C = \begin{bmatrix} 1 & -1 & 3 \\ 2 & -1 & 10 \\ 3 & -5 & 1 \end{bmatrix}.$$

Let us row reduce the matrices to row canonical form.

$$A=\begin{bmatrix} 1 & -2 & -1 \\ 3 & -4 & 5 \end{bmatrix} \xrightarrow{R_2-3R_1} \begin{bmatrix} 1 & -2 & -1 \\ 0 & 2 & 8 \end{bmatrix} \xrightarrow{R_1+R_2} \begin{bmatrix} 1 & 0 & 7 \\ 0 & 2 & 8 \end{bmatrix} \xrightarrow{\frac{1}{2}R_2} \begin{bmatrix} 1 & 0 & 7 \\ 0 & 1 & 4 \end{bmatrix}.$$

$$B = \begin{bmatrix} 1 & -1 & 2 \\ 2 & 3 & -1 \end{bmatrix} \xrightarrow{R_2-2R_1} \begin{bmatrix} 1 & -1 & 2 \\ 0 & 5 & -5 \end{bmatrix} \xrightarrow{5R_1+5R_2} \begin{bmatrix} 5 & 0 & 10 \\ 0 & 5 & -5 \end{bmatrix}$$

$$\xrightarrow[\frac{1}{5}R_2]{\frac{1}{5}R_1} \begin{bmatrix} 1 & 0 & 2 \\ 0 & 1 & -1 \end{bmatrix}.$$

$$C = \begin{bmatrix} 1 & -1 & 3 \\ 2 & -1 & 10 \\ 3 & -5 & 1 \end{bmatrix} \xrightarrow[R_3-3R_1]{R_2-R_1} \begin{bmatrix} 1 & -1 & 3 \\ 0 & 1 & 4 \\ 0 & -2 & -8 \end{bmatrix} \xrightarrow{R_3+2R_2} \begin{bmatrix} 1 & -1 & 3 \\ 0 & 1 & 4 \\ 0 & 0 & 0 \end{bmatrix}.$$

$$\xrightarrow{R_1+R_2} \begin{bmatrix} 1 & 0 & 7 \\ 0 & 1 & 4 \\ 0 & 0 & 0 \end{bmatrix}$$

∵ the nonzero rows of the reduced form of A & that of C are same, A & C have the same row space whereas B has a different row space.

Ex.4 Determine whether the following matrices have the same column space.

$$A = \begin{bmatrix} 1 & 3 & 5 \\ 1 & 4 & 3 \\ 1 & 1 & 9 \end{bmatrix}, B = \begin{bmatrix} 1 & 2 & 3 \\ -2 & -3 & -4 \\ 7 & 12 & 17 \end{bmatrix}.$$

Matrices A & B have the same column space if & only if the transpose A^T & B^T have the same row space. Thus let us reduce A^T & B^T to the row canonical form.

$$A^T = \begin{bmatrix} 1 & 1 & 1 \\ 3 & 4 & 1 \\ 5 & 3 & 9 \end{bmatrix} \xrightarrow[R_3-5R_1]{R_2-3R_1} \begin{bmatrix} 1 & 1 & 1 \\ 0 & 1 & -2 \\ 0 & -2 & 4 \end{bmatrix} \xrightarrow{R_3+2R_2} \begin{bmatrix} 1 & 1 & 1 \\ 0 & 1 & -2 \\ 0 & 0 & 0 \end{bmatrix}$$

$$\xrightarrow{R_1-R_2} \begin{bmatrix} 1 & 0 & 3 \\ 0 & 1 & -2 \\ 0 & 0 & 0 \end{bmatrix}.$$

$$B^T = \begin{bmatrix} 1 & -2 & 7 \\ 2 & -3 & 12 \\ 3 & -4 & 17 \end{bmatrix} \xrightarrow[R_3-3R_1]{R_2-2R_1} \begin{bmatrix} 1 & -2 & 7 \\ 0 & 1 & -2 \\ 0 & 2 & -4 \end{bmatrix} \xrightarrow{R_3-2R_1} \begin{bmatrix} 1 & -2 & 7 \\ 0 & 1 & -2 \\ 0 & 0 & 0 \end{bmatrix}$$

$$\xrightarrow{R_1+2R_2} \begin{bmatrix} 1 & 0 & 3 \\ 0 & 1 & -2 \\ 0 & 0 & 0 \end{bmatrix}.$$

$\because$ nonzero rows of A^T & B^T are same. Therefore A^T, B^T are of the same row space i.e. A & B are of the same column space.

Ex.5 Find (i) a basis for the row space, (ii) a basis for the column space, (iii) the rank of the matrix

$$A = \begin{bmatrix} 1 & -3 & 2 & 2 & 1 \\ 0 & 3 & 6 & 0 & -2 \\ 2 & -3 & -2 & 4 & 4 \\ 3 & -3 & 6 & 6 & 3 \\ 5 & -3 & 10 & 10 & 5 \end{bmatrix}.$$

(i) Let us reduce the matrix A to the row reduced echelon form. So,

$$A = \begin{bmatrix} 1 & -3 & 2 & 2 & 1 \\ 0 & 3 & 6 & 0 & -2 \\ 2 & -3 & -2 & 4 & 4 \\ 3 & -3 & 6 & 6 & 3 \\ 5 & -3 & 10 & 10 & 5 \end{bmatrix} \underset{\substack{R_4-3R_1 \\ R_5-5R_1}}{\overset{R_3-2R_1}{\approx}} \begin{bmatrix} 1 & -3 & 2 & 2 & 1 \\ 0 & 3 & 6 & 0 & -2 \\ 0 & 3 & -6 & 0 & 2 \\ 0 & 6 & 0 & 0 & 0 \\ 0 & 12 & 0 & 0 & 0 \end{bmatrix}$$

$$\underset{\approx}{\tfrac{1}{6}R_4 \leftrightarrow R_2} \begin{bmatrix} 1 & -3 & 2 & 2 & 1 \\ 0 & 1 & 0 & 0 & 0 \\ 0 & 3 & -6 & 0 & 2 \\ 0 & 3 & 6 & 0 & -2 \\ 0 & 12 & 0 & 0 & 0 \end{bmatrix} \underset{\substack{R_4-3R_2 \\ R_5-12R_2}}{\overset{R_3-3R_2}{\approx}} \begin{bmatrix} 1 & -3 & 2 & 2 & 1 \\ 0 & 1 & 0 & 0 & 0 \\ 0 & 0 & -6 & 0 & 2 \\ 0 & 0 & 6 & 0 & -2 \\ 0 & 0 & 0 & 0 & 0 \end{bmatrix} \overset{R_4+R_3}{\approx}$$

$$\begin{bmatrix}1 & -3 & 2 & 2 & 1\\0 & 1 & 0 & 0 & 0\\0 & 0 & -6 & 0 & 2\\0 & 0 & 0 & 0 & 0\\0 & 0 & 0 & 0 & 0\end{bmatrix} \underset{\approx}{\tfrac{1}{2}R_3} \begin{bmatrix}1 & -3 & 2 & 2 & 1\\0 & 1 & 0 & 0 & 0\\0 & 0 & -3 & 0 & 1\\0 & 0 & 0 & 0 & 0\\0 & 0 & 0 & 0 & 0\end{bmatrix} \underset{\approx}{R_1+3R_2} \begin{bmatrix}1 & 0 & 2 & 2 & 1\\0 & 1 & 0 & 0 & 0\\0 & 0 & -3 & 0 & 1\\0 & 0 & 0 & 0 & 0\\0 & 0 & 0 & 0 & 0\end{bmatrix}$$

$$\underset{\approx}{R_1-R_3} \begin{bmatrix}1 & 0 & 5 & 2 & 0\\0 & 1 & 0 & 0 & 0\\0 & 0 & -3 & 0 & 1\\0 & 0 & 0 & 0 & 0\\0 & 0 & 0 & 0 & 0\end{bmatrix}.$$

∴ the basis for the row space is

$$\{(1, 0, 5, 2, 0), (0, 1, 0, 0, 0), (0, 0, -3, 0, 1)\}.$$

(ii) Let us reduce the matrix A^T to the row reduced echelon form. So,

$$A^T = \begin{bmatrix}1 & 0 & 2 & 3 & 5\\-3 & 3 & -3 & -3 & -3\\2 & 6 & -2 & 6 & 10\\2 & 0 & 4 & 6 & 10\\1 & -2 & 4 & 3 & 5\end{bmatrix} \begin{matrix}R_2+3R_1\\ \approx\\ R_3-2R_1\\ R_4-2R_1\\ R_5-R_1\end{matrix} \begin{bmatrix}1 & 0 & 2 & 3 & 5\\0 & 3 & 3 & 6 & 12\\0 & 6 & -6 & 0 & 0\\0 & 0 & 0 & 0 & 0\\0 & -2 & 2 & 0 & 0\end{bmatrix} \begin{matrix}\tfrac{1}{3}R_2\\ \approx\\ \tfrac{1}{6}R_3\\ \tfrac{1}{2}R_5\end{matrix}$$

$$\begin{bmatrix}1 & 0 & 2 & 3 & 5\\0 & 1 & 1 & 2 & 4\\0 & 1 & -1 & 0 & 0\\0 & 0 & 0 & 0 & 0\\0 & -1 & 1 & 0 & 0\end{bmatrix} \underset{\approx}{R_2 \leftrightarrow R_3} \begin{bmatrix}1 & 0 & 2 & 3 & 5\\0 & 1 & -1 & 0 & 0\\0 & 1 & 1 & 2 & 4\\0 & 0 & 0 & 0 & 0\\0 & -1 & 1 & 0 & 0\end{bmatrix} \begin{matrix}R_3-R_2\\ \approx\\ R_5+R_2\end{matrix}$$

$$\begin{bmatrix}1 & 0 & 2 & 3 & 5\\0 & 1 & -1 & 0 & 0\\0 & 0 & 2 & 2 & 4\\0 & 0 & 0 & 0 & 0\\0 & 0 & 0 & 0 & 0\end{bmatrix} \underset{\approx}{\tfrac{1}{2}R_3} \begin{bmatrix}1 & 0 & 2 & 3 & 5\\0 & 1 & -1 & 0 & 0\\0 & 0 & 1 & 1 & 2\\0 & 0 & 0 & 0 & 0\\0 & 0 & 0 & 0 & 0\end{bmatrix} \underset{\approx}{R_1-2R_3}$$

$$\begin{bmatrix}1 & 0 & 0 & 1 & 1\\0 & 1 & -1 & 0 & 0\\0 & 0 & 1 & 1 & 2\\0 & 0 & 0 & 0 & 0\\0 & 0 & 0 & 0 & 0\end{bmatrix}.$$

∴ basis for the column space is

$$\left\{\begin{bmatrix}1\\0\\0\\1\\1\end{bmatrix},\begin{bmatrix}0\\1\\-1\\0\\0\end{bmatrix},\begin{bmatrix}0\\0\\1\\1\\2\end{bmatrix}\right\}.$$

N.B. Maximum possible linear independence is preferable.
Note the last two operations in (i) and last operation in (ii).

(iii) Rank A = dim.(row(A)) = dim.(col(A)) =3 from (i) and (ii).

Exercises eleven

1. Let V be the set of all ordered pairs of real numbers. Define addition and scalar multiplication by

$$(a, b) \circledR (c, d) = (a + c, b + d) \text{ and } r \copyright (a, b) = (a, rb)$$

respectively. Show that V is not a vector space.

2. Consider the differential equation $y'' + 5y' + 6y = 0$. A solution is a real valued function f satisfying the equation. Let V be the set of all solutions to the given differential equation. Define addition, (f ® g) and scalar multiplication, (c©f) by

$$(f \circledR g)(t) = f(t) + g(t)$$

$$\text{and} \quad (c \copyright f)(t) = g(cf)(t)$$

respectively. Prove that V is a vector space.

[**Hints :** $(c©f)''(t) + 5(c©f)'(t) + 6(c©f)(t)$
$= c[f''(t) + 5f(t) + 6f(t)] = c.0 = 0.$
∴ (c©f)∈ V. If f and g both ∈ V, then
$(f®g)''(t) + 5(f®g)'(t) + 6(f®g)(t)$
$= [f''(t) + 5f(t) + 6f(t)]® [g(t) + 5g(t) + 6g(t)]$
$= 0®0 = 0.$
∴ (f ® g)∈ V. Therefore, closure properties are satisfied.
If $c = -1$ then additive inverse (–f)∈ V and (f+(–f)) = 0 ∈ V. Similarly, show other axioms and reach the final decision]

3. Let M_2 be the set of all 2×2 matrices with addition and scalar multiplication

$$\begin{bmatrix}a_{11} & a_{12}\\a_{21} & a_{22}\end{bmatrix}+\begin{bmatrix}b_{11} & b_{12}\\b_{21} & b_{22}\end{bmatrix}=\begin{bmatrix}a_{11}+b_{11} & a_{12}+b_{12}\\a_{21}+b_{21} & a_{22}+b_{22}\end{bmatrix} \text{ and } \alpha\begin{bmatrix}a_{11} & a_{12}\\a_{21} & a_{22}\end{bmatrix}=\begin{bmatrix}0 & 0\\0 & 0\end{bmatrix}$$

respectively. Show that M_2 under the defined operations does not constitute a vector space.

[**Hints :** Note that scalar multiplication is not the usual scalar multiplications of matrix theory.
Note that M_2 satisfies all the conditions of sec.11.2 defn. 1 except (c)(4) and $4(m_1)$ of defn.2 of sec. 11.2]

4. Let $S_2 = \{(x_1, x_2) : x_1, x_2$ are real $\}$ i.e. S_2 be the set of all ordered pairs of real numbers with addition $(a_1, a_2) + (b_1, b_2) = (a_1 + b_1, a_2 - b_2)$ and scalar multiplication $\alpha\,(a_1, a_2) = (\alpha a_1, \alpha a_2)$, $\forall$ real α. Show that S_2 does not constitute a vector space.

[(**Hints :** $3(a_1)$, $3(a_2)$, $4(m_3)$ of defn.2 of sec 11.3 do not hold]

5. Let S be the set of all 2×2 matrices of the form

$$\begin{bmatrix} x & 0 \\ 0 & x \end{bmatrix},$$

x is an arbitrary real number. Show that S is a vector space under the standard operations defined on matrices.

6. Consider the following subsets of the 3-tuple space $R^3(R)$:—

(i) $W_2 = \{(0, x_2, x_3) : x_2, x_3 \in R\}$,

(ii) $W_3 = \{(x_1, 0, x_3) : x_1, x_3 \in R\}$,

(iii) $W_4 = \{(x_1, 0, 0) : x_1 \in R\}$

(iv) $W_5 = \{(0, x_2, 0) : x_2 \in R\}$ and

(v) $W_6 = \{(0, 0, x_3) : x_3 \in R\}$.

By testing the validity of the condition required for a subspace of vector space, it is easy to show that the each of the above subsets is a subspace of $R^3(R)$.

7. Prove that xy, yz, zx planes are subspaces of the space $V^3(R)$.

8. Prove that x, y & z axes are subspaces of the space $V^3(R)$.

9. Let V be a vector space and $\mathbf{u}, \mathbf{v} \in V$. Show that there is one and only one $\mathbf{x} \in V$ such that $\mathbf{u} + \mathbf{x} = \mathbf{v}$. Find $\mathbf{x} \in IR^5$ where $\mathbf{u} + \mathbf{x} = \mathbf{v}$ with

$$\mathbf{u} = (2, -1, 0, 3, 6) \text{ and } \mathbf{v} = (0, 1, 2, -1, -2).$$

[**Ans. x** =(–2, 2, 2, –4, –8)]

10. Explain why the following are linearly dependent sets of vectors (solve this problem by inspection) :

(a) $\mathbf{u}_1 = (1, 2)$ and $\mathbf{u}_2 = (-3, -6)$ in R_2.

(b) $\mathbf{u}_1 = (2, 3)$, $\mathbf{u}_2 = (-5, 8)$ and $\mathbf{u}_3 = (6, 1)$ in R_2.

(c) $p_1 = 2 + 3x - x^2$ and $p_2 = 6 + 9x - 3x_2$ in P_2

(d) $A = \begin{bmatrix} 1 & 3 \\ 2 & 0 \end{bmatrix}$ and $B = \begin{bmatrix} -1 & -3 \\ -2 & 0 \end{bmatrix}$ in $M_{2\times2}$.

[**Ans :** (a) $\mathbf{u}_2$ is a scalar multiple of $\mathbf{u}_1$,
(b) Vectors are linearly independent as number of vectors is greater than the number of components in each vector,
(c) p_2 is a scalar multiple of p_1,
(d) B is a scalar multiple of A]

11. Let V be the vector space of all real valued functions defined on the entire real line. Which of the following sets of vectors in V are linearly dependent / independent ?

(a) $2, 4\sin^2 x, \cos^2 x$; (b) $x, \cos x$;
(c) $1, \sin x, \sin 2x$; (d) $\cos 2x, \sin^2 x, \cos^2 x$;
(e) $(1+x)^2 \cdot x^2 + 2x, 3$; (f) $0, x, x^2$.

[**Ans.** (a) dependent ; (b) independent; (c) independent; (d) dependent ; (e) dependent ; (f) dependent]

12. Let $\mathbf{v}_1, \mathbf{v}_2, \mathbf{v}_3$ be vectors in R_3. Each one of them has their initial points at the origin. In each part determine whether the three vectors lie on the same line or not.

(a) $\mathbf{v}_1 = (3, -6, 9)$, $\mathbf{v}_2 = (2, -4, 6)$, $\mathbf{v}_3 = (1, 1, 1)$;
(b) $\mathbf{v}_1 = (2, -1, 4)$, $\mathbf{v}_2 = (4, 2, 3)$, $\mathbf{v}_3 = (2, 7, -6)$;
(c) $\mathbf{v}_1 = (4, 6, 8)$, $\mathbf{v}_2 = (2, 3, 4)$, $\mathbf{v}_3 = (-2, -3, -4)$.

[**Ans.** (a), (b) do not lie on the same line; (c) lie on the same line]

13. If f, g and h are vectors in vector space V of real valued functions defined for all real x and are twice differentiable, then prove that the Wronskian of f, g and h defined by

$$w(x) = \begin{vmatrix} \mathbf{f}(x) & \mathbf{g}(x) & \mathbf{h}(x) \\ \mathbf{f}'(x) & \mathbf{g}'(x) & \mathbf{h}'(x) \\ \mathbf{f}''(x) & \mathbf{g}''(x) & \mathbf{h}''(x) \end{vmatrix}$$

is not zero for the linear independence of those vectors. Hence show that

(a) $1, x, e^x$; (b) $\sin x, \cos x, x\sin x$;
(c) e^x, xe^x, x^2e^x ; (d) $1, x, x^2$ are linearly independent.

[**Hints :** w(x) is not identically zero]

14. Show that the span of $\{\alpha, \beta\}$ is the plane $x + y - z = 0$ where

$$\alpha = \begin{bmatrix} 1 \\ 0 \\ 1 \end{bmatrix}, \beta = \begin{bmatrix} -1 \\ 2 \\ 1 \end{bmatrix}.$$

15. Let W be a subspace of $V^4(R)$ generated by the vectors

$$(1, -2, 5, -3), (2, 3, 1, -4) \ \& \ (3, 8, -3, -5).$$

Find a basis and the dimension of W.

[**Ans** . Basis, S = {(1, –2, 5, –3), (0, 7, –9, 2)}; dim.=2]

16. Show that the vectors $\alpha_1 = (1, 0, -1)$, $\alpha_2 = (1, 2, 1)$, $\alpha_3 = (0, -3, 2)$ forms a basis for $V_3(R)$. Express each of the standard basis vector as a linear combination of $\alpha_1, \alpha_2, \alpha_3$.

[**Hints :** show that α_1, α_2 and α_3 are L.I. & generates $V_3(R)$. Then express each basis vectors as L.C. of $\alpha_1, \alpha_2, \alpha_3$. Thus **Ans.** $(1, 0, 0) = -\frac{7}{10}\alpha_1 - \frac{5}{10}\alpha_2 + \frac{1}{5}\alpha_3$, etc.]

17. Determine whether or not each of the following forms a basis of R_3 :

(i) (1, 2, –1) & (0, 3, 1) [**Ans.** No]
(ii) (2, 4, –3), (0, 1, 1) & (0, 1, –1) [**Ans.** Yes]
(iii) (1, 5, –6), (2, 1, 8), (3, –1, 4) & (2, 1, 1) [**Ans.** No]
(iv) (1, 3, –4), (1, 4, –3) & (2, 3, –11) [**Ans.** No]

18. Find a basis and the dimension of the subspace W of R^4 generated by

(i) (1, 4, –1, 3), (2, 1, –3, –1) & (0, 2, 1, –5) [**Ans.** dim.=2]
(ii) (1, –4, –2, 1), (1, –3, –1, 2) & (3, –8, –2, 7) . [**Ans.** dim.=2]

19. Find a basis and the dimension of the solution space W of each homogeneous system:

(i) $\left.\begin{array}{l} x+2y-2z+2s-t=0 \\ x+2y-z+3s-2t=0 \\ 2x+4y-7z+s+t=0 \end{array}\right\}$;

(ii) $\left.\begin{array}{l} x+2y-z+3s-4t=0 \\ 2x+4y-2z-s+5t=0 \\ 2x+4y-2z+4s-2t=0 \end{array}\right\}$

[**Ans.** (i) dim.W= 3, basis is

$\{(2, -1, 0, 0, 0), (4, 0, 1, -1, 0), (3, 0, 1, 0, 1)\}$];

(ii) dim.W=2, basis is $\{(2, -1, 0, 0, 0), (1, 0, 1, 0, 0)\}$]

20. Find a basis of $V^3(R)$ which contains the vectors $(1, -1, 1)$ and $(2, 1, 0)$.

[**Ans.** a basis of V^3 (R) is $(1, -1, 1)$, $(0, 3, -2)$ and $(0, 0, 1)$]

21. Find a homogeneous system whose solution set W is generated by

$$\{(1, -2, 0, 3), (1, -1, -1, 4), (1, 0, -2, 5)\}.$$

[**Ans.** $2x + y + z = 0, 5x + y - t = 0$]

22. Let W_1 and W_2 be the following subspaces of $V^4(R)$ such that

$W_1 = \{(a, b, c, d) : b + c + d = 0\}$ and

$W_2 = \{(a, b, c, d) : a + b = 0, c = 2d\}$.

Find the dimension and a basis of (i) W_1, (ii) W_2, (iii) $W_1 \cap W_2$.

[**Ans.** (i) dim.$W_1 = 3$, basis is $\{\alpha_1, \alpha_2, \alpha_3\}$,

$\alpha_1 = (1, 0, 0, 0), \alpha_2 = (0, -1, 1, 0), \alpha_3 = (0, -1, 0, 1)$

(ii) dim.$W_2 = 2$, basis is $\{\alpha_1, \alpha_2\}$,

$\alpha_1 = (-1, 1, 0, 0), \alpha_2 = (0, 0, 2\ 1)$

(iii) dim. $W_1 \cap W_2 = 1$, basis is $\{(3, -3, 2, 1)\}$]

23. Consider the following elements of the vector space M_2 of all 2×2 matrices:

$$A_1 = \begin{bmatrix} 1 & 1 \\ 1 & 1 \end{bmatrix}, A_2 = \begin{bmatrix} 1 & -1 \\ 1 & 0 \end{bmatrix}, A_3 = \begin{bmatrix} 1 & 0 \\ 0 & 1 \end{bmatrix}, A_4 = \begin{bmatrix} 0 & 1 \\ -1 & 0 \end{bmatrix} \text{ and } A_5 = \begin{bmatrix} 1 & 2 \\ 3 & 4 \end{bmatrix}.$$

(i) Show that the set $\{A_1, A_2, A_3, A_4, A_5\}$ spans M_2.
(ii) Show that A_5 is a linear combination of A_1, A_2, A_3, A_4.
(iii) Show that M_2 is also spanned by the set $\{A_1, A_2, A_3, A_4\}$.
(iv) Find a basis and the dimension of M_2.

[**Ans.** basis = $\{A_1, A_2, A_3, A_4\}$ and dim. $M_2 = 4$]

24. Prove that the row rank = column rank = 2 for the matrix

$$A = \begin{bmatrix} 5 & 6 & 7 \\ 7 & 3 & 4 \end{bmatrix}.$$

Find the dimension of solution space of $A\mathbf{x} = 0$. Hence verify that $s + c = n$.

[**Ans.** dim.(solution space) =1]

25. If

$$A = \begin{bmatrix} 1 & 2 \\ 3 & 4 \end{bmatrix}, B = \begin{bmatrix} -1 & -2 \\ 3 & 4 \end{bmatrix}$$

then prove that rank (A + B) < rank A + rank B.

[**Hints :** rank A= 2 =rank B. But rank (A + B) = 1]

26. Show that the following set of vectors is a basis for $M_{2\times 2}$:

$$\begin{bmatrix} 3 & 6 \\ 3 & -6 \end{bmatrix}, \begin{bmatrix} 0 & -1 \\ -1 & 0 \end{bmatrix}, \begin{bmatrix} 0 & -8 \\ -12 & -4 \end{bmatrix}, \begin{bmatrix} 1 & 0 \\ -1 & 2 \end{bmatrix}.$$

27. Let V be space spanned by $\mathbf{v}_1 = \cos^2 x$, $\mathbf{v}_2 = \sin^2 x$, $\mathbf{v}_3 = \cos 2x$.

(a) Show that $S=\{\mathbf{v}_1, \mathbf{v}_2, \mathbf{v}_3\}$ is not a basis for V.

(b) Find a basis for V.

(c) What is the dimension of V.

[**Hints :** (a) $\mathbf{v}_3 = \mathbf{v}_1 - \mathbf{v}_2$, so $\mathbf{v}_1, \mathbf{v}_2, \mathbf{v}_3$ are not linearly independent.
Ans. (b) Basis = $\{\mathbf{v}_1, \mathbf{v}_2\}$ or $\{\mathbf{v}_2, \mathbf{v}_3\}$, & (c) dim. V = 2]

28. Determine bases for the following subspaces of R^3.

(a) The plane $3x - 2y + 5z = 0$

(b) The plane $x - y = 0$

(c) The line,

$$\left.\begin{aligned} x &= 2t \\ y &= -t \\ z &= 4t \end{aligned}\right\}, -\infty < t < \infty$$

(d) All vectors of the form (a, b, c) where $b = a + c$.

[**Hints :** Each system is in echelon form. Note the number free variables. Obtain bases for each case giving particular values to free variables.
Ans. (a) (2, 3, 0), (–1, 1, 1) ; (b) (1, 1, 0), (1, 1, 5) ;
(c) (2, –1, 4) ; (d) (1, 2, 1), (1, 0, –1)]

29. Let $\{\mathbf{v}_1, \mathbf{v}_2, \mathbf{v}_3\}$ be a basis for a vector space V. Show that $\{\mathbf{u}_1, \mathbf{u}_2, \mathbf{u}_3\}$ is also a basis where

$$\mathbf{u}_1 = \mathbf{v}_1,\ \mathbf{u}_2 = \mathbf{v}_1 + \mathbf{v}_2,\ \mathbf{u}_3 = \mathbf{v}_1 + \mathbf{v}_2 + \mathbf{v}_3.$$

30. Show that the only subspaces of R^3 are zero subspace, lines through origin planes through origin and R^3 itself.

[**Hints :** (dim.of subspace of R^3) ≤ 3. So subspaces of R^3 must be of dimensions 0,1,2, 3 respectively. Hence subspace of R^3 are zero subspace, line through origin, planes through origin & R^3 itself]

31. Find **w** if $(\mathbf{w})_s = (6, -1, 4)$ and $S = \{\mathbf{v}_1, \mathbf{v}_2, \mathbf{v}_3\}$ where

$$\mathbf{v}_1 = (1, 0, 0),\ \mathbf{v}_2 = (2, 2, 0),\ \mathbf{v}_3 = (3, 3, 3).$$

[**Ans.** $\mathbf{w} = (16, 10, 12)$]

32. Find **q** if $(\mathbf{q})_s = (3, 0, 4)$ and $S = \{p_1, p_2, p_3\}$ where $p_1 = 1$, $p_2 = x$, $p_3 = x^2$.

[**Ans.** $\mathbf{q} = 3 + 4x^2$. **Hints**. $a + bx + cx_2 = 3p_1 + 0.p_2 + 4p_3$]

33. Find B if $(B)_s = (-8, 7, 6, 3)$ and $S = \{A_1, A_2, A_3, A_4\}$ with

$$A_1 = \begin{bmatrix} -1 & 1 \\ 0 & 0 \end{bmatrix}, A_2 = \begin{bmatrix} 1 & 1 \\ 0 & 0 \end{bmatrix}, A_3 = \begin{bmatrix} 0 & 0 \\ 1 & 0 \end{bmatrix}, A_4 = \begin{bmatrix} 0 & 0 \\ 0 & 1 \end{bmatrix}.$$

[**Hints :** $B = -8A_1 + 7A_2 + 6A_3 + 3A_4$ and

$B=\begin{bmatrix} a & b \\ c & d \end{bmatrix}$. **Ans.** $B=\begin{bmatrix} 15 & -1 \\ 6 & 3 \end{bmatrix}$]

34. Show that $R_1 + R_2$ is the linear space spanned by $\{\mathbf{x}_1, \cdots, \mathbf{x}_m, \mathbf{y}_1, \cdots, \mathbf{y}_p\}$, where $\{\mathbf{x}_1, \cdots, \mathbf{x}_m\}$ and $\{\mathbf{y}_1, \cdots, \mathbf{y}_p\}$ are any bases for R_1 and R_2. Show also that $\dim.(R_1 + R_2) = m + p - \dim.(R_1 \cap R_2)$.

35. Find the coordinate vector and coordinate matrix for A relative to $S = \{A_1, A_2, A_3, A_4\}$ where

$$A=\begin{bmatrix} 2 & 0 \\ -1 & 3 \end{bmatrix}, A_1=\begin{bmatrix} -1 & 1 \\ 0 & 0 \end{bmatrix}, A_2=\begin{bmatrix} 1 & 1 \\ 0 & 0 \end{bmatrix}, A_3=\begin{bmatrix} 0 & 0 \\ 1 & 0 \end{bmatrix}, A_4=\begin{bmatrix} 0 & 0 \\ 0 & 1 \end{bmatrix}.$$

[**Hints :** augmented matrix is

$$\begin{matrix} \text{vec}A_1 & \text{vec}A_2 & \text{vec}A_3 & \text{vec}A_4 & \text{vec}A \end{matrix}$$
$$\left[\begin{array}{cccc|c} -1 & 1 & 0 & 0 & 2 \\ 0 & 0 & 1 & 0 & -1 \\ 1 & 1 & 0 & 0 & 0 \\ 0 & 0 & 0 & 0 & 3 \end{array}\right] \text{ and } [A]_s = \begin{bmatrix} -1 \\ 1 \\ -1 \\ 3 \end{bmatrix}]$$

36. Consider the bases

$$B = \left\{\begin{bmatrix} 2 \\ 1 \\ 1 \end{bmatrix}, \begin{bmatrix} 2 \\ -1 \\ 1 \end{bmatrix}, \begin{bmatrix} 1 \\ 2 \\ 1 \end{bmatrix}\right\} \text{ and } B' = \left\{\begin{bmatrix} 3 \\ 1 \\ -5 \end{bmatrix}, \begin{bmatrix} 1 \\ 1 \\ -3 \end{bmatrix}, \begin{bmatrix} -1 \\ 1 \\ 2 \end{bmatrix}\right\}.$$

(a) Find the transition matrix from B-basis to B′-basis.

(b) Compute the coordinate matrix $[\mathbf{w}]_B$ where

$$\mathbf{w} = \begin{bmatrix} -5 \\ 8 \\ -5 \end{bmatrix} \text{ and hence compute } [\mathbf{w}]_{B'} \text{ by using (a).}$$

(c) Check the result in (b) by computing $[\mathbf{w}]_{B'}$ directly.

(d) Find the transition matrix from B′-basis to B-basis.

(e) Verify that the product of (a) and (d) is unit matrix.

[**Ans.** (a) ☐ ; (b) $[\mathbf{w}]_B = \begin{bmatrix} 9 \\ -9 \\ -9 \end{bmatrix}$; $[\mathbf{w}]_{B'} = \begin{bmatrix} -\frac{7}{2} \\ \frac{23}{2} \\ 6 \end{bmatrix}$;

(d) $\begin{bmatrix} \frac{35}{2} & \frac{19}{2} & -\frac{13}{2} \\ -\frac{19}{2} & -\frac{11}{2} & \frac{7}{2} \\ -13 & -7 & 5 \end{bmatrix}$]

37. Consider the bases $B = \{p_1, p_2\}$ and $B' = \{q_1, q_2\}$ where $p_1 = 6 + 2x$, $p_2 = 10 + 2x$, $q_1 = 2$, $q_2 = 3 + 2x$.

(a) Find the transition matrix from B to B′.

(b) Compute the coordinate matrix $[p]_B$ where $p = -4 + x$ and hence compute $[p]_{B'}$ using (a).

(c) Check the result by computing $[p]_{B'}$ directly.

(d) Find the transition matrix from B' to B.

(e) Verify that the product of (a) & (d) is unit matrix.

[**Hints:** (a) augmented matrix is

$$\begin{array}{cc} q_1 & q_2 & p_1 & p_2 \end{array}$$
$$\left[\begin{array}{cc|cc} 2 & 3 & 6 & 10 \\ 0 & 2 & 2 & 2 \end{array}\right].$$

Because fundamental bases of polynomial p_1 are 1 and x, so $q_1 = 2.1 + 0.x$, $q_2 = 3.1 + 2.x$, $p_1 = 6.1+2.x$, $p_2 = 10.1+2.x$.

Ans. (a) $\begin{bmatrix} -\frac{3}{4} & \frac{7}{9} \\ \frac{1}{3} & -\frac{1}{6} \end{bmatrix}$, (b) $[p]_B = \begin{bmatrix} 1 \\ -1 \end{bmatrix}$, $[p]_{B'} = \begin{bmatrix} -\frac{11}{4} \\ \frac{1}{2} \end{bmatrix}$, (d) $\begin{bmatrix} -\frac{2}{9} & \frac{7}{9} \\ \frac{1}{3} & -\frac{1}{6} \end{bmatrix}$]

38. Let V be the space spanned by $f_1 = \sin x$ and $f_2 = \cos x$.

(a) Show that $g_1 = 2\sin x + \cos x$ and $g_2 = 3\cos x$ form a basis for V.

(b) Find the transition matrix from $B=\{f_1, f_2\}$ to $B' = \{g_1, g_2\}$.

(c) Compute the coordinate matrix $[h]_B$, where $h = 2\sin x - 5\cos x$ and hence obtain $[h]_{B'}$ from (b).

(d) Check your work by computing $[h]_{B'}$ directly.

(e) Find the transition matrix from B' to B.

[**Hints :** (a) augmented matrix is

$$\begin{array}{cc} g_1 & g_2 & f_1 & f_2 \end{array}$$
$$\left[\begin{array}{cc|cc} 2 & 0 & 1 & 0 \\ 1 & 3 & 0 & 1 \end{array}\right].$$ **Ans.** (b) $\begin{bmatrix} \frac{1}{2} & 0 \\ -\frac{1}{6} & \frac{1}{3} \end{bmatrix}$,

(c) $[h]_B = \begin{bmatrix} 2 \\ -5 \end{bmatrix}$, $[h]_{B'} = \begin{bmatrix} 1 \\ -2 \end{bmatrix}$; (e) $\begin{bmatrix} 2 & 0 \\ 1 & 3 \end{bmatrix}$]

39. Let $S = \{2t^2 + t,\ t^2 + 3,\ t\}$ and $T = \{t^2 + 1,\ t-2,\ t + 3\}$ be ordered bases for polynomial P_2. Let

$\alpha = 8t^2 - 4t + 6$ and $\beta = 7t^2 - t + 9$.

(a) Find coordinate matrices of α and β with respect to the basis S.

(b) What is the transition matrix from S-basis to T-basis?

(c) Find coordinate matrices of α and β with respect to the basis T using result in (b).

(d) Find the coordinate matrices of α and β with respect to T directly.

(e) Find the transition matrix from T-basis to S-basis.

(f) Find the coordinate matrices of α and β with respect to S using result in (e). Compare the answers with those of (a).

[Ans.(a) $[\alpha]_s = \begin{bmatrix} 3 \\ 2 \\ -7 \end{bmatrix}$; $[\beta]_s = \begin{bmatrix} 2 \\ 3 \\ -3 \end{bmatrix}$. **Hints :** $[\alpha]_s = \begin{bmatrix} a \\ b \\ c \end{bmatrix}$.

So, $\alpha = a(2t^2 + t) + b(t^2 + 3) + ct$. Solve for a, b, c.

OR construct the augmented matrix

$$\begin{matrix} p_1 & p_2 & p_3 & \alpha_s \end{matrix}$$
$$\left[\begin{array}{ccc|c} 0 & 3 & 0 & 6 \\ 1 & 0 & 1 & -4 \\ 2 & 1 & 0 & 8 \end{array}\right] \approx \left[\begin{array}{ccc|c} 0 & 1 & 0 & 2 \\ 1 & 0 & 0 & 3 \\ 0 & 0 & 1 & -7 \end{array}\right] \approx \left[\begin{array}{ccc|c} 1 & 0 & 0 & 3 \\ 0 & 1 & 0 & 2 \\ 0 & 0 & 1 & 7 \end{array}\right] \text{ and } [\alpha]_s = \begin{bmatrix} 3 \\ 2 \\ 7 \end{bmatrix};$$

(b) $[p]_S^T = \begin{bmatrix} 2 & 1 & 0 \\ 1 & -\frac{2}{5} & \frac{3}{5} \\ 0 & \frac{2}{5} & \frac{2}{5} \end{bmatrix}$; (c) $[\alpha]_T = \begin{bmatrix} 8 \\ -2 \\ -2 \end{bmatrix}$; $[\beta]_T = \begin{bmatrix} 7 \\ -1 \\ 0 \end{bmatrix}$; (d) same as (c) ;

(e) $[p]_T^S = \begin{bmatrix} \frac{1}{3} & \frac{1}{3} & -\frac{1}{2} \\ \frac{1}{3} & -\frac{2}{3} & 1 \\ -\frac{1}{3} & \frac{2}{3} & \frac{3}{2} \end{bmatrix}$;

(f) $[\alpha]_s = [p]_T^S [\alpha]_T = \begin{bmatrix} 3 \\ 2 \\ -7 \end{bmatrix}$ and $[\beta]_s = [p]_S^T [\beta]_T = \begin{bmatrix} 2 \\ 3 \\ -3 \end{bmatrix}$, same as (a)]

40. Let

$$S = \left\{\begin{bmatrix} 1 & 1 \\ 0 & 0 \end{bmatrix}, \begin{bmatrix} 0 & 0 \\ 1 & 0 \end{bmatrix}, \begin{bmatrix} 0 & 0 \\ 0 & 1 \end{bmatrix}, \begin{bmatrix} 1 & 0 \\ 0 & 0 \end{bmatrix}\right\} \text{ and}$$

$$T = \left\{\begin{bmatrix} 1 & 0 \\ 0 & 1 \end{bmatrix}, \begin{bmatrix} 0 & 1 \\ 1 & 0 \end{bmatrix}, \begin{bmatrix} 0 & 2 \\ 0 & 1 \end{bmatrix}, \begin{bmatrix} 0 & 0 \\ 1 & 1 \end{bmatrix}\right\}$$

be ordered bases for R^2.

Let $\alpha = \begin{bmatrix} 1 & 1 \\ 1 & 1 \end{bmatrix}$ and $\beta = \begin{bmatrix} 1 & 2 \\ -2 & 1 \end{bmatrix}$. Follow the directions of exercise above.

[Ans. (a) $[\alpha]_s = \begin{bmatrix} 1 \\ 1 \\ 1 \\ 0 \end{bmatrix}$ $[\beta]_s = \begin{bmatrix} 2 \\ -2 \\ 1 \\ -1 \end{bmatrix}$; (b) $[p]_S^T = \begin{bmatrix} 1 & 0 & 0 & 1 \\ \frac{1}{3} & \frac{2}{3} & -\frac{2}{3} & 0 \\ \frac{1}{3} & -\frac{1}{3} & \frac{1}{3} & 0 \\ -\frac{1}{3} & \frac{1}{3} & \frac{2}{3} & 0 \end{bmatrix}$;

(c) $[\alpha]_T = \begin{bmatrix} 1 \\ \frac{1}{3} \\ \frac{1}{3} \\ \frac{2}{3} \end{bmatrix}$, $[\beta]_T = \begin{bmatrix} 1 \\ -\frac{4}{3} \\ \frac{5}{3} \\ -\frac{2}{3} \end{bmatrix}$; (d) Same as (c) ; (e) $[p]_T^S = \begin{bmatrix} 0 & 1 & 2 & 0 \\ 0 & 1 & 0 & 1 \\ 0 & 0 & 1 & 1 \\ 1 & -1 & -2 & 0 \end{bmatrix}$ (f) Same as (a)]

41. Find the dimension of the vector space spanned by

$(1,-2, 3, 1)$ & $(1,1,-2, 3)$.

[**Ans.** dim = 1]

42. Let U be the subspace of R^5 generated by

$\{(1, 3, -2, 2, 3), (1, 4, -3, 4, 2), (2, 3, -1, -2, 9)\}$

and let W be the subspace generated by

$\{(1, 3, 0, 2, 1), (1, 5, -6, 6, 3), (2, 5, 3, 2, 1)\}$.

Find a basis & the dimension of (i) U + W, (ii) $U \cap W$.

[**Ans. & Hints :** (i) Form 5×5 matrix with 6 vectors & reduce to echelon form. Nonzero rows form a basis. Basis of U + W is $\{(1, 3, -2, 2, 3), (0, 1, -1, 2, -1), (0, 0, 2, 0, -2)\}$; dim.(U + W) = 3.
(ii) $\{(1, 4, -3, 4, 2)\}$ is a basis of $(U \cap W)$ & dim.$(U \cap W) = 1$]

43. Let U and W be the subspaces of R^5 generated by

$\{(1, -1, -1, -2, 0), (1, -2, -2, 0, -3), (1, -1, -2, -2, 1)\}$

and $\{(1, -2, -3, 0, -2), (1, -1, -3, 2, -4), (1, -1, -2, 2, -5)\}$ respectively.

(i) Find two homogeneous systems whose solution spaces are U and W respectively.

(ii) Find a basis & the dimension of $U \cap W$.

[**Ans.** (i) $\begin{cases} 3x + 4y - z \quad - t = 0 \\ 4x + 2y + \quad s \quad = 0 \end{cases}$ and $\begin{cases} 4x + 2y - \quad s \quad = 0 \\ 9x + 2y + z + t = 0 \end{cases}$;

(ii) basis is $\{(1, -2, -5, 0, 0), (0, 0, 1, 0, -1)\}$. dim$(U \cap W) - 2$]

44. What subspace of R^3 is spanned by the vectors (1, 0, 0) & (0, 1, 0)?

[**Ans. & hints :** Let $L(S) = x(1, 0, 0) + y(0, 1, 0) = (x, y, 0)$ for all $x, y \in R$
$\therefore$ the required subspace spanned by the given vectors is $W = \{ (x, y, 0)$ for all $x, y \in R\}$]

45. Let W be the subspace of R^3 spanned by the set $S = \{(1, -1, 0), (-2, 1, 2). (-1, -1, 4)\}$. Show that the subset $S_1 = \{(1, -1, 0), (-2, 1, 2)\}$ spans W.

46. Find a basis for the subspace W of R^4 generated by $S = \{\mathbf{v}_1, \mathbf{v}_2, \mathbf{v}_3, \mathbf{v}_4, \mathbf{v}_5\}$ where $\mathbf{v}_1 = (1, 0, -1, 1)$, $\mathbf{v}_2 = (-2, 0, 2, -2)$, $\mathbf{v}_3 = (1, 1, 1, 1)$, $\mathbf{v}_4 = (1, -1, -3, 1)$ & $\mathbf{v}_5 = (1, -1, 1, -1)$.

[**Ans.**$\{ \mathbf{v}_1, \mathbf{v}_2, \mathbf{v}_3 \}$]

47. Find a basis for vector space V generated by the set

$S=\{(1, 0, 1, 0), (0, -1, 1, 2), (2, -1, 3, 2), (1, 1, 1, 1)\}$.

[**Ans.** basis = $\{(1, 0, 1, 0), (0, -1, 1, 2), (1, 1, 1, 1)\}$]

48. Determine whether $A\mathbf{x} = \mathbf{b}$ is consistent or not through the consideration that $\mathbf{b}$ lies in the column space of A where

(i) $A = \begin{bmatrix} 1 & 3 \\ 4 & -6 \end{bmatrix}$, $\mathbf{b} = \begin{bmatrix} -1 \\ 10 \end{bmatrix}$; (ii) $A = \begin{bmatrix} 1 & -4 \\ 2 & -8 \end{bmatrix}$, $\mathbf{b} = \begin{bmatrix} 0 \\ 1 \end{bmatrix}$;

(iii) $A = \begin{bmatrix} 1 & -1 & 1 \\ 1 & 1 & -1 \\ -1 & -1 & 1 \end{bmatrix}, \mathbf{b} = \begin{bmatrix} 2 \\ 0 \\ 0 \end{bmatrix}$.

[**Hints :** Verify that b is the linear combination of the columns of A. **Ans .** (i) & (ii) are consistent while (ii) is not]

49. Find a subset of the vectors that form a basis for the vector space spanned by the vectors. Hence express each $\mathbf{v}_i$ that is not in the basis as a linear combination of the basis vectors .

(i) $\mathbf{v}_1 = (1, -1, 5, 2)$, $\mathbf{v}_2 = (-2, 3, 1, 0)$, $\mathbf{v}_3 = (4, -5, 9, 4)$, $\mathbf{v}_4 = (0, 4, 2, -3)$, $\mathbf{v}_5 = (-7, 18, 2, -8)$;

(ii) $\mathbf{v}_1 = (1, 0, 1, 1)$, $\mathbf{v}_2 = (-3, 3, 7, 1)$, $\mathbf{v}_3 = (-1, 3, 9, 3)$, $\mathbf{v}_4 = (-5, 3, 5, -1)$.

[**Ans.** (i) $\{\mathbf{v}_1, \mathbf{v}_2, \mathbf{v}_4\}$, $\mathbf{v}_3 = 2\mathbf{v}_1 - \mathbf{v}_2$, $\mathbf{v}_5 = -\mathbf{v}_1 + 3\mathbf{v}_2 + 2\mathbf{v}_4$;
(ii) $\{\mathbf{v}_1, \mathbf{v}_2\}$, $\mathbf{v}_3 = 2\mathbf{v}_1 + \mathbf{v}_2$, $\mathbf{v}_4 = -2\mathbf{v}_1 + \mathbf{v}_2$.]

50. In the vector space , $V_3(R)$, let $\alpha_1 = (1, 2, 1)$, $\alpha_2 = (3, 1, 5)$, $\alpha_3 = (3, -4, 7)$. Prove that the subspaces spanned by $S = \{\alpha_1, \alpha_2\}$ and $T = \{\alpha_1, \alpha_2, \alpha_3\}$ are the same. i.e. $< S > = < T >$.

51. Show that the *xy* plane $W = \{(a, b, 0)\}$ in R^3 is generated by **u** and **v** where (i) $\mathbf{u} = (1, 2, 0)$ & $\mathbf{v} = (0, 1, 0)$; (ii) $\mathbf{u} = (2, -1, 0)$ & $\mathbf{v} = (1, 3, 0)$.

52. Find one vector in R^3 which generates the intersection of U & W where U is the *xy* plane :
$U = \{(a, b, 0)\}$ and W is the space generated by the vectors
$(1, 2, 3)$ & $(1, -1, 1)$.

[**Ans.** (2, –5, 0)]

53. Let $U = \{(1, 1, 1), (0, -2, 3)\}$. Find a subspace $W \in V^3$ such that $V^3 = U \oplus W$.

[**Ans.** $W = \{(a, b, c)\}$ such that $a \neq 3b + 2c$]

54. If $V_3 = U \oplus W$ and $\mathbf{x} = \mathbf{y} + \mathbf{z}$ such that vectors $\mathbf{x} \in V$, $\mathbf{y} \in U$ and $\mathbf{z} \in W$ then **y** is called the *projection* of **x** on U and **z** is called the *projection* of **x** on W.

If $V_3 = U \oplus W$ where $U = \{(1, 0, 3), (2, -1, 6)\}$, $W = L\{(1, 1, 1)\}$ find projection of $\mathbf{x} = (4, 1, -2)$ on U and on W

[**Ans .** Projection on U = –15(1, 0, 3) + 6 (2,– 1, 6) = (–3, –6, –9) ;
projection on W = 7(1, 1, 1) = (7, 7, 7)]

55. Which of the following are subspaces of $M_{2\times2}$, a vector space of all 2×2 matrices.

(a) all matrices of the form $\begin{bmatrix} a & b \\ c & d \end{bmatrix}$, *a, b, c, d* are integers ;

(b) matrtices as in (a) with $a + d = 0$;

(c) all 2×2 matrices A such that $A = A^T$;

(d) all 2×2 matrices A such that det A = 0. [**Ans .** (b) and (c)]

56. Determine which of the following are subspaces of P_3 .

(a) all polynomials $a_0 + a_1x + a_2x^2 + a_3x^3$ with $a_0 \neq 0$;

(b) as in (a) with $a_0 + a_1 + a_2 + a_3 = 0$;

(c) as in (a) with a_0, a_1, a_2, a_3 are integer

(d) all polynomials of the form $a_0 + a_1x$ where a_0 and a_1 are real numbers.

[**Ans.**(a), (b), (d)]

57. Express each one of (a) $5 + 9x + 5x^2$; (b) $2 + 6x^2$; (c) 0 ; (d) $2 + 2x + 3x^2$ as linear combinations of $p_1 = 2 + x + 4x^2$, $p_2 = 1 - x + 3x^2$ and $p_3 = 3 + 2x + 5x^2$.

[**Ans.** (a) $5 + 9x + 5x^2 = 5p_1 - 4p_2 + p_3$;

(b) $2+6x^2 = 4p_1-2p_3$
(c) $0 = 0p_1 - 0p_2 + 0p_3$;
(d) $2 + 2x + 3x^2 = \frac{1}{2}p_1 - \frac{1}{2}p_2 + \frac{1}{2}p_3$]

58. Determine which of the following lie in the space spanned by

$$f = \cos^2 x \quad \text{and} \quad g = \sin^2 x.$$

(a) $\cos 2x$, (b) $3 + x^2$, (c) 1, (d) $\sin x$.

[**Ans.** (a), (c)]

59. Determine if the following polynomials span P_2

$p_1 = 1+2x-x^2$, $p_2 = 3 + x^2$, $p_3 = 5 + 4x - x^2$ and $p_4 = -2 + 2x-2x^2$

[**Ans .** Do not span P_2]

60. Which of the following

$$\text{(a)} \begin{bmatrix} 6 & 3 \\ 0 & 8 \end{bmatrix}; \text{ (b)} \begin{bmatrix} -1 & 7 \\ 5 & 1 \end{bmatrix}; \text{ (c)} \begin{bmatrix} 0 & 0 \\ 0 & 0 \end{bmatrix}; \text{ (d)} \begin{bmatrix} 6 & 1 \\ -8 & -8 \end{bmatrix}$$

are linear combinations of

$$A = \begin{bmatrix} 1 & 2 \\ -1 & 3 \end{bmatrix}, B = \begin{bmatrix} 0 & 1 \\ 2 & 4 \end{bmatrix} \text{ and } C = \begin{bmatrix} 4 & -2 \\ 0 & -2 \end{bmatrix}.$$

[**Ans.** (a), (b), (c)]

GRE Sample Test

(Graduate Record Examination (GRE) sample test)

Directions: Each of the questions or incomplete statements below is followed by five suggested answers or completions .In each case, select the one that is the best of the choices offered . Readers are requested to set up identical questions and practice for GRE test.

T1. The solution set for the equation $\begin{vmatrix} 2 & 3 & x \\ 2 & 1 & x^2 \\ 6 & 7 & 3 \end{vmatrix} = 0$ is

(A) Empty , (B) $\{0\}$, (C) $\{1\}$, (D) $\{1,-3\}$, (E) $\{\sqrt{3}, -\sqrt{3}\}$. [**Ans.** (D)]

T2. Each of the two sets of data D_1, D_2 , D_3 is divided into categories C_1 , C_2 , C_3 and the following table is devised to the number of data in each category.

	C_1	C_2	C_3
D_1			
D_2			
D_3			

Furthermore there are three systems of weights W_1, W_2 ,W_3 each of which assigns a score for each detum in each category and the following table is devised to show score assigned to each category under each system of weights

	W_1	W_2	W_3
C_1			
C_2			
C_3			

Let x_{ij} be the number of elements of D_i that are put into C_j .Let y_{ij} be the weight assigned C_i under the system W_j . If $X=[x_{ij}]$ and $Y=[y_{ij}]$, $i, j=1,2,3$, which of the following would be the matrix of entries showing the total score of set D_i under W_j ?
(A) X + Y , (B) XY , (C) YX , (D) XYX^{-1} , (E) YXY^{-1} . [**Ans.** B]

T3. Which of the following is equal to the product $P^{-1}\begin{bmatrix} 1 & 2 \\ 3 & 4 \end{bmatrix}P$ for some invertible 2×2 matrix P ?

(A) $\begin{bmatrix} 3 & 4 \\ 1 & 2 \end{bmatrix}$, (B) $\begin{bmatrix} 2 & 2 \\ 3 & 2 \end{bmatrix}$, (C) $\begin{bmatrix} 2 & 0 \\ 0 & 0 \end{bmatrix}$, (D) $\begin{bmatrix} 4 & 3 \\ 2 & 1 \end{bmatrix}$, (E) $\begin{bmatrix} 2 & 1 \\ 4 & 3 \end{bmatrix}$.

[**Ans.** D. Because similar matrices have identical eigenvalues]

T4. If $A = \begin{bmatrix} 2 & 1 \\ 3 & -1 \end{bmatrix}$ and $I = \begin{bmatrix} 1 & 0 \\ 0 & 1 \end{bmatrix}$,which of the following is zero ?

(A) A^2-A-5I , (B) A^2+A-5I , (C) A^2+A-I , (D) A^2-4I , (E) $A^2-3A+5I$.

[**Ans.** A. Because characteristic equation $(2-\lambda)(-1-\lambda)-3=0$ i.e.$\lambda^2-\lambda-5=0$. By Caylay-Hamilton theorem , matrix satisfies the characteristic equation.]

T5. Geometrically what is represented by detA=0 if

$$A(x, y)=A=\begin{bmatrix} x & y & 1 \\ x_1 & y_1 & 1 \\ x_2 & y_2 & 1 \end{bmatrix}$$

and (x_1, y_1) and (x_2, y_2) are two different points in the plane ?

(A) Straight line, (B) A point , (C) Both (A) and (B), (D) A plane , (E) Both (B) and (D).
[**Ans.** A]

T6. If A_1 is of rank r_1 and A_2 is of rank r_2 , the rank of $\begin{bmatrix} A_1 & 0 \\ 0 & A_2 \end{bmatrix}$ is

(A) $r_1 + r_2$, (B) r_2 , (C) $r_1 - r_2$, (D) r_1 , (E) 0 .
[**Ans.** A. Because number of nonzero rows are $r_1 + r_2$.]

T7. If T is the linear transformation mapping the vectors (1, 0, 0), (0, 1, 0) and (0, 0, 1) to the vectors (2, 2, 1), (1, 1, 2) and (0, 0, 1) respectively , which of the following is the image of the vector (2, 3, 4) under T ?
(A) (7, 7, 12) , (B) (7, 7, 11), (C) (6, 6, 11), (D) (3, 3 ,4) , (E) (7 , 7, 9).
[**Ans.** A. Follow w.o.ex.1 under sec.12.7]

T8. Which of the following is the standard matrix for linear transformation

$$T\begin{bmatrix} x_1 \\ x_2 \\ x_3 \\ x_4 \end{bmatrix} = \begin{bmatrix} 7x_1 + 2x_2 - x_3 + x_4 \\ x_2 + x_3 \\ -x_1 \end{bmatrix}.$$

(A) $\begin{bmatrix} 7 & 2 & -1 & 1 \\ 0 & 1 & 1 & 0 \\ -1 & 0 & 0 & 0 \end{bmatrix}$, (B) $\begin{bmatrix} 1 & 0 & 0 & 0 \\ 0 & 1 & 0 & 0 \\ 0 & 0 & 1 & 0 \\ 0 & 0 & 0 & 1 \end{bmatrix}$, (C) $\begin{bmatrix} 7 & 2 & -1 & 1 \\ 0 & 1 & 0 & 0 \\ -1 & 0 & 0 & 0 \end{bmatrix}$,

(D) $\begin{bmatrix} 1 & 0 & 0 \\ 0 & 1 & 0 \\ 0 & 0 & 1 \end{bmatrix}$, (E) $\begin{bmatrix} 1 & 7 & 2 & -1 \\ 0 & 0 & 1 & 1 \\ 0 & -1 & 0 & 0 \end{bmatrix}$.

[**Ans.** A. Note w.o. ex.1 under sec.12.9.2]

T9. Which is the true statement defining function $f : A \rightarrow B$ by the following table

x	f_1	f_2	f_3	f_4
1	a	a	b	b
2	a	b	a	b

(A) f_1 and f_4 are one-one, (B) f_1 and f_4 are onto, (C) f_2 and f_3 are neither one-one nor onto, (D) f_2 and f_4 are both one-one and onto, (E) All f_1, f_2, f_3 and f_4 are one-one.
[**Ans.** D. Note prob2 exercises twelve]

T10. Suppose A= $\begin{bmatrix} a & 0 & 0 \\ 0 & b & 0 \\ 0 & 0 & c \end{bmatrix}$ and B is similar to A. Which one of the following is detB ?

(A) detB = ab , (B) detB = 0, (C) detB cannot be determined, (D) detB = abc, (E) Both (A) and (B). [**Ans.** D]

T11. Note the following statements

I. The transpose of a diagonal matrix is a diagonal matrix.

II. Every scalar matrix is Hermitian.

III. The product of two diagonal matrices of the same order is a diagonal matrix.

IV. Every upper triangular block matrix is an upper triangular matrix. Which one of the following is true ?

(A) I, (B) II & III , (C) II & IV, (D) I & IV, (E) I & III. [**Ans.** E]

T12. Determine the nullity of

(i) $T:R^5 \rightarrow R^7$ has rank 3 ; (ii) T: $P_4 \rightarrow P_3$ has rank 1; and (iii) The range of $T:R^6 \rightarrow R^3$ is R^3. Check the correct one from the following :

(A) Nullity of (i) is 4, (B) Nullity of (ii) is 2, (C) Nullity of (iii) is 2,

(D) Nullity of (i), (ii) and (iii) are 2, 4 and 3 respectively , (E) Nullity of (i) is 3 and that of (ii) is 2. [**Ans.** (i) 5 – 3 = 2 ,(ii) 5 – 1 = 4 , (iii) 6 – 3 = 3. D]

T13. Let A be a 5×7 matrix with rank 4. The solution of the dimension space of $A\mathbf{x} = \mathbf{0}$ is (A) 7, (B) 5, (C) 3, (D) 2, (E) 4.

[**Ans.** dim.solution space = nullity(T) = n – rank(A) = 7 – 4 = 3. C]

T14. Let V be any vector space and let T:V→V be defined by T(**v**) =6**v**. Check the best one of the following :

(A) Kernel T={0} and range of T=V , (B) Kernel T={6} and range of T=T(**v**),
(C) Kernel T={1} and range of T=V , (D) Kernel T={4} and range of T= **v** ,
(E) Kernel T={2} and range of T= 6. [**Ans.**A. Since Ker(T) = {0} & R(T) = V]

T15. The eigenvalues of A^{-1} are (A) 2,3 ; (B) $\frac{1}{2}, \frac{1}{3}$; (C) 2 , $\frac{1}{2}$; (D) 3 , $\frac{1}{3}$;

(E) 3, $\frac{1}{2}$ where A= $\begin{bmatrix} 1 & -1 \\ 2 & 4 \end{bmatrix}$.

[**Ans.** B. Since eigenvalues of A^{-1} are reciprocals to those of A]

T16. Let $T : R^3 \rightarrow R^3$ be defined by

$T\begin{bmatrix} x \\ y \\ z \end{bmatrix} = \begin{bmatrix} 2x+y-z \\ x+2y+z \\ x-y-z \end{bmatrix}$. Let $B_1 = \left\{ \begin{bmatrix} 1 \\ 1 \\ 1 \end{bmatrix}, \begin{bmatrix} 1 \\ 1 \\ 0 \end{bmatrix}, \begin{bmatrix} 1 \\ 0 \\ 0 \end{bmatrix} \right\}$ and $\left\{ \begin{bmatrix} 0 \\ 0 \\ 1 \end{bmatrix}, \begin{bmatrix} 0 \\ 1 \\ 1 \end{bmatrix}, \begin{bmatrix} 1 \\ 0 \\ 1 \end{bmatrix} \right\}$ be the ordered bases

of R^3. The transition matrix P from B_1 to A_1 is

(A) $\begin{bmatrix} 1 & 1 & 1 \\ 1 & 1 & 0 \\ 1 & 0 & 0 \end{bmatrix}$, (B) $\begin{bmatrix} 0 & 0 & 1 \\ 0 & 1 & 0 \\ 1 & 1 & 1 \end{bmatrix}$, (C) $\begin{bmatrix} -1 & -5 & -3 \\ 1 & 3 & 2 \\ 1 & 0 & 0 \end{bmatrix}$,

(D) $\begin{bmatrix} -1 & -2 & -1 \\ 1 & 1 & 0 \\ 1 & 1 & 1 \end{bmatrix}$, (E) $\begin{bmatrix} 1 & 1 & 1 \\ -1 & 0 & -1 \\ 0 & -1 & 1 \end{bmatrix}$ [**Ans.** D. note sec.13.20.1]

T17. Let $A = \begin{bmatrix} 1 & 3 \\ 2 & 6 \end{bmatrix}$. Then $A^n = 7^{n-1}\begin{bmatrix} 1 & 3 \\ 2 & 6 \end{bmatrix}$ is true when n must be

(A) positive integer, (B) $n = 0$, (C) any number, (D) rational,(E) $n = -1$.[**Ans.** A]

T18. Let $A=\begin{bmatrix} \lambda & 1 \\ 0 & \lambda \end{bmatrix}$. Then JCF of A^{-1} is

(A) $\begin{bmatrix} \frac{1}{\lambda} & -\frac{1}{\lambda^2} \\ 0 & \frac{1}{\lambda} \end{bmatrix}$, (B) $\begin{bmatrix} \frac{1}{\lambda} & 1 \\ 0 & \frac{1}{\lambda} \end{bmatrix}$, (C) $\begin{bmatrix} \frac{1}{\lambda} & 0 \\ 0 & \frac{1}{\lambda} \end{bmatrix}$, (D) $\begin{bmatrix} \frac{1}{\lambda} & 0 \\ \frac{1}{\lambda^2} & \frac{1}{\lambda} \end{bmatrix}$, (E) $\begin{bmatrix} \frac{1}{\lambda} & \frac{1}{\lambda^2} \\ 0 & \frac{1}{\lambda} \end{bmatrix}$. [**Ans.** B. Ex.2 exercises 14]

T19. Note the product $\begin{bmatrix} 2 & 1 \\ 1 & 1 \\ 2 & 1 \end{bmatrix} = \begin{bmatrix} \frac{2}{3} & -\frac{\sqrt{2}}{6} \\ \frac{1}{3} & \frac{\sqrt{2}}{3} \\ \frac{2}{3} & -\frac{\sqrt{2}}{3} \end{bmatrix} \begin{bmatrix} 3 & \frac{5}{3} \\ 0 & \frac{\sqrt{2}}{3} \end{bmatrix}$.

(A) It is a QR factorization, (B) It is a LU factorization, (C) It is a full rank factorization, (D) It is a polar decomposition, (E) It is a spectral decomposition.

[**Ans.** A . Ex.18 of exercises 14]

T20. Note matrices $A=\begin{bmatrix} 1 & 2 \\ 3 & 6 \end{bmatrix}$, $B = \begin{bmatrix} 0 & 1 \\ 0 & 7 \end{bmatrix}$.

(A) B is JCF of A, (B) B is Schur triangular form of A, (C) B is a diagonal form of A , (D) B is both Schur triangular form & diagonal form of A , (E) B is Smith canonical form of A. [**Ans.** D .Ex.35 of exercises 14]

T21. Note the matrices $A=\begin{bmatrix} 2 & 1 \\ 0 & 1 \end{bmatrix}$, $B = \begin{bmatrix} -2 & -1 \\ 2 & 1 \end{bmatrix}$, $C = \begin{bmatrix} 1 & 0 \\ 0 & 2 \end{bmatrix}$, $D = \begin{bmatrix} 2 & 0 \\ 0 & 1 \end{bmatrix}$. Check the correct statement of (A) A is similar to B, (B) A is similar to both C and D, (C) B is similar to C, (D) B is similar to both C and D, (E) C is similar to both A and B. [**Ans.** B. Note Ex.51 of exercises 13]

T22. Note the matrices

$A=\begin{bmatrix} 1 & -1 \\ 0 & 2 \end{bmatrix}$, $B = \begin{bmatrix} -3 & 4 \\ 0 & -1 \end{bmatrix}$, $C = \begin{bmatrix} 0 & 5 \\ 2 & -9 \end{bmatrix}$, $X_1 = \begin{bmatrix} 1 & 0 \\ -2 & -1 \end{bmatrix}$, $X_2 = \begin{bmatrix} 1 & 1 \\ -2 & -1 \end{bmatrix}$.

The general solution of the matrix equation $AX + XB = C$ is
(A) $X=X_1$, (B) $X=X_2$, (C) $X = aX_1 + bX_2$, (D) $X = aX_1$, (E) $X = bX_2$.

[**Ans.** C . Note Ex. 63 of exercises 13]

T23. The value of $\begin{bmatrix} 3 & 4 \\ 1 & 3 \end{bmatrix}^n$ is

(A) $\begin{bmatrix} \frac{1}{2}(5^n+1) & (5^n-1) \\ \frac{1}{4}(5^n-1) & \frac{1}{2}(5^n+1) \end{bmatrix}$, (B) $\begin{bmatrix} \frac{1}{2}(5^n+1) & (5^n-1) \\ \frac{1}{4}(5^n-1) & (5^n+1) \end{bmatrix}$, (C) $\begin{bmatrix} \frac{1}{2}(5^n+1) & (5^n+1) \\ \frac{1}{4}(5^n-1) & \frac{1}{2}(5^n+1) \end{bmatrix}$,

(D) $\begin{bmatrix} \frac{1}{2}(5^n+1) & (5^n-1) \\ \frac{1}{4}(5^n+1) & \frac{1}{2}(5^n+1) \end{bmatrix}$,

(E) $\begin{bmatrix} \frac{1}{2}(5^n-1) & (5^n-1) \\ \frac{1}{4}(5^n-1) & \frac{1}{2}(5^n+1) \end{bmatrix}$. [**Ans.** A. Note Ex.28 of exercises 14]

T24. Let A=$\begin{bmatrix} 1 & 2 \\ 1 & 2 \end{bmatrix}$. Then

(A) $A^{-1}=\begin{bmatrix} 1 & 1 \\ 1 & 0 \end{bmatrix}$, (B) $A^{+}=\begin{bmatrix} -1 & -1 \\ -2 & -2 \end{bmatrix}$, (C) $A^{+}=\begin{bmatrix} -1 & 1 \\ -2 & -2 \end{bmatrix}$,

(D) $A^{+}=\begin{bmatrix} -1 & -1 \\ -2 & 2 \end{bmatrix}$, (E) $A^{-1}=\begin{bmatrix} -1 & -1 \\ -2 & -2 \end{bmatrix}$.

[**Ans.** B. Note that given matrix is singular and has no ordinary inverse]

T25. If S is a plane in Euclidean 3-space containing (0, 0, 0), (2, 0, 0) and (0, 0, 1), then S is the (A) *xy*-plane , (B) *xz*-plae , (C) *yz*-plane , (D) plane $y-z=0$, (E) $x+2y-2z=0$. [**Ans.** B]

T26. In order to send an undetected message to an agent in the field , each letter in the message is replaced by the number of its position in the alphabet and that number is entered in a matrix M. Thus, for example , DEAD becomes the matrix M=$\begin{bmatrix} 4 & 5 \\ 1 & 4 \end{bmatrix}$. In order to further avoid detection, each message with four letters is sent to the agent encoded as MC, where C $=\begin{bmatrix} 2 & -1 \\ 1 & 1 \end{bmatrix}$. If the agent receives the matrix $\begin{bmatrix} 51 & -3 \\ 31 & -8 \end{bmatrix}$, then the message is

(A) RUSH , (B) COME , (C) ROME , (D) CALL, (E) Not uniquely determined by the information given. [**Ans.** C. Since

$$MC=\begin{bmatrix} 51 & -3 \\ 31 & -8 \end{bmatrix} \text{ or } M=\begin{bmatrix} 51 & -3 \\ 31 & -8 \end{bmatrix}C^{-1}=\begin{bmatrix} 51 & -3 \\ 31 & -8 \end{bmatrix}\frac{1}{3}\begin{bmatrix} 1 & 1 \\ -1 & 2 \end{bmatrix}=\begin{bmatrix} 18 & 15 \\ 13 & 5 \end{bmatrix}$$]

T27. Let * be the binary operation on the rational numbers given by $a*b=a+b+2ab$. Which of the the following are true ?

I. * is commutative .

II. There is a rational number that is a *-identity.

III.Every rational number has a *-inverse.

(A) I only , (B) II only , (C) I and II only , (D) I and III only, (E) I , II and III.

[**Ans.** C]

T28. A group G in which $(ab)^2=a^2b^2$ for all a, b in G is necessarily (A) finite , (B) cyclic, (C) of order two , (D) abelian , (E) none of the previous.

[**Ans.** D]

T29. Let f be a function such that the graph of f is a semicircle S with end points $(a, 0)$ and $(b, 0)$ where $a<b$. The graph of $y=3f(x)$ is a

(A) translation of S , (B) semicircle with radius three times that of S , (C) subset of an ellipse, (D) subset of a parabola , (E) subset of a hyperbola.

[**Ans.** B. Since S : $(x-a)(x-b)+y^2=0$ or $\left(x-\frac{a+b}{2}\right)^2+y^2=\left(\frac{a-b}{2}\right)^2$ and $y=f(x)=\sqrt{-(x-a)(x-b)}$. Now $y=3f(x)=3\sqrt{-(x-a)(x-b)}$ simplifies to $\left\{3\left(x-\frac{a+b}{2}\right)\right\}^2+y^2=\left\{3\left(\frac{a-b}{2}\right)\right\}^2$ which shows the truth of the answer]

T30. The dimension of the subspace spanned by the real vectors

$$\begin{bmatrix}1\\1\\0\\0\end{bmatrix},\begin{bmatrix}2\\2\\0\\0\end{bmatrix},\begin{bmatrix}0\\1\\0\\0\end{bmatrix},\begin{bmatrix}2\\0\\0\\3\end{bmatrix},\begin{bmatrix}1\\-2\\0\\8\end{bmatrix},\begin{bmatrix}0\\0\\0\\0\end{bmatrix}$$ is (A) 2, (B) 3, (C) 4, (D) 5, (E) 6.

[**Ans.** B. $\begin{bmatrix}1&2&0&2&1&0\\1&2&1&0&-2&0\\0&0&0&0&0&0\\0&0&0&3&0&0\end{bmatrix}\approx\begin{bmatrix}1&2&0&0&-\frac{13}{3}&0\\0&0&1&0&\frac{7}{3}&0\\0&0&0&1&\frac{8}{3}&0\\0&0&0&0&0&0\end{bmatrix}$]

T31. The rank of the matrix $\begin{bmatrix}1&2&3&4&5\\6&7&8&9&10\\11&12&13&14&15\\16&17&18&19&20\\21&22&23&24&25\end{bmatrix}$ is

(A) 1, (B) 2, (C) 3, (D) 4, (E) 5.

[**Ans.** B. Since the given matrix is equivalent to $\begin{bmatrix}1&1&0&0&0\\6&1&0&0&0\\11&1&0&0&0\\16&1&0&0&0\\21&1&0&0&0\end{bmatrix}$]

T32. If M is the matrix $\begin{bmatrix}0&1&0\\0&0&1\\1&0&0\end{bmatrix}$, then M^{100} is

(A) $\begin{bmatrix}0&1&0\\0&0&1\\1&0&0\end{bmatrix}$, (B) $\begin{bmatrix}0&1&0\\0&0&1\\1&0&0\end{bmatrix}$, (C) $\begin{bmatrix}1&0&0\\0&1&0\\0&0&1\end{bmatrix}$, (D) $\begin{bmatrix}0&0&0\\0&0&0\\0&0&0\end{bmatrix}$, (E) None of the previous.

[**Ans.** A. Because M is an elelementary matrix, it remains unchanged on exponentiation]

T33. In the *xyz*-space, the degree measure of the angle between the rays

$z=x\geq 0, y=0$ and $z=y\geq 0, x=0$ is

(A) 0°, (B) 30°, (C) 45°, (D) 60°, (E) 90°.

[**Ans.** D. Since the symmetric form of the rays are $\frac{x}{1}=\frac{y}{0}=\frac{z}{1}$ and $\frac{x}{0}=\frac{y}{1}=\frac{z}{1}$ respectively. So

$$\cos\theta=\frac{1.0+0.1+1.1}{\sqrt{1^2+0^2+1^2}\sqrt{0^2+1^2+1^2}}=\frac{1}{2}=\cos 60^\circ$$]

T34. Let V be the set of all real polynomials p(x). Let transformations T, S be defined on V by T : p(x) $\rightarrow$ xp(x) and S : p(x) $\rightarrow$ p$'$(x)= $\frac{d}{dx}p(x)$ and interpret (ST)(p(x)) as S(T(p(x))). Which of the following is true ?

(A) ST = 0 , (B) ST = T, (C) ST=TS , (D) ST − TS is the identity map of V onto itself, (E) ST − TS is the identity map of V onto itself.

[**Ans.** D. Since TS = TS (p(x)) = T(p′ (x)) = x p′ (x) and ST = ST(p(x)) = S(p(x)) = (x p(x))′ = x p′ (x) + p(x). ∴ ST − TS = p(x) = identity of V]

T35. Which of the following is the larger of the eigenvalues of the matrix $\begin{bmatrix} 5 & 1 \\ 1 & 5 \end{bmatrix}$? (A) 4, (B) 5 , (C) 6 , (D) 7, (E) 8.

[**Ans.** C. Since $(5-\lambda)^2 - 1 = 0$ gives $\lambda = 4, 6$.]

T36. Let V be the vector space, under the usual operations, of real polynomials that are of degree at most 3. Let W be the subspace of all polynomials p(x) in V such that p(0) = p(1) = p(−1) =0 . Then dim V + dim W is

(A) 4 , (B) 5 , (C) 6, (D) 7, (E) 8 .

[**Ans.** B. V : $p(x) = ax^3 + bx^2 + cx + d$ is of dimension 4. $p(0) = d = 0$, $p(1) = a + b + c$,
$p(-1) = -a + b - c = 0 \Rightarrow b = a + c$.
$\therefore$ W : $p(x) = ax^3 - ax = ax(x^2 - 1)$ is of dimension 1.
Hence dimV + dim W = 4 + 1 = 5]

T37. Let I ≠ A ≠ −I , where I is the identity matrix and A is a real 2×2 matrix . If a = A^{-1}, then the trace of A is

(A) 1 , (B) 1 , (C) 0 , (D) −1 , (E) −2 .

[**Ans.** C. Let A=$\begin{bmatrix} a & b \\ c & d \end{bmatrix}$,$A^{-1} = \frac{1}{ad-bc}\begin{bmatrix} d & -c \\ -b & a \end{bmatrix}$. So A = $A^{-1} \Rightarrow$

$a = \frac{d}{ad-bc}$ and $d = \frac{a}{ad-bc}$ $\Rightarrow \frac{a}{d} = \frac{d}{a} \Rightarrow a^2 - d^2 = 0 \Rightarrow a + d = 0$]

T38. In the matrix equation $X^2 = \begin{bmatrix} 0 & 1 \\ 0 & 0 \end{bmatrix}$, X is to be a 2×2 matrix with real elements , which one of the following is true of X ?

(A) X = $\begin{bmatrix} 0 & 1 \\ 0 & 0 \end{bmatrix}$, (B) X = $\begin{bmatrix} 0 & -1 \\ 0 & 0 \end{bmatrix}$, (C) X = $\begin{bmatrix} 0 & 1 \\ 1 & 0 \end{bmatrix}$, (D) X = $\begin{bmatrix} 0 & -1 \\ -1 & 0 \end{bmatrix}$,

(E) No such matrix exists.

[**Ans.** E. Take X = $\begin{bmatrix} a & c \\ b & d \end{bmatrix}$. Form equations $a^2 + bc = 0$, $ad + bd = 0$,
$ac + cd = 1$, $bc + d^2 = 0$ which reach to a contradiction]

T39. Solve the equation for the matrix X : $\begin{bmatrix} 0 & 2 \\ 2 & 0 \end{bmatrix} X = \begin{bmatrix} 1 & 2 \\ 3 & 4 \end{bmatrix}$.

(A) $\begin{bmatrix} \frac{3}{2} & 2 \\ \frac{1}{2} & 1 \end{bmatrix}$, (B) $\begin{bmatrix} 4 & 3 \\ 2 & 1 \end{bmatrix}$, (C) $\begin{bmatrix} 0 & 1 \\ \frac{2}{3} & 0 \end{bmatrix}$, (D) $\begin{bmatrix} 3 & 1 \\ \frac{2}{3} & 0 \end{bmatrix}$, (E) $\begin{bmatrix} 1 & \frac{1}{2} \\ 2 & \frac{3}{2} \end{bmatrix}$.

[**Ans.** E. Inverse of $\begin{bmatrix} 0 & 2 \\ 2 & 0 \end{bmatrix}$ is $\begin{bmatrix} 0 & \frac{1}{2} \\ \frac{1}{2} & 0 \end{bmatrix}$. So X = $\begin{bmatrix} 1 & 2 \\ 3 & 4 \end{bmatrix}\begin{bmatrix} 0 & \frac{1}{2} \\ \frac{1}{2} & 0 \end{bmatrix}$]

T40. If T is linear transformation mapping the vectors (0, 0, 1), (0, 1, 0) & (1 , 0, 0) into the vectors (0, 0, 1), (−1, −1, 1) & (2, 2, 1) respectively . What is the image of the vector (3, 4, 5) under T ?

(A) (1, 2, 3) , (B) (2, 3, 4), (C) (2, 5, 2) , (D) (3, 4, 7) , (E) (2, 2, 12).

[**Ans.** E. Proceed as in Ch.12]

T41. If a set has n elements, how many possible subsets are there, if we include the universal set and the empty set as subsets ?

(A) 2^n , (B) 2^{n-1} , (C) n^2, (D) $2^n - 1$, (E) $2^n + 1$. [**Ans.** Each element may either be chosen for a subset or rejected .We can make this decision in $2.2...2 = 2^n$ ways and hence this is the number of different subsets.]

T49. If P & Q are intersecting sets and Q' is the suplement of Q, then $(P \cap Q) \cup (P \cap Q'))$ =

(A) P' , (B) Q' , (C) Q , (D) P, (E) $(P \cap Q)'$.

[**Ans.** D. It is seen from the Venn diagram that the union of the $(P \cap Q')$ and $(P \cap Q)$ make up the set P]

T50. Topology deals mainly with those properties of configurations which are invariant under

(A) conformal mapping

(B) continuous, one to one transformation

(C) Euclidean transformation

(D) contact transformation

(E) polar reciprocation.

[**Ans.** B. Since topology deals essentially with deformations, it is not concerned with preservation of metric properties of figure or angle preservation. If the distance between two points in a surface tends to zero then distance between two mapped points must approach zero. This type of transformation is best described by (B)]

T51. Let r and s be the linear transformations with respective matrices R and S. Then the mapping r.s is a linear transformation whose matrix is

(A) R + S, (B) R – S, (C) R.S , (D) R/S, (E) none of these.

[**Ans.** C. Note sec.12.11]

T52. If a set X consists of m and a set Y consists of n elements then the total number of subsets of the Cartesian product X×Y is

(A) mn , (B) 2^{m+n} ,(C) $m + n$, (D) $2mn$,(E) $2^m + 2^n$. [**Ans.** The Cartesian product consists of the ordered pairs (x, y) & so there are mn elements in this set. Including the empty set and the universal set, there are 2^{mn} subsets]

T53. Which system is an example of a group ?

(A) The set of nonnegative integers under ordinary multiplication

(B) The set of nonnegative reals under ordinary multiplication

(C) The set of positive reals under ordinary multiplication

(D) The set of nonnegative rationals under multiplication

(E) The set consisting of the four complex fourth roots of unity under ordinary addition.

[**Ans.** C. In (A) inverse element of 0 does not exist. In (B) , the inverse elements of + ve reals do not exist. In (D), there is no inverse of the 0 element. In (E) the identity element 0 is missing. (C) satisfies all requirements: inverse element, identity element, associative law and closure]

T53. The set of 2×2 matrices fails to satisfy the requirements of a group under matrix multiplication because

(A) not every element has an inverse

(B) the set lacks an identity element

(C) the associative law is not satisfied

(D) the commutative law is not satisfied

(E) the closure law is not satisfied.

[**Ans.** A. The nonsingular matrix elements do not have inverse. All other requirements are satisfied except (D) and it is not necessary for a group]

T54. Consider the system of linear equations

$$\begin{aligned}3x+2y-5z&=3\\2x-6y+kz&=9\\5x-4y-\ z&=5\end{aligned}$$

If this system is consistent, find k

(A) 2, (B) −2 , (C) 4 , (D) 3, (E) 6. [**Ans** . C]

T55. A collection of objects, group G = $\{p, q, r, s\}$ is given with the binary operation * defined by the table

*	p	q	r	s
p	p	q	r	s
q	q	p	s	r
r	r	s	p	q
s	s	r	q	p

From the table, find the value of $(q * r)*(r*s)$

(A) p, (B) q , (C) r , (D) s , (E) none of those .

[**Ans.** C.From the table $q * r = s$ and $r*s = q$ then $s*q = r$]

T56. Which of the properties are (is) invariant under topological transformation?

I. Length of arc.
II. Degree of connectivity of a domain.
III.Cross-ratio.
IV.Genus of surface.

(A) I & II only , (B) II & III only, (C) IV only , (D) II & IV only , (E) III & IV only.

[**Ans.** D.Length of arc and crossratio are not preserved topologically. Degree of connectivity and genus of surface are preserved]

T57. Any finite set of points (nonempty) is of dimension

(A) −1 , (B) 0, (C) 1, (D) 2, (E) 3. [**Ans.** B. The empty set is considered of dimension −1.Then a point set of dimension 0 if it is not of dimension −1 (it contains at least one point)and if each point of S can be enclosed within an arbitrarily small region whose boundary intersects S in a set of dimension −1 .Thus the given set is of dimension 0]

T58. Find the matrix M^{-1} which is the inverse of the matrix M= $\begin{bmatrix}1&0\\3&2\end{bmatrix}$

(A) $\begin{bmatrix}3&0\\-2&2\end{bmatrix}$, (B) $\begin{bmatrix}1&0\\3&2\end{bmatrix}$,(C) $\begin{bmatrix}2&3\\1&0\end{bmatrix}$,(D) $\begin{bmatrix}1&0\\-\frac{2}{3}&\frac{1}{3}\end{bmatrix}$,(E) $\begin{bmatrix}\frac{2}{3}&\frac{1}{3}\\1&0\end{bmatrix}$.

[**Ans.** D.M is nonsingular]

T59. If the norm in a vector space is defined by $\|(x_1,x_2,x_3,x_4)\| = \sqrt{-x_1^2-x_2^2-x_3^2+x_4^2}$ then the space is a

(A) Minkowski space, (B) Function space , (C) Sobolev space , (D) Hilbert space , (E) Banach space . [**Ans.** A. Note sec. 16.2.10]

Symbol	Meaning
$[:::::],[:::::]_{m\times n}$	matrix , $m\times n$ matrix
$A,B,C,..X...;0_{m\times n}$	matrix ; $m\times n$ zero matrix
a, b, c....x, y...u,v...etc.	vectors
$\|A\|, \|B\|, \|C\|$, detA ; a_{ij} ; A_{ij} ; M_{ij} ; S , T ; A_{ij}	determinant ; *ij*th element of matrix A ; *ij*th cofactor of matrix A; minor of a_{ij} ; set ; *ij*th block of matrix A
${}^{n}_{m}A$; ${}_{m}A_{n}$; $A_{m\times n} \equiv [a_{ij}]_{m\times n}$; $A_{(m,n)} \equiv [a_{ij}]_{m\times n}$; $A_{(m,n)} \equiv (a_{ij})_{(m,n)}$; $A_{(m,n)} \equiv \lVert a_{ij} \rVert_{(m,n)}$ etc.	an $m\times n$ matrix A
$A_{i\cdot}$ (or A_{i*})	*i*th row of the matrix A
$A_{\cdot j}$ (or A_{*j})	*j*th column of the matrix A
R-matrix	real matrix
C-matrix	complex matrix
Mod A	modulus or absolute value of matrix A
diag A	diagonal of matrix A
Tr(A), tr(A)	trace of matrix A
P_n ,T_y etc.	normal stress
P_s	shearing stress
T_{xy} , T_{xz} etc.	stress
$\mathbf{e}_i$	standard unit vectors
s.u.	standard unit
e or **1**	one vector
$J_{m\times n}$ or $1_{m\times n}$	j-matrix or unity matrix
E_{ij}	matrix unit
U_{ij} (or U_i)	rectangular (or square) upper trapezoidal matrix
L_{ij} (or L_i)	rectangular (or square) lower trapezoidal matrix
i	current
$\dot{x}$	differential coefficient of x
A^T , $\tilde{A}$	transpose of matrix A

$\overline{A}$	complex conjugation (conjugation) of matrix A
$A^*, (\overline{A})^T$	conjugate transpose (Hermitian) of matrix A
A^{-1}	inverse of matrix A
$\otimes$	Kronecker product
$\oplus$ (or [illegible])	Kronecker sum
$\lVert \cdots \rVert$	matrix or vector norm
S_n	scalar matrix of order n
I_n	unit matrix of order n
$>$ (or $<$)	greater (or less) than
$\geq$ (or $\leq$)	greater (or less) than equal to
$\lvert :::::: \rvert$	determinant
$\lvert \cdot \rvert$	absolute value or modulus
–‖–	source of power
_/\/\/_	resistor
$\dfrac{\partial f_i}{\partial f_j}$ and $\dfrac{\partial^2 f}{\partial x_i \partial x_j}$	first and second partial derivatives used in the definition of Jacobian and Hessian repectively
δ_{ij}	Kronecker delta
I_s	unit matrix with unity along secondary diagonal
$\sigma_x, \sigma_y, \sigma_z$	Pauli spin matrices
${}^n cC_r$ (or ${}^n rC_r$)	combination of n things in column (or row) taken r at a time
$A(\alpha,\beta)$	submatrix with α rows and β columns of the matrix A
$A^{[2]}, A^{[3]}, A^{[k]}$	Kronecker square , cube , of matrix A , kth Kronecker power of matrix A
$A*B$	Jordan product of A and B
$A\times B$	Lie product of A and B
$F\Theta G$	Khatri-Rao product of matrices F and G
$A \dagger B$	Hadamard product of matrices A and B
TR and BL	complementary blocks along principal diagonal
BR and TL	complementary blocks along secondary

Symbol	Meaning
	diagonal
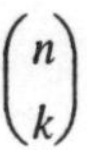	combination of n different things taken k at a time
LU	product of lower and upper triangular matrices
cofA	cofactor matrix of A
adjA or A^A	adjoint/adjugate matrix
Δ	determinant of a matrix
A_R (or A_L)	right (or left) inverse of nonsingular matrix A
Δa_{ij}	change or disturbance of an element a_{ij}
TIM	time independendent matrices
$\Rightarrow$	implies
Π	product of some factors

REFERENCES

BOOKS

1. Davis, P. J, *The Mathematics of Matrices: A First Book of Matrix Theory and Linear Algebra*, Gim-Blaisdell, 1965.

2. Ortega, James M., *Matrix theory: A Second Course*, Plenum Press, second printing 1989.

3. Goldberg, Jack L., *Matrix theory with Applications,* McGraw-Hill, Inc., 1991.

4. Steinberg, D. I., *Computational Matrix Algebra,* McGraw-Hill Book company, 1974.

5. Hohn, F. E., *Elementary Matrix Algebra,* The Macmillan Company, 3/e, 1973.

6. Perlis, S., *Theory of Matrices*, Addison-Wesley Publishing Company, Inc. Third printing, 1958.

7. Reza, Fazlollah, *Linear Spaces in Engineering,* Ginn and Co., 1971.

8. Joshi, A. W., *Matrices & Tensors in Physics* , Wiley Eastern Limited, 2/e, first Reprint 1985.

9. Lewis, David W., *Matrix Theory,* World Scientific, 1991.

10. Paige, L. J. & Swift, J. D., *Elements of Linear Algebra,* Ginn and Co., 1961.

11. Anton, Howard, *Elementary Linear Algebra,* John Wiley & sons, Inc. 1981.

12. Gupta, P. P. & Sharma, S. K., *Linear Algebra,* S. Chand & Co. Ltd., 1982.

13. Peddy, J. N. & Rasmüssen, M. L., *Advanced Engineering Analysis,* John Wiley & Sons Inc., 1982.

14. Pedoe, Danel, *A Geometric Introduction to Linear algebra,* John Wiley & Sons Inc., Third Printing, 1967.

15. Nering, Evar D., *Elementary Linear Algebra,* W. B. Saunders Co., 1974.

16. Hollingsworth, C. A., *Vectors, Matrices & Group Theory for Scientists & Engineers,* McGraw- Hill Book Company, 1967.

17. Anton, H. & Kolman, B., *Aplied Finite Mathematics,* Academic Press, Inc., 1974.

18. Smith, G. D., *Numerical solution of Partial Differential Equations: Finite Difference Methods,* Clarendon Press, 3/e, Reprint, 1992.

19. Türnbull, H. W. & Aitken, A. C., *An Introduction to The Theory of Canonical Matrices,* Blackie & Son Limited, Reprint, 1950.

20. Kolman, B., *Elementary Linear Algebra,* Macmillan Publishing Co. Inc., 1970

21. Basilevsky, A., *Applied Matrix Algebra in the Statistical Sciences,* North- Holland, 1983.

22. Ledermann, W. and Vajda, S., *Handbook of Applicable Mathematics, vol.* I: *Algebra,* John Wiley & Sons Ltd., 1980.

23. Campbell, Hugh G., *Linear Algebra with Applications,* Meredith Corporation, 1971.

24. Rice, John R., *Matrix Computations & Mathematical Software,* McGraw-Hill Book Company, 2nd print, 1985.

25. Pipes, Louis A., *Applied Mathematics for Engineers and Physicists,* McGraw-Hill Book Co., 3/e, 1970.

26. Finkbeiner, Daniel T., II, *Introduction to Matrices and Linear Transformations,* W. H. Freeman and Co., 2/e, 1966.

27. Turnbull, H. W., *Determinants, Matrices and Invariants,* Blackie & Son Ltd, 2/e, Reprint, 1948.

28. Aitken, A. C., *Determinants and Matrices,* Oliver & Boyd Ltd., 9/e, Reprint, 1964.

29. Demidovich, B. P. & Maron, I. A., *Computational Mathematics,* Mir Publishers, 1970.

30. Williams, G., *Computational Linear Algebra with Models,* Allyn and Bacon, Inc., 2/e, 1978.

31. Hadley, G., *Linear Algebra,* Addison-Wiley Publishing Co. Inc., Third printing, 1969.

32. Franklin, J. N., *Matrix Theory,* Prentice-Hall, Inc., 1968.

33. Graham, A., *Kronecker Products and Matrix Calculus with Applications,* Ellis Horwood Limited, 1981.

34. Magnus, J. R. and Neudecker, H., Matrix Differential Calculus with Applications in Statistics and Econometrics, John Wiley & Sons Ltd., 1988.

35. Magid, A. R., *Applied Matrix Models: A second course in Linear Algebra with computer Applications,* John Wiley & Sons, Inc., 1985.

36. Leon, S. J., *Linear Algebra with Applications,* Macmillan, 1980.

37. Hamilton, A. G., *Linear Algebra,* Cambridge University Press, 1989.

38. Mirsky, W., *An Introduction to Linear Algebra,* Oxford University Press, 1955.

39. Amundson, N.R., *Mathematical Methods in Chemical Engineering Matrices & their Application,* Prentice-Hall, 1966.

40. Pipes, Louis A., *Matrix Methods for Engineering,* Prentice-Hall Inc., 1963.

41. Pestel, E.C. and Lackie, F. A., *Matrix Methods in Elastomechanics,* McGraw-Hill Book Co., Inc., 1963

42. Bellman, R., *Introduction to Matrix Analysis*, McGraw-Hill Book Co., 2/e, 1970.

43. Ben-Israel, A. and Grevilla, T. N. E., *Generalized Inverses: Theory and Applications,* John Willey & Sons, 1974.

44. Searle, S. R., *Linear Models*, John Willey & Sons, Inc., 1971.

45. Horn, R. A. & Johnson, C. R., *Matrix Analysis,* Cambridge University Press, 1985.

46. Lancaster, P., *Theory of Matrices,* Academic Press, 1969.

47. Ball,W. W. R. and Coxeter, H. S. M., *Mathematical Recreations & Essays,* Macmillan and Co. Ltd, 11/e, 1939.

48. Ogata, K., *State Space Analysis of Control Systems,* Prentice- Hall, Inc., 1967.

49. Frazer, R. A., Duncan, W. J. and Collar, A. R., *Elementary Matrices and Some Applications to Dynamics and Differential Equations,* Cambridge University Press, Macmillan company, 1947.

50. Faddeev, D. K. and Faddeeva, V. N., *Computational Methods of Linear Algebra,* W. H. Freeman and Co., 1963.

51. Stein, F. M., *Introduction to Matrices and Determinants,* Wadsworth Publishing Co., Inc., 1967.

52. Eves, H., *Elementary Matrix Theory,* Allyn and Bacon, Inc ., 1966.

53. Brinkmann, H. W. and Klotz, E. A., *Linear Algebra and Analytic Geometry*, Addison-Wesley Publishing Company, 1971.

54. Venit, S. and Bishop, W., *Elementary Linear Algebra,* Prindle, Weber & Schmidt, 1981.

55. Wilkinson, J. H., *The Algebraic Eigenvalue Problem,* Oxford University Press, Reprint, 1992.

56. Halmos, P. R., *Finite Dimensional Vector spaces,* D. Van Nostrand company, Inc., 1958.

57. Vidyasagar, M., *Nonlinear System Analysis*, Prentice-Hall Inc., 1978.

58. Jamshedi, M. and Malek-Zavarei, M., *Linear Control Systems,* Pergamon Press, 1986.

59. Mac Daffee, C . C ., *The Theory of Matrices,* Chelsea Publishing Company, 5/e, 1946.

60. Desoer, C. A. and Vidyasagar, M., *Feedback Systems: Input-Output Properties,* cademic Press, 1975.

APERS

1. Johnson, C. R., *Positive Definite Matrices,* Am. Math. Monthly, vol.77, pp. 259-264, arch, 1970.

2. Levine, J and Naleikhan, H. M., *On the Construction of Involutory Matrices,* Am. Math. onthly, vol. 69, pp. 267-272, 1962.

3. Sherman J. and Morrison W. J., *Adjustment of An Inverse Matrix Corresponding to a hange in one Element of a Given Matrix,* Ann. Math. Statistics, vol. 21, pp. 124-127, 1949.

4. Graybill, F. A., Meyer, C. D. and Painter, R. J., *Note on the Computation of the eneralized Inverse of a Matrix,* SIAM Rev, vol. 8, No.4, pp. 522-524, 1966.

5. Barnett, S., *Matrix Differential Equations and Kronecker products,* SIAM J. Appl. Math., ol.24, No.1, pp. 1-5, 1973.

6. Decell, H. P. Jr., *An Application of the Cayley-Hamilton theorem to generalized Matrix nversion,* SIAM Rev, vol. 7, No.4, pp. 526- 528, 1965.

7. Cline, R. E., *Note on the Generalized Inverse of the Product of Matrices,* SIAM Rev, ol.6,No.1, pp. 57-58, 1964.

8. Penrose, R., *A Generalized Inverse for Matrices,* Proc. Camb. Phil. Soc., vol. 51, pp. 406-13, 1955.

9. Cline, R. E., *Note on an Extension of the Moore-Penrose Inverse,* Linear Algebra and its pplications, vol. 40, pp. 19-23, 1981.

10. Dwyer, P. S., *Some Applications of Matrix Derivatives in Multivariate Analysis,* Am. tatist Assoc. J., vol. 62, pp. 607- 625, 1967.

11. Councilman, S., *Eigenvalues and Eigenvectors of N-Matrices,* Am. Math. Monthly, vol. 3, pp. 392-393, 1986.

12. Kaufman, I., *The Inversion of the Vandermode Matrix,* IEEE trans, Auto. Control, vol. Ac 14 , pp. 774-777, December, 1969.

13. Tou, J. T., *Determination of the Inverse Vandermode Matrix,* IEEE Auto. Control, vol. AC-9, pp. 314, 1964

14. Macon, N. and Spitzbant, A., *Inverses of Vandermode Matrices,* Am. Math. Monthly, vol.65, pp. 95-100, 1958.

15. Chidambara, M. R., *On the Numerical Inversion of a Recurrent Problem: The Vandermode Matrix,* IEEE Auto, Control vol. AC-10, pp. 492-495, October, 1965.

16. Gerstein, L. J., *A New Algorithm for Computing the Rank of a Matrix,* Am. Math. Monthly, vol. 95, pp. 950-952, December, 1988.

17. Wilde, A. C., *Transformation Leaving the Determinant of Circulant Matrices Invariant,* Am. Math. Monthly, vol. 93, pp. 356 -361, 1986.

18. Ablow, C. M. and Brenner, J. L., *Roots and Canonical Forms for Circular Matrices,* Trans, Am. Math. Soc., vol. 107, 1963.

19. Lancaster, H. O., *The Helmert Matrices,* Am. Math. Montly, vol. 72, pp. 4-12, 1965.

20. Butson, A. T., *Generalized Hadamard Matrices,* Proc. Am. Math. Soc., vol. 13, pp. 894-898, 1962.

21. Golomb, S. W. and Baumert, L. D., *The Search for Hadamard Matrices,* Am. Math. Monthly, vol.70, pp. 12 -17, 1963.

22. Hedayat, A. & Walls, W. D., *Hadamard Matrices and their Applications,* Ann. Statist., vol. 6, No. 6, pp. 1184-1238, 1978.

23. Hazra, A. K., *An overview of the applications of matrices,* presented in Regional Mathematics Workshop in Dhaka, Bangladesh, 1989.

24. Hazra, A. K., *Matrix Partitioning Technique to Derive Abel's Identity for nth order Ordinary Differential Equations (ODE),* Bangladesh J. Sci. Res., vol.12, No. 1, pp. 61-67, 1994.

25. Hazra, A. K., *Applications of Sylvester's Dialytic Elimination*, Proceedings of 10th Mathematical Conference of Bangladesh, p 117-120 , 1996.

26. Hazra, A. K., *On The Evolutionary Trends In Geometry,* J. Math. Math. Sciences, vol.10, pp. 33-40, 1995.

27. Hazra, A. K., *Increasing the Order and a New Expansion Method of Determinant,* to appear.

28. Hazra, A. K., *Properties and Uses of Cofactor Matrix,* to appear.

29. Hazra, A. K., *A Construction of Magic Squares,* to appear.

30. Hazra, A. K., *Modal matrix- its construction and use,* Journal of Bangladesh Academy of Science, Vol.23, No.1, p 57-65, 1999.

31. Brewer, J. W., *Kronecker Products and Matrix Calculus in System theory,* IEEE Trans, Circuits And Systems, vol. CAS 25, No. 9, 1978.

32. Brewer, J. W., *The Derivative of the Exponential Matrix with Respect to a Matrix,* IEEE Trans. Auto. Control, vol. AC 22, No. 4, 1977.